THEORETICAL KINEMATICS

O. BOTTEMA

Professor Emeritus
University of Technology, Delft

B. ROTH

Professor
Stanford University, Stanford, CA

DOVER PUBLICATIONS, INC.
NEW YORK

Published in Canada by General Publishing Company, Ltd., 30 Lesmill Road, Don Mills, Toronto, Ontario.
Published in the United Kingdom by Constable and Company, Ltd.

This Dover edition, first published in 1990, is an unabridged and corrected republication of the work originally published in 1979 by North-Holland Publishing Company, Amsterdam, as Volume 24 of the North-Holland Series in Applied Mathematics and Mechanics.

Manufactured in the United States of America
Dover Publications, Inc., 31 East 2nd Street, Mineola, N.Y. 11501

Library of Congress Cataloging-in-Publication Data

Bottema, O. (Oene)
Theoretical kinematics / O. Bottema, B. Roth.
p. cm.
Reprint, corrected. Originally published: Amsterdam ; New York : North-Holland Pub. Co., 1979.
Includes bibliographical references.
ISBN 0-486-66346-9
1. Kinematics. I. Bottema, O. (Oene). II. Roth, Bernard. III. Title.
QA841.B66 1990
531′.112—dc20 90-31416
CIP

To Ferdinand Freudenstein

Friend and colleague; in appreciation for having first brought us together, and for many outstanding contributions to kinematics.

PREFACE

This book deals with the science of kinematics. Formally, kinematics is that branch of mechanics which treats the phenomenon of motion without regard to the cause of the motion. In kinematics there is no reference to mass or force; the concern is only with relative positions and their changes.

We have used the adjective *theoretical* in our title in order to distinguish our subject from *applied* kinematics, which deals with the application of kinematics: to mechanical contrivances, to the theory of machines, and to the analysis and synthesis of mechanisms. Much of what is written herein could be used to study mechanical devices. However, our aim is broader: what we give is the development of the theory independent of any particular application, a presentation of the subject as a fundamental science in its own right. By this, we hope to make these results equally accessible to other fields. This is important because our science touches on many areas: *everything that moves has kinematical aspects.* In recent years interesting applications of kinematics have, for example, been made in areas as diverse as: animal locomotion, art, biomechanics, geology, robots and manipulators, space mechanics, structural chemistry, and surgery.

Theoretical kinematics is a large subject and it is impossible to treat it completely in a single volume. In this book we restrict ourselves to what might be called the kinematics of spaces. We consider unbounded infinite spaces which contain idealized mathematical elements embedded in them. We mainly treat the linear elements (points, lines, planes) and study the geometric properties which arise as the space moves. Essentially we are dealing with aspects of what mathematicians call transformation geometry. We consider those transformations in Euclidean spaces such that all distances remain fixed. Often we will find it convenient to add the elements at infinity to our space; this clarifies the results and simplifies the terminology and analytical development.

A characteristic feature of this book is the principle of starting with general concepts and problems and then specializing gradually to more simple cases.

Following this idea, Chapters I and II are written in terms of n-dimensional Euclidean spaces. Then we specialize to the important case $n = 3$, the major topic of the book, which could be called *rigid body kinematics of three dimensions.* We think of our mathematical space as representing an unbounded, rigid, material body.

The word *displacement* is used in kinematics in a rather special way. It implies that we have no interest in how a motion actually proceeds: we consider only the position before and the position after the motion. If the distances moved by any of the points are finite we say that we have a *finite* displacement, or simply a displacement. Also of interest is the limit of a finite displacement if the displaced position approaches the undisplaced one; these infinitesimal displacements play an important role in the theory, they are the subject of *instantaneous* kinematics. Our Chapters III, IV, V deal with two, three, and four (or more) positions theory, both for finite and instantaneous displacements. Chapter VI studies motion proper: continuous sets of displacements. In accordance with our principle further specialization follows: spherical kinematics is dealt with in Chapter VII, planar kinematics in Chapter VIII. To illustrate the foregoing theory Chapter IX studies a selection of special motions; this chapter, more than any other borders on the subject of applied kinematics. The last four chapters are mutually independent, each developes the fundamentals of an aspect of kinematics not previously dealt with in the book. In Chapter X we consider motions which depend upon two or more parameters. Chapter XI returns to planar kinematics and studies it by a new technique: planar positions are mapped on to the points of a three-dimensional space with an adapted metric. The final two Chapters, XII and XIII, are respectively concerned with the kinematics associated with various geometries in addition to the Euclidean one and with special mathematical methods.

In writing this book we have tried to expand the theory and to fill various gaps. We have also derived some old results in new ways. We believe this book to be the first on theoretical kinematics written in the English language. It addresses itself to the researcher in the field and to the advanced student.

The treatment in the text is mainly analytical. However, geometric interpretations are given, and in places synthetic reasoning is also applied. With regard to the necessary mathematical tools, it is essential that the reader understands the basics of homogeneous coordinates of points and planes in order to be able to study what happens to the elements at infinity; when we deal with lines in space we use the homogeneous Plücker coordinates, defined in the text. Furthermore, we have, in the main, limited ourselves to the mathematics of algebraic geometry, vector and matrix algebra and elementary calculus. Special mathematical methods (such as quaternions and dual numbers) are reserved for the final chapter, where they are briefly introduced as subjects in their own right.

The reader should note that in order to limit the length of the text we have presented some material without proof; it is generally set in small print and denoted "Example". Although mainly written in the form of exercises, we regard this material as an essential part of the text.

We wish to thank the publishers for their courtesy and patience. Thanks are also due to Joel Yudken for his careful typing of the manuscript, and to the National Science Foundation for financial support of the research which led to the development of some of the results contained herein.

O.B.
B.R.

TABLE OF CONTENTS

CHAPTER I

EUCLIDEAN DISPLACEMENTS

1. Displacements

A *transformation* $T(P \rightarrow P')$ of Euclidean n-space E^n is defined by giving each point P an image point P' such that a one-to-one correspondence between P and P' exists: each point P' is the image of exactly one point P. Hence $P' \rightarrow P$ is also a transformation, the inverse of T, denoted by T^{-1}. If $T_1(P \rightarrow P')$ and $T_2(P' \rightarrow P'')$ are two transformations $P \rightarrow P''$ is a transformation, called the *product* T_2T_1. Obviously $(T_3T_2)T_1$ is equal to $T_3(T_2T_1)$: the multiplication of transformations is associative. There is a unit transformation $I(P \rightarrow P)$. It follows from all this that the transformations of E^n form a *group*.

If $T(P \rightarrow P')$ has the property that for all point pairs P, R, the distance $P'R'$ is equal to the distance PR the transformation is called a *Euclidean displacement*. In this book kinematics is essentially the study of Euclidean displacements, and, unless otherwise stated, a displacement is Euclidean. If D is a displacement, D^{-1} is a displacement; if D_1 and D_2 are displacements the same holds for D_2D_1; I is a displacement. Therefore: the displacements form a group. It is a sub-group of the group of transformations and it consists of those which leave the distance of any two points *invariant*.

If $D(P \rightarrow P')$ and P, R, S are collinear points, from $P'R' = PR$, $P'S' = PS$, $(R'S' = RS)$, it follows that P', R', S' are collinear. Therefore: a displacement transforms a line l, into a line l'; D is a *collineation*. By the same reasoning we conclude that the points of a plane or, in general, the points of a linear subspace E^m of E^n are transformed into those of a linear subspace $E^{m'}$ of the same dimension.

If l, m are parallel lines, which means that they are in a plane E^2 and have no point of intersection, l' and m' are in the plane $E^{2'}$ and they have no common point because D is a transformation. Thus parallelism of lines is maintained by displacements. A displacement is an *affine* transformation. If E^k and E^m are parallel subspaces so are $E^{k'}$ and $E^{m'}$.

If P, R, S are the vertices of a triangle the images P', R', S' form a triangle with equal corresponding sides and thus with equal corresponding angles. Therefore: displacements are angle-preserving (often called conformal) transformations. In particular: right angles correspond to right angles. This implies that not only the distance between two points but also the distance between a point and a linear subspace, and the distance between two parallel subspaces, are invariant under displacements.

We define an n-dimensional reference space Σ^n and regard this space as *fixed* (in the sense that its points are invariant under all displacements). We take an arbitrary point O, in Σ^n, as the origin, and associate with each point $P, R, S, \ldots$ of E^n its *position vector* $\boldsymbol{P} = OP$, $\boldsymbol{R} = OR$, $\boldsymbol{S} = OS, \ldots$ in Σ^n. For the displaced space with points $P', R', S', \ldots$ in $E^{n\prime}$ we have position vectors $\boldsymbol{P}' = OP'$, $\boldsymbol{R}' = OR'$, $\boldsymbol{S}' = OS', \ldots$ in Σ^n. For points in $E^{n\prime\prime}$ we have $\boldsymbol{P}'', \boldsymbol{R}'', \boldsymbol{S}'', \ldots$ in Σ^n, and so on. For the vectors in Σ^n we have the usual definition for the scalar (or so-called dot) product of two vectors. Geometrically, the scalar product of two vectors, say $\boldsymbol{v}_1$ and $\boldsymbol{v}_2$, is the scalar $v_1 v_2 \cos\theta$; v_i being the length of $\boldsymbol{v}_i$, and θ the angle between $\boldsymbol{v}_1$ and $\boldsymbol{v}_2$. It is easy to show that it is commutative and distributive; $\boldsymbol{v}_i \cdot \boldsymbol{v}_i$ or $\boldsymbol{v}_i^2$ for short, is the square of the length of $\boldsymbol{v}_i$.

A displacement $\mathrm{D}(P \to P')$ is seen to be a transformation such that $(\boldsymbol{P}' - \boldsymbol{R}')^2 = (\boldsymbol{P} - \boldsymbol{R})^2$ for all point pairs P, R. Obviously the transformation $\boldsymbol{P}' = \boldsymbol{P} + \boldsymbol{d}$, $\boldsymbol{d}$ being a fixed vector, is a displacement. It is a special one, called a *translation*. If P, R are two points, $P'R'$ has not only the same length as PR but the line $P'R'$ is also parallel to PR. I is a translation. If D is a translation D^{-1} is a translation. If D_1 and D_2 are the translations $\boldsymbol{P}' = \boldsymbol{P} + \boldsymbol{d}_1$, $\boldsymbol{P}'' = \boldsymbol{P}' + \boldsymbol{d}_2$, then $\mathrm{D}_2\mathrm{D}_1$ is a translation and $\mathrm{D}_2\mathrm{D}_1 = \mathrm{D}_1\mathrm{D}_2$. Therefore the translations form a commutative group, which is a subgroup of the displacements group. $\boldsymbol{d}$ is called the vector of the translation. It is independent of the choice of origin.

If for a certain displacement D a point P coincides with its image P' this point is called a fixed point or a double point of D. All points are fixed points for the displacement I. A translation, different from I, has no fixed points.

A displacement for which a point O is a fixed point is called a *rotation* about O. I is a rotation about O. If R is a rotation about O then so is R^{-1}; if R_1 and R_2 are rotations about O, $\mathrm{R}_2\mathrm{R}_1$ and $\mathrm{R}_1\mathrm{R}_2$ are rotations about O. Thus the rotations about O constitute a subgroup of the displacements group. It is not commutative, for in general $\mathrm{R}_2\mathrm{R}_1$ and $\mathrm{R}_1\mathrm{R}_2$ are different displacements, as we shall see.

If D is a displacement, O an arbitrary point, O' its image, S the translation with vector $\boldsymbol{d} = \overrightarrow{OO'}$ then $\mathrm{S}^{-1}\mathrm{D}$ is clearly a rotation R about O. From

$S^{-1}D = R$ it follows $D = SR$; *a displacement is the product of a rotation (about an arbitrarily chosen point) and a translation.*

2. Orthogonal matrices

We consider a rotation R about the origin O. Let $U_1, U_2, \ldots, U_n$ be n linearly independent unit vectors (in E^n, measured in Σ^n); then an arbitrary vector P may be written in a unique way as a linear function of U_i, namely as

$$P = (P \cdot U_1)U_1 + (P \cdot U_2)U_2 + \cdots + (P \cdot U_n)U_n = \sum_{i=1}^{n} (P \cdot U_i)U_i,$$

or $P = (P \cdot U_i)U_i$, if we make use of the summation convention for repeated indices. If P' is the image of P and U'_i $(i = 1, 2, \ldots, n)$ that of U_i, we have $P' = (P' \cdot U'_i)U'_i$. But from the definition of a displacement the relation $P' \cdot U'_i = P \cdot U_i$ is true for all i. Therefore $P' = (P \cdot U_i)U'_i$. This means that the image P' of a general point P is known if the images U'_i of the set U_i are given for the rotation.

It is clear that if the U_i are given the set U'_i cannot be arbitrary, because the relation $U_i \cdot U_j = U'_i \cdot U'_j$ must hold for all i and j. In order to obtain simple conditions it seems wise to make use of Cartesian sets: let U_i be n mutually orthogonal unit vectors. Then it is necessary that the unit vectors U'_i are also mutually orthogonal. On the other hand this is also sufficient to insure that the transformation is a displacement: if P and R are two points, we have $P' - R' = \{(P \cdot U_i) - (R \cdot U_i)\}U'_i$ and therefore $(P' - R')^2 = \Sigma_i\{(P - R) \cdot U_i\}^2$, in view of $U'_i \cdot U'_j = \delta_{ij}$ ($\delta_{ij} = 0$ for $i \neq j$ and $\delta_{ij} = 1$ for $i = j$), and this is furthermore equal to $(P - R)^2$. The conclusion is: a rotation about O is fully determined if the ordered set of mutually orthogonal unity vectors U'_i is known as the image of a given set U_i of such vectors $(i = 1, \ldots, n)$. So the problem to determine analytically the rotations about O is reduced to that of finding all such sets U'_i. Any U'_i can be expressed in terms of U_k; indeed we have $U'_i = (U'_i \cdot U_k)U_k$ or, if we denote $U'_i \cdot U_k$ by a_{ik} we have $U'_i = a_{ik}U_k$. The orthogonality relations for U'_i are seen to be $U'_i \cdot U'_j = \delta_{ij}$ or $a_{ik}a_{jk} = \delta_{ij}$ in view of $U_i \cdot U_j = \delta_{ij}$. They are $\frac{1}{2}n(n + 1)$ in number and express as many conditions for the n^2 numbers $U'_i \cdot U_k$, that is for the cosines of the angles between the U and the U' axes.

The matrix $\| a_{ik} \|$ with a_{ik} in the ith row and the kth column is called an orthogonal matrix. If the basis vectors U_i are given this matrix determines the rotation; indeed $U'_i = a_{ik}U_k$ and $P' = (P \cdot U_i)U'_i$ and thus $P' = a_{ik}(P \cdot U_i)U_k$, the right hand side being a double sum over the repeated indices i and k. The symbol $\boldsymbol{P}' = \mathbf{A}\boldsymbol{P}$ is often introduced for this relation, where $\mathbf{A}$ is an abbreviation for the matrix $\| a_{ik} \|$, $\boldsymbol{P}$ is the column matrix representation of the vector, and $\boldsymbol{P}'$ is also in column matrix form. The transformation $\boldsymbol{P} \rightarrow \boldsymbol{P}'$ is thus expressed as a matrix multiplication.

To our rotation R (essentially given by $\boldsymbol{U}_i \rightarrow \boldsymbol{U}'_i$, that is by $\boldsymbol{U}'_i = a_{ik}\boldsymbol{U}_k$) belongs the inverse rotation R^{-1}, expressed by $\boldsymbol{U}_i = a'_{ik}\boldsymbol{U}'_k$, where $\| a'_{ik} \|$ is the inverse of the matrix $\| a_{ik} \|$, which because of our (1, 1) correspondence cannot be singular. In view of $a_{ik} = \boldsymbol{U}'_i \cdot \boldsymbol{U}_k$ and analogously $a'_{ik} = \boldsymbol{U}_i \cdot \boldsymbol{U}'_k$ it follows $a'_{ik} = a_{ki}$. Therefore $\mathbf{A}$ *has the property that its inverse is identical with its transpose*: $\mathbf{A}^{-1} = \mathbf{A}^{\mathrm{T}}$, and it is easily seen that this property characterizes the *orthogonal* matrices.

If Δ is the determinant of $\mathbf{A}$, Δ^{-1} is that of $\mathbf{A}^{-1}$ and Δ is that of $\mathbf{A}^{\mathrm{T}}$; therefore $\Delta^2 = 1$ and thus $\Delta = \pm 1$. Both cases exist: for I we have $\Delta = 1$; for the orthogonal matrix $a_{11} = -1$, $a_{ii} = 1$ $(i \neq 1)$, $a_{ik} = 0$ $(i \neq k)$, $\Delta = -1$. If R_1 and R_2 are two rotations about O, to which the matrices $\mathbf{A} = \| a_{ik} \|$ and $\mathbf{B} = \| b_{ik} \|$ belong, then from $\boldsymbol{U}'_i = a_{ik}\boldsymbol{U}_k$ and $\boldsymbol{U}''_i = b_{ik}\boldsymbol{U}'_k$ it follows that $\boldsymbol{U}''_i = b_{ik}a_{kl}\boldsymbol{U}_l$, or $\boldsymbol{U}''_i = c_{il}\boldsymbol{U}_l$ with $c_{il} = b_{ik}a_{kl}$; that is, if $\mathbf{C} = \| c_{ik} \|$ we have $\mathbf{C} = \mathbf{BA}$ from the multiplication rule for matrices. It is well-known that the multiplication of matrices is not commutative. This implies that *the product of two rotations is* also *not commutative.*

3. The eigenvalues of an orthogonal matrix

We have seen that a rotation R about O is accompanied by an orthogonal matrix $\mathbf{A} = \| a_{ik} \|$. It is clear that this matrix depends on the way we have described the rotation analytically and this was done by means of the arbitrarily chosen set of Cartesian basis vectors $\boldsymbol{U}_i$. Suppose that we take another orthogonal Cartesian set $\boldsymbol{V}_i$ and that by our rotation it transforms to $\boldsymbol{V}'_i$. Applying the general formula $\boldsymbol{P}' = a_{ik}(\boldsymbol{P} \cdot \boldsymbol{U}_i)\boldsymbol{U}_k$ for $\boldsymbol{P} = \boldsymbol{V}_j$ we obtain $\boldsymbol{V}'_j = a_{ik}(\boldsymbol{V}_j \cdot \boldsymbol{U}_i)\boldsymbol{U}_k = q_{ji}a_{ik}\boldsymbol{U}_k = b_{jk}\boldsymbol{U}_k$; q_{ji} are the elements of an orthogonal matrix $\mathbf{Q}$ and b_{jk} those of $\mathbf{B} = \mathbf{QA}$. Furthermore $\boldsymbol{U}_k = (\boldsymbol{U}_k \cdot \boldsymbol{V}_l)\boldsymbol{V}_l$ in which $\boldsymbol{U}_k \cdot \boldsymbol{V}_l = q_{lk}$, so that $\boldsymbol{V}'_j = b_{jk}q_{lk}\boldsymbol{V}_l = b_{jk}q'_{kl}\boldsymbol{V}_l$ where $q'_{kl} = q_{lk}$ is the (k, l)-element of $\mathbf{Q}^{\mathrm{T}}$ and also $\mathbf{Q}^{-1}$; $b_{jk}q'_{kl} = a'_{jl}$, however, is the (j, l)-element of $\mathbf{BQ}^{-1}$, that is $\mathbf{QAQ}^{-1}$. The result is that the rotation R is expressed by $\boldsymbol{V}'_j = a'_{jl}\boldsymbol{V}_l$ or in general by $\boldsymbol{P}' = \mathbf{A}'\boldsymbol{P}$ in which $\mathbf{A}' = \mathbf{QAQ}^{-1}$. The conclusion is: if a rotation R about O is described by an orthogonal matrix $\mathbf{A}$ in a certain reference system $\boldsymbol{U}_i$ then in any system it is described by a matrix $\mathbf{A}' = \mathbf{QAQ}^{-1}$, $\mathbf{Q}$ being itself an orthogonal matrix whose elements are the cosines between the bases of the two systems. Making use of the terminology of matrix algebra we may say that $\mathbf{A}'$ and $\mathbf{A}$ are equivalent within the group of orthogonal matrices, which we denote by $\mathbf{A}' \cong \mathbf{A}$. It is easily seen that this concept is symmetric ($\mathbf{A}' \cong \mathbf{A}$ implies $\mathbf{A} \cong \mathbf{A}'$) and transitive (from $\mathbf{A}_1 \cong \mathbf{A}$ and $\mathbf{A}_2 \cong \mathbf{A}$ it follows $\mathbf{A}_1 \cong \mathbf{A}_2$). Summing up: *a rotation* R *about O corresponds to a class of mutually equivalent orthogonal matrices.* In each reference system the rotation is described by a matrix which is a member of the class. The class as such is independent of the reference system.

It is clear that a property of a matrix $\mathbf{A}$ which describes the rotation is only of interest if it is independent of the choice of the reference system, which means that $\mathbf{A}$ shares that property with all its equivalent matrices. A property common to all members of the class is called an *invariant.*

We have seen that the determinant Δ of an orthogonal matrix is either 1 or -1. Obviously equivalent matrices $\mathbf{A}$ and $\mathbf{A}' = \mathbf{QAQ}^{-1}$ must have equal determinants. Therefore all members of a class have $\Delta = 1$ or all have $\Delta = -1$. Hence rotations correspond to the number 1 or to the number -1. The first are called direct rotations, the others are indirect. Kinematics proper deals with direct rotations only.

For a (direct) orthogonal matrix $\mathbf{A}$ we consider $F(\lambda) = |\mathbf{A} - \lambda\mathbf{I}| = 0$ which is an equation of the nth degree for the scalar λ, the characteristic equation of $\mathbf{A}$. If we replace $\mathbf{A}$ by an equivalent matrix $\mathbf{A}'$, such that $\mathbf{A}' = \mathbf{QAQ}^{-1}$, $\mathbf{Q}$ being an orthogonal matrix, we have $\mathbf{A}' - \lambda\mathbf{I} = \mathbf{QAQ}^{-1} - \lambda\mathbf{I} = \mathbf{Q}(\mathbf{A} - \lambda\mathbf{I})\mathbf{Q}^{-1}$, for matrix multiplication is distributive and $\mathbf{QIQ}^{-1} = \mathbf{I}$. Therefore $|\mathbf{A}' - \lambda\mathbf{I}| = |\mathbf{A} - \lambda\mathbf{I}|$: the characteristic equations of two equivalent matrices are identical. If we write it out:

$$F(\lambda) = (-\lambda)^n + c_1(-\lambda)^{n-1} + c_2(-\lambda)^{n-2} + \cdots + c_{n-1}(-\lambda) + 1 = 0, \tag{3.1}$$

the coefficients c_k $(k = 1, \ldots, n-1)$ are obviously invariants of $\mathbf{A}$ and the same holds for the zeros λ_j $(j = 1, \ldots, n)$ of $F(\lambda)$, the eigenvalues of $\mathbf{A}$. c_1 is equal to Σa_{ii}, the trace of $\mathbf{A}$; c_k is the sum of the $\binom{n}{k}$ central minors of $\mathbf{A}$ of the kth order. But in view of the fundamental relations $\mathbf{A}^{-1} = \mathbf{A}^{\mathrm{T}}$, $|\mathbf{A}| = 1$, the elements of $\mathbf{A}$ are equal to their own minors and each minor of the kth order is equal to its complementary minor of order $(n-k)$. This implies $c_k = c_{n-k}$ $(k = 1, \ldots, n-1)$. Hence (3.1) is a reciprocal equation: if λ_j is a root λ_j^{-1} is a root.

Assume that λ_k is a root. This implies that the system of homogeneous linear equations with unknowns X_i

$$\lambda_k X_j = a_{ji} X_i \quad (j = 1, 2, \ldots, n) \tag{3.2}$$

has a non-zero solution. That means that there is a point P, different from O, with the image point P', such that $\boldsymbol{P}' = \lambda_k \boldsymbol{P}$. In view of the fundamental property of a displacement $(\overrightarrow{OP})^2 = (\overrightarrow{OP'})^2$ and therefore $(\overrightarrow{OP})^2 = \lambda_k^2(\overrightarrow{OP})^2$.

There are two cases: if λ_k is real then P is a real point, hence $\lambda_k^2 = 1$ and a real eigenvalue of $\mathbf{A}$ is either equal to 1 or to -1. When $\lambda_k = \pm 1$, eq. (3.2) yields a unique, real line, which is invariant under the rotation, called the axis of the rotation. If λ_k is imaginary $\lambda_k^2 \neq 1$, hence $(\overrightarrow{OP})^2 = 0$, P and P' are imaginary and they lie on the same null-line (i.e., isotropic line) through O. This line coincides with its image and is therefore invariant for the rotation. Suppose that $\lambda = \rho e^{i\varphi}$ $(\rho > 0, \varphi \neq 0, \text{mod. } \pi)$ is a non-real eigenvalue and $\boldsymbol{P}$ an eigenvector, which means $\boldsymbol{P}' = (\rho e^{i\varphi})\boldsymbol{P}$. Then if $\boldsymbol{P} = \boldsymbol{q} + i\boldsymbol{h}$, $\boldsymbol{q}$ and $\boldsymbol{h}$ being real vectors, $(\boldsymbol{P})^2 = (\boldsymbol{P}')^2$ implies $\boldsymbol{q}^2 = \boldsymbol{h}^2 = k^2$ and $\boldsymbol{q} \cdot \boldsymbol{h} = 0$. But as the characteristic equation has real coefficients $\rho e^{-i\varphi}$ is another eigenvalue and we have $\boldsymbol{R}' = (\rho e^{-i\varphi})\boldsymbol{R}$, $\boldsymbol{R} = \boldsymbol{q} - i\boldsymbol{h}$. The plane W through $\boldsymbol{P}$ and $\boldsymbol{R}$ contains two different invariant lines and it is therefore an invariant plane (i.e., all the image points of W also lie in W). But W is obviously a real plane for it contains the real vectors $\boldsymbol{q} = \frac{1}{2}(\boldsymbol{P} + \boldsymbol{R})$ and $\boldsymbol{h} = -\frac{1}{2}i(\boldsymbol{P} - \boldsymbol{R})$, which are furthermore orthogonal and of the same length k. Moreover if $\boldsymbol{q}'$ corresponds to $\boldsymbol{q}$,

$$\boldsymbol{q}' = \tfrac{1}{2}(\boldsymbol{P}' + \boldsymbol{R}') = \tfrac{1}{2}\rho(e^{i\varphi}\boldsymbol{P} + e^{-i\varphi}\boldsymbol{R}) = \tfrac{1}{2}\rho\{e^{i\varphi}(\boldsymbol{q} + i\boldsymbol{h}) + e^{-i\varphi}(\boldsymbol{q} - i\boldsymbol{h})\}$$
$$= \rho(\boldsymbol{q}\cos\varphi - \boldsymbol{h}\sin\varphi),$$

similarly $\boldsymbol{h}' = \rho(\boldsymbol{q}\sin\varphi + \boldsymbol{h}\cos\varphi)$. From $(\boldsymbol{q}')^2 = (\boldsymbol{q})^2$ it follows however that $\rho = 1$. Therefore: all eigenvalues λ_k of an orthogonal matrix have the property $|\lambda_k| = 1$; if λ_k is real it is either $+1$ or -1; if it is imaginary $\lambda_k = e^{i\varphi}$ and the conjugate value $e^{-i\varphi}$ (which is equal to λ_k^{-1}) is also an eigenvalue in accordance with the fact that the characteristic equation is a reciprocal one. Furthermore from the foregoing $\boldsymbol{q}' \cdot \boldsymbol{q} = \boldsymbol{h}' \cdot \boldsymbol{h} = k^2 \cos\varphi$, hence the displacement within the invariant plane W is a planar rotation about O with the rotation angle φ.

Let $e^{i\varphi_1}, e^{-i\varphi_1}$ and $e^{i\varphi_2}, e^{-i\varphi_2}$ be two different pairs of conjugate imaginary eigenvalues, and W_1 and W_2 the real invariant planes corresponding to them. W_1 contains two conjugate imaginary vectors $\boldsymbol{P}$ and $\boldsymbol{R}$ such that $\boldsymbol{P}' = (e^{i\varphi_1})\boldsymbol{P}$ and $\boldsymbol{R}' = (e^{-i\varphi_1})\boldsymbol{R}$; W_2 contains vectors $\boldsymbol{S}$ and $\boldsymbol{T}$ such that $\boldsymbol{S}' = (e^{i\varphi_2})\boldsymbol{S}$, $\boldsymbol{T}' = (e^{-i\varphi_2})\boldsymbol{T}$. Then $\boldsymbol{P}' \cdot \boldsymbol{S}' = (e^{i(\varphi_1+\varphi_2)})\boldsymbol{P} \cdot \boldsymbol{S}$ since $\boldsymbol{P}' \cdot \boldsymbol{S}' = \boldsymbol{P} \cdot \boldsymbol{S}$ and $e^{i(\varphi_1+\varphi_2)} \neq 1$ the conclusion is $\boldsymbol{P} \cdot \boldsymbol{S} = 0$. In the same way we prove $\boldsymbol{P} \cdot \boldsymbol{T} = 0$, $\boldsymbol{R} \cdot \boldsymbol{S} = 0$, $\boldsymbol{R} \cdot \boldsymbol{T} = 0$. Hence the scalar product $(k_{11}\boldsymbol{P} + k_{12}\boldsymbol{R}) \cdot (k_{21}\boldsymbol{S} + k_{22}\boldsymbol{T}) = 0$ for any set of numbers k_{ij}. But that means: each line in W_1 is orthogonal to each line in W_2: the two planes are (absolutely) orthogonal. In the same way it is proved that *an invariant line, corresponding to a real eigenvalue, is orthogonal to all invariant planes.*

Example 1. Show that for $n = 3$ the characteristic equation of $\mathbf{A} = \| a_{ij} \|$ reads $\lambda^3 - c\lambda^2 + c\lambda - 1 = 0$, with $c = a_{11} + a_{22} + a_{33}$. The eigenvalues are $\lambda = 1$ and the roots of $\lambda^2 - (c-1)\lambda + 1 = 0$; for the rotation angle φ we have $\varphi = \arccos[(c-1)/2]$. Show that the direction of the rotation axis is given by $(a_{23} + a_{32})^{-1} : (a_{31} + a_{13})^{-1} : (a_{12} + a_{21})^{-1}$, as well as $(a_{32} - a_{23}) : (a_{13} - a_{31}) : (a_{21} - a_{12})$.

4. Standard representations of a displacement

A direct rotation for which all eigenvalues are different will be called general. We introduce now a distinction which is fundamental for kinematics and which arises from the theory developed so far, namely n is even and n is odd. If n is even, $n = 2m$, the equation $F(\lambda) = 0$ has (in general) no real root; there are m pairs of conjugate imaginary roots: $e^{\pm i\varphi_k}$ $(k = 1, 2, \ldots, m)$. Hence there are m real invariant planes, all mutually orthogonal. We denote the n Cartesian coordinates of E^n by $X_1, Y_1, \ldots, X_m, Y_m$ and choose our frame such that the invariant plane W_k corresponding to φ_k has the equations $X_i = Y_i = 0$, $i \neq k$. Hence the general rotation about O can be expressed in the canonical form

$$(4.1) \qquad \begin{aligned} X'_k &= X_k \cos\varphi_k - Y_k \sin\varphi_k, \\ Y'_k &= X_k \sin\varphi_k + Y_k \cos\varphi_k, \quad k = 1, \ldots, m, \end{aligned}$$

that means: *the general rotation is built up and determined by m planar rotations in m mutually orthogonal planes.*

If n is odd, $n = 2m + 1$, we have one real eigenvalue $\lambda = 1$, and m pairs of conjugate imaginary $e^{\pm i\varphi_k}$, $k = 1, \ldots, m$. The first corresponds to a real invariant line l, the others to m real invariant planes W_k. We denote the coordinates by $X_1, Y_1, \ldots, X_m, Y_m, Z$, give l the equations $X_i = Y_i = 0$ and W_k the equations $X_i = Y_i = Z = 0$ $(i \neq k)$. The rotation is then given by

$$(4.2) \qquad \begin{aligned} X'_k &= X_k \cos\varphi_k - Y_k \sin\varphi_k, \\ Y'_k &= X_k \sin\varphi_k + Y_k \cos\varphi_k, \quad k = 1, \ldots, m, \\ Z' &= Z. \end{aligned}$$

For n even there is in general no fixed point different from O, for n odd we have always at least a line l of fixed points, the "axis" of the rotation. In each $(n-1)$-dimensional space orthogonal to l and intersecting it at O' a general even dimensional rotation takes place about O'.

We do not want to discuss systematically the special cases which occur if the eigenvalues are not all different. In the canonical representation some φ_k may be the same without changing the rotation essentially; the most special case in this respect is that for which all φ_k are equal, the equal-angled rotation.

Other special cases occur if $\lambda = 1$ or $\lambda = -1$ is a root for $n = 2m$; in both cases it is a double root. The consequence is the same as putting $\varphi_k = 0$ or $\varphi_k = \pi$. In the first case all points of the corresponding plane W_k are fixed points, in the second we have a half-turn in W_k. If for instance $\varphi_1 = \varphi_2 = 0$, all points of the 4-space $X_i = Y_i = 0$ ($i \neq 1$, $i \neq 2$) are fixed points. Similar special cases appear for odd values of n.

We have seen earlier that a displacement can always be considered as the product of a rotation about a point O, chosen arbitrarily, and a translation. If O is replaced by a point O' and the frame at the latter is taken parallel to that at O, we see at once that the rotation part in both cases is essentially the same; therefore the coefficients c_i, the roots λ_k, and the angles φ_k are characteristic for the displacement. This means that a displacement can be expressed analytically by

$$\begin{aligned} X'_k &= X_k \cos\varphi_k - Y_k \sin\varphi_k + a_k \\ Y'_k &= X_k \sin\varphi_k + Y_k \cos\varphi_k + b_k \end{aligned} \tag{4.3}$$

$k = 1, 2, \ldots, m$, for $n = 2m$. Here a_k, b_k are the components of the translation vector. For $n = 2m + 1$ the representation reads

$$\begin{aligned} X'_k &= X_k \cos\varphi_k - Y_k \sin\varphi_k + a_k \\ Y'_k &= X_k \sin\varphi_k + Y_k \cos\varphi_k + b_k \\ Z' &= Z + c_0 \end{aligned} \tag{4.4}$$

$k = 1, \ldots, m$; a_k, b_k, c_0, the components of the translation vector, vary with the choice of origin. We have for both cases

$$\boldsymbol{P}' = \mathbf{A}\boldsymbol{P} + \boldsymbol{d} \tag{4.5}$$

$\mathbf{A}$ represents an orthogonal matrix (or more precisely: stands for a class of equivalent orthogonal matrices); $\boldsymbol{P}'$, $\boldsymbol{P}$ and $\boldsymbol{d}$ are column matrices.

The question whether a general displacement has fixed points is now seen to be answered differently for $n = 2m$ and for $n = 2m + 1$. In the first case the determinant of the n linear equations $X'_k = X_k$, $Y'_k = Y_k$ is unequal to zero (because $\lambda = 1$ is *not* an eigenvalue) so these equations have a unique solution: in E^{2m} a displacement has in general *one* fixed point and it is therefore a mere rotation; hence (4.1) is the standard representation for this case. For $n = 2m + 1$ the determinant arising from $X'_k = X_k$, $Y'_k = Y_k$, $Z' = Z$ is zero and no solution exists: in E^{2m+1} a displacement has in general no fixed point. *There is, however, an invariant direction* l. Hence any $(n-1)$-space perpendicular to it is displaced parallel to itself. But such a space has an even number of dimensions and its own displacement has an invariant point. We take this to be $X_k = Y_k = 0$ $(k = 1, \dots, m)$. Hence the canonic form of a general displacement in E^{2m+1} reads

(4.6) $\quad X'_k = X_k \cos\varphi_k - Y_k \sin\varphi_k, \quad Y'_k = X_k \sin\varphi_k + Y_k \cos\varphi_k, \quad Z' = Z + c_0$

and the displacement is seen to be *a (generalized) screw displacement*, built up by a rotation about axis $\mathrm{l}(X_k = Y_k = 0)$ and a translation parallel to l.

5. Cayley's formula for an orthogonal matrix

A rotation about the origin O is given by

(5.1) $$\boldsymbol{P}' = \mathbf{A}\boldsymbol{P}$$

$\mathbf{A}$ being an orthogonal matrix. As all distances are invariant for a displacement we have $(\overrightarrow{OP})^2 = (\overrightarrow{OP'})^2$, hence

$$\boldsymbol{P}' \cdot \boldsymbol{P}' - \boldsymbol{P} \cdot \boldsymbol{P} = 0$$

or

(5.2) $$(\boldsymbol{P}' - \boldsymbol{P}) \cdot (\boldsymbol{P}' + \boldsymbol{P}) = 0$$

for any $\boldsymbol{P}$. This means that $\boldsymbol{f} = \boldsymbol{P}' - \boldsymbol{P}$ and $\boldsymbol{g} = \boldsymbol{P}' + \boldsymbol{P}$ are orthogonal vectors. We have (with $\boldsymbol{f}$, $\boldsymbol{g}$, and $\boldsymbol{P}$ in column matrix form):

(5.3) $$\boldsymbol{f} = (\mathbf{A} - \mathbf{I})\boldsymbol{P}, \quad \boldsymbol{g} = (\mathbf{A} + \mathbf{I})\boldsymbol{P}, \quad \boldsymbol{f} \cdot \boldsymbol{g} = 0.$$

We exclude the special case that -1 is an eigenvalue of $\mathbf{A}$: then $\mathbf{A} + \mathbf{I}$ is a non-singular matrix and we have

$$\boldsymbol{P} = (\mathbf{A} + \mathbf{I})^{-1}\boldsymbol{g},$$

and therefore

$$\boldsymbol{f} = (\mathbf{A} - \mathbf{I})(\mathbf{A} + \mathbf{I})^{-1}\boldsymbol{g}.$$

Putting

(5.4) $$(\mathbf{A} - \mathbf{I})(\mathbf{A} + \mathbf{I})^{-1} = \mathbf{B}$$

it follows that

(5.5) $$f = \mathbf{B}g.$$

If $\mathbf{B} = \| b_{ik} \|$ and if g_i are the components of g then $f \cdot g = 0$ implies

$$\sum_{i,k} (b_{ik} + b_{ki}) g_i g_k = 0$$

for any vector g. The conclusion is that $b_{ik} + b_{ki} = 0$ for all i, k. *Hence* $\mathbf{B}$ *is a skew matrix.* From (5.4) it follows

$$\mathbf{A} - \mathbf{I} = \mathbf{B}(\mathbf{A} + \mathbf{I})$$

or

(5.6) $$(\mathbf{I} - \mathbf{B})\mathbf{A} = \mathbf{I} + \mathbf{B}.$$

It is well-known that if $\mathbf{B}$ is a real skew matrix we have $|\mathbf{B}| \geq 0$, hence $|\mathbf{B} + \lambda \mathbf{I}|$ is a polynomial in λ with non-negative coefficients and thus with no real zero's different from $\lambda = 0$. This implies that $|\mathbf{B} - \mathbf{I}| \neq 0$. The conclusion is that *any orthogonal matrix*, for which -1 is not an eigenvalue, *may be written as*

(5.7) $$\mathbf{A} = (\mathbf{I} - \mathbf{B})^{-1}(\mathbf{I} + \mathbf{B})$$

in which $\mathbf{B}$ *is skew*; (5.7) is Cayley's formula.

On the other hand, if in (5.7) $\mathbf{B}$ is any skew matrix we have $(\mathbf{I} - \mathbf{B})\mathbf{A} = \mathbf{I} + \mathbf{B}$ and therefore $\mathbf{A}^T(\mathbf{I} - \mathbf{B})^T = (\mathbf{I} + \mathbf{B})^T$, or $\mathbf{A}^T(\mathbf{I} + \mathbf{B}) = \mathbf{I} - \mathbf{B}$; hence $\mathbf{A}\mathbf{A}^T = (\mathbf{I} - \mathbf{B})^{-1}(\mathbf{I} + \mathbf{B})(\mathbf{I} - \mathbf{B})(\mathbf{I} + \mathbf{B})^{-1}$ or, as the second and third factor may be interchanged, $\mathbf{A}\mathbf{A}^T = \mathbf{I}$ which implies that $\mathbf{A}$ is an orthogonal matrix. The set of *orthogonal matrices*, (for which -1 is not an eigenvalue) *is identical with the set* $(\mathbf{I} - \mathbf{B})^{-1}(\mathbf{I} + \mathbf{B})$, $\mathbf{B}$ *being skew.*

Cayley has shown that the orthogonal matrices with an eigenvalue -1 may be derived from (5.7) by a limit procedure.

We have found that there exists a $(1, 1)$ correspondence between orthogonal matrices and skew matrices. If $\mathbf{B} = \| b_{ik} \|$ is skew, $b_{ii} = 0$ and $b_{ki} = -b_{ik}$ for $i > k$. Hence $\mathbf{B}$ is determined if we know the elements b_{ik} for $i > k$, that is, all elements to one side of the main diagonal. Their number is $\frac{1}{2}n(n-1)$. If $\mathbf{B}$ is known $\mathbf{A}$ may be found by formula (5.7). Hence the theorem: *any orthogonal matrix in* E^n *may be written as a function of* $\frac{1}{2}n(n-1)$ *parameters.*

Example 2. Show that for $n = 2m$ and for $n = 2m + 1$ the elements of an orthogonal matrix $\mathbf{A}$ in E^n are rational functions of degree $2m$ of the parameters.

Example 3. Consider the case $n = 2$. If $\mathbf{B} = \begin{Vmatrix} 0 & -a \\ a & 0 \end{Vmatrix}$, show that $\mathbf{A} = \Delta^{-1}\begin{Vmatrix} 1-a^2 & -2a \\ 2a & 1-a^2 \end{Vmatrix}$, with $\Delta = 1 + a^2$. Determine $\mathbf{A}$ if $a = \tan(\varphi/2)$. Show that the exceptional case follows from $a \to \infty$ and that it represents a half-turn.

Example 4. Consider the case $n = 3$. If

$$\mathbf{B} = \begin{Vmatrix} 0 & -b_3 & b_2 \\ b_3 & 0 & -b_1 \\ -b_2 & b_1 & 0 \end{Vmatrix}$$

derive by (5.7) the matrix $\mathbf{A}$. Show that by introducing the homogeneous parameters c_i so that $b_i = c_i/c_0$ we obtain

$$\mathbf{A} = \Delta^{-1}\begin{Vmatrix} c_0^2 + c_1^2 - c_2^2 - c_3^2 & 2(-c_0c_3 + c_1c_2) & 2(c_0c_2 + c_1c_3) \\ 2(c_0c_3 + c_2c_1) & c_0^2 - c_1^2 + c_2^2 - c_3^2 & 2(-c_0c_1 + c_2c_3) \\ 2(-c_0c_2 + c_3c_1) & 2(c_0c_1 + c_3c_2) & c_0^2 - c_1^2 - c_2^2 + c_3^2 \end{Vmatrix},$$

with $\Delta = c_0^2 + c_1^2 + c_2^2 + c_3^2$

6. A displacement as the product of two reflections

We shall prove that a displacement in E^n can always be written as the product of two displacements of a simple type, called reflections. Again we must distinguish between the cases n is even and n is odd.

Let $n = 2m$. We consider in E^{2m} a linear subspace V of dimension m and derive from it a displacement by the following procedure. If P is any point, Q its orthogonal projection on V, P' the point on the extended line $\overrightarrow{PQ}$ such that $\overrightarrow{QP'} = \overrightarrow{PQ}$, then $(P \to P')$ is a displacement. That is easily verified by elementary geometry. The displacement is called the *reflection* into V.

If X_i, Y_i $(i = 1, \dots, m)$ are Cartesian coordinates in Σ^n, V can be given the equations $Y_i = 0$ for all i, and then the reflection is expressed by

$$(6.1) \qquad X_i' = X_i, \quad Y_k' = -Y_k \quad (i, k = 1, \dots, m);$$

it is a direct displacement if m is even, indirect if m is odd. The square of a reflection is the unity displacement. We consider two m-spaces V_1 and V_2 in E^{2m}; $\mathfrak{R}_i$ is the reflection with respect to V_i. $\mathrm{D} = \mathfrak{R}_2\mathfrak{R}_1$ is clearly a displacement and moreover direct. In the general case V_1 and V_2 have one (finite) point of intersection O; it is clear that O is a fixed point of D: *the product of two reflections is a rotation.* There are several special cases: the intersection of V_1 and V_2, may be a k-dimensional subspace $(k = 1, 2, \dots, m-1)$. Another case is that for which they have their point of intersection at infinity: V_1 and V_2 are parallel and D is a translation; the direction of which is parallel to the E^m absolutely orthogonal to both V_1 and V_2, and the magnitude is determined by the relative position of V_1 and V_2.

In the general case, when V_1 and V_2 intersect at one point O, n-dimensional Euclidean geometry teaches that they give rise to m, generally different, angles α_k describing their mutual position. We state without proof that the m rotation angles φ_i of the rotation D satisfy the conditions $\varphi_i = 2\alpha_i$. It may be shown moreover that any rotation about O is the product of two reflections with respect to suitably chosen m dimensional spaces V_1 and V_2 through O, and, moreover, in ∞^m ways.

In E^{2m+1} we have similar theorems. Again a reflection takes place with respect to an m-space. If V_1 and V_2 are two m-spaces, $\mathfrak{R}_i$ the reflection in V_i then $\mathrm{D} = \mathfrak{R}_2\mathfrak{R}_1$ is a (direct) displacement, but as V_1 and V_2 have in general no intersection, it is not a rotation. Here again it may be shown that a general displacement may be written (in ∞^{m+1} ways) as the product of two reflections. We return to these theorems later on for $n = 2$ and $n = 3$, for which cases important consequences may be derived.

7. Indirect displacements

We consider the displacement of E^n:

$$P^* = \mathbf{A}P + d \tag{7.1}$$

where $\mathbf{A}$ is an orthogonal matrix with determinant -1; let $\mathbf{A}$ be general (all eigenvalues mutually different). It is easily verified that for all values of n, either even or odd, one of the eigenvalues is -1. If the point P is displaced to P^* their midpoint M satisfies

$$M = \tfrac{1}{2}(P^* + P) = \tfrac{1}{2}(\mathbf{A} + \mathbf{I})P + \tfrac{1}{2}d \tag{7.2}$$

but -1 being an eigenvalue of $\mathbf{A}$ the determinant $|\mathbf{A} + \mathbf{I}|$ is equal to zero. Hence for any P the coordinates of M satisfy the same linear equation. This implies that all mid-points M lie in a Σ^{n-1}, the subspace of E^n coincident with Σ^{n-1} is denoted by U^{n-1}. Obviously, if P is in U^{n-1} the same holds for P^*; hence U^{n-1} is an invariant $(n-1)$-space for the displacement. Without any loss of generality we may suppose the equation of U^{n-1} to be $X_n = 0$. Then we have $X'_n = -X_n$ and the displacement of U^{n-1} in itself is seen to be one with an orthogonal matrix with determinant 1 and therefore a direct displacement. The conclusion is: *the displacement* (7.1) *is the product* (with arbitrary order of its factors) *of the reflection into* U^{n-1} *and a direct displacement of* E^n which is essentially a $(n-1)$-dimensional displacement induced by that in U^{n-1}. A $(n-1)$-space parallel to U^{n-1} is reflected into U^{n-1}.

For further discussion the parity of n is of interest.

Example 5. Show that for $n = 2m$ one of the eigenvalues of $\mathbf{A}$ is 1; for $n = 2m + 1$ those different from -1 are imaginary.

Example 6. Investigate the cases $n = 2$ and $n = 3$; determine the invariant points; show that for $n = 2$ a standard representation for an indirect displacement is $X_1^* = X_1 + d$, $X_2^* = -X_2$ and for $n = 3$: $X_1^* = X_1 \cos\varphi - X_2 \sin\varphi$, $X_2^* = X_1 \sin\varphi + X_2 \cos\varphi$, $X_3^* = -X_3$.

8. Transformations and operators

The equation for a general displacement (4.5) is of fundamental importance in kinematics, and as such will appear repeatedly in this book. The basic interpretation we use is:

$$\boldsymbol{P}' = \mathbf{A}\boldsymbol{P} + \boldsymbol{d} \tag{8.1}$$

represents two operators $\mathbf{A}$ and $\boldsymbol{d}$ which change the n-dimensional position vector $\boldsymbol{P}$ into the n-dimensional position vector $\boldsymbol{P}'$. We take the point P in a system E^n and measure all quantities relative to a reference system in Σ^n. All quantities in (8.1) are measured in a coordinate system fixed in Σ^n. However, in many applications it will be more useful to measure $\boldsymbol{P}$ in E^n. To accomplish this we select parallel coordinate systems in E^n and Σ^n with origins o and O respectively. We substitute $\boldsymbol{P} = \boldsymbol{p} + \overrightarrow{Oo}$, and let $\boldsymbol{d}$ now represent the total displacement of point o (that is, $\boldsymbol{d} + \mathbf{A}(\overrightarrow{Oo}) \rightarrow \boldsymbol{d}$). The resulting equation can be written as follows

$$\boldsymbol{P} = \mathbf{A}\boldsymbol{p} + \boldsymbol{d}, \tag{8.2}$$

with the understanding that here $\boldsymbol{p}$ is measured in E^n while all other quantities are still measured in Σ^n. (We have dropped the prime from the left-hand-side, but here $\boldsymbol{P}$ has the same meaning as $\boldsymbol{P}'$ does in (8.1).

(8.2) suggests another interpretation for the operators $\mathbf{A}$ and $\boldsymbol{d}$: If we have a vector $\boldsymbol{p}$ measured in an arbitrary system E^n, then $\mathbf{A}$ and $\boldsymbol{d}$ represent the elements of the coordinate transformation which transform the coordinate system of E^n to the coordinate system of Σ^n. The only difference associated with the displacement or the coordinate transformation interpretation has to do with the algebraic sign associated with the sense of the rotation and translation. In this book we will henceforth interpret $\mathbf{A}$ and $\boldsymbol{d}$ as displacement operators unless explicitly stated otherwise.

Example 7. Consider a two dimensional system and show by elementary geometry that if

$$\mathbf{A} = \begin{Vmatrix} \cos\phi & -\sin\phi \\ \sin\phi & \cos\phi \end{Vmatrix} \quad \text{and} \quad \boldsymbol{d} = \begin{Vmatrix} a \\ b \end{Vmatrix}$$

represent the displacement of the plane E when it is rotated by an angle ϕ *counter-clockwise* about the origin of the system and then translated a, b, the coordinate transformation interpretation for this physical situation would imply that the coordinate axes of E^2 have been translated $-a$, $-b$ and rotated *clockwise* by an angle ϕ.

We will often deal with a series of displacements, i.e., we will start with E^n in one position say E_1^n and then displace it to another, say E_2^n, and then say E_3^n, all relative to Σ^n. To deal with this situation we will employ the nomenclature, for the displacement from position i to position j:

$$\boldsymbol{P}_j = \mathbf{A}_{ij}\boldsymbol{P}_i + \boldsymbol{d}_{ij} \tag{8.3}$$

where all quantities are measured in Σ^n.

It is frequently convenient to define the operators $\mathbf{A}_{ij}$ and $\boldsymbol{d}_{ij}$ relative to a secondary reference system Σ_k^n. The result might be inferred from our discussion in Section 3, however we will develop it here in its more general context. We can rewrite (8.3) as

$$\boldsymbol{P}_j = \mathbf{D}_{ij}\boldsymbol{P}_i \tag{8.4}$$

if we use homogeneous coordinates for $\boldsymbol{P}_j$ and $\boldsymbol{P}_i$ and define $\mathbf{D}_{ij}$ such that

$$\mathbf{D}_{ij} = \left\Vert \begin{array}{cccc|c} & \mathbf{A}_{ij} & & & \boldsymbol{d}_{ij} \\ \hline 0 & 0 & 0 \cdots & 0 & 1 \end{array} \right\Vert \tag{8.5}$$

If the system Σ^n is itself displaced by $\mathbf{D}_{0k}$ relative to another reference system Σ_k^n, we have from (8.4)

$$\mathbf{D}_{0k}\boldsymbol{P}_j = \mathbf{D}_{0k}(\mathbf{D}_{ij}\boldsymbol{P}_i) = \mathbf{D}_{0k}\mathbf{D}_{ij}\boldsymbol{P}_i$$

and this can be written as

$$\mathbf{D}_{0k}\boldsymbol{P}_j = (\mathbf{D}_{0k}\mathbf{D}_{ij}\mathbf{D}_{0k}^{-1})(\mathbf{D}_{0k}\boldsymbol{P}_i). \tag{8.6}$$

Hence if we set

$${}_k\mathbf{D}_{ij} = \mathbf{D}_{0k}\mathbf{D}_{ij}\mathbf{D}_{0k}^{-1} \tag{8.7}$$

we have

$${}_k\boldsymbol{P}_j = {}_k\mathbf{D}_{ik}({}_k\boldsymbol{P}_i). \tag{8.8}$$

That is to say (8.4) may be written in any system, say Σ_k^n, provided the operator $\mathbf{D}_{ij}$ defined in Σ^n is replaced by ${}_k\mathbf{D}_{ij}$ computed from (8.7). (A transformation of the form (8.7) is known as a *similarity transformation.*) It should be noted that we have used only the displacement operator interpretation of $\mathbf{D}_{0k}$ in this development.

Example 8. Show that if E_1^n is displaced by $\mathbf{D}_{12}$ to E_2^n and then by $\mathbf{D}_{23}$ to E_3^n, the same final position is obtained if E_1^n is first displaced by $\mathbf{D}_{23}$ and then by $\mathbf{D}_{23}\mathbf{D}_{12}\mathbf{D}_{23}^{-1}$. That is, if $\mathbf{D}_{12}$ is itself "displaced" by $\mathbf{D}_{23}$ the displacements commute.

CHAPTER II

INSTANTANEOUS KINEMATICS

1. Definition of a motion

We have seen that a general displacement in n-space may be given analytically by

$$P = \mathbf{A}p + d \tag{1.1}$$

in which $\boldsymbol{P}$ and $\boldsymbol{p}$ are the position vectors, represented by column matrices, of a point P in the fixed space Σ^n and the moving space E^n respectively; $\mathbf{A}$ is an orthogonal matrix and $\boldsymbol{d}$ a translation vector. If $\mathbf{A}$ and $\boldsymbol{d}$ are functions of a parameter t, which may be identified with the time, (1.1) gives us a continuous series of displacements, called a *motion*. Any point P of the moving space E^n describes a curve, its *path*, in the fixed space Σ^n. We shall deal with the motion concept later on in detail, here we consider a special feature.

We consider a certain position of the moving space, given by $t = 0$ and restrict our study to the properties of the motion for the limit case $t \to 0$. For any function $\mathrm{F}(t)$ we shall denote $\mathrm{d}^k \mathrm{F}/\mathrm{d}t^k$ at $t = 0$ by F_k. We have then: $\boldsymbol{P}_0$ yields the position vector of the point P (in the zero position), $\boldsymbol{P}_1$ is its velocity vector (which implies the tangent to the path), $\boldsymbol{P}_k$ for $k = 2$ is its acceleration, from which follows the curvature of its path, and so on. As all theorems deal with one particular instant during the motion this subject is called *instantaneous kinematics*. It has developed into an important branch of our science. Of course it is easier to treat than the kinematics of motion proper. Moreover it gives rise to attractive results, which have furthered our insight into general motion, to which it is an approximation comparable with the Taylor expansion of a function of one variable. We obtain information about a motion, at least during a certain period, if we know its characteristics for one instant. The value of the information is of course dependent on how much instantaneous knowledge is available, or to be more exact, to which order the derivatives of $\boldsymbol{P}$ are known. If $\boldsymbol{P}_k$ are given for $k = 0, 1, 2, \ldots, m$, our subject is instantaneous kinematics of the mth order.

2. The Taylor expansion of an orthogonal matrix

$\mathbf{A}$ and $\boldsymbol{d}$ being functions of t means that the n^2 elements of $\mathbf{A}$ and the n elements of $\boldsymbol{d}$ are functions of t in the usual sense. Without any loss of generality we may suppose that for $t = 0$ the origins in E^n and Σ^n coincide, so that $\mathbf{A}_0 = \mathbf{I}$, $\boldsymbol{d}_0 = 0$. The elements of $\mathbf{A}$ may be expanded in Taylor series and this gives rise to the following expansion

$$\mathbf{A}(t) = \mathbf{I} + \mathbf{A}_1 t + \frac{1}{2!}\mathbf{A}_2 t^2 + \frac{1}{3!}\mathbf{A}_3 t^3 + \cdots \tag{2.1}$$

in which $\mathbf{A}_k$ $(k = 1, 2, \ldots)$ is a constant matrix. $\mathbf{A}(t)$ is orthogonal, which means that for all values of t we must have

$$\mathbf{A}\mathbf{A}^{\mathrm{T}} = \mathbf{I}$$

or

$$\left(\mathbf{I} + \mathbf{A}_1 t + \frac{1}{2!}\mathbf{A}_2 t^2 + \cdots\right)\left(\mathbf{I} + \mathbf{A}_1^{\mathrm{T}} t + \frac{1}{2}\mathbf{A}_2^{\mathrm{T}} t^2 + \cdots\right) = \mathbf{I} \tag{2.2}$$

for all t. Hence on the left-hand-side the coefficient of t^k must vanish for $k = 1, 2, \ldots$. This gives us

$$\mathbf{A}_1 + \mathbf{A}_1^{\mathrm{T}} = 0$$

$$\mathbf{A}_2 + 2\mathbf{A}_1\mathbf{A}_1^{\mathrm{T}} + \mathbf{A}_2^{\mathrm{T}} = 0$$

$$\mathbf{A}_3 + 3\mathbf{A}_2\mathbf{A}_1^{\mathrm{T}} + 3\mathbf{A}_1\mathbf{A}_2^{\mathrm{T}} + \mathbf{A}_3^{\mathrm{T}} = 0$$

and in general

$$\sum_{i=0}^{k} \binom{k}{i} \mathbf{A}_{k-i}\, \mathbf{A}_i^{\mathrm{T}} = 0, \quad k = 1, 2, \ldots \tag{2.3}$$

with $\mathbf{A}_0 = \mathbf{A}_0^{\mathrm{T}} = \mathbf{I}$.

From the first equation it follows that $\mathbf{A}_1$ is *skew*; we write $\mathbf{A}_1 = \mathbf{B}_1$ and therefore $\mathbf{A}_1^{\mathrm{T}} = -\mathbf{B}_1$, $\mathbf{B}_1$ denoting an (arbitrary) skew matrix. Then we have $\mathbf{A}_1\mathbf{A}_1^{\mathrm{T}} = -\mathbf{C}_2 = -\mathbf{B}_1^2$, $\mathbf{C}_2$ being *symmetric*. The second equation leads to $\mathbf{A}_2 + \mathbf{A}_2^{\mathrm{T}} = 2\mathbf{C}_2$; hence if $\mathbf{A}_2 = \mathbf{C}_2 + \mathbf{B}_2$ we have $\mathbf{A}_2^{\mathrm{T}} = \mathbf{C}_2 + \mathbf{B}_2^{\mathrm{T}}$ and therefore $\mathbf{B}_2 + \mathbf{B}_2^{\mathrm{T}} = 0$, which means that $\mathbf{B}_2$ must be *skew*. The next step gives $\mathbf{A}_3 + \mathbf{A}_3^{\mathrm{T}} = 3(\mathbf{B}_2\mathbf{B}_1 + \mathbf{B}_1\mathbf{B}_2) = 2\mathbf{C}_3$, $\mathbf{C}_3$ being *symmetric*. Therefore $\mathbf{A}_3 = \mathbf{C}_3 + \mathbf{B}_3$, $\mathbf{A}_3^{\mathrm{T}} = \mathbf{C}_3 + \mathbf{B}_3^{\mathrm{T}}$ and $\mathbf{B}_3$ is *skew*.

By continuing the procedure we arrive (by induction) at the following result: *the expansion of an orthogonal matrix* $\mathbf{A}(t)$ *in the neighborhood of the unit matrix reads*

$$\mathbf{A}(t) = \mathbf{I} + (\mathbf{C}_1 + \mathbf{B}_1)t + \frac{1}{2!}(\mathbf{C}_2 + \mathbf{B}_2)t^2 + \frac{1}{3!}(\mathbf{C}_3 + \mathbf{B}_3)t^3 + \cdots \tag{2.4}$$

in which $\mathbf{B}_k$ $(k = 1, 2, \ldots)$ is an arbitrary *skew* matrix and $\mathbf{C}_k$ $(k = 1, 2, \ldots)$ is a *symmetric* matrix which depends on $\mathbf{B}_1, \mathbf{B}_2, \ldots, \mathbf{B}_{k-1}$ and is given by the recursion formula

$$\mathbf{C}_k = \frac{-1}{2} \sum_{i=1}^{k-1} \binom{k}{i} (\mathbf{C}_{k-i} + \mathbf{B}_{k-i})(\mathbf{C}_i - \mathbf{B}_i), \tag{2.5}$$

with $\mathbf{C}_1 = 0$. The expansion up to the mth order depends on the m skew matrices $\mathbf{B}_1, \mathbf{B}_2, \ldots, \mathbf{B}_m$.

Example 1. Show that $\mathbf{C}_2 = \mathbf{B}_1^2$, $\mathbf{C}_3 = \frac{3}{2}(\mathbf{B}_1\mathbf{B}_2 + \mathbf{B}_2\mathbf{B}_1)$, $\mathbf{C}_4 = -3\mathbf{B}_1^4 + 3\mathbf{B}_2^2 + 2(\mathbf{B}_1\mathbf{B}_3 + \mathbf{B}_3\mathbf{B}_1)$.
Example 2. If for $n = 2$ we have $\mathbf{B}_k = \left\| \begin{smallmatrix} 0 & -b_k \\ b_k & 0 \end{smallmatrix} \right\|$, determine $\mathbf{C}_2, \mathbf{C}_3, \mathbf{C}_4, \mathbf{C}_5$.

In view of the fundamental property $\mathbf{A}\mathbf{A}^{\mathrm{T}} = \mathbf{I}$, the expansion of the *inverse* matrix follows at once from (2.4):

$$\mathbf{A}^{-1}(t) = \mathbf{I} + (\mathbf{C}_1 - \mathbf{B}_1)t + \frac{1}{2!}(\mathbf{C}_2 - \mathbf{B}_2)t^2 + \cdots \tag{2.6}$$

Therefore: if $\mathbf{B}_k$ are the skew matrices determining a motion those of the inverse motion are $\tilde{\mathbf{B}}_k = -\mathbf{B}_k$.

So far we have only dealt with the rotation part of the motion (1.1). The translation part is much simpler for it is fully determined by the vector $\boldsymbol{d}(t)$. If the origins in E^n and Σ^n coincide and therefore $\boldsymbol{d}_0 = 0$ the expansion for a general motion is seen to be

$$\boldsymbol{P} = \boldsymbol{p} + \{\mathbf{B}_1\boldsymbol{p} + \boldsymbol{d}_1\}t + \tfrac{1}{2}\{(\mathbf{C}_2 + \mathbf{B}_2)\boldsymbol{p} + \boldsymbol{d}_2\}t^2 + \tfrac{1}{6}\{(\mathbf{C}_3 + \mathbf{B}_3)\boldsymbol{p} + \boldsymbol{d}_3\}t^3 + \cdots \tag{2.7}$$

and we have in particular

$$\boldsymbol{P}_k = (\mathbf{C}_k + \mathbf{B}_k)\boldsymbol{p} + \boldsymbol{d}_k \tag{2.8}$$

which for $k = 1$ gives us the velocity, for $k = 2$ the acceleration and for $k = m$ the $(m-1)^{\text{th}}$ order acceleration of any point P of E^n in the zero position.

As a counterpart of (2.7) we can also derive the expansion for the inverse motion. If we solve for $\boldsymbol{p}$ from (1.1) we obtain $\boldsymbol{p} = \mathbf{A}^{-1}(\boldsymbol{P} - \boldsymbol{d})$ and therefore for $\boldsymbol{P} = 0$:

$$\boldsymbol{p} = -\{\mathbf{I} - \mathbf{B}_1 t + \tfrac{1}{2}(\mathbf{C}_2 - \mathbf{B}_2)t^2 + \cdots\}\{\boldsymbol{d}_1 t + \tfrac{1}{2}\boldsymbol{d}_2 t^2 + \cdots\},$$

from which it follows, using the tilda ($\tilde{}$) to denote the inverse motion,

$$\tilde{\boldsymbol{d}}_0 = 0, \quad \tilde{\boldsymbol{d}}_1 = -\boldsymbol{d}_1, \quad \tilde{\boldsymbol{d}}_2 = -\boldsymbol{d}_2 + 2\mathbf{B}_1\boldsymbol{d}_1, \quad \tilde{\boldsymbol{d}}_3 = -\boldsymbol{d}_3 + 3\mathbf{B}_1\boldsymbol{d}_2 - 3(\mathbf{C}_2 - \mathbf{B}_2)\boldsymbol{d}_1$$

and in general

$$\tilde{d}_k = -\sum_{i=0}^{k-1} \binom{k}{i} (\mathbf{C}_i - \mathbf{B}_i) \boldsymbol{d}_{k-i}. \tag{2.9}$$

3. The angular velocity matrix

Although (2.7) gives us a complete account of the behavior of the moving space in the neighborhood of $t = 0$, the apparatus developed so far is not always the most suitable for the study of instantaneous kinematics. A different approach shall be given now. We restrict ourselves for the time being to rotational motion, given by

$$\boldsymbol{P} = \mathbf{A}(t)\boldsymbol{p}. \tag{3.1}$$

From this follows the velocity vector of a point P of the moving space:

$$\dot{\boldsymbol{P}} = \dot{\mathbf{A}}(t)\boldsymbol{p}. \tag{3.2}$$

In the next step, which is essential for the new method, we eliminate $\boldsymbol{p}$ from (3.1) and (3.2) and use instead the position vector $\boldsymbol{P}$, of the point, in the fixed space. We obtain

$$\dot{\boldsymbol{P}} = \dot{\mathbf{A}}\mathbf{A}^{-1}\boldsymbol{P}. \tag{3.3}$$

(3.2) may be considered as the velocity as experienced by an observer who belongs to the moving space E^n, while (3.3) is that registered by a person at rest in the fixed space Σ^n who sees the moving point passing. The two concepts are well-known in continuum mechanics where they are denoted, respectively, as the Lagrangian and the Eulerian view-point. The matrix $\dot{\mathbf{A}}\mathbf{A}^{-1}$ which appears in (3.3) has an important property: Differentiating $\mathbf{A}\mathbf{A}^{\mathrm{T}} = \mathbf{I}$ we obtain

$$\dot{\mathbf{A}}\mathbf{A}^{\mathrm{T}} + \mathbf{A}(\mathbf{A}^{\mathrm{T}})^{\cdot} = 0. \tag{3.4}$$

In view of the general relation $(\mathbf{M}_1\mathbf{M}_2)^{\mathrm{T}} = \mathbf{M}_2^{\mathrm{T}}\mathbf{M}_1^{\mathrm{T}}$ and the obvious property $(\mathbf{M}^{\mathrm{T}})^{\cdot} = (\dot{\mathbf{M}})^{\mathrm{T}}$, the second term of (3.4) is equal to $\mathbf{A}(\mathbf{A}^{\mathrm{T}})^{\cdot} = (\mathbf{A}^{-1})^{\mathrm{T}}(\dot{\mathbf{A}})^{\mathrm{T}} = (\dot{\mathbf{A}}\mathbf{A}^{-1})^{\mathrm{T}}$; since the first term is $\dot{\mathbf{A}}\mathbf{A}^{-1}$ we conclude that this product is a *skew* matrix. We denote it by $\mathbf{\Omega}$, which means that (3.3) may be written

$$\dot{\boldsymbol{P}} = \mathbf{\Omega}(t)\boldsymbol{P} \tag{3.5}$$

where the skew matrix $\mathbf{\Omega}$ is defined by

$$\mathbf{\Omega}(t) = \dot{\mathbf{A}}(t)\mathbf{A}^{-1}(t). \tag{3.6}$$

This matrix shall play an important part in the theory; it is called the *angular velocity matrix.*

Example 3. If $\mathbf{A} = \|a_{ij}\|$ and $\mathbf{\Omega} = \|\Omega_{ij}\|$ show that $\Omega_{ij} = -\Omega_{ji} = \Sigma_{k=1}^{n} \dot{a}_{ik}a_{jk}$.

If instead of (3.1) we consider the general motion $\boldsymbol{P} = \mathbf{A}\boldsymbol{p} + \boldsymbol{d}$, we have $\dot{\boldsymbol{P}} = \dot{\mathbf{A}}\boldsymbol{p} + \dot{\boldsymbol{d}}$, or after eliminating $\boldsymbol{p}$:

$$\dot{\boldsymbol{P}} = \mathbf{\Omega}(\boldsymbol{P} - \boldsymbol{d}) + \dot{\boldsymbol{d}}. \tag{3.7}$$

This means that the combination $(\mathbf{\Omega}, \dot{\boldsymbol{d}})$, which consists of a skew matrix and a vector, determines by its values at a certain moment, the velocity of all points (i.e., the so-called velocity *distribution*) at that moment.

From (3.7) we may answer the question whether there are, at a certain instant, points with velocity zero. To discuss this problem we must distinguish two cases: n is even or n is odd.

If $n = 2m$ the skew matrix $\mathbf{\Omega}$ is (restricting ourselves to the general case) not singular; hence $\mathbf{\Omega}^{-1}$ exists. Then from (3.7) and $\dot{\boldsymbol{P}} = 0$ it follows that

$$\boldsymbol{P} = \boldsymbol{d} - \mathbf{\Omega}^{-1}\dot{\boldsymbol{d}}. \tag{3.8}$$

Hence in $2m$-space there is in general at any moment one point with zero velocity. It is called the *pole* at that moment or the *instantaneous rotation center.* The formula (3.8) gives its position in the fixed space. From $\boldsymbol{P} = \mathbf{A}\boldsymbol{p} + \boldsymbol{d}$ it follows that its position in the moving space is given by

$$\boldsymbol{p} = -(\dot{\mathbf{A}})^{-1}\dot{\boldsymbol{d}}. \tag{3.9}$$

If in (3.8) and (3.9) t is considered variable, these formulas represent the locus of the pole in the fixed and the moving space; these curves are called the *polhodes* or the *centrodes* of the motion.

We restrict ourselves to the zero position, take the origins in E^n and Σ^n coinciding (which implies $\boldsymbol{d}_0 = 0$) and located at the pole (which gives us by (3.8) or (3.9) that $\boldsymbol{d}_1 = 0$). Then from (3.7):

$$\boldsymbol{P}_1 = \mathbf{\Omega}_0\boldsymbol{P}_0 = \mathbf{\Omega}_0\boldsymbol{p} \tag{3.10}$$

represents the instantaneous velocity distribution with respect to the chosen origin, in $2m$-space. To write $\mathbf{\Omega}_0$ in a standard form we make use of the same Cartesian frame chosen in Chapter I (in connection with (4.1)). If φ_k are functions of t and if $\dot{\varphi}_k(0)$ is denoted by ω_k we obtain

$$\Omega_0 = \begin{Vmatrix} 0 & -\omega_1 & 0 & 0 & \cdot & 0 & 0 \\ \omega_1 & 0 & 0 & 0 & \cdot & 0 & 0 \\ 0 & 0 & 0 & -\omega_2 & \cdot & 0 & 0 \\ 0 & 0 & \omega_2 & 0 & \cdot & 0 & 0 \\ \cdot & \cdot & \cdot & \cdot & \cdot & \cdot & \cdot \\ 0 & \cdot & \cdot & \cdot & \cdot & 0 & -\omega_m \\ 0 & \cdot & \cdot & \cdot & \cdot & \omega_m & 0 \end{Vmatrix} \tag{3.11}$$

which expresses the velocity distribution by means of the angular velocities ω_k $(k = 1, 2, \ldots, m)$ of the m invariant planes through O.

In a space with an odd number of dimensions the situation is different and less symmetric. A skew $(2m+1) \times (2m+1)$ matrix $\boldsymbol{\Omega}$ is always singular. This implies that the equation $\dot{\boldsymbol{P}} = 0$ which follows from (3.7) has in general no solution: in $(2m+1)$-space there is in general no point with zero velocity. To discuss this problem we consider first a motion with $\boldsymbol{d} = 0$. Then $\dot{\boldsymbol{P}} = 0$ if $\boldsymbol{\Omega P} = 0$, $\boldsymbol{\Omega}$ being singular; the conclusion is (supposing that the rank of $\boldsymbol{\Omega}$ is not less than $2m$) that there exists a line l through the origin all of whose points have a zero velocity. This implies that any point not on l has a velocity vector perpendicular to l; in any E^{2m} perpendicular to l any point has a velocity vector lying in this space; the velocity distribution in such a space is of the type described before. The $2m$-space as a whole does not move. Now we add a translation part $\boldsymbol{d}$ as the motion of the origin. There are two cases: $\boldsymbol{d}_1$ has a non-zero component parallel to l, which is the general case, or $\boldsymbol{d}_1$ is perpendicular to l, this being a special case. In the general case the velocity of a point P has two components: one is the sum of the rotation and translation in the $2m$-space through P perpendicular to l and the second is parallel to l, the latter being the same for all points P. Furthermore, if we choose the origin at the pole of the $2m$-space the line l will pass through it, and the rotation and translation in the $2m$-space becomes simply a pure rotation.

Example 4. Derive from formula (4.2) in Chapter I a standard representation of the velocity distribution in a $2m+1$-space.

The theorems on the velocity distribution are obviously limit cases of those considered in Chapter I, where two *finitely* separated positions of a moving space were studied. In the situation at hand the two positions coincide but the direction of approach is known; they are said to be *infinitesimally* separated, sometimes also denoted by the term "two consecutive positions".

We have defined $\mathbf{\Omega}$ as $\dot{\mathbf{A}}\mathbf{A}^{-1}$. From (2.4) and (2.6) it follows

$$\dot{\mathbf{A}}(t) = \mathbf{B}_1 + (\mathbf{C}_2 + \mathbf{B}_2)t + \tfrac{1}{2}(\mathbf{C}_3 + \mathbf{B}_3)t^2 + \cdots$$

$$\mathbf{A}^{-1}(t) = \mathbf{I} - \mathbf{B}_1 t + \tfrac{1}{2}(\mathbf{C}_2 - \mathbf{B}_2)t^2 + \cdots \tag{3.12}$$

Hence the series

$$\mathbf{\Omega}(t) = \mathbf{\Omega}_0 + \mathbf{\Omega}_1 t + \tfrac{1}{2}\mathbf{\Omega}_2 t^2 + \cdots \tag{3.13}$$

has its coefficients given by

$$\mathbf{\Omega}_k = \sum_{i=1}^{k+1} \binom{k}{i-1} (\mathbf{C}_i + \mathbf{B}_i)(\mathbf{C}_{k-i+1} - \mathbf{B}_{k-i+1}). \tag{3.14}$$

These formulas give us the Taylor expansion of the angular velocity matrix, in terms of $\mathbf{B}_1, \mathbf{B}_2, \ldots$.

Example 5. Show that

$$\mathbf{\Omega}_0 = \mathbf{B}_1, \qquad \mathbf{\Omega}_1 = \mathbf{B}_2, \qquad \mathbf{\Omega}_2 = -\mathbf{B}_1^3 + \tfrac{1}{2}(\mathbf{B}_1\mathbf{B}_2 - \mathbf{B}_2\mathbf{B}_1) + \mathbf{B}_3,$$

$$\mathbf{\Omega}_3 = -\tfrac{3}{2}(\mathbf{B}_1^2\mathbf{B}_2 + \mathbf{B}_2\mathbf{B}_1^2) - 3\mathbf{B}_1\mathbf{B}_2\mathbf{B}_1 + (\mathbf{B}_1\mathbf{B}_3 - \mathbf{B}_3\mathbf{B}_1) + \mathbf{B}_4.$$

$\tilde{\mathbf{\Omega}}$, the angular velocity matrix of the inverse motion, is $(\mathbf{A}^{-1})\mathbf{A}$ and it is easy to see that it follows from $\mathbf{\Omega}$ by writing $-\mathbf{B}_k$ instead of $\mathbf{B}_k$ for any index k. Therefore, we obtain $\tilde{\mathbf{\Omega}}_k = \Sigma_{i=1}^{k+1}\binom{k}{i-1}(\mathbf{C}_i - \mathbf{B}_i)(\mathbf{C}_{k-i+1} + \mathbf{B}_{k-i+1})$ here $\mathbf{C}_i$ is the matrix analogous to (2.5). We have in particular

$$\tilde{\mathbf{\Omega}}_0 = -\mathbf{\Omega}_0, \qquad \tilde{\mathbf{\Omega}}_1 = -\mathbf{\Omega}_1, \qquad \tilde{\mathbf{\Omega}}_2 + \mathbf{\Omega}_2 = \mathbf{B}_1\mathbf{B}_2 - \mathbf{B}_2\mathbf{B}_1$$

$$\tilde{\mathbf{\Omega}}_3 + \mathbf{\Omega}_3 = 2(\mathbf{B}_1\mathbf{B}_3 - \mathbf{B}_3\mathbf{B}_1) \tag{3.15}$$

4. The case $n = 3$

All results derived so far are valid in a space with an arbitrary number of dimensions. For the remainder of this chapter, we shall confine ourselves to the case $n = 3$, called *spatial* kinematics (the case $n = 2$, planar kinematics, is an important subcase, which will as a matter of fact also be developed autonomously in Chapter VIII).

The restriction to three-dimensional space is of course in the first place justified by its relation to the physical world. It is remarkable, however, that the analytics contain a certain elegance which does not exist for higher values of n.

For $n = 3$ the number of independent elements of the skew matrix $\mathbf{\Omega}$ is equal to *three* and if we write it as follows

$$\mathbf{\Omega} = \begin{Vmatrix} 0 & -\Omega_Z & \Omega_Y \\ \Omega_Z & 0 & -\Omega_X \\ -\Omega_Y & \Omega_X & 0 \end{Vmatrix}, \tag{4.1}$$

we may fix a correspondence between $\mathbf{\Omega}$ and the vector with Cartesian components $(\Omega_X, \Omega_Y, \Omega_Z)$ which we shall denote by $\boldsymbol{\Omega}$. If (p_x, p_y, p_z) are the elements of the column vector $\boldsymbol{p}$ we have

$$\mathbf{\Omega}\boldsymbol{p} = \begin{Vmatrix} \Omega_Y p_z - \Omega_Z p_y \\ \Omega_Z p_x - \Omega_X p_z \\ \Omega_X p_y - \Omega_Y p_x \end{Vmatrix}.$$

Therefore we have an equivalence between the matrix multiplication and the vector product:

$$\mathbf{\Omega}\boldsymbol{p} = \boldsymbol{\Omega} \times \boldsymbol{p}, \tag{4.2}$$

and this makes it possible to treat spatial kinematics by means of vector algebra. For $n > 3$ this is not true: a skew matrix (with $\frac{1}{2}n(n-1)$ elements) does not correspond to a vector and we need instead tensor algebra. For $n = 2$, $\mathbf{\Omega}$ reduces essentially to a scalar.

The vector corresponding to the 3×3 skew matrix $\mathbf{B}_k$ will be denoted by $\boldsymbol{b}_k$.

We mention some rules for this equivalent vector. They can easily be verified by means of the well-known formula

$$\boldsymbol{v}_1 \times (\boldsymbol{v}_2 \times \boldsymbol{v}_3) = (\boldsymbol{v}_1 \cdot \boldsymbol{v}_3)\boldsymbol{v}_2 - (\boldsymbol{v}_1 \cdot \boldsymbol{v}_2)\boldsymbol{v}_3. \tag{4.3}$$

We have

$$\mathbf{B}_1\mathbf{B}_2\boldsymbol{p} = \boldsymbol{b}_1 \times (\boldsymbol{b}_2 \times \boldsymbol{p}) = (\boldsymbol{b}_1 \cdot \boldsymbol{p})\boldsymbol{b}_2 - (\boldsymbol{b}_1 \cdot \boldsymbol{b}_2)\boldsymbol{p}. \tag{4.4}$$

Example 6. Show that $(\mathbf{B}_1\mathbf{B}_2 - \mathbf{B}_2\mathbf{B}_1)\boldsymbol{p} = (\boldsymbol{b}_1 \times \boldsymbol{b}_2) \times \boldsymbol{p}$.
Example 7. Show that $\mathbf{B}^2\boldsymbol{p} = -\boldsymbol{b}^2\boldsymbol{p} + (\boldsymbol{b} \cdot \boldsymbol{p})\boldsymbol{b}$, $\mathbf{B}^3\boldsymbol{p} = -\boldsymbol{b}^2(\boldsymbol{b} \times \boldsymbol{p})$, and more generally $\mathbf{B}^{2k+1}\boldsymbol{p} = (-1)^k\boldsymbol{b}^{2k}(\boldsymbol{b} \times \boldsymbol{p})$, $\mathbf{B}^{2k}\boldsymbol{p} = (-1)^{k-1}\boldsymbol{b}^{2k-2}(-\boldsymbol{b}^2\boldsymbol{p} + (\boldsymbol{b} \cdot \boldsymbol{p})\boldsymbol{b})$.

Obviously the multiplication of a column matrix (i.e., a vector) by a matrix which is given as a polynomial of skew matrices can always be written by means of vector and scalar products. This is especially true for the matrices $\mathbf{C}_k$ of Section 2, which, as we have seen, are polynomials of $\mathbf{B}_1, \mathbf{B}_2, \ldots, \mathbf{B}_{k-1}$.

This implies that the expansion (2.7) may be written in terms of the arbitrary vectors $\boldsymbol{b}_k$ and $\boldsymbol{d}_k$. By means of (4.4), Example 1, and Example 7 we obtain

$$
(4.5)\qquad \begin{aligned} \boldsymbol{P} = \boldsymbol{p} + \{\boldsymbol{b}_1 \times \boldsymbol{p} + \boldsymbol{d}_1\}t + \tfrac{1}{2}\{-\boldsymbol{b}_1^2\boldsymbol{p} + (\boldsymbol{b}_1 \cdot \boldsymbol{p})\boldsymbol{b}_1 + \boldsymbol{b}_2 \times \boldsymbol{p} + \boldsymbol{d}_2\}t^2 \\ + \tfrac{1}{6}\{\tfrac{3}{2}(\boldsymbol{b}_1 \cdot \boldsymbol{p})\boldsymbol{b}_2 + \tfrac{3}{2}(\boldsymbol{b}_2 \cdot \boldsymbol{p})\boldsymbol{b}_1 - 3(\boldsymbol{b}_1 \cdot \boldsymbol{b}_2)\boldsymbol{p} + \boldsymbol{b}_3 \times \boldsymbol{p} + \boldsymbol{d}_3\}t^3 + \cdots \end{aligned}
$$

In the same way the series (3.13) may be converted:

$$
(4.6)\qquad \begin{aligned} \boldsymbol{\Omega} = \boldsymbol{b}_1 + \boldsymbol{b}_2 t + \tfrac{1}{2}\{\boldsymbol{b}_1^2\boldsymbol{b}_1 + \tfrac{1}{2}\boldsymbol{b}_1 \times \boldsymbol{b}_2 + \boldsymbol{b}_3\}t^2 \\ + \tfrac{1}{6}\{\tfrac{9}{2}(\boldsymbol{b}_1 \cdot \boldsymbol{b}_2)\boldsymbol{b}_1 + \tfrac{3}{2}\boldsymbol{b}_1^2\boldsymbol{b}_2 + \boldsymbol{b}_1 \times \boldsymbol{b}_3 + \boldsymbol{b}_4\}t^3 + \cdots . \end{aligned}
$$

This implies that $\boldsymbol{b}_1$ is the angular velocity vector at the zero position and $\boldsymbol{b}_2$ the angular acceleration; thus a kinematical interpretation of these two vectors has been derived.

We return now to (3.7). Obviously in three-dimensional space it may be written

$$
(4.7)\qquad \dot{\boldsymbol{P}} = \boldsymbol{\Omega} \times (\boldsymbol{P} - \boldsymbol{d}) + \dot{\boldsymbol{d}}.
$$

If for $t = 0$ the origins in E^3 and Σ^3 coincide, we have $\boldsymbol{P}_0 = \boldsymbol{p}$ and $\boldsymbol{d}_0 = 0$, and therefore

$$
(4.8)\qquad \boldsymbol{P}_1 = \boldsymbol{\Omega}_0 \times \boldsymbol{p} + \boldsymbol{d}_1.
$$

If instead of O we choose the common origin at O' such that $\overrightarrow{OO'} = \boldsymbol{e}$, (using primes to denote the new position vectors) we have $\boldsymbol{P} = \boldsymbol{P}' + \boldsymbol{e}$, $\boldsymbol{p} = \boldsymbol{p}' + \boldsymbol{e}$ and thus

$$
(4.9)\qquad \boldsymbol{P}_1' = \boldsymbol{\Omega}_0 \times \boldsymbol{p}' + \boldsymbol{d}_1 + \boldsymbol{\Omega}_0 \times \boldsymbol{e}
$$

which shows that $\boldsymbol{\Omega}_0$ is independent of the chosen origin. This is not true for the translation part: we have $\boldsymbol{d}_1' = \boldsymbol{d}_1 + \boldsymbol{\Omega}_0 \times \boldsymbol{e}$.

Formula (4.7) confirms that there is in general no point with zero velocity: since $\boldsymbol{\Omega} \times (\boldsymbol{P} - \boldsymbol{d})$ is orthogonal to $\boldsymbol{\Omega}$ this would require $\boldsymbol{\Omega}$ and $\dot{\boldsymbol{d}}$ to be mutually orthogonal. This latter condition is necessary and also sufficient. Let $\boldsymbol{\Omega} \cdot \dot{\boldsymbol{d}} = 0$; consider the point $\boldsymbol{p} = \boldsymbol{\Omega}_0^{-2}(\boldsymbol{\Omega}_0 \times \boldsymbol{d}_1)$, its velocity is according to (4.8) equal to $\boldsymbol{\Omega}_0^{-2}\boldsymbol{\Omega}_0 \times (\boldsymbol{\Omega}_0 \times \boldsymbol{d}_1) + \boldsymbol{d}_1$ and this is after reduction by means of (4.3) seen to be zero.

In the general case, when no such a point exists, we may ask for those points with velocity *parallel* to $\boldsymbol{\Omega}$. Such points satisfy

$$
(4.10)\qquad \boldsymbol{\Omega} \times (\boldsymbol{P} - \boldsymbol{d}) + \dot{\boldsymbol{d}} = \sigma\boldsymbol{\Omega},
$$

σ being a scalar. Scalar multiplication of both sides by $\boldsymbol{\Omega}$ yields

$$
(4.11)\qquad \sigma = \boldsymbol{\Omega}^{-2}(\boldsymbol{\Omega} \cdot \dot{\boldsymbol{d}})
$$

and (4.10) reads

(4.12) $$\boldsymbol{\Omega} \times (\boldsymbol{P} - \boldsymbol{d}) = -(\dot{\boldsymbol{d}} - \sigma\boldsymbol{\Omega}) = -\dot{\boldsymbol{d}}'$$

$\dot{\boldsymbol{d}}'$ being the component of $\dot{\boldsymbol{d}}$ orthogonal to $\boldsymbol{\Omega}$, thus $\boldsymbol{\Omega} \cdot \dot{\boldsymbol{d}}' = 0$. It is well-known from vector algebra that the equation $\boldsymbol{a} \times \boldsymbol{x} = \boldsymbol{b}$, with $\boldsymbol{a} \cdot \boldsymbol{b} = 0$ has the set of solutions $\boldsymbol{x} = -\boldsymbol{a}^{-2}(\boldsymbol{a} \times \boldsymbol{b}) + \mu\boldsymbol{a}$, μ being an arbitrary scalar. Hence the solutions of (4.12) are

(4.13) $$\boldsymbol{P} = \boldsymbol{d} + \boldsymbol{\Omega}^{-2}(\boldsymbol{\Omega} \times \dot{\boldsymbol{d}}) + \mu\boldsymbol{\Omega}$$

which means that there exists at any instant a line s in space, parallel to $\boldsymbol{\Omega}$, which is the locus of points whose velocity is parallel to $\boldsymbol{\Omega}$. If S is any point on s we have from (4.12): $\boldsymbol{\Omega} \times (\boldsymbol{S} - \boldsymbol{d}) = \sigma\boldsymbol{\Omega} - \dot{\boldsymbol{d}}$ and then from (4.7)

(4.14) $$\dot{\boldsymbol{P}} = \boldsymbol{\Omega} \times (\boldsymbol{P} - \boldsymbol{S}) + \sigma\boldsymbol{\Omega}$$

which expresses that at any instant the velocity distribution is identical with that of a screw motion, with s as its axis, with angular velocity $\boldsymbol{\Omega}$ and translation velocity $\sigma\boldsymbol{\Omega}$. The scalar σ being the ratio of the linear and the angular velocity, is known as the pitch of the screw motion. It must be understood that here, unlike in Chapter I, s, $\boldsymbol{\Omega}$, and σ belong to a certain instant (in other words that they are functions of t).

The screw axis s is given by (4.13): all $\boldsymbol{P}$ from this equation are position vectors $\boldsymbol{S}$, μ being a parameter. This is not the most suitable way to handle it. We shall make use of the method by which in line geometry the position of a line is described: One takes an arbitrary vector $\boldsymbol{Q}$ along the line and determines the moment $\boldsymbol{Q}'$ of $\boldsymbol{Q}$ with respect to the origin, this is the vector $\boldsymbol{S} \times \boldsymbol{Q}$ where $\boldsymbol{S}$ is the position vector of any point on the line. It is well-known that the line is uniquely determined by the ordered pair of (free) vectors $\boldsymbol{Q}$ and $\boldsymbol{Q}'$; the first gives it geometric direction and the other contains the location. Obviously all pairs $k\boldsymbol{Q}$, $k\boldsymbol{Q}'$, k being an arbitrary scalar, unequal to zero, describe one and the same line. Moreover the two vectors must satisfy the relation $\boldsymbol{Q} \cdot \boldsymbol{Q}' = 0$. The components of $\boldsymbol{Q}$ and $\boldsymbol{Q}'$ are the homogeneous Cartesian Plücker coordinates of the line. We shall call $\boldsymbol{Q}$, $\boldsymbol{Q}'$ its Plücker vectors.

Example 8. If (x_1, x_2, x_3, x_4) and (y_1, y_2, y_3, y_4) are the homogeneous Cartesian coordinates of two different points on a line and $x_i y_j - x_j y_i$ is denoted by q_{ij}, show that $\boldsymbol{Q} = (q_{41}, q_{42}, q_{43})$, $\boldsymbol{Q}' = (q_{23}, q_{31}, q_{12})$.

Example 9. Alternatively, if the second Plücker vector is $\boldsymbol{Q}' = \boldsymbol{Q} \times \boldsymbol{S}$, show that $\boldsymbol{Q}'$ remains as $\boldsymbol{Q}' = (q_{23}, q_{31}, q_{12})$ provided that now $\boldsymbol{Q} = (q_{14}, q_{24}, q_{34})$. (The nomenclature $\boldsymbol{Q}(Q_1, Q_2, Q_3)$, $\boldsymbol{Q}'(Q_1', Q_2', Q_3')$ will frequently be used instead of the double subscripts, in which case we will note which system is indicated.)

Example 10. Show that the lines $(\boldsymbol{Q}_1, \boldsymbol{Q}_1')$ and $(\boldsymbol{Q}_2, \boldsymbol{Q}_2')$ are parallel if $\boldsymbol{Q}_1 \times \boldsymbol{Q}_2 = 0$ and perpendicular if $\boldsymbol{Q}_1 \cdot \boldsymbol{Q}_2 = 0$.

Example 11. Show that the line passes through the origin if $\boldsymbol{Q}' = 0$ and conversely.

If two lines intersect, their common point being S, we may write $\boldsymbol{Q}'_1 = \boldsymbol{S} \times \boldsymbol{Q}_1$, $\boldsymbol{Q}'_2 = \boldsymbol{S} \times \boldsymbol{Q}_2$ and therefore

$$\boldsymbol{Q}'_1 \cdot \boldsymbol{Q}_2 = (\boldsymbol{S} \times \boldsymbol{Q}_1) \cdot \boldsymbol{Q}_2 = (\boldsymbol{Q}_1 \times \boldsymbol{Q}_2) \cdot \boldsymbol{S},$$

$$\boldsymbol{Q}'_2 \cdot \boldsymbol{Q}_1 = (\boldsymbol{S} \times \boldsymbol{Q}_2) \cdot \boldsymbol{Q}_1 = (\boldsymbol{Q}_2 \times \boldsymbol{Q}_1) \cdot \boldsymbol{S},$$

which implies

$$\boldsymbol{Q}'_1 \cdot \boldsymbol{Q}_2 + \boldsymbol{Q}'_2 \cdot \boldsymbol{Q}_1 = 0. \tag{4.15}$$

Example 12. Show that from (4.15) it follows that the two lines are either intersecting or parallel, which means that they are coplanar.

The Plücker vectors of the screw axis s are given by $\boldsymbol{Q} = \boldsymbol{\Omega}$ and $\boldsymbol{Q}' = \boldsymbol{S} \times \boldsymbol{\Omega}$, $\boldsymbol{S}$ being a point on s and therefore satisfying (4.13). Hence

$$\boldsymbol{Q} = \boldsymbol{\Omega}, \qquad \boldsymbol{Q}' = \dot{\boldsymbol{d}} - \boldsymbol{\Omega} \times \boldsymbol{d} - \sigma\boldsymbol{\Omega}, \tag{4.16}$$

σ being given by (4.11).

The velocity distribution at a certain instant is determined by the screw-axis s, the vector $\boldsymbol{\Omega}$ along s, and the pitch σ. It may be represented by a pair of ordered vectors $\boldsymbol{L}$, $\boldsymbol{M}$ defined as follows

$$\boldsymbol{L} = \boldsymbol{\Omega}, \qquad \boldsymbol{M} = \boldsymbol{Q}' + \sigma\boldsymbol{\Omega}, \tag{4.17}$$

$\boldsymbol{Q}'$ being the second Plücker vector of the line s. $\boldsymbol{L}$ and $\boldsymbol{M}$ determine the distribution (or "the infinitesimal screw motion") uniquely. Indeed, we have $\boldsymbol{\Omega} = \boldsymbol{L}$ and $\boldsymbol{L} \cdot \boldsymbol{M} = \sigma \boldsymbol{L}^2$ and hence $\sigma = (\boldsymbol{L} \cdot \boldsymbol{M})/\boldsymbol{L}^2$. Furthermore $\boldsymbol{Q}' = \boldsymbol{M} - \sigma\boldsymbol{L}$, which gives us the position of s. From (4.14) it follows then (in view of $\boldsymbol{S} \times \boldsymbol{\Omega} = \boldsymbol{Q}'$):

$$\dot{\boldsymbol{P}} = \boldsymbol{L} \times \boldsymbol{P} + \boldsymbol{M}. \tag{4.18}$$

Example 13. If $\boldsymbol{L} \cdot \boldsymbol{M} = 0$ the screw motion is a pure rotation.

The (independent) vectors $\boldsymbol{L}$ and $\boldsymbol{M}$ may be called the Ball vectors of the infinitesimal screw. Sir Robert Ball, in his classic treatise (BALL [1900]) deals with geometrical properties of sets of infinitesimal screws and their applications to dynamics.

5. Canonical systems

In this section we shall introduce special coordinate systems in order to simplify the study of instantaneous kinematics.

First of all we suppose, as we did before, that the origins of the fixed and the moving space coincide in the zero position. This implies $\boldsymbol{d}_0 = 0$ and $\boldsymbol{P}_0 = \boldsymbol{p}$.

Furthermore we take this common origin on the screw axis s_0 of the zero position. The consequences are: $\boldsymbol{d}_1 = \sigma_0\boldsymbol{\Omega}_0$, $\boldsymbol{P}_1 = \boldsymbol{\Omega}_0 \times \boldsymbol{p} + \sigma_0\boldsymbol{\Omega}_0$, the Plücker vectors of s_0 are $\boldsymbol{Q}_0 = \boldsymbol{\Omega}_0$, $\boldsymbol{Q}'_0 = 0$.

For a definite choice of O on s_0 we determine the common perpendicular n of s_0 and $\mathrm{s}(t)$ in its limit position for $t \to 0$. In order to study the behavior of $\mathrm{s}(t)$ for small values of t we expand its Plücker vectors in a power series, restricting ourselves to first order terms. From (4.16) it follows

$$(5.1)\qquad \begin{aligned} \boldsymbol{Q} &= \boldsymbol{\Omega}_0 + \boldsymbol{\Omega}_1 t,\\ \boldsymbol{Q}' &= (\boldsymbol{d}_1 - \boldsymbol{\Omega}_0 \times \boldsymbol{d}_0 - \sigma_0\boldsymbol{\Omega}_0) + (\boldsymbol{d}_2 - \boldsymbol{\Omega}_0 \times \boldsymbol{d}_1 - \boldsymbol{\Omega}_1 \times \boldsymbol{d}_0 - \sigma_0\boldsymbol{\Omega}_1 - \sigma_1\boldsymbol{\Omega}_0)t\\ &= (\boldsymbol{d}_2 - \sigma_0\boldsymbol{\Omega}_1 - \sigma_1\boldsymbol{\Omega}_0)t. \end{aligned}$$

Let $\boldsymbol{N}$, $\boldsymbol{N}'$ be the Plücker vectors of n. As n is perpendicular to s_0 and $\mathrm{s}(t)$ we have

$$(5.2)\qquad \boldsymbol{N} = \boldsymbol{\Omega}_0 \times (\boldsymbol{\Omega}_0 + \boldsymbol{\Omega}_1 t) = (\boldsymbol{\Omega}_0 \times \boldsymbol{\Omega}_1)t.$$

We exclude the special case $\boldsymbol{\Omega}_0 \times \boldsymbol{\Omega}_1 = 0$, which would mean that s_0 and $\mathrm{s}(t)$ are, to the first order, parallel.

s_0 and n intersect, according to (4.15), if

$$(5.3)\qquad \boldsymbol{\Omega}_0 \cdot \boldsymbol{N}' = 0.$$

The condition that n and $\mathrm{s}(t)$ intersect reads

$$(\boldsymbol{\Omega}_0 + \boldsymbol{\Omega}_1 t)\cdot \boldsymbol{N}' + (\boldsymbol{d}_2 - \sigma_0\boldsymbol{\Omega}_1 - \sigma_1\boldsymbol{\Omega}_0)\cdot \boldsymbol{N}t = 0,$$

or in view of (5.2) and (5.3), dividing $\boldsymbol{N}$ and $\boldsymbol{N}'$ by t:

$$(5.4)\qquad \boldsymbol{\Omega}_1 \cdot \boldsymbol{N}' + \boldsymbol{d}_2 \cdot (\boldsymbol{\Omega}_0 \times \boldsymbol{\Omega}_1) = 0.$$

The limit of the common perpendicular n is implicitly determined by (5.4), by $\boldsymbol{N} = (\boldsymbol{\Omega}_0 \times \boldsymbol{\Omega}_1)$, and by $\boldsymbol{N}\cdot\boldsymbol{N}' = 0$. It intersects s_0 at a point S_0 which is called the striction point on s_0 with respect to the ruled surface which is the locus of $\mathrm{s}(t)$. We choose the origin at S_0, which implies $\boldsymbol{N}' = 0$. Hence (5.4) gives us as a consequence

$$(5.5)\qquad \boldsymbol{d}_2 \cdot (\boldsymbol{\Omega}_0 \times \boldsymbol{\Omega}_1) = 0$$

which means that $\boldsymbol{d}_2$ is linearly dependent on $\boldsymbol{\Omega}_0$ and $\boldsymbol{\Omega}_1$.

We shall call S_0 the canonical origin of the zero position. Summing up what we have done so far, we draw the following conclusions. We started by expressing the motion of E^3 with respect to Σ^3 in the neighborhood of a zero position by means of a Taylor expansion, the coefficients being functions of two sets of arbitrarily chosen vectors: $\boldsymbol{b}_k$ $(k = 1, 2, \ldots)$ and $\boldsymbol{d}_k$ $(k = 0, 1, 2, \ldots)$. The first set was related to only the rotational part of the motion. When we changed our choice of origin to the canonical origin the first set remained the same, which could have been expected because we have seen that the pertinent vector $\boldsymbol{\Omega}(t)$ is independent of the origin. The second set does however depend upon our choice of origin. For the canonical system we obtained three relations:

$$\boldsymbol{d}_0 = 0, \qquad \boldsymbol{d}_1 = \sigma_0 \boldsymbol{\Omega}_0, \qquad \boldsymbol{d}_2 = \lambda \boldsymbol{\Omega}_0 + \mu \boldsymbol{\Omega}_1, \tag{5.6}$$

σ_0, λ, μ being constant scalars. The vectors $\boldsymbol{d}_3, \boldsymbol{d}_4, \ldots$ are still arbitrary.

Up to second order terms we have thus the following configuration (Fig. 1): $\boldsymbol{d}_1$ and $\boldsymbol{\Omega}_0$ are vectors along the same line, $\boldsymbol{d}_2$ is on the plane spanned by $\boldsymbol{\Omega}_0$ and $\boldsymbol{\Omega}_1$, $\boldsymbol{\Omega}_0 \times \boldsymbol{\Omega}_1$ is perpendicular to this plane (and along the common perpendicular n).

As seen from (4.5) and (4.6) the series for $\boldsymbol{P}$ and $\boldsymbol{\Omega}$ start with simple expressions but the higher order terms are complicated. Therefore the study

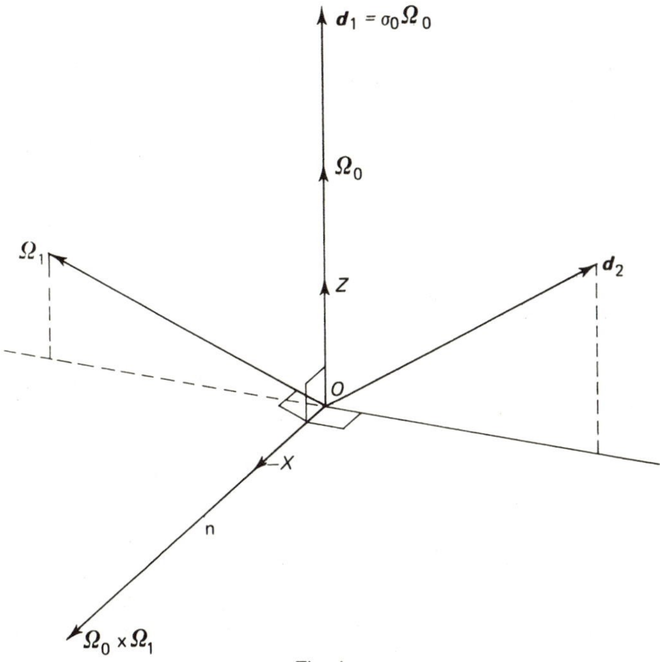

Fig. 1.

of general instantaneous kinematics is restricted for practical reasons to three or four terms at most. This is also the justification for the introduction of a notation whereby each $\boldsymbol{b}_i$ $(i = 1, 2, 3, 4)$ is given a special symbol:

$$\boldsymbol{b}_1 = \boldsymbol{\omega}, \qquad \boldsymbol{b}_2 = \boldsymbol{\varepsilon}, \qquad \boldsymbol{b}_3 = \boldsymbol{\gamma}, \qquad \boldsymbol{b}_4 = \boldsymbol{\kappa}. \tag{5.7}$$

We have then for (4.6):

$$\begin{gathered} \boldsymbol{\Omega}_0 = \boldsymbol{\omega}, \qquad \boldsymbol{\Omega}_1 = \boldsymbol{\varepsilon}, \qquad \boldsymbol{\Omega}_2 = \omega^2\boldsymbol{\omega} + \tfrac{1}{2}(\boldsymbol{\omega} \times \boldsymbol{\varepsilon}) + \boldsymbol{\gamma}, \\ \boldsymbol{\Omega}_3 = \tfrac{9}{2}(\boldsymbol{\omega} \cdot \boldsymbol{\varepsilon})\boldsymbol{\omega} + \tfrac{3}{2}\omega^2\boldsymbol{\varepsilon} + \boldsymbol{\omega} \times \boldsymbol{\gamma} + \boldsymbol{\kappa}. \end{gathered} \tag{5.8}$$

Obviously $\boldsymbol{\omega}$ and $\boldsymbol{\varepsilon}$ are the angular velocity and acceleration vectors of the zero position.

Example 14. Show that

$$(\Omega^2)_0 = \omega^2, \quad (\Omega^2)_1 = 2\boldsymbol{\omega} \cdot \boldsymbol{\varepsilon}, \quad (\Omega^2)_2 = 2(\omega^4 + \varepsilon^2 + \boldsymbol{\omega} \cdot \boldsymbol{\gamma}),$$

$$(\Omega^2)_3 = 18\omega^2(\boldsymbol{\omega} \cdot \boldsymbol{\varepsilon}) + 6\boldsymbol{\varepsilon} \cdot \boldsymbol{\gamma} + 2\boldsymbol{\omega} \cdot \boldsymbol{\kappa}.$$

Until now we have used vectors to represent the position of a point, its velocity, its acceleration, etc. If we want to introduce Cartesian frames it is natural to choose O_{XYZ} in Σ^3 and o_{xyz} in E^3 coinciding for $t = 0$. Considering Fig. 1 it seems suitable to use a frame simply connected with the configuration. We take O_Z along $\boldsymbol{\Omega}_0$; the plane of $\boldsymbol{\Omega}_0$, $\boldsymbol{\Omega}_1$ and $\boldsymbol{d}_2$ as $X = 0$, and O_X along $-\boldsymbol{\Omega}_0 \times \boldsymbol{\Omega}_1$. Then we have:

$$\begin{gathered} \boldsymbol{P} = (X, Y, Z), \quad \boldsymbol{p} = (x, y, z), \quad \boldsymbol{b}_1 = \boldsymbol{\Omega}_0 = (0, 0, \omega), \quad \boldsymbol{b}_2 = \boldsymbol{\Omega}_1 = (0, \varepsilon_Y, \varepsilon_Z), \\ \boldsymbol{d}_1 = (0, 0, \sigma_0\omega), \qquad \boldsymbol{d}_2 = (0, \mu\varepsilon_Y, \lambda\omega + \mu\varepsilon_Z), \end{gathered} \tag{5.9}$$

and for (4.5) we obtain

$$\begin{aligned} X_0 &= x, & X_1 &= -\omega y, & X_2 &= -\omega^2 x - \varepsilon_Z y + \varepsilon_Y z, \\ Y_0 &= y, & Y_1 &= \omega x, & Y_2 &= \varepsilon_Z x - \omega^2 y + \mu\varepsilon_Y, \\ Z_0 &= z, & Z_1 &= \sigma_0\omega, & Z_2 &= -\varepsilon_Y x + \lambda\omega + \mu\varepsilon_Z, \end{aligned}$$

$$\begin{aligned} X_3 &= -3\omega\varepsilon_Z x - \gamma_Z y + \gamma_Y z + d_{3_X}, \\ Y_3 &= \gamma_Z x - 3\omega\varepsilon_Z y + (\tfrac{3}{2}\omega\varepsilon_Y - \gamma_X)z + d_{3_Y}, \\ Z_3 &= -\gamma_Y x + (\tfrac{3}{2}\omega\varepsilon_Y + \gamma_X)y + d_{3_Z}. \end{aligned} \tag{5.10}$$

These formulas may be considered as fundamental for the study of instantaneous kinematics (up to the third order). As the frame is canonical the constants appearing in (5.10) are characteristic of the motion, they are called its (kinematical) instantaneous invariants: For the first order there are two, ω

and σ_0, which determine the velocity distribution. There are four more for the second order, ε_Y, ε_X, λ, μ, which together with ω determine the acceleration distribution. Third order invariants ($\gamma_X, \gamma_Y, \gamma_Z, d_{3_X}, d_{3_Y}, d_{3_Z}$) are six in number, and the same holds obviously for any order >3.

Formulas like (5.10) were first derived, for spatial kinematics, by VELDKAMP [1967a].

Example 15. If the dimensional symbol for a length is L, and for time T, show that for ω we have T^{-1}, for $\varepsilon_Y, \varepsilon_Z$: T^{-2}, for $\gamma_X, \gamma_Y, \gamma_Z$: T^{-3}, for σ_0: LT^{-1}, for λ: LT^{-1}, for μ: L, for $d_{3_X}, d_{3_Y}, d_{3_Z}$: LT^{-3}.

For a canonical frame the expansion (4.6) reduces to

$$\begin{aligned} \Omega_X &= \tfrac{1}{2}(-\tfrac{1}{2}\omega\varepsilon_Y + \gamma_X)t^2 + \tfrac{1}{6}(-\omega\gamma_Y + \kappa_X)t^3 + \cdots \\ \Omega_Y &= \varepsilon_Y t + \tfrac{1}{2}\gamma_Y t^2 + \tfrac{1}{6}(\tfrac{3}{2}\omega^2\varepsilon_Y + \omega\gamma_X + \kappa_Y)t^3 + \cdots \\ \Omega_Z &= \omega + \varepsilon_Z t + \tfrac{1}{2}(\omega^3 + \gamma_Z)t^2 + \tfrac{1}{6}(6\omega^2\varepsilon_Z + \kappa_Z)t^3 + \cdots \end{aligned} \tag{5.11}$$

In order to derive the instantaneous invariants for the inverse motion we remark that for a 3-dimensional space the formulas (3.15) read

$$\begin{gathered} \tilde{\boldsymbol{\Omega}}_0 = -\boldsymbol{\Omega}_0, \qquad \tilde{\boldsymbol{\Omega}}_1 = -\boldsymbol{\Omega}_1, \qquad \tilde{\boldsymbol{\Omega}}_2 + \boldsymbol{\Omega}_2 = \boldsymbol{\Omega}_0 \times \boldsymbol{\Omega}_1, \\ \tilde{\boldsymbol{\Omega}}_3 + \boldsymbol{\Omega}_3 = 2\boldsymbol{\Omega}_0 \times \boldsymbol{\Omega}_2, \end{gathered} \tag{5.12}$$

and (2.9) gives us

$$\begin{gathered} \tilde{\boldsymbol{d}}_0 = 0, \qquad \tilde{\boldsymbol{d}}_1 = -\boldsymbol{d}_1, \qquad \tilde{\boldsymbol{d}}_2 = -\boldsymbol{d}_2 + 2\boldsymbol{\Omega}_0 \times \boldsymbol{d}_1, \\ \tilde{\boldsymbol{d}}_3 = -\boldsymbol{d}_3 + 3\boldsymbol{\Omega}_0 \times \boldsymbol{d}_2 + 3\Omega_0^2 \boldsymbol{d}_1 - 3(\boldsymbol{\Omega}_0 \cdot \boldsymbol{d}_1)\boldsymbol{\Omega}_0 + 3\boldsymbol{\Omega}_1 \times \boldsymbol{d}_1. \end{gathered} \tag{5.13}$$

Hence, if the frame for the direct motion is canonical, we have $\tilde{\boldsymbol{d}}_2 = -\boldsymbol{d}_2$. This implies that the inverse and the direct motion have the same screw axis and the same canonical origin, and that furthermore the canonical frame $O_{\tilde{X}\tilde{Y}\tilde{Z}}$ follows from O_{XYZ} by reflection into O_X. This implies that the instantaneous invariants of the inverse motion (with respect to its own canonical frame) of the first and the second order are the same as for the direct motion. From (5.12) and (5.13) it follows that this is not true for higher order invariants.

Example 16. Determine the third order invariants of the inverse motion in terms of those of the direct motion.

6. Geometrical invariants

We have defined the motion of a space E by giving a continuous set of its positions with respect to the fixed space Σ; we associated with each position a value of the parameter t, which was meant to be the time, as measured by a clock. It is of course possible for two motions M_1 and M_2 to have the same positions of E but have different time schemes associated with them. In this case M_1 and M_2 have different properties: for instance at the same position the velocity and the acceleration of a certain point of E will in general be different for the two motions. On the other hand it is obvious that there are other characteristics which are the same for them: the path of any point is identical for the two motions and it follows that the path's geometrical properties (tangents, curvatures, etc.) are the same for both. These remarks give rise to a certain classification in kinematics: there are properties, to be called *time-independent*, which regardless of the time scheme remain unaltered provided only that the same set of positions is travelled through; their study is the subject of *geometric kinematics.* Other properties are *time-dependent*: they are related to the particular time scheme of the motion.

Example 17. Show that the instantaneous screw axis and the pitch are time-independent concepts; if the time scheme is changed the angular velocity vector is multiplied by a scalar.

In geometric kinematics the parameter actually being made use of is irrelevant and we are free to choose it in a way to simplify the analytic description of the motion. The possibilities are, however, rather restricted. Because the parameter is a scalar the freedom of choice can only enable us to normalize a scalar function related to the motion, while the motion itself is a phenomenon mainly described by vectorial means.

We cannot use some of the scalars which characterize the lower order properties of a motion: the pitch cannot be used because it is immune to parameter transformation; if we take the length of $\boldsymbol{d}$ we exclude the important spherical motions which require $\boldsymbol{d} = 0$. The only acceptable choice seems to be to normalize the length Ω of the angular velocity vector, which is in accordance with the developments of this chapter where we supposed $\boldsymbol{\Omega}$ to be unequal to zero. Therefore we introduce for geometric kinematics a parameter, say ϕ, such that $\boldsymbol{\Omega}(\phi)$ is normalized so that $\Omega = 1$. As a consequence we have the infinite series of relations

$$(6.1) \qquad (\Omega^2)_0 = 1, \qquad (\Omega^2)_k = 0 \quad \text{if } k > 0.$$

Considering Example 14 it follows that

$$\omega = 1, \qquad \boldsymbol{\omega} \cdot \boldsymbol{\varepsilon} = 0, \qquad \boldsymbol{\omega} \cdot \boldsymbol{\gamma} = -(1 + \varepsilon^2), \qquad \boldsymbol{\omega} \cdot \boldsymbol{\kappa} = -3\boldsymbol{\varepsilon} \cdot \boldsymbol{\gamma}, \ldots$$

which implies

$$(6.2) \qquad \varepsilon_Z = 0, \qquad \varepsilon_Y = \varepsilon, \qquad \gamma_Z = -(1 + \varepsilon^2), \qquad \kappa_Z = -3\varepsilon\gamma_Y, \ldots$$

The formulas (5.10) are now reduced to:

$$
\begin{aligned}
&X_0 = x, && X_1 = -y, && X_2 = -x + \varepsilon z, && X_3 = (1+\varepsilon^2)y + \gamma_Y z + d_{3_X}, \ldots \\
&Y_0 = y, && Y_1 = x, && Y_2 = -y + \mu\varepsilon, && Y_3 = -(1+\varepsilon^2)x + (\tfrac{3}{2}\varepsilon - \gamma_X)z + d_{3_Y}, \ldots \\
&Z_0 = z, && Z_1 = \sigma_0, && Z_2 = -\varepsilon x + \lambda, && Z_3 = -\gamma_Y x + (\tfrac{3}{2}\varepsilon + \gamma_X)y + d_{3_Z}, \ldots
\end{aligned}
\tag{6.3}
$$

These basic formulas contain the geometric instantaneous invariants: one of the first order (σ_0), three of the second ($\varepsilon, \lambda, \mu$), five of order three ($\gamma_X, \gamma_Y, d_{3_X}, d_{3_Y}, d_{3_Z}$), and obviously again five for any higher order.

Example 18. Determine the expansion (5.11) for the case where the parameter has been normalized.
Example 19. If $(\boldsymbol{Q}, \boldsymbol{Q}')$ are the Plücker vectors of the screw axis determine $\boldsymbol{Q}_0, \boldsymbol{Q}'_0, \boldsymbol{Q}_1, \boldsymbol{Q}'_1$.
Example 20. Show that $\boldsymbol{d}_1 = (0, 0, \sigma_0)$, $\boldsymbol{d}_2 = (0, \mu\varepsilon, \lambda)$.
Example 21. Show that $\sigma_1 = \lambda$.
Example 22. Prove that if the parameter is normalized for a motion it is normalized for the inverse motion as well.

Restricting ourselves to second order terms, it follows from (6.3):

$$
\begin{aligned}
X &= x - yt + \tfrac{1}{2}(-x + \varepsilon z)t^2, \qquad Y = y + xt + \tfrac{1}{2}(-y + \mu\varepsilon)t^2, \\
Z &= z + \sigma_0 t + \tfrac{1}{2}(-\varepsilon x + \lambda)t^2,
\end{aligned}
\tag{6.4}
$$

or, solving for x, y, z:

$$
\begin{aligned}
x &= X + Yt + \tfrac{1}{2}(-X - \varepsilon Z)t^2, \qquad y = Y - Xt + \tfrac{1}{2}(-Y - \mu\varepsilon)t^2, \\
z &= Z - \sigma_0 t + \tfrac{1}{2}(\varepsilon X - \lambda)t^2.
\end{aligned}
\tag{6.5}
$$

Let a, b, c, d be the homogeneous coordinates of a plane of E^3 and A, B, C, D those of its variable position in Σ^3. Substituting (6.5) into $ax + by + cz + d = 0$ and rearranging the left-hand-side we obtain

$$
\begin{aligned}
&A_0 = a, && A_1 = -b, && A_2 = -a + \varepsilon c, \\
&B_0 = b, && B_1 = a, && B_2 = -b, \\
&C_0 = c, && C_1 = 0, && C_2 = -\varepsilon a, \\
&D_0 = d, && D_1 = -\sigma_0 c, && D_2 = -\lambda c - \mu\varepsilon b,
\end{aligned}
\tag{6.6}
$$

as the dual counterpart of (6.3).

Example 23. Determine A_3, B_3, C_3, D_3 in terms of a, b, c, d and the instantaneous invariants up to the third order.

The number of data which needs to be taken into consideration in order to study the kth order properties of geometric instantaneous kinematics is 1 for $k = 1$, 4 for $k = 2$, 9 for $k = 3$, and in general $(5k - 6)$ for $k > 1$. Time-dependent instantaneous kinematics requires 2 numbers for $k = 1$, and $6k - 7$ for $k > 1$. It is obvious that although our method is applicable for all k, practical reasons restrict the investigation to a low order. It may be said that the study of general spatial kinematics has been concentrated mainly on second order properties and that the knowledge about this restricted subject is satisfactory but by no means complete. Higher order investigations have rarely been made systematically except for special cases such as spherical and planar motion.

The results of this chapter will be applied in Chapter VI to continuous kinematics, and also in Chapter III–V to m-positions theory ($m = 2, 3, 4$); the special case of m consecutive (or infinitesimally separated) positions is identical with instantaneous kinematics of the $(m - 1)^{\text{th}}$ order. This approach will furnish us rather simple examples for concepts and theorems of general $(m - 1)^{\text{th}}$ order theory.

CHAPTER III

TWO POSITIONS THEORY

1. The displacement

We consider two positions of a three dimensional space E with respect to a fixed space Σ. The positions of E will be denoted by E_1 and E_2 respectively. The positions in Σ of points $A, B, \ldots$ of E will be $A_1, B_1, \ldots$ when E is at E_1, and $A_2, B_2, \ldots$ when E is at E_2; those of lines $\mathrm{l}, \mathrm{m}, \ldots$ are $\mathrm{l}_1, \mathrm{m}_1, \ldots$ and $\mathrm{l}_2, \mathrm{m}_2, \ldots$; the two positions in Σ of planes $\alpha, \beta, \ldots$ of E are $\alpha_1, \beta_1, \ldots$ and $\alpha_2, \beta_2, \ldots$.

A_1 and A_2 are called homologous points, l_1 and l_2 are homologous lines, α_1 and α_2 homologous planes. In this chapter we study relationships between homologous elements. Our methods are mainly algebraic; the subject was originally developed using (mainly) geometrical reasoning (by Chasles [1831], Brisse [1875], Schoenflies [1892], and others).

As shown in Chapter I, Section 4, the transformation $E_1 \rightarrow E_2$ can always be generated by a unique screw displacement in Σ. If s is the screw axis, the displacement is the product (in an arbitrary order) of a rotation of angle ϕ about s and a translation of distance d parallel to s.

In order to describe the displacement analytically we take a Cartesian frame O_{XYZ} in Σ such that O_Z coincides with s, the origin O and the axes O_X and O_Y are arbitrary. In E we take a Cartesian frame o_{xyz} such that the two frames coincide in the first position. Hence if the coordinates of A in E are (x, y, z) we have for A_1 and A_2 respectively

$$X_1 = x, \qquad Y_1 = y, \qquad Z_1 = z \tag{1.1}$$

and

$$X_2 = x \cos\phi - y \sin\phi, \qquad Y_2 = x \sin\phi + y \cos\phi, \qquad Z_2 = z + d. \tag{1.2}$$

Let h be the line $X \sin\gamma - Y \cos\gamma = Z - a = 0$, intersecting the screw axis orthogonally at the point $H(0, 0, a)$. The projection of (X, Y, Z) on h is seen to be

$$(X\cos\gamma + Y\sin\gamma)\cos\gamma,\ (X\cos\gamma + Y\sin\gamma)\sin\gamma,\ a$$

and its reflection with respect to h is

$$X' = X\cos 2\gamma + Y\sin 2\gamma,\quad Y' = X\sin 2\gamma - Y\cos 2\gamma,\quad Z' = -Z + 2a, \tag{1.3}$$

which verifies that the reflection is a direct displacement; namely, the half turn about h. Let R_i $(i = 1,2)$ be the reflection with respect to h_i: $X\sin\gamma_i - Y\cos\gamma_i = Z - a_i = 0$. For the product R_2R_1 we obtain

$$\begin{aligned} X'' &= X'\cos 2\gamma_2 + Y'\sin 2\gamma_2 = X\cos 2(\gamma_2-\gamma_1) - Y\sin 2(\gamma_2-\gamma_1),\\ Y'' &= X'\sin 2\gamma_2 - Y'\cos 2\gamma_2 = X\sin 2(\gamma_2-\gamma_1) + Y\cos 2(\gamma_2-\gamma_1),\\ Z'' &= -Z' + 2a_2 = Z + 2(a_2 - a_1), \end{aligned} \tag{1.4}$$

which when compared with (1.2) proves that R_2R_1 is the screw displacement with $\phi = 2(\gamma_2-\gamma_1)$, $d = 2(a_2 - a_1)$; the rotation angle is twice the angle between h_1 and h_2, the translation is twice the distance H_1H_2.

On the other hand *the screw displacement may be written in* ∞^2 *many ways as the product, in the correct order, of two reflections with respect to lines intersecting the screw axis orthogonally*. The parameters γ_i and a_i for one of the two lines $(\mathrm{h}_1, \mathrm{h}_2)$ may be chosen arbitrarily. We shall make use of this decomposition of a screw displacement in Chapter IV.

2. Homologous points

Pairs of homologous points A_1 and A_2 are given by (1.1) and (1.2). The midpoint M of A_1A_2 plays an important part in the theory. Its coordinates are seen to be

$$X_M = \tfrac{1}{2}x(1+\cos\phi) - \tfrac{1}{2}y\sin\phi,\qquad Y_M = \tfrac{1}{2}x\sin\phi + \tfrac{1}{2}y(1+\cos\phi),$$
$$Z_M = z + \tfrac{1}{2}d \tag{2.1}$$

and its distance ρ to the screw axis is

$$\rho^2 = X_M^2 + Y_M^2 = (x^2+y^2)\cos^2(\tfrac{1}{2}\phi). \tag{2.2}$$

On the other hand we have for the distance between A_1 and A_2

$$(\overrightarrow{A_1A_2})^2 = 4(x^2+y^2)\sin^2(\tfrac{1}{2}\phi) + d^2,$$

and therefore, if $\phi \neq \pi$,

$$(\overrightarrow{A_1A_2})^2 = 4\rho^2\tan^2(\tfrac{1}{2}\phi) + d^2. \tag{2.3}$$

The formulas (2.1) show that M is obtained from A_1 by a linear transformation, which is non-singular if $\phi \neq \pi$. In that case the inverse transformation exists and we find: *an arbitrary point M is the midpoint of a unique pair of homologous points*, namely

$$(2.4)\qquad \begin{aligned} &A_1\colon X_M + Y_M \tan\tfrac{1}{2}\phi, \quad -X_M \tan\tfrac{1}{2}\phi + Y_M, \quad Z_M - \tfrac{1}{2}d, \\ &A_2\colon X_M - Y_M \tan\tfrac{1}{2}\phi, \quad X_M \tan\tfrac{1}{2}\phi + Y_M, \quad Z_M + \tfrac{1}{2}d. \end{aligned}$$

If, however, $\phi = \pi$, which means that the rotation part of the screw displacement is a half turn, all points M are on the screw axis and each is the midpoint of ∞^2 pairs.

3. The complex of lines joining homologous points

Supposing that the screw displacement is not a pure rotation, which means $d \neq 0$, it has no finite fixed points, and therefore A_1 and A_2 are distinct and the line A_1A_2 is well-defined. We investigate the locus of the line joining A_1 and A_2, the so-called join A_1A_2. The Plücker coordinates of a join follow from (1.1) and (1.2). If the Plücker vectors are denoted as $\boldsymbol{Q}(Q_1, Q_2, Q_3)$, $\boldsymbol{Q}'(Q_1', Q_2', Q_3')$ and satisfy the fundamental relation

$$(3.1)\qquad Q_1Q_1' + Q_2Q_2' + Q_3Q_3' = 0,$$

we obtain, (using the definition given in Chapter II, Example 9),

$$(3.2)\qquad \begin{aligned} Q_1 &= x(1-\cos\phi) + y\sin\phi, & Q_1' &= -xz\sin\phi + yz(1-\cos\phi) + yd \\ Q_2 &= -x\sin\phi + y(1-\cos\phi), & Q_2' &= -xz(1-\cos\phi) - yz\sin\phi - xd \\ Q_3 &= -d, & Q_3' &= (x^2+y^2)\sin\phi, \end{aligned}$$

which gives the set of joins in terms of the three parameters x, y, z. To eliminate these we use Q_1/Q_3, Q_2/Q_3 and Q_3'/Q_3 which leads to three equations for x and y only. The first two are linear; they give

$$(3.3)\qquad \begin{aligned} 2Q_3x &= d(-Q_1 + Q_2\cot\tfrac{1}{2}\phi), \\ 2Q_3y &= d(-Q_1\cot\tfrac{1}{2}\phi - Q_2), \end{aligned}$$

and thus

$$4Q_3^2(x^2+y^2)\sin^2(\tfrac{1}{2}\phi) = d^2(Q_1^2 + Q_2^2).$$

From this it follows, if $\phi \neq \pi$,

$$(3.4)\qquad d(Q_1^2 + Q_2^2) + 2Q_3Q_3'\tan\tfrac{1}{2}\phi = 0.$$

Hence *the joins* A_1A_2 *belong to a quadratic line complex* C. On the other hand, if $(\boldsymbol{Q}, \boldsymbol{Q}')$ is a line satisfying (3.4) and if $Q_3 \neq 0$, the x, y coordinates of A_1 follow from (3.3). The coordinate z may be found from Q_1'/Q_2' which gives us a linear equation for z. The derivation is however somewhat simpler if we calculate the coordinates of A_1A_2 by means of (2.4), supposing $\phi \neq \pi$. We obtain then

$$(3.5) \qquad \begin{aligned} Q_1 &= 2Y_M \tan\tfrac{1}{2}\phi, & Q_1' &= -2X_MZ_M \tan\tfrac{1}{2}\phi + Y_Md \\ Q_2 &= -2X_M \tan\tfrac{1}{2}\phi, & Q_2' &= -X_Md - 2Y_MZ_M \tan\tfrac{1}{2}\phi \\ Q_3 &= -d, & Q_3' &= 2(X_M^2 + Y_M^2)\tan\tfrac{1}{2}\phi, \end{aligned}$$

which may be verified by means of (2.1) and (3.2). Eliminating X_M, Y_M, Z_M we have at once (3.4) and moreover

$$(3.6) \qquad \begin{aligned} X_M &= \tfrac{1}{2}Q_3^{-1}(Q_2d \cot\tfrac{1}{2}\phi), \qquad Y_M = \tfrac{1}{2}Q_3^{-1}(-Q_1d \cot\tfrac{1}{2}\phi), \\ Z_M &= \tfrac{1}{2}Q_1^{-1}(Q_2d \cot\tfrac{1}{2}\phi - 2Q_2') = \tfrac{1}{2}Q_2^{-1}(-Q_1d \cot\tfrac{1}{2}\phi + 2Q_1') \\ &= \tfrac{1}{2}(Q_3Q_3')^{-1}(Q_1Q_2' - Q_2Q_1')d \cot\tfrac{1}{2}\phi, \end{aligned}$$

the three answers for Z_M being equal in view of (3.1) and (3.4). It follows from this that each line satisfying (3.4) is a join A_1A_2, the midpoint of the two homologous points being given by (3.6). The complex C is therefore indeed the locus of the lines joining A_1 and A_2. It is a complex in Σ.

As may be seen from its equation (3.4), the complex C is invariant for any rotation about s and for any translation parallel to s: it has complete screw symmetry about s. This follows also from the fact that the origin O and the axes O_X, O_Y have been chosen arbitrarily and that the equation of C does not depend on these choices.

The lines of a quadratic complex passing through a given point are the generators of a quadratic cone. Those points for which the cone is degenerate constitute the singularity surface of the complex. It is well-known that this surface is of the fourth order. To determine it in our case we use homogeneous coordinates, and intersect the cone with vertex at $T(X_0, Y_0, Z_0, W_0)$ with a plane, $Z = 0$ say. All the lines on the cone are the join of T with the point (X, Y, O, W). We derive the condition that these lines belong to C. We have

$$Q_1 = X_0W - W_0X, \quad Q_2 = Y_0W - W_0Y, \quad Q_3 = Z_0W, \quad Q_3' = X_0Y - Y_0X,$$

and it follows that the conic in $Z = 0$ has the equation

$$(3.7) \qquad \begin{aligned} &W_0^2(X^2 + Y^2) - 2(X_0W_0 + aY_0Z_0)XW \\ &\quad - 2(Y_0W_0 - aX_0Z_0)YW + (X_0^2 + Y_0^2)W^2 = 0, \end{aligned}$$

with $a = d^{-1}\tan\frac{1}{2}\phi$. Its discriminant reads $a^2(X_0^2 + Y_0^2)Z_0^2W_0^2$. If $Z_0 = 0$ the vertex T is in the plane $Z = 0$ and the cone is in general not degenerate. Therefore: the singularity surface of C consists of the plane at infinity (counted twice since $W^2 = 0$) and the two isotropic planes $X + iY = 0$, $X - iY = 0$ through the screw axis. The configuration is of course screw symmetric about s.

If the singularity surface of a quadratic complex degenerates into four planes it is called tetrahedral. Hence C *is a tetrahedral complex*. It is a special type within its class, because two of the four planes coincide. If T is on s the cone degenerates into the isotropic planes through s. If T is a point of the plane V at infinity (that is if $W_0 = 0$) the cone consists of the plane V and the plane $-2aZ_0(Y_0X - X_0Y) + (X_0^2 + Y_0^2)W = 0$, parallel to s. For a general point T (3.7) is a circle, which means that all planes orthogonal to s intersect the cones of the complex in circles.

4. Normal planes

We have investigated the lines A_1A_2 in Σ which join two homologous points. Another geometric entity related to A_1, A_2 and of importance for the theory of displacements is the normal plane β: the plane through the midpoint M of A_1,A_2 and perpendicular to the line A_1A_2.

The direction of A_1A_2 in terms of the coordinates of M follows from (2.4) and is seen to be

$$2Y_M\tan\tfrac{1}{2}\phi : -2X_M\tan\tfrac{1}{2}\phi : -d. \tag{4.1}$$

Therefore the equation of the normal plane through $M(X_M, Y_M, Z_M, W_M)$ reads

$$X(2Y_M\tan\tfrac{1}{2}\phi) - Y(2X_M\tan\tfrac{1}{2}\phi) - ZW_Md + WZ_Md = 0. \tag{4.2}$$

The coordinates, U_i, of the normal plane are seen to be linear functions of the coordinates of M;

$$\begin{aligned} U_1 &= 2Y_M\tan\tfrac{1}{2}\phi, & U_2 &= -2X_M\tan\tfrac{1}{2}\phi, \\ U_3 &= -W_Md, & U_4 &= Z_Md, \end{aligned} \tag{4.3}$$

which means that the relation $M \rightarrow$ β is a correlation; as M and β are incident it is called a null-correlation. The transformation matrix of (4.3) is skew:

$$\left\|\begin{matrix} 0 & 2\tan\frac{1}{2}\phi & 0 & 0 \\ -2\tan\frac{1}{2}\phi & 0 & 0 & 0 \\ 0 & 0 & 0 & -d \\ 0 & 0 & d & 0 \end{matrix}\right\| \tag{4.4}$$

β is the so-called null-plane of M. Inversely each plane $\beta(U_1, U_2, U_3, U_4)$ has a null-point:

$$X_M : Y_M : Z_M : W_M = -U_2 d\cot\tfrac{1}{2}\phi : U_1 d\cot\tfrac{1}{2}\phi : 2U_4 : -2U_3. \tag{4.5}$$

A null-correlation is always associated with a linear complex of lines: those passing through any point and lying in the null-plane of that point.* To determine the equation of the linear complex associated with our null-correlation we join an arbitrary point to the points at infinity of its null-plane, that is we join the points

$$X_M, \quad Y_M, \quad Z_M, \quad W_M$$

and

$$W_M X d, \quad W_M Y d, \quad 2(Y_M X - X_M Y)\tan\tfrac{1}{2}\phi, \quad 0,$$

X_M, Y_M, Z_M, W_M, X and Y all being arbitrary. We have

$$Q_3 = -2W_M(Y_M X - X_M Y)\tan\tfrac{1}{2}\phi, \qquad Q_3' = W_M(X_M Y - Y_M X)d,$$

and therefore the equation of the complex, which we shall denote by Γ reads

$$dQ_3 - 2Q_3'\tan\tfrac{1}{2}\phi = 0. \tag{4.6}$$

Γ has complete screw symmetry about O_Z; the screw axis is its principal axis. We shall return to this linear complex when we study the midpoint line.

* For any linear complex the null-planes of the points of a line l pass through a line l_0; l and l_0 are conjugate lines of the complex, the relation being involutory.

If $l(\boldsymbol{Q}, \boldsymbol{Q}')$ is any line it passes through the points $(0, Q_3', -Q_2', Q_1)$ and $(-Q_3', 0, Q_1', Q_2)$, the null-planes of which, according to (4.3), are $(2Q_3'\tan\frac{1}{2}\phi, 0, -Q_1 d, -Q_2' d)$ and $(0, 2Q_3'\tan\frac{1}{2}\phi, -Q_2 d, Q_1' d)$. Their line of intersection $l_0(\boldsymbol{Q}_0, \boldsymbol{Q}_0')$ is given by

$$\begin{aligned} Q_{0_1} &= Q_1, & Q'_{0_1} &= Q'_1 \\ Q_{0_2} &= Q_2, & Q'_{0_2} &= Q'_2 \\ Q_{0_3} &= kQ'_3, & Q'_{0_3} &= k^{-1}Q_3. \end{aligned} \tag{4.7}$$

with $k = 2d^{-1}\tan\frac{1}{2}\phi$.

The screw axis is conjugate to the line at infinity perpendicular to it.

5. Homologous planes

A plane of E is given by the equation $u_1x + u_2y + u_3z + u_4 = 0$, its plane coordinates being u_i $(i = 1, \dots, 4)$. From the screw displacement

$$X = x\cos\phi - y\sin\phi, \qquad Y = x\sin\phi + y\cos\phi, \qquad Z = z + d,$$

it follows

$$x = X\cos\phi + Y\sin\phi, \qquad y = -X\sin\phi + Y\cos\phi, \qquad z = Z - d.$$

Hence the plane u_i has in Σ the equation

$$(u_1\cos\phi - u_2\sin\phi)X + (u_1\sin\phi + u_2\cos\phi)Y + u_3Z + (-u_3d + u_4) = 0.$$

Therefore, if U_i are the plane coordinates in Σ, planes are transformed as follows

$$\begin{aligned} &U_1 = u_1\cos\phi - u_2\sin\phi, \qquad U_2 = u_1\sin\phi + u_2\cos\phi,\\ &U_3 = u_3, \qquad U_4 = -u_3d + u_4. \end{aligned} \tag{5.1}$$

If, as before, the frames of E_1 and Σ coincide the two homologous planes α_1 and α_2, associated with respectively E_1 and E_2, are given in Σ by

$$\begin{aligned} &\alpha_1\text{:}\quad u_1, \quad u_2, \quad u_3, \quad u_4\\ &\alpha_2\text{:}\ u_1\cos\phi - u_2\sin\phi,\ u_1\sin\phi + u_2\cos\phi,\ u_3,\ -u_3d + u_4. \end{aligned} \tag{5.2}$$

Equation (5.2) for homologous planes is the equation analogous to (1.1) and (1.2) for homologous points.

The angles of α_1 and α_2 with s are equal $([\pi/2] - \arccos[u_3/(u_1^2 + u_2^2 + u_3^2)^{1/2}]$ in this case, as could be expected). If γ denotes this angle, we find that the angle ψ between α_1 and α_2 is

$$\sin\tfrac{1}{2}\psi = \cos\gamma\,\sin\tfrac{1}{2}\phi, \tag{5.3}$$

α_1 and α_2 coincide only for the two isotropic planes through s and for the plane at infinity, these three planes being the fixed planes of the displacement; α_1 and α_2 are only parallel if they are perpendicular to s. In general α_1 and α_2 have a line of intersection: (α_1, α_2). We investigate the set of these lines. Their Plücker coordinates follow from (5.2) and are seen to be

$$\begin{aligned} Q_1 &= u_3[-u_1\sin\phi + u_2(1-\cos\phi)],\\ Q_1' &= u_1u_4(1-\cos\phi) + u_2u_4\sin\phi - u_1u_3d,\\ Q_2 &= u_3[-u_1(1-\cos\phi) - u_2\sin\phi], \end{aligned}$$

$$
\begin{aligned}
(5.4) \qquad Q_2' &= -u_1u_4\sin\phi + u_2u_4(1-\cos\phi) - u_2u_3d,\\
Q_3 &= (u_1^2+u_2^2)\sin\phi,\\
Q_3' &= -u_3^2 d,
\end{aligned}
$$

which is comparable to (3.2). From this it follows

$$
\begin{aligned}
(Q_1^2+Q_2^2) &= 2u_3^2(u_1^2+u_2^2)(1-\cos\phi),\\
Q_3Q_3' &= -u_3^2(u_1^2+u_2^2)d\sin\phi,
\end{aligned}
$$

and the result of eliminating u_i is therefore

$$
(5.5) \qquad d(Q_1^2+Q_2^2)+2Q_3Q_3'\tan\tfrac{1}{2}\phi = 0,
$$

that is again the equation of C. Hence *the set of joins of homologous points A_1, A_2 coincides with the set of the lines of intersection* (α_1, α_2) *of homologous planes.* That the two loci must be identical may be seen directly as follows. Let A_1 and A_2 be two homologous points. A_1 coincides with a point B_2 of E_2, homologous with B_1 of E_1. On the other hand A_2 coincides with a point C_1 of E_1, the homologous point of which is C_2. Since the lines A_1C_1 and A_2B_2 coincide they are the intersection of the planes through $A_1B_1C_1$ and $A_2B_2C_2$. The conclusion is that the plane $B_2A_2C_2$ of E_2 is homologous with the plane $B_1A_1C_1$ of E_1; the intersection of these homologous planes is A_1A_2.

We have arrived at the conclusion that *on any line of the complex* C *lie two homologous points A_1 and A_2 and that two homologous planes α_1 and α_2 pass through it.* A line of C may be given by the parameters X_M, Y_M, Z_M, the coordinates of the midpoint M on it. The Plücker vectors of the line are then determined by (3.5) and the two homologous points A_1 and A_2 by (2.4). In order to find the two homologous planes through it we proceed as follows. α_1 contains B_1, A_1, A_2 and it passes therefore through the following three points as well: the midpoint M of A_1A_2 (with coordinates X_M, Y_M, Z_M), the point at infinity of A_1A_2 (with coordinates $2Y_M\tan\frac{1}{2}\phi$, $-2X_M\tan\frac{1}{2}\phi$, $-d$, 0, in view of (2.4)) and the midpoint of B_1B_2 (which is the point in E_1 to which M taken as a point in E_2 is homologous, such that its coordinates are $X_M\cos\phi + Y_M\sin\phi$, $-X_M\sin\phi + Y_M\cos\phi$, $Z_M - d$). For the plane α through these points we find the coordinates

$$
(5.6) \qquad
\begin{gathered}
(X_M - Y_M\cot\tfrac{1}{2}\phi)d, \quad (X_M\cot\tfrac{1}{2}\phi + Y_M)d, \quad -2(X_M^2+Y_M^2),\\
(X_M^2+Y_M^2)(2Z_M-d).
\end{gathered}
$$

α_2 is the plane through A_1A_2 and the midpoint of C_1C_2, which is the point in E_2 homologous with M taken as a point in E_1, and thus $(X_M\cos\phi - Y_M\sin\phi$,

$X_M \sin\phi + Y_M \cos\phi,\ Z_M + d)$. The coordinates of α_2 are therefore found by replacing ϕ and d in (5.6) by $-\phi$ and $-d$, and we obtain

$$(5.7)\qquad \begin{gathered}(-X_M - Y_M \cot\tfrac{1}{2}\phi)d,\quad (X_M \cot\tfrac{1}{2}\phi - Y_M)d,\quad -2(X_M^2 + Y_M^2),\\ (X_M^2 + Y_M^2)(2Z_M + d).\end{gathered}$$

It may be verified by means of (3.5) that α_1 and α_2 as given by equations (5.6) and (5.7) do intersect in the join A_1A_2 and hence are homologous.

6. The midpoint plane

If A_1, B_1, C_1 are three non-collinear points of a plane α_1 any point P_1 of α_1 is linearly dependent on them; P_2 on α_2 is linearly dependent on A_2, B_2, C_2. Therefore the midpoint of P_1P_2 is linearly dependent on the midpoints of A_1A_2, B_1B_2 and C_1C_2. Hence, if we consider all points P_1 on α_1, *the locus of the midpoints of the displacement vectors P_1P_2 is a plane.* We shall call it the midpoint plane of the homologous planes α_1 and α_2; it will be denoted by μ and its coordinates by $\mu_1, \mu_2, \mu_3, \mu_4$.

Let α_1 be the plane (u_1, u_2, u_3, u_4) and suppose for the time being that no coordinate is zero. Then if

$$(6.1)\qquad A_1 = (-u_4u_1^{-1}, 0, 0),\qquad B_1 = (0, -u_4u_2^{-1}, 0),\qquad C_1 = (0, 0, -u_4u_3^{-1})$$

are three linearly independent points of α_1, their homologous points are

$$(6.2)\qquad \begin{aligned} A_2 &= (-u_4u_1^{-1}\cos\phi_1, -u_4u_1^{-1}\sin\phi_1, d_1),\\ B_2 &= (u_4u_2^{-1}\sin\phi_1, -u_4u_2^{-1}\cos\phi_1, d_1),\\ C_2 &= (0, 0, -u_4u_3^{-1} + d_1).\end{aligned}$$

The homogeneous coordinates of the midpoints of A_1A_2, B_1B_2, C_1C_2 (with $\phi_1 = 2\phi$, $d_1 = 2d$) follow from (6.1) and (6.2):

$$(6.3)\qquad \begin{aligned} &(-u_4\cos^2\phi, -u_4\sin\phi\cos\phi, u_1d, u_1),\\ &(u_4\sin\phi\cos\phi, -u_4\cos^2\phi, u_2d, u_2),\\ &(0, 0, -u_4 + u_3d, u_3).\end{aligned}$$

Hence (in view of $u_4^2\cos\phi \neq 0$) *we obtain for the coordinates of the midpoint plane* μ:

$$(6.4)\qquad \begin{gathered}\mu_1 = u_1\cos\phi - u_2\sin\phi,\qquad \mu_2 = u_1\sin\phi + u_2\cos\phi,\\ \mu_3 = u_3\cos\phi,\qquad \mu_4 = (-u_3d + u_4)\cos\phi,\end{gathered}$$

it can be easily verified that the same formulas hold for the planes α_1 excluded so far.

From (6.4) it follows that, if μ is given, the planes α_1 and α_2 are respectively

$$\alpha_1: \mu_1 \cos\phi + \mu_2 \sin\phi,\ -\mu_1 \sin\phi + \mu_2 \cos\phi,\ \mu_3/\cos\phi,\ (\mu_3 d + \mu_4)/\cos\phi,$$

(6.5)

$$\alpha_2: \mu_1 \cos\phi - \mu_2 \sin\phi,\ \mu_1 \sin\phi + \mu_2 \cos\phi,\ \mu_3/\cos\phi,\ (-\mu_3 d + \mu_4)/\cos\phi.$$

The conclusion is: *any plane* μ, *in space, is the midpoint plane of one pair of homologous planes* α_1, α_2. A pair of homologous planes may be represented by its midpoint plane in the same way as a pair of homologous points is by its midpoint (2.4.)

From (6.5) follow the planes $\beta = \alpha_1 + \alpha_2$, $\beta' = \alpha_1 - \alpha_2$:

$$\beta: \mu_1 \cos\phi,\ \mu_2 \cos\phi,\ \mu_3/\cos\phi,\ \mu_4/\cos\phi,$$

(6.6)

$$\beta': \mu_2 \sin\phi,\ -\mu_1 \sin\phi,\ 0,\ \mu_3 d/\cos\phi.$$

β and β' pass through the intersection of α_1 and α_2; they are harmonically conjugated with respect to α_1, α_2 and they are moreover mutually orthogonal. *Hence* β *and* β' *are the bisector planes of* α_1, α_2.

Example 1. Show that β' is parallel to the screw axis s.
Example 2. If μ is orthogonal to s show that β coincides with μ, and β' coincides with the plane at infinity.

The coordinates of the intersection of α_1, α_2 in terms of μ_i follow from (6.5), or more easily from (6.6). We obtain

$$Q_1 = \mu_1\mu_3 \tan\phi, \qquad Q_2 = \mu_2\mu_3 \tan\phi, \qquad Q_3 = -(\mu_1^2 + \mu_2^2)\sin\phi\cos\phi,$$

(6.7)

$$Q_1' = \mu_1\mu_3 d - \mu_2\mu_4 \tan\phi, \qquad Q_2' = \mu_2\mu_3 d + \mu_1\mu_4 \tan\phi,$$

$$Q_3' = \mu_3^2 d/\cos^2\phi.$$

Hence, eliminating μ_i, the locus of the intersection of α_1, α_2 is seen to be

$$d(Q_1^2 + Q_2^2) + Q_3 Q_3' \tan\phi = 0, \tag{6.8}$$

which is the equation of the same tetrahedral complex C we found before (5.5).

Summing up some results of this section and of Sections 3 and 5, we obtain the following situation. Any *point* M *of* Σ *is the midpoint of a displacement vector* $\overrightarrow{A_1A_2}$. The line A_1A_2 belongs to the complex C. Through it pass two homologous planes α_1, α_2, their midpoint plane being μ. Obviously μ passes through M and it is determined by it. The relationship between μ and M is seen to be

$$X_M = d_1\mu_2\mu_3^2, \quad Y_M = -d_1\mu_1\mu_3^2, \quad Z_M = \mu_4(\mu_1^2+\mu_2^2)\sin\phi_1, \tag{6.9}$$
$$W_M = -\mu_3(\mu_1^2+\mu_2^2)\sin\phi_1.$$

Conversely we obtain

$$\mu_1 = d_1 Y_M W_M^2, \quad \mu_2 = -d_1 X_M W_M^2, \quad \mu_3 = W_M(X_M^2 + Y_M^2)\sin\phi_1, \tag{6.10}$$
$$\mu_4 = -Z_M(X_M^2+Y_M^2)\sin\phi_1.$$

(6.9) and (6.10) indicate that *there exists a birational cubic relationship between the midpoint M and the midpoint plane* μ (incident with one another) *belonging to the same displacement line.*

7. Homologous lines

Two points (x, y, z, w) and (x', y', z', w') of a line l are transformed by the screw displacement into

$$x\cos\phi - y\sin\phi,\ x\sin\phi + y\cos\phi,\ z + dw,\ w$$

$$x'\cos\phi - y'\sin\phi,\ x'\sin\phi + y'\cos\phi,\ z' + dw',\ w',$$

which enables us to calculate the Plücker coordinates in Σ of the line $\mathrm{l}(\boldsymbol{Q}, \boldsymbol{Q}')$ in terms of its coordinates $\mathrm{l}(\boldsymbol{q}, \boldsymbol{q}')$ in E.

$$\begin{aligned}
Q_1 &= q_1\cos\phi - q_2\sin\phi,\\
Q_1' &= q_1'\cos\phi - q_2'\sin\phi + q_1 d\sin\phi + q_2 d\cos\phi,\\
Q_2 &= q_1\sin\phi + q_2\cos\phi,\\
Q_2' &= q_1'\sin\phi + q_2'\cos\phi - q_1 d\cos\phi + q_2 d\sin\phi,\\
Q_3 &= q_3,\\
Q_3' &= q_3'.
\end{aligned}\tag{7.1}$$

Hence two homologous lines l_1 and l_2 in Σ are given by

$$\begin{aligned}
&\mathrm{l}_1\colon q_1,\quad q_2,\quad q_3,\quad q_1',\quad q_2',\quad q_3'\\
&\mathrm{l}_2\colon q_1\cos\phi - q_2\sin\phi,\quad q_1\sin\phi + q_2\cos\phi,\quad q_3,\\
&\qquad q_1'\cos\phi - q_2'\sin\phi + q_1 d\sin\phi + q_2 d\cos\phi,\\
&\qquad q_1'\sin\phi + q_2'\cos\phi - q_1 d\cos\phi + q_2 d\sin\phi,\quad q_3'.
\end{aligned}\tag{7.2}$$

This relationship for homologous lines l_1 and l_2 is analogous to (1.1), (1.2) for homologous points and (5.2) for homologous planes. If γ is the angle of l_1 (and of l_2) with the screw axis, the angle δ between l_1 and l_2 is given by

$$\sin\tfrac{1}{2}\delta = \sin\gamma \sin\tfrac{1}{2}\phi. \tag{7.3}$$

The distance D between two lines $(\boldsymbol{Q}_1, \boldsymbol{Q}'_1)$ and $(\boldsymbol{Q}_2, \boldsymbol{Q}'_2)$ is given by the formula (see SOMMERVILLE [1934])

$$D = \mathfrak{T}\mathfrak{N}^{-1}, \tag{7.4}$$

in which

$$\mathfrak{T} = \boldsymbol{Q}_1 \cdot \boldsymbol{Q}'_2 + \boldsymbol{Q}_2 \cdot \boldsymbol{Q}'_1, \qquad \mathfrak{N} = |\boldsymbol{Q}_1 \times \boldsymbol{Q}_2|. \tag{7.5}$$

Applying this to (7.2) we find for two homologous lines l_1 and l_2

$$\mathfrak{T} = (q_1^2 + q_2^2)d \sin\phi + 4q_3q'_3 \sin^2(\tfrac{1}{2}\phi), \tag{7.6}$$

and

$$\mathfrak{N}^2 = 4\sin^2(\tfrac{1}{2}\phi)(q_1^2 + q_2^2)\{(q_1^2 + q_2^2)\cos^2(\tfrac{1}{2}\phi) + q_3^2\}. \tag{7.7}$$

Those lines l_1 which are intersected by their homologous line l_2 satisfy $\mathfrak{T} = 0$, or

$$d(q_1^2 + q_2^2) + 2q_3q'_3 \tan(\tfrac{1}{2}\phi) = 0. \tag{7.8}$$

This equation has already appeared in this chapter: as that of the lines in Σ joining two homologous points (3.4), and as that of the lines of intersection of two homologous planes (5.5) (and also (6.8)). It follows that (7.8) represents a tetrahedral complex C_1 in E_1, which coincides with the complex C in Σ; the homologous configuration of C_1 is the complex C_2 which also coincides with C.

The identity of the three sets of lines may be shown directly. Let l_1 and l_2 be two homologous lines intersecting at the point R; this is a point of E_2, A_2 say, which is the point homologous to A_1 on l_1, hence line l_1 coincides with the join A_1A_2. On the other hand R is a point B_1 of E_1, its homologous point B_2 is on l_2 and therefore the line l_2 coincides with the join B_1B_2. Inversely if we have a join A_1A_2 and consider it as l_1 its homologous line l_2 passes through A_2 intersecting l_1 at this point.

From this argument it follows, that through any point R of Σ pass two homologous lines; we now determine the plane $\alpha(R)$ containing them. Let R be the point (X, Y, Z, W). Considered as a point A_1 of E_1 it has the homologous point

$$A_2\colon X\cos\phi - Y\sin\phi,\ X\sin\phi + Y\cos\phi,\ Z + dW,\ W,$$

considered as a point B_2 it is the homologous point of

$$B_1\colon X\cos\phi + Y\sin\phi,\ -X\sin\phi + Y\cos\phi,\ Z - dW,\ W.$$

The plane $\alpha(R)$ passes through R, A_2 and B_1 and therefore through R and the points

$$X\cos\phi,\quad Y\cos\phi,\quad Z,\quad W$$

and

$$Y\sin\phi,\quad -X\sin\phi,\quad -dW,\quad 0.$$

Hence its plane coordinates are

$$\begin{aligned} &U_1 = dYW^2, \qquad U_2 = -dXW^2, \qquad U_3 = (X^2+Y^2)W\sin\phi,\\ &U_4 = -(X^2+Y^2)Z\sin\phi. \end{aligned} \tag{7.9}$$

This establishes a cubic relationship between any point R and the plane $\alpha(R)$ through it, which contains the two homologous lines intersecting at R. The relationship is birational for from (7.9) follows the inverse relation

$$\begin{aligned} &X = dU_2U_3^2, \qquad Y = -dU_1U_3^2, \qquad Z = (U_1^2+U_2^2)U_4\sin\phi,\\ &W = -(U_1^2+U_2^2)U_3\sin\phi. \end{aligned} \tag{7.10}$$

Hence *each plane of Σ contains two intersecting homologous lines.*

8. The linear complex Γ'

Equations (7.9) establish that each point R of Σ is associated with the plane $\alpha(R)$ through R. The direction of n, the line normal to $\alpha(R)$ at R, is:

$$dYW\colon -dXW\colon (X^2+Y^2)\sin\phi,$$

hence the Plücker coordinates of n are

$$\begin{aligned} Q_1 &= -dYW^2,\\ Q_1' &= (X^2+Y^2)Y\sin\phi + dXZW,\\ Q_2 &= dXW^2,\\ Q_2' &= -(X^2+Y^2)X\sin\phi + dYZW,\\ Q_3 &= -(X^2+Y^2)W\sin\phi,\\ Q_3' &= -d(X^2+Y^2)W, \end{aligned} \tag{8.1}$$

which show that the lines n belong to a linear complex Γ' with the equation

$$dQ_3 - Q_3' \sin\phi = 0. \tag{8.2}$$

If a line satisfies (8.2) the point R may be found by inverting (8.1). We obtain

$$X = -Q_2Q_3', \qquad Y = Q_1Q_3', \qquad Z = (-Q_1Q_2' + Q_2Q_1'), \qquad W = Q_1^2 + Q_2^2, \tag{8.3}$$

showing that each line of Γ' is associated with one point R on it.

Γ' has complete screw symmetry about s. It is different from Γ given by (4.6); their intersection is the congruence of lines $Q_3 = 0$, $Q_3' = 0$ intersecting the screw axis orthogonally.

9. The midpoint line

If l_1 and l_2 are homologous lines, any point A_1 on l_1 has its homologous point A_2 on l_2. The restriction that A_1 be on l_1 only means that the coordinates of A_1 are linear functions of a parameter λ. The same is true for those of A_2; hence the coordinates of the midpoint M of A_1A_2 are linear functions of λ. Therefore: the locus of the midpoints is a line m, the midpoint line of l_1 and l_2. Since any point M is the midpoint of a unique point-pair A_1, A_2, any line m is the midpoint line of a unique pair l_1, l_2. This follows from (2.4) which can also be used to show that if Q_i, Q_i' are the coordinates of m, those of $l_1(\boldsymbol{P}, \boldsymbol{P}')$ are

$$\begin{aligned}
P_1 &= Q_1 + Q_2 \tan\tfrac{1}{2}\phi, \\
P_1' &= Q_1' + Q_2' \tan\tfrac{1}{2}\phi - \tfrac{1}{2}d(-Q_1 \tan\tfrac{1}{2}\phi + Q_2), \\
P_2 &= -Q_1 \tan\tfrac{1}{2}\phi + Q_2, \\
P_2' &= -Q_1' \tan\tfrac{1}{2}\phi + Q_2' + \tfrac{1}{2}d(Q_1 + Q_2 \tan\tfrac{1}{2}\phi), \\
P_3 &= Q_3, \\
P_3' &= Q_3'[1 + \tan^2(\tfrac{1}{2}\phi)].
\end{aligned} \tag{9.1}$$

Those of l_2 are obtained by replacing ϕ and d in (9.1) by $-\phi$ and $-d$.

Giving m a certain orientation implies an orientation on l_1 and l_2. If M is the point (X_M, Y_M, Z_M) on m, and $(\cos\alpha, \cos\beta, \cos\gamma)$ the direction of m, the points A_1 on l_1 are by virtue of (2.4)

$$\begin{aligned}
&X_M + \lambda\cos\alpha + (Y_M + \lambda\cos\beta)\tan\tfrac{1}{2}\phi, \\
&-(X_M + \lambda\cos\alpha)\tan\tfrac{1}{2}\phi + (Y_M + \lambda\cos\beta), \\
&Z_M + \lambda\cos\gamma - \tfrac{1}{2}d.
\end{aligned}$$

Hence the direction of l_1 is given by

$$(9.2)\qquad \cos\alpha + \cos\beta \tan\tfrac{1}{2}\phi,\ -\cos\alpha \tan\tfrac{1}{2}\phi + \cos\beta,\ \cos\gamma,$$

and that of l_2 by

$$(9.3)\qquad \cos\alpha - \cos\beta \tan\tfrac{1}{2}\phi,\ \cos\alpha \tan\tfrac{1}{2}\phi + \cos\beta,\ \cos\gamma.$$

For the angle ψ_i between the oriented lines m and l_i we obtain

$$(9.4)\qquad \cos\psi_i = (1 + \tan^2(\tfrac{1}{2}\phi)\sin^2\gamma)^{-1/2}$$

for $i = 1, 2$. Hence the oriented line m makes equal angles with the oriented lines l_1 and l_2.

The direction cosines of the oriented line joining A_1 and A_2 are seen to be

$$-2\tan\tfrac{1}{2}\phi(Y_M + \lambda\cos\beta)/a,\quad 2\tan\tfrac{1}{2}\phi(X_M + \lambda\cos\alpha)/a,\quad d/a,$$

where a is the distance A_1A_2. For the angle η between A_1A_2 and m we obtain

$$(9.5)\qquad \cos\eta = (1/a)\{2\tan\tfrac{1}{2}\phi(-Y_M\cos\alpha + X_M\cos\beta) + d\cos\gamma\},$$

which implies that $a\cos\eta$ is independent of λ and leads to the conclusion that *the projections of all joins A_1A_2 onto the midpoint line* m *are of equal length for all points on* l_1 *and* l_2.

We have in particular: if any join A_1A_2 intersects m orthogonally all joins A_1A_2 do so. This occurs when the right-hand side of (9.5) is zero. For the line m, through X_M, Y_M, Z_M and with direction $\cos\alpha$, $\cos\beta$, $\cos\gamma$ we have

$$Q_3 = -\cos\gamma,\qquad Q_3' = X_M\cos\beta - Y_M\cos\alpha.$$

Hence $\cos\eta = 0$ if

$$dQ_3 - 2Q_3'\tan\tfrac{1}{2}\phi = 0,$$

but this is identical to (4.6). Therefore: all joins A_1A_2 intersect m orthogonally if m is a line of the linear complex Γ. In this situation the endpoints of the common perpendicular of l_1, l_2 are a pair of homologous points and all points A_1 of l_1 may be displaced to their position A_2 on l_2 by a half-turn (i.e., a rotation of π radians) about m.

This property is a special case of the more general statement: *Any line* l_1 *may be transformed into its homologous line* l_2 *by a rotation about* m′ *which is the line conjugate to the midpoint line* m *with respect to* Γ. The proof follows: If line m is arbitrary all normal planes of A_1A_2 are null-planes of points M on m with respect to the null correlation associated with Γ. Therefore (as shown in

Section 4) all normal planes pass through one line m′, conjugate to m with respect to Γ. Let A_1, B_1 be two points on l_1 and A_2, B_2 the points homologous to them. Since line m′ lies in the plane normal to the join A_1A_2, it is always possible to determine a plane α′ perpendicular to m′ which contains A_1 and A_2. Similarly B_1 and B_2 are in a plane β′ parallel to α′ (Fig. 2). As $A_1B_1 = A_2B_2$ their projections A_1B_1' and A_2B_2' on α′ are equal. Therefore we have in α′ the points A_1, A_2, B_1', B_2' such that $A_1B_1' = A_2B_2'$. Planar kinematics implies that there is a rotation in α′ such that A_1, B_1' are transformed into A_2, B_2'. As the normal lines in α', of joins A_1A_2 and $B_1'B_2'$, pass through the intersection T' of m′ and α′, this point must be the rotation center. Therefore there is a rotation in space about m′ such that A_1, B_1 are transformed into A_2 and B_2, which implies that all points of l_1 are brought into their homologous positions. If m belongs to Γ the conjugate line m′ coincides with it and we have the special case of the rotation being a half-turn, as mentioned in the foregoing.

What are the loci of l_1 and l_2 if m belongs to Γ? If the coordinates of m are Q_i, Q_i' two points of m are $(0, Q_3', -Q_2', Q_1)$ and $(-Q_3', 0, Q_1', Q_2)$. The corresponding points of l_1 are according to (2.4) $(Q_3'\tan\frac{1}{2}\phi,$ $Q_3', -Q_2' - \frac{1}{2}dQ_1, Q_1)$ and $(-Q_3', Q_3'\tan\frac{1}{2}\phi, Q_1' - \frac{1}{2}dQ_2, Q_2)$. Hence for the Plücker coordinates (q_i, q_i') of l_1 we have $q_3 : q_3' = Q_3\cos^2(\frac{1}{2}\phi) : Q_3'$; m belongs to Γ if $dQ_3 - 2Q_3'\tan\frac{1}{2}\phi = 0$ and therefore l_1 satisfies $dq_3 - q_3'\sin\phi = 0$, but that is the equation of Γ′. Thus we have proved that *when the midpoint line of*

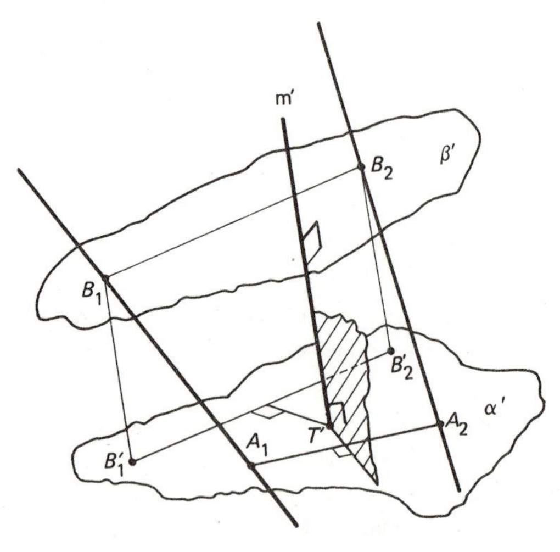

Fig. 2.

l_1 *and* l_2 *belongs to* Γ (which implies that l_1 and l_2 are interchanged by a half-turn about m) *the lines* l_1 *and* l_2 *belong to linear complexes* Γ_1' *and* Γ_2' *in* E_1 *and* E_2 *respectively, both coinciding with the complex* Γ' *in* Σ.

10. The line-bisectors

The *spatial distance* of two lines l_1, l_2 is defined as the pair of numbers D, δ, where D stands for the length of their common perpendicular ($D \geq 0$) and δ is their angle ($0 \leq \delta \leq \pi$); the latter is an ambiguous concept, δ can be replaced by $\pi - \delta$.

Let O be the midpoint of the common perpendicular of l_1 and l_2, l_1' and l_2' are the lines through O parallel to l_1 and l_2, b and b′ are the bisectors of l_1', l_2' (Fig. 3, where O is taken as the origin of a Cartesian frame, with O_Z along the common perpendicular and O_X, O_Y along b and b′). We define b and b′ to be *the bisectors of* l_1, l_2.

If $\tan(\frac{1}{2}\delta) = u$ the lines l_1 and l_2 are represented by $y + ux = z + \frac{1}{2}D = 0$ and $y - ux = z - \frac{1}{2}D = 0$. This implies that l_1 and l_2 are interchanged by the transformation $X = x$, $Y = -y$, $Z = -z$, that is the half-turn about O_X. But each line n which intersects O_X orthogonally is invariant for this transformation. The conclusion is that any line n has equal spatial distances to l_1 and l_2; in the same way we prove that each line n′ intersecting O_Y orthogonally has this property. Hence *any line which intersects orthogonally one of the bisectors* b *or* b′, of l_1 and l_2, *has equal spatial distances to* l_1 *and* l_2. There are therefore two linear congruences of such lines.

Example 3. Show that the sets n and n′ and two other linear congruences are together the locus of lines with equal distances to l_1 and l_2, (for instance by making use of the formulas for the angle and the distance of two lines given by their Plücker vectors).
Example 4. Show that there are sixteen lines each with equal spatial distances to three given lines l_1, l_2, l_3.
Example 5. Consider the special case of the three lines being coplanar or concurrent.

The definition of D, δ has the disadvantage of being ambiguous. (We have the same situation for two lines in plane geometry.) It may be removed if we deal with orientated or *directed* lines, sometimes called *spears.* If l_1 and l_2 are given a direction, the parallel lines l_1' and l_2' are also directed (Fig. 4.), their angle δ has a well defined sense and so has the spatial distance D, δ of l_1 and l_2. The two bisectors b and b′ are no longer equivalent: we can distinguish between the *internal bisector*, b say, and the *external bisector* b′. (b and b′ are

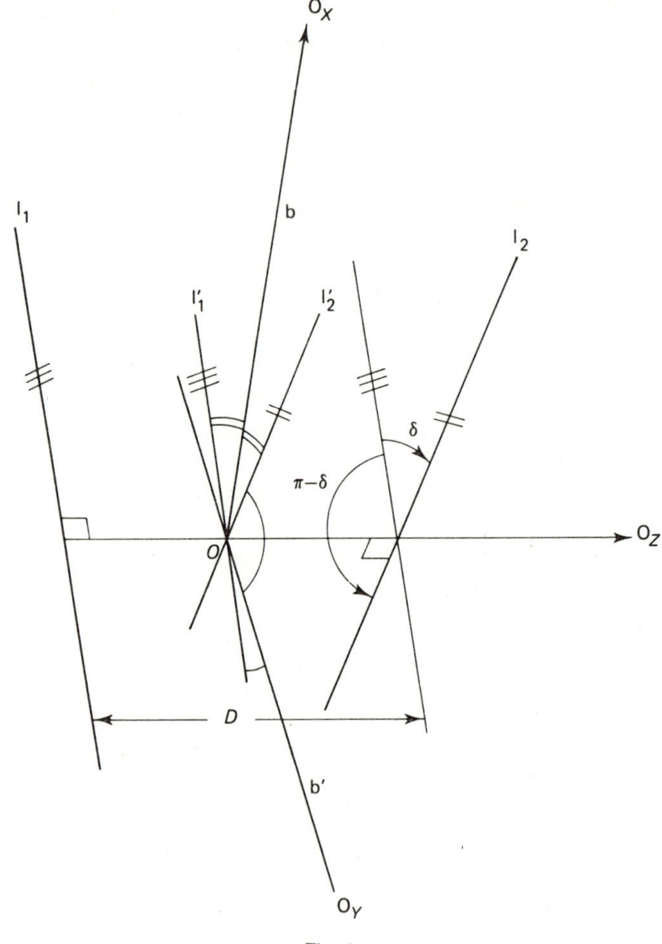

Fig. 3.

considered as undirected lines.) The spatial distance for two directed lines changes in general if we change the direction of one of them, but it remains the same if the directions of both are replaced by the opposite directions.

Let l_1 and l_2 be two directed lines and b, taken as O_X in the figures, be their internal bisector; let furthermore n be a line intersecting O_X orthogonally and let a direction be taken on it. If l_1 is reflected into O_X it coincides with the directed line l_2, but the direction of n is reversed. If on the other hand n′ is a

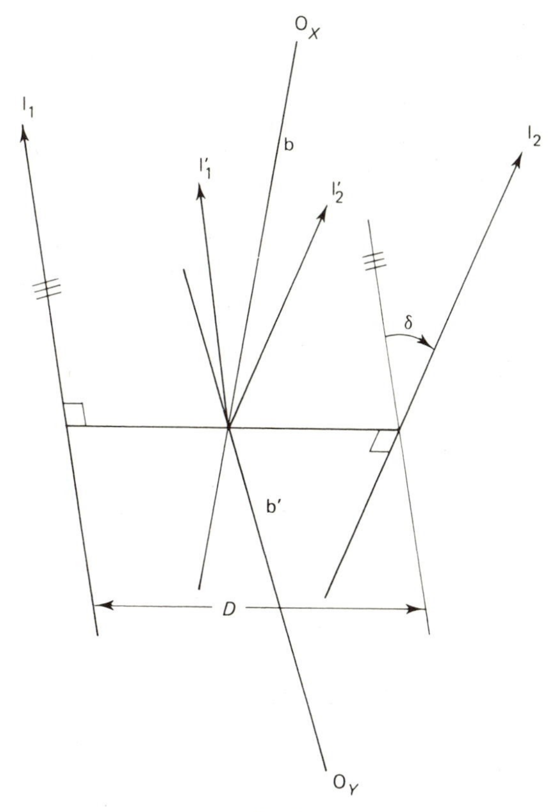

Fig. 4.

line orthogonally intersecting O_Y the analogous argument shows that n′ has (with both of its directions) equal distances to l_1 and l_2. The conclusion is: *the locus of lines with equal distances to two directed lines is the congruence of lines intersecting their external bisector orthogonally.*

A consequence, (which will be elaborated upon in Chapter IV, Section 7) is: if three directed lines l_1, l_2, l_3 are given there is one line l_c with equal distances to all three; it is the common perpendicular of the external bisectors of the pairs (l_2, l_3), (l_3, l_1) and (l_1, l_2); l_c will be called the central line of l_1, l_2, l_3.

Example 6. Show that l_c remains the same if the direction of each line l_i is reversed.
Example 7. Determine l_c if the (directed) lines l_i are coplanar.
Example 8. Determine l_c if the lines l_i pass through one point.

Let s_{12}, taken as the Z-axis, be the axis of a screw displacement with angle ϕ $(0 \leq \phi \leq 2\pi)$ and translation distance d $(d \geq 0)$. A directed line l_1 is transformed into the directed line l_2 (Fig. 5); $A_1'A_1$ and $A_2'A_2$, both equal to p, are respectively the (directed) common perpendiculars of l_1 and s_{12}, and of l_2 and s_{12}. The internal bisector of $A_1'A_1$ and $A_2'A_2$ is taken as the X-axis; A_1B_1 and A_2B_2, both of length q are direction vectors on l_1 and l_2.

The screw displacement is given by

$$X = x\cos\phi - y\sin\phi, \qquad Y = x\sin\phi + y\cos\phi, \qquad Z = z + d. \tag{10.1}$$

Furthermore we have for $A_1 = (x_1, y_1, z_1)$:

$$x_1 = X_1 = p\cos\tfrac{1}{2}\phi, \qquad y_1 = Y_1 = -p\sin\tfrac{1}{2}\phi, \qquad z_1 = Z_1 = -\tfrac{1}{2}d, \tag{10.2}$$

which is transformed by (10.1) into $A_2 = (X_2, Y_2, Z_2)$ with

$$X_2 = p\cos\tfrac{1}{2}\phi, \qquad Y_2 = p\sin\tfrac{1}{2}\phi, \qquad Z_2 = \tfrac{1}{2}d. \tag{10.3}$$

If the angle between A_1B_1 and O_Z is equal to α, we have $B_1 = (x_1 + x_1', y_1 + y_1', z_1 + z_1')$, with

$$\begin{aligned} x_1' = X_1' = q\sin\alpha\sin\tfrac{1}{2}\phi, \qquad y_1' = Y_1' = q\sin\alpha\cos\tfrac{1}{2}\phi, \\ z_1' = Z_1' = q\cos\alpha. \end{aligned} \tag{10.4}$$

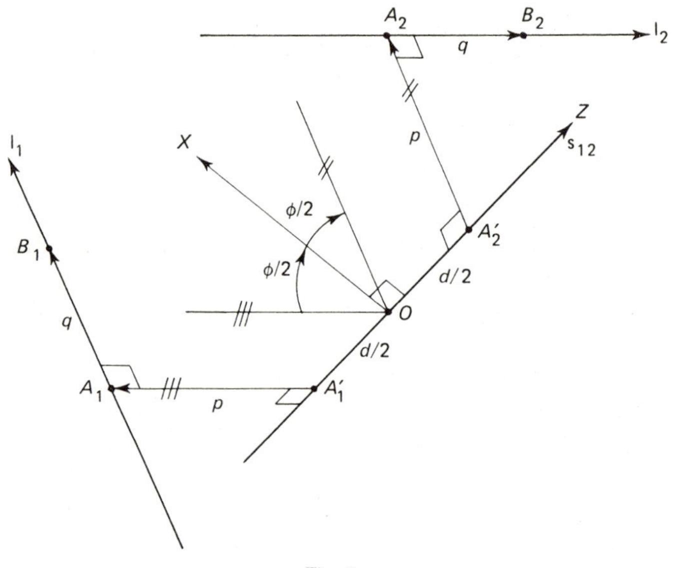

Fig. 5.

B_1 is by (10.1) transformed into $B_2 = (X_2 - X_1', Y_2 + Y_1', Z_2 + Z_1')$. The reflection into O_X is given by $X' = x$, $Y' = -y$, $Z' = -z$. It transforms A_1 into A_2 and the line l_1 into l_2 but with the opposite orientation. Each line n which intersects O_X orthogonally is invariant but its direction is also reversed. *Hence any line* n *orthogonally intersecting* O_X (with arbitrary direction) *has equal distances to the directed lines* l_1 *and* l_2.

Example 9. Show that the external bisector of the directed lines l_1 and l_2 coincides with the internal bisector of the directed perpendiculars $A_1'A_1$ and $A_2'A_2$.
Example 10. Consider the special cases $p = 0$ and $d = 0$.

11. The instantaneous case

Instantaneous kinematics up to the first order may be considered a limit case of two positions theory. We suppose that $\phi \to 0$ and $d \to 0$, but that $d/\phi \to \sigma$, the pitch of the instantaneous screw. Two homologous points A_1 and A_2 coincide, but their join has as limit position the tangent to the orbit described by the moving point A. The locus of these lines is still a quadratic complex C, the equation of which is the limit case of (3.4) and therefore reads

$$\sigma(Q_1^2 + Q_2^2) + Q_3 Q_3' = 0. \tag{11.1}$$

Furthermore the linear complexes Γ and Γ' coincide in the instantaneous case, the limit of both (4.6) and (8.2) being

$$\sigma Q_3 - Q_3' = 0. \tag{11.2}$$

We mention only one application. If a line moves continuously it generates a ruled surface. A line which is intersected by the next consecutive generator is called a torsal line of the surface. Hence the theorem: *those lines of a moving space which are at a certain moment torsal lines of their trajectory surface belong to the quadratic complex* (11.1).

The formulas (11.1) and (11.2) could also have been derived by means of the first order results of Chapter II (6.3) and (6.6). The Plücker coordinates of the line joining (x, y, z) to its consecutive point follow from the matrix

$$\begin{Vmatrix} x & y & z & 1 \\ -y & x & \sigma_0 & 0 \end{Vmatrix},$$

and they are therefore

$$Q_1 = y, \qquad Q_2 = -x, \qquad Q_3 = -\sigma_0,$$
$$Q_1' = -xz + \sigma_0 y, \qquad Q_2' = -yz - \sigma_0 x, \qquad Q_3' = x^2 + y^2;$$

eliminating x, y, z the locus of these lines is the quadratic complex C with the equation $\sigma_0(Q_1^2 + Q_2^2) + Q_3 Q_3' = 0$.

Example 11. Show from Chapter II (6.6) that C is also the locus of the intersection of two consecutive homologous planes.

Example 12. If (x_i, y_i, z_i), $i = 1, 2$, are two distinct points on a line l, show that l and its consecutive homologous line intersect if the points $(x_1, y_1, z_1, 1)$, $(x_2, y_2, z_2, 1)$, $(-y_1, x_1, \sigma_0, 0)$ and $(-y_2, x_2, \sigma_0, 0)$ are coplanar; show that the locus of the lines l with this property is also the complex C, in accordance with the general two positions theory.

12. Another method to derive the screw displacement equations

So far we have used special coordinate systems in order to simplify our equations. Sometimes however it is desirable to express the screw transformation in terms of a general coordinate system. We have already given one form of the general rotation matrix in Chapter I, Example 4, and will develop it further in Chapter VI. Here we derive another useful (albeit, less elegant) form. Before deriving the general formulas we will develop a very useful equation known as *Rodrigues' formula for a general screw displacement.* We obtain the formula as follows: we take point P_1 and consider its displacement to position P_2 in Σ. We consider its displacement as one whereby P_1 first rotates about the screw axis s to position P_2^r and then translates parallel to s from P_2^r to position P_2. It follows that P_1 and P_2^r lie in the same plane normal to the screw axis. We take S_P as the point where s cuts this plane, P_m the point where the perpendicular bisector of chord $P_1P_2^r$ cuts the chord, and the vectors $\boldsymbol{r}_1$ and $\boldsymbol{r}_2$ as respectively the position vectors of P_1 and P_2^r relative to S_P. Then from Fig. 6 it follows that

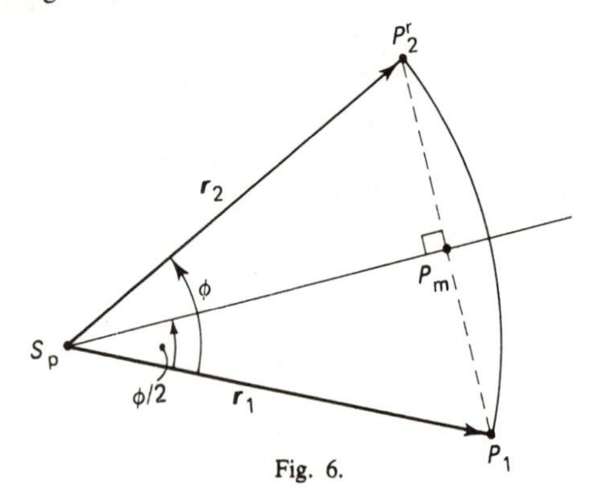

Fig. 6.

$$\tan(\phi/2) = P_m P_1 / P_m S_p = |\boldsymbol{r}_2 - \boldsymbol{r}_1| / |\boldsymbol{r}_1 + \boldsymbol{r}_2|, \tag{12.1}$$

where ϕ is the rotation angle of the displacement about the screw axis. If we introduce the unit vector $\boldsymbol{s}$ along the screw axis (with its positive sense defined in accordance with the right-hand rule relative to the rotation angle ϕ), we may write the vector expression

$$\boldsymbol{r}_2 - \boldsymbol{r}_1 = (\tan(\phi/2))\boldsymbol{s} \times (\boldsymbol{r}_2 + \boldsymbol{r}_1) \tag{12.2}$$

which follows from (12.1) as far as magnitude is concerned, and in direction from the figure.

(12.2) is Rodrigues' formula for a planar displacement (Chapter VIII) with the origin at the center of the displacement.

If we consider the origin of coordinates located elsewhere on the screw axis say at point S_0 then the position vectors of P_1 and P_2^r are $\boldsymbol{R}_1$ and $\boldsymbol{R}_2$, as shown in Fig. 7, and we have $\boldsymbol{r}_1 = \boldsymbol{R}_1 - (S_0 S_P)\boldsymbol{s}$, $\boldsymbol{r}_2 = \boldsymbol{R}_2 - (S_0 S_P)\boldsymbol{s}$ which when substituted into (12.2) yields

$$\boldsymbol{R}_2 - \boldsymbol{R}_1 = (\tan(\phi/2))\boldsymbol{s} \times (\boldsymbol{R}_2 + \boldsymbol{R}_1) \tag{12.3}$$

which is Rodrigues' formula for a spherical displacement (Chapter VII) about an axis passing through the origin of coordinates.

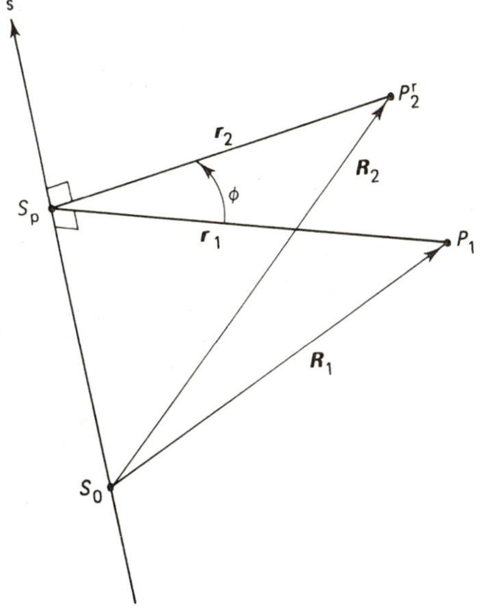

Fig. 7.

If the origin of coordinates, O, does not lie on the screw axis, we take $\boldsymbol{S}_0$ as the position vector from O to S_0, and $\boldsymbol{P}_1$ and $\boldsymbol{P}_2^r$ respectively as the position vectors of P_1 and P_2^r from O. We have $\boldsymbol{R}_1 = \boldsymbol{P}_1 - \boldsymbol{S}_0$, $\boldsymbol{R}_2 = \boldsymbol{P}_2^r - \boldsymbol{S}_0$ which when substituted into (12.3) yields

$$\boldsymbol{P}_2^r - \boldsymbol{P}_1 = (\tan(\phi/2))\boldsymbol{s} \times (\boldsymbol{P}_2^r + \boldsymbol{P}_1 - 2\boldsymbol{S}_0), \tag{12.4}$$

which is Rodrigues' formula for general spherical displacement measured relative to any arbitrary origin O.

For a general spatial displacement we must add the screw translation, d, which carries P_2^r to position P_2. Thus we substitute $\boldsymbol{P}_2^r = \boldsymbol{P}_2 - d\boldsymbol{s}$ into (12.4), the result is

$$\boldsymbol{P}_2 - \boldsymbol{P}_1 = (\tan(\phi/2))\boldsymbol{s} \times (\boldsymbol{P}_2 + \boldsymbol{P}_1 - 2\boldsymbol{S}_0) + d\boldsymbol{s}, \tag{12.5}$$

which is Rodrigues' formula for a general screw displacement.

Formulas (12.2)–(12.5) were developed by Rodrigues [1840] and used by him to solve various problems dealing with the resultant of a series of displacements. Since they involve $\boldsymbol{P}_2$ and $\boldsymbol{P}_1$ on both sides of the equal sign, they usually require further manipulation. An alternative development starts with the fact that

$$\boldsymbol{r}_2 = \boldsymbol{r}_1 \cos\phi + \boldsymbol{s} \times \boldsymbol{r}_1 \sin\phi, \tag{12.6}$$

which follows from the planar figure (Fig. 6) if we drop the normal from P_2^r to $\boldsymbol{r}_1$, and keep in mind $|\boldsymbol{r}_1| = |\boldsymbol{r}_2|$. Now substituting $\boldsymbol{r}_2 = \boldsymbol{R}_2 - (\boldsymbol{R}_2 \cdot \boldsymbol{s})\boldsymbol{s}$, $\boldsymbol{r}_1 = \boldsymbol{R}_1 - (\boldsymbol{R}_1 \cdot \boldsymbol{s})\boldsymbol{s}$ and using the facts that $\boldsymbol{s} \times \boldsymbol{r}_1 = \boldsymbol{s} \times \boldsymbol{R}_1$ and $\boldsymbol{s} \cdot \boldsymbol{R}_1 = \boldsymbol{s} \cdot \boldsymbol{R}_2$ yields

$$\boldsymbol{R}_2 = \boldsymbol{R}_1 \cos\phi + \boldsymbol{s} \times \boldsymbol{R}_1 \sin\phi + (\boldsymbol{R}_1 \cdot \boldsymbol{s})\boldsymbol{s}(1 - \cos\phi). \tag{12.7}$$

Substituting

$$\cos\phi = (1 - \tan^2(\phi/2))/(1 + \tan^2(\phi/2)),$$

$$\sin\phi = 2(\tan(\phi/2))/(1 + \tan^2(\phi/2))$$

and rearranging (12.7) we get

$$\boldsymbol{R}_2 = \boldsymbol{R}_1 + \boldsymbol{s} \times \boldsymbol{R}_1 [2(\tan(\phi/2))/(1 + \tan^2(\phi/2))] + [2\tan^2(\phi/2)/(1 + \tan^2(\phi/2))][(\boldsymbol{R}_1 \cdot \boldsymbol{s})\boldsymbol{s} - \boldsymbol{R}_1]$$

but since $(\boldsymbol{R}_1 \cdot \boldsymbol{s})\boldsymbol{s} - \boldsymbol{R}_1 = \boldsymbol{s} \times (\boldsymbol{s} \times \boldsymbol{R}_1)$ we have

$$\boldsymbol{R}_2 = \boldsymbol{R}_1 + [2(\tan(\phi/2))/(1 + \tan^2(\phi/2))]\boldsymbol{s} \times (\boldsymbol{R}_1 + \tan(\phi/2)\boldsymbol{s} \times \boldsymbol{R}_1). \tag{12.8}$$

If we substitute $\boldsymbol{R}_1 = \boldsymbol{P}_1 - \boldsymbol{S}_0$, $\boldsymbol{R}_2 = \boldsymbol{P}_2 - d\boldsymbol{s} - \boldsymbol{S}_0$ we get from (12.7)

$$\begin{aligned}\boldsymbol{P}_2 = (\boldsymbol{P}_1 - \boldsymbol{S}_0)\cos\phi + \boldsymbol{s}\times(\boldsymbol{P}_1 - \boldsymbol{S}_0)\sin\phi \cdot \\ + [(\boldsymbol{P}_1 - \boldsymbol{S}_0)\cdot\boldsymbol{s}]\boldsymbol{s}(1-\cos\phi) + \boldsymbol{S}_0 + d\boldsymbol{s},\end{aligned} \tag{12.9}$$

and

$$\begin{aligned}\boldsymbol{P}_2 = \boldsymbol{P}_1 + (2(\tan(\phi/2))/(1+\tan^2(\phi/2)))\boldsymbol{s} \\ \times[\boldsymbol{P}_1 - \boldsymbol{S}_0 + \tan(\phi/2)\boldsymbol{s}\times(\boldsymbol{P}_1 - \boldsymbol{S}_0)] + d\boldsymbol{s}.\end{aligned} \tag{12.10}$$

Equations (12.9) and (12.10) each give equivalent vector forms of the general displacement equation, one being in terms of the full angle and the other the half-angle. In both cases the point P_1 is brought to position P_2 by a screw displacement ϕ, d about an axis with direction $\boldsymbol{s}$ passing through a point S_0. If the screw axis passes through the origin of coordinates in Σ, all the terms with $\boldsymbol{S}_0$ disappear from (12.9) and (12.10).

Example 13. Show that the *dyadic* form of (12.9) is

$$\boldsymbol{P}_2 = \mathbf{R}\cdot(\boldsymbol{P}_1 - \boldsymbol{S}_0) + \boldsymbol{S}_0 + d\boldsymbol{s}$$

where the rotation dyadic $\mathbf{R}$ is

$$\mathbf{R} = \mathbf{U}\cos\phi - \boldsymbol{s}\times\mathbf{U}\sin\phi + \boldsymbol{ss}(1-\cos\phi);$$

$\mathbf{U}$ is the unit dyadic. (For definitions see, for example, WILLS [1931].)

If we expand (12.9) in terms of the scalar components $\boldsymbol{P}_2(X_2, Y_2, Z_2)$, $\boldsymbol{P}_1(X_1, Y_1, Z_1)$, $\boldsymbol{S}_0(S_{0X}, S_{0Y}, S_{0Z})$, $\boldsymbol{s}(s_X, s_Y, s_Z)$, we get the matrix equation

$$\left\|\begin{matrix} X_2 \\ Y_2 \\ Z_2 \end{matrix}\right\| = \mathbf{A}\left\|\begin{matrix} X_1 \\ Y_1 \\ Z_1 \end{matrix}\right\| + \left\|\begin{matrix} d_1 \\ d_2 \\ d_3 \end{matrix}\right\| \tag{12.11}$$

where the elements of $\mathbf{A}$ are a_{ij}:

$$\begin{aligned}
a_{11} &= (s_X^2 - 1)(1-\cos\phi) + 1, \\
a_{12} &= s_X s_Y(1-\cos\phi) - s_Z\sin\phi, \\
a_{13} &= s_X s_Z(1-\cos\phi) + s_Y\sin\phi, \\
a_{21} &= s_Y s_X(1-\cos\phi) + s_Z\sin\phi, \\
a_{22} &= (s_Y^2 - 1)(1-\cos\phi) + 1, \\
a_{23} &= s_Y s_Z(1-\cos\phi) - s_X\sin\phi, \\
a_{31} &= s_Z s_X(1-\cos\phi) - s_Y\sin\phi, \\
a_{32} &= s_Z s_Y(1-\cos\phi) + s_X\sin\phi, \\
a_{33} &= (s_Z^2 - 1)(1-\cos\phi) + 1,
\end{aligned} \tag{12.12}$$

and

$$
\begin{aligned}
d_1 &= ds_X - S_{0_X}(a_{11}-1) - S_{0_Y}a_{12} - S_{0_Z}a_{13},\\
d_2 &= ds_Y - S_{0_X}a_{21} - S_{0_Y}(a_{22}-1) - S_{0_Z}a_{23},\\
d_3 &= ds_Z - S_{0_X}a_{31} - S_{0_Y}a_{32} - S_{0_Z}(a_{33}-1).
\end{aligned} \tag{12.13}
$$

Hence (12.11) is a description of the displacement in terms of the screw parameters: ϕ (the rotation angle), d (the translation distance), s (the unit direction along the screw), S_0 (a vector from the origin of coordinates to any point on the screw axis). At first it might seem as though we have eight rather than six parameters. However s must satisfy $s \cdot s = 1$ and so it represents only two parameters. Similarly since S_0 is any point on the screw axis the vector S_0 only requires two parameters, hence one might set say $S_{0_Z} = 0$ or, what is more usual, take S_0 as the foot of the normal from the origin, in which case $S_0 \cdot s = 0$. The signs of ϕ, d are related to s by the right-hand-screw rule.

Example 14. Show that the a_{ij} of (12.12) are the elements of an orthogonal matrix.

Example 15. Show that (12.12) and the form of **A** given in Chapter I, Example 4 are identical if $c_0 = \cos\frac{1}{2}\phi$, $c_1 = s_X \sin\frac{1}{2}\phi$, $c_2 = s_Y \sin\frac{1}{2}\phi$, $c_3 = s_Z \sin\frac{1}{2}\phi$. This follows directly if we use half-angles in (12.12): $a_{11} = 2(s_X^2 - 1)\sin^2\frac{1}{2}\phi + 1$, $a_{12} = 2\sin\frac{1}{2}\phi(s_X s_Y \sin\frac{1}{2}\phi - s_Z \cos\frac{1}{2}\phi)$, $a_{13} = 2\sin\frac{1}{2}\phi(s_X s_Z \sin\frac{1}{2}\phi + s_Y \cos\frac{1}{2}\phi)$, $a_{21} = 2\sin\frac{1}{2}\phi(s_X s_Y \sin\frac{1}{2}\phi + s_Z \cos\frac{1}{2}\phi)$, etc.

Example 16. Show that if matrix **A** is given, the equivalent screw rotation can be determined from the relation

$$\phi = \arccos[(a_{11} + a_{22} + a_{33} - 1)/2]$$

and the axis direction from

$$s_X = (a_{32} - a_{23})/(2\sin\phi), \qquad s_Y = (a_{13} - a_{31})/(2\sin\phi),$$
$$s_Z = (a_{21} - a_{12})/(2\sin\phi).$$

Example 17. Show that (12.11) also follows from Rodrigues' formula (12.5). (Here, after substituting the scalar components, we must solve a system of three linear equations for X_2, Y_2, Z_2 and then use $\tan(\phi/2) = (1-\cos\phi)/\sin\phi$ — the process is rather lengthy.)

Example 18. Show that the line with direction l_1 (in both E_1 and Σ) is displaced by the screw displacement into the line with direction l_2, in Σ, as given by $l_2 = \mathbf{A}l_1$ with **A** defined by (12.12) or its equivalent.

Example 19. If we take an infinitesimally small displacement then $\phi \to \Delta\phi$, $d \to \Delta d$, $P_2 \to P + \Delta P$, $P_1 \to P$, where Δ denotes a first order infinitesimal. Using this limiting process and then dividing by Δt, show that Rodrigues' formula (12.5) yields

$$\dot{P} = \omega s \times (P - S_0) + \dot{d}s, \quad \text{with } \omega = \dot{\phi}.$$

Also show that (12.11) yields

$$\dot{P} = \mathbf{B}P + D$$

where **B** is skew and has elements $b_{11} = 0$, $b_{12} = -\omega s_Z$, $b_{13} = \omega s_Y$, etc., and D has elements $d_1 = \dot{d}s_X + \omega(S_{0_Y}s_Z - S_{0_Z}s_Y)$, etc.

Equation (12.5) can be used to determine the screw parameters: Assume we are given two positions of E in terms of the positions of three non-collinear points say P, Q, R. We write (12.5) twice: once for point P and once for point Q. Subtracting the Q equation from the P equation yields

$$(\boldsymbol{P}_2 - \boldsymbol{Q}_2) - (\boldsymbol{P}_1 - \boldsymbol{Q}_1) = (\tan(\phi/2))\boldsymbol{s} \times [(\boldsymbol{P}_2 - \boldsymbol{Q}_2) + (\boldsymbol{P}_1 - \boldsymbol{Q}_1)]. \tag{12.14}$$

Forming the cross product with $[(\boldsymbol{R}_2 - \boldsymbol{Q}_2) - (\boldsymbol{R}_1 - \boldsymbol{Q}_1)]$ yields

$$\begin{aligned} \tan(\phi/2)\boldsymbol{s} = \{&[(\boldsymbol{R}_2 - \boldsymbol{Q}_2) - (\boldsymbol{R}_1 - \boldsymbol{Q}_1)] \times [(\boldsymbol{P}_2 - \boldsymbol{Q}_2) - (\boldsymbol{P}_1 - \boldsymbol{Q}_1)]\}/ \\ &\{[(\boldsymbol{R}_2 - \boldsymbol{Q}_2) - (\boldsymbol{R}_1 - \boldsymbol{Q}_1)] \cdot [(\boldsymbol{P}_2 - \boldsymbol{Q}_2) + (\boldsymbol{P}_1 - \boldsymbol{Q}_1)]\}. \end{aligned} \tag{12.15}$$

In obtaining (12.15) we have made use of the fact that $[(\boldsymbol{R}_2 - \boldsymbol{Q}_2) - (\boldsymbol{R}_1 - \boldsymbol{Q}_1)]$ is perpendicular to s, which is obvious if one substitutes $\boldsymbol{R}$ for $\boldsymbol{P}$ in (12.14).

If we operate on (12.5) with $\boldsymbol{s} \times$ and also set $\boldsymbol{s} \cdot \boldsymbol{S}_0 = 0$ we have for the normal, $\boldsymbol{S}_{0_n}$, to the screw from the origin,

$$\boldsymbol{S}_{0_n} = \tfrac{1}{2}[\boldsymbol{P}_1 + \boldsymbol{P}_2 + (s \times (\boldsymbol{P}_2 - \boldsymbol{P}_1)/\tan(\phi/2)) - s \cdot (\boldsymbol{P}_2 + \boldsymbol{P}_1)s]. \tag{12.16}$$

Finally operating on (12.5) with $\boldsymbol{s} \cdot$ yields

$$d = \boldsymbol{s} \cdot (\boldsymbol{P}_2 - \boldsymbol{P}_1). \tag{12.17}$$

From (12.15) we obtain ϕ and s, from (12.16) $\boldsymbol{S}_0$, and from (12.17) d. Hence the screw is completely determined from three non-collinear points.

Example 20. Show that if the displacement is such that (at least) one point of E remains fixed in Σ (i.e., it is either spherical (Chapter VII) or planar (Chapter VIII) we may set $\boldsymbol{Q}_1 = \boldsymbol{Q}_2 = 0$ in (12.15).

Example 21. Show that for the limiting case described in Example 19 formula (12.15) yields

$$\omega\boldsymbol{s} = ((\dot{\boldsymbol{R}} - \dot{\boldsymbol{Q}}) \times (\dot{\boldsymbol{P}} - \dot{\boldsymbol{Q}}))/((\dot{\boldsymbol{R}} - \dot{\boldsymbol{Q}}) \cdot (\boldsymbol{P} - \boldsymbol{Q})),$$

(12.16) yields

$$\boldsymbol{S}_{0_n} = \boldsymbol{P} + ((s \times \dot{\boldsymbol{P}})/\omega) - s \cdot \boldsymbol{P}s,$$

and (12.17) yields $\dot{d} = \boldsymbol{s} \cdot \dot{\boldsymbol{P}}$, where $\dot{\boldsymbol{P}}$, $\dot{\boldsymbol{Q}}$, $\dot{\boldsymbol{R}}$ are the velocities of points P, Q, and R.

Example 22. Discuss the special cases for which (12.15) and the limiting case given in the previous example become indeterminate. (See for example STIELTJES [1884].)

The theorem: *a general displacement in three-space can be represented by a unique screw displacement* is usually referred to as Chasles' theorem. (Although, both Mozzi and Cauchy seem to have preceeded Chasles with this result.) We mention two important special cases: When $d = 0$ the displacement is a *pure rotation*. This causes no special difficulty, all the screw parameters are still unique and remain defined by the equations of this section. However, if $\phi = 0$ the displacement is a *pure translation* and the screw axis is no longer a unique line: any line parallel to a displacement vector can be taken as the screw axis. The displacement of every point is exactly the same, and so the displacement vector of any one point uniquely defines the magnitude of the translation, d, and the direction of the screw, $\mathbf{s}$. Even though the direction of the screw is uniquely defined, its location is completely arbitrary.

An alternative, which is analogous to a well-known result in planar kinematics, is to choose the line at infinity in the planes normal to the displacement vector. The axis then is unique, and the vector of translation $d\mathbf{s}$ is now regarded as the result of an infinitesimal pure rotation, i.e. $\phi \to 0$, about an infinitely distant axis.

Example 23. Verify that the foregoing discussion also applies to the limiting case described in Example 19.

CHAPTER IV

THREE POSITIONS THEORY

1. The screw triangle

We consider three positions E_1, E_2, E_3 of a moving space E with respect to a fixed space Σ.

The displacement $E_1 \rightarrow E_2$ may be performed by the screw motion $s_{12}(\phi_{12}, d_{12})$ about the line s_{12}, ϕ_{12} and d_{12} being the rotation angle and the translation distance respectively. The displacement $E_2 \rightarrow E_3$ takes place by means of the screw motion $s_{23}(\phi_{23}, d_{23})$. As a line is determined by four parameters and a screw displacement therefore by six, a three positions configuration needs twelve pieces of data. $s_{12}(\phi_{12}, d_{12})$ and $s_{23}(\phi_{23}, d_{23})$ describe the three positions completely, hence $s_{31}(\phi_{31}, d_{31})$ must be determined by them. Assuming for the time being that the two screw displacements are general and that s_{12} and s_{23} are skew, we proceed as follows.

According to Chapter III, Section 1 any screw motion $s(\phi, d)$ may be expressed as the product of two line reflections; the two lines intersect s orthogonally, their distance being $\frac{1}{2}d$ and their angle $\frac{1}{2}\phi$. One of the two lines may be chosen arbitrarily, the other is then uniquely determined. Decomposing $s_{12}(\phi_{12}, d_{12})$ we take as the second reflection that with respect to the common perpendicular n_2 of s_{12} and s_{23}; let n_1 be the other line. (See Fig. 8.) Furthermore we decompose $s_{23}(\phi_{23}, d_{23})$ in a similar way, now choosing n_2 as the first line; let n_3 be the second one. The displacement $E_1 \rightarrow E_3$ can take place by the series of four reflections into respectively n_1, n_2, n_2, n_3, which reduces to the reflections in n_1 and n_3 respectively. The result of the latter is a screw motion about the common perpendicular s_{13} of n_1 and n_3, ϕ_{13} and d_{13} being twice the angle and the distance of these lines. For the sake of symmetry we will measure from n_3 to n_1 thereby obtaining $s_{31}(\phi_{31}, d_{31})$ instead of $s_{13}(\phi_{13}, d_{13})$.

Our construction has not only given us a geometrical method to find the product of two screw displacements but shows moreover that the configura-

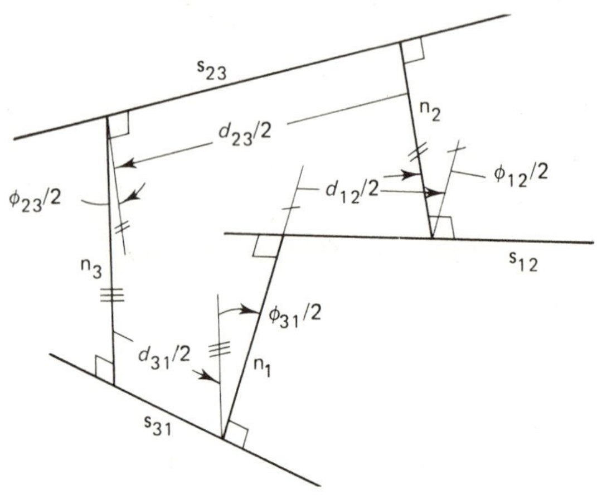

Fig. 8.

tion of three positions is completely given when the three axes s_{23}, s_{31}, s_{12} are known; the numerical data ϕ_{ij}, d_{ij} follow from the angle and distance of the two common perpendiculars n_i, n_j of the axis s_{ij}. The description of three positions by means of three lines is not only relatively simple, but moreover attractively symmetric because all three positions are dealt with equivalently. Note that the figure of three lines depends on 12 data as is expected.

If A_1 is a point of E_1 successive reflection in n_1 and n_2 gives us A_2, by successive reflection of A_1 in n_1 and n_3 we obtain A_3. Hence if A^* is the reflection of A_1 in n_1, it is also that of A_2 in n_2 and of A_3 in n_3. Therefore a triplet of three homologous points A_1, A_2, A_3 may always be found by reflecting an arbitrary point A^* of Σ into n_1, n_2, n_3 respectively; A^* is the *basic point* or the *fundamental point* of the triplet. In the same manner three homologous lines (or planes) may be obtained by reflecting an arbitrary basic line (or plane) into n_1, n_2, n_3.

The configuration of the three screw axes s_{23}, s_{31}, s_{12} and their common perpendiculars n_1, n_2, n_3 is called the *screw triangle*[†]; s_{23}, s_{31}, s_{12} are its "*vertices*" and n_1, n_2, n_3 their opposite "*sides*". Several cases of degeneration of the screw triangle appear if the screw displacements are not all general or if their axes are not all skew lines. So, for instance, if s_{23} is the axis of a pure rotation n_2 and n_3 intersect; if the displacement is a translation, n_2 and n_3 are parallel. Two screw axes may have a point of intersection or may be parallel

[†]The name originates with ROTH [1967a], the concept with HALPHEN [1882]. The geometry of Fig. 8 represents Halphen's theorem on the composition of two general displacements.

or coincide. An important situation is that for which the three lines s_{ij} have a common perpendicular. Some of the special cases will be dealt with later on.

General expressions relating the elements of the screw triangle can be obtained as follows:

The screw triangle, shown in Fig. 9, is defined so that the distance between screws s_{ij} and s_{jk} is a_j and the angle between them is α_j. The distance between normals n_i and n_k is $d_{ik}/2$, and the angle between them is $\phi_{ik}/2$. The triangle is shown as a set of directed vectors so that the positive sense of $\mathbf{s}_{ik}$ is from n_i toward n_k and that of $\mathbf{n}_j$ is from s_{ij} toward $\mathbf{s}_{jk}$ where j, k are in the cyclic order 1, 2, or 2, 3, or 3, 1, and angles are measured in the sense given by the right-hand screw rule. (See Figs. 9, 10, 11.)

With these conventions, if we take the spherical triangle with edges parallel to $\mathbf{s}_{12}$, $\mathbf{s}_{23}$, $\mathbf{s}_{31}$ we obtain from the cosine law

$$\cos(\phi_{31}/2) = \cos(\phi_{12}/2)\cos(\phi_{23}/2) - \cos\alpha_2 \sin(\phi_{12}/2)\sin(\phi_{23}/2) \tag{1.1}$$

and from the sine law

$$\sin(\phi_{31}/2) = \frac{\sin\alpha_2}{\sin\alpha_3}\sin(\phi_{12}/2). \tag{1.2}$$

These expressions are equally valid for any cyclic permutation of indices.

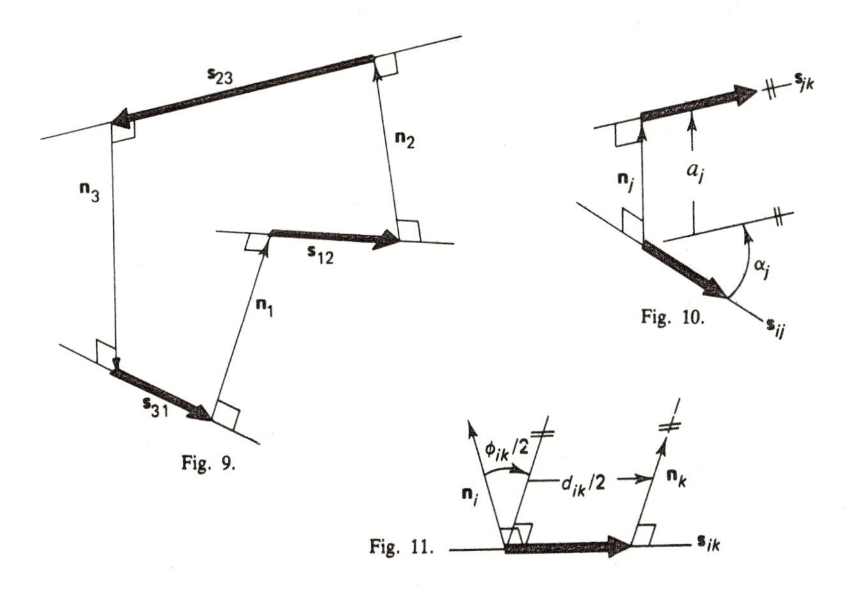

Fig. 9.

Fig. 10.

Fig. 11.

The several other forms of the spherical cosine law can be used to develop alternative expressions. Another approach is to use unit vectors $\mathbf{s}_{ij}$ and $\mathbf{n}_i$, respectively parallel to the screw axes and their normals. We have:

$$\mathbf{n}_i = \frac{\mathbf{s}_{ji} \times \mathbf{s}_{ik}}{|\mathbf{s}_{ji} \times \mathbf{s}_{ik}|} \tag{1.3}$$

then

$$\mathbf{s}_{31} \tan(\phi_{31}/2) = \frac{\mathbf{n}_3 \times \mathbf{n}_1}{(\mathbf{n}_3 \cdot \mathbf{n}_1)}$$

yields

$$\tan(\phi_{31}/2) = \frac{\mathbf{s}_{12} \cdot (\mathbf{s}_{23} \times \mathbf{s}_{31})}{(\mathbf{s}_{23} \times \mathbf{s}_{31}) \cdot (\mathbf{s}_{31} \times \mathbf{s}_{12})}. \tag{1.4}$$

Since the three screws and their three normals form a closed polygon, with lengths $d_{ij}/2$ and a_j respectively, we have from the projection onto s_{31}:

$$\frac{d_{31}}{2} = \frac{d_{12}}{2} \cos\alpha_1 + \frac{d_{23}}{2} \cos\alpha_3 - a_2 \sin\alpha_1(\sin(\phi_{12}/2)). \tag{1.5}$$

In terms of vectors, we can obtain an expression for $d_{31}/2$ as follows: let $\mathbf{A}_{ij}$ be the position vector of an arbitrary point on s_{ij} and let $\mathbf{I}^j_{ij}$ be the position vector of the point where n_j intersects s_{ij}. For purposes of derivation we define the unit vector normal to s_{12} and n_1 as $\mathbf{v}$.

$$\mathbf{v} = \frac{\mathbf{s}_{12} \times (\mathbf{s}_{31} \times \mathbf{s}_{12})}{|\mathbf{s}_{31} \times \mathbf{s}_{12}|}. \tag{1.6}$$

From the definition of $\mathbf{v}$ it follows that $\mathbf{v} \cdot (\mathbf{A}_{12} - \mathbf{I}^1_{12}) = 0$, and $\mathbf{v} \cdot (\mathbf{I}^1_{12} - \mathbf{I}^1_{31}) = 0$. Hence

$$\begin{aligned}
\mathbf{v} \cdot (\mathbf{A}_{12} - \mathbf{A}_{31}) &= \mathbf{v} \cdot \{(\mathbf{A}_{12} - \mathbf{I}^1_{12}) + (\mathbf{I}^1_{12} - \mathbf{I}^1_{31}) + (\mathbf{I}^1_{31} - \mathbf{A}_{31})\} \\
&= \mathbf{v} \cdot (\mathbf{I}^1_{31} - \mathbf{A}_{31}) \\
&= \mathbf{v} \cdot \mathbf{s}_{31}(\mathbf{I}^1_{31} - \mathbf{A}_{31}) \cdot \mathbf{s}_{31}.
\end{aligned}$$

If we substitute for $\mathbf{v}$ from (1.6) it follows that the distance between points I^1_{31} and A_{31} is

$$(\mathbf{I}^1_{31} - \mathbf{A}_{31}) \cdot \mathbf{s}_{31} = \frac{\mathbf{s}_{12} \times (\mathbf{s}_{31} \times \mathbf{s}_{12})}{1 - (\mathbf{s}_{31} \cdot \mathbf{s}_{12})^2} \cdot (\mathbf{A}_{12} - \mathbf{A}_{31}).$$

By replacing $\mathbf{n}_1$ with $-\mathbf{n}_3$, and $\mathbf{s}_{12}$ with $\mathbf{s}_{23}$ we obtain an analogous expression for the distance between I^3_{31} and A_{31}. Since

$$\frac{d_{31}}{2} = \{(\boldsymbol{I}^1_{31} - \boldsymbol{A}_{31}) - (\boldsymbol{I}^3_{31} - \boldsymbol{A}_{31})\} \cdot \boldsymbol{s}_{31}$$

it follows that

$$\frac{d_{31}}{2} = \frac{\boldsymbol{s}_{12} \times (\boldsymbol{s}_{31} \times \boldsymbol{s}_{12})}{1 - (\boldsymbol{s}_{31} \cdot \boldsymbol{s}_{12})^2} \cdot (\boldsymbol{A}_{12} - \boldsymbol{A}_{31}) - \frac{\boldsymbol{s}_{23} \times (\boldsymbol{s}_{31} \times \boldsymbol{s}_{23})}{1 - (\boldsymbol{s}_{31} \cdot \boldsymbol{s}_{23})^2} \cdot (\boldsymbol{A}_{23} - \boldsymbol{A}_{31}). \tag{1.7}$$

Expressions (1.4), (1.5) and (1.7) each give rise to three (two additional) equations if we cyclically permute the indices. Given any two screw displacements, it is possible to determine the third displacement by use of these equations. For example if we know s_{12} and s_{23}, s_{31} can be obtained as follows: We permute (1.4) so that it gives us expressions for $\tan(\phi_{12}/2)$ and $\tan(\phi_{23}/2)$. We solve these for $\boldsymbol{s}_{31}$ and obtain

$$\boldsymbol{s}_{31} = k_1(\boldsymbol{s}_{12} \tan(\phi_{12}/2) + \boldsymbol{s}_{23} \tan(\phi_{23}/2) + \boldsymbol{s}_{23} \times \boldsymbol{s}_{12} \tan(\phi_{12}/2) \tan(\phi_{23}/2)) \tag{1.8}$$

where

$$\begin{aligned} k_1 = -1/\{&\tan^2(\phi_{12}/2) + \tan^2(\phi_{23}/2) + 2(\boldsymbol{s}_{12} \cdot \boldsymbol{s}_{23}) \tan(\phi_{12}/2) \tan(\phi_{23}/2) \\ &+ (\boldsymbol{s}_{12} \times \boldsymbol{s}_{23})^2 \tan^2(\phi_{23}/2)\}^{\frac{1}{2}}. \end{aligned} \tag{1.9}$$

Upon substituting (1.8) into (1.4) we have

$$\tan(\phi_{31}/2) = \frac{1}{k_1(1 - (\boldsymbol{s}_{12} \cdot \boldsymbol{s}_{23}) \tan(\phi_{12}/2) \tan(\phi_{23}/2))}. \tag{1.10}$$

If we permute (1.7) so as to obtain expressions for $d_{12}/2$ and $d_{23}/2$, we can solve these two equations for $\boldsymbol{A}_{31}$. We obtain, if we set $\boldsymbol{A}_{31} \cdot \boldsymbol{s}_{31} = 0$,

$$\boldsymbol{A}_{31_n} = [1/(\boldsymbol{s}_{12} \cdot (\boldsymbol{s}_{31} \times \boldsymbol{s}_{23}))]\{k_2(\boldsymbol{s}_{31} \times \boldsymbol{s}_{23}) + k_3(\boldsymbol{s}_{12} \times \boldsymbol{s}_{31})\} \tag{1.11}$$

where $\boldsymbol{A}_{31_n}$ is the vector from the origin *normal* to s_{31}; so that $\boldsymbol{A}_{31} = \boldsymbol{A}_{31_n} + \lambda \boldsymbol{s}_{31}$, λ being an arbitrary parameter. Here

$$\begin{aligned} k_2 = \boldsymbol{A}_{12} \cdot [\boldsymbol{s}_{31} \times (\boldsymbol{s}_{12} \times \boldsymbol{s}_{31})] + ((\boldsymbol{s}_{12} \times \boldsymbol{s}_{31})^2/(\boldsymbol{s}_{12} \times \boldsymbol{s}_{23})^2)[\boldsymbol{s}_{23} \times (\boldsymbol{s}_{12} \times \boldsymbol{s}_{23})] \cdot (\boldsymbol{A}_{23} - \boldsymbol{A}_{12}) \\ - (d_{12}/2)(\boldsymbol{s}_{12} \times \boldsymbol{s}_{31})^2 \end{aligned}$$

and k_3 may be obtained from k_2 by interchanging subscripts 12 and 23 and changing the sign on the last term. Substituting (1.8) and (1.11) into (1.7) yields

$$\begin{aligned} (d_{31}/2) = k_1[&(\boldsymbol{A}_{12} - \boldsymbol{A}_{23}) \cdot (\boldsymbol{s}_{12} \times \boldsymbol{s}_{23}) \tan(\phi_{12}/2) \tan(\phi_{23}/2) \\ &+ (d_{12}/2)(\tan(\phi_{12}/2) + (\boldsymbol{s}_{12} \cdot \boldsymbol{s}_{23}) \tan(\phi_{23}/2)) \\ &+ (d_{23}/2)((\boldsymbol{s}_{12} \cdot \boldsymbol{s}_{23}) \tan(\phi_{12}/2) + \tan(\phi_{23}/2))]. \end{aligned} \tag{1.12}$$

This completes our description of screw s_{31} which is entrirely determined by (1.8), (1.10), (1.11), and (1.12) if we know $s_{12}(\phi_{12}, d_{12})$ and $s_{23}(\phi_{23}, d_{23})$.

Use the geometry of the screw triangle to show that:

Example 1. The product of two general displacements is itself a general displacement.

Example 2. The product of two rotations is a general displacement unless the rotation axes intersect, in which case the product is a pure rotation and the point of intersection is a fixed point.

Example 3. The product of two rotations about parallel axes is a pure rotation.

Example 4. The product of a translation and a general displacement has the same rotation as the given general displacement. (This follows if, for example, when $\phi_{23} = 0$, s_{12} and s_{13} are parallel.)

Example 5. The product of equal and opposite rotations about parallel axes is a translation normal to the direction of the axes.

Example 6. The effect of reversing the order of two general displacements is to reverse the sense of the resultant screw and to displace it by reflecting it (i.e. a half-turn) about n_2. Hence, two displacements commute only when they are such that the resultant screw (i.e. s_{13}) intersects their common normal (n_2). However, the magnitude of the product (i.e., $|d_{13}|$, $|\phi_{13}|$) is invariant to the order.*

Example 7. Show that if $\phi_{23} = 0$, a derivation similar to the general case yields:

$$\tan(\phi_{13}/2) = \tan(\phi_{12}/2) = (\boldsymbol{A}_{31} - \boldsymbol{A}_{12}) \cdot (\boldsymbol{s}_{31} \times (\boldsymbol{s}_{23} \times \boldsymbol{s}_{31}))/((\boldsymbol{A}_{31} - \boldsymbol{A}_{12}) \cdot (\boldsymbol{s}_{23} \times \boldsymbol{s}_{31}))$$

$$(d_{31}/2) = (\boldsymbol{s}_{31} \cdot \boldsymbol{A}_{23} + (\boldsymbol{s}_{31} \cdot \boldsymbol{s}_{23})\boldsymbol{s}_{23} \cdot (\boldsymbol{A}_{23} - \boldsymbol{A}_{31}))/(1 - (\boldsymbol{s}_{31} \cdot \boldsymbol{s}_{23})^2) - k_3$$

where k_3 is arbitrary.

In the foregoing we have used a double subscript notation for the screws and their displacement parameters. A simpler notation follows if we denote s_{ij} by s_k, ϕ_{ij} by ϕ_k, and d_{ij} by d_k. In which case we replace $\boldsymbol{s}_{23}$, $\boldsymbol{s}_{31}$, $\boldsymbol{s}_{12}$ by $\boldsymbol{s}_1$, $\boldsymbol{s}_2$, $\boldsymbol{s}_3$ respectively. The common perpendiculars maintain the same subscripts as before. Hence, the perpendicular of s_2, s_3 is n_1, that of s_3, s_1 is n_2, and the one between s_1, s_2 is n_3. The following section is written in terms of this single subscript notation.

2. The plane at infinity

The plane V at infinity is as a whole invariant for any displacement. Therefore, if A is a point of E in V, the three homologous points A_1, A_2, A_3 are in V.

The screw axes s_1, s_2, s_3 intersect V at the points P_1, P_2, P_3, which in the general case are the vertices of a triangle. The perpendiculars n_1, n_2, n_3 intersect V at N_1, N_2, N_3. As s_1 is perpendicular to n_2 and n_3 the line N_2N_3 is the polar line of P_1 with respect to the isotropic conic Ω; N_3N_1 and N_1N_2 are

* We have used s_{31} heretofore in order to make the equations symmetrical. The screws s_{31} and s_{13} both have the same axis, they differ only in sense: ϕ_{13}, d_{13} are measured from n_1 to n_3 while ϕ_{31}, d_{31} are measured from n_3 to n_1.

the polar lines of P_2 and P_3. Hence $P_1P_2P_3$ and $N_1N_2N_3$ are reciprocally polar triangles with respect to Ω; N_1, N_2, N_3 are the poles of P_2P_3, P_3P_1, P_1P_2; one triangle determines the other one.

It is well-known from plane projective geometry that P_1N_1, P_2N_2 *and* P_3N_3 *pass through one point* H, or in other words that the two triangles are in Desarguesian position (Fig. 12). This may be proved also by making use of space geometry in the following way. Take an arbitrary point O, and let s_i', n_i' pass through O, parallel to s_i, n_i $(i = 1, 2, 3)$. The statement comes to this: in the trihedron $Os_1's_2's_3'$ the plane through s_1' perpendicular to the plane $s_2's_3'$, and the two similar planes pass through one line, a known theorem of solid geometry. A proof may be given by vector algebra. If $\boldsymbol{s}_i$ is a vector along s_i',

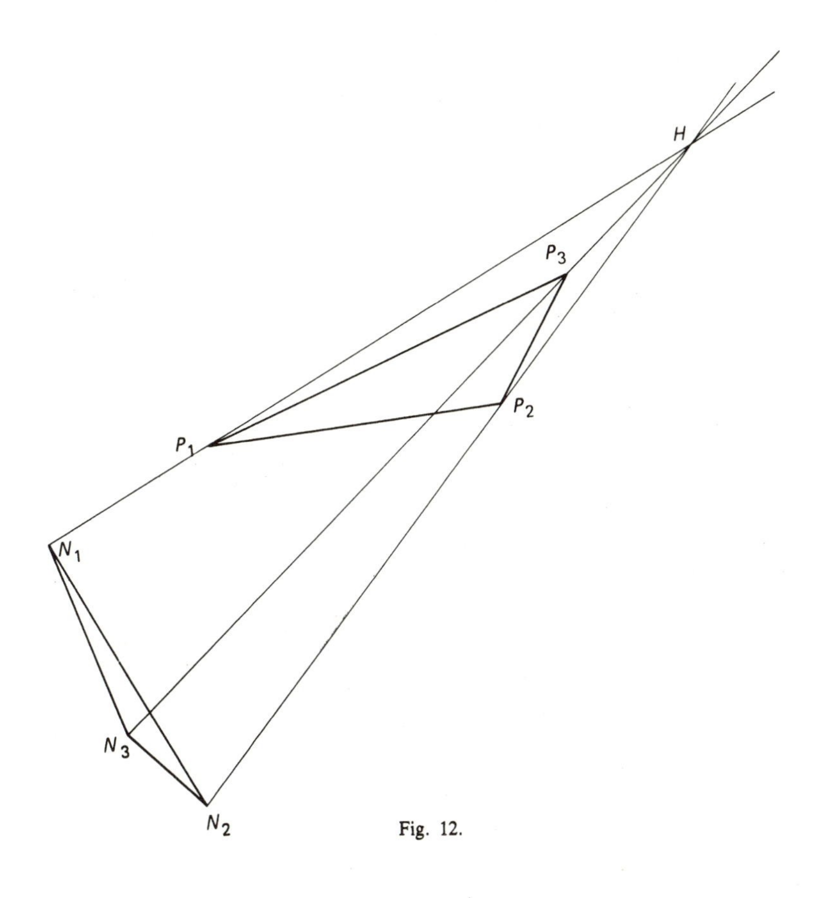

Fig. 12.

vectors along n_1', n_2', n_3' are given by $s_2 \times s_3$, $s_3 \times s_1$, $s_1 \times s_2$ and the theorem follows from the fact that $s_1 \times (s_2 \times s_3)$, $s_2 \times (s_3 \times s_1)$, $s_3 \times (s_1 \times s_2)$ are linearly dependent.

In order to describe a triad of homologous points A_1, A_2, A_3 in V we remark that each point of V corresponds to a direction in space, parallel lines having the same direction. If in space we consider a basic line a^* its three related homologous lines a_1, a_2, a_3 are found as the reflections of a^* into n_1, n_2, n_3 respectively. In as far as directions concern us, all lines may be replaced by parallel lines all passing through O, denoted by primes. Thus the direction of a_1 is found by reflecting $a^{*\prime}$ into n_1' and so on. But $a^{*\prime}$ and n_1' are in one plane and therefore a_1' is also in this plane. This means that A_1 is on the line N_1A^*, A^* being the point of a^* at infinity. The reflection into n_1' is a displacement and therefore a linear transformation in space; hence it induces a linear transformation in V. As N_1 and all points of its polar line are fixed points and as moreover the transformation is involutory, A_1 is seen to be that point on N_1A^* which is harmonic with A^* with respect to the point N_1 and the intersection of N_1A^* and the polar line. This projective transformation in V will be called reflection into the point N_1 or, with the same justification, reflection into the polar line P_2P_3 of N_1. Hence: *any three homologous points in* V *are found as reflections of an arbitrary basic point* A^* *into the three points* N_i (or into the sides of the triangle $P_1P_2P_3$).

We have dwelt on what takes place in V for several reasons. The metric in V is not Euclidean; the absolute being a non-degenerate and imaginary conic Ω, we have dealt with kinematics in plane *elliptic* geometry. A systematic treatment of kinematics in spaces other than Euclidean ones is given in Chapter XII. Here however, since kinematics in V appears naturally it serves as a good introductory example. A second reason for looking at V is that Euclidean kinematics can not be understood without asking what is going on at infinity. A third justification is the fact that kinematics in V is strongly related to (if not identical with) *spherical kinematics*, to be treated in Chapter VII, and that *plane Euclidean kinematics* (Chapter VIII) may be considered as a limiting case of the latter.

3. The planes through three homologous points

The preceding sections have given us some insights about the triads $(A_1A_2A_3)$ of homologous points, each of which is related to its own basic point A^*. The points A_i will in general be linearly independent; hence they determine a plane α. We now consider the relationship between A^* and α. If A^* is in V this is trivial: all points of V correspond to the plane V itself.

Let A^* be a finite point and its rectangular coordinates be (x_1, x_2, x_3). Then those of A_i, as we know, are linear functions of x_i. Hence, introducing homogeneous coordinates, we may construct a matrix of three rows and four columns the elements of which are the coordinates of a triad of homologous points:

$$\|L_{i1} \quad L_{i2} \quad L_{i3} \quad x_4\|, \quad (i = 1,2,3) \tag{3.1}$$

in which L_{ik} is a homogeneous linear function of x_1, x_2, x_3, x_4. The coordinates U_i of the plane α are proportional to the four determinants of (3.1). We obtain

$$U_1 = F_1x_4, \qquad U_2 = F_2x_4, \qquad U_3 = F_3x_4, \qquad U_4 = G, \tag{3.2}$$

in which F_i $(i = 1,2,3)$ are quadratic polynomials and G is a cubic polynomial in x_i. If $x_4 = 0$ (3.2) gives us the plane V, so that (3.2) expresses the relationship between A^* and α for all cases.

Let, on the other hand, α be an arbitrary plane; if it will contain a point A_i then A^* must be in the plane α_i, the reflection of α into n_i. The three planes α_i (which are a triad of homologous planes) have in general one point of intersection A^* and its three reflections are all in α. Hence: *a plane α contains in general one triad of homologous points.*

If U_i are the coordinates of a plane and U'_i those of its reflection into a line, then U'_1, U'_2, U'_3 are linear functions of U_1, U_2, U_3, while U'_4 is a linear function of U_1, U_2, U_3, U_4. The coordinates of a triad α_i can therefore be arranged in a 3×4 matrix:

$$\|L_{i1} \quad L_{i2} \quad L_{i3} \quad L_{i4}\|, \quad i = 1,2,3, \tag{3.3}$$

in which the linear functions L_{ij} $(i,j = 1,2,3)$ depend on U_1, U_2, U_3 only. We obtain for the coordinates of A^*:

$$x_1 = G_1 + D_1U_4, \quad x_2 = G_2 + D_2U_4, \quad x_3 = G_3 + D_3U_4, \quad x_4 = G_4, \tag{3.4}$$

in which G_i are cubic polynomials and D_i quadratic polynomials of U_1, U_2, U_3.

(3.2) and its inverse (3.4) show that *there exists a birational cubic relationship between a basic point A^* and the plane α through the triad of homologous points associated with it*; *it is the same as that between a basic plane α^* and the point of intersection of the triad of homologous planes associated with it.*

4. Collinear homologous points

The cubic relationship obviously has singularities. Indeed if A^* is such that A_1, A_2, A_3 are one line l the plane α is not unique: A^* is associated with all planes of the pencil through l. Inversely, if α is such that α_1, α_2, α_3 pass through one line l′ it is associated with all points on l′.

If A_1, A_2, A_3 are collinear the rank of the matrix (3.1) must be less than three (and inversely a rank of less than three implies collinearity). Suppose first that A^* is a finite point: which implies $x_4 \neq 0$. No two rows of (3.1) can be linearly dependent because a general displacement has no finite fixed point. In view of the fourth column the three rows are linearly dependent if the last row is equal to the sum of the first and the second, multiplied by λ and $1-\lambda$ respectively, in which λ is an arbitrary number unequal to ∞, 0 or 1. This gives rise to three homogeneous linear equations for x_i:

$$L_{31} = \lambda L_{11} + (1-\lambda)L_{21}, \qquad L_{32} = \lambda L_{12} + (1-\lambda)L_{22},$$
$$L_{33} = \lambda L_{13} + (1-\lambda)L_{23}, \tag{4.1}$$

the solutions of which are

$$x_1 : x_2 : x_3 : x_4 = C_1(\lambda) : C_2(\lambda) : C_3(\lambda) : C_4(\lambda), \tag{4.2}$$

C_i being a polynomial of the third degree in λ. Hence *the locus of the finite points A^* for which A_1, A_2, A_3 are collinear is a twisted cubic*, which we denote by c. All parameter values λ, unequal to ∞, 0, 1 correspond to finite points on c. Hence $\lambda = \infty$, $\lambda = 0$, $\lambda = 1$ correspond to the three intersections of c and V. From (4.1) it follows that for $\lambda = \infty$ A_1 and A_2 coincide, for $\lambda = 0$ A_2 and A_3 coincide, for $\lambda = 1$ A_3 and A_1 conicide. Therefore the points $\lambda = \infty$, $\lambda = 0$, $\lambda = 1$ of c are the points P_3, P_1, P_2 respectively, these points being the only real ones for which two points A_i coincide. The conclusion is: *the twisted cubic* c *passes through the points at infinity of the screw axes* s_i. It is therefore a cubic hyperbola.

Suppose now that A^* is a point of V, $x_4 = 0$. According to (3.1) the three points A_i are collinear if

$$|L_{i1} \quad L_{i2} \quad L_{i3}| = 0; \tag{4.3}$$

in this determinant $x_4 = 0$ is substituted in L_{ij} which is now a linear function in just x_1, x_2, x_3. Obviously (4.3) represents *a cubic* c′ *in* V.

If two of the three homologous points, A_2 and A_3 say, coincide then A_1, A_2, A_3 are collinear and thus A^* is on c′. There are six cases for which this situation takes place. If A^* coincides with P_1 its reflections A_2 into P_1P_3 and

A_3 into P_1P_2 both coincide with P_1; therefore P_1 is a point of c′ and so are P_2 and P_3. Furthermore let I_1 and I_1' be the intersections of N_2N_3 and Ω (which means that they are the isotropic points of any plane perpendicular to s_1). These two points are interchanged by reflection into N_2 as well as into N_3. Thus if A^* is at I_1 then A_2 and A_3 coincide at I_1', which means that I_1 is on c′. *The plane cubic* c′ *passes through the points at infinity of the screw axes and through the isotropic points of the planes perpendicular to these axes.* In general c′ is determined by these nine points.

It may be shown, as follows, that c′ passes through three more points of the configuration. If m is a line in V on which three homologous points A_1, A_2, A_3 lie then obviously the three reflections m_1, m_2, m_3 of m pass through a point A^* on c′. Now take N_2N_3 as line m; its reflection m_2 into N_2 is m itself, its reflection m_3 into N_3 is again m itself, its reflection m_1 into N_1 passes through the intersection T_1 of m and P_2P_3 because all points of P_2P_3 are fixed points of this reflection. Hence T_1 is a point of c′ and so is the intersection T_2 of N_3N_1 and P_3P_1, and similarly the intersection T_3 of N_1N_2 and P_1P_2.

T_1, T_2, T_3 are the third intersections of c′ and P_2P_3, P_3P_1, and P_1P_2 respectively. Moreover, according to Desargues' theorem they are on one line h. (Fig. 13.)

The curve c′ plays a part in plane elliptic kinematics. Its meaning for common spatial kinematics is obviously as follows: if three homologous lines l_i are parallel to one plane the basic line of the triad is parallel to a generator of a cubic cone having c′ as its intersection with the plane V at infinity.

Summing up we have: *the complete locus of the basic point* A^* *such that* A_1, A_2, A_3 *are collinear consists of the twisted cubic* c *and the plane cubic* c′. The locus of A_i is a twisted cubic c_i and the plane cubic c_i'.

The triad of homologous points associated with a point A^* is on one line if A^* is on c or on c′. For such a point A^* the plane α is not uniquely determined; corresponding to it is not one plane but a pencil of planes. This means that *the points on* c *and* c′ *are the singular points of the cubic relationship* (3.2). Hence for such points the expressions U_i given by (3.2) are all zero. Therefore the varieties $U_1 = 0$, $U_2 = 0$, $U_3 = 0$, $U_4 = 0$ contain the curves c and c′. In view of the expressions for U_1, U_2, U_3 it follows that the quadrics $F_1 = 0$, $F_2 = 0$, $F_3 = 0$ pass through c, and the cubic surface $G = 0$ passes through c and c′. We mention some consequences. The planes through a point $B(b_1, b_2, b_3, b_4)$ satisfy the equation $b_1U_1 + b_2U_2 + b_3U_3 + b_4U_4 = 0$; *the locus of the points associated with these planes is a cubic surface through* c *and* c′. If $b_4 = 0$, that is if the planes are all parallel to a certain line the cubic surface degenerates into V and a quadric. Furthermore we may understand

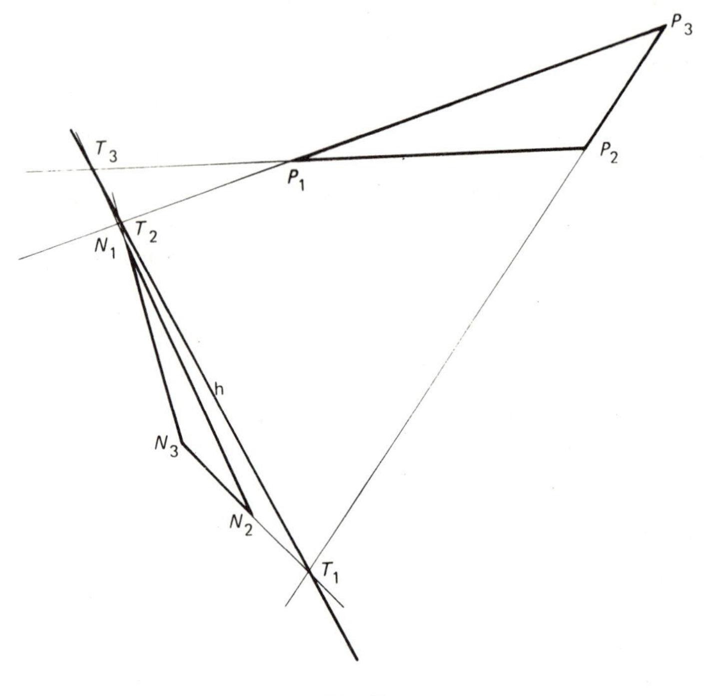

Fig. 13.

now more clearly why the inverse of (3.2) is a unique point. Let $U_1 : U_2 : U_3 : U_4 = k_1 : k_2 : k_3 : k_4$ be a given plane. The corresponding point satisfies $k_2U_1 - k_1U_2 = 0$, $k_3U_1 - k_1U_3 = 0$, $k_4U_1 - k_1U_4 = 0$ (or three similar equations if $k_1 = 0$). The finite points satisfying the first two are on the intersection of two quadrics through c; it consists therefore of c and one of its chords. The third equation represents a cubic surface through c; its third intersection with the chord is the unique non-singular point corresponding to the given plane. Singular *planes* of the cubic relationship will be dealt with later (Section 6).

5. Lines through three homologous points

There are ∞^1 points A^* for which A_1, A_2, A_3 are on one line; hence there are ∞^1 such lines. Their locus will now be considered. It consists of two sets, lines associated with points of c and those associated with c′. The first are lines in space generating a ruled surface F, the others are in V enveloping a curve of a certain class.

Let A^* be a point of c; its reflections A_i into n_i are on a line l. Hence the midpoints B_i of A^*A_i are on a line l′ parallel to l, B_i is the projection of A^* on n_i. Each line l′ intersects n_1, n_2, n_3, it is a transversal of these three lines and therefore belongs to the regulus of such transversals which is the hyperboloid Q through n_i; the lines n_i belong to the other regulus. The first conclusion reads: each line l is parallel to a line of the regulus of l′, or in other words all lines l have their points at infinity on the conic k, the intersection of Q and V. Each line l′ gives rise to a line l; indeed the three planes through B_i perpendicular to n_i have a point of intersection A^* from which l follows. Hence there is one line l through any point of k and a 1,1 relationship, without singularities, between the points of k and the points A^* on c. From this it follows that the points of k and c are projectively related. On k lie the three special points N_i; the line l′ through N_1 is parallel to n_1, B_1 is at infinity, hence A^* is at infinity and coincides with P_1. Therefore in the projectivity between k and c the points N_i on k correspond to the points P_i on c $(i = 1,2,3)$; the projectivity is determined by the three pairs.

The points A_1 are the reflections of A^* into n_1; their locus is the twisted cubic c_1, the reflection of c into n_1. This implies a projectivity between the points A_1 of c_1 and those of k. The curve c_1 passes through P_2, P_3 and the point P_1' which is the reflection of P_1 into N_1. In the projectivity between k and c_1 N_1, N_2, N_3 correspond to P_1', P_2, P_3. We have arrived at the following conclusion: *the lines* l *join corresponding points of two projective curves, the twisted cubic* c_1 *and the conic* k.

The coordinates of the points of c_1 are cubic polynomials of a parameter λ; we introduce a parameter on k such that corresponding points are denoted by the same value of λ; the coordinates of the point of k are quadratic functions of λ. Therefore the Plücker coordinates of the lines l are of the *fifth* degree in λ. *The locus* F *of the lines* l *in space, on which three homologous points lie, is a rational ruled surface of the fifth order.*

Three generators of F are in V, the line N_1P_1' (which is the same line as N_1P_1) and the lines N_2P_2, N_3P_3. These three lines pass through the point H. Thus *the intersection of* F *and* V *is a degenerate curve of the fifth order*; *it consists of the conic* k *and the three lines N_iP_i through H.*

c_1 is a twisted cubic on F and so are c_2 and c_3, the reflections of c into n_2 and n_3; c_1 passes through P_1', P_2, P_3, c_2 through $P_1P_2'P_3$, c_3 through $P_1P_2P_3'$; two curves c_i have one common point at infinity.

Ruled surfaces of the fifth order have been investigated and projectively classified by Cayley, by Schwarz and more recently by Edge. Our F is a (metrically specialized) surface of type II in Schwarz's classification. We return to some of its properties later on (Section 10).

Considering now triads of collinear homologous points A_1, A_2, A_3 in V, we know that their basic point A^* is on the plane cubic c′. In order to determine the set of lines m on which these triads lie we remark that they must be such that their reflections m_i in N_i pass through one point, the basic point. But in view of the complete duality in V between points and lines the determination of homologous lines passing through one point is essentially the same as the determination of homologous points on one line. Hence *the set of lines on which three homologous points of* V *lie is the curve* C′ *of the third class*, polar with c′ with respect to Ω. The lines N_2N_3, N_3N_1, N_1N_2 belong to the set; so do the lines N_1P_1, N_2P_2, N_3P_3, passing through H (which is dual to the line h through T_1, T_2, T_3) and the three pairs of tangents of Ω through P_1, P_2, P_3 (dual to the six isotropic points of c′). *The complete locus of lines through three homologous points consists of* F *and* C′; *it is of the eighth order.*

6. The intersection of three homologous planes

We know from Sections 3 and 4 that their exists a birational cubic relationship between a basic point A^* and the plane α through the three reflections A_1, A_2, A_3 of A^* into n_i. If A^* is on c the points A_i are on one line l; A^* corresponds then to the planes of the pencil through l, it is a singular point of the relationship. In general one triad of homologous planes passes through an arbitrary point A^*, namely that consisting of the three reflections α_1, α_2, α_3 of the plane α corresponding to A^*. If A^* is on c there are ∞^1 triads of homologous planes through it.

We consider now the dual problem. In general three homologous planes α_1, α_2, α_3 have one point of intersection; it is the point A^* corresponding to the basic plane α of α_i. We consider now those triads of homologous planes which have a common line of intersection; their basic plane α, is a singular plane of the cubic relationship for it does not correspond to one point but to a line of points A^*. For such a singular plane $\alpha(U_1, U_2, U_3, U_4)$ the expressions x_i of (3.4) are all zero, or in other words the matrix (3.3) has rank two. It is a 3×4 matrix the elements of which are homogeneous linear functions of U_i. Our problem is a special case of a more general one. Let a $p \times q$ matrix be given, the elements of which are linear functions of n variables x_i. One may ask for which points (with coordinates x_i) the matrix has a given rank; this variety is called a *determinantal locus*. The subject was extensively treated by ROOM [1938], who has derived general formulas for the order of such a variety. We give a direct proof for our case. If the matrix (3.3) has rank two its four 3×3 determinants must be zero. If D_1' and D_2' are the determinants obtained by omitting the first and the second column respectively, $D_1'=0$ and $D_2'=0$ define cubic varieties and their intersection has order 9. But to arrive at the determinantal locus we must leave out all those points for which the 3×2 matrix of the first and second column has rank one. Following the same procedure we find that they are on a variety of order $4-1=3$. Hence our locus has order six.

The variety of singular planes is a developable (*or torse*) T *of class six*. A developable is a set of ∞^1 planes. Its class being six means that six planes pass through an arbitrary point.

Singular *points* are those for which the matrix (3.1) has rank two. Due to its special properties we have treated it somewhat differently, but we remark that in this case the variety also has order six, even though it degenerates into two cubic varieties, the curves c and c′.

In the case of T no degeneration takes place (as we shall see) but it has a special relation to the plane V at infinity. First of all V belongs to it; indeed for $U_1=U_2=U_3$ the first three columns of (3.3) are all zero. But furthermore, of the six planes of T through an arbitrary point of V three always coincide with V. Hence: V *is a triple plane of* T.

A general plane contains one triad of homologous points; in a singular plane there are ∞^1 (in V where are ∞^2). Let A_1, A_2, A_3 and B_1, B_2, B_3 be two triads in a plane α. Hence any point M_1 on A_1B_1 has its homologous M_2 on A_2B_2 and M_3 on A_3B_3. Hence the triads M_1, M_2, M_3 of homologous points in the singular plane α are all inscribed in three (homologous) lines l_i. The lines M_1M_2 join congruent point rows on l_1 and l_2, they are the tangents of a parabola; the lines M_1M_3 are the tangents of a second parabola. The two parabolas have (besides l_1 and the line at infinity) two common tangents. Therefore: among the triads in α, there are two collinear ones. On the other hand: if a plane contains two collinear triads, it is singular. Collinear triads lie on a generator of F. Hence: *the developable* T *of singular planes consists of the planes through two intersecting generators of the fifth order ruled surface* F.

We have generated F by joining corresponding points on the projectively related curves c_1 and k. Let $c_{1_i}(\lambda)$ be the coordinates of a point of c_1, and $k_i(\lambda)$ ($i=1,2,3,4$) the corresponding homogeneous coordinates of a point of k. The generator λ_1 and the generator λ intersect if

$$
(6.1)\qquad \begin{vmatrix} c_{1_1}(\lambda_1) & c_{1_2}(\lambda_1) & c_{1_3}(\lambda_1) & c_{1_4}(\lambda_1) \\ k_1(\lambda_1) & k_2(\lambda_1) & k_3(\lambda_1) & k_4(\lambda_1) \\ c_{1_1}(\lambda) & c_{1_2}(\lambda) & c_{1_3}(\lambda) & c_{1_4}(\lambda) \\ k_1(\lambda) & k_2(\lambda) & k_3(\lambda) & k_4(\lambda) \end{vmatrix} = 0.
$$

By subtracting the third row from the first, and the fourth from the second it is seen that among the five roots of (6.1) $\lambda = \lambda_1$ appears twice. Hence: *a generator of* F *is intersected by three other generators.* Therefore three planes of T pass through any generator of F. If l is a generator and L its point at infinity, three finite planes of T pass through L and we see again that the other three coincide with V.

If l is a generator and l′ an intersecting one the locus of the point of intersection of l and l′ is the double curve d of F. *The developable* T *of singular planes consists of the planes tangent to* F *at the points of its double curve.*

On each generator l there are three points of d; if l and l′ intersect at M the plane through l and l′ being tangent to F is tangent to d at M; this plane has four more intersections with d, two on l and two on l′. *Hence the order of the double curve* d *is six.*

We remark finally that the points at infinity of the three homologous lines of a singular plane are obviously homologous points and they lie on one line. The conclusion is: the planes of T intersect V in the lines of C′, the set is of the third class.

7. Three homologous lines

If a basic line l* is reflected into n_i we obtain a triad of homologous lines l_1, l_2, l_3. We consider the special triads with the property that all three lines mutually intersect. In Chapter III we have shown that l_1 and l_2 intersect if they both belong to a (special) tetrahedral complex. Therefore we could answer our question by considering the intersections of three such complexes. There is a more simple method however if we keep in mind that if three lines mutually intersect they are in one plane or pass through one point.

Let l_1, l_2, l_3 be in one plane α. Then for any point M_1 on l_1 the homologous points M_2 and M_3 are on l_2 and l_3 respectively. Hence α contains an infinity of triads of homologous points and it is therefore singular; on the other hand a singular plane contains one triad of homologous lines. *The planes of three coplanar homologous lines are the singular planes.* There are ∞^1 singular planes; hence the locus of the lines l_i of a coplanar triad is a ruled surface R_i and that of its basic lines l^* is a ruled surface R^*.

Any singular plane α contains two intersecting generators a and b of the ruled surface F. On a there are three homologous points A_1, A_2, A_3 and on b three homologous points B_1, B_2, B_3. The homologous lines in α are therefore A_1B_1, A_2B_2 and A_3B_3. But A_1 and B_1 for instance are both on the twisted cubic c_1; thus l_1 is a chord of c_1 and similar for l_2 and l_3; l^* is a chord of c.

If L_i is the point at infinity of l_i the three points L_1, L_2, L_3 are a homologous triad of points and moreover are collinear. Hence the locus of L_i is the plane cubic c_i' and the result is: *the ruled surface* R_i *consists of those lines which are chords of* c_i *and intersect* c_i'. The generators of R^* are the lines l^* intersecting c' and intersecting c twice.

Any generator a of F is met by three other generators b, b′ and b″. Therefore through the point A_1 on a pass three generators of R_1, joining A_1 to B_1, B_1', B_1''. *Hence* c_1 *is a threefold curve on* R_1.

Through a point outside a twisted cubic there passes always one of its chords. The conclusion is: *through any point of the curve* c_1' *passes one generator of* R_1, *and therefore* c_1' *is a directrix of* R_1.

To determine the order of R_1 we may proceed as follows. There is an involutory 3,3 correspondence on c_1, the join of two corresponding points being a generator of R_1. The order of R_1 is the number of generators intersecting an arbitrary line m in space. Through a point A_1 on c_1 pass two chords intersecting m; in fact, they join A_1 to the two remaining intersections A', A'' of c_1 and the plane through A_1 and m. Therefore m induces on c_1 an involutory 2,2 correspondence. A p, p correspondence and a q, q have pq pairs in common. Consequently six generators of R_1 intersect m. *The order of* R_1 *is six* and the same holds for R_i and R^*.

The twisted cubic c_1 passes through the points P_1', P_2 and P_3 at infinity and so does the plane cubic c_1'. In the 3,3 correspondence P_1' and P_2, P_1' and P_3, P_2 and P_3 are corresponding points. Hence the lines $P_1'P_2$, $P_1'P_3$, P_2P_3 are generators of R_1, they are after all the only chords of c_1 in V and they intersect c_1'. The intersection of R_1 and V consists of c_1' and these three chords, and it is accordingly of order six.

The problem to determine *triads of homologous lines* l_1, l_2, l_3 *passing through one point A* is easily dealt with in view of our former results. Indeed it means that in the inverse displacement three homologous points are on one line. According to Sections 4 and 5, the locus of A is a twisted cubic and the loci of l_i are ruled surfaces of the fifth order.

In the same way, using the inverse displacements, the results of this section could give the same information about *three homologous planes passing through one line* as was obtained in Section 6.

In general the triad of homologous lines l_1, l_2, l_3 is skew, i.e., none of the lines intersect. The simplest surface which contains them is either a hyperboloid of one sheet or a hyperbolic paraboloid. Such surfaces are of little interest to us since they do not lead to important properties associated with three positions of a line. Instead we consider a special line congruence defined by the homologous triad l_1, l_2, l_3; this will lead us to an important correspondence between lines in the moving and fixed space.

We ask ourselves: if we consider only the positions of the lines l_i (and not their points), does there exist a single line in Σ which could be the screw axis for displacements $l_1 \to l_2$, $l_2 \to l_3$, and $l_3 \to l_1$? We will show that the answer is affirmative, and that each homologous set l_1, l_2, l_3 defines a unique line l_c, in Σ, which we call its axis. Since we do not restrict the pitch of the screw along l_c, the displacement $l_i \to l_j$ will follow by rotating and translating l_i about l_c using different pitches (i.e., ratios of rotations to translations) for each displacement. Since l_c acts as a screw axis the directed distance and angle between l_c and l_i is the same in each position $i = 1, 2, 3$. This means we may think of l_i as being coupled to l_c by a rigid bar with joints at each end which allow turning and sliding (see Fig. 14). It is clear that any l_i coupled to a fixed line l_c in this

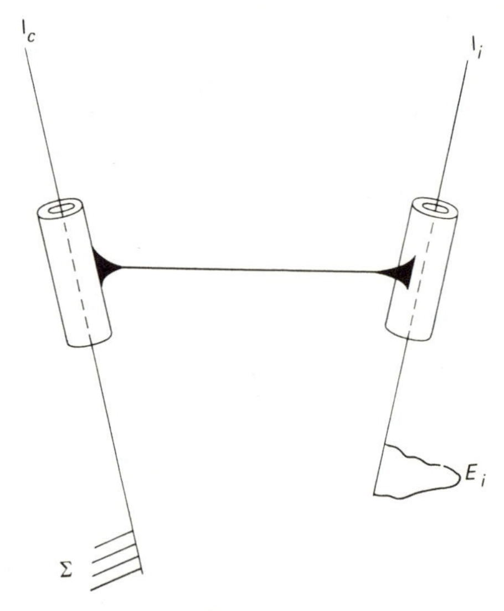

Fig. 14.

way has two freedoms: it may be rotated about and translated along l_c. Hence, the locus of the three lines l_i is such that they belong to a line congruence. Just as a circle in a plane is the locus of all points at a fixed distance from a point in that plane, this congruence in 3-space is the locus of all lines at a fixed angle and distance from a line in that space. The line l_c is referred to as the *central axis* of the congruence. *The three directed lines* l_i *completely and uniquely define the congruence and the central axis* l_c. Conversely, if we invert, holding E_i fixed, l_i is the central axis for the congruence defined by the positions of l_c. It then follows that l_i (as well as the basic line l^* corresponding to l_i) is in 1,1 reciprocal correspondence with axis l_c.

In order to formulate the foregoing analytically we define the Plücker vectors $(\boldsymbol{l}_i, \boldsymbol{l}'_i)$ for the directed line l_i so that $\boldsymbol{l}_i$ is parallel to l_i and, furthermore, if A_i and B_i are two distinct points on l_i, the vector $\boldsymbol{l}_i$ is directed from A_i toward B_i (for all values of i). The moment vector $\boldsymbol{l}'_i$ is defined by $\boldsymbol{l}'_i = \boldsymbol{a}_i \times \boldsymbol{l}_i$ where $\boldsymbol{a}_i$ is the position vector of a point A_i on l_i. Similarly we have the Plücker vectors $(\boldsymbol{l}_c, \boldsymbol{l}'_c)$ for a directed line l_c in Σ. Letting Θ_i and D_i denote respectively the scalar angle and distance between l_i and l_c, it follows that

$$\boldsymbol{l}_i \cdot \boldsymbol{l}_c = |\boldsymbol{l}_i||\boldsymbol{l}_c| \cos \Theta_i \tag{7.1}$$

$$\boldsymbol{l}'_i \cdot \boldsymbol{l}_c + \boldsymbol{l}'_c \cdot \boldsymbol{l}_i = |\boldsymbol{l}_i||\boldsymbol{l}_c| D_i \sin \Theta_i. \tag{7.2}$$

We point out in passing that, because of the definition of moment vector, the sign convention is such that when $D_i \sin \Theta_i$ is negative (positive) the right hand rule applied to the cross product of line vectors along l_i and l_c — in any order — is such that the thumb points from the first toward (away from) the second line of the cross product.

From (7.1) the condition $\cos\Theta_1 = \cos\Theta_2 = \cos\Theta_3$, assuming $|\boldsymbol{l}_i| =$ constant (usually $\boldsymbol{l}_i^2 = 1$), yields

$$(\boldsymbol{l}_j - \boldsymbol{l}_1) \cdot \boldsymbol{l}_c = 0 \quad j = 2, 3. \tag{7.3}$$

The two equations (7.3) represent necessary, but not sufficient, conditions for $\Theta_1 = \Theta_2 = \Theta_3$.

Since for a general displacement $1 \to j$, $\boldsymbol{l}_j - \boldsymbol{l}_1$ can be written as a homogeneous linear function of the coordinates of l^* (the basic line l^* is given by Plücker vectors $\boldsymbol{l}^*$, $\boldsymbol{l}'^*$), it follows that if we know s_{12}, s_{31} (7.3) represents the 1,1 reciprocal correspondence between the directions of l^* and l_c. That is to say we may choose $\boldsymbol{l}^*$ and uniquely determine the directions of l_c, or choose $\boldsymbol{l}_c$ and uniquely determine the directions of l^*. It also follows from (7.3) that the singular directions of this correspondence are the screws: If $\boldsymbol{l}_c$ is taken parallel to either s_{12}, s_{23}, or s_{31} there is a single infinity of corresponding directions l^* all of which are parallel to one plane, while if l^* is taken parallel to either s^*_{12}, s^*_{23}, or s^*_{31} the corresponding l_c are all parallel to a plane. We point out that this 1,1 correspondence depends only upon the directions of the screws and the rotations about them, it in no way contains their positions or translation. Hence, we may say it depends only upon the *spherical properties* of the displacement.

From (7.2) the condition $D_1 \sin \Theta_1 = D_2 \operatorname{Sin} \Theta_2 = D_3 \sin \Theta_3$, assuming $|l_i| =$ constant, yields

$$(7.4) \qquad (l'_j - l'_1) \cdot l_c + l'_c \cdot (l_j - l_1) = 0 \quad j = 2, 3.$$

If (7.3) is satisfied then (7.4) is sufficient to guarantee that $D_1 = D_2 = D_3$ and $\Theta_1 = \Theta_2 = \Theta_3$. Recalling that l'_i $(i = 1, 2, 3)$ is a linear non-homogeneous function of l'^* (it is also a function of l^*), it is obvious that if l^* and l_c are known, (7.4) may be regarded as two non-homogeneous linear equations in l'^* and l'_c.

For any pair of directions l^*, l_c which satisfy the directional correspondence, equations (7.4) together with the fundamental relationships

$$(7.5) \qquad l^* \cdot l'^* = 0, \qquad l_c \cdot l'_c = 0$$

are sufficient to determine the l'^*, l'_c correspondence. Clearly this is also a reciprocal 1,1 correspondence: if we choose l'^* we can use (7.4) and (7.5) to uniquely determine l'_c, or choosing l'_c we can determine l'^*.

We have thus proved the existence, in general, of a 1,1 reciprocal correspondence of basic lines l^* and fixed lines l_c, such that the fixed l_c is a screw axis for the displacement of lines l_i $(i = 1, 2, 3)$. The foregoing proof also provides the outline of a computational procedure for determining corresponding pairs l^*, l_c.

We determine the order of the correspondence between the basic line and the central axis by means of an example. We take three mutually orthogonal screws with axes

$$s_{23}\colon\ Y = b, Z = -c; \qquad s_{31}\colon\ X = -a, Z = c; \qquad s_{12}\colon\ X = a, Y = -b.$$

It follows then that the normals between the screws are

$$n_1\colon\ Y = -b, Z = c; \qquad n_2\colon\ X = a, Z = -c; \qquad n_3\colon\ X = -a, Y = b.$$

If a basic line, l^*, is denoted by its Plücker coordinates p_{ij}, then the corresponding l_i must have the components:

$$\mathrm{l}_1\colon (p_{14},\ -p_{24},\ -p_{34}),\ (p_{23}-2bp_{34}-2cp_{24},\ -p_{31}-2cp_{14},\ -p_{12}-2bp_{14}),$$

$$\mathrm{l}_2\colon (-p_{14},\ p_{24},\ -p_{34}),\ (-p_{23}-2cp_{24},\ p_{31}-2cp_{14}-2ap_{34},\ -p_{12}-2ap_{24}),$$

$$\mathrm{l}_3\colon (-p_{14},\ -p_{24},\ p_{34}),\ (-p_{23}-2bp_{34},\ -p_{31}-2ap_{34},\ p_{12}-2ap_{24}-2bp_{14}).$$

Now we can apply the formulas (7.3)–(7.5). If we denote the Plücker coordinates of l_c by c_{ij}, we have from (7.3)

$$-p_{14}c_{14}+p_{24}c_{24}=0$$

and

$$-p_{14}c_{14}+p_{34}c_{34}=0,$$

from which it follows

$$c_{14}\colon c_{24}\colon c_{34}=p_{24}p_{34}\colon p_{34}p_{14}\colon p_{14}p_{24}.$$

Similarly, (7.4) yields

$$(-p_{23}+bp_{34})c_{14}+(p_{31}-ap_{34})c_{24}-ap_{24}c_{34}+bp_{14}c_{34}-p_{14}c_{23}+p_{24}c_{31}=0,$$

and

$$(-p_{23}+cp_{24})c_{14}+cp_{14}c_{24}-ap_{34}c_{24}+p_{12}c_{34}-ap_{24}c_{34}-p_{14}c_{23}+c_{12}p_{34}=0,$$

while (7.5) gives

$$c_{31}c_{24}+c_{12}c_{34}+c_{23}c_{14}=0.$$

We have now three non-homogeneous linear equations for c_{23}, c_{31}, c_{12}, which may be solved in terms of $c_{14}, c_{24}, c_{34}, a, b, c$. The common denominator is seen to be $\mathfrak{N}=p_{24}^2p_{34}^2+p_{34}^2p_{14}^2+p_{14}^2p_{24}^2$. Multiplying all six numbers c_{ij} by $\mathfrak{N}$ we obtain

$$c_{14}=p_{24}p_{34}\mathfrak{N},\qquad c_{24}=p_{34}p_{14}\mathfrak{N},\qquad c_{34}=p_{14}p_{24}\mathfrak{N},$$

$$c_{23}=p_{14}[-ap_{14}(p_{24}^2+p_{34}^2)^2+bp_{24}p_{34}^2(p_{34}^2+p_{14}^2)+cp_{34}p_{24}^2(p_{14}^2+p_{24}^2)$$
$$-p_{24}p_{34}(p_{24}^2+p_{34}^2)p_{23}+p_{14}p_{34}^3p_{31}+p_{14}p_{24}^3p_{12}]$$

and cyclically for c_{31} and c_{12}.

Hence the relationship between l^* and l_c is seen to be a sextic correspondence for this case. Following a similar procedure it is possible to show that this relationship is in fact also sextic for any general set of three displacements.

Example 8. Show that, in general, the results are as follows: From (7.3) it follows $c_{14} : c_{24} : c_{34} = Q_1 : Q_2 : Q_3$ and $c_{23}=S_1/\mathfrak{F}$, $c_{31}=S_2/\mathfrak{F}$, $c_{12}=S_3/\mathfrak{F}$, where Q_i, S_i, and $\mathfrak{F}$ are homogeneous functions of the second, the sixth and fourth degree respectively. Hence the required result has the form

$$c_{14} = Q_1\tilde{\mathfrak{F}}, \qquad c_{24} = Q_2\tilde{\mathfrak{F}}, \qquad c_{34} = Q_3\tilde{\mathfrak{F}}, \qquad c_{23} = S_1, \qquad c_{31} = S_2, \qquad c_{12} = S_3.$$

A natural next step is to inquire if any special relationship exists between the screw axis for the displacement of E and the central axis l_c of the congruence determined by l_1, l_2, l_3. The results follow directly from the screw triangle geometry: since displacement $\mathrm{l}_i \to \mathrm{l}_j$ may be obtained by a screw displacement about l_c, all possible configurations of E_j must follow from screwing E_i first about l_c and then about a screw axis coincident with l_j. Hence we may regard l_c as s_{12}, l_j as s_{23}, and s_{ij} as s_{13} in the screw triangle (Section 1). The result, as shown in Fig. 15, is the theorem: *the normal* n_i, *between* l_i *and* s_{ij}, *and the normal* n_c, *between* l_c *and* s_{ij}, *are related so that the distance between them equals* $d_{ij}/2$ *and the angle between them is* $\phi_{ij}/2$, both measured about s_{ij} in the sense of n_i to n_c with n_i and n_c extended if necessary so that $\phi_{ij}/2 \leq \pi$.

Applying this theorem to two screws say s_{12}, s_{13} yields another method to determine the l_i, l_c correspondence: If we choose a line l_1, the normal n_c is determined by screwing the normal between l_1 and s_{12} about s_{12} by $(\phi_{12}/2, d_{12}/2)$. The normal n_c', see Fig. 16, is obtained by screwing the normal between l_1 and s_{13} about s_{13} by $(\phi_{13}/2, d_{13}/2)$. The axis l_c corresponding to l_1 is the common normal to n_c, n_c'.

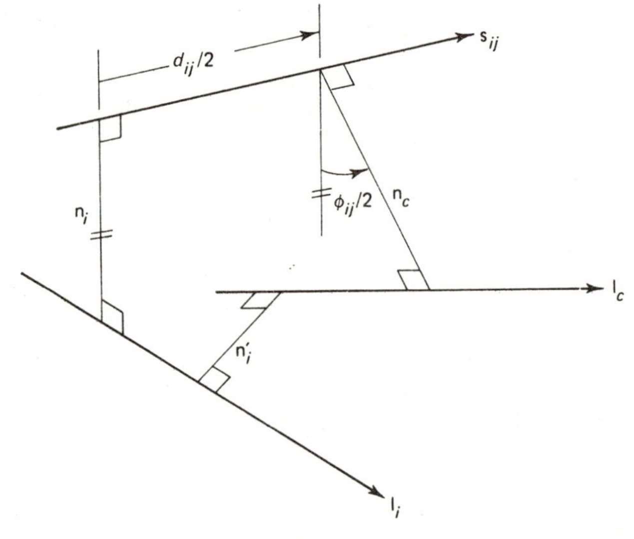

Fig. 15.

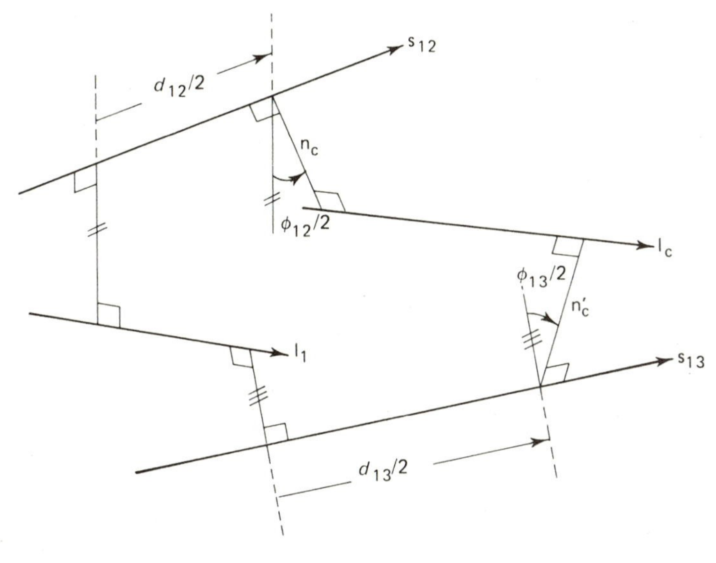

Fig. 16.

Similarly if we start with l_c and determine n_c and n'_c, screwing n_c and n'_c backwards along s_{12} and s_{13} yields the lines for which l_1 is the common normal.

Returning to Fig. 15, for the screw triangle composed of "screws" s_{ij}, l_c and l_i, it follows that the angle between n'_i and n_c must be $\alpha_{ij}/2$ and the distance between them $c_{ij}/2$, measured from n'_i to n_c, where α_{ij} and c_{ij} are the screw displacements of line l_i about axis l_c. Analogously, if n'_j is the normal between l_j (not shown) and l_c, the angle from n_c to n'_j is $\alpha_{ij}/2$ and the distance is $c_{ij}/2$. We will make use of these properties in Chapter V, Section 8.

If we consider two sets of homologous triads l_1, l_2, l_3 and g_1, g_2, g_3 we have two different central axes l_c and g_c. The following theorem follows: *the angle and distance between the normals from l_i and g_i to s_{ij} are equal respectively to the angle and distance between the normals from l_c and g_c to s_{ij}*. The proof of this follows from Fig. 17. Since $\angle p_i p_c = \angle n_i n_c = \phi_{ij}/2$, dist $|p_i p_c| =$ dist $|n_i n_c| = d_{ij}/2$, and all the angles and distances are measured about s_{ij}, it follows if $\angle n_i p_c = \alpha$ and dist $|n_i p_c| = d$, that $\angle p_i n_i = (\phi_{ij}/2) - \alpha = \angle p_c n_c$, and dist $|p_i n_i| = (d_{ij}/2) - d =$ dist $|p_c n_c|$.

Example 9. Show that the basic lines l^*, g^* corresponding to l_1, l_2, l_3 and g_1, g_2, g_3, respectively, may be used in the above theorems instead of l_i and g_i if the sense of the angle and distance between normals is reversed. So for example if the normals from l_c and g_c to s_{ij} measure (from the

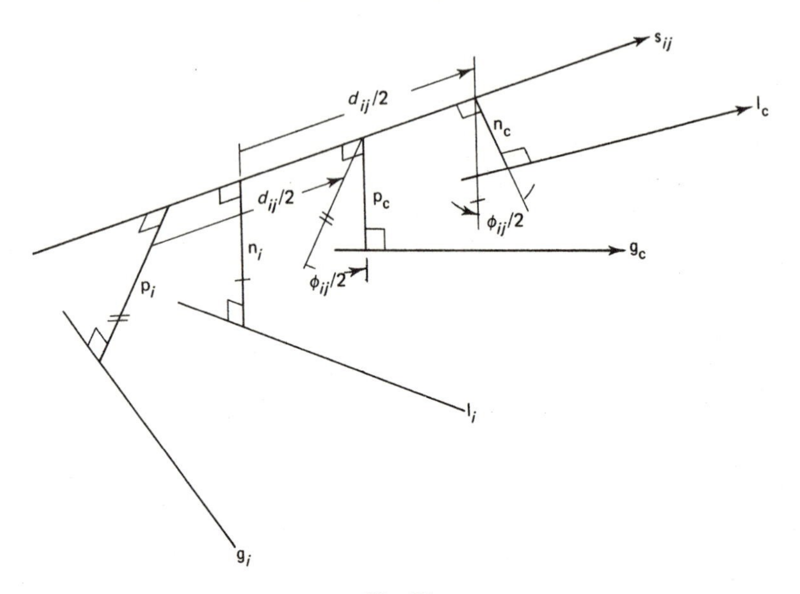

Fig. 17.

l_c to the g_c normal) β, D, the normals from l^* and g^* to s_{ij} measure (from l^* to the g^* normal) $-\beta$, $-D$; the sign for angle β and distance D is determined by the screw s_{ij}.

Example 10. The image screws are similarly related to l_i and l_c as the screws. Specifically, show that the angle and distance between the normals from l_i and l_c to s^i_{jk} are the directed angle $\phi_{jk}/2$ and distance $d_{jk}/2$.

Example 11. Show that if n_i is the normal between l_i and s_{ij}, and n_c is the normal from l_c to s_{ij}, then the angle and distance between n_i and n_c is $-\phi_{ij}/2$ and $-d_{ij}/2$ measured from n_i to n_c about s_{ij}.

In Chapter III, Section 10 we introduced the concept of an internal line-bisector together with the notion of a directed line. These can be applied to the foregoing. We use this to restate our results in the following way: The three homologous lines l_1, l_2, l_3 are obtained as reflections of the basic line l^* into the sides n_1, n_2, n_3 of the screw triangle. If l^* is a directed line it induces directions on l_i $(i = 1,2,3)$, and therefore a unique central line l_c, of l_i, is defined; if the direction of l^* is reversed the same holds for all l_i and hence we have the same central line.

We consider the screw axis s_{12} (which we also denote as s_3) and its directed common perpendiculars $A^*_3A^*$, $A^1_3A_1$, $A^2_3A_2$ with l^*, l_1, l_2 respectively. (Fig. 18.) We denote the directed line $A^*_3A^*$ by m_3. The lines n_1 and n_2 also intersect s_{12} orthogonally; let the points of intersection with s_{12} be N_1 and N_2.

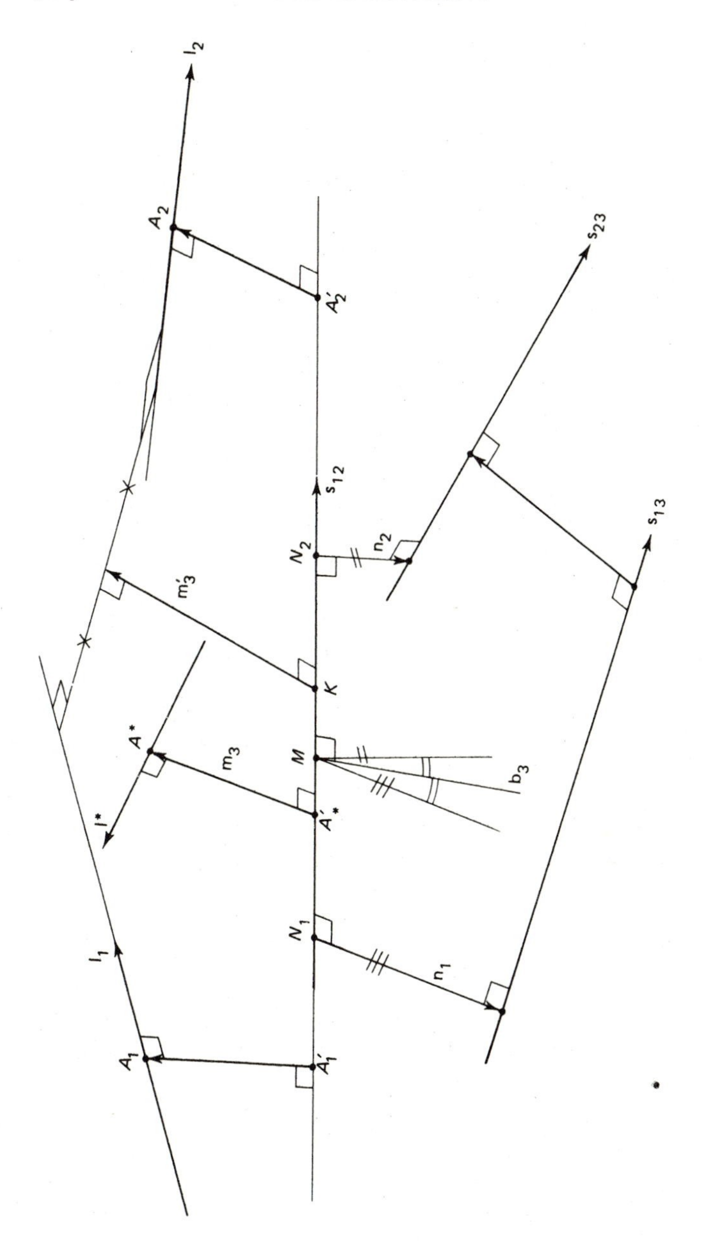

Fig. 18.

We give n_1 and n_2 an orientation; that on n_1 will be from N_1 to the intersection of n_1 and s_{13}, and similarly for n_2. The internal bisector b_3 of n_1 and n_2 could be called an internal bisector of the screw triangle; it passes through the midpoint M of N_1, N_2.

The internal bisector of $A_1'A_1$ and $A_2'A_2$ passes through the midpoint K of $A_1'A_2'$.

We have $MN_1 + MN_2 = 0$, $N_1A_1' + N_1A_*' = 0$, $N_2A_2' + N_2A_*' = 0$. Hence $MA_1' = MN_1 + N_1A_1' = MN_1 - N_1A_*' = 2(MN_1) - MA_*'$ and analogously $MA_2' = 2(MN_2) - MA_*'$. Therefore $MK = \frac{1}{2}(MA_1' + MA_2') = -MA_*'$, which means that K is found by reflecting A_*' into M.

On the other hand, if we project n_1, n_2, $A_*'A^*$, $A_1'A_1$ and $A_2'A_2$, all perpendicular to s_{12}, on a plane orthogonal to s_{12} (Fig. 19), the same argument can be used for the angles (i.e., $\angle b_3n_1 + \angle b_3n_2 = 0$, $\angle n_1, A_1'A_1 + \angle n_1m_3 = 0$, etc.). We conclude that the bisector of $A_1'A_1$ and $A_2'A_2$ is the reflection of $A_*'A^*$ into the bisector of n_1, n_2. The conclusion is: *the bisector* m_3' *of* l_1, l_2 *coincides with the reflection of the common perpendicular* m_3 *of* s_3 *and* l^* *into the bisector* b_3 *of the screw triangle.* We shall call m_3' the isogonally conjugated line of m_3 with respect to the "vertex" s_3 of the screw triangle.

Our result is the following theorem: *if* l^* *is the basic line of the three homologous lines* l_1, l_2, l_3, *if furthermore* m_i *is the common perpendicular of* l^* *and* s_i, *and if* m_i' *is the isogonal conjugate of* m_i *with respect to vertex* s_i *of the screw triangle, then the line* l_c *of* l_1, l_2, l_3 (which are given a direction in accordance with an arbitrarily chosen direction on l^*) *is the common perpendicular of the three lines* m_i'.

Example 12. Show that the relationship between the lines l^* and l_c is involutory.

Example 13. Show that the following geometrical theorem holds: if a spatial triangle is given with vertices s_i, if furthermore m_i is a line intersecting s_i orthogonally ($i = 1,2,3$) and if m_i' is isogonally conjugated to m_i, then: if m_i have a common perpendicular p, the lines m_i' have a common perpendicular p′ (the lines p and p′ are said to be isogonally conjugate with respect to the triangle).

An interesting special case of the l^*, l_c correspondence occurs when the three homologous lines l_1, l_2, l_3 have a common normal. Clearly l_c is the common normal, and $\Theta_i = \pi/2$, $D_i = 0$ ($i = 1,2,3,4$). Taking $\Theta_i = \pi/2$, we have $\cos\Theta_i = 0$, thus (7.1) and (7.3) yields

$$\boldsymbol{l}_i \cdot \boldsymbol{l}_c = 0 \quad i = 1,2,3. \tag{7.6}$$

If, in Σ, we take the components of $\boldsymbol{l}_i$ as (l_i, m_i, n_i), (7.6) will have non-trivial solutions if

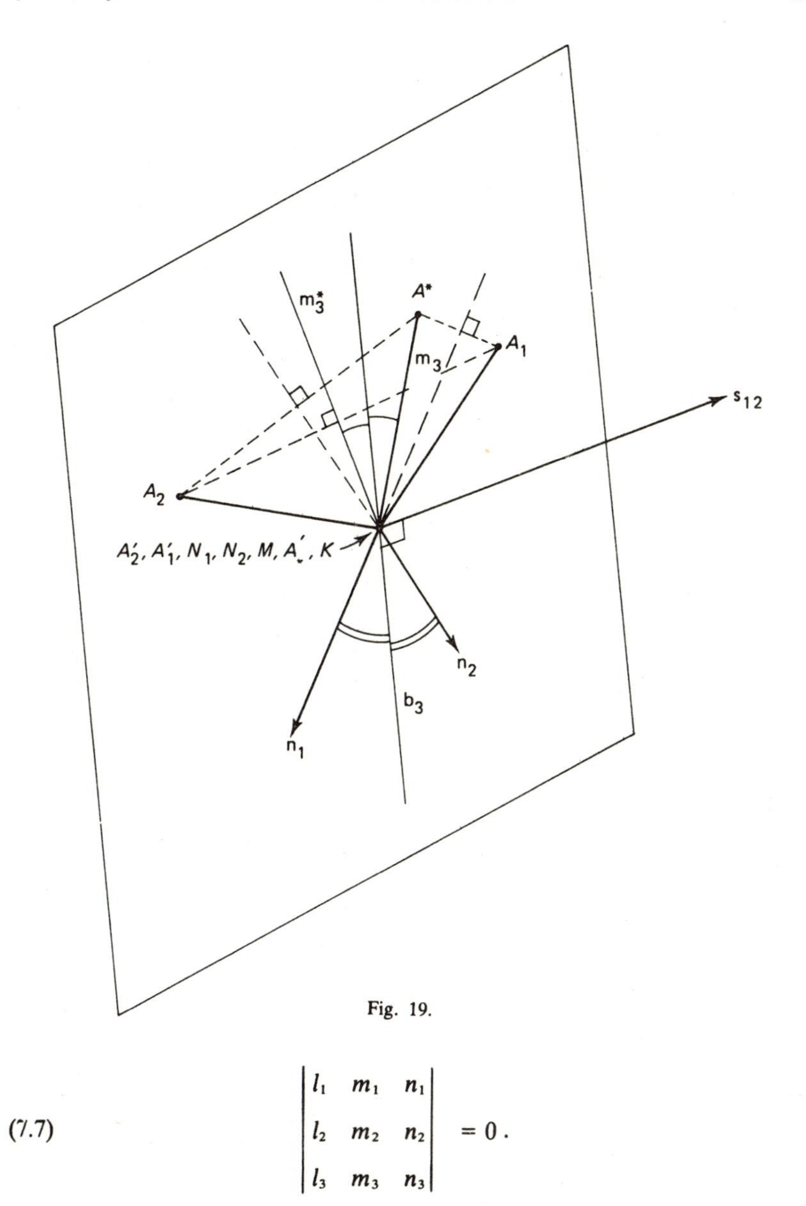

Fig. 19.

$$(7.7) \qquad \begin{vmatrix} l_1 & m_1 & n_1 \\ l_2 & m_2 & n_2 \\ l_3 & m_3 & n_3 \end{vmatrix} = 0 .$$

Since each term can be written as a linear homogeneous function of the components in E_i of l_i (or l^*), (7.7) is the equation of a cubic cone in E_i (or

E^*), the directions of its generators give the directions for those l_i for which three positions have a common normal. Alternatively, by writing all $\boldsymbol{l}_i$ as functions of $\boldsymbol{l}^*$ (or one of the $\boldsymbol{l}_i$), (7.6) yields a cubic cone in Σ whose elements are parallel to $\boldsymbol{l}_c$. The cubic cone of $\boldsymbol{l}_c$ has all its generators normal to the corresponding generators of the cubic cone of $\boldsymbol{l}_i$.

Similarly $D_i = 0$ yields from (7.2) and (7.4)

$$\boldsymbol{l}'_i \cdot \boldsymbol{l}_c + \boldsymbol{l}'_c \cdot \boldsymbol{l}_i = 0 \quad i = 1,2,3. \tag{7.8}$$

If we are to find a solution for, say, $\boldsymbol{l}'_c$, since (7.8) and the fundamental relationship $\boldsymbol{l}_c \cdot \boldsymbol{l}'_c = 0$ yield four equations, the system must have a rank of three. Hence:

$$\begin{vmatrix} \boldsymbol{l}'_1 \cdot \boldsymbol{l}_c & l_1 & m_1 & n_1 \\ \boldsymbol{l}'_2 \cdot \boldsymbol{l}_c & l_2 & m_2 & n_2 \\ \boldsymbol{l}'_3 \cdot \boldsymbol{l}_c & l_3 & m_3 & n_3 \\ 0 & l_c & m_c & n_c \end{vmatrix} = 0. \tag{7.9}$$

For any $\boldsymbol{l}_i$ and $\boldsymbol{l}_c$ this is a linear equation in $\boldsymbol{l}'_i$ or $\boldsymbol{l}'^*$ (recalling that all $\boldsymbol{l}'_i$ can be written as linear functions of any one of them, or of $\boldsymbol{l}'^*$). Taken together with

$$\boldsymbol{l}'_i \cdot \boldsymbol{l}_i = 0 \qquad (\text{or } \boldsymbol{l}'^* \cdot \boldsymbol{l}^* = 0) \tag{7.10}$$

we have a single infinity of lines $\boldsymbol{l}'_i$ (or $\boldsymbol{l}'^*$) corresponding to each pair of directions $\boldsymbol{l}_i, \boldsymbol{l}_c$. A similar result follows from (7.8) if we solve for $\boldsymbol{l}'_i$ (or $\boldsymbol{l}'^*$) instead of $\boldsymbol{l}'_c$. To summarize: the locus of lines l_i, such that their homologous triads each have a single common normal, is a line congruence (whose directions are given by the cubic cone (7.7), and location by the single infinity of vectors determined by (7.9) and (7.10)). The infinitesimal displacement case of this congruence will be studied in Chapter XIII, Section 5; we will prove it is a (7.3) congruence. The locus of common normals l_c is a corresponding congruence in Σ.

There are several other special cases of interest, these include the further restrictions that the displacement of l_i along l_c is a pure translation, the displacement of E about l_i is a pure rotation, and combinations of these. The analysis of these cases is rather involved, see ROTH [1968], and TSAI and ROTH [1972]. We conclude by remarking, without proof, that there are always two triads of homologous lines, and only two such triads, such that a set of general displacements of E, given by s_{12}, s_{23}, may be obtained by a set of pure rotations about l_c and l_i, see TSAI and ROTH [1973], and VELDKAMP [1967a].

8. The circle axes

Three homologous points A_1, A_2, A_3 are in general the vertices of a triangle. Of interest is its circle axis, that is the line m through the circumcenter of the triangle and orthogonal to its plane; m is the locus of points at equal distances to A_1, A_2 and A_3. It is the intersection of the normal planes β_1, β_2, β_3 of A_2A_3, A_3A_1, A_1A_2 respectively. We know (from Chapter III, Section 4) that β_2 and β_3 are each associated to A_1 by a non-singular correlation. Hence β_2 and β_3 are corresponding planes of a spatial projectivity. The locus of their line of intersection is a tetrahedral complex C_1. To determine its four singular planes we determine when planes β_2 and β_3 coincide: one such singular β_2 plane is perpendicular to A_1A_3 and passes through the midpoint M_{13} of A_1 and A_3; β_3 is perpendicular to A_1A_2 and passes through the midpoint M_{12} of A_1 and A_2. If β_2 and β_3 coincide with one plane β, the points A_1, A_2, A_3 must be on one line l. However, as M_{12} and M_{13} are different points (because A_2 and A_3 are different) β must also go through l; since it is perpendicular to l as well, it is an isotropic plane, passing through a tangent t, of the isotropic conic Ω. The conclusion is: *the locus of the circle axes* m *is a tetrahedral complex* C_1; *the faces of its tetrahedron are isotropic planes.*

As a tetrahedral complex is quadratic the lines m of C_1 at infinity are the tangents of a conic k′. Such lines are the intersection of two (and therefore of three) parallel normal planes. Two normal planes however are parallel if and only if A_1, A_2, A_3 are on one line l. We know from the preceding section that these lines l have their point at infinity on the conic k, the intersection of V and the hyperboloid $Q(n_1, n_2, n_3)$. *Hence the conics* k *and* k′ *in* V *are polar reciprocals with respect to the isotropic conic* Ω. As[†] N_1, N_2, N_3 are on k it follows that k′ is an inscribed conic of the triangle $P_1P_2P_3$. The lines at infinity on the faces of the tetrahedron are the common tangents of Ω and k′. They are two pairs of conjugate imaginary lines. The faces of the tetrahedron are two pairs of conjugate imaginary planes; its vertices are two pairs of conjugate imaginary points; two opposite edges are real, the other two pairs are conjugate imaginary; the characteristic cross ratio of the tetrahedral complex is a real number.

From any basic point A^* follow three homologous points A_1, A_2, A_3 and to these there corresponds in general one circle axis m; we shall say that the line m is associated with the point A^* and conversely A^* is associated with m. Let B^* be a point on m; then we have

[†]Here, N_i and P_i are the points in V defined in Section 2.

(8.1) $$|B^* A_1| = |B^* A_2| = |B^* A_3|.$$

By reflection into n_1 the points B^* and A_1 are transformed into B_1 and A^* respectively. Hence $|B^* A_1| = |B_1 A^*|$; similarly $|B^* A_2| = |B_2 A^*|$ and $|B^* A_3| = |B_3 A^*|$. From (8.1) it then follows that

$$|A^* B_1| = |A^* B_2| = |A^* B_3|,$$

and we have therefore the theorem: *if B^* is on the circle axis associated with A^* then A^* is on the circle axis associated with B^*.*

If A^* is on the twisted cubic c the homologous points A_1, A_2, A_3 are collinear; the planes β_1, β_2, β_3 are parallel and the circle axis associated with A^* is a line of the plane V at infinity. The locus of these lines consists of the tangents of the conic k′. If B^* is any point of V two of these tangents m′ and m″ pass through it; they are associated with two points $A^{*\prime}$ and $A^{*\prime\prime}$ on c. In view of our theorem the circle axis associated with B^* passes through both $A^{*\prime}$ and $A^{*\prime\prime}$, which means that it is their join. Hence *the circle axis associated with a point at infinity is a chord of* c. Conversely, any chord of c is associated with a point of V: namely, the intersection point of the two lines in V which are the circle axes associated with the end-points of the chord. It then follows that *all chords of* c *are lines of the tetrahedral complex* C_1 and the tangents of c are the circle axes associated with the points of k′. The cone of the complex associated with a point of c is the (quadratic) cone of the chords through it.

If the planes β_1, β_2, β_3 coincide they are of course parallel, which means that A^* is on c. Therefore there must be four points, T_i say, $i = 1, 2, 3, 4$, of c that are not associated with a single axis but with all lines of a plane γ_i; these four planes are the faces of the tetrahedron of C_1. T_i is associated with every line of γ_i; in view of our theorem the circle axis associated with any point of γ_i passes through T_i, which implies that ∞^2 axes pass through T_i. The conclusion is that *T_i is a vertex of the tetrahedron.*

Summing up we have: *the locus of the circle axes of the triads of homologous points is a tetrahedral complex* C_1; *the vertices of its tetrahedron are on the twisted cubic* c; *its faces are tangent to the isotropic conic Ω.*

If we disregard questions of reality we may introduce a projective coordinate system with T_i as fundamental points; let x_i be point coordinates and p_{ij} line coordinates in this system. If the point A^* and the line m are associated the coordinates of m are quadratic functions of the coordinates of A^*: this follows from the correspondence described at the beginning of this section. We know that m is undetermined if A^* coincides with a point T_i. This means that every p_{ij} is zero if any three of the four coordinates x_i are zero; in other words p_{ij} are functions of the products $x_k x_l$ $(k \neq l)$. On the other hand every point of the plane $x_i = 0$ has its associated line through T_i; it follows from this that the line associated with any point $(x_1, x_2, 0, 0)$ on the edge T_1T_2 say (different from T_1 and T_2) is the opposite edge $x_1 = x_2 = 0$, this is the line for which $p_{34} = 1$ and all other coordinates are zero. From all this it follows that the relationship between the point $A^*(x_i)$ and the associated line $\mathrm{m}(p_{ij})$, for a suitable choice of the unit point, reads

$$\begin{aligned} p_{14} &= k_1x_2x_3, & p_{23} &= x_1x_4 \\ p_{24} &= k_2x_3x_1, & p_{31} &= x_2x_4 \\ p_{34} &= k_3x_1x_2, & p_{12} &= x_3x_4 \end{aligned} \tag{8.2}$$

in which the coefficients k_i satisfy $k_1 + k_2 + k_3 = 0$ in view of $p_{14}p_{23} + p_{24}p_{31} + p_{34}p_{12} = 0$ (the fundamental relation for line coordinates). From (8.2), the locus of the circle axes is given by the equation

$$k_2p_{14}p_{23} - k_1p_{24}p_{31} = 0, \tag{8.3}$$

or the equivalent ones

$$k_3p_{24}p_{31} - k_2p_{34}p_{12} = 0, \qquad k_1p_{34}p_{12} - k_3p_{14}p_{23} = 0, \tag{8.4}$$

or more symmetrically by

$$(k_2 - k_3)p_{14}p_{23} + (k_3 - k_1)p_{24}p_{31} + (k_1 - k_2)p_{34}p_{12} = 0, \tag{8.5}$$

which is a standard form for the equation of a tetrahedral complex. From (8.2) if follows

$$x_1 : x_2 : x_3 : x_4 = k_2p_{12}p_{23} : k_2p_{12}p_{31} : k_2p_{12}^2 : p_{23}p_{24}, \tag{8.6}$$

which by means of (8.3) and (8.4) may be written in other, equivalent forms. The conclusion is: a line m of the complex C_1 is (in general) associated with *one* basic point A^*; their relationship is birational.

The line $\mathrm{m}(p_{ij})$ is the intersection of the planes

$$p_{23}x_1 + p_{31}x_2 + p_{12}x_3 = 0 \quad \text{and} \quad p_{34}x_2 + p_{42}x_3 + p_{23}x_4 = 0. \tag{8.7}$$

Hence if the point (x'_i) is on the line associated with (x_i) we have in view of (8.2) and (8.7):

$$\begin{aligned} 0 &= x_1x'_1 + x_2x'_2 + x_3x'_3 = k_3x_2x'_2 - k_2x_3x'_3 + x_4x'_4 \\ &= k_1x_3x'_3 - k_3x_1x'_1 + x_4x'_4 = k_2x_1x'_1 - k_1x_2x'_2 + x_4x'_4. \end{aligned} \tag{8.8}$$

It follows from this that conversely (x_i) is on the line associated with (x'_i); this provides an analytical verification of the theorem given at the beginning of this section. Furthermore it is seen from (8.8) that m is the polar line of its associated point with respect to the pencil of quadrics to which the four cones

$$(8.9)\qquad \begin{aligned} &x_1^2+x_2^2+x_3^2=0, && k_3x_2^2-k_2x_3^2+x_4^2=0,\\ &k_1x_3^2-k_3x_1^2+x_4^2=0, && k_2x_1^2-k_1x_2^2+x_4^2=0, \end{aligned}$$

belong. This then leads us to conclude that: a point lies on the line associated with it if it lies on all quadrics of the pencil. Hence *the locus of the basic point A^* which has equal distances to the corresponding homologous points A_1, A_2, A_3 is a biquadratic space curve of the first kind.*

9. The radius of a circle

For any planar triangle the radius, r, of the circumscribed circle is given by $r=\frac{1}{2}(a/\sin\alpha)$ where α is an interior angle and a is the length of the opposite side. Taking $\boldsymbol{A}_1$, $\boldsymbol{A}_2$, $\boldsymbol{A}_3$ as the position vectors, in Σ, of the homologous set A_1, A_2, A_3 we have, from the above formula: the radius of the circle through a triple of homologous points is

$$(9.1)\qquad r=\tfrac{1}{2}\{|(\boldsymbol{A}_3-\boldsymbol{A}_2)|\,|(\boldsymbol{A}_3-\boldsymbol{A}_1)|\,|(\boldsymbol{A}_2-\boldsymbol{A}_1)|\}/|(\boldsymbol{A}_3-\boldsymbol{A}_1)\times(\boldsymbol{A}_2-\boldsymbol{A}_1)|.$$

If the points A_i $(i=1,2,3)$ are known this expression can be used to compute the radius from their position vectors $\boldsymbol{A}_i$.

By squaring this expression we can remove the absolute value restrictions and obtain the rationalized equation:

$$(9.2)\qquad 4r^2[(\boldsymbol{A}_3-\boldsymbol{A}_1)\times(\boldsymbol{A}_2-\boldsymbol{A}_1)]^2-(\boldsymbol{A}_3-\boldsymbol{A}_2)^2(\boldsymbol{A}_3-\boldsymbol{A}_1)^2(\boldsymbol{A}_2-\boldsymbol{A}_1)^2=0$$

since all the $\boldsymbol{A}_i$ $(i=1,2,3)$ can be expressed as linear functions of $\boldsymbol{A}^*$ (or if we wish one of the $\boldsymbol{A}_i$), this equation represents a sixth order surface embedded in Σ (or E_i), and is the locus of all A^* (or A_i) for which the corresponding A_i $(i=1,2,3)$ lie on a circle of radius r.

There are two special questions of interest: namely, how small and how large can r be?

As r approaches infinity this surface degenerates into the singular quartic surface $[(\boldsymbol{A}_3-\boldsymbol{A}_1)\times(\boldsymbol{A}_2-\boldsymbol{A}_1)]^2=0$ and the plane at infinity counted twice. The only real points on these surfaces lie on the twisted cubic c_i and the plane cubic c_i' (discussed in Section 4).

In order to find the minimum value of r we introduce the points π_i, which denotes the point where the normal n_i bisects the line connecting A_iA^*. In this way we define three points π_i $(i=1,2,3)$ such that they lie on n_i, and have the important property that the circle through π_1, π_2, π_3 is always one-half as large as the circle through A_1, A_2, A_3. With this definition we have

$$\boldsymbol{\pi}_i = \boldsymbol{P}_i + \mu_i \boldsymbol{n}_i$$

where $\boldsymbol{\pi}_i$ is the position vector to π_i, $\boldsymbol{P}_i$ the position vector to a point on n_i, $\boldsymbol{n}_i$ a unit vector parallel to n_i, and μ_i is a scalar parameter (Fig. 20). It then follows that

$$\boldsymbol{A}_j - \boldsymbol{A}_i = 2(\boldsymbol{\pi}_j - \boldsymbol{\pi}_i) = 2(\boldsymbol{P}_j - \boldsymbol{P}_i + \mu_j \boldsymbol{n}_j - \mu_i \boldsymbol{n}_i). \tag{9.3}$$

Equation (9.2) can be written as an explicit function of μ_i if we substitute (9.3) into (9.2). Now, operating on (9.2) with $\mathrm{d}/\mathrm{d}\mu_i$ and setting $\mathrm{d}r/\mathrm{d}\mu_i = 0$ we find the conditions for stationary values of r. The details are rather lengthy (they are given by BOTTEMA et al. [1970]), the conclusions are relatively simple: The minimum radius $r_{\min}$ is always twice the radius of the smallest circle which can be inscribed in the skew lines n_i ($i = 1, 2, 3$).

There are two distinct cases: if d_{ij} is the largest of the three translations (along the screw axes s_{12}, s_{23}, s_{31}), then we have either $r_{\min} = d_{ij}/2$ or $r_{\min} > d_{ij}/2$.

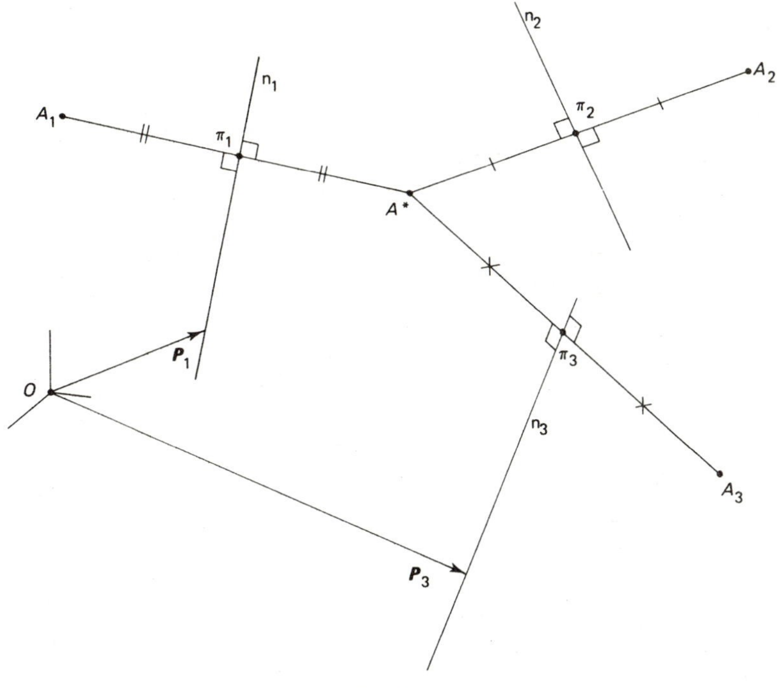

Fig. 20.

If $r_{\min} = d_{ij}/2$, the screw axis is collinear with a diameter of "the" smallest circle, and there are two smallest circles. Their sets of homologous points A_k, B_k ($k = 1, 2, 3$) lie on circles of radii $d_{ij}/2$ with centers on s_{ij}; A_i, A_j, B_i, B_j all lie on s_{ij} as shown in Fig. 21. Also, the center of each circle coincides with the base point of the other.

As long as the distance of n_k from π_c (shown in Fig. 21) is less than $d_{ij}/2$ there will be two such minimum radius circles. If this distance equals $d_{ij}/2$ the circles coincide. However, when n_k is at a greater distance, $r_{\min} > d_{ij}/2$ and we have the second case. In this case the minimum radius points A_i are characterized by their basic point, A^*, and center point, A_c, coinciding. In addition the center of their π_i-circle, π_c, also coincides with A^* and A_c. In general there will be at most 12 circles with the property that they touch all three n_i and yet do not have a screw axis as a diameter. All of these will define π_i points (where they touch the n_i) which in turn will define sets of homologous A_i, since their A_c, A^*, π_c points all coincide. Each such set, of homologous A_i, defines a circle with the property that its radius is stationary with respect to the radii associated with all neighboring points. The smallest of these stationary radii will be the minimum radius, $r_{\min}$, in the case when $r_{\min} > d_{ij}/2$.

10. An oblique coordinate system

In the analytical sections of this chapter we have used orthogonal coordinate systems. It is possible to reduce the complexity of some of the formulas by introducing a special oblique system.

Three positions are completely determined by the three screw axes, s_i, or by

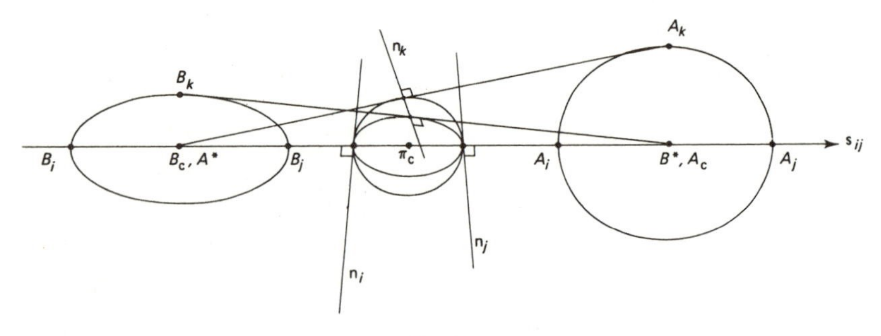

Fig. 21.

their common perpendiculars, n_i. To describe this configuration in a suitable manner we introduce an oblique (or affine) coordinate system in the following way. Let V_{ij} be the plane through n_i parallel to n_j $(i \neq j)$. As V_{ij} and V_{ji} are parallel the six planes are the faces of a parallelepiped (Fig. 22). We take its center O as the origin. Furthermore we orient n_1 so that its positive direction is from its intersection with s_2 to that with s_3, and continue in the same way, cyclically, for the positive directions of n_2 and n_3. The axes of our system O_{xyz} pass through O parallel to the oriented lines n_i. Hence if the edges of the parallelepiped are $2a$, $2b$, $2c$ the lines n_i are given by

$$(10.1) \quad n_1\colon\ y = b, z = -c; \qquad n_2\colon\ z = c, x = -a; \qquad n_3\colon\ x = a, y = -b.$$

It follows from the formula for the distance of any point from the origin that the equation of the isotropic conic Ω reads

$$(10.2) \qquad x^2 + y^2 + z^2 + 2(yz \cos\alpha + zx \cos\beta + xy \cos\gamma) = w = 0,$$

in which α is the angle between n_2 and n_3, β the angle between n_3 and n_1, γ the angle between n_1 and n_2.

The polar line of $N_1 = (1, 0, 0, 0)$ with respect to Ω is

$$(10.3) \qquad x + y \cos\gamma + z \cos\beta = w = 0$$

and all planes through it are orthogonal to n_1. It follows from this that s_1 is

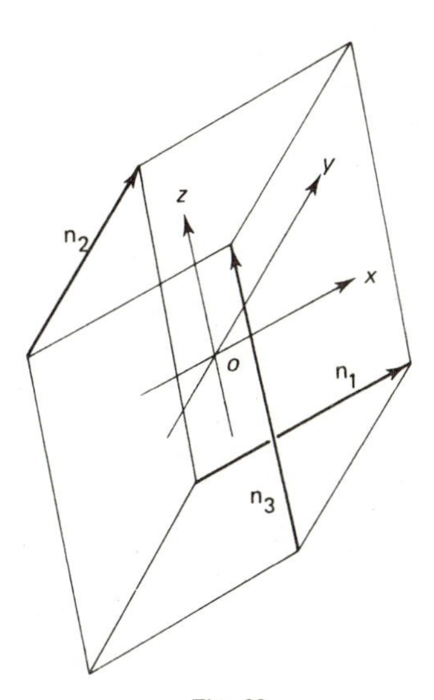

Fig. 22.

parallel to the intersection of the planes $x \cos\gamma + y + z \cos\alpha = 0$ and $x \cos\beta + y \cos\alpha + z = 0$. Hence the direction of s_1 reads (according to a well-known formula from spherical trigonometry)

$$s_1\colon\ \sin\alpha, -\sin\beta\cos\gamma', -\sin\gamma\cos\beta', \tag{10.4}$$

α', β', γ' being the angles between the faces of the trihedron, the sides of which are α, β, γ.

We shall make use of a second coordinate system O_{XYZ}, with the same origin, and directions such that

$$\begin{aligned} X &= x + y\cos\gamma + z\cos\beta,\\ Y &= x\cos\gamma + y + z\cos\alpha,\\ Z &= x\cos\beta + y\cos\alpha + z. \end{aligned} \tag{10.5}$$

or inversely

$$\Re x = X\sin^2\alpha + Y(\cos\alpha\cos\beta - \cos\gamma) + Z(\cos\alpha\cos\gamma - \cos\beta),$$

and cyclically for y and z,

with $\Re = 1 + 2\cos\alpha\cos\beta\cos\gamma - \cos^2\alpha - \cos^2\beta - \cos^2\gamma$. Obviously the axes O_X, O_Y and O_Z are parallel to s_1, s_2, s_3. Three homologous points A_i are the reflections of the basic point A^* into n_i. Let A^* have coordinates (x^*, y^*, z^*) and (X^*, Y^*, Z^*) in the two systems; A_i' is its projection on n_i.

The plane through A^* orthogonal to n_1 has the equation

$$X = X^*, \tag{10.6}$$

and for its intersection A_1' with n_1 we obtain

$$A_1'\colon\ x = X^* - b\cos\gamma + c\cos\beta, \qquad y = b, \qquad z = -c \tag{10.7}$$

or, what is the same thing

$$\begin{aligned} A_1'\colon\ x &= x^* + (y^* - b)\cos\gamma + (z^* + c)\cos\beta,\\ y &= b, \qquad z = -c. \end{aligned} \tag{10.8}$$

For A_2' and A_3' we have similar formulas.

A_1' is the midpoint of A^* and A_1. Hence, from (10.8),

$$\begin{aligned} A_1\colon\ x &= x^* + 2(y^* - b)\cos\gamma + 2(z^* + c)\cos\beta,\\ y &= 2b - y^*, \qquad z = -2c - z^*, \end{aligned} \tag{10.9}$$

or in the O_{XYZ}-system

$$A_1:\ X = X^*,$$

$$(10.10)\qquad Y = 2X^* \cos\gamma - Y^* + 2\sin\gamma(b\sin\gamma - c\sin\beta\cos\alpha'),$$

$$Z = 2X^*\cos\beta - Z^* - 2\sin\beta(c\sin\beta - b\sin\gamma\cos\alpha'),$$

with similar expressions for A_2 and A_3.

11. Applications

If A_1, A_2, A_3 are collinear then the same holds for A'_1, A'_2, A'_3. Making use of (10.7) the conclusion is that the basic point A^* must be such that the matrix

$$\left\|\begin{matrix} X^* - b\cos\gamma + c\cos\beta & b & -c & 1 \\ -a & Y^* - c\cos\alpha + a\cos\gamma & c & 1 \\ a & -b & Z^* - a\cos\beta + b\cos\alpha & 1 \end{matrix}\right\|$$

(11.1)

has rank two. Introducing three homogeneous parameters μ_1, μ_2, μ_3 such that $\mu_1 + \mu_2 + \mu_3 = 0$, and making use of homogeneous coordinates, we obtain

$$\mathrm{X}^* = \mu_2\mu_3\{\mu_1(b\cos\gamma - c\cos\beta) + a(\mu_2 - \mu_3)\}$$

$$\mathrm{Y}^* = \mu_3\mu_1\{\mu_2(c\cos\alpha - a\cos\gamma) + b(\mu_3 - \mu_1)\}$$

$$(11.2)\qquad \mathrm{Z}^* = \mu_1\mu_2\{\mu_3(a\cos\beta - b\cos\alpha) + c(\mu_1 - \mu_2)\}$$

$$\mathrm{W}^* = \mu_1\mu_2\mu_3,$$

where $X^* = \mathrm{X}^*/\mathrm{W}^*$, $Y^* = \mathrm{Y}^*/\mathrm{W}^*$, $Z^* = \mathrm{Z}^*/\mathrm{W}^*$. Hence the locus of A^* such that A_1, A_2, A_3 are collinear is a twisted cubic, as we found before. Its intersections with the plane at infinity follow from $\mu_1 = 0$, $\mu_2 = 0$ and $\mu_3 = 0$ respectively; they are therefore the points (1, 0, 0, 0), (0, 1, 0, 0) and (0, 0, 1, 0), which are the points at infinity of the screw axes s_i, as is expected. If A_1, A_2, A_3 are collinear they are on a line m; in order to find the locus of m we determine first the locus of the lines m′ on which the three collinear points A'_1, A'_2, A'_3 lie. Making use of (10.7) and similar expressions for the coordinates of A'_2, A'_3, and substituting for X^*, Y^*, Z^* the ratios of X^*, Y^*, Z^*, with W^* as given by (11.2) we obtain for three collinear points A'_i:

$$A'_1:\ x = a(\mu_2 - \mu_3),\ y = b\mu_1,\ z = -c\mu_1,\ w = \mu_1,$$

$$(11.3)\qquad A'_2:\ x = -a\mu_2,\ y = b(\mu_3 - \mu_1),\ z = c\mu_2,\ w = \mu_2,$$

$$A'_3:\ x = a\mu_3,\ y = -b\mu_3,\ z = c(\mu_1 - \mu_2),\ w = \mu_3.$$

Hence the Plücker coordinates of m′ in the $O_{x,y,z}$-system are

$$p_1 = a\mu_2\mu_3, \qquad p_2 = b\mu_3\mu_1, \qquad p_3 = c\mu_1\mu_2,$$
$$(11.4) \qquad p_1' = bc\mu_1^2, \qquad p_2' = ca\mu_2^2, \qquad p_3' = ab\mu_3^2.$$

Any line m′ intersects n_1, n_2 and n_3 and is therefore a generator of the regulus determined by them. The equation of the corresponding hyperboloid is, as seen from (10.1):

$$(11.5) \qquad ayz + bzx + cxy + abcw^2 = 0.$$

If we transform from O_{xyz} to O_{XYZ}, the Plücker coordinates in the latter system being P_i, P_i', we obtain for the lines m′ instead of (11.4):

$$P_1 = a\mu_2\mu_3 + b\mu_3\mu_1\cos\gamma + c\mu_1\mu_2\cos\beta,$$
$$(11.6) \qquad P_2 = a\mu_2\mu_3\cos\gamma + b\mu_3\mu_1 + c\mu_1\mu_2\cos\alpha,$$
$$P_3 = a\mu_2\mu_3\cos\beta + b\mu_3\mu_1\cos\alpha + c\mu_1\mu_2,$$

$$P_1' = (bc\mu_1^2\sin\alpha - ca\mu_2^2\sin\beta\cos\gamma' - ab\mu_3^2\sin\gamma\cos\beta')\sin\alpha,$$
$$(11.7) \qquad P_2' = (-bc\mu_1^2\sin\alpha\cos\gamma' + ca\mu_2^2\sin\beta - ab\mu_3^2\sin\gamma\cos\alpha')\sin\beta,$$
$$P_3' = (-bc\mu_1^2\sin\alpha\cos\beta' - ca\mu_2^2\sin\beta\cos\alpha' + ab\mu_3^2\sin\gamma)\sin\gamma.$$

We obtain the line $m^*(\boldsymbol{Q}, \boldsymbol{Q}')$ doubling the distance of (points on) line m′ from point A^*. Hence

$$Q_1 = P_1W^*, \qquad Q_2 = P_2W^*, \qquad Q_3 = P_3W^*,$$
$$(11.8) \qquad Q_1' = 2P_1'W^* - Z^*P_2 + Y^*P_3, \qquad Q_2' = 2P_2'W^* - X^*P_3 + Z^*P_1,$$
$$Q_3' = 2P_3'W^* - Y^*P_1 + X^*P_2.$$

Parametric expressions for the locus of the lines m then follow directly from substituting (11.2) and (11.6) into (11.8). The result is

$$Q_1 = \mu_1\mu_2\mu_3(a\mu_2\mu_3 + b\mu_3\mu_1\cos\gamma + c\mu_1\mu_2\cos\beta),$$
$$(11.9) \qquad Q_1' = \mu_1[2\mu_2\mu_3P_1' - \mu_2\{\mu_3(a\cos\beta - b\cos\alpha) + (\mu_1 - \mu_2)c\}P_2$$
$$+ \mu_3\{\mu_2(c\cos\alpha - a\cos\gamma) + (\mu_3 - \mu_1)b\}P_3]$$

and Q_2, Q_3, Q_2', Q_3' follow by cyclic permutation. As P_i, P_i', given by (11.6) and (11.7) are quadratic functions of μ_1, μ_2, μ_3, it follows from (11.9) that the Plücker coordinates of m are of the fifth order in the parameters. Hence the locus of the lines on which there lie three homologous points is a ruled surface F of the fifth order, which is as we found before.

We derive from (11.9) the intersection of F and the plane at infinity. If $\mu_1\mu_2\mu_3 \neq 0$ the corresponding lines m do not lie in the plane at infinity; the direction of any such line is given by $Q_1 : Q_2 : Q_3$ and for this case the locus of the intersections of F with V is the conic k. In the O_{xyz} system we use p_1, p_2, p_3 to denote the directions of m, and since m is parallel to m′ we have $p_1 : p_2 : p_3 = a\mu_2\mu_3 : b\mu_3\mu_1 : c\mu_1\mu_2$. It follows that the equation for k is $ayz + bzx + cxy = w = 0$, in accordance with (11.5).

If $\mu_1\mu_2\mu_3 = 0$ the corresponding lines, m, are in the plane at infinity; there are three such lines. For $\mu_1 = 0$, we have from (11.8) and (11.6), keeping in mind that $W^* = Y^* = Z^* = 0$:

$$(11.10) \qquad Q_1 = Q_2 = Q_3 = Q_1' = 0, \qquad Q_2' : Q_3' = -\cos\beta : \cos\gamma;$$

hence it is the line m_1 joining $L_1 = (1, 0, 0, 0)$ to $N_1 = (1, \cos\gamma, \cos\beta, 0)$. In a similar manner we could get the other two lines. The three lines L_iN_i pass through $H = (\cos\beta\cos\gamma, \cos\gamma\cos\alpha, \cos\alpha\cos\beta, 0)$. All this is in accordance with Section 5.

Until now we dealt with finite homologous points. If $A^* = (x^*, y^*, z^*, 0)$ is in V we know that A_1 is found as follows. The polar line of $N_1(1, 0, 0, 0)$ is $X = 0$. Hence the join N_1A^* is

$$(11.11) \qquad x = x^* - X^* + \lambda, \qquad y = y^*, \qquad z = z^*, \qquad w = 0.$$

$\lambda = \infty$ gives N_1, $\lambda = 0$ the intersection with the polar line, and $\lambda = X^*$ gives A^*. The reflection of A^* is therefore given by $\lambda = -X^*$. Hence

$$A_1\!: \; x = x^* - 2X^*, \; y = y^*, \; z = z^*, \; w = 0,$$

and similarly

$$(11.12) \qquad \begin{aligned} &A_2\!: \; x = x^*, \; y = y^* - 2Y^*, \; z = z^*, \; w = 0, \\ &A_3\!: \; x = x^*, \; y = y^*, \; z = z^* - 2Z^*, \; w = 0. \end{aligned}$$

The three homologous points are on one line if A^* satisfies

$$\begin{vmatrix} x - 2X & y & z \\ x & y - 2Y & z \\ x & y & z - 2Z \end{vmatrix} = 0,$$

or

$$(11.13) \qquad xYZ + yZX + zXY - 2XYZ = 0,$$

which is therefore the equation of the plane cubic c′ we have met in Section 5.

It passes through the points at infinity of the screw axes; for L_1, for instance, we have $Y^* = Z^* = 0$. This curve has a simple equation in the XYZ-system: substituting (10.5) into (11.13) yields c':

$$\begin{aligned}&XYZ(1-\cos^2\alpha-\cos^2\beta-\cos^2\gamma)\\&+X(Y^2+Z^2)(\cos\beta\cos\gamma-\cos\alpha)\\&+Y(Z^2+X^2)(\cos\gamma\cos\alpha-\cos\beta)\\&+Z(X^2+Y^2)(\cos\alpha\cos\beta-\cos\gamma)=0.\end{aligned} \tag{11.14}$$

In view of the duality of the metrics in V and in accordance with our remarks at the end of Section 5, the equation of the locus of the lines at infinity on which three homologous points are situated reads, in the O_{xyz}-system

$$\begin{aligned}&u_1u_2u_3(1-\cos^2\alpha-\cos^2\beta-\cos^2\gamma)+u_1(u_2^2+u_3^2)(\cos\beta\cos\gamma-\cos\alpha)\\&+u_2(u_3^2+u_1^2)(\cos\gamma\cos\alpha-\cos\beta)\\&+u_3(u_1^2+u_2^2)(\cos\alpha\cos\beta-\cos\gamma)=0.\end{aligned} \tag{11.15}$$

12. Three mutually orthogonal screw axes

Until now we have supposed that the screw axes s_i determining the three positions of a space are three lines in general position. As we have remarked before, there are several special cases. If the three s_i pass through one point we deal with spherical kinematics, if they are mutually parallel we have essentially the case of plane kinematics. Both subjects will be considered separately in Chapters VII and VIII.

In this section we consider another special case, namely that in which the three axes s_i are mutually orthogonal. The kinematic meaning of this is obviously that the transformation of any position into another is a screw displacement the rotational part of which is a half turn. Furthermore n_i is parallel to s_i. The parallelepiped considered in Section 10 is rectangular. The oblique coordinate systems O_{XYZ} and O_{xyz} coincide and they are now Cartesian systems. In the formulas of Sections 10 and 11 we have $\cos\alpha = \cos\beta = \cos\gamma = 0$. With these simplifictions this special case is particularly suitable for illustrating the general theory.

From the formulas (10.6)–(10.8) it follows that if $A^*(X^*, Y^*, Z^*)$ is the basic point, its projections A_1', A_2', A_3' on n_i are

$$(X^*, b, -c), \quad (-a, Y^*, c), \quad (a, -b, Z^*) \tag{12.1}$$

and we have from (10.10), for the three homologous points

$$\begin{aligned} A_1 &= (X^*, -Y^* + 2b, -Z^* - 2c), \\ A_2 &= (-X^* - 2a, Y^*, -Z^* + 2c), \\ A_3 &= (-X^* + 2a, -Y^* - 2b, Z^*). \end{aligned} \tag{12.2}$$

Furthermore it follows from (11.2) that the locus of A^* such that A_1, A_2, A_3 are collinear is the twisted cubic

$$\begin{aligned} X^* &= a\mu_2\mu_3(\mu_2 - \mu_3), \qquad Y^* = b\mu_3\mu_1(\mu_3 - \mu_1), \\ Z^* &= c\mu_1\mu_2(\mu_1 - \mu_2), \qquad W^* = \mu_1\mu_2\mu_3. \end{aligned} \tag{12.3}$$

The fifth order ruled surface F on whose generators three collinear homologous points lie follows from (11.6), (11.7) and (11.9); for $\mu_1\mu_2\mu_3 \neq 0$ we obtain

$$\begin{aligned} Q_1 &= a\mu_2\mu_3, \qquad Q_2 = b\mu_1\mu_3, \qquad Q_3 = c\mu_1\mu_2, \\ Q_1' &= -bc\mu_1^2, \qquad Q_2' = -ca\mu_2^2, \qquad Q_3' = -ab\mu_3^2. \end{aligned} \tag{12.4}$$

The conclusion is that the surface F is *degenerate* and consists of a quadric regulus and three pencils of lines in V. The regulus is a hyperboloid; from (11.4) and (11.5) it follows that its equation is

$$aYZ + bZX + cXY + abcW^2 = 0. \tag{12.5}$$

We remark furthermore that the curve c′ of the third order and the set of lines C′ of the third class are trivial in this case, their equations being $XYZ = 0$ and $u_1u_2u_3 = 0$ respectively.

So far we have only applied this special case to the results of the general theory of Section 11. We shall now develop for this case an analytic treatment of the tetrahedral complex of circle axes — we have not done this for the general case in view of its complexity.

From (12.2) it follows that the direction of the line A_2A_3 is given by

$$2a : (-Y^* - b) : (Z^* - c), \tag{12.6}$$

and the midpoint of A_2A_3 by

$$-X^*, -b, c. \tag{12.7}$$

Hence the plane β_1 through this point orthogonal to A_2A_3 has the coordinates

$$\begin{aligned} &2aW^*, \quad -Y^* - bW^*, \quad Z^* - cW^*, \\ &2aX^* - b(Y^* + bW^*) - c(Z^* - cW^*), \end{aligned} \tag{12.8}$$

and those of β_2 are

$$\text{(12.9)}\qquad \begin{gathered} X^* - aW^*, \quad 2bW^*, \quad -Z^* - cW^*, \\ -a(X^* - aW^*) + 2bY^* - c(Z^* + cW^*). \end{gathered}$$

The line of intersection of planes β_1 and β_2 gives us the coordinates of the circle axis associated with the basic point A^*:

$$\text{(12.10)}\qquad \begin{aligned} Q_1 &= Y^*Z^* + cY^*W^* - bZ^*W^* + 3bc\,W^{*2}, \\ Q_2 &= Z^*X^* + aZ^*W^* - cX^*W^* + 3ca\,W^{*2}, \\ Q_3 &= X^*Y^* + bX^*W^* - aY^*W^* + 3ab\,W^{*2}, \\ Q_1' &= X^*(-2aX^* + bY^* + cZ^*) + (b^2 - c^2)X^*W^* + 3a(bY^* - cZ^*)W^* \\ &\quad + a(2a^2 - b^2 - c^2)W^{*2}, \\ Q_2' &= Y^*(-2bY^* + cZ^* + aX^*) + (c^2 - a^2)Y^*W^* + 3b(cZ^* - aX^*)W^* \\ &\quad + b(2b^2 - c^2 - a^2)W^{*2}, \\ Q_3' &= Z^*(-2cZ^* + aX^* + bY^*) + (a^2 - b^2)Z^*W^* + 3c(aX^* - bY^*)W^* \\ &\quad + c(2c^2 - a^2 - b^2)W^{*2}. \end{aligned}$$

These formulas give us a representation of the tetrahedral complex C_1 in terms of the homogeneous parameters X^*, Y^*, Z^*, W^*. The line coordinates are quadratic polynomials in these parameters as is expected.

If A^* is on the twisted cubic c we obtain its associated circle axis by substituting (12.3) into (12.10), which gives us

$$\text{(12.11)}\qquad \begin{gathered} Q_1 = Q_2 = Q_3 = 0, \quad Q_1' = 8a\mu_2\mu_3 G, \quad Q_2' = 8b\mu_3\mu_1 G, \\ Q_3' = 8c\mu_1\mu_2 G \end{gathered}$$

with

$$\text{(12.12)}\qquad G = a^2\mu_2^2\mu_3^2 + b^2\mu_3^2\mu_1^2 + c^2\mu_1^2\mu_2^2.$$

If $G \neq 0$ the basic point A^* has one associated circle axis, lying in V. The locus of these lines is given by

$$\text{(12.13)}\qquad Q_1' : Q_2' : Q_3' = a\mu_2\mu_3 : b\mu_3\mu_1 : c\mu_1\mu_2,$$

which implies that it is the set of lines satisfying

$$\text{(12.14)}\qquad aU_2U_3 + bU_3U_1 + cU_1U_2 = 0,$$

or the tangents of the conic k′:

$$\text{(12.15)}\qquad a^2X^2 + b^2Y^2 + c^2Z^2 - 2bcYZ - 2caZX - 2abXY = 0.$$

If, however, $G = 0$ the associated circle axis of A^* is undetermined. Hence the vertices T_i of the tetrahedron are those points of c which satisfy

$$a^2\mu_2^2\mu_3^2 + b^2\mu_3^2\mu_1^2 + c^2\mu_1^2\mu_2^2 = 0. \tag{12.16}$$

In view of $\mu_1 + \mu_2 + \mu_3 = 0$ this equation is equivalent to

$$\begin{aligned} g(\mu_1, \mu_2) = b^2\mu_1^4 + 2b^2\mu_1^3\mu_2 \\ + (a^2 + b^2 + c^2)\mu_1^2\mu_2^2 + 2a^2\mu_1\mu_2^3 + a^2\mu_2^4 = 0. \end{aligned} \tag{12.17}$$

For the invariants I and J of the biquadratic form g we obtain

$$6I = (a^2 + b^2 + c^2)^2, \qquad 36J = 54a^2b^2c^2 - (a^2 + b^2 + c^2)^3, \tag{12.18}$$

which gives us for the discriminant

$$R = J^2 - (1/6)I^3 = (1/12)a^2b^2c^2\{27a^2b^2c^2 - (a^2 + b^2 + c^2)^3\}, \tag{12.19}$$

which implies that R is only equal to zero if $a = b = c$. The conclusion is: if a, b and c are not all equal, the equation (12.16) has four different roots and the four vertices of the tetrahedron are distinct. The equality of a, b, and c occurs only when all three screws have equal translations.

If (12.16) is satisfied the three planes β_i coincide; their common coordinates follow from (12.8) or (12.9) after substituting (12.3) and are seen to be

$$U_1 = a\mu_2\mu_3, \qquad U_2 = b\mu_3\mu_1, \qquad U_3 = c\mu_1\mu_2, \tag{12.20}$$

$$\begin{aligned} U_4 = (\mu_3^2\mu_1/\mu_2)b^2 - (\mu_2^2\mu_1/\mu_3)c^2 = (\mu_1^2\mu_2/\mu_3)c^2 - (\mu_3^2\mu_2/\mu_1)a^2 \\ = (\mu_2^2\mu_3/\mu_1)a^2 - (\mu_1^2\mu_3/\mu_2)b^2. \end{aligned} \tag{12.21}$$

From (12.20) and (12.16) it follows that the plane is indeed isotropic.

The biquadratic equation (12.17) can always be solved, but the roots are complicated functions of a, b, c. If two of these lengths are equal a certain simplification takes place. The solution is simple if $a = b = c$; we obtain two double roots

$$\mu_1 : \mu_2 : \mu_3 = 1 : w_1 : w_1^2, \qquad \mu_1 : \mu_2 : \mu_3 = 1 : w_2 : w_2^2, \tag{12.22}$$

in which w_1 and w_2 are the imaginary cubic roots of unity. Two pairs of vertices coincide, the two points being (recalling (12.3)):

$$X = Y = Z = \pm ia\sqrt{3}. \tag{12.23}$$

They lie on the same diagonal of the rectangular parallelepiped which is a cube in this special case. The faces of the tetrahedron coincide in pairs; the conic k and Ω have two pairs of coinciding intersections.

Returning to the case of arbitrary a, b, c we seek the equation $f(Q_1, Q_2, Q_3, Q'_1, Q'_2, Q'_3) = 0$ of the tetrahedral complex C_1. This follows if we can eliminate the parameters X*, Y*, Z*, W* from the set of equations (12.10). f is a quadratic function of its six variables and we know in view of (12.11) that the agregate of terms without Q_i must be $f_1 = aQ'_2Q'_3 + bQ'_3Q'_1 + cQ'_1Q'_2$. Substituting in f_1 the terms without W* we obtain†

$$f_1 = 5abc \sum \mathrm{Y}^{*2}\mathrm{Z}^{*2} - 2\sum ab^2\mathrm{Y}^{*3}\mathrm{Z}^* - 2\sum ac^2\mathrm{Y}^*\mathrm{Z}^{*3}$$

$$+\sum a(a^2 - b^2 - c^2)\mathrm{X}^{*2}\mathrm{Y}^*\mathrm{Z}^*.$$

The second term on the right can only be compensated by

$$f_2 = -aQ'_1(bQ_2 + cQ_3) - bQ'_2(cQ_3 + aQ_1) - cQ'_3(aQ_1 + bQ_2)$$

$$= 2\sum ab^2\mathrm{Y}^{*3}\mathrm{Z}^* + 2\sum a^2c\mathrm{X}^{*3}\mathrm{Y}^* - 2abc\sum \mathrm{Y}^{*2}\mathrm{Z}^{*2}$$

$$-\sum a(b^2 + c^2)\mathrm{X}^{*2}\mathrm{Y}^*\mathrm{Z}^*.$$

Continuing in this manner we may build up f step by step. Omitting the rather lengthy necessary algebra we obtain at last the following equation of the tetrahedral complex C_1 of circle axes

$$\begin{aligned} &aQ'_2Q'_3 + bQ'_3Q'_1 + cQ'_1Q'_2 + aQ'_1(aQ_1 - bQ_2 - cQ_3) \\ &\quad + bQ'_2(-aQ_1 + bQ_2 - cQ_3) + cQ'_3(-aQ_1 - bQ_2 + cQ_3) \\ &\quad + -3abc(Q_1^2 + Q_2^2 + Q_3^2) \\ &\quad + (a^2 + b^2 + c^2)(aQ_2Q_3 + bQ_3Q_1 + cQ_1Q_2) = 0. \end{aligned} \tag{12.24}$$

13. Three screw axes with a common perpendicular

Continuing our discussions of special cases in three positions theory we consider the case when the three lines s_1, s_2, s_3 (generally supposed to be

† Here Σ implies a summation of three terms. The second two are cyclically obtained starting from the given one. So for example

$$\sum ab^2\mathrm{X}^{*3}\mathrm{Z}^* \equiv ab^2\mathrm{X}^{*3}\mathrm{Z}^* + bc^2\mathrm{Y}^{*3}\mathrm{X}^* + ca^2\mathrm{Z}^{*3}\mathrm{Y}^*.$$

mutually skew) have a single common perpendicular n. This means that n_1, n_2, n_3 all coincide. Hence the homologous points A_i, since they are the reflections of the basic point A^* into n_i, coincide and we have no displacements at all. Therefore, this special configuration of the three lines s_i can only have a meaning when we deal with *infinitesimal* displacements.

To discuss the situation we start with two infinitesimal displacements S_{23}, S_{31} (further on denoted by S_1 and S_2) and ask for a third S_{12} (denoted by S_3) such that the result of the three is no displacement at all. The condition can readily be found if we make use of the Ball vectors (Chapter II, Section 4) of a infinitesimal displacement. If $(\boldsymbol{L}_i, \boldsymbol{M}_i)$ corresponds to S_i the velocity $\boldsymbol{v}_i$ of an arbitrary point A with position vector $\boldsymbol{p}$ is given by

$$\boldsymbol{v}_i = \boldsymbol{L}_i \times \boldsymbol{p} + \boldsymbol{M}_i, \tag{13.1}$$

and hence the condition for no displacement, written in terms of velocity, reads

$$\boldsymbol{L}_1 \times \boldsymbol{p} + \boldsymbol{M}_1 + \boldsymbol{L}_2 \times \boldsymbol{p} + \boldsymbol{M}_2 + \boldsymbol{L}_3 \times \boldsymbol{p} + \boldsymbol{M}_3 = 0. \tag{13.2}$$

Since this must be valid for any $\boldsymbol{p}$, this implies

$$\boldsymbol{L}_1 + \boldsymbol{L}_2 + \boldsymbol{L}_3 = 0, \tag{13.3}$$

$$\boldsymbol{M}_1 + \boldsymbol{M}_2 + \boldsymbol{M}_3 = 0. \tag{13.4}$$

If a line n with Plücker vectors $\boldsymbol{n}$, $\boldsymbol{n}'$ is the common perpendicular of s_1 and s_2 we have

$$\boldsymbol{n} \cdot \boldsymbol{L}_1 = 0, \qquad \boldsymbol{n} \cdot \boldsymbol{L}_2 = 0, \tag{13.5}$$

in view of the orthogonality, and

$$\boldsymbol{n} \cdot (\boldsymbol{M}_1 - \sigma_1 \boldsymbol{L}_1) + \boldsymbol{n}' \cdot \boldsymbol{L}_1 = 0, \qquad \boldsymbol{n} \cdot (\boldsymbol{M}_2 - \sigma_2 \boldsymbol{L}_2) + \boldsymbol{n}' \cdot \boldsymbol{L}_2 = 0, \tag{13.6}$$

because n intersects s_1 and s_2. Substituting (13.5) into (13.6) we obtain

$$\boldsymbol{n} \cdot \boldsymbol{M}_1 + \boldsymbol{n}' \cdot \boldsymbol{L}_1 = 0, \qquad \boldsymbol{n} \cdot \boldsymbol{M}_2 + \boldsymbol{n}' \cdot \boldsymbol{L}_2 = 0. \tag{13.7}$$

From (13.3), (13.4), (13.5) and (13.7) we derive

$$\boldsymbol{n} \cdot \boldsymbol{L}_3 = 0, \qquad \boldsymbol{n} \cdot \boldsymbol{M}_3 + \boldsymbol{n}' \cdot \boldsymbol{L}_3 = 0, \tag{13.8}$$

but these imply that n intersects s_3 orthogonally. Hence *the three axes* s_i *have a common perpendicular* and therefore they build the configuration which was the starting point of this section.

Obviously the infinitesimal screw motion S_3 is determined by S_1 and S_2. To investigate the situation we take the O_X-axis along the common perpendicular n. Let s_i intersect O_X at $(a_i, 0, 0)$, and let α_i be the angle between $\boldsymbol{\omega}_i$ and O_Y (Fig. 23), measured from O_Y to $\boldsymbol{\omega}_i$ according to the right-hand rule about axis O_X. $\boldsymbol{\omega}_i$ is the angular velocity vector along s_i, its scalar magnitude is ω_i.

First of all we have from (13.3):

$$\boldsymbol{\omega}_1 + \boldsymbol{\omega}_2 + \boldsymbol{\omega}_3 = 0, \tag{13.9}$$

which gives us $\boldsymbol{\omega}_3$ in terms of $\boldsymbol{\omega}_1$ and $\boldsymbol{\omega}_2$ and therefore the direction of s_3. In particular, with $\alpha_1 - \alpha_2 = \beta$,

$$\omega_3^2 = \omega_1^2 + \omega_2^2 + 2\omega_1\omega_2 \cos\beta. \tag{13.10}$$

If $\boldsymbol{e}$ is the unit vector along O_X we have

$$\boldsymbol{M}_i = a_i\boldsymbol{e} \times \boldsymbol{\omega}_i + \sigma_i\boldsymbol{\omega}_i,$$

which implies in view of (13.4)

$$\boldsymbol{e} \times [a_1\boldsymbol{\omega}_1 + a_2\boldsymbol{\omega}_2 - a_3(\boldsymbol{\omega}_1 + \boldsymbol{\omega}_2)] + \sigma_1\boldsymbol{\omega}_1 + \sigma_2\boldsymbol{\omega}_2 - \sigma_3(\boldsymbol{\omega}_1 + \boldsymbol{\omega}_2) = 0, \tag{13.11}$$

a vector equation in terms of the two unknown scalars a_3 and σ_3. We obtain explicit expressions for these as follows: As $\boldsymbol{\omega}_i \cdot \boldsymbol{e} = 0$, we have

$$\boldsymbol{\omega}_i \times (\boldsymbol{e} \times \boldsymbol{\omega}_j) = (\boldsymbol{\omega}_i \cdot \boldsymbol{\omega}_j)\boldsymbol{e},$$

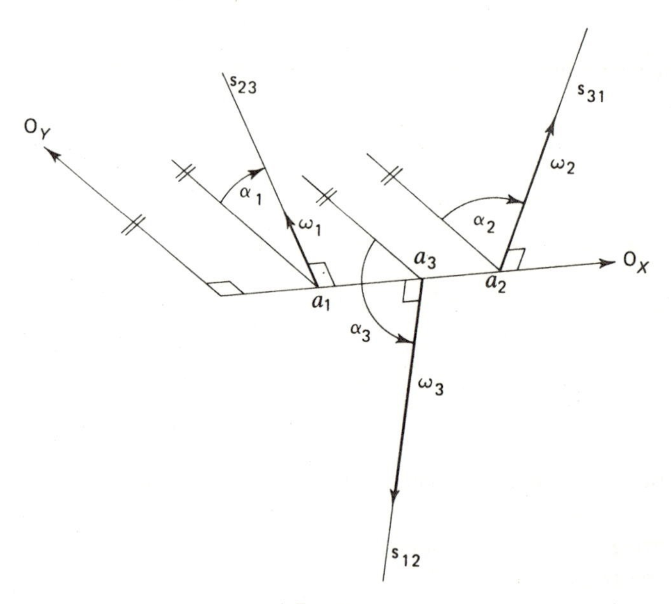

Fig. 23.

and therefore, after multiplying (13.11) vectorially by $\boldsymbol{\omega}_1$ and $\boldsymbol{\omega}_2$ respectively, we obtain

$$(a_1 - a_3)\omega_1^2 + (a_2 - a_3)\omega_1\omega_2 \cos\beta - (\sigma_2 - \sigma_3)\omega_1\omega_2 \sin\beta = 0,$$

$$(a_2 - a_3)\omega_2^2 + (a_1 - a_3)\omega_1\omega_2 \cos\beta - (\sigma_3 - \sigma_1)\omega_1\omega_2 \sin\beta = 0,$$

and by addition

$$(13.12) \qquad a_3\omega_3^2 = a_1\omega_1^2 + a_2\omega_2^2 + \omega_1\omega_2\{(a_1 + a_2)\cos\beta + (\sigma_1 - \sigma_2)\sin\beta\}.$$

As $\boldsymbol{e} \times (\boldsymbol{e} \times \boldsymbol{\omega}) = -\boldsymbol{\omega}$, if we multiply (13.11) vectorially be $\boldsymbol{e}$, it may be written as

$$(13.13) \qquad \boldsymbol{e} \times \{\sigma_1\boldsymbol{\omega}_1 + \sigma_2\boldsymbol{\omega}_2 - \sigma_3(\boldsymbol{\omega}_1 + \boldsymbol{\omega}_2)\} - \{a_1\boldsymbol{\omega}_1 + a_2\boldsymbol{\omega}_2 - a_3(\boldsymbol{\omega}_1 + \boldsymbol{\omega}_2)\} = 0.$$

Since this is of the same form as (13.11) it follows that the same argument that produced (13.12) yields

$$(13.14) \quad \sigma_3\omega_3^2 = \sigma_1\omega_1^2 + \sigma_2\omega_2^2 + \omega_1\omega_2\{(\sigma_1 + \sigma_2)\cos\beta - (a_1 - a_2)\sin\beta\}.$$

Hence the theorem: *Instantaneous displacements* S_{23}, S_{31}, S_{12} (or S_1, S_2, S_3) *are specified by the rotation vectors* $\boldsymbol{\omega}_i$, *the pitches* σ_i *and the points* $(a_i, 0, 0)$, $i = 1, 2, 3$. If S_1 and S_2 are given, S_3 is determined by (13.9), (13.12) and (13.14).

It follows in particular that if S_1 and S_2 are pure rotations ($\sigma_1 = \sigma_2 = 0$), S_3 is in general *not* a pure rotation. It is only a pure rotation in the case when $(a_1 - a_2)\sin\beta = 0$, hence s_1 and s_2 must either intersect or be parallel. In the latter case ($\beta = 0$), s_3 is parallel to s_1 and s_2 and we have furthermore the scalar relationships

$$(13.15) \qquad \omega_1 + \omega_2 + \omega_3 = 0, \qquad a_1\omega_1 + a_2\omega_2 + a_3\omega_3 = 0.$$

This special case is known as the Aronhold–Kennedy theorem. If s_1 and s_2 intersect ($a_1 = a_2 = a$) we have $a_3 = a$: s_1, s_2, s_3 pass through one point. To summarize: S_3 can only be a pure rotation if S_1 and S_2 are pure rotations which are either parallel or intersect. These cases correspond respectively to planar and spherical displacements.

In general three positions theory we have met the fundamental theorem that three screw axes determine the three screw displacements completely. In the special case at hand the question arises whether the three axes s_i of instantaneous screws (having a common perpendicular) determine the displacements. In other words: can we find the scalars ω_i and σ_i if the directions of $\boldsymbol{\omega}_i$ and the coordinates a_i $(i = 1, 2, 3)$ are given. The equations (13.3) and (13.4), which must be satisfied, read

$$(13.16) \qquad \boldsymbol{\omega}_1 + \boldsymbol{\omega}_2 + \boldsymbol{\omega}_3 = 0,$$

(13.17) $$\boldsymbol{e} \times \{a_1\boldsymbol{\omega}_1 + a_2\boldsymbol{\omega}_2 + a_3\boldsymbol{\omega}_3\} + \sigma_1\boldsymbol{\omega}_1 + \sigma_2\boldsymbol{\omega}_2 + \sigma_3\boldsymbol{\omega}_3 = 0.$$

Obviously, if $(\omega_1, \omega_2, \omega_3; \sigma_1, \sigma_2, \sigma_3)$ is a solution, $(k_1\omega_1, k_1\omega_2, k_1\omega_3;\ \sigma_1 + k_2, \sigma_2 + k_2, \sigma_3 + k_2)$ with arbitrary constants k_1 and k_2 is also a solution. Hence ω_i can at most be determined to within a multiplicative constant and σ_i to within an additive constant. We shall now prove that the ratios of ω_i and the differences $\sigma_i - \sigma_j$ follow uniquely from the given data: From the orientation of the lines s_i the angles α_i are known; (13.16) gives us

(13.18) $$\begin{aligned} \omega_1 \cos\alpha_1 + \omega_2 \cos\alpha_2 + \omega_3 \cos\alpha_3 &= 0, \\ \omega_1 \sin\alpha_1 + \omega_2 \sin\alpha_2 + \omega_3 \sin\alpha_3 &= 0. \end{aligned}$$

Hence, if $\alpha_i - \alpha_j = \beta_k$ where i, j, k take on the values 1, 2, 3 in cyclic order

(13.19) $$\omega_1 : \omega_2 : \omega_3 = \sin\beta_1 : \sin\beta_2 : \sin\beta_3,$$

i.e., the scalar values ω_i are proportional to the sines of the angles between the screw axes.

Furthermore we have $\boldsymbol{\omega}_i = (0, \omega_i \cos\alpha_i, \omega_i \sin\alpha_i)$ and $\boldsymbol{e} \times \boldsymbol{\omega}_i = (0, -\omega_i \sin\alpha_i, \omega_i \cos\alpha_i)$; hence (13.17) is equivalent to the two equations

(13.20) $$\begin{aligned} &-a_1\omega_1 \sin\alpha_1 - a_2\omega_2 \sin\alpha_2 - a_3\omega_3 \sin\alpha_3 + \sigma_1\omega_1 \cos\alpha_1 \\ &\quad + \sigma_2\omega_2 \cos\alpha_2 + \sigma_3\omega_3 \cos\alpha_3 = 0, \\ &a_1\omega_1 \cos\alpha_1 + a_2\omega_2 \cos\alpha_2 + a_3\omega_3 \cos\alpha_3 + \sigma_1\omega_1 \sin\alpha_1 \\ &\quad + \sigma_2\omega_2 \sin\alpha_2 + \sigma_3\omega_3 \sin\alpha_3 = 0. \end{aligned}$$

Eliminating σ_1, making use of (13.19) and using the fact that $\beta_1 = -(\beta_2 + \beta_3)$, we obtain

(13.21) $$\sigma_2 - \sigma_3 = (a_2 - a_1)\cot\beta_3 + (a_3 - a_1)\cot\beta_2,$$

and similarly

(13.22) $$\sigma_3 - \sigma_1 = (a_3 - a_2)\cot\beta_1 + (a_1 - a_2)\cot\beta_3,$$

(13.23) $$\sigma_1 - \sigma_2 = (a_1 - a_3)\cot\beta_2 + (a_2 - a_3)\cot\beta_1.$$

The three (linearly dependent) equations (13.21), (13.22), (13.23) determine σ_i, up to an additive constant, in terms of the given data.

In the general theory of three positions attention has been given to the case of three collinear homologous points. The corresponding problem in our special case requires determining the locus of those points for which the velocity $\boldsymbol{v}_1$ due to S_1 is proportional to $\boldsymbol{v}_2$. If (X, Y, Z) are the coordinates of a point it follows from (13.1), making use of our coordinate system, that

$$\mathbf{v}_i = \{-Y\sin\alpha_i + Z\cos\alpha_i,\ (X-a_i)\sin\alpha_i + \sigma_i\cos\alpha_i,$$
$$-(X-a_i)\cos\alpha_i + \sigma_i\sin\alpha_i\}\omega_i. \tag{13.24}$$

Hence the locus satisfies the two equations

$$((X-a_1)\sin\alpha_1 + \sigma_1\cos\alpha_1)/((X-a_2)\sin\alpha_2 + \sigma_2\cos\alpha_2)$$
$$= ((X-a_1)\cos\alpha_1 - \sigma_1\sin\alpha_1)/((X-a_2)\cos\alpha_2 - \sigma_2\sin\alpha_2)$$
$$= (-Y\sin\alpha_1 + Z\cos\alpha_1)/(-Y\sin\alpha_2 + Z\cos\alpha_2). \tag{13.25}$$

The first equality leads to the quadratic equation

$$X^2 + \{-(a_1+a_2) + (\sigma_1-\sigma_2)\cot\beta_3\}X + a_1a_2 + \sigma_1\sigma_2$$
$$+ (a_1\sigma_2 - a_2\sigma_1)\cot\beta_3 = 0. \tag{13.26}$$

It is left to the reader to show that (13.26) is invariant for cyclical permutation of S_1, S_2, S_3. If we substitute in turn the two roots of (13.26) into the second equation resulting from (13.25), we obtain both times a linear homogeneous equation for Y and Z. The conclusion is that the locus consists of two lines l_1 and l_2 intersecting O_X orthogonally. As the line l_3 at infinity of the O_{YZ}-plane ($X = W = 0$) also satisfies (13.25), the complete locus consists of three lines of which two are in general skew, while the third intersects the other two. The twisted cubic of the general case is in this case therefore degenerate. We remark that l_1 and l_2 may be either real or imaginary. Moreover they depend on the configuration of the three axes s_1, s_2, s_3 and on the values of σ_i (and not only on their differences) as well.

14. The instantaneous case

We consider now three consecutive positions as a special case of the theory of three positions, which means that we deal with (geometrical) instantaneous kinematics of the second order. According to (6.3) of Chapter II the pertinent formulas are

$$\begin{aligned} X_0 &= x, & X_1 &= -y, & X_2 &= -x + \varepsilon z, \\ Y_0 &= y, & Y_1 &= x, & Y_2 &= -y + \mu\varepsilon, \\ Z_0 &= z, & Z_1 &= \sigma_0, & Z_2 &= -\varepsilon x + \lambda, \end{aligned} \tag{14.1}$$

where O_{XYZ} and o_{xyz} are the coinciding canonical frames and σ_0, ε, λ, μ the instantaneous invariants up to the second order.

The plane U through three homologous positions reduces to the osculating plane to the path of the moving point. If $U = (A, B, C, D)$ we obtain from (14.1)

$$
\begin{aligned}
A &= -\varepsilon x^2 + \lambda x + \sigma_0 y - \mu\varepsilon\sigma, \\
B &= -\varepsilon xy - \sigma_0 x + \lambda y + \sigma_0 \varepsilon z, \\
C &= x^2 + y^2 - \varepsilon xz - \mu\varepsilon y, \\
D &= \varepsilon x(x^2 + y^2 + z^2) - (x^2 + y^2)(z + \lambda) - \varepsilon(\sigma_0 - \mu) yz + \mu\varepsilon\sigma_0 x.
\end{aligned}
\tag{14.2}
$$

The coordinates of the osculating plane are cubic functions of the coordinates of the corresponding point.

If three consecutive positions of a point P are collinear it passes through an inflection point of its path. The locus of such points P is the *inflection curve*. P satisfies the condition that its velocity and acceleration vectors coincide. Hence

$$-x + \varepsilon z = -y\tau, \qquad -y + \mu\varepsilon = x\tau, \qquad -\varepsilon x + \lambda = \sigma_0 \tau,$$

and making use of homogeneous coordinates we obtain

$$
\begin{aligned}
x &= \varepsilon(-\sigma_0\tau + \lambda), \qquad y = \varepsilon(\sigma_0\tau^2 - \lambda\tau + \mu\varepsilon^2), \\
z &= -\sigma_0\tau^3 + \lambda\tau^2 - (\mu\varepsilon^2 + \sigma_0)\tau + \lambda, \qquad w = \varepsilon^2,
\end{aligned}
\tag{14.3}
$$

which represents a twisted cubic c, τ being the parameter. It has three coinciding points at infinity (for $\tau = \infty$) and it is therefore a *parabolic* cubic. The point at infinity is $(0, 0, 1, 0)$, on the screw axis, in accordance with the general case wherein the cubic passes through the points at infinity of the three screw axes.

An inflection tangent joins a point (x, y, z, w) on c and the point at infinity $(-y, x, \sigma_0, 0)$. It follows from (14.3) that the locus of the latter is a conic k. The Plücker coordinates of the inflection tangent are polynomials in τ of the fifth order and the locus of the tangents is therefore a quintic ruled surface F, in accordance with the general case.

Example 14. Derive from (14.2) the osculating plane corresponding to a point on the screw axis, and especially to the origin.

Example 15. Determine the equation of the conic k and show that it is in general not degenerate.

Example 16. Determine the generators of F as functions of τ.

Example 17. Verify the general properties of F for this case; determine in particular the complete intersection of F and the plane at infinity.

CHAPTER V

FOUR AND MORE POSITIONS

1. Choice of screw axes

In Chapter III we saw that the relative displacement between two arbitrary positions depends upon six parameters. Furthermore, a convenient set of such parameters is given by the elements of a screw displacement: the four coordinates of the line collinear with the screw axis, the rotation angle, and the displacement along the axis. Clearly then we may choose any line in 3-space and find a double infinity of possible displacements which have this line as their screw axis. Every displacement however generally determines a unique screw axis.

In Chapter IV we have seen that, provided we exclude pure translational displacements, *three* separate positions have their relative configurations completely determined by the set of three lines coinciding with the axes of the displacement — screws s_{12}, s_{23}, s_{31}. In other words, a general set of three lines, no two of which are parallel, uniquely defines the relative displacements between the three positions 1, 2, and 3. Each ordered set of three lines defines a different ordered set of relative displacements, and reciprocally, each ordered set of three positions defines a unique ordered set of three lines. A bi-rational correspondence of this kind between ordered sets of lines and displacements of a body exists only for three positions. One way to account for this is to compare the number of parameters associated with a set of lines to the number associated with a corresponding set of displacements: Displacements $1 \to 2$ and $2 \to 3$ depend upon $2 \times 6 = 12$ parameters. Clearly these 12 parameters are sufficient to determine the entire configuration defined by three positions 1, 2, 3. The three lines (coinciding with s_{23}, s_{31}, s_{12}) depend upon $3 \times 4 = 12$ parameters. We see then that the relative displacements between three positions, and a configuration of three lines both depend upon 12 parameters. This then accounts for the fact that any three lines may be used as axes and that they uniquely and entirely determine the displacements.

For four positions, there are six screw axes (s_{12}, s_{13}, s_{14}, s_{23}, s_{24}, s_{34}) associated with the relative displacements. However, the three displacements $1 \rightarrow 2$, $2 \rightarrow 3$, $3 \rightarrow 4$ determine the four positions, so the entire configuration depends upon $3 \times 6 = 18$ parameters. This also implies that we are free to choose only four and one-half (since $18/4 = 4.5$) and not all six of the screw-axis lines. Because a configuration of six arbitrary lines involves $6 \times 4 = 24$ arbitrary parameters, it cannot generally coincide with the six screw axes associated with four positions (which are determined by only 18 arbitrary parameters). In the next section we show which sets of six lines can in fact be sets of four position axes.

2. The configuration of the screw axes

In the previous section we have seen that it is possible to arbitrarily specify at most 4.5 axes; this means we can choose four axes arbitrarily and in addition we have two free parameters at our disposal. Once the four axes and two additional parameters are specified, we expect the four positions, and hence all six axes, to be determined. There are in fact two different cases depending upon which four axes are chosen arbitrarily:

Case 1. Three axes form a screw triangle. If three of the axes refer to the same three positions they form a screw triangle of the type discussed in Chapter IV. s_{ij} and s_{ji} are coinciding lines and the same triangle follows from either ordering of i and j. Hence there are actually only four possible screw triangles: $s_{12}s_{23}s_{13}$, $s_{12}s_{24}s_{14}$, $s_{13}s_{34}s_{14}$, and $s_{23}s_{34}s_{24}$. If any three of our four lines belong to such a set, they uniquely and completely determine the three positions to which they correspond. The two remaining parameters may be taken as the remaining freedoms in the choice of a fifth axis which is already restricted (as we shall soon see) to intersect a fixed line at a right angle. However, it will turn out that these two free parameters are most simply taken as the angular and linear displacements (ϕ, d) associated with the fourth screw axis (i.e., the one that does not belong to the screw triangle). The sixth axis follows uniquely once any five axes are known. We now illustrate this with arguments which rely on the screw triangle geometry developed in Chapter IV.

We take four arbitrary lines as, for example, the axes s_{12}, s_{23}, s_{13}, s_{34}. Since s_{12}, s_{23}, s_{13} form a screw triangle, the screw displacements (ϕ_{12}, d_{12}), (ϕ_{13}, d_{13}), and (ϕ_{23}, d_{23}) are also known. If we now use the remaining two parameters to specify (ϕ_{34}, d_{34}), enough is known about triangle $s_{14}s_{13}s_{34}$ to determine axis s_{14} and displacements (ϕ_{14}, d_{14}). These follow uniquely from the geometry of the screw triangle. Similarly, from triangle $s_{23}s_{34}s_{24}$, the axis s_{24} and associated displacements are now known. Alternatively, if l_3 is the normal between s_{23} and s_{34}, screwing line l_3 about axis s_{23} by an amount $(-\phi_{23}/2, -d_{23}/2)$ yields the line l_2 which intersects s_{23} at a right angle and (because $s_{23}s_{34}s_{24}$ form a screw triangle) also s_{24} at a right angle. Since l_2 is uniquely determined by the original choice of four lines, it follows that s_{24} may be any line which intersects a known line (l_2) at a right angle. The locus of all possible lines s_{24} is then clearly a linear line congruence with axis at l_2. (Such a congruence is also called a (1,1) congruence, since only one line of the congruence lies in an arbitrarily situated plane and only one line passes through an arbitrary point.) If we select one line of the congruence to be s_{24} we use up our two free parameters, and the remaining parts of the configuration of six axes and their associated screw displacements follow uniquely. Choosing a line of the s_{24} congruence is equivalent to choosing (ϕ_{34}, d_{34}). Conversely, we now see that if we choose (ϕ_{34}, d_{34}) as our two free parameters, the effect is the same as selecting a line of the linear congruence for s_{24}. Clearly a similar argument could be used to first determine that s_{14} must be a line of a linear congruence and then show that s_{24} follows uniquely. The conclusion is: *if four screw axes are chosen such that three belong to the same screw triangle, the remaining two axes are members of linear congruences. There are two freedoms remaining in the system, and if these are used to specify one member of either congruence, the fifth and sixth axes follow uniquely.*

Case 2. No three of the four specified axes belong to the same screw triangle. This is the more symmetric and by far the more interesting of the two cases. If the axes are paired so that common normals are drawn between axes with common subscripts we have a closed configuration composed of eight lines as shown in Fig. 24. Only four of these lines are independent, the other four being common normals; such a figure can be considered to be a *spatial quadrilateral.* The common normals are drawn only between axes with a common subscript. Two such axes which have a subscript in common are said to be *adjacent.* Axes that do not have a common subscript are said to be *complimentary.* There are three pairs of complimentary axes $s_{12}s_{34}$, $s_{13}s_{24}$, $s_{14}s_{23}$, and every spatial quadrilateral with non-adjacent axes formed from two such pairs is called a *complimentary screw quadrilateral.* Fig. 24 shows the general scheme of such a quadrilateral. The only restriction on the subscripts is that $i \neq j \neq k \neq l$. This means there are three different possible configurations depending upon if i, j, k, l are 1, 2, 3, 4; 1, 2, 4, 3; or 1, 3, 2, 4. This follows from the fact that there are three sets of complimentary axes taken two at a time, any other permutation of 1, 2, 3, 4 yields one of these configurations.

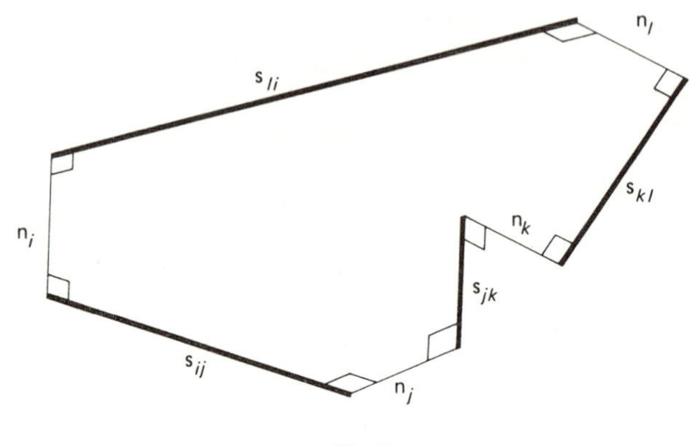

Fig. 24.

If we select four axes so that no three form a screw triangle, then the four must form a complimentary screw quadrilateral. The question remains what are the restrictions on the choice of the remaining two axes. Using the notation of Fig. 24, we seek to determine s_{ik} and s_{jl}. The answer again lies in the property of the screw triangle. Taking first s_{ik} we note that s_{ik} forms screw triangles with axes s_{ij}, s_{jk} as well as s_{li}, s_{kl}. If we draw the normals in these two triangles we have lines i_1, k_1 and i_2, k_2 as shown in Fig. 25. From the geometry of screw triangle s_{ij}, s_{ik}, s_{jk} it follows that, regardless of where s_{ik} actually is, the angle and distance between i_1, k_1 must be respectively $(\phi_{ik}/2, d_{ik}/2)$, measured from i_1 to k_1. Similarly triangle s_{il}, s_{ik}, s_{kl} requires that the angle and distance between i_2, k_2 also be $(\phi_{ik}/2, d_{ik}/2)$, measured from i_2 to k_2; which means that the angle and distance between i_1, k_1 must equal respectively the angle and distance between i_2, k_2. In addition, since lines i_1, k_1, i_2, k_2 are all normal to and intersect the same line (s_{ik}) it follows that the angle and distance between i_1, i_2, measured from i_1 toward i_2, will equal respectively the angle and distance between k_1, k_2, measured from k_1 toward k_2. We now have two restrictions on the location of s_{ik}, these restrictions are in terms of the equality of two angles and two distances and in no way imply knowledge of the actual values of these angles or distances.

We conclude: *all possible axes* s_{ik} *must the situated so that the angle and distance between normals to them from one pair of adjacent axes are equal to respectively the angle and distance between normals from the other pair of adjacent axes.* If any one pairing of adjacent axes satisfies this condition the other pairings will also satisfy this condition.

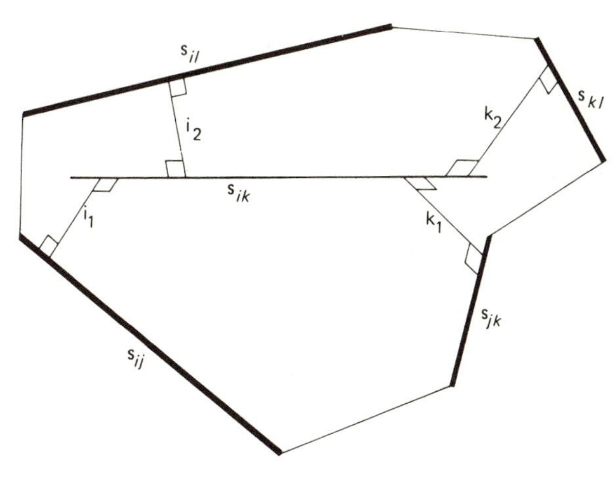

Fig. 25.

This then implies that the locus of s_{ik} will be a line congruence, since it is a line and its locus depends upon two scalar conditions. In the next three sections we derive the equations and properties of this congruence. In Sections 3 and 4 we deal only with the rotational aspects; in Section 5 we introduce the condition on the distances.

3. The screw cone

Analytic expressions for the equality of the angle between i_1, k_1 and the angle between i_2, k_2 are most easily obtained by first considering only the angles, and not the distances, between the lines. This is facilitated by using the spherical indicatrix of the quadrilateral. In other words, since we are only interested in angles, we can represent any line by a vector parallel to it and passing through the origin of coordinates. We use, therefore, a set of intersecting vectors $\mathbf{s}_{ij}$, $\mathbf{s}_{ik}$, $\mathbf{s}_{kl}$, $\mathbf{s}_{li}$ parallel to the corresponding axes s_{ij}, s_{ik}, s_{kl}, s_{li}. From triangle s_{ij}, s_{jk}, s_{ik} we have then

$$((\mathbf{s}_{ij} \times \mathbf{s}_{ik}) \times (\mathbf{s}_{jk} \times \mathbf{s}_{ik}))/((\mathbf{s}_{ij} \times \mathbf{s}_{ik}) \cdot (\mathbf{s}_{jk} \times \mathbf{s}_{ik})) = \hat{\mathbf{s}}_{ik} \tan(\phi_{ik}/2), \tag{3.1}$$

where $\hat{\mathbf{s}}_{ik}$ is the unit vector parallel to axis s_{ik}. Equation (3.1) may be expanded and simplified to:

$$\begin{aligned} &((\mathbf{s}_{ij} \times \mathbf{s}_{jk}) \cdot \mathbf{s}_{ik})/[(\mathbf{s}_{ij} \cdot \mathbf{s}_{jk})(\mathbf{s}_{ik} \cdot \mathbf{s}_{ik}) - (\mathbf{s}_{ij} \cdot \mathbf{s}_{ik})(\mathbf{s}_{jk} \cdot \mathbf{s}_{ik})] = \\ &= (\tan(\phi_{ik}/2))/|\mathbf{s}_{ik}|. \end{aligned} \tag{3.2}$$

Repeating for triangle s_{li}, s_{kl}, s_{ik} we get a similar expression in which s_{li} replaces s_{ij}, and s_{kl} replaces s_{jk}. Equating both of these expressions and clearing the denominator yields

$$\begin{aligned}(3.3)\qquad &[(s_{ij}\times s_{jk})\cdot s_{ik}][(s_{li}\cdot s_{kl})(s_{ik}\cdot s_{ik})-(s_{li}\cdot s_{ik})(s_{kl}\cdot s_{ik})] \\ &-[(s_{li}\times s_{kl})\cdot s_{ik}][(s_{ij}\cdot s_{jk})(s_{ik}\cdot s_{ik})-(s_{ij}\cdot s_{ik})(s_{jk}\cdot s_{ik})]=0.\end{aligned}$$

In (3.3), s_{ik} is the only unknown. Since it appears as a homogeneous cubic function, the directions of all possible axes s_{ik} must be parallel to the rulings of a cubic cone. This equation defines the *only* possible directions for an axis for which $\angle i_1, k_1 = \angle i_2, k_2$. From (3.3) it follows that s_{ij}, s_{jk}, s_{li}, s_{kl} are also rulings of this same cubic cone. By the symmetry of our argument it follows that the sixth axis s_{jl} must also be parallel to one of the rulings of this same cone. Hence all six axes are parallel to rulings of the same cubic cone. It may also be shown that all three complimentary screw quadrilaterals associated with a given set of six axes define the same cubic cone. It seems logical then to name such a cone: *the screw cone.*

4. The correspondence between axis direction and rotation angle

We now explore some of the properties of the screw cone by asking the following question: Assuming a complimentary screw quadrilateral is known and axis s_{ik} unknown, if a given rotation angle ϕ_{ik} is specified, how many of the cone directions (i.e., screw cone rulings) are possible directions for s_{ik}?

From (3.2) it follows that any arbitrary value for ϕ_{ik} requires s_{ik} to be parallel to the directions of the fourth-order cone:

$$(4.1)\qquad [(s_{ij}\times s_{jk})\cdot s_{ik}]^2 s_{ik}^2-\tan^2(\phi_{ik}/2)[(s_{ij}\cdot s_{jk})(s_{ik}^2)-(s_{ij}\cdot s_{ik})(s_{jk}\cdot s_{ik})]^2=0.$$

This cone intersects the screw cone in $4\times 3=12$ lines. However the directions s_{ij} and s_{jk} are double lines of (4.1) so they count as four of the twelve intersections. Since these four are spurious, we have in general at most eight directions for any value of $\tan^2(\phi_{ik}/2)$; four for $+\phi_{ik}$ and four for $-\phi_{ik}$. Hence, *for a given value of the rotational angle ϕ_{ik} there are at most four directions for* s_{ik}. By permutation of subscripts in (4.1) we can see that there are at most 24 (i.e., 6×4) directions on the screw cone — four for each screw — for which the corresponding rotation angle has a given value.

There are two special cases: $\phi_{ik}=\pi$ and $\phi_{ik}=2\pi$. When $\phi_{ik}=\pi$ the fourth order cone degenerates into the quadric cone

$$(4.2)\qquad (s_{ij}\cdot s_{jk})(s_{ik}^2)-(s_{ij}\cdot s_{ik})(s_{jk}\cdot s_{ik})=0$$

counted twice. We still have four directions, the difference in this case is that the $\phi_{ik} = +\pi$ and $\phi_{ik} = -\pi$ directions coincide, and so the eight directions for $\pm\phi_{ik}$ become the same four lines counted twice.

When $\phi_{ik} = 2\pi$, the fourth order cone (4.1) degenerates into the imaginary quadric $s_{ik}^2 = 0$, and the plane

$$(4.3) \qquad (s_{ij} \times s_{jk}) \cdot s_{ik} = 0$$

counted twice. This plane (4.3) cuts the screw cone in one real line other than s_{ij} and s_{jk}. The conclusion is then, *when $\phi_{ik} = 2\pi$ there is only one possible direction for* s_{ik} and this direction is the same for $+2\pi$ and -2π. We will call such directions π-directions, since $\phi_{ik}/2 = \pm\pi$ for such axes. The π-direction can be determined most simply if we realize that using triangle s_{li}, s_{kl}, s_{ik} in (3.2) would result in (4.3) being

$$(4.4) \qquad (s_{li} \times s_{kl}) \cdot s_{ik} = 0.$$

The π-direction is then obtained as the intersection of the planes (4.3) and (4.4). From these equations it follows clearly that the π-direction is normal to the normals to s_{ij}, s_{jk} and s_{li}, s_{kl} respectively.

If $\phi_{ik} \to 0$ the corresponding direction approaches a π-direction, since $\tan(\phi_{ik}/2) \to 0$. *The π-direction is the only possible direction for* s_{ik} *when ϕ_{ik} is either $\pm 2\pi$ or ± 0.* There are six lines on the screw cone which represent π-directions, one for each screw.

5. The distance condition

In Section 2 we saw that axis s_{ik} must satisfy a condition which was phrased in terms of the respective equivalence of two angles and two distances. We have so far been dealing with only the angle condition. We turn now to the distance part of the restriction. We proceed by determining the coordinates of the normals to s_{ik} from the relevant screw axes, and then deriving expressions for the distances between the normals. This problem was first solved by ROTH [1967a], here we give a new derivation. In Chapter XIII, Section 3 we illustrate how to derive this same result using a process called *dualization*.

Using screw axes s_{ij}, s_{jk}, and s_{ik} we will develop an expression for $(d_{ik}/2)\sec^2(\phi_{ik}/2)$. Recalling, from Section 2 and Fig. 25, that $d_{ik}/2$ is the distance between i_1 and k_1, and that $\phi_{ik}/2$ is the angle between i_1 and k_1 (both measured from i_1 towards k_1), we first determine i_1 and k_1. Since i_1 is the common perpendicular between s_{ij} and s_{ik} we define it in terms of the line vectors along these screws. If the line vectors along s_{ij} and s_{ik} are given by the Plücker vectors $(\boldsymbol{s}_{ij}, \boldsymbol{s}'_{ij})$ and $(\boldsymbol{s}_{ik}, \boldsymbol{s}'_{ik})$ respectively we have for the Plücker vectors $(\boldsymbol{i}_1, \boldsymbol{i}'_1)$ along i_1

$$\boldsymbol{i}_1 = (\boldsymbol{s}_{ik} \times \boldsymbol{s}_{ij})/|\boldsymbol{s}_{ik} \times \boldsymbol{s}_{ij}| \tag{5.1}$$

and

$$\boldsymbol{i}'_1 = [\boldsymbol{s}_{ik} \times \boldsymbol{s}'_{ij} + \boldsymbol{s}'_{ik} \times \boldsymbol{s}_{ij} + (\boldsymbol{i}_1 D_i \boldsymbol{s}_{ik} \cdot \boldsymbol{s}_{ij})]/|\boldsymbol{s}_{ik} \times \boldsymbol{s}_{ij}| \tag{5.2}$$

where $D_i = (\boldsymbol{s}_{ik} \cdot \boldsymbol{s}'_{ij} + \boldsymbol{s}'_{ik} \cdot \boldsymbol{s}_{ij})/|\boldsymbol{s}_{ik} \times \boldsymbol{s}_{ij}|$; the D_i term will drop out and does not affect our final result.

The Plücker vectors $(\boldsymbol{k}_1, \boldsymbol{k}'_1)$ along line k_1 are

$$\boldsymbol{k}_1 = (\boldsymbol{s}_{ik} \times \boldsymbol{s}_{jk})/|\boldsymbol{s}_{ik} \times \boldsymbol{s}_{jk}| \tag{5.3}$$

and

$$\boldsymbol{k}'_1 = [\boldsymbol{s}_{ik} \times \boldsymbol{s}'_{jk} + \boldsymbol{s}'_{ik} \times \boldsymbol{s}_{jk} + (\boldsymbol{k}_1 D_k \boldsymbol{s}_{ik} \cdot \boldsymbol{s}_{jk})]/|\boldsymbol{s}_{ik} \times \boldsymbol{s}_{jk}| \tag{5.4}$$

where $D_k = (\boldsymbol{s}_{ik} \cdot \boldsymbol{s}'_{jk} + \boldsymbol{s}'_{ik} \cdot \boldsymbol{s}_{jk})/|\boldsymbol{s}_{ik} \times \boldsymbol{s}_{jk}|$; the D_k term will not appear in our final result. The moment between lines i_1 and k_1 is $(d_{ik}/2)\sin(\phi_{ik}/2)$, so it follows that

$$\boldsymbol{i}_1 \cdot \boldsymbol{k}'_1 + \boldsymbol{i}'_1 \cdot \boldsymbol{k}_1 = -(d_{ik}/2)\sin(\phi_{ik}/2). \tag{5.5}$$

Similarly

$$\boldsymbol{i}_1 \times \boldsymbol{k}'_1 + \boldsymbol{i}'_1 \times \boldsymbol{k}_1 = [(d_{ik}/2)\cos(\phi_{ik}/2)]\hat{\boldsymbol{s}}_{ik} + \sin(\phi_{ik}/2)\boldsymbol{s}'_{ik}/|\boldsymbol{s}_{ik}|. \tag{5.6}$$

Since the angle between i_1 and k_1 is $\phi_{ik}/2$ we also have

$$\boldsymbol{i}_1 \times \boldsymbol{k}_1 = \sin(\phi_{ik}/2)\hat{\boldsymbol{s}}_{ik} \tag{5.7}$$

$$\boldsymbol{i}_1 \cdot \boldsymbol{k}_1 = \cos(\phi_{ik}/2). \tag{5.8}$$

Forming the product of (5.6) and (5.8) and substracting from it the product of (5.5) and (5.7), and then dividing the result by the square of (5.8) yields

$$\begin{aligned} &\{(\boldsymbol{i}_1 \times \boldsymbol{k}'_1 + \boldsymbol{i}'_1 \times \boldsymbol{k}_1)(\boldsymbol{i}_1 \cdot \boldsymbol{k}_1) - (\boldsymbol{i}_1 \cdot \boldsymbol{k}'_1 + \boldsymbol{i}'_1 \cdot \boldsymbol{k}_1)(\boldsymbol{i}_1 \times \boldsymbol{k}_1)\}/(\boldsymbol{i}_1 \cdot \boldsymbol{k}_1)^2 = \\ &\quad = [(d_{ik}/2)\sec^2(\phi_{ik}/2)]\hat{\boldsymbol{s}}_{ik} + \tan(\phi_{ik}/2)\boldsymbol{s}'_{ik}/|\boldsymbol{s}_{ik}|. \end{aligned} \tag{5.9}$$

If we substitute into the left-hand side eqs. (5.1), (5.2), (5.3), and (5.4), and then operate on (5.9) with the scalar product of $\boldsymbol{s}_{ik}$ we obtain,

$$(b_1 c_1 - a_1 d_1)/c_1^2 = (d_{ik}/2)\sec^2(\phi_{ik}/2) \tag{5.10}$$

where

$$
\begin{aligned}
(5.11)\qquad & a_1 = (\mathbf{s}_{ij} \times \mathbf{s}_{jk}) \cdot \mathbf{s}_{ik} \\
& b_1 = (\mathbf{s}'_{ij} \times \mathbf{s}_{jk}) \cdot \mathbf{s}_{ik} + (\mathbf{s}_{ij} \times \mathbf{s}'_{jk}) \cdot \mathbf{s}_{ik} + (\mathbf{s}_{ij} \times \mathbf{s}_{jk}) \cdot \mathbf{s}'_{ik} \\
& c_1 = (\mathbf{s}_{ij} \cdot \mathbf{s}_{jk})(\mathbf{s}_{ik} \cdot \mathbf{s}_{ik}) - (\mathbf{s}_{ij} \cdot \mathbf{s}_{ik})(\mathbf{s}_{jk} \cdot \mathbf{s}_{ik}) \\
& d_1 = (\mathbf{s}'_{ij} \cdot \mathbf{s}_{jk})(\mathbf{s}_{ik} \cdot \mathbf{s}_{ik}) + (\mathbf{s}_{ij} \cdot \mathbf{s}'_{jk})(\mathbf{s}_{ik} \cdot \mathbf{s}_{ik}) \\
& \qquad - (\mathbf{s}'_{ij} \cdot \mathbf{s}_{ik})(\mathbf{s}_{jk} \cdot \mathbf{s}_{ik}) - (\mathbf{s}_{ij} \cdot \mathbf{s}'_{ik})(\mathbf{s}_{jk} \cdot \mathbf{s}_{ik}) \\
& \qquad - (\mathbf{s}_{ij} \cdot \mathbf{s}_{ik})(\mathbf{s}'_{jk} \cdot \mathbf{s}_{ik}) - (\mathbf{s}_{ij} \cdot \mathbf{s}_{ik})(\mathbf{s}_{jk} \cdot \mathbf{s}'_{ik}).
\end{aligned}
$$

Repeating this entire development using screw triangle s_{li}, s_{kl}, s_{ik} (we now work with lines i_2 and k_2 instead of i_1 and k_1, see Fig. 25), we find that

$$(5.12)\qquad (b_2c_2 - a_2d_2)/c_2^2 = (d_{ik}/2)\sec^2(\phi_{ik}/2)$$

where a_2, b_2, c_2, d_2 correspond respectively to a_1, b_1, c_1, d_1 and are obtained by substituting subscripts *li* for *ij*, and *kl* for *jk* in (5.11). Equating (5.10) and (5.12) yields

$$(5.13)\qquad c_2^2(b_1c_1 - a_1d_1) - c_1^2(b_2c_2 - a_2d_2) = 0.$$

In terms of this notation, the screw cone equation (3.3) is

$$(5.14)\qquad a_1c_2 - a_2c_1 = 0.$$

If a line has a direction which satisfies (5.14) then, provided $c_1 \neq 0$, $c_2 \neq 0$, it will satisfy (5.13) iff

$$(5.15)\qquad b_1c_2 - b_2c_1 + d_2a_1 - d_1a_2 = 0.$$

We now have the rather simple result: *only lines which belong to the congruence defined by equations* (5.14) *and* (5.15) *can be screw axis* s_{ik}. Eq. (5.15) in effect is the translational displacement (i.e., d_{ik}) condition, but it is only valid for lines which are parallel to rulings of the screw cone.

6. The screw congruence

It remains for us to study eq. (5.15) in order to better understand the properties of the congruence of screw axes.

The Plücker vector-coordinates of the line which is the axis s_{ik} are $\mathbf{s}_{ik}$, $\mathbf{s}'_{ik}$. Eq. (5.15) is homogeneous in these coordinates: each term either contains cubic terms in $\mathbf{s}_{ik}$, or quadratic terms in $\mathbf{s}_{ik}$ multiplied by linear terms in $\mathbf{s}'_{ik}$. Hence (5.15) represents a *cubic line complex*. The properties of interest to us become most apparent if we substitute for $\mathbf{s}'_{ik}$

(6.1) $$\mathbf{s}'_{ik} = \mathbf{p}_{ik} \times \mathbf{s}_{ik},$$

where $\mathbf{p}_{ik}$ is a position vector from the origin of coordinates to any point on axis s_{ik}. Now, if a direction $\mathbf{s}_{ik}$ is chosen, eq. (5.15) (with (6.1)) may be written as

(6.2) $$(\mathbf{p}_{ik} \times \mathbf{s}_{ik}) \cdot \mathbf{K} + k_1 = 0.$$

Here $\mathbf{K}$ and k_1 are respectively a vector and scalar, both of these are known since they are determined by $\mathbf{s}_{ik}$, $\mathbf{s}_{ij}$, $\mathbf{s}_{jk}$, $\mathbf{s}_{il}$, $\mathbf{s}_{lk}$. Hence, each direction $\mathbf{s}_{ik}$ of the screw cone generally determines a different plane; these planes contain all possible axes s_{ik}. The result is then: given the direction of s_{ik}, eq. (6.2) represents an infinity of coplanar lines, all parallel to $\mathbf{s}_{ik}$, any one of which may be the screw axis s_{ik}. This situation is illustrated in Fig. 26 where one of the elements of the screw cone has been selected (shown in heavy), and this in turn determines the plane of parallel lines from which the actual axis is chosen (shown in heavy).

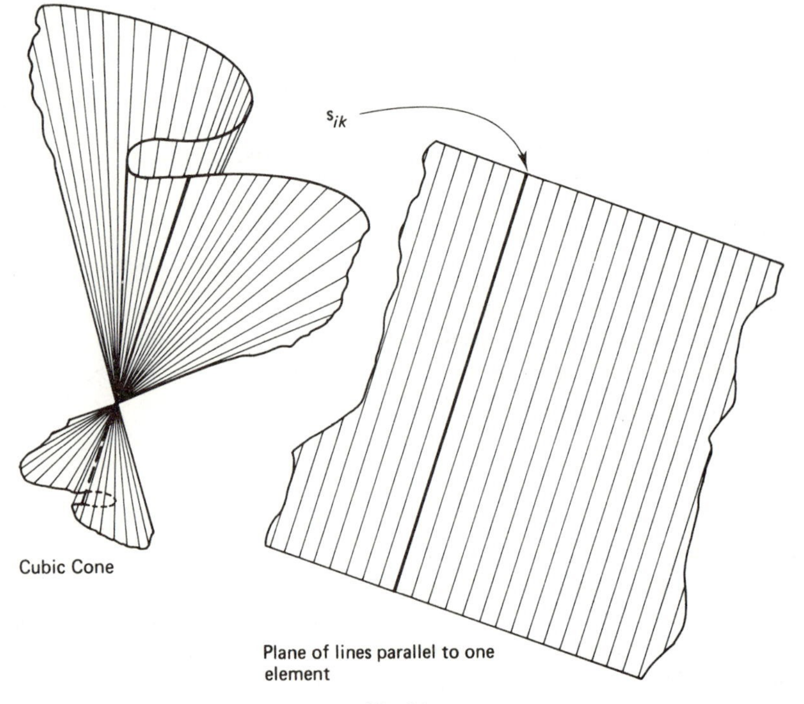

Fig. 26.

Equation (6.2) allows us to determine the order of the screw congruence: By fixing $\boldsymbol{p}_{ik}$ we determine the restriction on all axes which pass through a point (for which $\boldsymbol{p}_{ik}$ is the position vector). For any general value of $\boldsymbol{p}_{ik}$, (6.2) represents a cubic cone of $\boldsymbol{s}_{ik}$ with vertex at the origin. Hence all possible screw axes through an arbitrary point are given by the directions common to (6.2) (with $\boldsymbol{p}_{ik}$ given) and the screw cone, provided we exclude those directions for which $c_1 = 0$ or $c_2 = 0$. Two such cones intersect, in general, in nine lines (counting both real and imaginary intersections). We conclude then that through a general point, given by $\boldsymbol{p}_{ik}$, there are nine possible directions for s_{ik}. Which is to say that *nine lines of the congruence pass through a general point*, this means its order is nine.

In order to determine the class of the congruence, we consider an arbitrary plane P. Any plane P_0, parallel to P and passing through the apex of the screw cone, intersects the cone in three lines. These three lines determine three possible directions for s_{ik}. Substituting these in turn into (6.2) we obtain three planes each of which cuts P in a different line. Hence *three lines of the congruence lie in an arbitrary plane*. The class is therefore three, and *the screw congruence is a* (9,3) *congruence.*

This congruence is the intersection of two cubic complexes, both having the plane V at infinity as a singularity. One complex is that of all lines parallel to the generators of the screw cone; it consists of all lines intersecting a cubic curve k in V. Obviously every line in V not only belongs to the complex but it must be counted thrice because it intersects k three times. Hence this complex has V (i.e., the lines of V) as a three-fold manifold.

The other cubic complex is given by (5.15) (and (5.11)). In its equation the components of $\boldsymbol{s}'_{ik}$ appear (only) linearly. This means that the lines of V are double lines of this complex. Hence the intersection of the two complexes contains every line of V $3 \times 2 = 6$ times. Their complete intersection, in general a (9,9) congruence, is therefore degenerated into a (0,6) and the required $(9,9) - (0,6) = (9,3)$ congruence.

Example 1. Show that the cubic complex given by (5.15) is built up of ∞^2 pencils of parallel lines, one in each direction. (Hint: The lines, of the complex, through a point are the rulers of a cubic cone. If the point is in V this cone degenerates into a pencil in V, counted twice, and a third pencil in space.)

Example 2. Confirm that the class of the screw congruence is three by use of the following argument: consider an arbitrary plane W and its line of intersection n with V. The lines of the complex (5.15) in W are the tangents of a (rational) curve c_3 of class 3, and obviously n is a double tangent of this curve. The line n has three intersections S_i $(i = 1, 2, 3)$ with k. We must consider the tangents to c_3 through S_i. Of the three tangents through, say S_1, two coincide and only one is not in V. Hence in W there appear three lines of the congruence, which confirms that its class is three.

Choosing the position of the axis s_{ik} (i.e., $\boldsymbol{p}_{ik}$ in (6.2)) determines the displacement parameter d_{ik} as well. Hence $\boldsymbol{p}_{ik}$ and d_{ik} are related; we examine this correspondence: If we choose d_{ik} then, for every direction $\boldsymbol{s}_{ik}$, equation (5.10) yields a linear equation in $\boldsymbol{s}'_{ik}$ and therefore, by virtue of (6.1), in $\boldsymbol{p}_{ik}$, i.e., the point p_{ik} lies in a plane. The intersection of this plane and the plane given by (6.2) is a line. To each displacement d_{ik} there therefore corresponds one location for each direction $\boldsymbol{s}_{ik}$. Thus for any d_{ik} there is an infinite number of possible axes s_{ik}; one corresponding to each direction of the screw cone.

Once the fifth axis is chosen, cases 1 and 2 (see Section 2) become identical and clearly the sixth axis follows uniquely. So if we choose four arbitrary axes, which define a complimentary screw quadrilateral, and use the two remaining freedoms to choose one line from the (9,3) congruence (as the fifth axis), the set of displacements associated with four positions is uniquely defined. For a given set of four positions all three complimentary screw quadrilaterals define the same congruence; all six screws belong to this same congruence.

We have already seen that there exist six so-called π-directions. Corresponding to each such direction is a line which is a common normal between two opposite sides of a complimentary screw quadrilateral (a *side* is the normal between two adjacent screws). These six *π-lines* belong to the congruence, and have the additional property that the displacement along them can only be null.

From the principle of kinematic inversion it follows that the six screw axes (s_{12}, s_{13}, s_{14}, s^1_{23}, s^1_{24}, s^1_{34}) in the moving system E_1 form analogous configurations to the foregoing. In fact there are image-screw triangles and image-screw congruences associated with the moving system. All the relationships between the axes are analogous to those in the fixed system Σ. These configurations of axes move with the system and exist in each of the four positions E_1, E_2, E_3, E_4. Viewed from Σ these configurations are all different, however, the relative positions of the axes are the same in all positions E_i, and we can speak of the moving configuration. The equations for the congruence, of image screw axes, in E_i follow from (5.14) and (5.15) if we replace $\boldsymbol{s}_{ij}$, $\boldsymbol{s}_{jk}$, $\boldsymbol{s}_{ik}$, $\boldsymbol{s}_{li}$, $\boldsymbol{s}_{kl}$, $\boldsymbol{s}'_{ij}$, $\boldsymbol{s}'_{jk}$, $\boldsymbol{s}'_{ik}$, $\boldsymbol{s}'_{li}$, $\boldsymbol{s}'_{kl}$ by $\boldsymbol{s}^{i}_{ij}$, $\boldsymbol{s}^{i}_{jk}$, $\boldsymbol{s}^{i}_{ik}$, $\boldsymbol{s}^{i}_{li}$, $\boldsymbol{s}^{i}_{kl}$, $\boldsymbol{s}'^{i}_{ij}$, $\boldsymbol{s}'^{i}_{jk}$, $\boldsymbol{s}'^{i}_{ik}$, $\boldsymbol{s}'^{i}_{li}$, $\boldsymbol{s}'^{i}_{kl}$. The superscript i indicates the values we would get under the inverse displacement when Σ is in the position corresponding to E_i.

Although the idea of obtaining the image screws by use of inverse displacements is conceptually simple, for actual computations and graphical constructions the following procedure is often more useful: To obtain the image screws in E_i, we take the coordinate system in E_i congruent with the system in Σ. This means that all i-subscripted vectors coincide with their image vectors (i.e., $s^i_{ij} = s_{ij}$, $s^i_{ik} = s_{ik}$, $s^i_{li} = s_{li}$, $s'^i_{ij} = s'_{ij}$, $s'^i_{ik} = s'_{ik}$, $s'^i_{li} = s'_{li}$). Each of the remaining image vectors can be obtained by a reflection. For example, by reflecting s_{jk} into the normal between s_{ij} and s_{ik} (subscript i because we want the image in E_i, subscripts j and k because we are dealing with s_{jk}) we obtain s^i_{jk} whose coordinates are s^i_{jk}, s'^i_{jk}.

Example 3. Show that for a given set of four positions, the screw-axis congruence intersects the congruence of image screw-axes in (at most) nine (real) lines. Show that in general only three of these are screws for the given set of positions.

Example 4. Under what conditions will more than three of the above mentioned intersections be screw axes?

Example 5. Show that in general no more than three screws can be perpendicular to a given direction.

Example 6. Verify that if four screws are perpendicular to one direction the screw congruence degenerates (into a linear and quadratic congruence, or three linear ones). Does this imply any degeneracy in the image congruence?

Example 7. Show that when the screw axes are concurrent and the translation along each screw is null, the screw congruence degenerates into a ruling which is the screw cone. (This is the case called spherical kinematics, see Chapter VII). Similarly, for the image congruence show we have the same degeneracy.

Example 8. By limit reasoning, deduce that when the point of concurrency (referred to in the previous example) is at infinity the screw cone becomes a cubic cylinder. (The planes normal to the generators of this cylinder intersect it in the circular-cubics developed in the chapter on planar kinematics, Chapter VIII). Similarly, for the image congruence deduce the analogous result.

7. Homologous points

A point A of E has four positions A_1, A_2, A_3, A_4 in the fixed space Σ. The coordinates X_i, Y_i, Z_i of A_i are linear functions of the coordinates x, y, z of A. The condition for the coplanarity of A_i $(i = 1, 2, 3, 4)$ reads $|X_i\ Y_i\ Z_i\ 1| = 0$. Developing this 4×4 determinant we obtain: *the locus of a point A in E such that its four homologous positions are in one plane of Σ is a cubic surface* H. Strictly speaking the locus consists of H and the plane at infinity of E all points of which have of course coplanar positions. This is confirmed if we make use of homogeneous coordinates.

In Chapter IV we have shown that the locus of A such that three positions are collinear is a twisted cubic c. If three of the four points A_i $(i = 1, 2, 3, 4)$ are collinear, the four points are coplanar. Hence if c_1 is the twisted cubic determined by the positions E_2, E_3, E_4; c_2 is the cubic determined by positions 3, 4, 1; and so on, we obtain: *the four twisted cubics* c_i *are curves on* H.

The cubic c_1 intersects the plane at infinity of E in the points P_{23}, P_{34}, P_{42} on s_{23}, s_{34}, s_{42} respectively. From this it follows that the plane cubic which is the intersection of H and the plane at infinity passes through the six points P_{ij}.

There are in general no finite points A of E such that all four A_i $(i = 1,2,3,4)$ are collinear, because A must satisfy four conditions in this case. (This is the same as saying that in general the four c_i do not have a common finite point). However if A is at infinity so are all A_i, and there are only two conditions in this case. It then follows, as we shall show in Chapter VII on spherical kinematics, that there are six such points in the plane at infinity.

The four positions A_i $(i = 1,2,3,4)$ of a point $A(x, y, z, w)$ of the moving space E are in general the vertices of a tetrahedron T in the fixed space Σ. Let the circumcenter of T be the point M; it is the intersection of the perpendicular bisector planes of the edges. M may be found as the intersection of the normal planes of A_1A_2, A_2A_3 and A_3A_4. We know from Chapter III that the coordinates of these planes are linear functions of x, y, z, w. Hence the four homogeneous coordinates of M are cubic functions of x, y, z, w.

The points M in the fixed space and A in the moving space have the property that $|MA_1| = |MA_2| = |MA_3| = |MA_4|$, this constant distance being the radius of the circumsphere of T. If we interchange the roles of the two spaces, that is if we consider the inverse motion, this implies that A is the circumcenter of the four positions of M with respect to E. This means that its coordinates are cubic functions of those of M. The conclusion is: there exists a birational cubic relationship between A and M.

Analytically, the conditions $|MA_1| = |MA_2| = |MA_3| = |MA_4|$ may be written as $(MA_j)^2 - (MA_1)^2 = 0$, $j = 2,3,4$ which in terms of position vectors $\boldsymbol{M}$ and $\boldsymbol{A}_i$ (in Σ, from the origin of coordinates to points M and A_i, respectively) yields:

$$2\boldsymbol{M}\cdot(\boldsymbol{A}_j - \boldsymbol{A}_1) - (\boldsymbol{A}_j^2 - \boldsymbol{A}_1^2) = 0, \quad j = 2,3,4. \tag{7.1}$$

Equation (7.1) is the equation of the perpendicular bisector plane of A_jA_1. If the four positions are known, then corresponding to any point A the coordinates of center $M(M_X, M_Y, M_Z, M_W)$ follow from (7.1) as:

$$M_X : M_Y : M_Z : M_W = G_X : G_Y : G_Z : H \tag{7.2}$$

where

$$\begin{aligned} G_X &= |(\boldsymbol{A}_j^2 - \boldsymbol{A}_1^2), (Y_j - Y_1), (Z_j - Z_1)| \\ G_Y &= |(X_j - X_1), (\boldsymbol{A}_j^2 - \boldsymbol{A}_1^2), (Z_j - Z_1)| \\ G_Z &= |(X_j - X_1), (Y_j - Y_1), (\boldsymbol{A}_j^2 - \boldsymbol{A}_1^2)| \\ H &= |(X_j - X_1), (Y_j - Y_1), (Z_j - Z_1)| \end{aligned} \qquad j = 2,3,4$$

are all (3×3) determinants (with the jth row elements shown) whose elements may each be written (using (8.2) of Chapter I) as linear homogeneous functions of the coordinates of A, i.e., (x, y, z, w). Similarly, using the inverse motion (7.2) gives us the coordinates of A as the ratio of two cubic functions of the coordinates of M. The cubic nature of this (1,1) correspondence between A and M implies that points on an nth order surface in E correspond to points on a $3n^{th}$ order surface in Σ, and vice versa.*

In particular it follows that the locus of the points A corresponding to the points M of a plane is a cubic surface. If the locus of M is the plane at infinity the locus of the corresponding points A is the cubic surface H considered in the first paragraph of this section and given by eq. (7.2) as $H = 0$. On the other hand all points A at infinity correspond to those of a cubic surface embedded in Σ. When M is at infinity the corresponding point A lies on H (embedded in E) and has its four homologous positions A_i $(i = 1, 2, 3, 4)$ on a plane in Σ with the directions (G_X, G_Y, G_Z).

The circumcenter M is determined as the intersection of the normal planes of A_1A_2, A_1A_3, A_1A_4 (i.e., (7.1)) the coordinates of which are linear in x, y, z, w. It may be that these three planes have a line l in common, which implies that any point of l has equal distance to the four points A_i; obviously this can only occur if the A_i $(i = 1, 2, 3, 4)$ are coplanar and also lie on a circle. The analytic condition is that the rank of the 4×3-matrix of the coordinates, of the three normal planes (i.e., (7.1)), is only 2. The necessary and sufficient conditions for the rank of eq. (7.1) to be 2 are that $G_X = 0$, $G_Y = 0$, $G_Z = 0$, $H = 0$ and that at least one (2×2) minor of the system have a non-zero determinant. Alternatively, a rank of 2 is guaranteed by $G_Z = 0$, $H = 0$ provided their common (2×2) determinants do not all vanish. We use this latter condition in our analysis:

In general two cubic surfaces such as G_Z and H (represented here by equations $G_Z = 0$, $H = 0$) intersect in a ninth order curve, however in this case the curve degenerates into two curves of lower order: a cubic I_Z and a sextic κ. The cubic curve I_Z is spurious for our purposes, since it is the locus of points for which $G_Z = 0$, $H = 0$, but $G_X \neq 0$, $G_Y \neq 0$. The cubic follows from the two common 2×2-minors of G_Z and H; setting them to zero we have

* If we take the centers $M(M_X, M_Y, M_Z, M_W)$ on an nth order surface $f(M_X^n, M_Y^n, M_Z^n, M_W^n, M_X^{n-1}, \ldots) = 0$ and substitute (7.2) into this equation, we obtain the corresponding locus of A's: $f(G_X^n, G_Y^n, G_Z^n, H^n, G_X^{n-1}, \ldots) = 0$ which when expanded yields $f(x^{3n}, y^{3n}, z^{3n}, w^{3n}, x^{3n-1}, \ldots) = 0$, a surface of order $3n$. Similarly if we start with the moving points $A(x, y, z, w)$ on an nth order surface $g(x^n, y^n, z^n, w^n, x^{n-1}, \ldots) = 0$ and use the inverse motion, the corresponding centers are on a surface $g(M_X^{3n}, M_Y^{3n}, M_Z^{3n}, M_W^{3n}, M_X^{3n-1}, \ldots) = 0$ of order $3n$.

$$\begin{vmatrix} X_2 - X_1 & Y_2 - Y_1 \\ X_3 - X_1 & Y_3 - Y_1 \end{vmatrix} = 0, \qquad \begin{vmatrix} X_2 - X_1 & Y_2 - Y_1 \\ X_4 - X_1 & Y_4 - Y_1 \end{vmatrix} = 0.$$

These represent two quadratic surfaces which intersect in the cubic curve I_Z and the line $X_2 - X_1 = 0$, $Y_2 - Y_1 = 0$. In general, G_Z and H will not be zero for points on this line, but must vanish for points on the cubic I_Z. So we have, by our method of analysis, introduced an extraneous cubic I_Z. The remaining part of the intersection of $H = 0$, $G_Z = 0$ is a sextic κ. Hence the theorem: *the locus of the points A such that the four homologous points A_i $(i = 1, 2, 3, 4)$ are on a circle is a curve κ of the sixth order lying on the surface* H *embedded in E.*

κ is a singular curve of the cubic relationship between A and M: any point of κ corresponds to all points of a line l; the locus of all such lines l is a ruled surface (of "singular" lines) in Σ.

Conversely, there is a singular sixth order curve in Σ which is the locus of the points M which correspond to the points of a line, and all such lines form a ruled surface in E.

Example 9. Verify that a plane of E contains six singular points of the correspondence, and that the cubic surface in Σ corresponding to this plane contains six singular lines l.

A curve such as κ, the points of which are singular with regard to a given (1,1) correspondence, is called the "fundamental curve" of the correspondence. It is easy to show that κ has a genus of 3 if we apply the condition $2(p' - p) = (n' - n)(N_1 + N_2 - 4)$ (see Semple and Roth [1949, pg. 91]), which relates the orders n and n', and genera p and p' of the two curves which together are the complete intersection of two surfaces of order N_1 and N_2; we have: $N_1 = 3$ $(H = 0)$, $N_2 = 3$ $(G_Z = 0)$; $n = 3$, $p = 0$ (since I_Z is a twisted cubic), $n' = 6$ (κ is a sextic).

If a point of E corresponds to a plane in Σ the rank of the 4×3-matrix given by eq. (7.1) is only 1. This requires that all the (2×2) determinants vanish, but this is impossible under a general displacement. Hence there is no point A which generally corresponds to a plane, and κ is the only singular locus, of A, in regard to the (A, M) reciprocal cubic correspondence.

8. Homologous planes and lines

The four homogeneous coordinates of a plane in E are linear functions of those in Σ. Therefore if four positions of E with respect to Σ are given, the locus of the planes in E such that each quadruple of homologous planes passes through one point is the set of tangent planes to a surface of the fourth order. It has four subsets, each being a cubic torse, such that in three of the four positions the homologous planes pass through one line.

The Plücker coordinates of a line l in E are linear functions of those in Σ. This enables us to make some remarks on quadruples of homologous lines satisfying certain conditions. For instance the quadruple may be parabolic, which means that its two transversals coincide. The analytic condition is the vanishing of a determinant of order four the elements of which are quadratic functions of the coordinates of the four lines. Hence the locus of the lines in E such that their homologous quadruple is parabolic is a complex of order eight. If four lines lie on a hyperboloid their coordinates satisfy three conditions. Hence the locus of lines in E such that their homologous quadruples are on a hyperboloid is a ruled surface.

In Chapter IV, Section 7 we saw that a homologous triad l_1, l_2, l_3 defines a line congruence with central axis l_c. This congruence is the locus of all lines at a given distance and angle with l_c. In this section we ask which homologous quadruples l_i, $i = 1,2,3,4$, lie on one such congruence; i.e., which lines l_i, $i = 1,2,3,4$, may be displaced by screw displacements about a single line l_c in Σ. We proceed as in Chapter IV, Section 7 except now we add an additional position, i.e, we require

$$(l_j - l_1) \cdot l_c = 0 \quad j = 2,3,4 \tag{8.1}$$

and

$$(l'_j - l'_1) \cdot l_c + l'_c \cdot (l_j - l_1) = 0 \quad j = 2,3,4. \tag{8.2}$$

The result, after we analyze these equations, will be that all l (or l_i) with these properties lie on a line congruence. Similarly all corresponding l_c lie on a corresponding congruence. Thus there are double infinites of quadruples l_i and axes l_c for which $\theta_1 = \theta_2 = \theta_3 = \theta_4$ and $D_1 = D_2 = D_3 = D_4$. We will see that these congruences are congruent with the (9,3) congruence of screw axes defined in Section 5, l_c lies on the screw axis congruence in Σ, and l_i on the image screw axis congruence in E_i.

The analysis of (8.1) is as follows: if we take the components (measured in Σ) of l_i as l_i, m_i, n_i we require that

$$\begin{vmatrix} l_2 - l_1 & m_2 - m_1 & n_2 - n_1 \\ l_3 - l_1 & m_3 - m_1 & n_3 - n_1 \\ l_4 - l_1 & m_4 - m_1 & n_4 - n_1 \end{vmatrix} = 0. \tag{8.3}$$

Since we can write each term as a linear homogeneous function of the coordinates in E of $\boldsymbol{l}$ (or one of the $\boldsymbol{l}_i$), (8.3) represents a cubic cone the generators of which give the direction for l (or l_i). The corresponding directions $\boldsymbol{l}_c$ then follow from (8.1). Alternatively we can write the condition that the rank of (8.1), as an equation for $\boldsymbol{l}$ (or $\boldsymbol{l}_i$), be 2, and show that the directions l_c are also parallel to the generators of a cubic cone. Clearly, there is a (1,1) reciprocal correspondence between the generators of the two cones.

From (8.2) and $\boldsymbol{l}'_c \cdot \boldsymbol{l}_c = 0$ we have four equations for $\boldsymbol{l}'_c$, the condition that the rank of this system should be three gives a linear equation in $\boldsymbol{l}'$ (or $\boldsymbol{l}'_i$). This together with $\boldsymbol{l}' \cdot \boldsymbol{l} = 0$ (or $\boldsymbol{l}'_i \cdot \boldsymbol{l}_i = 0$) yields a single infinity of values for $\boldsymbol{l}'$ (or $\boldsymbol{l}'_i$) for each pair of directions $\boldsymbol{l}$, $\boldsymbol{l}_c$. For any one of these $\boldsymbol{l}'$ (or $\boldsymbol{l}'_i$) vectors, the corresponding $\boldsymbol{l}'_c$ follows from (8.2). Alternatively we could proceed in an exactly analogous manner and find that there are an infinity of moment vectors $\boldsymbol{l}'_c$ which satisfy (8.2) for every pair of directions $\boldsymbol{l}$, $\boldsymbol{l}_c$. The conclusion is then, as we stated above, there are a double infinity of l, l_c; l and l_c belong respectively to line congruences fixed in E and Σ.

We will now show that the axes l_c satisfy the condition given at the end of Section 2, and therefore l_c belong to the (9,3) screw axis congruence. We again introduce a complimentary screw quadrilateral, say, s_{12}, s_{23}, s_{34}, s_{41}, and consider first screws s_{12}, s_{23}. We know from Chapter IV, Section 7, that if n_c is the normal between s_{12} and l_c, n_2 the normal between l_2 and l_c, n'_c the normal between s_{23} and l_c, see Fig. 27, then the angle from n_c to n_2 is $\alpha_{12}/2$ and the

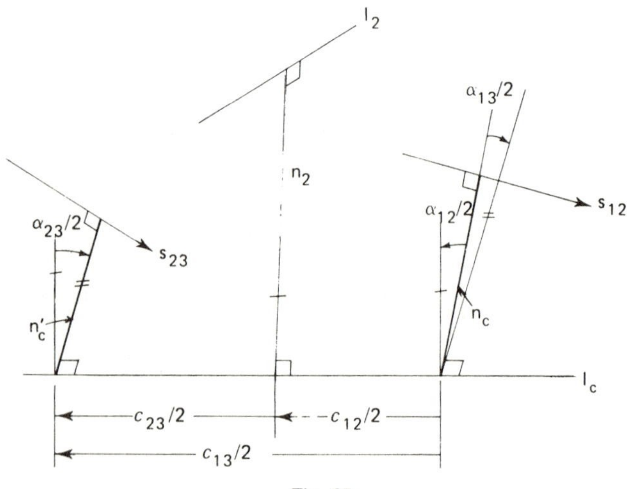

Fig. 27.

distance from n_c to n_2 is $c_{12}/2$. Similarly, the angle between n_2 and n_c' is $\alpha_{23}/2$ and the distance is $c_{23}/2$. Since n_c, n_2, n_c' all are normal to l_c it follows that the angle between n_c and n_c' is $\alpha_{12}/2+\alpha_{23}/2=\alpha_{13}/2$ and the distance is $c_{12}/2+c_{23}/2=c_{13}/2$.

In an analogous way it follows that the angle between n_c'' and n_c''' (respectively, the normals between l_c and s_{41}, and l_c and s_{34}) is $-\alpha_{41}/2-\alpha_{34}/2=\alpha_{13}/2$ and the distance between them is $-c_{41}/2-c_{34}/2=c_{13}/2$, see Fig. 28.

Hence l_c must be located so that the normals to it from adjacent sides of the complimentary screw quadrilateral are at equal angles and distance i.e., $\angle n_c n_c' = \angle n_c'' n_c'''$ and distance $n_c n_c'$ = distance $n_c'' n_c'''$. This is precisely the condition for a line to belong to the (9,3) screw congruence which we have already studied in some detail. Similarly, we can by inversion show that l_i must belong to the image screw congruence. Hence all the properties which we developed for the screw congruences are directly applicable to the l_i, l_c congruences defined by equations (8.1) and (8.2).

As a special case of the foregoing we seek out those lines l_i, $i=1,2,3,4$, which are intersected orthogonally by a single line. Clearly l_c is the line that cuts the l_i orthogonally; we require $\theta_1=\theta_2=\theta_3=\theta_4=\pi/2$, $D_1=D_2=D_3=D_4=0$. The solution can be obtained by using the equations developed in Chapter IV, Section 7 where we considered the same problem for three positions. If we set $i=1,2,3,4$ in (7.6) and (7.8) of Chapter IV we find that there are, at most, six sets of homologous quadruples l_i which have the sought

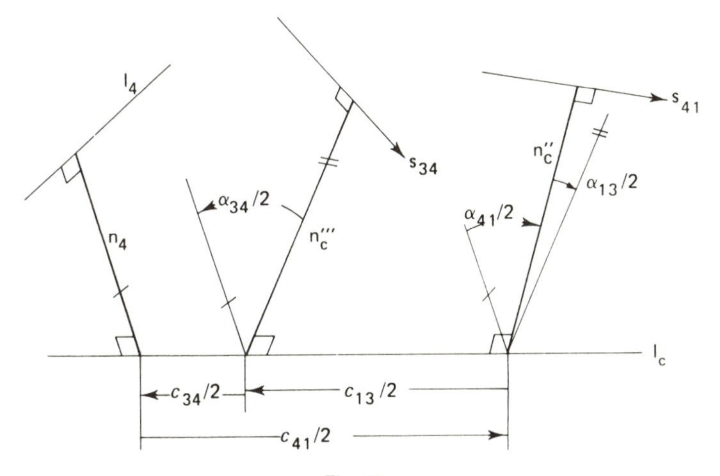

Fig. 28.

after property. This follows directly from the fact that in addition to (7.7) of Chapter IV we now have to satisfy a second cone-of-directions which is obtained by replacing the subscript 3 by 4 in (7.7) of Chapter IV. These two cubic cones intersect in at most 9 real lines, but three of these correspond to the screw axis directions and are spurious. Hence there are at most 6 possible directions. Now, equation (7.8) of Chapter IV yields an additional linear equation in $\boldsymbol{l}'_i$ which can be obtained by replacing the subscript 3 by 4 in (7.9) of Chapter IV. The result is then that we have one moment vector $\boldsymbol{l}'_i$ for each one of the 6 directions $\boldsymbol{l}$ and hence at most 6 homologous quadruples with the desired property.

9. A special case of four positions

As an illustration of the preceding general theory we consider a special case. The six screw axes s_{ij} are in general skew lines; we suppose here that they are the edges of a tetrahedron $B_1\, B_2\, B_3\, B_4$ (with faces β_i) such that s_{12} coincides with $B_3\, B_4$ and so on (Fig. 29). It is easy to verify that the edges are a compatible sextuple for they satisfy the conditions derived in Section 5.

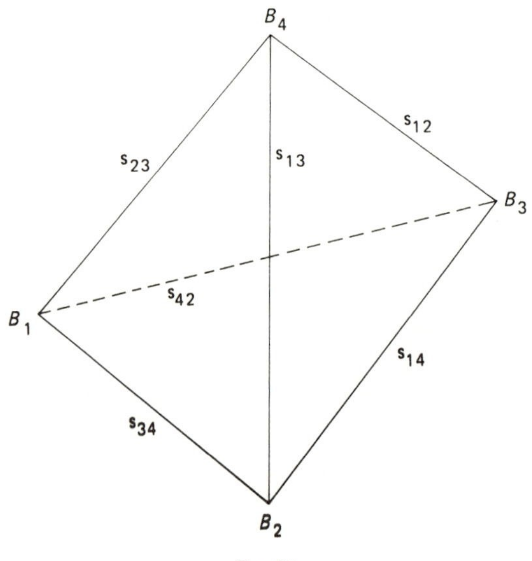

Fig. 29.

Indeed, the common perpendicular of s_{23}, s_{34} say, is the normal at B_1 on the face β_3, and so on. The conclusion is that the six displacements are all pure rotations and the rotation angle ϕ_{ij} is twice the dihedral angle between the two faces through the edge s_{ij}.

If A_i $(i = 1, 2, 3, 4)$ is a set of homologous points, A_2 is the position taken by A_1 after a rotation about s_{12} by angle ϕ_{12}. In other words, we can obtain A_2 from A_1 by reflecting A_1 in β_1 and then reflecting the resulting image, A^*, into β_2. Hence any quadruple of homologous points consists of the reflections of an arbitrary point A^* in the faces of the tetrahedron. A^* will be called the fundamental (also sometimes cardinal, or base, or representative) point of A_1, A_2, A_3, A_4. If A^* lies in β_i it coincides with A_i; if A^* is on the edge s_{ij} it coincides with A_i and A_j; if A^* coincides with B_1 say, it coincides with A_2, A_3, A_4.

A_1 and A_2 are the reflections of A^* into β_1 and β_2 respectively. Hence A^*, A_1, A_2 are in a plane U, perpendicular to s_{12}, intersecting β_1, β_2 along l_1, l_2 and the edge s_{12} at O. (Fig. 30.) It follows that $|OA_1| = |OA^*| = |OA_2|$; OA_1A_2 is an isosceles triangle with O as vertex. The perpendicular bisector of A_1A_2 in U coincides with the bisector OP of the angle A_1OA_2. Hence $\angle l_2OP = \angle A^*Ol_1$. (Here, l_i is the perpendicular bisector of A_iA^*.) The normal plane

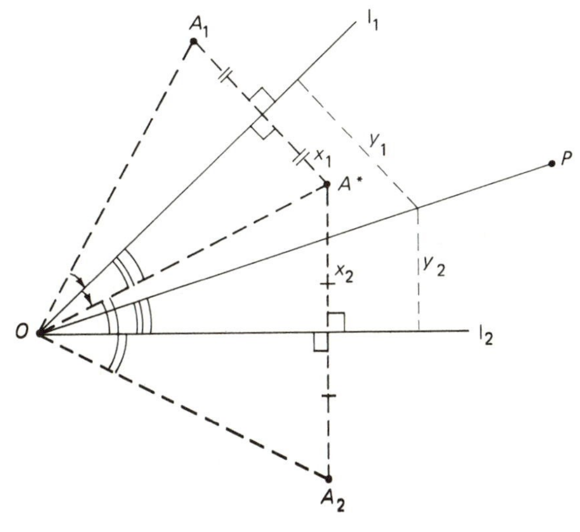

Fig. 30.

of A_1A_2 passes through OP and s_{12}. The conclusion is: the normal plane of A_1A_2 coincides with the plane we obtain if the plane through A^* and s_{12} is reflected into the bisector plane of the edge s_{12}. (The bisector plane of an edge contains the edge and bisects the dihedral angle between the faces which meet at the edge.) The angles of the normal plane of A_1A_2 with the faces β_1, β_2 are the same as those of the plane A^*s_{12} with β_2 and β_1.

The six normal planes of A_iA_j pass through one point M, the circumcenter of the homologous points A_i $(i = 1,2,3,4)$. We have proved the theorem: if A^* is an arbitrary point and if we reflect the plane A^*s_{ij} into the bisector plane of s_{ij} the six planes so constructed pass through one point M. We have established a correspondence between the fundamental point A^* of a quadruple of homologous points and their circumcenter M. It is called the isogonal correspondence with respect to the tetrahedron. Obviously it is involutory: if M is conjugate to A^*, then A^* is conjugate to M.

If A^* lies in the face β_i it follows from the construction that M always coincides with B_i. Conversely any point of β_i is conjugate to B_i. Hence the correspondence has singularities and it can not be linear. Its analytic form can easily be obtained: If x_i is the distance of A^* to the face β_i, it follows from Fig. 30 that the distances y_1, y_2 from any point on OP (to β_1 and β_2 respectively) are related to the x_i as follows: $y_1 : y_2 = x_2 : x_1$. Hence, since M is on OP, the distances of M to the faces are inversely proportional to those of A^*. If we introduce homogeneous coordinates, such that the coordinates of a point are proportional to its distances to the faces β_i, the relationship between $M(y_i)$ and $A^*(x_i)$ is given by

$$y_1 : y_2 : y_3 : y_4 = x_2x_3x_4 : x_3x_4x_1 : x_4x_1x_2 : x_1x_2x_3, \tag{9.1}$$

and conversely (if $x_1x_2x_3x_4 \neq 0$),

$$x_1 : x_2 : x_3 : x_4 = y_2y_3y_4 : y_3y_4y_1 : y_4y_1y_2 : y_1y_2y_3. \tag{9.2}$$

Hence there exists a cubic relationship between A^* and M; as there is a linear relation between A^* and A_1 say, this is in accordance with the general results of Section 7.

The point M is in the plane at infinity if $\Sigma \mathfrak{b}_i y_i = 0$, where $\mathfrak{b}_i$ stands for the area of the face β_i. Therefore: the locus of the point A^* such that the four homologous points A_i are in one plane has the equation

$$\mathfrak{b}_1x_2x_3x_4 + \mathfrak{b}_2x_3x_4x_1 + \mathfrak{b}_3x_4x_1x_2 + \mathfrak{b}_4x_1x_2x_3 = 0, \tag{9.3}$$

which represents a cubic surface H*; its reflection in β_1 say, is the surface H considered in Section 7.

H* is Cayley's cubic surface; it has double points at B_i and the six edges of the tetrahedron are on it. Any plane of points M corresponds, by the isogonal relationship, to a cubic surface of the same type.

Three homologous points, A_1, A_2, A_3 say, are collinear only if they are on one of the edges through B_4. Similarly, A^* must lie on an edge if three of the A_i are collinear, and the four cubic curves c_i of Section 7 each degenerate into three concurrent lines, lying on H.

The four points A_i lie on a circle if two A_i coincide. Hence such a point A_k is on edge s_{kl} or s_{jk}. Similarly A^* must lie on an edge for this condition. The conclusion is that the sextic curve κ in this case degenerates into the six lines which are the edges of the tetrahedron formed by reflecting tetrahedron B_1, B_2, B_3, B_4 into face β_1. These lines all lie on H. Similarly, the edges of the original tetrahedron form the degenerate sextic κ^* on H* which is the locus of all fundamental points with corresponding A_i $(i = 1, 2, 3, 4)$ on circles.

10. More than four positions

For n positions we have $[n(n-1)]/2$ screw axes and $[n(n-1)(n-2)]/6$ screw triangles. There are $6(n-1)$ independent parameters associated with the description of these positions (relative to each other), and $[4n(n-1)]/2$ parameters associated with the screw axes. It follows that the screw axes are not all independent (whenever $n > 3$). In fact, the axes require $2(n-3)(n-1)$ more parameters than do the positions, and therefore, in general, the number of independent axes is at most $[n(n-1)]/2 - [2(n-3)(n-1)]/4 = [3(n-1)]/2$. When n is odd the number of independent axes is an integer; whilst for n even, the number is given by the quotient, and the remainder of 2 implies that there are two additional free parameters. For $n > 4$ not all sets of $3(n-1)/2$ axes are independent. We illustrate these concepts with the case $n = 5$.

Example 10. Verify that the above is consistent with our results for three ($n = 3$) and four positions ($n = 4$).

For five positions we have ten screw axes but only six of these are independent. Furthermore, we know from the study of four positions that, in general, six arbitrary axes cannot belong to a set of four positions, and that there are at most four arbitrary axes for any set of four positions. It will be recalled (Section 1) that there are two different cases according to whether any three of the four axes belong to the same screw triangle.

If three axes belong to one screw triangle, the other three must also belong to one screw triangle, and the two triangles must have only one position in common. All such sets of two triangles uniquely define the five positions. All ten screws follow by successive applications of the screw triangle geometry.

Example 11. Verify that the six axes of triangles s_{12}, s_{13}, s_{23}, and s_{14}, s_{15}, s_{45} are independent, whereas those of s_{12}, s_{13}, s_{23} and s_{14}, s_{15}, s_{24} or s_{12}, s_{14}, s_{24} are not.

Example 12. Verify that the ten axes can be obtained from the screw triangle geometry as follows: if we choose axes which form triangles s_{12}, s_{13}, s_{23} and s_{14}, s_{15}, s_{45}, then ϕ_{1j}, d_{1j} and ϕ_{1k}, d_{1k} follow from these triangles. Now considering position 1, j, k, these parameters and axes s_{1j}, s_{1k} uniquely determine axes s_{jk} ($j = 2, 3$; $k = 4, 5$; $j \neq k$).

The second case occurs when no three of the six axes belong to one screw triangle: the only independent axes are members of two independent complimentary-screw-quadrilaterals. Such screw quadrilaterals always have two screw axes in common. For example, axes s_{12}, s_{13}, s_{34}, s_{24}, s_{35}, s_{25} are independent and they form the two quadrilaterals s_{12}, s_{13}, s_{34}, s_{24} and s_{12}, s_{13}, s_{35}, s_{25} which have axes s_{12} and s_{13} in common. We know that each of these complimentary-screw-quadrilaterals determines a (9,3) congruence (Section 6). There is always a seventh screw axis (as yet undetermined) which is a member of both (9,3) congruences. (For the present example this axis is s_{23}.)

This seventh axis follows from the intersection of the two congruences: we consider two sets of four positions (1, 2, 3, 4 and 1, 2, 3, 5 in this case). Then they each define one screw cone; the two (cubic) screw cones have $3 \times 3 = 9$ generators in common. Parallel to each of these nine generators there is one plane of each congruence. Each pair of planes intersect in one line common to both congruences, and hence there are nine such lines* (not necessarily all real). Two of these are s_{12}, s_{13}; any one of the remaining seven may be s_{23} (there must always be at least one real line).

Once the seventh axis is known, two of the remaining axes (s_{14} and s_{15}, in this case) follow uniquely since each represents the sixth axis for one of the sets of four positions. Finally, the tenth axis (s_{45}, in this case) is uniquely determined by any one of three screw triangles.

In general, these are the only possibilities for five positions. More than five positions can be analyzed similarly, the screw triangle and complimentary-screw-quadrilateral are the basic building blocks in all cases.

We turn now to homologous points. For five positions, a point A of E has five positions A_1, A_2, A_3, A_4, A_5 in the fixed space Σ. From Section 7 it follows that the condition for coplanarity of A_i $(1, 2, \ldots, 5)$ requires that the 5×4 matrix $\|X_i\, Y_i\, Z_i\, 1\|$ has a rank of 3. Thus the coplanar points are those points with coordinates such that

* Of the 90 lines predicted by Halphen's theorem (for the number of lines common to two congruences), 81 are at infinity.

(10.1) $$|X_i \quad Y_i \quad Z_i \quad 1| = 0 \quad (i = 1,2,3,4)$$
and

(10.2) $$|X_i \quad Y_i \quad Z_i \quad 1| = 0 \quad (i = 1,2,3,5)$$

provided not all the common 3×3 determinants of (10.1) and (10.2) vanish. Since (10.1) represents the cubic surface H, it follows that (10.2) also represents a cubic surface, say H_5. Furthermore, we know that both surfaces must contain the twisted cubic c_4, which is the locus of all points for which positions A_1, A_2, A_3 are collinear. c_4 is the locus of all points for which the common 3×3 determinants vanish. The result is then: *points A with five coplanar positions are located on a sixth order curve*, κ^1, which is the intersection of H and H_5 excluding c_4. The same analysis as used for κ (Section 7) shows that the genus of κ^1 is 3.

For six positions we have the additional condition $|X_i \; Y_i \; Z_i \; 1| = 0$ $(i = 1,2,3,6)$ which implies a third cubic surface H_6 also containing c_4. There are 10 points common to H_6 and κ^1 excluding those on c_4. This follows from the formula $I = n(N_1 + N_2 - 4) - (2p - 2)$, given by Semple and Roth [1949], in which I is the number of intersections of two nonsingular curves, with order and genus (n, p) and (n^1, p^1), which form the complete intersection of two surfaces of orders N_1 and N_2. Here $I = 3(3+3-4) - (2 \cdot 0 - 2) = 8$; of the 18 intersections of H_6 and κ^1 8 are on c_4, and so $18 - 8 = 10$ are the ones we seek. Hence the theorem: *There are generally* 10 *points* (not necessarily all real) *in E with six homologous positions in one plane of* Σ. Strictly speaking we must also add the plane at infinity to all the loci of coplanar points. Points at infinity remain in their plane for any number of positions, but there are generally no finite points with more than six positions on one plane.

Example 13. Show that there are generally no points with five homologous positions on one circle. Hint: two sixth order curves κ (for positions 1, 2, 3, 4) and say κ_5 (for positions 1, 2, 3, 5) will generally not have any common points.

Equation (7.1) gives the condition that a point A of E has a constant distance in four positions from a fixed point M of Σ. We now determine the loci of points A for which this condition is valid for five, six and seven homologous positions. We can interpret this question as asking for those points in E which have five, six or seven homologous positions on a sphere (of radius $|MA_j|$) in Σ.

For five positions we write (7.1) with $j = 2,3,4,5$. This means that M can exist only if the four equations are dependent. If we substitute $A_j = (X_j, Y_j, Z_j, W_j)$ and $M = (M_X, M_Y, M_Z, M_W)$, then the required condition is the vanishing of the 4×4 determinant:

(10.3) $|2(X_j - X_1), 2(Y_j - Y_1), 2(Z_j - Z_1), (A_j^2 - A_1^2)| = 0, \quad j = 2, 3, 4, 5.$

Each element is a linear homogeneous function of the coordinates of A, i.e., (x, y, z, w), and therefore (10.3) represents a fourth order surface, G_5, embedded in E. Hence, *the locus of all points with five homologous positions on a sphere is a fourth order surface.* Similarly, the locus of corresponding centers M is a fourth order surface embedded in Σ. This can easily be seen if we write (10.3) in terms of the coordinates of M rather than A, or by studying the inverse displacements. For six positions we have $j = 2, 3, 4, 5, 6$ in (7.1). Hence a unique M will exist only if the rank of the 5×4 matrix

(10.4) $\|2(X_j - X_1), 2(Y_j - Y_1), 2(Z_j - Z_1), (A_j^2 - A_1^2)\| \quad j = 2, 3, 4, 5, 6$

is equal to 3. This condition is satisfied by those points with coordinates given both by (10.3), i.e., by G_5, and by the 4×4 determinant obtained by using the subscript 6 instead of 5 in the last row of (10.3). This latter determinant yields G_6 which is the fourth order surface containing all the points A with positions A_1, A_2, A_3, A_4, A_6 on a sphere. The intersection of G_5 and G_6 gives the required locus provided we exclude those points for which the rank of the matrix (10.4), with $j = 2, 3, 4$ is 2. In Section 7 we saw that a rank of 2 corresponds to a sextic κ (of genus 3) which is the locus of all points A with positions A_1, A_2, A_3, A_4 on a circle. Since κ is embedded in both G_6 and G_5, their intersection obviously contains it. Hence G_6 and G_5 intersect in κ and a $4 \times 4 - 6 = 10$ order curve, k. The genus of k is eleven, this follows from $2(p^1 - p) = (n^1 - n)(N_1 + N_2 - 4)$ (see Section 7) if we set $N_1 = N_2 = 4$, $n = 6$, $n^1 = 10$, $p = 3$. Our result is: *the locus of all points A, in E, with six homologous positions on a sphere in Σ is a tenth order curve, of genus eleven, embedded in E.* The corresponding centers M are located on an analogous curve, of order ten and genus eleven, embedded in Σ.

For seven positions we have $j = 2, 3, 4, 5, 6, 7$ in (10.4), and require that its rank is again 3. Hence we have to find those points on k which intersect G_7, G_7 being the determinant given by (10.3) with $j = 7$ instead of 5 in the last row. The result follows from the formula $I = n(N_1 + N_2 - 4) - (2p - 2)$ introduced earlier in this section. Here we have $I = 6(4 + 4 - 4) - (2 \cdot 3 - 2) = 20$. Hence of the $4 \times 10 = 40$ intersections of G_7 and k, 20 lie on the intersections of k and κ and must be discounted (since for points on κ the rank of (10.4) with $j = 2, 3, 4$ is 2), we have then $40 - 20 = 20$ points. The conclusion is *there are generally twenty points (not necessarily all real) A, in E, such that their seven homologous positions A_i $(i = 1, \ldots, 7)$ lie on a sphere in Σ.*

Many of these results for points on planes, lines, circles and spheres were originally obtained by SCHOENFLIES [1886] using synthetic (as opposed to analytic) methods. Our treatment follows that given by ROTH [1967b]. It is of course possible to obtain additional results for other loci, such as points with several positions on a cylinder or a helix, and to find analogous results for the instantaneous case as well as mixed finite and infinitesimally separated displacements. A general theory has been developed by CHEN and ROTH [1969a, b].

If we study planes in E we can obtain the results for five and six positions by extending our four position analysis. We have, for a plane U, with coordinates in Σ $\mathrm{U}_j(u_{1j}, u_{2j}, u_{3j}, u_{4j})$ and passing through $M(M_X, M_Y, M_Z, M_W)$, $u_{1j}M_X + u_{2j}M_Y + u_{3j}M_Z + u_{4j}M_W = 0$ $j = 1,2,3,4,5,6$. The methods are exactly the same as for points. The results follow more immediately by inversion: the planes of E with five homologous positions passing through a point of Σ are such that the intersection point (of each homologous set of five planes) is on a sixth order space curve of genus three. This curve is the curve κ^1 for the inverse displacements. Similarly there are ten (not necessarily real) planes with their six homologous positions through a point of Σ. These fixed points are the ten planar points of the inverse displacement. Of course the plane at infinity also satisfies these conditions.

The lines l in E which have four positions on the congruence defined by l_1, l_2, l_3 and the central axis l_c (see Chapter IV, Section 7) were shown to have their four positions on a (9,3) line congruence, Section 8. To study five positions we add another position to equation (8.1) and (8.2) i.e., we now have $j = 2,3,4,5$.

The analysis follows the same lines of reasoning we used for the points: the five position results are given by the intersection of the results of two four position problems if we disregard the singular elements. If we consider $j = 2,3,4$ and then $j = 2,3,5$ in (8.1) and (8.2), the conclusion is that the lines we seek are the intersections of two four-position (9,3) image-screw-congruences. They intersect in nine lines but three are singular: s_{12}, s_{13}, s_{23}^1. Hence, there are at most six lines l, in E, with five homologous positions at a constant distance and angle from a line l_c in Σ. Similarly if we seek the lines l_c, we intersect two (9,3) screw congruences and, discounting the singular elements (in this case s_{12}, s_{13}, s_{23}), obtain: *there are at most six lines* l_c *in* Σ *which are central axes for five positions of* E. A detailed treatment of this problem has been given by ROTH [1967a]. The instantaneous case yields approximately the same result: the singular element is always the screw axis counted three times (see CHEN and ROTH [1969b]). There are generally no lines with more than five positions on a l,l_c congruence. In all this lines at infinity are not counted.

11. The instantaneous case

We have already mentioned some results which apply when the positions are considered to be infinitesimally separated. Here we expand our discussion. For m infinitesimally separated positions we are dealing with instantaneous kinematics up to order $(m-1)$. The basic formulas come from Chapter II, (4.5). For the canonical system in the case of geometrical instantaneous kinematics we have from (6.3) of Chapter II:

$$\begin{aligned} &X_0 = x, \quad X_1 = -y, \quad X_2 = -x + \varepsilon z, \quad X_3 = (1+\varepsilon^2)y + \gamma_Y z + d_{3_X}, \\ &Y_0 = y, \quad Y_1 = x, \quad Y_2 = -y + \mu\varepsilon, \quad Y_3 = -(1+\varepsilon^2)x + (\tfrac{3}{2}\varepsilon - \gamma_X)z + d_{3_Y}, \\ &Z_0 = z, \quad Z_1 = \sigma_0, \quad Z_2 = -\varepsilon x + \lambda, \quad Z_3 = -\gamma_Y x + (\tfrac{3}{2}\varepsilon + \gamma_X)y + d_{3_Z}. \end{aligned} \tag{11.1}$$

These results enable us to explicitly develop all instantaneous geometric properties up to the third order. Following the development given in Chapter II, fourth, fifth and higher order formulas could easily be obtained. However, as we get five new geometric invariants with each increase in order, we limit ourselves here to explicit formulas for the third order properties. Higher order properties will be given implicitly.

Example 14. Show that for five consecutive positions the additional terms are

$$\begin{aligned} X_4 &= (1+\varepsilon^2)x + 3\varepsilon\gamma_Y y + (2\gamma_X + \kappa_Y)z + d_{4_X}, \\ Y_4 &= -3\varepsilon\gamma_Y x + (1+4\varepsilon^2)y + (2\gamma_Y - \kappa_X)z + d_{4_Y}, \\ Z_4 &= (2\gamma_X - \kappa_Y)x + (2\gamma_Y + \kappa_X)y - 3\varepsilon^2 z + d_{4_Z}, \end{aligned}$$

where κ_X, κ_Y, d_{4_X}, d_{4_Y}, d_{4_Z} are the fourth order invariants.

Example 15. Obtain the expressions for X_5, Y_5, Z_5 and X_6, Y_6, Z_6.

If a point has four consecutive positions in its osculating plane it passes through a position where its path has zero torsion. Such points must satisfy the condition

$$\begin{vmatrix} X_0 & Y_0 & Z_0 & 1 \\ X_1 & Y_1 & Z_1 & 0 \\ X_2 & Y_2 & Z_2 & 0 \\ X_3 & Y_3 & Z_3 & 0 \end{vmatrix} = 0. \tag{11.2}$$

If we substitute from (11.1) and expand this determinant the result is a cubic polynomial in terms of x, y, z. This is the limit case of the cubic surface H discussed in Section 7. If we use homogeneous coordinates it is obvious that the plane at infinity also satisfies this condition.

Example 16. Show that the cubic terms of H are $(x^2+y^2)(\gamma_Y x - (\frac{3}{2}\varepsilon + \gamma_X)y) + \frac{3}{2}\varepsilon^2 xyz + \varepsilon xz(\gamma_Y x - \gamma_X y)$.
Example 17. Discuss the special cases $\varepsilon = 0$, or $\gamma_X = 0$, or $\gamma_Y = 0$ and combinations of these.
Example 18. Show that the origin of coordinates lies on H only if $\mu = 0$, or $\varepsilon = 0$, or $\sigma_0 = 0$, or $d_{3_X} = 0$.

Obviously the inflection curve c ((14.3) of Chapter IV) lies on the cubic surface H. In general, no finite points can have undulations in their paths; and so no finite points of H have all four positions on a line. However, there are six points at infinity which are at undulations of their paths (as will be shown in Chapter VII, Section 4 and Chapter XII, Section 4).

Example 19. Show that the inflection curve c may be obtained from

$$\begin{vmatrix} X_1 & Z_1 \\ X_2 & Z_2 \end{vmatrix} = 0 \quad \text{and} \quad \begin{vmatrix} Y_1 & Z_1 \\ Y_2 & Z_2 \end{vmatrix} = 0,$$

if we exclude the line $Z_1 = Z_2 = 0$.
Example 20. Show that an undulation point would require in addition to the conditions of Example 19,

$$\begin{vmatrix} X_1 & Z_1 \\ X_3 & Z_3 \end{vmatrix} = 0 \quad \text{and} \quad \begin{vmatrix} Y_1 & Z_1 \\ Y_3 & Z_3 \end{vmatrix} = 0$$

(excluding $Z_1 = Z_3 = 0$), and that this is generally impossible for a finite point.

Four infinitesimally separated positions of a point in E determine the sphere in Σ, called the osculating sphere, on which they lie. If $M(M_X, M_Y, M_Z, M_W)$ are the homogeneous coordinates of the center point of the osculating sphere of the point $A(x, y, z, w)$, it is necessary that these coordinates satisfy:

$$(11.3) \quad (X-(M_X/M_W))^2 + (Y-(M_Y/M_W))^2 + (Z-(M_Z/M_W))^2 = R^2$$

where R is the radius of the sphere. If we take the first three derivatives of (11.3) and evaluate them at the zero position we obtain

$$M_X X_1 + M_Y Y_1 + M_Z Z_1 - M_W(X_0X_1 + Y_0Y_1 + Z_0Z_1) = 0$$

$$M_X X_2 + M_Y Y_2 + M_Z Z_2 - M_W[(X_0X_2 + Y_0Y_2 + Z_0Z_2) + (X_1^2 + Y_1^2 + Z_1^2)]$$

$$(11.4)$$

$$M_X X_3 + M_Y Y_3 + M_Z Z_3 - M_W[(X_0X_3 + Y_0Y_3 + Z_0Z_3) + 3(X_1X_2 + Y_1Y_2 + Z_1Z_2)] = 0.$$

Example 21. Show that (11.4) can be obtained by substituting the coordinates of the four positions of A into (11.3) and then subtracting the equation for the first position from the one for the second, the ones for the first and second from the equation for the third position, and the first three from the fourth.

If we substitute (11.1) we obtain from (11.4):

$$
\begin{aligned}
& -M_X y + M_Y x + M_Z \sigma_0 w - M_W \sigma_0 z = 0 \\
& M_X(-x+\varepsilon z) + M_Y(-y+\mu\varepsilon w) + M_Z(-\varepsilon x + \lambda w) \\
& \quad - M_W(\mu\varepsilon y + \lambda z + \sigma_0^2 w) = 0 \\
(11.5) \qquad & M_X((1+\varepsilon^2)y + \gamma_Y z + d_{3_X} w) + M_Y(-(1+\varepsilon^2)x + (\tfrac{3}{2}\varepsilon - \gamma_X)z + d_{3_Y} w) \\
& \quad + M_Z(-\gamma_Y x + (\tfrac{3}{2}\varepsilon + \gamma_X)y + d_{3_X} w) \\
& \quad - M_W((3\varepsilon(\mu - \sigma_0) + d_{3_X})x + d_{3_Y} y + d_{3_Z} z + 3\sigma_0\lambda w) = 0.
\end{aligned}
$$

The equations are linear in M_X, M_Y, M_Z, M_W and x, y, z, w and define a reciprocal cubic correspondence between them. The results are similar to the finite displacement case (7.2). That is:

$$(11.6) \qquad M_X : M_Y : M_Z : M_W = G_X : G_Y : G_Z : H$$

where G_X, G_Y, G_Z, H are each cubic polynomials of x, y, z, w formed from the 3×3 determinants of the coefficients of (11.5). With these new definitions it immediately follows that the results of Section 7 are also valid for the infinitesimal case. Most important is the result that *the locus of all points with four positions on a circle is the sextic* κ which is the common intersection of $H = 0$, $G_X = 0$, $G_Y = 0$, $G_Z = 0$.

Example 22. Modify the arguments used in Section 7 for the present case and verify that κ does indeed have the same order in this limit case.
Example 23. Investigate the behavior of κ at infinity.

For five positions we find the coplanar points from (11.2), which is H, and

$$(11.7) \qquad \begin{vmatrix} X_1 & Y_1 & Z_1 \\ X_2 & Y_2 & Z_2 \\ X_4 & Y_4 & Z_4 \end{vmatrix} = 0$$

which, if we substitute from (11.1) and Example 14, yields a cubic surface. This is the limit case of the surface we have called H_5 (Section 10). The surfaces H and H_5 intersect in c and a sixth order curve which is the limit of κ^1. Thus, *all points with five infinitesimally separated positions on a plane is a curve* κ^1 *embedded in E.*

Example 24. Show that there are generally no finite points with five infinitesimally separated positions on a circle.

Continuing in the same way if we use

$$\begin{vmatrix} X_1 & Y_1 & Z_1 \\ X_2 & Y_2 & Z_2 \\ X_5 & Y_5 & Z_5 \end{vmatrix} = 0 \tag{11.8}$$

we obtain the surface corresponding to H_6 in Section 10. The result is: *there are generally* 10 *points with six infinitesimally separated positions on a plane.* There are in general no points with seven positions on a plane.

Example 25. Modify the arguments of Section 10 for the present case.

The points which remain on their osculating sphere for a fifth position satisfy (11.4) and also

$$\begin{aligned} M_X X_4 + M_Y Y_4 + M_Z Z_4 - M_W [(X_0 X_4 + Y_0 Y_4 + Z_0 Z_4) + 3(X_2^2 + Y_2^2 + Z_2^2) \\ + 4(X_1 X_3 + Y_1 Y_3 + Z_1 Z_3)] = 0. \end{aligned} \tag{11.9}$$

If we form the determinant of the coefficients of M_X, M_Y, M_Z, M_W in (11.4) and (11.9), and substitute (11.1) and Example 14 we have a fourth degree polynomial in x, y, z, w. This is the equation of the fourth order surface which is the limit of G_5 introduced in Section 10.

For six positions we have in addition to (11.4) and (11.9):

$$\begin{aligned} M_X X_5 + M_Y Y_5 + M_Z Z_5 - M_W [(X_0 X_5 + Y_0 Y_5 + Z_0 Z_5) \\ + 10(X_2 X_3 + Y_2 Y_3 + Z_2 Z_3) \\ + 5(X_1 X_4 + Y_1 Y_4 + Z_1 Z_4)] = 0. \end{aligned} \tag{11.10}$$

Using (11.4) and (11.10) we obtain a new fourth order surface which is analogous to G_6 of Section 10.

For seven positions, (11.4) and

$$\begin{aligned} M_X X_6 + M_Y Y_6 + M_Z Z_6 - M_W (X_0 X_6 + Y_0 Y_6 + Z_0 Z_6) \\ + 6(X_1 X_5 + Y_1 Y_5 + Z_1 Z_5) + 15(X_2 X_4 + Y_2 Y_4 + Z_2 Z_4) \\ + 10(X_3^2 + Y_3^2 + Z_3^2) = 0 \end{aligned} \tag{11.11}$$

yield a fourth order surface analogous to G_7 of Section 10.

Using reasoning analogous to that in Section 10, it is easy to show that the locus of the points with six positions on a sphere is a tenth order curve. For seven infinitesimal positions there are 20 points (not necessarily real) which have all seven positions on their osculating spheres.

Example 26. Verify that the finite position discussion of Section 10 applies to the infinitesimal case.

In general the finite position results are directly transferable to the infinitesimal position case. The major difference is that the several screw axes of the finite case tend to amalgamate and the screw axis for the first order (or the point on it at infinity) tends to become a multiple line (or point) in cases where the finite axes (or their points at infinity) are members of the given locus.

We complete this section with one further example.

In Chapter IV, Section 7 and in Section 8 of this chapter we treated the line congruence defined by three positions of a moving line being at a fixed angle and distance from a central axis l_c. For the infinitesimal case we have for the directions the equivalent of (8.1):

$$\boldsymbol{L}_1 \cdot \boldsymbol{l}_c = 0,$$

$$\boldsymbol{L}_2 \cdot \boldsymbol{l}_c = 0,$$

$$\boldsymbol{L}_3 \cdot \boldsymbol{l}_c = 0.$$

Here $\boldsymbol{l}_c$ is the direction vector of l_c and $\boldsymbol{L}_i$ (L_i, M_i, N_i) is the ith derivative of the direction of the moving line as measured in Σ. Hence, the moving line must satisfy

$$\begin{vmatrix} L_1 & M_1 & N_1 \\ L_2 & M_2 & N_2 \\ L_3 & M_3 & N_3 \end{vmatrix} = 0. \tag{11.12}$$

for four infinitesimally separated positions. Here a moving line with directions l, m, n in E has its geometric derivatives given by (using (11.1)):

$$\begin{aligned} &L_0 = l, && L_1 = -m, && L_2 = -l + \varepsilon n, && L_3 = (1+\varepsilon^2)m + \gamma_Y n, \\ &M_0 = m, && M_1 = l, && M_2 = -m, && M_3 = -(1+\varepsilon^2)l \\ & && && && \qquad + (\tfrac{3}{2}\varepsilon - \gamma_X)n, \\ &N_0 = n, && N_1 = 0, && N_2 = -\varepsilon l, && N_3 = -\gamma_Y l + (\tfrac{3}{2}\varepsilon + \gamma_X)m. \end{aligned} \tag{11.13}$$

Hence, (11.12) is a cubic cone with the screw axis (i.e., $l = m = 0$, $n = 1$) as a double line.

For five positions we add the condition

$$\begin{vmatrix} L_1 & M_1 & N_1 \\ L_2 & M_2 & N_2 \\ L_4 & M_4 & N_4 \end{vmatrix} = 0, \tag{11.14}$$

where L_4, M_4, N_4 can be easily obtained from Example 14. (11.14) is a cubic cone; it passes through the screw axis and its tangent plane at the screw coincides with one of those of (11.12). Hence, (11.12) and (11.14) have the screw axis as a triple intersection, they also have six other generally distinct lines of intersection. These six lines give the directions for the moving lines which in five infinitesimally separated positions remain at a constant angle from some fixed line.

For the location condition we use equations analogous to (8.2):

$$\begin{aligned} &\boldsymbol{L}_1' \cdot \boldsymbol{l}_c + \boldsymbol{l}_c' \cdot \boldsymbol{L}_1 = 0, \\ (11.15) \qquad &\boldsymbol{L}_2' \cdot \boldsymbol{l}_c + \boldsymbol{l}_c' \cdot \boldsymbol{L}_2 = 0, \\ &\boldsymbol{L}_3' \cdot \boldsymbol{l}_c + \boldsymbol{l}_c' \cdot \boldsymbol{L}_3 = 0 \end{aligned}$$

and for a fifth position we add

$$(11.16) \qquad \boldsymbol{L}_4' \cdot \boldsymbol{l}_c + \boldsymbol{l}_c' \cdot \boldsymbol{L}_4 = 0$$

where $\boldsymbol{l}_c'$ and $\boldsymbol{L}_i'$ are the moment vectors of respectively l_c and the ith order representations of the moving line. The components of $\boldsymbol{L}_i'$ are obtained from (11.1) and (11.13) using $\boldsymbol{L}' = \boldsymbol{X} \times \boldsymbol{L}$ and $\boldsymbol{l}' = \boldsymbol{x} \times \boldsymbol{l}$, where $\boldsymbol{X}$ and $\boldsymbol{x}$ are position vectors in Σ and E respectively. For any line with direction and moment components in E given by (l, m, n), (l', m', n') we have:

$$(11.17) \qquad \begin{aligned} &L_0' = l', \quad L_1' = -\sigma_0 m - m', \quad L_2' = -l' + \mu\varepsilon n - 2\sigma_0 l \\ &\qquad\qquad\qquad\qquad\qquad\qquad + \varepsilon n' - \lambda m, \ldots, \\ &M_0' = m', \quad M_1' = \sigma_0 l + l', \quad M_2' = -m' - 2\sigma_0 m + \lambda l, \ldots, \\ &N_0' = n', \quad N_1' = 0, \quad N_2' = -\varepsilon l' - \mu\varepsilon l, \ldots . \end{aligned}$$

(As will be seen in Chapter VI, (8.1), this is the same as describing a line in terms of Plücker coordinates p_{41}, p_{42}, p_{43}, p_{23}, p_{31}, p_{12}.)

With these equations we can now solve for the locations of the moving lines. We do not repeat the details since the arguments and results given in connection with (8.2) and at the end of Section 10 are applicable to this case also.

Detailed discussion of the theory and equations presented in this section, together with many other examples can be found in CHEN and ROTH [1969a, b].

CHAPTER VI

CONTINUOUS KINEMATICS

1. Displacements in three-dimensional space

In Chapters III–V we have dealt with a finite number of positions of a space E with respect to a fixed space Σ. We consider now a continuous series of positions of E with respect to Σ. If the position of E depends on m parameters we shall say that E has a motion with *m degrees of freedom*. The most important case is $m = 1$, i.e., motions with (only) one parameter, say τ. We must distinguish between geometric, i.e. time-independent, motions (in which case we are only interested in those properties which are invariant under transformations of the parameter) and time-dependent motions, in which case τ stands for the *time* t.

In order to describe a motion we will make use of the analytic apparatus developed in Chapter I, restricting ourselves to the case $n = 3$: kinematics in three-dimensional space. We first reconsider the case of finite displacements, this time developed in such a way as to lead naturally to continuous displacements (i.e., motions): If $\boldsymbol{P}$ and $\boldsymbol{p}$ are the position vectors of a point P, in Σ and E respectively, we have (Chapter I, eq. (8.2))

$$\boldsymbol{P} = \mathbf{A}\boldsymbol{p} + \boldsymbol{d}, \tag{1.1}$$

$\mathbf{A}$ being an orthogonal matrix and $\boldsymbol{d}$ the displacement vector of the origin of E. In Chapter I (eq. (5.7)) we have shown that $\mathbf{A}$ can always be expressed as

$$\mathbf{A} = (\mathbf{I} - \mathbf{B})^{-1}(\mathbf{I} + \mathbf{B}), \tag{1.2}$$

$\mathbf{B}$ being a *skew* matrix. We apply this general theorem to the case $n = 3$. The skew matrix $\mathbf{B}$ is then

$$\mathbf{B} = \begin{Vmatrix} 0 & -b_3 & b_2 \\ b_3 & 0 & -b_1 \\ -b_2 & b_1 & 0 \end{Vmatrix}, \tag{1.3}$$

where b_i are three real numbers. From (1.3) it follows that

$$\mathbf{I}-\mathbf{B} = \begin{Vmatrix} 1 & b_3 & -b_2 \\ -b_3 & 1 & b_1 \\ b_2 & -b_1 & 1 \end{Vmatrix},$$

$$|\mathbf{I}-\mathbf{B}| = \Delta = 1 + b_1^2 + b_2^2 + b_3^2,$$

$$(\mathbf{I}-\mathbf{B})^{-1} = \Delta^{-1} \begin{Vmatrix} 1+b_1^2 & b_1b_2-b_3 & b_1b_3+b_2 \\ b_1b_2+b_3 & 1+b_2^2 & b_2b_3-b_1 \\ b_1b_3-b_2 & b_2b_3+b_1 & 1+b_3^2 \end{Vmatrix},$$

and

$$\mathbf{I}+\mathbf{B} = \begin{Vmatrix} 1 & -b_3 & b_2 \\ b_3 & 1 & -b_1 \\ -b_2 & b_1 & 1 \end{Vmatrix}.$$

Hence,

$$\mathbf{A} = (\mathbf{I}-\mathbf{B})^{-1}(\mathbf{I}+\mathbf{B})$$

$$= \Delta^{-1} \begin{Vmatrix} 1+b_1^2-b_2^2-b_3^2 & 2(b_1b_2-b_3) & 2(b_1b_3+b_2) \\ 2(b_2b_1+b_3) & 1-b_1^2+b_2^2-b_3^2 & 2(b_2b_3-b_1) \\ 2(b_3b_1-b_2) & 2(b_3b_2+b_1) & 1-b_1^2-b_2^2+b_3^2 \end{Vmatrix}, \tag{1.4}$$

which gives us $\mathbf{A}$ expressed explicitly in terms of the three parameters b_i (called the Rodrigues parameters).

Example 1. If $\mathbf{A} = \|a_{ik}\|$ is an orthogonal matrix and $\mathbf{I} = \|\delta_{ik}\|$, it follows that $\Sigma_k\, a_{ik}a_{jk} = \delta_{ij}$; show that (1.4) satisfies these relations.

Example 2. Referring to Chapter III, (12.12) show that $b_1 = s_X \tan\frac{1}{2}\phi$, $b_2 = s_Y \tan\frac{1}{2}\phi$, $b_3 = s_Z \tan\frac{1}{2}\phi$.

The characteristic equation for the matrix $\mathbf{A}$ is determined by evaluating $|\mathbf{A}-\lambda\mathbf{I}| = 0$, the result is*

* Instead of expanding the rather complex coefficients of the lower order terms it is sufficient to determine the coefficients of λ^3 and λ^2 (which are simple to obtain) and then make use of the fact that this equation must be reciprocal (Chapter I, Section 3).

$$-\lambda^3 + \Delta^{-1}(3 - b^2)\lambda^2 - \Delta^{-1}(3 - b^2)\lambda + 1 = 0 \tag{1.5}$$

where $b^2 = \Sigma b_i^2$ (in which case $\Delta = 1 + b^2$). This characteristic equation can be factored as follows

$$(\lambda - 1)\{\lambda^2 + 2((b^2 - 1)/(b^2 + 1))\lambda + 1\} = 0,$$

and so its three roots $\lambda_0, \lambda_1, \lambda_2$ are:

$$\lambda_0 = 1$$

$$\lambda_{1,2} = (1 - b^2 \pm 2ib)/(1 + b^2) = \cos\phi \pm i\sin\phi \tag{1.6}$$

(provided $\phi = 2\arctan(b)$).

The three roots correspond to the three lines through O which are invariant under the rotation; the real root λ_0 corresponds to the rotation axis l, the complex roots $\lambda_{1,2}$ to the isotropic lines in the plane perpendicular to l. As the eigenvector corresponding to λ_0 turns out to be (b_1, b_2, b_3), in view of Chapter I, Section 3 we have: *the direction of the rotation axis of* (1.4) *is* (b_1, b_2, b_3) *and the rotation angle* ϕ *satisfies* $\tan\frac{1}{2}\phi = b$. This gives us a geometrical interpretation of the parameters b_i.

Example 3. Show that the inverse rotation of (1.4) has the parameters $-b_i$.

The general displacement (1.1) depends on the three parameters b_i of $\mathbf{A}$ and the components d_i of $\boldsymbol{d}$. This implies that the unrestricted motion of E has *six* degrees of freedom.

We shall now determine the screw axis s of (1.1). Since the screw axis is parallel to the rotation axis of $\mathbf{A}$, if $\boldsymbol{x}$ is the position vector of a point on s its displaced position $\boldsymbol{X}$ is given by $\boldsymbol{X} = \boldsymbol{x} + \mu\boldsymbol{b}$, μ being a constant scalar. For a point on s equation (1.1) yields

$$(\mathbf{A} - \mathbf{I})\boldsymbol{x} = \mu\boldsymbol{b} - \boldsymbol{d}, \tag{1.7}$$

which is equivalent to three (linear) scalar equations for the coordinates x_i of $\boldsymbol{x}$. But $(\mathbf{A} - \mathbf{I})$ is a singular matrix and the equations have therefore ∞^1 solutions. As $(\mathbf{A} - \mathbf{I})\boldsymbol{b} = 0$ is a condition for which the equations are dependent, the components of the right-hand side of (1.7) in the direction of $\boldsymbol{b}$ must vanish. This is only the case if $\mu b^2 - \boldsymbol{b}\cdot\boldsymbol{d} = 0$, or $\mu = (\boldsymbol{b}\cdot\boldsymbol{d})/b^2$, which implies that the translation component of the screw motion is given by $((\boldsymbol{b}\cdot\boldsymbol{d})/b^2)\boldsymbol{b}$; hence its length is equal to the component of $\boldsymbol{d}$ in the direction $\boldsymbol{b}$. Of the three equations only two are independent. The first and the second are

$$\begin{aligned} &-2(b_2^2 + b_3^2)x_1 + 2(b_1b_2 - b_3)x_2 + 2(b_1b_3 + b_2)x_3 - (\mu b_1 - d_1)\Delta = 0, \\ &2(b_1b_2 + b_3)x_1 - 2(b_3^2 + b_1^2)x_2 + 2(b_2b_3 - b_1)x_3 - (\mu b_2 - d_2)\Delta = 0, \end{aligned} \tag{1.8}$$

which are the equations of two planes, the intersection of which is the screw axis; they enable us to compute its Plücker coordinates and therefore the Plücker vectors $\boldsymbol{S}$ and $\boldsymbol{S}'$ of s. After some algebra (including use of the condition $\mu b^2 = \boldsymbol{b} \cdot \boldsymbol{d}$) we obtain

$$\boldsymbol{S} = 2\boldsymbol{b}, \qquad \boldsymbol{S}' = \boldsymbol{b} \times \boldsymbol{d} - \boldsymbol{d} + \mu \boldsymbol{b}, \tag{1.9}$$

which together with the rotation angle ϕ, defined by $\tan \phi/2 = b$, and the translation distance $(\boldsymbol{b} \cdot \boldsymbol{d})/b$ gives a complete description of the screw displacement in terms of the six parameters b_i and d_i.

Example 4. Show that (1.9) satisfy the fundamental relation $\boldsymbol{S} \cdot \boldsymbol{S}' = 0$.
Example 5. Discuss (1.9) for the case $\boldsymbol{b} = 0$ and for $\boldsymbol{d} = 0$.

2. Study's soma

The matrix (1.4) obtains a more elegant form if we let $b_i = c_i/c_0$; $c_0 \neq 0$:

$$\Delta_1 \mathbf{A} = \begin{Vmatrix} c_0^2 + c_1^2 - c_2^2 - c_3^2 & 2(c_1c_2 - c_0c_3) & 2(c_1c_3 + c_0c_2) \\ 2(c_2c_1 + c_0c_3) & c_0^2 - c_1^2 + c_2^2 - c_3^2 & 2(c_2c_3 - c_0c_1) \\ 2(c_3c_1 - c_0c_2) & 2(c_3c_2 + c_0c_1) & c_0^2 - c_1^2 - c_2^2 + c_3^2 \end{Vmatrix} \tag{2.1}$$

where $\Delta_1 = c_0^2 + c_1^2 + c_2^2 + c_3^2$.

Thus we have $\mathbf{A}$ as a homogeneous function of four parameters c_0, c_1, c_2, c_3. For (1.6) we obtain $\lambda_{1,2} = [2c_0^2 - \Delta_1 \pm 2c_0(c_0^2 - \Delta_1)^{1/2}]/\Delta_1$. Formula (1.4) is only valid if no characteristic root is equal to -1. As we remarked (Chapter I, Section 5) an expression for $\mathbf{A}$ in this case was obtained by Cayley by means of a limit procedure. We see that $\lambda_{1,2} = -1$ if $c_0 = 0$, which has been excluded so far. Therefore we can also accept the possibility for c_0 to equal the number zero (which means that b_i tend to infinity, their ratios being constant). The conclusion is that (2.1) is a representation of an orthogonal matrix in all cases, provided that the c_i are not all zero. We can *normalize* the parameters such that $\Delta_1 = 1$, in which case we have that $c_0 = \cos \phi/2$; for $c_0 = 0$ the rotation is a half turn. For $c_0 = 1$, $c_1 = c_2 = c_3 = 0$ we have the unity transformation $\mathbf{I}$; the inverse rotation of (c_0, c_1, c_2, c_3) is $(-c_0, c_1, c_2, c_3)$. The numbers c_i are called the *Euler* parameters of the rotation. They are also useful in the representation of a displacement by means of *quaternions*, a subject to be treated in Chapter XIII.

The general displacement of E, given by (1.1), depends on six parameters; in other words, the "geometry", for which a position of E is an element, is six-dimensional. Such a position was called by STUDY [1903] a *soma*. As coordinates of a soma we could make use of the six numbers b_i, d_i but then we exclude the half turns. This exception disappears if we apply the four homogeneous parameters c_i and the three non-homogeneous d_i, but this is of course not a very attractive method. *Study* therefore introduced in soma space instead of d_i four homogeneous numbers g_i $(i = 0, 1, 2, 3)$ defined as follows

$$
\begin{aligned}
g_0 &= \qquad\qquad d_1c_1 + d_2c_2 + d_3c_3, \\
g_1 &= -d_1c_0 \qquad\quad + d_3c_2 - d_2c_3, \\
g_2 &= -d_2c_0 - d_3c_1 \qquad\quad + d_1c_3, \\
g_3 &= -d_3c_0 + d_2c_1 - d_1c_2 \qquad\quad ,
\end{aligned} \tag{2.2}
$$

and we represent a position of E by the *eight* homogeneous coordinates $(c_0, c_1, c_2, c_3;\ g_0, g_1, g_2, g_3)$. From (2.2) it follows that they are not independent; indeed one has the *fundamental relation*

$$c_0g_0 + c_1g_1 + c_2g_2 + c_3g_3 = 0. \tag{2.3}$$

If (c_i, g_i) are given satisfying (2.3) the position is uniquely determined: the rotational part of the displacement depends on the c_i only, and the d_i follow from (2.2) as:

$$
\begin{aligned}
\Delta_1 d_1 &= g_0c_1 - g_1c_0 + g_2c_3 - g_3c_2, \\
\Delta_1 d_2 &= g_0c_2 - g_2c_0 + g_3c_1 - g_1c_3, \\
\Delta_1 d_3 &= g_0c_3 - g_3c_0 + g_1c_2 - g_2c_1,
\end{aligned} \tag{2.4}
$$

(which is easy to check by direct substitution).

There exists an analogy between the soma coordinates $(c_i\,; g_i)$ $(i = 0, 1, 2, 3)$ and the Plücker coordinates (q_i, q'_i) $(i = 1, 2, 3)$ of a line in three-dimensional space. As a counterpart of the Plücker vectors $(\boldsymbol{q}\,; \boldsymbol{q}')$ of a line we could introduce two "Study vectors" (in four-dimensional space) $\boldsymbol{c}$ and $\boldsymbol{g}$ with components c_i and g_i respectively. The fundamental relations have the same form: $\boldsymbol{q} \cdot \boldsymbol{q}' = 0$ and $\boldsymbol{c} \cdot \boldsymbol{g} = 0$. We must be aware, however, that the four numbers c_i can not all be zero. In the corresponding case $q_1 = q_2 = q_3 = 0$ we have a line in the plane at infinity. Study has introduced singular (or improper) soma's to make the analogy complete.

We now derive some implications of the representation of a displacement by the pair of vectors $(\boldsymbol{c}\,; \boldsymbol{g})$. The formula (1.1) gives the *identity* displacement if $\mathbf{A} = \mathbf{I}$, $\boldsymbol{d} = 0$; hence its coordinates are: $c_0 = 1$, $c_i = 0$ $(i = 1, 2, 3)$, $g_i = 0$ $(i = 1, 2, 3, 4)$.

If the displacement is a *translation* we have $\mathbf{A} = \mathbf{I}$, that is $c_1 = c_2 = c_3 = 0$, from which it follows, by virtue of (2.2), that $g_0 = 0$ and from (2.4) (if $c_0 = 1$) that the translation vector reads $\boldsymbol{d} = -(g_1, g_2, g_3)$.

If the displacement is a *rotation* it follows that vector $\boldsymbol{d}$ is orthogonal to the direction of the axis of rotation (c_1, c_2, c_3). Hence, in view of (2.2), for a rotation $g_0 = 0$.

The inverse displacement $\tilde{D}$ of the displacement D given by (1.1) has the equation

$$\boldsymbol{p} = \mathbf{A}^{-1}(\boldsymbol{P} - \boldsymbol{d}). \tag{2.5}$$

Keeping in mind that $\mathbf{A}^{-1} = \mathbf{A}^{\mathrm{T}}$ we obtain $\boldsymbol{p} = \tilde{\mathbf{A}}\boldsymbol{P} + \tilde{\boldsymbol{d}}$ if we let

$$\tilde{\mathbf{A}} = \mathbf{A}^{\mathrm{T}}, \qquad \tilde{\boldsymbol{d}} = -\mathbf{A}^{\mathrm{T}}\boldsymbol{d}. \tag{2.6}$$

From (1.5) it follows that $\tilde{c}_0 = -c_0$, $\tilde{c}_i = c_i$ $(i = 1, 2, 3)$. Furthermore we find after some algebra, by means of (2.1) and (2.4):

$$\begin{aligned} \Delta_1\tilde{d}_1 &= -g_0c_1 + g_1c_0 + g_2c_3 - g_3c_2 \\ \Delta_1\tilde{d}_2 &= -g_0c_2 + g_2c_0 + g_3c_1 - g_1c_3 \\ \Delta_1\tilde{d}_3 &= -g_0c_3 + g_3c_0 + g_1c_2 - g_2c_1 \end{aligned} \tag{2.7}$$

and then, comparing with (2.4):

$$\tilde{g}_0 = -g_0, \quad \tilde{g}_i = g_i \quad (i = 1, 2, 3). \tag{2.8}$$

Hence the inverse displacement of $(c_0, c_1, c_2, c_3; g_0, g_1, g_2, g_3)$ reads $(-c_0, c_1, c_2, c_3; -g_0, g_1, g_2, g_3)$, which may give an idea of the elegance of this concept.

Several other questions follow naturally: what is the screw axis and pitch of a displacement?: what are the Study vectors of the product of two displacements?; and so on. It may be shown, however, that this new apparatus can more efficiently be handled by other algebraic means, *viz.* quaternions and dual numbers. We shall do so in Chapter XIII. One such result (given without proof) is: In general two displacements have no point in common*, if they have there is a common line. The condition for the displacements $(\boldsymbol{c}; \boldsymbol{g})$ and $(\boldsymbol{c}'; \boldsymbol{g}')$ to have a common point (and line) is:

$$\boldsymbol{c} \cdot \boldsymbol{g}' + \boldsymbol{c}' \cdot \boldsymbol{g} = 0, \tag{2.9}$$

which is a complete analog of the condition for the intersection of two lines given by their Plücker vectors.

* By two displacements having a point in common we mean that $\boldsymbol{p}' = \boldsymbol{p}''$ when $(\boldsymbol{c}, \boldsymbol{q})$ brings $\boldsymbol{p} \to \boldsymbol{p}'$ and $(\boldsymbol{c}', \boldsymbol{q}')$ brings $\boldsymbol{p} \to \boldsymbol{p}''$.

3. Eulerian angles

In Chapter I we have, by Cayley's method, derived a general expression for an $n \times n$-orthogonal matrix. In Section 1 we have applied this to $n = 3$ and obtained the formula (1.4) with three (or four homogeneous) parameters.

In Chapter III (eq. (12.12)) an equivalent form was derived in terms of the screw parameters. Another, more simple but less symmetric formula has been given by Euler in terms of three angles named after him. As shown in Fig. 31, two Cartesian frames O_{XYZ} and O_{xyz} with the same origin are given. The plane O_{XY} and O_{xy} intersect in a directed line d (called the line of nodes). We denote the angle θ between O_Z and O_z, and the angle ξ between O_X and d; let ψ be the angle between d and O_x. The positive sense of each angle is

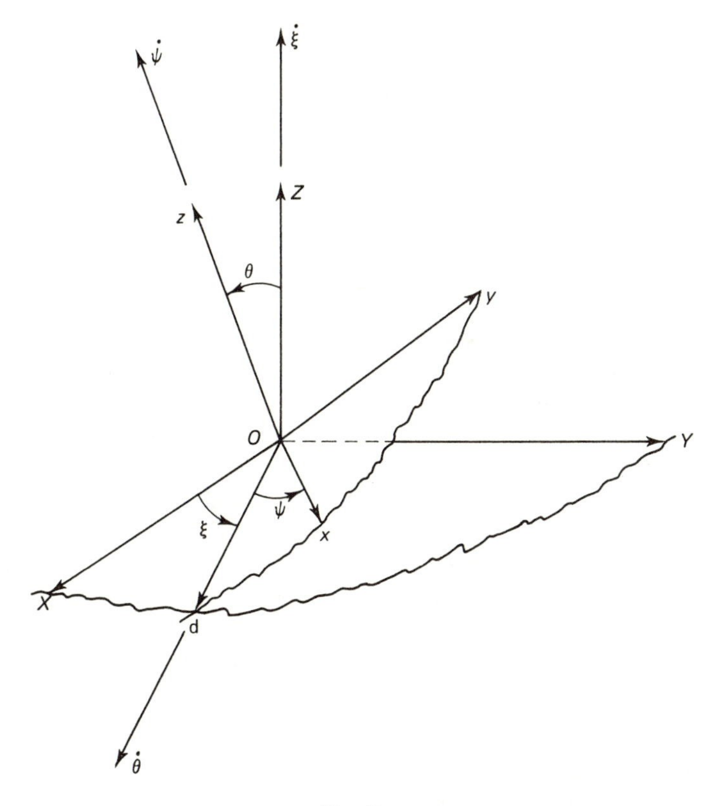

Fig. 31.

according to the right-hand-rule.* θ is measured from O_Z to O_z with d as axis. ξ from O_X to d about the Z axis, and ψ from d to O_x with O_z as axis. If we take $0 \leq \xi < \pi$ the system is uniquely defined.

The three angles θ, ξ, ψ determine the position of O_{xyz} with respect to O_{XYZ}. If we consider unit vectors along O_x, O_y, O_z and calculate their components along O_X, O_Y, O_Z by elementary geometry, we obtain the transformation matrix which gives the coordinates X, Y, Z of any point as linear homogeneous functions of its coordinates x, y, z:

$$(3.1)\quad \mathbf{A} = \left\| \begin{matrix} \cos\psi\cos\xi & -\sin\psi\cos\xi & \sin\xi\sin\theta \\ \quad -\sin\psi\sin\xi\cos\theta & \quad -\cos\psi\sin\xi\cos\theta & \\ \cos\psi\sin\xi & -\sin\psi\sin\xi & -\cos\xi\sin\theta \\ \quad +\sin\psi\cos\xi\cos\theta & \quad +\cos\psi\cos\xi\cos\theta & \\ \sin\psi\sin\theta & \cos\psi\sin\theta & \cos\theta \end{matrix} \right\|$$

It may be worth mentioning that EULER [1770], more interested in numbers than in geometrical reasoning, derived his formula in a purely algebraic way: If $\mathbf{A} = \| a_{ik} \|$, in a formal way he set $a_{33} = \cos\theta$; in view of $a_{31}^2 + a_{32}^2 = a_{13}^2 + a_{23}^2 = 1 - \cos^2\theta$, he continued by taking $a_{31} = \sin\theta\sin\alpha$, $a_{32} = \sin\theta\cos\alpha$, $a_{13} = \sin\theta\sin\beta$, and $a_{23} = \sin\theta\cos\beta$. To determine the elements a_{11}, a_{12}, a_{21} and a_{22} he made use of the property that each is equal to its minor, which furnishes four linear equations for the unknowns.

Example 6. Given the matrix (2.1) determine the Eulerian angles.

We have until now determined the angular position of a moving space in several ways. We have the numbers b_i or c_i, or the screw parameters, or the Eulerian angles. All of these concepts seem complicated in contrast to the simple way the velocity distribution of E can be described by a single vector $\boldsymbol{\Omega}$, with its three components $\Omega_X, \Omega_Y, \Omega_Z$. In a natural way the question arises whether it is possible to obtain a given position of E from a zero position by rotating it successively about O_X, O_Y and O_Z with finite angles. Unfortunately this is not an acceptable method because *whenever the axes are embedded in the fixed system* the composition of two finite rotations is complicated and not even commutative. We shall prove that the introduction of three rotation angles $\alpha_X, \alpha_Y, \alpha_Z$ about O_X, O_Y, O_Z such that $\dot{\alpha}_X = \Omega_X$, $\dot{\alpha}_Y = \Omega_Y$, $\dot{\alpha}_Z = \Omega_Z$ is fundamentally impossible. We make use of the Eulerian angles θ, ξ, ψ. In O_{XYZ}

* We use the angle ξ instead of the more usual ϕ in order to avoid confusion with the screw rotation angle ϕ. For a discussion of alternative definitions of Euler angles see GOLDSTEIN [1950, pg. 108].

the angular velocity $\dot\theta$ is represented by the vector $(\dot\theta\cos\xi, \dot\theta\sin\xi, 0)$, $\dot\xi$ by $(0,0,\dot\xi)$, $\dot\psi$ by $(\dot\psi\sin\theta\sin\xi, -\dot\psi\sin\theta\cos\xi, \dot\psi\cos\theta)$, as follows from Fig. 31.

Hence the following relations between the differentials should hold (the parameter τ being irrelevant):

$$\begin{aligned} \mathrm{d}\alpha_X &= \cos\xi\,\mathrm{d}\theta & &+ \sin\theta\sin\xi\,\mathrm{d}\psi, \\ \mathrm{d}\alpha_Y &= \sin\xi\,\mathrm{d}\theta & &- \sin\theta\cos\xi\,\mathrm{d}\psi, \\ \mathrm{d}\alpha_Z &= & \mathrm{d}\xi &+ \cos\theta\,\mathrm{d}\psi, \end{aligned} \tag{3.2}$$

and the question is whether this sytem can be integrated so that $\alpha_X, \alpha_Y, \alpha_Z$ are determined as functions of θ, ξ, ψ. From the third equation it follows that $\partial\alpha_Z/\partial\theta = 0$, $\partial\alpha_Z/\partial\psi = \cos\theta$, but in view of $\partial^2\alpha_Z/\partial\theta\,\partial\psi = \partial^2\alpha_Z/\partial\psi\,\partial\theta$ this leads to a contradiction. Hence (3.2) is a non-integrable, or using the usual term for similar situations in dynamics: (3.2) represents a non-holonomic system. It is impossible from (3.2) to express $\alpha_X, \alpha_Y, \alpha_Z$ in terms of the Eulerian angles. The differentials $\Omega_X\,\mathrm{d}t$, $\Omega_Y\,\mathrm{d}t$, $\Omega_Z\,\mathrm{d}t$ have been called non-holonomic velocity parameters (see for example HAMEL [1912]). Dividing (3.2) by $\mathrm{d}t$ gives the components of the angular velocity vector in the O_{XYZ} system.

There is occasionally some confusion on this point, especially since (3.1) is often obtained by three successive rotations, but in this case the axes are O_z, d, and O_Z and not the fixed system O_X, O_Y, O_Z.

Example 7. Show that if

$$\mathbf{A}_1 = \left\|\begin{matrix} \cos\xi & -\sin\xi & 0 \\ \sin\xi & \cos\xi & 0 \\ 0 & 0 & 1 \end{matrix}\right\|, \quad \mathbf{A}_2 = \left\|\begin{matrix} 1 & 0 & 0 \\ 0 & \cos\theta & -\sin\theta \\ 0 & \sin\theta & \cos\theta \end{matrix}\right\|, \quad \mathbf{A}_3 = \left\|\begin{matrix} \cos\psi & -\sin\psi & 0 \\ \sin\psi & \cos\psi & 0 \\ 0 & 0 & 1 \end{matrix}\right\|$$

then (3.1) is given by $\mathbf{A} = \mathbf{A}_1\mathbf{A}_2\mathbf{A}_3$.

Example 8. Show by using (8.7) of Chapter I that if the displacement of E is a rotation $\mathbf{R}_\xi$ about O_Z, followed by rotation $\mathbf{R}_\theta$ about the line of nodes d, and then a rotation $\mathbf{R}_\psi$ about O_z, then (using the notation of Example 7) $\mathbf{R}_\xi = \mathbf{A}_1$, $\mathbf{R}_\theta = \mathbf{A}_1\mathbf{A}_2\mathbf{A}_1^{-1}$, $\mathbf{R}_\psi = (\mathbf{A}_1\mathbf{A}_2)\mathbf{A}_3(\mathbf{A}_1\mathbf{A}_2)^{-1}$ and therefore $\mathbf{A} = \mathbf{R}_\psi\mathbf{R}_\theta\mathbf{R}_\xi = \mathbf{A}_1\mathbf{A}_2\mathbf{A}_3$. Alternatively, show if we rotate about axes fixed in Σ the order is: ψ about O_Z, θ about O_X, ξ about O_Z.

Example 9. Show that by introducing two new systems E', E'' a model of the foregoing could be constructed if system E is hinged to E' along axis O_z, E' is hinged to E'' along axis d, and E'' is hinged to Σ along O_Z. If rotation axes O_z, d, and O_Z are concurrent at O the model completely describes the displacement of E relative to Σ as given by (3.1). Using this model and the method of Example 8 prove that the three rotations done in any order yield exactly the same final position of E. Hence such finite rotations do commute.

Example 10. Show that the angular velocity vector in O_{xyz} has the components

$$\begin{aligned} \Omega_x &= \dot\xi\sin\theta\sin\psi + \dot\theta\cos\psi \\ \Omega_y &= \dot\xi\sin\theta\cos\psi - \dot\theta\sin\psi \\ \Omega_z &= \dot\xi\cos\theta + \dot\psi. \end{aligned}$$

We will return to the question of determining suitable analytical descriptions for the displacements in Chapter XIII where we will list additional methods. Of the methods developed in Sections 1–3, the results of Section 1 will be used in Chapter X, and the others in Chapter XIII. It should, however, be pointed out that the Eulerian angles have to date had relatively little application in kinematics other than in areas associated with rigid body dynamics.

4. One-parameter-motions

A displacement of the space E with respect to Σ is given by

$$\boldsymbol{P} = \mathbf{A}\boldsymbol{p} + \boldsymbol{d} = T(\boldsymbol{p}), \tag{4.1}$$

T being a linear, non-homogeneous function, containing six independent parameters. If these are functions of m variables $(m = 1, 2, \ldots 5)$ E has a motion with m degrees of freedom. For the time being we restrict ouselves to the case $m = 1$; cases $m > 1$ are considered in Chapter X. The variable will be denoted by τ; differentiation with respect to τ will be represented by a dot.

If we make use of Section 1 any motion of E can be described by giving the six parameters b_i, d_i as functions of τ; the same holds for Section 2 if the Study vectors (satisfying $\boldsymbol{c} \cdot \boldsymbol{g} = 0$) are known as functions of τ. The six functions contain the complete information about the motion and in principle they give sufficient information to enable one to answer any question concerning the motion. So, for instance, if we fix τ, (4.1) gives the position of any point of E at that moment. If we choose $\boldsymbol{p}$ we obtain, with τ as parameter, the path of P; if moreover $\boldsymbol{p}$ varies we get the set of ∞^3 paths. In a similar way we may study the ruled surface generated by a line of E (and the system of ∞^4 surfaces generated by the set of all lines), the system of ∞^1 planes (the developable or torse) which is the locus of the positions of a plane of E during the motion, or, to give another example, the (cyclic) surface generated by a circle of E.

It must be said that general theorems for these kinds of "global" problems or "problems in the large" are almost nonexistent, probably because they are either trivial or too complicated. Anyhow, theoretical kinematics did not develop in this direction, but restricted itself mainly to two features of the general properties of a motion. In one of them, one does not deal with six arbitrary functions, but one considers a specified motion (or a class of such motions), defined either analytically or, more often, geometrically. All studies of this type deal with examples of special motions, which for some reason are interesting, usually because of their technical applicability. They will be treated in Chapter IX, on special motions.

The other line of research has a different view point. If the motion is given we know the loci generated by a geometrical element of E such as a point, a line, a curve, and a plane. Moreover, we know the differential-geometrical properties of these loci. So for example, for the path of a point we know its tangents, osculating planes, its curvature, and its torsion. It will be seen that there exist general theorems of these local characteristics (covariant configurations and associated numbers), which as a rule state certain relationships between local properties of the whole set of moving points for a fixed value of τ. These theorems deal with instantaneous kinematics and their study, which will be the contents of the next sections, therefore makes use of the developments given in Chapter II.

5. The axodes

In Chapter II we have derived the following concepts for instantaneous kinematics. The angular velocity (Chapter II, (3.6)):

$$\mathbf{\Omega} = \dot{\mathbf{A}}\mathbf{A}^{-1}; \tag{5.1}$$

the pitch (Chapter II, (4.11)):

$$\sigma = (\boldsymbol{\Omega} \cdot \dot{\boldsymbol{d}})\Omega^{-2}; \tag{5.2}$$

and the Plücker vectors of the instantaneous screw axis (Chapter II, (4.16)):

$$\boldsymbol{Q} = \boldsymbol{\Omega}, \qquad \boldsymbol{Q}' = \dot{\boldsymbol{d}} - \boldsymbol{\Omega} \times \boldsymbol{d} - \sigma\boldsymbol{\Omega}. \tag{5.3}$$

In order to express these quantities in terms of the parameters introduced in this chapter we take $\mathbf{A}$ as given by (1.4), calculate $\dot{\mathbf{A}}$, and keeping in mind that $\mathbf{A}^{-1} = \mathbf{A}^{\mathrm{T}}$ we obtain, after some algebra,

$$\mathbf{\Omega} = 2\Delta^{-1} \left\| \begin{matrix} 0 & -b_1\dot{b}_2 + b_2\dot{b}_1 - \dot{b}_3 & -b_1\dot{b}_3 + b_3\dot{b}_1 + \dot{b}_2 \\ b_1\dot{b}_2 - b_2\dot{b}_1 + \dot{b}_3 & 0 & -b_2\dot{b}_3 + b_3\dot{b}_2 - \dot{b}_1 \\ b_1\dot{b}_3 - b_3\dot{b}_1 - \dot{b}_2 & b_2\dot{b}_3 - b_3\dot{b}_2 + \dot{b}_1 & 0 \end{matrix} \right\|$$

$$= 2\Delta^{-1}(\mathbf{B}\dot{\mathbf{B}} - \dot{\mathbf{B}}\mathbf{B} + \dot{\mathbf{B}}) \tag{5.4}$$

or, if we convert to vectors

$$\boldsymbol{\Omega} = 2\Delta^{-1}(\boldsymbol{b} \times \dot{\boldsymbol{b}} + \dot{\boldsymbol{b}}). \tag{5.5}$$

Furthermore from (5.2) it follows

$$\sigma = \tfrac{1}{2}\Delta((\boldsymbol{b} \times \dot{\boldsymbol{b}} + \dot{\boldsymbol{b}}) \cdot \dot{\boldsymbol{d}})/(b^2\dot{b}^2 - (\boldsymbol{b} \cdot \dot{\boldsymbol{b}})^2), \tag{5.6}$$

and from (5.3)

$$\begin{aligned} \boldsymbol{Q} &= 2\Delta^{-1}(\boldsymbol{b} \times \dot{\boldsymbol{b}} + \dot{\boldsymbol{b}}), \\ \boldsymbol{Q}' &= \dot{\boldsymbol{d}} - 2\Delta^{-1}(\boldsymbol{b} \times \dot{\boldsymbol{b}} + \dot{\boldsymbol{b}}) \times \boldsymbol{d} \\ &\quad - ((\boldsymbol{b} \times \dot{\boldsymbol{b}} + \dot{\boldsymbol{b}}) \cdot \dot{\boldsymbol{d}}(\boldsymbol{b} \times \dot{\boldsymbol{b}} + \dot{\boldsymbol{b}}))/(b^2\dot{b}^2 - (\boldsymbol{b} \cdot \dot{\boldsymbol{b}})^2). \end{aligned} \tag{5.7}$$

These formulas could also have been derived from (1.9) by a limit procedure.

If we transform the parameter by substituting $\tau = f(\tau_1)$, all first derivatives are (at the same instant) multiplied by the same factor. From this it follows that the scalar σ and the screw axis s are geometrical invariants of the motion (while $\boldsymbol{\Omega}$ is not). If τ varies, $\sigma(\tau)$ is a variable function during the motion, and s(τ) represents a ruled surface S′ in the fixed space Σ, which will be called the (fixed) axode of the motion. As a counterpart there exists a "moving" axode which is the locus, S, of s considered as a line in the moving space *E*. *The fixed axode of the continuous motion* $(\boldsymbol{b}(\tau), \boldsymbol{d}(\tau))$ *is given by* (5.7), *which expresses the Plücker coordinates of its generators as functions of the parameter* τ. A similar representation of the moving axode may be found by considering the inverse motion.

$\sigma(\tau)$ and s(τ) are instantaneous invariants of the first order. At any instant the velocity of a moving point is the same as that of a screw motion about s, with angular velocity Ω and translation component $\sigma\Omega$.

At any instant the moving axode S has the generator s in common with S′. To further investigate the relative position of S with respect to S′, instantaneously, we need instantaneous invariants of higher order, as developed in Chapter II, Section 6. As we are restricting ourselves here to local properties, we make use of the special coordinate system introduced in Chapter II, Section 5 to study the geometric properties of the canonical system. For $\tau = 0$, the position under consideration, the two Cartesian frames coincide, hence $\boldsymbol{b} = 0$ and $\boldsymbol{d} = 0$, $\Delta = 1$. It follows from (5.5) that $\boldsymbol{\Omega}_0 = 2\dot{\boldsymbol{b}}$. Furthermore, from (5.5) and in view of $\dot{\Delta} = 2\boldsymbol{b} \cdot \dot{\boldsymbol{b}}$

$$\dot{\boldsymbol{\Omega}} = 2\Delta^{-2}\{\Delta(\boldsymbol{b} \times \ddot{\boldsymbol{b}} + \ddot{\boldsymbol{b}}) - 2(\boldsymbol{b} \times \dot{\boldsymbol{b}} + \dot{\boldsymbol{b}})(\boldsymbol{b} \cdot \dot{\boldsymbol{b}})\},$$

and therefore (since $\boldsymbol{b} = 0$ and $\Delta = 1$)

$$\boldsymbol{\Omega}_1 = 2\ddot{\boldsymbol{b}}. \tag{5.8}$$

Differentiating $\dot{\boldsymbol{\Omega}}$ and putting $\tau = 0$ we obtain

(5.9) $$\boldsymbol{\Omega}_2 = 2(\dddot{\boldsymbol{b}} + \dot{\boldsymbol{b}} \times \ddot{\boldsymbol{b}} - 2\dot{b}^2\dot{\boldsymbol{b}}).$$

Comparing all this with (5.8) of Chapter II we have

(5.10) $$\boldsymbol{\omega} = 2\dot{\boldsymbol{b}}, \qquad \boldsymbol{\varepsilon} = 2\ddot{\boldsymbol{b}}, \qquad \boldsymbol{\gamma} = 2(\dddot{\boldsymbol{b}} - 4\dot{b}^2\dot{\boldsymbol{b}}),$$

so that the instantaneous invariants $\boldsymbol{\omega}, \boldsymbol{\varepsilon}, \boldsymbol{\gamma}$ are determined as functions of the motion parameters b_i. If we substitute $\boldsymbol{\Omega}_0 = 2\dot{\boldsymbol{b}}$ into $\boldsymbol{d}_1 = \sigma_0\boldsymbol{\Omega}_0$, and also (5.8) and (5.9) into (5.1) of Chapter II, we find the pitch σ_0 and the Plücker vectors $(\boldsymbol{Q}, \boldsymbol{Q}')$ of the screw axis s (and their derivatives) as functions of the parameters (b_i, d_i), of the motion, and their derivatives.

All these formulas simplify considerably if we normalize the parameter and make use of the canonical coordinate frame introduced in Chapter II, Section 6.

Example 11. Show that at $\tau = 0$ a canonical frame yields $\boldsymbol{b} = 0$, $\boldsymbol{d} = 0$, $\dot{\boldsymbol{b}} = (0, 0, \frac{1}{2})$, $\dot{\boldsymbol{d}} = (0, 0, \sigma_0)$, $\ddot{\boldsymbol{b}} = (0, \frac{1}{2}\varepsilon, 0)$, $\ddot{\boldsymbol{d}} = (0, \mu\varepsilon, \lambda)$.

Only the first order instantaneous properties of a motion are represented by the screw axis and the pitch σ_0. It follows from this that first order geometrical properties of the axodes depend on the invariants of the motion up to the second order, that is (using the nomenclature of Chapter II, Section 6) on $\sigma_0, \varepsilon, \lambda, \mu$, and that, in general, m-th order properties of S′ and S are functions of the motion invariants up to the $(m + 1)$-th order. The first order properties of s, taken as a generator of the axode S′, are the tangent plane of S′ at each point of s and the point on s which is (in the limit) the foot of the common perpendicular between s and its neighboring generator on S′; in the geometry of ruled surface this latter point is called the striction point of s, we have taken it as the origin O of our canonic frame.

Now we have seen in Chapter II, Section 5 that the second order invariants of a motion are essentially the same as those of the inverse motion (the difference in sign arises from the different orientation of the respective canonic frames). This implies that first order properties of s taken as either a generator of S′ or of S are identical, which means that the common generator s has the same striction point on S′ and S, and moreover that the tangent planes to S′ and to S, at any point on s, coincide. The axodes S′ and S are tangent to one another at every point of their common generator. The situation where two ruled surfaces are so related with respect to one another is expressed by saying that they are *raccording* along s.

It follows from Chapter II, Sections 5 and 6 that the Plücker vectors, up to the first order, of the screw axis are

$$(5.11) \qquad \boldsymbol{Q} = \{0, \varepsilon\tau, 1\}, \qquad \boldsymbol{Q}' = \{0, (\mu - \sigma_0)\varepsilon\tau, 0\},$$

hence its direction is $\{0, \varepsilon\tau, 1\}$ and its intersection with the plane O_{XY} reads $\{p_{31}, p_{32}, 0, p_{34}\}$ or $\{(\sigma_0 - \mu)\varepsilon\tau, 0, 0\}$. Therefore if we characterize a point on s by its distance u along s to the plane O_{XY}, the axodes S′ and S (in the first order neighborhood of s in terms of the parameters τ and u) are represented by

$$(5.12) \qquad X = (\sigma_0 - \mu)\varepsilon\tau, \qquad Y = \varepsilon\tau u, \qquad Z = u, \qquad W = 1.$$

Hence, up to this order they coincide with the hyperbolic paraboloid

$$(5.13) \qquad (\sigma_0 - \mu)YW - XZ = 0,$$

their raccording quadric. The invariant $\delta = (\mu - \sigma_0)$ which appears in this equation is a well-known quantity in the theory of ruled surfaces. It follows immediately from (5.11) that it is the limit of the ratio of the distance and angle between two consecutive generators; it is denoted by the (not very fortunately chosen) term: distribution parameter (German: Drall). Obviously the common generator has the same δ on S′ and S.

The tangent plane to (5.13), and therefore to S′ and S, at the point $(0, 0, z_0, 1)$ has the equation

$$(5.14) \qquad \delta Y + z_0 X = 0;$$

hence δ determines the relationship between a point on s and the corresponding tangent plane through s; this could be taken as a definition of the number δ. The tangent plane of the axodes at the striction point ($z_0 = 0$) is $Y = 0$; the common asymptotic tangent plane is $X = 0$. This gives us a new interpretation of the canonical frame. It is a well known property of any ruled surface that the tangent plane at the striction point is orthogonal to the asymptotic tangent plane. This may be used to define the striction point as that point along the ruling at which the tangent plane is orthogonal to the asymptotic tangent plane for the same ruling.

Example 12. Show that the raccording quadric (5.13) is an orthogonal hyperbolic paraboloid.
Example 13. Show that the normal to S′ and S at the striction point coincides with the axis O_Y of the canonical frame.

Summing up, the position of S with respect to S′ is such that they have a common generator s, the striction points of each surface on s coincide at O, their tangent planes at O coincide. This determines the position of S and therefore of E with respect to Σ. Since the δ's of s are the same on both surfaces the axodes are in raccordance in view of (5.14).

This situation yields, at any instant, the following geometric description for the motion of E with respect to Σ: In the spaces E and Σ the ruled surfaces S and S′, respectively, have a (1, 1) correspondence between their generators s and s′. The surfaces are not arbitrary: for two corresponding generators the number δ must be the same. The motion of E is generated by the following motion of S: at any instant a pair of corresponding generators coincide and the two surfaces are in raccordance along this common line (which determines the position of E completely). The motion is built up of a series of infinitesimal screw displacements with successive common generators as axes. This yields a special combination of rolling and sliding of the two ruled surfaces, which is called "schroten" in German (Dutch: schrooien; French: virer, viration), whereby at any instant the rolling and sliding are taking place about a common axis. The locus of the striction points on successive generators of S′ is called its striction curve k′; in the same way we have a striction curve k on S. Obviously at any instant the curves k′ and k intersect at the instantaneous striction point. The curves k and k′ depend on the second order instantaneous invariants of the motion ($\sigma_0, \varepsilon, \lambda, \mu$). Hence their first order differential-geometric properties, such as their tangents and their arc-elements, depend on the third order invariants, that is on the set ($\sigma_0, \varepsilon, \lambda, \mu, \gamma_X, \gamma_Y, d_{3_X}, d_{3_Y}, d_{3_Z}$), introduced in Chapter II, Section 6. We know that these are in general not the same as for the inverse motion. A consequence is that, for instance, k and k′ are in general not tangent at their common point.

The representation of a motion by successive infinitesimal screw displacements of surfaces is theoretically interesting, but in the general case does not seem to lead to easily manipulatable results. Some examples of axodes will be given in Chapter IX: in the main, however, applications are restricted to such special cases as spherical and planar kinematics (Chapters VII and VIII). The relationships between the local properties of the axodes (and their striction curves) and the higher order instantaneous invariants do not seem to have been developed.

There are of course many special cases. An important one appears when $\sigma_0 = 0$: the instantaneous motion is then a rotation which means (using an equivalent terminology) the motion is, at the instant, tangent to a rotation. If, moreover, $\sigma_1 = 0$, which implies $\lambda = 0$, the motion osculates a rotation. It is possible that during a continuous motion σ is permanently zero; this special motion will be treated in Chapter IX. Another special case is $\delta = \mu - \sigma_0 = 0$; this implies that, in view of (5.14), all tangent planes along s coincide with the plane $X = 0$, which means that S′ (and S) are locally developable. If $\delta = 0$ for every generator, the axodes S′ and S are both developable surfaces and their cuspidal curves (the tangents of which are the generators) coincide with the striction curves k′ and k.

If at every instant the screw axis in Σ and the angular velocity Ω and the pitch σ are given, the motion of E is implicitly determined if we know its initial position. It may in principle be found by a procedure of integration; a method for doing this was given by DARBOUX [1887]. He showed, however, that the analytical treatment leads to a differential equation of the Riccati type which implies that the solution cannot be given (except in special cases) by quadratures.

6. The point-paths

In this section we investigate the differential geometrical properties of the paths of the moving points. If, for the canonic frame, (x, y, z) and (X, Y, Z) are the coordinates, of a point, in E and Σ respectively, we have according to Chapter II, (6.3), up to the second order:

$$\begin{aligned} X &= x, & X_1 &= -y, & X_2 &= -x + \varepsilon z, \\ Y &= y, & Y_1 &= x, & Y_2 &= -y + \mu\varepsilon, \\ Z &= z, & Z_1 &= \sigma_0, & Z_2 &= -\varepsilon x + \lambda. \end{aligned} \tag{6.1}$$

The *tangent* to the path is a first order concept. Making use of homogeneous coordinates, its Plücker coordinates are the minors of the matrix

$$\begin{Vmatrix} x & y & z & w \\ -y & x & \sigma_0 w & 0 \end{Vmatrix}, \tag{6.2}$$

and are seen to be

$$\begin{aligned} &p_{14} = yw, \quad p_{24} = -xw, \quad p_{34} = -\sigma_0 w^2, \\ &p_{23} = \sigma_0 yw - xz, \quad p_{31} = -\sigma_0 xw - yz, \quad p_{12} = x^2 + y^2. \end{aligned} \tag{6.3}$$

They are quadratic functions of the point coordinates.

From (6.3) it follows that the instantaneous tangents satisfy the equation

$$p_{12}p_{34} + \sigma_0(p_{14}^2 + p_{24}^2) = 0, \tag{6.4}$$

which represents the same tetrahedral complex as in Chapter III, (11.1). If p_{ij} are the coordinates of a line of the complex of instantaneous tangents, i.e., if they satisfy (6.4), we may solve for the point (x, y, z, w) from (6.3) obtaining

(6.5) $$x : y : z : w = \sigma_0 p_{14} p_{24} : -\sigma_0 p_{14}^2 : p_{34}(p_{13} + \sigma_0 p_{24}) : p_{14} p_{34}$$

or, eliminating σ_0

(6.6) $$x : y : z : w = -p_{12} p_{24} : p_{12} p_{14} : (p_{13} p_{14} + p_{23} p_{24}) : (p_{14}^2 + p_{24}^2).$$

Hence any line of the complex is in general the tangent of only one of its points. There exists a birational relationship between the points of E and their instantaneous tangents; (6.3) and (6.6) are both quadratic.

Example 14. This relationship has some singularities. Show from (6.3) that the only point for which the tangent is undetermined is $(0, 0, 1, 0)$, i.e., the point Z at infinity of the screw axis. Show that, although (6.5) and (6.6) are meaningless, in this case the point corresponding to a line at infinity ($p_{14} = p_{24} = p_{34} = 0$), which is always a line of the complex, is $(-p_{12}p_{23}, p_{12}p_{13}, p_{13}^2 + p_{23}^2, 0)$; if the line passes through $Z(p_{12} = 0)$ it corresponds to Z unless $p_{13}^2 + p_{23}^2 = 0$ when it is undetermined.

Any line of the complex (6.4) corresponds in general to one point on it. Through an arbitrary point P_0 pass ∞^1 lines of the complex, they are the generators g of a quadratic cone. One of the generators, g_0 say, corresponds to P_0 itself. We now determine the locus of the points P corresponding to the generators g. An arbitrary plane through P_0 contains two generators, to each there corresponds a point; hence the plane contains three points of the locus, P_0 being one of them. The conclusion is: the locus of the points corresponding to the generators of a cone of the complex is a twisted cubic k_3, passing through its vertex. This curve can more simply be defined as the locus of the points with path-tangents which all pass through a given point of Σ.

Example 15. Show that g_0 is the tangent of k_3 at P_0.
Example 16. Show that the tangent plane of the cone (of the complex) through g_0 is the osculating plane of k_3 at P_0.

Analogously we may ask for the locus of the points corresponding to the lines of the complex lying in a given plane Λ. These lines are the tangents of a conic k and as each of them has *one* corresponding point on it we may expect the locus to be a straight line. This is easily verified analytically. The locus consists of the points for which the point and the path-tangent (or, in other words: the velocity vector) lies in Λ; they satisfy the two linear equations $\Lambda_1 x + \Lambda_2 y + \Lambda_3 z + \Lambda_4 = 0$ and $-\Lambda_1 y + \Lambda_2 x + \Lambda_3 \sigma_0 = 0$. Hence the locus asked for is a line m, corresponding to the plane Λ; we shall meet it again in Section 7. For the line-coordinates of m we obtain $p_{12} = \sigma_0 \Lambda_3^2$, $p_{34} = -(\Lambda_1^2 + \Lambda_2^2)$, $p_{14} = \Lambda_1 \Lambda_3$, $p_{24} = \Lambda_2 \Lambda_3$ and as they satisfy (6.4) the line m itself belongs to the complex and is therefore a tangent of the conic k. The tangent point is the corresponding point of m.

The normal plane at the point (x, y, z, w) of a path is normal to its tangent;

the latter has the direction $X : Y : Z = -y : x : \sigma_0 w$. Hence the equation of the normal plane is $N_1X + N_2Y + N_3Z + N_4W = 0$, with

$$N_1 = -y, \qquad N_2 = x, \qquad N_3 = \sigma_0 w, \qquad N_4 = -\sigma_0 z, \tag{6.7}$$

which verifies that the plane and the point are corresponding elements of a null-system. Hence the locus of all normals is a linear complex L and it is easy to derive that its equation is

$$p_{12} - \sigma_0 p_{34} = 0. \tag{6.8}$$

The osculating plane of a path, passing through three consecutive points, is a second order concept. For the point (x, y, z, w) of E its plane coordinates U_i in Σ follow, in view of (6.1) from the matrix

$$\left\| \begin{matrix} x & y & z & w \\ -y & x & \sigma_0 w & 0 \\ -x + \varepsilon z & -y + \mu\varepsilon w & -\varepsilon x + \lambda w & 0 \end{matrix} \right\| \tag{6.9}$$

and we obtain

$$\begin{aligned} U_1 &= (-\varepsilon x^2 + \lambda x w + \sigma_0 y w - \sigma_0 \varepsilon \mu w^2) w, \\ U_2 &= (-\varepsilon x y + \lambda y w - \sigma_0 x w + \sigma_0 \varepsilon z w) w, \\ U_3 &= (x^2 + y^2 - \varepsilon x z - \mu \varepsilon y w) w, \\ U_4 &= -z(x^2 + y^2 - \varepsilon(xz + \mu y w)) \\ &\quad + (x^2 + y^2)(\varepsilon x - \lambda w) + \sigma_0 w \varepsilon (\mu x w - y z). \end{aligned} \tag{6.10}$$

On the other hand it follows from (6.9) that

$$\begin{aligned} &U_1 x + U_2 y + U_3 z + U_4 w = 0, \\ &U_2 x - U_1 y + \sigma_0 U_3 w = 0, \\ &(-U_1 - \varepsilon U_3) x - U_2 y + \varepsilon U_1 z + (\mu \varepsilon U_2 + \lambda U_3) w = 0, \end{aligned} \tag{6.11}$$

which implies that x, y, z, w are cubic functions of U_i, and leads to the inverse expressions of (6.10). Hence a plane of Σ is in general the osculating plane to the path of one point of E. There exists a birational cubic relationship between the points of E and the planes of Σ; a point and its corresponding plane are incident.

We already know from Chapter IV that the relationship has singularities. If $w = 0$ the corresponding plane is $(0, 0, 0, 1)$; any point at infinity corresponds to the plane at infinity, and so this plane corresponds to each of its points. More interesting is the case when (6.9) has rank two. Then three consecutive points of a path are collinear: the path has an inflection point. This occurs if the second and the third row of (6.9) are proportional, that is if

$$(6.12) \qquad -x+\varepsilon z=-uy, \qquad -y+\mu\varepsilon w=ux, \qquad -\varepsilon x+\lambda w=\sigma_0 uw,$$

from which it follows

$$(6.13) \qquad \begin{aligned} x:y:z:w = \varepsilon(-\sigma_0 u+\lambda):\varepsilon(\sigma_0 u^2-\lambda u+\varepsilon^2\mu):\\ \{-\sigma_0 u^3+\lambda u^2-(\sigma_0+\varepsilon^2\mu)u+\lambda\}:\varepsilon^2, \end{aligned}$$

u being a parameter.

Hence the locus of those points of E which are at an inflection point of their paths is a twisted cubic, the instantaneous inflection curve c_3. It is a parabolic cubic since it osculates the plane at infinity (at Z when the parameter takes on the value $u=\infty$). The directions of the inflection tangent in the fixed space is (for $\varepsilon\neq 0$):

$$(6.14) \qquad X:Y:Z=-(\sigma_0 u^2-\lambda u+\varepsilon^2\mu):(-\sigma_0 u+\lambda):\sigma_0\varepsilon.$$

From (6.13) and (6.14) the Plücker coordinates of an inflection tangent follow; they are fifth order functions of u. Hence the locus of the inflection tangents is a quintic ruled surface F_5; all this is in accordance with the results in Chapter IV on three positions theory.

Example 17. Determine the tangent to c_3 at a general point and especially at Z.
Example 18. Determine the Plücker coordinates of the generators of F_5.
Example 19. Derive a representation of F_5 by means of two parameters u and v.
Example 20. Determine the intersection of F_5 and the plane at infinity.
Example 21. Investigate whether there are points on c_3 where the inflection tangent coincides with the tangent to c_3.
Example 22. Show that for $\sigma_0=0$ the curve c_3 degenerates into a line and a conic; determine the plane of the latter. Investigate F_5 for this case.

The normals at a point of a path are first order covariants. Their locus, the linear complex L depends only on the first order invariant σ_0. Now that we have derived the osculating plane U we are able to study the two special normals: the principal normal, lying in U, and the binormal, orthogonal to U.

The principal normal n_1 is the intersection of the normal plane (6.7) and the osculating plane (6.10). Hence the Plücker coordinates of n_1 are polynomials of degree four in the point coordinates (x, y, z, w).

The question arises whether an arbitrary normal is the principal normal for one or more of its points. A line p_{ij} is a normal if it satisfies (6.8), from which it follows in view of the fundamental relation that if $p_{34}\neq 0$ the normal n also satisfies

(6.15) $$\sigma_0 p_{34}^2 + p_{14}p_{23} + p_{24}p_{31} = 0.$$

By definition, the direction of n is $(p_{14} : p_{24} : p_{34})$ and its intersection with O_{XY} is $(p_{31}, p_{32}, 0, p_{34})$. This implies that a point P on n has the coordinates (supposing $p_{34} \neq 0$),

(6.16) $$X = p_{31} + rp_{14}, \qquad Y = p_{32} + rp_{24}, \qquad Z = rp_{34}, \qquad W = p_{34},$$

r being a parameter on n. The osculating plane of P follows by substitution of (6.16) into (6.10), keeping in mind that $X = x$, etc. The normal n and the plane U both pass through P. Hence, n will be in U (which implies that n is the principal normal of P) if one more point different from P, is in U. For this point we take $(p_{14}, p_{24}, p_{34}, 0)$ which obviously satisfies (6.16). The condition is then $U_1 p_{14} + U_2 p_{24} + U_3 p_{34} = 0$, which is clearly a quadratic equation for r. The conclusion is: a normal is in general the principal normal of two of its points. (STICHER [1972], BOTTEMA [1975b].)

A real point has a real principal normal, which implies that the roots of the quadratic equation are not always imaginary. On the other hand it is not sure that they are always real. One counter example is sufficient to disprove this.

Example 23. Consider the line $x = y$, $z = k$, which satisfies (6.8) and is therefore a normal. Show that the two points for which this line is the principal normal are imaginary if $k > (\lambda^2 - 2\sigma_0\mu\varepsilon^2)/2\sigma_0\varepsilon^2$.

The quadratic equation for r reads

(6.17) $$Ar^2 + Br + C = 0,$$

with

(6.18) $$\begin{aligned} A &= (-\varepsilon p_{14})(p_{14}^2 + p_{24}^2 + p_{34}^2) + p_{34}(p_{14}^2 + p_{24}^2) \\ B &= -2\varepsilon p_{31}(p_{14}^2 + p_{24}^2) - \varepsilon p_{24}p_{12}p_{34} + p_{34}q_1, \\ C &= -\varepsilon p_{31}(p_{31}p_{14} + p_{32}p_{24}) + p_{34}q_2, \end{aligned}$$

q_1 and q_2 denote quadratic functions of p_{ij}. If we calculate the discriminant of (6.17) it may be shown, after some algebra (using (6.15)), that this expression of degree six has p_{34}^2 as a factor. Since p_{34} was supposed to be unequal to zero, those normals for which the two points coincide satisfy an equation of degree four, representing a quartic line complex. As the normals are lines of L the conclusion is: the locus of the principal normals to the paths is a subset of the linear complex L of all normals; to L belongs a (4, 4) congruence which separates in L the lines which are principal normals from those which are not.

The normals in a given plane π of Σ are the lines of a pencil with vertex P_0, the null-point of π with respect to the linear complex L. On any line p of the pencil there are two points, real or imaginary, for which p is the principal normal. We now determine the locus c of these points. P_0 belongs to the locus for its principal normal is a line, p_0 say, of the pencil. On any line p, different from p_0, there are two points of c different from P_0. Hence the locus c is a curve of order three. Obviously p_0 is the tangent of c at P_0. A cubic curve has in general a class of six, which implies that the number of lines through P_0, tangent to c at a point different from P_0, is four. These lines separate, in the pencil, those lines intersecting c in two real points from those intersecting c in two imaginary points (different from P_0). They are the four lines of the (4, 4) congruence, mentioned above, lying in π (and passing through P_0).

Example 24. Let π be the plane $Z = z_0$, then $P = (0, 0, z_0)$. Show that c is the circular cubic given by the equation $(\lambda - \varepsilon x)(x^2 + y^2) - \sigma_0 \varepsilon (z_0 y + \mu x) = 0$. Two of the lines, through P, of the (4, 4) congruence are isotropic. Determine the other two and investigate whether they are real or imaginary.

The binormal n_2 at a point P of a path is perpendicular to the osculating plane. Hence its direction numbers are $U_1 : U_2 : U_3$ given by (6.10) and therefore they are quadratic functions of the coordinates of P. It follows that the Plücker coordinates of n_2 are cubic functions of x, y, z and w.

Example 25. Show that n_2 is undetermined if P is on the inflection curve. Show that for a point P at infinity the binormal coincides with PZ, provided P is not on the pair of imaginary lines $x^2 + y^2 = w = 0$, in this case n_2 is undetermined.

We now investigate the inverse relationship between the point P and its binormal. Any normal p_{ij} satisfies (6.15) and a point P on it is given by (6.16). We shall try to determine the parameter r such that the normal is the binormal of P, or in other words that it is orthogonal to the osculating plane of P. We know that the line is orthogonal to a line of the plane (the tangent of the path). Hence we need only one condition more to ascertain that it is the binormal. It follows that the two conditions $U_1 : U_2 : U_3 = p_{14} : p_{24} : p_{34}$ are dependent and we may restrict ourselves to $p_{24}U_1 - p_{14}U_2 = 0$. If we do so the quadratic term r^2 vanishes and making use of (6.15) the equation reads

$$(6.19)\qquad (p_{14}^2 + p_{24}^2)r - \varepsilon p_{13}p_{34} - \lambda p_{34}^2 + \mu\varepsilon p_{24}p_{34} + p_{31}p_{14} + p_{32}p_{24} = 0,$$

a linear equation for the parameter r. The conclusion is: a normal is in general the binormal of *one* of its points. (BOTTEMA [1975b]), or in other words: the locus of the binormals is the (complete) linear complex L.

If we solve for r from (6.19) and substitute it in (6.16), then after some reduction by means of (6.15) we obtain

$$
\begin{aligned}
x &= -\sigma_0 p_{24} p_{34} + p_{14}(\varepsilon p_{13} + \lambda p_{34} - \mu\varepsilon p_{24}),\\
y &= \sigma_0 p_{14} p_{34} + p_{24}(\varepsilon p_{13} + \lambda p_{34} - \mu\varepsilon p_{24}),\\
z &= p_{13} p_{14} + p_{23} p_{24} + p_{34}(\varepsilon p_{13} + \lambda p_{34} - \mu\varepsilon p_{24}),\\
w &= p_{14}^2 + p_{24}^2.
\end{aligned} \tag{6.20}
$$

The coordinates of the point for which the normal p_{ij} is the binormal are quadratic functions of p_{ij}. The relationship between a point and its binormal is birational.

Example 26. Verify that the point (6.20) is on the line p_{ij}.

The binormals in a given plane π of Σ are the lines of the pencil with vertex P_0, the null-point of π with respect to L. On any line lies one point P such that the line is the binormal of P. The locus of P, in plane π, is seen to be a conic K, passing through P_0; the binormal of P_0 itself is the tangent, at P_0, to K. The points of K at infinity have normals passing through P_0, which implies that these points are on the singular lines $x^2 + y^2 = w = 0$. Hence K is an ellipse.

Let $\mathfrak{B}$ be an arbitrary plane of Σ, with the equation $V_1 X + V_2 Y + V_3 Z + V_4 W = 0$. Its null-point P_0 is then $(\sigma_0 V_2, -\sigma_0 V_1, -V_4, V_3)$. If $P(x, y, z, w)$ is in $\mathfrak{B}$ we have for the line PP_0:

$$
p'_{14} = xV_3 - \sigma_0 w V_2, \qquad p'_{24} = yV_3 + \sigma_0 w V_1, \qquad p'_{34} = zV_3 + wV_4. \tag{6.21}
$$

The condition $p'_{14} : p'_{24} = U_1 : U_2$ which gave us (6.19) also gives us an equation for P. Substituting for U_i from (6.10) and keeping in mind that P lies in $\mathfrak{B}$, this condition can be reduced to

$$
\begin{aligned}
&V_3(x^2 + y^2) + (\sigma_0 V_1 + \lambda V_2 - \mu\varepsilon V_3) yw + (\lambda V_1 - \sigma_0 V_2 + \varepsilon V_4) xw\\
&\quad + \sigma_0 V_2 \varepsilon zw - \sigma_0 \mu\varepsilon V_2 w^2 = 0,
\end{aligned} \tag{6.22}
$$

which represents a paraboloid of revolution, passing through the singular lines $x^2 + y^2 = w = 0$. The conic K is the intersection of this quadric and the plane $\mathfrak{B}$.

Example 27. Determine the conic K in a plane orthogonal to the screw axis.
Example 28. Determine the conic K in a plane parallel to the screw axis.

We have thus obtained the properties of the triad to the path of any point, of the moving space, at a certain instant. Summing up: the tangent, the principal normal, and the binormal have Plücker coordinates which are polynomials of the point coordinates of degree two, four, and three respectively.

Example 29. Show that the direction numbers of the three lines are polynomials of degree one, three and two respectively. Write them out explicitly and check that they are mutually orthogonal.

In Chapter IV on three positions theory we introduced the circle axis, that is the axis of the circle through three homologous points. Its counterpart in instantaneous continuous kinematics is the curvature axis of the path of the moving point. The curvature axis is the line of intersection of two consecutive normal planes of the path. The normal plane at the point X, Y, Z of the curve $X(\tau), Y(\tau), Z(\tau)$ has the coordinates

$$\dot{X}, \quad \dot{Y}, \quad \dot{Z}, \quad -(X\dot{X}+Y\dot{Y}+Z\dot{Z}).$$

Hence the curvature axis is the intersection of this plane and the plane with the coordinates

$$\ddot{X}, \quad \ddot{Y}, \quad \ddot{Z}, \quad -(\dot{X}^2+\dot{Y}^2+\dot{Z}^2)-(X\ddot{X}+Y\ddot{Y}+Z\ddot{Z}).$$

In the instantaneous case, making use of (6.1), the two planes (for the point (x, y, z) of E) have the coordinates:

$$(6.24)\qquad \begin{aligned} &-y, \quad x, \quad \sigma_0, \quad -\sigma_0 z, \\ &-x+\varepsilon z, \quad -y+\mu\varepsilon, \quad -\varepsilon x+\lambda, \quad -\sigma_0^2-\mu\varepsilon y-\lambda z. \end{aligned}$$

From this the Plücker coordinates of the curvature axis follow:

$$(6.25)\qquad \begin{aligned} p_{14} &= -\varepsilon x^2+\lambda x+\sigma_0 y-\sigma_0\mu\varepsilon, \\ p_{24} &= -\varepsilon xy+\lambda y-\sigma_0 x+\sigma_0\varepsilon z, \\ p_{34} &= x^2+y^2-\varepsilon xz-\mu\varepsilon y, \\ p_{23} &= \mu\varepsilon y^2+\lambda yz-\sigma_0 xz+\sigma_0^2 y+\sigma_0\varepsilon z^2, \\ p_{31} &= -\mu\varepsilon xy+\sigma_0\varepsilon\mu z-\lambda xz-\sigma_0 yz-\sigma_0^2 x, \\ p_{12} &= \sigma_0(-\sigma_0^2-\mu\varepsilon y-\varepsilon xz). \end{aligned}$$

It is orthogonal to the osculating plane (6.10). The Plücker coordinates of the curvature axis are quadratic functions of the point coordinates. To every point of E there corresponds a curvature axis in Σ; hence at any instant the locus of these curvature axes is a complex of lines. Its equation would follow if we were able to eliminate (x, y, z) from (6.25). This seems too complicated, but we may at least determine the degree of the complex. A line of (6.25) is at infinity if $p_{14}=p_{24}=p_{34}=0$, this means that the osculating plane is not determined and we know that this is the case for the points on the inflection curve. If we substitute their coordinates from (6.13) into p_{23}, p_{31}, p_{12} we obtain expressions of order six in the parameter u. Hence the complex curve in the plane at infinity is a (rational) curve of class six. The conclusion is that the complex is of the sixth degree.

Example 30. Show that for a point of the inflection curve the two planes (6.24) are parallel.
Example 31. Show that the line at infinity of the plane $X = 0$ is the curvature axis of the point at infinity of the screw axis O_Z.

The center of curvature at a point of a path is the intersection of the curvature axis and the osculating plane. (It may be defined as the center of the circle through three consecutive points of the path.)

To determine the center of curvature M we note that the equation of the osculating plane may be written as

$$p_{14}X + p_{24}Y + p_{34}Z + U_4W = 0, \tag{6.26}$$

where p_{14}, p_{24}, p_{34} are given by (6.25) and U_4 by (6.10). A point on the curvature axis is

$$X = p_{31} + rp_{14}, \quad Y = p_{32} + rp_{24}, \quad Z = rp_{34}, \quad W = p_{34}, \tag{6.27}$$

where r is a parameter and $p_{34} \neq 0$. For the intersection of (6.26) and (6.27) we obtain

$$r = (p_{14}p_{13} + p_{24}p_{23} - p_{34}U_4)/(p_{14}^2 + p_{24}^2 + p_{34}^2).$$

If we substitute this into (6.27) it is easy to see that (after some reduction by means of the fundamental relation for p_{ij}), all four coordinates have the factor p_{34}. The result is that M is found to be

$$\begin{aligned} X &= p_{21}p_{24} + p_{31}p_{34} - p_{14}U_4, \\ Y &= p_{12}p_{14} + p_{32}p_{34} - p_{24}U_4, \\ Z &= p_{13}p_{14} + p_{23}p_{24} - p_{34}U_4, \\ W &= p_{14}^2 + p_{24}^2 + p_{34}^2. \end{aligned} \tag{6.28}$$

In view of (6.10) and (6.25) we obtain: the coordinates of the center of curvature are polynomials of degree five in the coordinates of the moving point.

If $p_{14} = p_{24} = p_{34} = 0$, that is if we deal with an inflection point, the curvature center is undetermined. Indeed: any point at infinity in the normal plane is a center.

Example 32. Verify that (6.28) satisfies (6.26).
Example 33. Show that the center of curvature of the origin O is on O_Y.

Example 34. Determine the center of curvature corresponding to a point on the screw axis and show that the locus of all such points is a twisted cubic.

From the symmetrical expressions (6.28) follows a rather simple formula for the distance from the origin to the center of curvature, $|OM|$. We have

$$X^2+Y^2+Z^2=\sum(p_{21}p_{24}+p_{31}p_{34})^2+U_4^2\cdot\sum p_{14}^2$$
$$=\sum p_{14}^2\cdot\sum p_{23}^2-(p_{14}p_{23}+p_{24}p_{31}+p_{34}p_{12})^2+U_4^2\cdot\sum p_{14}^2,$$

therefore, in view of the fundamental relation for p_{ij}, we obtain

$$(OM)^2=(p_{23}^2+p_{31}^2+p_{12}^2+U_4^2)/(p_{14}^2+p_{24}^2+p_{34}^2), \tag{6.29}$$

p_{ij} and U_4 being given by (6.25) and (6.10) as functions of the coordinates of the point P of which M is the center of curvature.

A more difficult, and as yet (it seems) unsolved, problem is to determine the point P of the moving space if the center of curvature M in Σ is given, or in other words to derive the inverse relationship of (6.28). An example shows that this is not rational: in general M does not correspond to *one* point P. Take M at the origin O: $X=Y=Z=0$. Then the corresponding point P must have its osculating plane and curvature axis both passing through O, hence $U_4=0$, $p_{23}=p_{31}=p_{12}=0$. Points satisfying the last three equations are for instance those of the line $y=-\sigma_0^2/\mu\varepsilon$, $z=0$. If we substitute this in $U_4=0$ we obtain a cubic equation for x, which means that there are at least three points P, real or imaginary, corresponding to M.

The radius of curvature ρ of the path could be found from (6.28) as the distance between P and M. The result follows more easily if we make use of the formula for ρ as it is derived in the differential geometry of space curves. The general equation for path curvature reads

$$\rho^2=(X_1^2+Y_1^2+Z_1^2)^3/[(X_2Y_1-X_1Y_2)^2+(Y_2Z_1-Y_1Z_2)^2+(Z_2X_1-Z_1X_2)^2], \tag{6.30}$$

so that we obtain

$$\rho^2=(x^2+y^2+\sigma_0^2)^3/(p_{14}^2+p_{24}^2+p_{34}^2), \tag{6.31}$$

p_{i4} being given by (6.25). The numerator is a polynomial of degree six, the denominator one of degree four, in terms of the coordinates of the point in E. The radius cannot be zero unless $\sigma_0=0$, which is as expected because in general there is no point instantaneously at rest: in general no path has a cusp. The radius is infinite only if $p_{14}=p_{24}=p_{34}=0$: the moving point is at an inflection of its path.

The principal normal, the binormal, and the curvature are second order properties of the path and of the motion; they depend on the four instantaneous invariants $\sigma_0, \varepsilon, \mu, \lambda$. Third order properties such as the torsion and the osculating sphere (through four consecutive points of the path) depend on five more invariants, denoted in Chapter II, (6.3), as $\gamma_X, \gamma_Y, d_{3_X}, d_{3_Y}, d_{3_Z}$. Although an apparatus to study these properties was completely developed in Chapter II, its general theory, involving nine constants, seems too complicated to insure reasonably concise results; even the second order theory, given in the preceding sections, was far from simple and left many open questions. This is the reason why we shall not continue these investigations in detail here. The general principles have already been developed and the results follow the comparable cases for finite displacements. To illustrate this we briefly consider the osculating sphere.

The osculating sphere of a space curve is the sphere through four consecutive points of the curve. Its center N is the intersection of three consecutive normal planes and is therefore a point on the curvature axis. If the curve is given by $X(\tau), Y(\tau), Z(\tau)$ the normal plane at the point has the coordinates

$$\dot{X}, \quad \dot{Y}, \quad \dot{Z}, \quad -(X\dot{X} + Y\dot{Y} + Z\dot{Z}),$$

and those of the two consecutive normal planes are

$$\ddot{X}, \quad \ddot{Y}, \quad \ddot{Z}, \quad -\sum X\ddot{X} - \sum \dot{X}^2,$$

and

$$\dddot{X}, \quad \dddot{Y}, \quad \dddot{Z}, \quad -\sum X\dddot{X} - 3\sum \dot{X}\ddot{X}.$$

If we apply this to a path in the zero position, substituting the formulas from Chapter II, (6.3) it is easy to verify that the fourth coordinate of each of these three planes is a linear (and not a quadratic) function of x, y, z. The homogeneous coordinates of the sphere center N are the 3×3 determinants of the matrix

$$\left\| \begin{matrix} -y & x & \sigma_0 & -\sigma_0 z \\ -x + \varepsilon z & -y + \mu\varepsilon & -\varepsilon x + \lambda & -\mu\varepsilon y - \lambda z - \sigma_0^2 \\ (1+\varepsilon^2)y & [-(1+\varepsilon^2)x & [-\gamma_Y x & -[3\varepsilon(\mu - \sigma_0)x \\ \quad + \gamma_Y z + d_{3_X} & \quad + ((3/2)\varepsilon - \gamma_X)z & \quad + ((3/2)\varepsilon + \gamma_X)y & \quad + 3\sigma_0\lambda + x d_{3_X} \\ & \quad + d_{3_Y}] & \quad + d_{3_Z}] & \quad + y d_{3_Y} + z d_{3_Z}] \end{matrix} \right\|$$

(6.32)

Hence coordinates of N are polynomials of degree three in the coordinates of the moving point. The locus of those points for which four consecutive positions are coplanar is a cubic surface; its equation is obtained by putting the first 3×3 determinant of (6.32) equal to zero. In the general case the radius of the osculating sphere is the distance $|NP|$.

7. The motion of a plane

In the preceding section we have studied the properties of the path of a point of E with respect to Σ. We now consider the motion of a plane of E. To develop the necessary apparatus we note that according to (6.1) we may write, up to the second order,

$$\begin{aligned} X &= x - y\tau + \tfrac{1}{2}(-x+\varepsilon z)\tau^2, \\ Y &= y + x\tau + \tfrac{1}{2}(-y+\mu\varepsilon)\tau^2, \qquad (7.1) \\ Z &= z + \sigma_0\tau + \tfrac{1}{2}(-\varepsilon x+\lambda)\tau^2. \end{aligned}$$

If we solve these equations for x, y, z, again up to the second order, we obtain

$$\begin{array}{llll} & x_0 = X, & x_1 = Y, & x_2 = -X-\varepsilon Z, \\ (7.2) & y_0 = Y, & y_1 = -X, & y_2 = -Y-\varepsilon\mu, \\ & z_0 = Z, & z_1 = -\sigma_0, & z_2 = \varepsilon X-\lambda. \end{array}$$

The formulas (7.2) can also be obtained by considering the instantaneous invariants of the inverse motion (Chapter II, (6.5)).

Let $ax+by+cz+d=0$ be the equation of a plane u of E; by means of (7.2) its position in Σ, up to the second order is given by the equation

$$(aX+bY+cZ+d)+(aY-bX-c\sigma_0)\tau$$
$$+\tfrac{1}{2}\{a(-X-\varepsilon Z)+b(-Y-\varepsilon\mu)+c(\varepsilon X-\lambda)\}\tau^2=0,$$

or $AX+BY+CZ+D=0$, A, B, C, D being functions of τ. We obtain

$$\begin{array}{llll} & A_0 = a, & A_1 = -b, & A_2 = -a+\varepsilon c, \\ & B_0 = b, & B_1 = a, & B_2 = -b, \\ (7.3) & C_0 = c, & C_1 = 0, & C_2 = -\varepsilon a, \\ & D_0 = d, & D_1 = -\sigma_0 c, & D_2 = -\mu\varepsilon b-\lambda c, \end{array}$$

as the dual counterpart of the formulas (6.1). During its motion the plane u coincides with a continuous set of ∞^1 planes in Σ; the configuration thus generated is called a developable or torse, the dual of the path of a moving point. The differential geometry of a torse is less developed than that of a curve for the simple reason that it is more difficult to visualize its geometrical properties.

We consider first order concepts. At any instant the moving plane has a line of intersection m with its consecutive plane; m is the dual of a path tangent, it is called the characteristic of the plane under consideration. We have met it already in the previous section, there it appeared as the locus of the points for which the path tangents lie in a plane. The characteristic of the plane $ax + by + cz + d = 0$ in the zero position is its intersection with the plane $-bx + ay - \sigma_0 c = 0$ and its coordinates are therefore

(7.4)
$$\begin{aligned} p_{14} &= -ac, & p_{23} &= -\sigma_0 ac + bd, \\ p_{24} &= -bc, & p_{31} &= -\sigma_0 bc - ad, \\ p_{34} &= a^2 + b^2, & p_{12} &= -\sigma_0 c^2. \end{aligned}$$

Eliminating a, b, c, d we obtain

(7.5)
$$p_{12}p_{34} + \sigma_0(p_{14}^2 + p_{24}^2) = 0$$

as the equation of the locus of all characteristics. It is the same as (6.4). Hence the locus of m is a tetrahedral line complex in Σ, coinciding with the locus of the path-tangents. This is as expected, in view of the results derived in Chapter III on two positions theory.

By means of (7.4) it follows that to any plane of E there corresponds in general one characteristic. Conversely, to any line of the complex there corresponds one plane, as seen by the inverse formulas of (7.4):

(7.6)
$$a : b : c : d = \sigma_0 p_{14} p_{24} : \sigma_0 p_{24}^2 : p_{12} p_{24} : p_{12}(\sigma_0 p_{14} - p_{23}).$$

Some special cases may be mentioned. To the plane at infinity ($a = b = c = 0$) corresponds any line in it, as is of course what should be expected. A plane is parallel to its consecutive plane if in (7.4) $p_{14} = p_{24} = p_{34} = 0$, that is if $a = b = 0$ (u is orthogonal to the screw axis), or if $c = 0$, $a^2 + b^2 = 0$ (u is parallel to one of the isotropic planes through the screw axis).

Example 35. Determine the characteristic of a plane through the screw axis.

In Section 6 we have derived that any line of the complex (7.5) corresponds to a point on it for which it is the path tangent; we see now that it corresponds also to a plane through it for which it is the characteristic.

Example 36. Show that the planes corresponding to the lines of the complex (7.5) through a point P form a pencil of planes passing through the path tangent of P.
Example 37. Show that the planes corresponding to the lines of the complex in a given plane U_0 are those of a torse of the third class of which U_0 itself is an element; this set consists of the osculating planes of a twisted cubic.

To any plane u corresponds a characteristic m, but because m belongs to the complex it has a point P on it for which it is the path tangent. This implies a relationship, between the plane u and the point P (and conversely), geometrically defined by the property: the path tangent of P coincides with the characteristic of u. The analytic expressions follow at once from (6.5) and (7.4):

$$(7.7) \quad x = \sigma_0 bc^2, \quad y = -\sigma_0 ac^2, \quad z = d(a^2+b^2), \quad w = -c(a^2+b^2),$$

and, respectively, from (6.3) and (7.6):

$$(7.8) \quad a = -\sigma_0 yw^2, \quad b = \sigma_0 xw^2, \quad c = -w(x^2+y^2), \quad d = z(x^2+y^2),$$

which implies that the relationship is birational and of degree three.

Example 38. Show that the point P corresponding to the plane u is the tangent point of the characteristic of u and the complex conic in U_0; describe analagously the plane corresponding to a given point.

To summarize the first order properties of a moving plane u and to help visualize the results, we note that the first order situation of a moving space is completely determined by the instantaneous screw axis s and pitch σ_0. Therefore by a suitable choice of the coordinate system any moving plane u (not parallel to s) may be given the equation $x = z \tan \alpha$, (that is $a = 1$, $b = 0$, $c = -\tan \alpha$, $d = 0$) so that α, the angle between u and s determines the situation. The characteristic m of u is its intersection with the plane (obtained from the first order terms of (7.3)): $-bx + ay - \sigma_0 c = 0$, which in this case becomes $y + \sigma_0 \tan \alpha = 0$. The point P on m defined by (7.7) is seen to be $(0, -\sigma_0 \tan \alpha, 0)$. In the plane U_0, coinciding with u, we introduce the Cartesian frame O_{YR_u}, O_{R_u} being the projection of the screw axis s on U_0 (Fig. 32). Then m is parallel to O_{R_u} and $OP = -\sigma_0 \tan \alpha$.

If Q is a variable point on m, its coordinates are $x = q \tan \alpha$, $y = -\sigma_0 \tan \alpha$, $z = q$, $w = 1$, q being a parameter. The path tangent l of Q joins Q to $(\sigma_0 \tan \alpha, q \tan \alpha, \sigma_0, 0)$; hence its line coordinates are

$$p_{14} = \sigma_0 \tan \alpha, \quad p_{24} = q \tan \alpha, \quad p_{34} = \sigma_0,$$

$$p_{23} = (\sigma_0^2 + q^2) \tan \alpha, \quad p_{31} = 0, \quad p_{12} = -(\sigma_0^2 + q^2) \tan^2 \alpha.$$

They of course satisfy (7.5). The line l intersects O_Y at

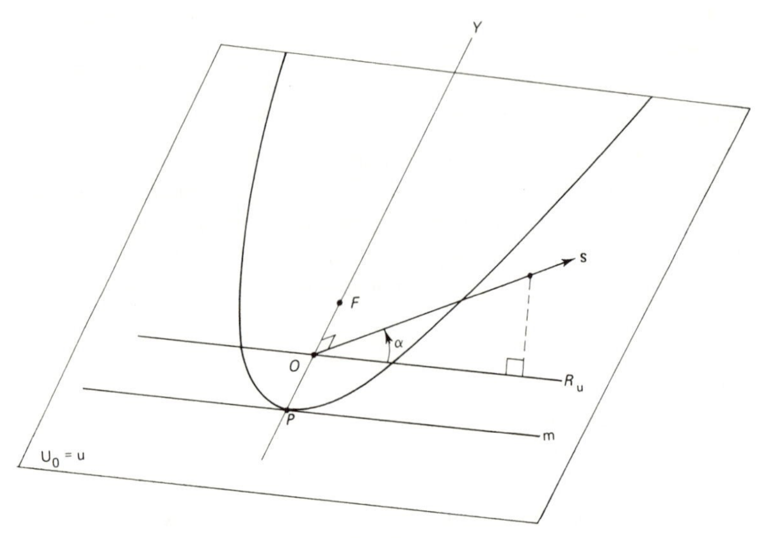

Fig. 32.

$\{0, -(\sigma_0^2 + q^2)\tan\alpha/\sigma_0, 0\}$ and O_{R_u} (or the plane O_{XZ}) at $\{(\sigma_0^2 + q^2)\tan\alpha/q, 0, (\sigma_0^2 + q^2)/q\}$; after some algebra it is seen that the equation of l in the O_{YR_u} frame reads

$$\sigma_0 y - q \tan\alpha\, r_u + (\sigma_0^2 + q^2)\tan\alpha = 0, \tag{7.9}$$

where r_u is the coordinate measured along O_{R_u}. This is a quadratic set of lines. Through any point of U_0, (y, r_u) pass two lines of the set, corresponding to the roots of the quadratic equation

$$q^2 \tan\alpha - q r_u \tan\alpha + \sigma_0(y + \sigma_0 \tan\alpha) = 0.$$

The envelope of the set is the locus of the points for which the two lines coincide. Hence its equation reads

$$r_u^2 \tan\alpha - 4\sigma_0(y + \sigma_0 \tan\alpha) = 0, \tag{7.10}$$

which represents therefore the point equation of the conic in which the complex (7.5) cuts U_0. It is a parabola p, which is in accordance with the fact that any line at infinity belongs to the complex (7.5). The vertex of p is the point P, the tangent at P is the characteristic m. Elementary analytic geometry shows that the focus F of p is given by $y = \sigma_0 \cot\alpha$, $r_u = 0$. It follows from (6.8) that the null-point of the plane $Ax + By + Cz + D = 0$ with respect to the linear complex L of normals is given by $x = B\sigma_0$, $y = -A\sigma_0$, $z = -D$, $w = C$. Hence the null-point of U_0 reads $(0, \sigma_0 \cot\alpha, 0, 1)$ and it coincides with F, a theorem due originally to SCHOENFLIES [1886]; F is therefore the (only) point of u for which the path tangent is perpendicular to u.

We now consider some second order properties of a moving plane u of E. As it moves, this plane generates a torse in Σ. Two consecutive positions of u define its characteristic line m; three consecutive positions of u define a point K at which they intersect, the locus of all points K is a curve k associated with the torse. The tangents of k are the characteristics, and the osculating planes of k are the planes of the torse. In the instantaneous position the coordinates of K are, in view of (7.3), for the plane (a, b, c, d), the determinants of the matrix

$$\left\| \begin{matrix} a & b & c & d \\ -b & a & 0 & -\sigma_0 c \\ -a+\varepsilon c & -b & -\varepsilon a & -\mu\varepsilon b - \lambda c \end{matrix} \right\| \tag{7.11}$$

or explicitly

$$\begin{aligned} X &= \mu\varepsilon abc + \lambda ac^2 - \varepsilon a^2 d + \sigma_0 bc(-\varepsilon a + c), \\ Y &= \varepsilon\mu b^2 c + \sigma_0 a^2 c - \sigma_0 ac^2 + \lambda bc^2 - \varepsilon abd + \sigma_0 \varepsilon c^3, \\ Z &= (a^2 + b^2)(-\mu\varepsilon b - \lambda c + d) - \sigma_0 \varepsilon bc^2 - \varepsilon acd, \\ W &= (a^2 + b^2)(\varepsilon a - c) + \varepsilon ac^2. \end{aligned} \tag{7.12}$$

Hence the coordinates of K are cubic polynomials in terms of the coordinates of the moving plane u.

Example 39. Show that, conversely, the coordinates (a, b, c, d) of u are cubic polynomials of the coordinates (X, Y, Z, W) of K.
Example 40. Show that the matrix (7.11) may be derived from (6.11) if (U_1, U_2, U_3, U_4) is changed into (a, b, c, d), (x, y, z, w) into (X, Y, Z, W), and the instantaneous invariants into those of the inverse motion.
Example 41. Show from (7.12) that the locus of K for a set of parallel planes is a straight line.

8. The motion of a line

We deal now with the motion of a line l of E, the Plücker coordinates of which are p_{ij}.

From (7.1) after some algebra, it follows that its coordinates P_{ij} in Σ are, up to the second order, given by

$$
\begin{aligned}
P_{14} &= p_{14} - p_{24}\tau + \tfrac{1}{2}(-p_{14} + \varepsilon p_{34})\tau^2, \\
P_{24} &= p_{24} + p_{14}\tau - \tfrac{1}{2}p_{24}\tau^2, \\
P_{34} &= p_{34} \qquad - \tfrac{1}{2}\varepsilon p_{14}\tau^2, \\
P_{23} &= p_{23} + (\sigma_0 p_{24} - p_{31})\tau + \tfrac{1}{2}(-p_{23} - \mu\varepsilon p_{34} + 2\sigma_0 p_{14} + \varepsilon p_{12} + \lambda p_{24})\tau^2, \\
P_{31} &= p_{31} + (p_{23} - \sigma_0 p_{14})\tau + \tfrac{1}{2}(-p_{31} + 2\sigma_0 p_{24} - \lambda p_{14})\tau^2, \\
P_{12} &= p_{12} \qquad + \tfrac{1}{2}(-\varepsilon p_{23} + \mu\varepsilon p_{14})\tau^2.
\end{aligned}
\tag{8.1}
$$

Our first question may be: when does a line intersect itself in consecutive positions. The condition $p_{14}P_{23} + p_{24}P_{31} + p_{34}P_{12} + p_{23}P_{14} + p_{31}P_{24} + p_{12}P_{34} = 0$ gives after substitution of (8.1) (the constant term and the linear term vanish):

$$p_{12}p_{34} + \sigma_0(p_{14}^2 + p_{24}^2) = 0. \tag{8.2}$$

Hence the lines of the tetrahedral complex given by (8.2) are those lines which at the zero position are intersected by their next consecutive position. We have met this same complex already two times, in Sections 6 and 7 respectively, but we must be aware that in the latter cases the complex was in Σ, while (8.2) lies in E. In the zero position the three coincide. We have dealt with this phenomenon in Chapter III on two positions theory.

Obviously the intersection point of a line l of (8.2) and its next consecutive position is the point given by (6.5): the point on l for which l is the path tangent. Indeed, the tangents of a path are the generators of a developable ruled surface, or, what is the same thing, a surface for which every two consecutive generators intersect.

Example 42. Verify (6.5) by using (8.1) and the coordinates of the intersection point of two intersecting lines p_{ij} and q_{ij}.

In general the line l of E describes in Σ a ruled surface R which is not developable. Two properties of the first order are of interest: the striction point, S, of l with respect to R and the distribution parameter of l on R. To simplify the analysis we note, as we did in Section 7 with respect to the first order displacement of a plane, that first order properties depend only on the screw axis s and the pitch σ_0. This enables us to introduce a suitable coordinate system for any line. Obviously we may restrict ouselves to a line l, intersecting the X-axis orthogonally at $S(a, 0, 0)$, and therefore given by $x = a$, $z = y\tan\beta$ (here, β is the angle between l and the X, Y plane). Its coordinates in Σ are

$$P_{14} = 0, \quad P_{24} = 1, \quad P_{34} = \tan\beta, \quad P_{23} = 0, \quad P_{31} = a\tan\beta, \quad P_{12} = -a; \tag{8.3}$$

and its next consecutive position $\mathrm{l}(\tau)$ follows from (8.1), up to the first order,

$$P_{14} = -\tau, \quad P_{24} = 1, \quad P_{34} = \tan\beta,$$
(8.4)
$$P_{23} = (\sigma_0 - a\tan\beta)\tau, \quad P_{31} = a\tan\beta, \quad P_{12} = -a.$$

The common perpendicular n' of l and $\mathrm{l}(\tau)$ has the direction numbers $(0, -\tan\beta, 1)$; if we intersect l with the plane through $\mathrm{l}(\tau)$ and the direction n', then for $\tau \to 0$ the point of intersection is S. Hence S is the striction point on l. In general: the striction point on a moving line l is the foot, on l, of the common perpendicular n between l and the screw axis s. The tangent plane to the ruled surface R at S is the plane through l parallel to s. The normal to R at S is the line n (Fig. 33).

A point P on l, such that $SP = u$, has the coordinates

$$x = a, \quad y = u\cos\beta, \quad z = u\sin\beta.$$

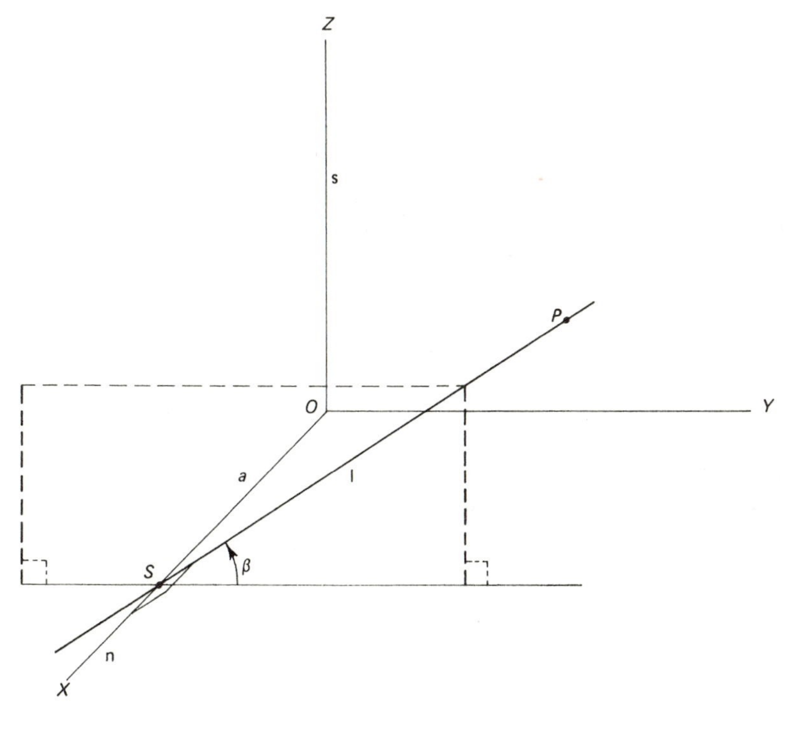

Fig. 33.

Its path is on R; the direction of the tangent to the path is $(-u\cos\beta, a, \sigma_0)$ and the tangent plane U to R at P(u) is seen to be

$$(8.5)\qquad (\sigma_0 - a\tan\beta)(X - aW) + u(-Y\sin\beta + Z\cos\beta) = 0.$$

For $u = 0$ we obtain the tangent plane U_0 at the striction point S, for $u = \infty$ the asymptotic tangent plane, orthogonal to the former.* The coordinates of U are linear functions of the distance $SP = u$. In other words, the pencil of tangent planes through l and the point row on l are projective sets, as is the case on any ruled surface.

It is easily seen that the angle γ between the tangent plane U, given by (8.5), and the plane U_0, satisfies $\tan\gamma = u/(\sigma_0 - a\tan\beta)$. The distribution parameter of l on R is defined by $u/\tan\gamma$.

Hence the distribution parameter δ of a line l, whose distance to the screw axis is a and whose angle with its normal plane is β, is given by

$$(8.6)\qquad \delta = \sigma_0 - a\tan\beta.$$

Example 43. If $\delta = 0$ the tangent plane at any point of l coincides with U_0; show, for instance by means of (8.3), that such a line l belongs to the complex (8.2); show geometrically that l is the path tangent of S.

If Q is a point on the path tangent of $P(u)$ on l, the locus of Q is given by

$$X = a - uv\cos\beta,\quad Y = u\cos\beta + av,\quad Z = u\sin\beta + \sigma_0 v,\quad W = 1,$$

which is therefore the representation, by means of the parameters u and v, of the raccording quadric of R along l. Eliminating u and v its equation is seen to be

$$(8.7)\qquad \delta^2(X - aW)W - (Y\tan\beta - Z)(\sigma_0 Y - aZ) = 0.$$

Example 44. Show that the quadric is a paraboloid and that for $\delta = 0$ it degenerates into two planes.

These same first order results for a moving line will now be developed for the case where the line is in a general position. With s along O_Z, if p_{ij} are the coordinates of l, the common perpendicular of l and the screw axis s has the direction $(-p_{24}, p_{14}, 0)$; the plane through s and this direction is $p_{14}X + p_{24}Y = 0$. This plane intersects l at its striction point S whose coordinates are

$$(8.8)\qquad X = p_{21}p_{24},\quad Y = p_{12}p_{14},\quad Z = p_{31}p_{41} + p_{32}p_{42},\quad W = p_{14}^2 + p_{24}^2.$$

The distance of S to the Z-axis is $a = p_{21}/(p_{14}^2 + p_{24}^2)^{1/2}$. For the angle β, which is the complement of the angle between l and O_Z, we obtain $\sin\beta =$

* It is a well-known result in the algebraic geometry of ruled surfaces that an asymptotic tangent plane along a ruling is orthogonal to the tangent plane at the striction point of the line.

$p_{43}/(p_{14}^2+p_{24}^2+p_{34}^2)^{1/2}$, which implies $\tan\beta = p_{43}/(p_{14}^2+p_{24}^2)^{1/2}$. Hence, from (8.6), it follows that the distribution parameter of the line p_{ij}, of E, on its generated ruled surface R, in Σ, is

$$\delta = \sigma_0 + p_{12}p_{34}/(p_{14}^2+p_{24}^2). \tag{8.9}$$

Example 45. Show that the locus of lines in E with a given δ is a tetrahedral complex and that for $\delta = 0$ it is (8.2).

Example 46. Show that the equation of the tangent plane of R at S reads $p_{24}X + p_{41}Y + p_{12}W = 0$.

Example 47. Show that the path tangent at S has the direction $\{p_{12}p_{14}, p_{12}p_{24}, -\sigma_0(p_{14}^2+p_{24}^2)\}$ and verify that it coincides with l only if l belongs to the complex (8.2).

The striction point S on a moving line l and its distribution parameter δ with respect to R are first order covariants of the motion. We consider now some second order properties.

One of these is the tangent at S of the striction curve on R. To determine it we consider the striction point, denoted by $S(\tau)$, on the line $\mathrm{l}(\tau)$, the first order consecutive line of l. It is the foot on $\mathrm{l}(\tau)$ of the common perpendicular $\mathrm{n}(\tau)$ between $\mathrm{l}(\tau)$ and the first order consecutive screw axis $\mathrm{s}(\tau)$ of s. According to (5.11) the coordinates of the latter are

$$P_{41}=0, \quad P_{42}=\varepsilon\tau, \quad P_{43}=1, \quad P_{23}=0, \quad P_{31}=(\mu-\sigma_0)\varepsilon\tau, \quad P_{12}=0. \tag{8.10}$$

The formulas (8.1) show that those of $\mathrm{l}(\tau)$ are

$$\begin{aligned} &P_{14}=p_{14}-p_{24}\tau, \quad P_{24}=p_{24}+p_{14}\tau, \quad P_{34}=p_{34},\\ &P_{23}=p_{23}+(\sigma_0p_{24}-p_{31})\tau, \quad P_{31}=p_{31}+(p_{23}-\sigma_0p_{14})\tau, \quad P_{12}=p_{12}, \end{aligned} \tag{8.11}$$

p_{ij} being the coordinates of l in E.

The line $\mathrm{n}(\tau)$ perpendicular to (8.10) and (8.11) has, to the first order, the direction

$$p_{24}+(p_{14}-\varepsilon p_{34})\tau, \quad -p_{14}+p_{24}\tau, \quad \varepsilon p_{14}\tau. \tag{8.12}$$

The plane through $\mathrm{s}(\tau)$ and the direction (8.12) reads

$$\{p_{14}-p_{24}\tau\}X+\{p_{24}+(p_{14}-\varepsilon p_{34})\tau\}Y+(\varepsilon p_{24}\tau)Z+\varepsilon(\sigma_0-\mu)p_{14}\tau=0. \tag{8.13}$$

The point $S(\tau)$ that we seek is the intersection of this plane and the line $\mathrm{l}(\tau)$. After a certain amount of algebra we obtain for the coordinates of $S(\tau)$:

$$\begin{aligned} &X=-p_{12}p_{24}N^{-1}-p_{14}B\tau, \quad Y=p_{12}p_{14}N^{-1}-p_{24}B\tau,\\ &Z=(p_{13}p_{14}+p_{23}p_{24})N^{-1}-(p_{34}B-\delta)\tau, \end{aligned} \tag{8.14}$$

with

(8.15) $$N = p_{14}^2 + p_{24}^2, \quad B = \{-p_{12} - \varepsilon p_{23} - \varepsilon(\sigma_0 - \mu)p_{14}\}N^{-1},$$

δ being the distribution parameter (8.9) of l.

For $\tau = 0$, (8.14) gives us the striction point S of (8.8). The direction of the tangent at S to the striction line follows from (8.14):

(8.16) $$-p_{14}B, \quad -p_{24}B, \quad -p_{34}B + \delta.$$

Example 48. Show that (8.14) is in the tangent plane of R at S, as should be expected.
Example 49. Show that the tangent at S to the striction curve coincides with l if l is a path tangent.

Other second order properties of a moving line deal with the curvature configuration of the ruled surface R at an arbitrary point P of l. Of the two principal curvatures at P one is zero, the other one—in the plane through P orthogonal to l—could be found by Euler's theorem because we are able to determine the curvature of another curve on R, *viz.* the path of P. We know, however, from (6.31) that this is expressed by a complicated function of the coordinates of P and for this reason we do not develop this line of reasoning any further.

There is another concept, associated with the generator of a ruled surface, which is studied in the differential geometry of lines. It may be defined by kinematical terminology. Consider the orthogonal triad T, with its origin at the striction point S and consisting of three lines through it: the generator l, the normal n at S on R, and the line orthogonal to both and hence lying in the tangent plane. This triad could be regarded as a kind of (first order) representation of the system containing the moving line l. The position of T is a function of τ and we may study the motion of T. This is obviouly different from the motion of E with respect to Σ: S is not a fixed point on l, the origin of T describes the striction curve which is not the path of a point of E. We consider the position T(0) of T for $\tau = 0$ and the consecutive position T(τ); the latter follows from T(0) by a screw motion about an axis the limit of which for $\tau \to 0$ will be defined as the "striction axis" of l and denoted by q. It is obviously a second order covariant of l.

To determine q we note that the motion of T may be considered as the result of the translation of S, which follows from (8.14), and a rotation about an axis q′ through S. The direction of l(τ) is given by (8.1) to be $\{(p_{12} - p_{24}\tau), (p_{24} + p_{14}\tau), p_{34}\}$; hence if $S(\tau)B_1(\tau)$ is a unitvector on l(τ), B_1 is a definite point of T. If $X(\tau)$, $Y(\tau)$, $Z(\tau)$ are the coordinates of $S(\tau)$, those of $B_1(\tau)$ are

$$X(\tau) + (p_{14} - p_{24}\tau)N_1^{-1/2}, \quad Y(\tau) + (p_{24} + p_{14}\tau)N_1^{-1/2}, \quad Z(\tau) + p_{34}N_1^{-1/2},$$

with $N_1 = p_{14}^2 + p_{24}^2 + p_{34}^2$. This implies that the relative velocity of B_1 with respect to S, at $\tau = 0$, is

$$\{-p_{24}N_1^{-1/2},\ p_{14}N_1^{-1/2},\ 0\}. \tag{8.17}$$

We take a unit vector $S(\tau)B_2(\tau)$ on the normal $\mathrm{n}(\tau)$; the relative coordinates of $B_2(\tau)$ with respect to $S(\tau)$ are in view of (8.12):

$$\{p_{24} + (p_{14} - \varepsilon p_{34})\tau\}N_2^{-1/2},\quad \{-p_{14} + p_{24}\tau\}N_2^{-1/2},\quad \varepsilon p_{14}\tau N_2^{-1/2},$$

with $N_2 = N - 2\varepsilon p_{24}p_{34}\tau$; from this it follows (after some algebra) that the relative velocity of B_2 with respect to S, at $\tau = 0$, is

$$p_{14}(N - \varepsilon p_{14}p_{34})N^{-3/2},\quad p_{24}(N - \varepsilon p_{14}p_{34})N^{-3/2},\quad \varepsilon p_{14}N^{-1/2}. \tag{8.18}$$

The rotation axis q′ through S is perpendicular to both (8.17) and (8.18), which implies that the direction of q′ is

$$-\varepsilon p_{14}^2,\quad -\varepsilon p_{14}p_{24},\quad N - \varepsilon p_{14}p_{34}, \tag{8.19}$$

from which it follows that q′ lies in the tangent plane. If ω' is the scalar angular velocity, of T, the vector $\boldsymbol{\omega}'$ has in view of (8.19), the components

$$\varepsilon p_{14}^2 N_3^{-1/2}\omega',\quad \varepsilon p_{14}p_{24}N_3^{-1/2}\omega',\quad (-N + \varepsilon p_{14}p_{34})N_3^{-1/2}\omega', \tag{8.20}$$

with

$$N_3 = \varepsilon^2 p_{14}^2 N_1 - 2\varepsilon p_{14}p_{34}N + N^2. \tag{8.21}$$

The condition that $\boldsymbol{\omega}' \times SB_1$ is equal to (8.17) (or $\boldsymbol{\omega}' \times SB_2$ equal to (8.19)) gives us

$$\omega' = -N_3^{-1/2}N^{-1}, \tag{8.22}$$

which implies that $\boldsymbol{\omega}'$ has the components

$$-\varepsilon p_{14}^2 N^{-1},\quad -\varepsilon p_{14}p_{24}N^{-1},\quad -(-N + \varepsilon p_{14}p_{34})N^{-1}. \tag{8.23}$$

Hence, the rotation part of the screw motion of T is now determined. It must be combined with the translation, which follows from the velocity of S. As the latter also lies in the tangent plane, the striction axis q, parallel to q′, will intersect the normal n. Hence the theorem: *the instantaneous screw axis* s, *the moving line* l *and its striction axis* q *have a common perpendicular.*

To fix the position of q we must determine its intersection Q with n. If S' is the intersection of n and s, and we set $S'Q = d$, the coordinates of Q are

$$-p_{24}dN^{-1/2},\quad p_{14}dN^{-1/2},\quad (p_{13}p_{14} + p_{23}p_{24})N^{-1}, \tag{8.24}$$

and its relative coordinates with respect to S:

$$(8.25) \qquad -p_{24}(d-p_{12}N^{-1/2})N^{-1/2}, \quad p_{14}(d-p_{12}N^{-1/2})N^{-1/2}, \quad 0.$$

The velocity of Q caused by the rotation is equal to $\boldsymbol{\omega}' \times \boldsymbol{r}$, where $\boldsymbol{\omega}'$ is given by (8.23) and the vector $\boldsymbol{r}$ by (8.25). If we add to this the translation velocity $(-p_{14}B, -p_{24}B, -p_{34}B+\delta)$ we obtain the complete velocity of Q. The condition that this has the direction of $\boldsymbol{\omega}'$ gives us an equation for d. We obtain

$$(8.26) \quad dN^{-1/2} = \{(NB - \varepsilon p_{14}\delta)N/((N-\varepsilon p_{14}p_{34})^2 + N\varepsilon^2 p_{14}^2)\} + p_{12}N^{-1},$$

in which $N = p_{14}^2 + p_{24}^2$; B is given by (8.15) and δ by (8.9). The coordinates of Q and hence the Plücker coordinates of the striction axis q follow from (8.24) and (8.26). They depend on the instantaneous invariants $\sigma_0, \varepsilon, \mu$.

Example 50. Show that the coordinates q_{ij} of q are polynomials of degree six in p_{ij}.
Example 51. Determine the pitch belonging to the axis q.

Our development is valid for any line l which is not parallel to the screw axis s. (For a line parallel to O_Z we have $p_{14} = p_{24} = 0$, hence $N = 0$; the striction point S on l is undetermined.) Our formulas become special for a line l which is the path tangent of a point P; for such a line, belonging to (8.2), we have $\delta = 0$ and $\sigma_0 N = -p_{12}p_{34}$.

Example 52. Modify the formulas (8.15), (8.16), (8.23), (8.26) for the case just mentioned.

Until now we have indicated a moving line 1 of E by its Plücker coordinates p_{ij} with respect to the canonical frame. Another way to do so, less symmetrically but more "geometrically", makes use of cylindrical coordinates. Let (r, h, ψ) be the coordinates of the striction point S and η the angle of l with the O_{XY}-plane; we exclude the case $\eta = \pi/2$. (Fig. 34.) Then we have

$$(8.27) \quad \begin{aligned} &p_{14} = -\sin\psi\cos\eta, \quad p_{24} = \cos\psi\cos\eta, \quad p_{34} = \sin\eta \\ &p_{23} = h\cos\psi\cos\eta - r\sin\psi\sin\eta, \\ &p_{31} = h\sin\psi\cos\eta + r\cos\psi\sin\eta, \\ &p_{12} = -r\cos\eta, \end{aligned}$$

from which it follows

$$(8.28) \quad \begin{aligned} &\delta = \sigma_0 - r\tan\eta, \quad N^{-1/2} = \cos\eta, \\ &B = \{r + \varepsilon(\sigma_0 - \mu)\sin\psi + \varepsilon r\sin\psi\tan\eta - h\varepsilon\cos\psi\}/\cos\eta. \end{aligned}$$

This implies that the direction of the tangent at S to the striction curve (8.16) is

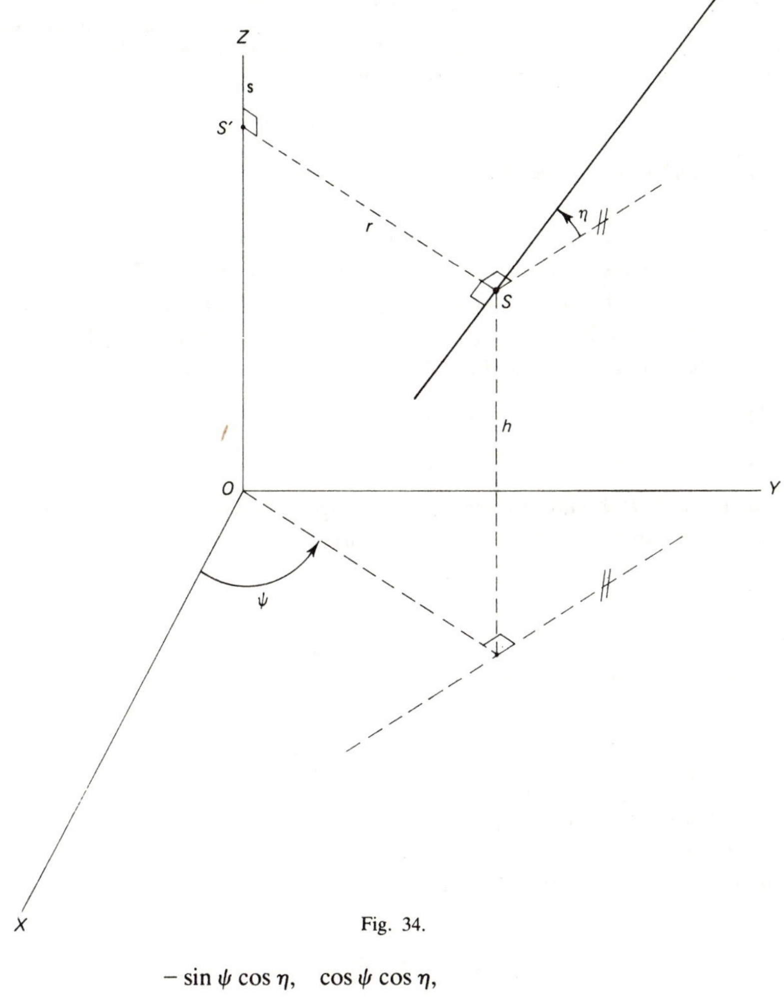

Fig. 34.

$$-\sin\psi\cos\eta,\quad \cos\psi\cos\eta,$$

$$(8.29)\qquad \begin{aligned}&\{\varepsilon(\sigma_0-\mu)\sin\eta\sin\psi+\varepsilon r\sin\eta\tan\eta\sin\psi\\&\quad -h\varepsilon\sin\eta\cos\psi+\sigma_0\cos\eta\}/\{r+\varepsilon(\sigma_0-\mu)\sin\psi\\&\qquad +\varepsilon r\sin\psi\tan\eta-h\varepsilon\cos\psi\}.\end{aligned}$$

Example 53. Verify that for $\delta = 0$ the direction (8.29) coincides with that of l.
Example 54. Show that the direction of the striction axis reads

$$-\varepsilon\sin^2\psi,\quad \varepsilon\sin\psi\cos\psi,\quad 1+\varepsilon\sin\psi\tan\eta.$$

Example 55. Show that parallel lines have parallel striction axes.
Example 56. Express the position of the intersection Q of the normal n and the striction axis (8.24, 8.26) by means of cylindrical coordinates. (It can be verified that this result is exactly the same as Disteli's equation ((9.16) and Chapter XIII, (5.16)).

9. A generalization of the Euler–Savary formula

In studying the motion of a line it is interesting to examine the limiting case of the (1, 1) correspondence discussed in Chapter IV, Section 7. This means we seek the line l in the moving system which, to within the second order, remains at a fixed distance D and angle θ from a given line l_c in Σ. Hence, if the Plücker vectors of l are $(\boldsymbol{l}, \boldsymbol{l}')$ and those of l_c are $(\boldsymbol{l}_c, \boldsymbol{l}'_c)$, we require that

$$\boldsymbol{l} \cdot \boldsymbol{l}_c = \cos\theta \tag{9.1}$$

$$\boldsymbol{l} \cdot \boldsymbol{l}'_c + \boldsymbol{l}' \cdot \boldsymbol{l}_c = D \sin\theta \tag{9.2}$$

be such that $\boldsymbol{l}_c$, $\boldsymbol{l}'_c$, θ, and D remain constant up to the second order. If τ is the motion parameter, we have for the first order

$$(\mathrm{d}\boldsymbol{l}/\mathrm{d}\tau) \cdot \boldsymbol{l}_c = 0 \tag{9.3}$$

$$(\mathrm{d}\boldsymbol{l}/\mathrm{d}\tau) \cdot \boldsymbol{l}'_c + (\mathrm{d}\boldsymbol{l}'/\mathrm{d}\tau) \cdot \boldsymbol{l}_c = 0, \tag{9.4}$$

and for the second order properties

$$(\mathrm{d}^2\boldsymbol{l}/\mathrm{d}\tau^2) \cdot \boldsymbol{l}_c = 0 \tag{9.5}$$

$$(\mathrm{d}^2\boldsymbol{l}/\mathrm{d}\tau^2) \cdot \boldsymbol{l}'_c + (\mathrm{d}^2\boldsymbol{l}'/\mathrm{d}\tau^2) \cdot \boldsymbol{l}_c = 0 \tag{9.6}$$

The Plücker coordinates of l in the canonical system are given by the vector components $\boldsymbol{l}(p_{14}, p_{24}, p_{34})$ and $\boldsymbol{l}'(p_{23}, p_{31}, p_{12})$; it follows that all the derivatives can be expressed in terms of p_{ij} and the motion invariants if we substitute eq. (8.1). In fact, (8.1) yields

$$\mathrm{d}\boldsymbol{l}/\mathrm{d}\tau = (-p_{24}, p_{14}, 0) \tag{9.7}$$

$$\mathrm{d}\boldsymbol{l}'/\mathrm{d}\tau = (\sigma_0 p_{24} - p_{31}, p_{23} - \sigma_0 p_{14}, 0) \tag{9.8}$$

which shows that $(\mathrm{d}\boldsymbol{l}/\mathrm{d}\tau, \mathrm{d}\boldsymbol{l}'/\mathrm{d}\tau)$ are the Plücker vectors of the common normal between l and the screw axis. From (9.3) and (9.4) it follows that this normal also intersects l_c at a right angle. The result is that *the screw axis and each conjugate line pair* $(\mathrm{l}, \mathrm{l}_c)$ *always have one common normal* as shown in Fig. 35. We call this common normal a *ray* and locate it relative to the canonical system by its intercept distance h, measured along the screw axis,

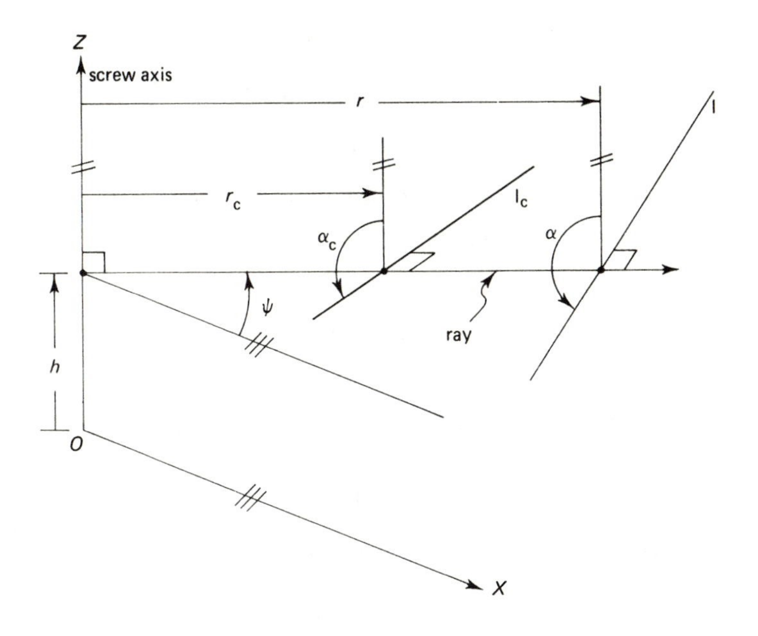

Fig. 35.

and the ray angle ψ, measured with respect to the X-axis of the canonical system. It should be noted that (9.7) and (9.8) are the Plücker coordinates of the ray in terms of the coordinates of l.

We introduce coordinates (r, α) and (r_c, α_c) which define the positions of l and l_c, respectively, along the ray. These coordinates are all measured relative to the screw axis as shown in the figure. By referring to the figure it is possible to express the Plücker coordinates of l in terms of the four coordinates (h, ψ), (r, α), we have:

$$p_{14} = \sin\alpha \sin\psi, \quad p_{24} = -\sin\alpha\cos\psi, \quad p_{34} = \cos\alpha \tag{9.9}$$

$$\begin{aligned} &p_{23} = -r\sin\psi\cos\alpha - h\sin\alpha\cos\psi, \\ &p_{31} = r\cos\psi\cos\alpha - h\sin\alpha\sin\psi, \quad p_{12} = r\sin\alpha. \end{aligned} \tag{9.10}$$

Example 57. Verify that (9.9) and (9.10) are the same as (8.27), if we substitute $\eta = \alpha + \pi/2$.

Similarly, if q_{ij} are the coordinates of l_c we have:

(9.11) $$q_{14} = \sin\alpha_c \sin\psi, \quad q_{24} = -\sin\alpha_c\cos\psi, \quad q_{34} = \cos\alpha_c$$

(9.12) $$q_{23} = -r_c\sin\psi\cos\alpha_c - h\sin\alpha_c\cos\psi,$$
$$q_{31} = r_c\cos\psi\cos\alpha_c - h\sin\alpha_c\sin\psi, \quad q_{12} = r_c\sin\alpha_c.$$

Substituting from eq. (8.1) into (9.5) yields:

(9.13) $$(-p_{14} + \varepsilon p_{34})q_{14} - p_{24}q_{24} - \varepsilon p_{14}q_{34} = 0.$$

Into this we substitute from (9.9) and (9.11) to obtain

(9.14) $$\cot\alpha - \cot\alpha_c = 1/(\varepsilon\sin\psi).$$

This can be considered as the *Spherical Euler–Savary equation.* It gives the correspondence between the inclination of l and l_c in terms of the ray angle and the second order invariant ε. For spherical motions (9.14) defines the second order, $(\mathrm{l}, \mathrm{l}_c)$ correspondence; for spatial motions it gives only the angular part of the correspondence. The locational aspects are obtained from eq. (9.6):

Into (9.6) we substitute from eq. (8.1) and obtain

(9.15) $$\begin{aligned}&(-p_{14} + \varepsilon p_{34})q_{23} - p_{24}q_{31} - \varepsilon p_{14}q_{12}\\ &\quad + (-p_{23} - \mu\varepsilon p_{34} + 2\sigma_0 p_{14} + \varepsilon p_{12} + \lambda p_{24})q_{14}\\ &\quad + (-p_{31} + 2\sigma_0 p_{24} - \lambda p_{14})q_{24} + (-\varepsilon p_{23} + \mu\varepsilon p_{14})q_{34} = 0.\end{aligned}$$

Substituting eqs. (9.9)–(9.12), and (9.14) we find that

(9.16) $$[r/\sin^2\alpha - r_c/\sin^2\alpha_c]\varepsilon\sin\psi = h\cot\psi - 2\sigma_0 + \mu.$$

Once the angles α and α_c are known, eq. (9.16) gives the correspondence between r and r_c in terms of the ray coordinates (h, ψ) and three invariants $\varepsilon, \mu, \sigma_0$. (Note that it is independent of one of the second order invariants, λ, which has dropped out.) Clearly (9.16) is linear in r and r_c, as expected.

Eqs. (9.14) and (9.16) together form the spatial analog of the Euler–Savary formula for planar motion (Chapter VIII, Section 9). They were first obtained by DISTELI [1914] using an analysis similar to the derivation given in Section 7.

The following convention governs the signs: (r, α) and (r_c, α_c) are according to the right-hand-screw rule with the thumb pointing along the ray; the sense of the ray is such that $0 \leqslant \psi < \pi$. (h, ψ) are defined with the thumb in the direction of the screw axis. It should be noted that ψ, which is the angle between the X-axis and the ray, is the compliment of the angle between the axode tangent plane (i.e. O_{xz}) and the (single) plane defined by the spherical images of l, l_c, and the screw axis.

10. Some general remarks

We make some general remarks on the four preceding sections.

In Section 6 we considered the locus of a moving point, its path, and the tangents and osculating planes of this path. In Section 7 we studied a moving plane, the set of its positions, its characteristics, and the curve which is the locus of the points of intersection of three consecutive planes. In both cases there appears a developable surface. However, it is obvious that the two configurations belong to essentially different concepts. In the first case the point is in E, the tangents and the osculating planes are in Σ; in the second the plane is in E, the characteristics and the enveloping curve are in Σ. It is clear that an object of the first type is even different from that of the second type under the inverse motion.

In Section 8 the moving element is a line. In general its locus is not a developable ruled surface, but if it is, the resulting configuration is also different from the two types just mentioned: the tangent belongs to E, the points of the cuspidal curve and its osculating planes are elements of Σ.

We add a remark about Sections 6, 8, and 9. In each a different curvature concept has been considered: In Section 6 the center of curvature of a point of a path and its radius of curvature; in Section 8 the striction axis and its pitch for a moving line which is a generator of a ruled surface; in Section 9 the center line which lies at a fixed distance and angle from a moving line.

In Chapter VIII, on plane kinematics, we shall see that there is for this special case a simple expression, the famous Euler–Savary formula, for the center of curvature M of the path of a moving point A. It contains of course instantaneous invariants up to the second order (only one in number in this case) and moreover A and M are collinear with the origin of the canonic frame. One has often tried to generalize this formula to spatial kinematics, but it is important to understand precisely which analog is being sought. If one remains bound by the notion of curvature the analog is far from satisfying as our results in Sections 6 and 8 clearly show. The best of the two is obviously that of Section 8; indeed, the moving line l, its striction axis q and the screw axis s have a common perpendicular which is a reasonable analog for the collinearity of A, M, and O in the planar case. But we must keep in mind that the curvature configuration of l does not only depend on the distance d as defined in Section 8, but also on the angle between q and s, and moreover on the pitch belonging to the axis q. That means that the analog if it is to be defined *only* in terms of curvature is incomplete *per se*, and that it does not seem to have been developed. On the other hand, substituting (8.9) into (8.26) removes the distribution parameter, and the analog seems much improved.

Using a different approach, we arrive at the same analog in Section 9. Here we identified line pairs for which the distance and angle between lines l in E and l_c in Σ remains constant to the second order of the motion of E. Each pair l, l_c is analogous to the point pair A, M of the planar curvature theory. The collinearity of A, M, and the canonic frame's origin is now given by the condition that l, l_c and s have a common perpendicular. This analog viewed in terms of "dualizing" the spherical analog of the planar theory seems to be a perfect one. This notion will be further developed when we discuss dual numbers in Chapter XIII, Section 5.

There is one other idea associated with the Euler–Savary equation which should not be confused with the former. There exists in plane kinematics a formula resembling the Euler–Savary one, but essentially different from it. We shall call it the *second* Euler–Savary formula. It deals with a relationship between the curvatures of the centrodes at their point of contact, the pole of the motion; it expresses the fact that these two curvatures, being obviously third order concepts, have a difference which depends on second order instantaneous invariants only. From this theorem follows a construction which enables one to find the center of curvature of any moving point if the pole and the centers of curvature of the centrodes are known. This is, of course, an inefficient theorem from the theoretical point of view, since it makes use of third order data to determine a second order covariant (this implies that it is a non-covertible construction: the knowledge of the center of curvature of any moving point does not enable one to derive the centrodes' centers of curvature), but it has its advantages in practice because special motions are often defined by their centrodes.

The problem to generalize the construction based on the second Euler–Savary formula to spatial kinematics was successfully solved by DISTELI [1914] (see also GARNIER [1954, 1956], DIZIOĞLU [1974], TÖLKE [1976], VELDKAMP [1976]). He developed a method to determine the striction axis of any moving line when the striction axes of the screw axis on the two axodes are known. It will not be reproduced here; we refer to the respective literature and to our Chapter XIII, Section 5, where we use dual numbers to obtain this result. DISTELI [1914] also derived spatial analogs of the Aronhold and Bobillier constructions of planar kinematics (Chapter VIII).

11. The velocity distribution

Until now we have dealt in this chapter with geometric kinematics, considering the motion of E independent of the time scale. This enabled us to make use of canonic frames and a normalized parameter satisfying $\omega = 1$.

We consider now time-dependent kinematics. Our analytical apparatus is given by Chapter II, (4.5), modified by the special choice of the origin, which implies that (5.6) and (5.8) of Chapter II hold. Hence we have up to the second order

$$(11.1)\qquad \begin{aligned} &\boldsymbol{P}_0 = \boldsymbol{p}, \qquad \boldsymbol{P}_1 = \boldsymbol{\omega} \times \boldsymbol{p} + \sigma_0 \boldsymbol{\omega}, \\ &\boldsymbol{P}_2 = -\omega^2 \boldsymbol{p} + \boldsymbol{\varepsilon} \times \boldsymbol{p} + (\boldsymbol{\omega} \cdot \boldsymbol{p})\boldsymbol{\omega} + \lambda \boldsymbol{\omega} + \mu \boldsymbol{\varepsilon}. \end{aligned}$$

If we introduce the Cartesian frame, with O_Z along $\boldsymbol{\omega}$, the O_{YZ} plane through $\boldsymbol{\omega}$ and $\boldsymbol{\varepsilon}$, O_X along $\boldsymbol{\varepsilon} \times \boldsymbol{\omega}$, we obtain $\boldsymbol{\omega} = (0, 0, \omega)$, $\boldsymbol{\varepsilon} = (0, \varepsilon_2, \varepsilon_3)$ and therefore

$$(11.2)\qquad \begin{array}{lll} X_0 = x, & X_1 = -\omega y, & X_2 = -\omega^2 x + \varepsilon_2 z - \varepsilon_3 y, \\ Y_0 = y, & Y_1 = \omega x, & Y_2 = \varepsilon_3 x - \omega^2 y + \mu \varepsilon_2, \\ Z_0 = z, & Z_1 = \sigma_0 \omega, & Z_2 = -\varepsilon_2 x + \lambda \omega + \mu \varepsilon_3, \end{array}$$

the columns of (11.2) give us the instantaneous position, velocity and acceleration respectively. For $\omega = 1$, $\varepsilon_2 = \varepsilon$, $\varepsilon_3 = 0$ the formulas (11.2) simplify to those for time independent motion.

Example 58. If L and T are the symbols for length and time respectively show that the dimensions of the coordinates and of σ_0 (the geometric pitch) are L, that of ω is T^{-1}, those of ε_2 and ε_3 are T^{-2}, λ has the dimension LT^{-1}, that of μ is L.

If v_x, v_y, v_z stand for the velocity components of the point (x, y, z) the "velocity distribution" of E is given by

$$(11.3)\qquad v_x = -\omega y, \quad v_y = \omega x, \quad v_z = \sigma_0 \omega.$$

It depends on the two parameters ω and σ_0, which we suppose both unequal to zero, this being the general case. We now derive some properties of the distribution: there is no point with velocity zero. The scalar value v of the velocity is given by $v^2 = (x^2 + y^2 + \sigma_0^2)\omega^2$; the locus of all points with the same given v is a cylindrical shell with axis O_Z; the velocity is minimum for points on O_Z.

The points whose velocity has a given direction (α, β, γ) satisfy $-y : x : \sigma_0 = \alpha : \beta : \gamma$; their locus is (if $\gamma \neq 0$) a line parallel to the screw axis; they all have the same v. More interesting is the locus of those points P for which the velocity is directed to or from a given finite point A. In view of the screw symmetry of the distribution we may without loss of generality take the point as $A = (a, 0, 0)$. Then the conditions are

$$(x - a) : -y = y : x = z : \sigma_0 = u,$$

u being a parameter. From this it follows that the homogeneous coordinates of P are

$$x : y : z : w = a : au : \sigma_0 u(1 + u^2) : 1 + u^2. \tag{11.4}$$

Hence the locus is (for $a \neq 0$) a twisted cubic. It passes through A (for $u = 0$), through the point at infinity on the screw axis (for $u = \infty$), and through the isotropic points of the O_{XY}-plane (for $1 + u^2 = 0$). On the portion of the cubic corresponding to $u > 0$ the velocity is directed from A, while for the portion $u < 0$ it is directed towards A. For v we have $v^2 = w^2\{\sigma_0^2 + a^2/(1 + u^2)\}$, which implies that, of all the points on the cubic curve, A has the maximum velocity. We have already met this cubic locus in Section 6; it lies on the (quadratic) cone of A with respect to the tetrahedral complex (6.4).

We now turn to investigate the velocities of the points on a line l. If l coincides with the screw axis s all points have equal velocity, directed along s; if l is parallel to s the velocity vectors are parallel and in one plane. For the general case we may restrict ourselves to the line $x = a$, $z = y \tan \nu$. An arbitrary point on it is $(a, u, u \tan \nu)$ and this point has as velocity vector $(-\omega u, \omega a, \sigma_0 \omega)$. The endpoint of the velocity vector has the coordinates $(a - \omega u, u + \omega a, u \tan \nu + \sigma_0 \omega)$, hence the locus of the endpoints of the velocity vectors for all points on l is a line l′. Since $v^2 = \omega^2(u^2 + a^2 + \sigma_0^2)$, the velocity is a minimum if $u = 0$, that is for the striction point of l.

Example 59. Show that as ω varies the locus of the endpoints of the velocity vectors of l is the quadric surface $(X - a)(\sigma_0 - a \tan \nu)^2 + (Z - Y \tan \nu)(\sigma_0 Y - aZ) = 0$.

Example 60. Show that this quadric is a hyperbolic paraboloid and that the two systems of generators are given by $u =$ constant and $\omega =$ constant.

Example 61. Show that the angle ψ between l and l′ satisfies $\cos \psi = (1 + \omega^2 \cos^2 \nu)^{-1/2}$.

Example 62. Show that l and l′ intersect if $\sigma_0 = a \tan \nu$; the paraboloid degenerates and l belongs to the line complex (8.2).

The direction of l is $(0, \cos \nu, \sin \nu)$, hence the projection on l of the velocity of a point P on l is $\omega(a \cos \nu + \sigma_0 \sin \nu)$, this is the same for all points. This projection is zero if $\tan \nu = -a/\sigma_0$. Obviously then, if one point of a line l has its velocity directed normal to l so do all points of l. This implies that all path normals have this property: the velocity of a point P is along the tangent to its path and therefore perpendicular to any normal. The conclusion is, in view of (6.8), that any line l of the linear complex of normals has the property that the velocity of every point on it is perpendicular to l.

Example 63. Prove directly that l belongs to the linear complex if $\tan \nu = -a/\sigma_0$.
Example 64. Determine the paraboloid of Example 59 if l is a path normal and show that it is orthogonal.

We consider now the velocity vectors of the points of a moving plane u of E.

In the general case u has an intersection with the screw axis and in view of the symmetry of the distribution u may be given the equation $x = z \tan \alpha$, as we did in Section 7. A point P in u is then $(z \tan \alpha, y, z)$ and its velocity vector reads $(-\omega y, \omega z \tan \alpha, \omega\sigma_0)$. The direction of a normal on u is $(\cos \alpha, 0, -\sin \alpha)$. Hence the velocity vector of P is normal to u if and only if $x = z = 0$, $y = \sigma_0 \cot \alpha$. This point P_0 is the null-point of u with respect to the linear complex of the path normals. Its velocity is $v = \omega\sigma_0/\sin \alpha$.

Points in u for which the velocity vector is in u satisfy $y + \sigma_0 \tan \alpha = 0$; hence, in view of Section 7 their locus is the characteristic m of u. The velocity vectors of these points coincide with the tangents of the conic, in u, of the tetrahedral complex (7.5).

Example 65. If u is parallel to the screw axis it may be given by $x = a$; the velocity of $P(a, y, z)$ of u is $(-\omega y, \omega a, \sigma_0\omega)$. Show that: points on the line $x = a$, $y = 0$ have their velocity vectors in u, all these vectors are parallel and of equal length, and that the line is the characteristic of u. Show also: that a point not on it has a non-zero velocity component perpendicular to u; that for all points of u the velocity component in u is the same, and that there are no points with a velocity normal to u.

The endpoints of the velocity vectors of the points of m are on the line $\mathrm{m}' = \{(\sigma_0\omega + q)\tan \alpha, (\omega q - \sigma_0)\tan \alpha, q + \sigma_0\omega)\}$, q being a parameter; it intersects m, in accordance with the fact that m belongs to the tetrahedral complex (7.5).

The endpoint of the velocity vector of an arbitrary point of u is given by

$$X = z \tan \alpha - \omega y, \quad Y = y + \omega z \tan \alpha, \quad Z = z + \omega\sigma_0,$$

y and z being parameters. This verifies that these endpoints are in a plane. It intersects u along a line which is the locus of the endpoints of the characteristic of u.

Example 66. Determine the equation of this plane.
Example 67. Investigate the set of these planes if ω is variable.
Example 68. Describe the velocity distribution of a moving space if $\omega = 0$.

12. The acceleration distribution

From the third column of (11.2) it follows that the acceleration vector $\boldsymbol{a}$ of the moving point (x, y, z) reads

(12.1) $$(-\omega^2 x - \varepsilon_3 y + \varepsilon_2 z,\ \varepsilon_3 x - \omega^2 y + \mu\varepsilon_2,\ -\varepsilon_2 x + \lambda\omega + \mu\varepsilon_3).$$

Any point with its acceleration directed along its path tangent has no normal acceleration and therefore its path curvature is zero; hence the locus of all such points is the inflection curve ((6.13) and also Chapter IV, (14.3)).

Example 69. Determine the inflection curve from the condition $\boldsymbol{a} = \kappa \boldsymbol{v}$, κ being a scalar. Show that it is the curve (6.13). Prove that it is a non time-dependent concept.

Example 70. Show that $\lambda\omega + \mu\varepsilon_3$ is the tangential, and $\mu\varepsilon_2$ the normal component of the acceleration of the origin. Prove that the curvature of the path of O is equal to $1/R = \mu\varepsilon_2/\sigma_0^2\omega^2$ and that the center of curvature M is $(0, R, 0)$; prove that R is time-independent; show that it is in accordance with formula (6.30).

The condition $\boldsymbol{a} = 0$ implies three linear equations for x, y, z. This suggests that in the general case (i.e., $\omega\varepsilon_2 \neq 0$) there is one point Γ, which we call the acceleration pole, with zero acceleration. From (12.1) it follows that Γ is the point

(12.2) $$x = (\lambda\omega + \mu\varepsilon_3)/\varepsilon_2, \quad y = \{(\lambda\omega + \mu\varepsilon_3)\varepsilon_3 + \mu\varepsilon_2^2\}/\omega^2\varepsilon_2,$$
$$z = \{(\lambda\omega + \mu\varepsilon_3)(\omega^4 + \varepsilon_3^2) + \mu\varepsilon_2^2\varepsilon_3\}/\omega^2\varepsilon_2^2.$$

Obviously Γ is a point on the inflection curve.

To study the acceleration distribution we introduce a Cartesian frame $\Gamma_{\xi\eta\rho}$, with its origin at the acceleration pole and axes parallel to those of O_{xyz}. From (12.1) it follows that if the origin has zero acceleration $\lambda = \mu = 0$, and the acceleration of a point $P = (\xi, \eta, \rho)$ is seen to be

(12.3) $$(-\omega^2\xi - \varepsilon_3\eta + \varepsilon_2\rho,\ \varepsilon_3\xi - \omega^2\eta,\ -\varepsilon_2\xi).$$

We remark that the acceleration distribution towards Γ depends only on ω and $\boldsymbol{\varepsilon}$, invariants of the rotational part of the motion. This is however not true for the position of Γ. As the three components of $\boldsymbol{a}$ are homogeneous linear functions of the coordinates, we obtain: all points on a line through Γ have parallel acceleration vectors; the magnitude $|\boldsymbol{a}|$ is proportional to the radius vector ΓP. We have $(\xi, \eta, \rho)\cdot(a_\xi, a_\eta, a_\rho) = -\omega^2(\xi^2 + \eta^2)$. Hence there is one line through Γ, the coordinate axis Γ_ρ, all points P of which have their acceleration vector orthogonal to ΓP; the acceleration has the direction Γ_ξ and its value is $\varepsilon_2\rho$.

Points P whose acceleration is directed along ΓP satisfy

(12.4) $$-\omega^2\xi - \varepsilon_3\eta + \varepsilon_2\rho = h\xi, \quad \varepsilon_3\xi - \omega^2\eta = h\eta, \quad -\varepsilon_2\xi = h\rho.$$

Eliminating the coordinates we obtain a cubic equation for the parameter h:

(12.5) $$h^3 + 2\omega^2 h^2 + (\omega^4 + \varepsilon_2^2 + \varepsilon_3^2)h + \omega^2\varepsilon_2^2 = 0.$$

Hence there are three lines l_1, l_2, l_3 through Γ (two of them may be imaginary) such that any point on l_i has its acceleration along l_i; these lines shall be called the (instantaneous) acceleration axes.

The coefficients of (12.5) are all positive; therefore, if a root of (12.5) is real it is negative. The conclusion is: if a line l_i is real the acceleration of any point on it is directed towards Γ.

From (12.3) it follows that $\boldsymbol{a}$ is equal to the sum of two vectors $\boldsymbol{a}_1 = (-\omega^2\xi, -\omega^2\eta, -\omega^2\rho)$ and $\boldsymbol{a}_2 = (-\varepsilon_3\eta + \varepsilon_2\rho, \varepsilon_3\xi, -\varepsilon_2\xi + \omega^2\rho)$. The first has the direction $P\Gamma$; hence a point on l_i satisfies $\boldsymbol{a}_2 = -h_1\boldsymbol{a}_1$, i.e.,

$$-\varepsilon_3\eta + \varepsilon_2\rho = h_1\xi\omega^2, \quad \varepsilon_3\xi = h_1\eta\omega^2, \quad -\varepsilon_2\xi + \omega^2\rho = h_1\rho\omega^2.$$

Eliminating ξ, η, ρ we obtain for the new parameter h_1

$$\omega^4h_1^3 - \omega^4h_1^2 + (\varepsilon_2^2 + \varepsilon_3^2)h_1 - \varepsilon_3^2 = 0, \tag{12.6}$$

which is of course equivalent to (12.5). The angular velocity vector $\boldsymbol{\omega}$ is $(0, 0, \omega)$, the angular acceleration $\boldsymbol{\varepsilon} = (0, \varepsilon_2, \varepsilon_3)$, and the angle α between them is given by $\cos\alpha = \varepsilon_3/\varepsilon$, with $\varepsilon^2 = \varepsilon_2^2 + \varepsilon_3^2$. If we introduce the non-negative number $k = \varepsilon^2/\omega^4$, we obtain for (12.6):

$$h_1^3 - h_1^2 + kh_1 - k\cos^2\alpha = 0. \tag{12.7}$$

After some algebra the discriminant of (12.7) is seen to be

$$D = 4k^2 + (27\cos^4\alpha - 18\cos^2\alpha - 1)k + 4\cos^2\alpha, \tag{12.8}$$

and the condition that the three acceleration axes are real and distinct is

$$D < 0; \tag{12.9}$$

if $D > 0$ one is real and two are imaginary. The examples $\cos\alpha = 0$, $k < \frac{1}{4}$, and $\cos^2\alpha = 1$, show respectively that both cases are possible.

Example 71. Show from (12.8) that a necessary condition for (12.9) to be valid is $\cos^2\alpha < 1/9$.
Example 72. Show that if the three acceleration axes are real, the trihedron with edges l_i (or the related spherical triangle) has the property that each angle is equal to, or the supplement of, the opposite side. (BOTTEMA [1965].)
Example 73. Show that there are two velocity and acceleration distributions corresponding to the same acceleration axes. (BOTTEMA [1965].)

From (12.3) it follows that the locus of those points P which each have an acceleration vector with the same given scalar magnitude $|\boldsymbol{a}|$ is the quadric Q with the equation

$$(\omega^4 + \varepsilon_2^2 + \varepsilon_3^2)\xi^2 + (\omega^4 + \varepsilon_3^2)\eta^2 + \varepsilon_2^2\rho^2 - 2\omega^2\varepsilon_2\xi\rho - 2\varepsilon_2\varepsilon_3\eta\rho = a^2. \tag{12.10}$$

Applying the theory of quadric surfaces to (12.10) we find that the characteristic equation (which follows from the quadratic terms of the left-hand side) reads

$$(12.11) \qquad s^3 - 2(\omega^4 + \varepsilon^2)s^2 + (\omega^4 + \varepsilon^2)^2 s - \omega^4 \varepsilon_2^4 = 0,$$

which shows, by virtue of Descartes' Rule of Signs, that the three roots s_1, s_2, s_3 (which are always real) are all positive. Hence Q is an ellipsoid. The roots must satisfy the equation

$$(s_1 + s_2 + s_3)^2 - 4(s_2 s_3 + s_3 s_1 + s_1 s_2) = 0,$$

and, as s_i is proportional to the square of the reciprocal of the semi-length of the ith principal axis of Q, we obtain for the reciprocals of the semi-axes lengths b_1, b_2, b_3:

$$b_1^4 + b_2^4 + b_3^4 - 2b_2^2 b_3^2 - 2b_3^2 b_1^2 - 2b_1^2 b_2^2 = 0,$$

or

$$(12.12) \qquad (b_1 + b_2 + b_3)(-b_1 + b_2 + b_3)(b_1 - b_2 + b_3)(b_1 + b_2 - b_3) = 0,$$

which shows that the ellipsoid Q has the following property: the reciprocal of the shortest axis is equal to the sum of the reciprocals of the lengths of the other two principal axes.

Example 74. Prove that Q is an ellipsoid of revolution if and only if $\varepsilon^2 = 2\omega^4$, $\varepsilon_3 = 0$.

Are there lines l such that a point on l has an acceleration along l? Any line may be given in terms of the equations:

$$(12.13) \qquad \xi = (p_{31} + p_{14}\tau)/p_{34}, \quad \eta = (p_{32} + p_{24}\tau)/p_{34}, \quad \rho = \tau.$$

p_{ij} being the Plücker coordinates of l with respect to the $\Gamma_{\xi\eta\rho}$-frame, $p_{34} \neq 0$. The acceleration vector of the point $P(\tau)$ follows from (12.3); for the condition we seek it must be equal to $(k_1 p_{14}, k_1 p_{24}, k_1 p_{34})$, which gives us three linear equations for τ and k_1. Eliminating τ and k_1 we obtain for the locus of the lines l with said property

$$(12.14) \qquad \begin{aligned} &(\omega^4 + \varepsilon_3^2)p_{12}p_{34} - \omega^2 \varepsilon_2 (p_{12}p_{14} + p_{32}p_{34}) \\ &\quad + \varepsilon_2 \varepsilon_3 (p_{24}p_{12} + p_{34}p_{31}) + \varepsilon_2^2 p_{24}p_{31} = 0. \end{aligned}$$

Example 75. Show that (12.14) represents a tetrahedral complex; the associated tetrahedron has Γ as the only finite vertex, the three edges through Γ are the acceleration axes l_1, l_2, l_3, the plane at infinity is the face opposite Γ.

Example 76. Show that the locus of the endpoints of the acceleration vectors of the points of any line l is a line l'. l and l' intersect if l belongs to the complex (12.14)

It is easy to see that on a moving line l there is in general one point whose acceleration is perpendicular to l.

Example 77. Determine this point for the line (12.13)

A plane of E can be specified by giving the coordinates ξ, η, ρ of its points P as linear functions of two parameters τ_1 and τ_2. Then, by (12.3), the components of the acceleration of P are linear functions of τ_1 and τ_2, and the same holds for the coordinates of the endpoint of $\boldsymbol{a}$. Hence, the locus of the endpoints of the acceleration vectors of the points of a moving plane u is a plane, u′ say. In general u and u′ have a line of intersection, l′. Hence the locus of those points of u which have their acceleration in u will be a line l.

Example 78. If (u_1, u_2, u_3, u_4) are the coordinates of the plane, those of u′ may be obtained by determining the acceleration of three points of u, for instance, its intersections with the coordinate axes. Show that the u'_i are linear functions of u_i. Determine the planes u such that u′ and u are parallel. Show that there are three planes u which coincide with u′: the planes through two acceleration axes; any point of such a plane has its acceleration in it.
Example 79. Show that in the plane u there is in general one point whose acceleration is perpendicular to u, *viz.* the point

$$\xi = -\omega^2\varepsilon_2 u_3 u_4/N,$$

$$\eta = -\varepsilon_2(\varepsilon_2 u_2 + \varepsilon_3 u_3)u_4/N,$$

$$\rho = -\{\varepsilon_2\varepsilon_3 u_2 - \varepsilon_2\omega^2 u_1 + (\omega^4 + \varepsilon_3^2)u_3\}u_4/N,$$

with $N = (\varepsilon_2 u_2 + \varepsilon_3 u_3)^2 + \omega^4 u_3^2$.

CHAPTER VII

SPHERICAL KINEMATICS

1. Spherical displacements

We have seen in Chapter I that for a general displacement of the three-space E with respect to the fixed space Σ there is no invariant point. This chapter deals with those special displacements for which a point, O, does not move. For the equation (Chapter I, (4.5)) of a displacement this means that, taking the fixed point O as the origin, the displacement of O represented by the vector $\boldsymbol{d}$, is identically zero. Hence a displacement is given by

$$P = \mathbf{A}p \tag{1.1}$$

in any coordinate system for which the origin coincides with the fixed point. $\mathbf{A}$ is an orthogonal matrix. A consequence is that the degree of freedom of E, which is six in the general case, reduces to three. Furthermore we know that for any such displacement not only the point O but any point of a certain line s through O is invariant. Hence any such displacement is a rotation about a line through O.* The direction of a line (through O) is determined by two data, and as the displacement is known if s and the rotation angle ϕ are given, we see once more that the degree of freedom of E is three.

For a point P, different from O, the distance $|OP| = R$ must be constant. This implies that the locus of the possible positions of P is the sphere $(O; R)$. Hence our displacements are such that any point moves on a sphere with center O. That is the reason why this special branch of our science is called spherical kinematics.

If $S_1 = (O, R_1)$ and $S_2 = (O, R_2)$ are two spheres, P_1 a point on S_1, and P_2 that intersection of OP_1 and S_2 for which OP_2/OP_1 is the positive number R_2/R_1, then any configuration of points P_2 is obviously directly similar to that of the corresponding points P_1. This implies that we may restrict ourselves essentially to the motion on one of the concentric spheres.

Another remark seems of interest: As P moves on its sphere, the line OP

* This result is known as Euler's theorem on rotations about a point.

moves about O, and its position is known if we know that of its intersection P' with the plane at infinity V. Hence a displacement of P corresponds to that of P'. Spherical kinematics is therefore related to the study of certain transformations in the plane V. We know that spatial displacements have the property that the isotropic conic Ω in V is—as a whole—invariant. The plane geometry, dealing with those linear transformations for which a non-degenerated conic is invariant, is known as a non-Euclidean geometry. If the conic is imaginary, as in our case, the geometry is called elliptic. Hence there is a strong relationship between kinematics in plane elliptic geometry and spatial spherical kinematics. The two concepts are essentially identical (or isomorphic, as mathematicians call them), since the point P' of V corresponds to one line OP' and conversely. This isomorphism is only valid if we study the displacement of lines. For points this is not true, since a point P' in V corresponds to two points situated diametrically opposite on the sphere. The analog would be complete if two such points would be identified with each P'. We remark that non-Euclidean kinematics has been the subject of systematic research, for instance by GARNIER [1951]. We shall deal with it in Chapter XII.

2. The lines joining homologous points

The n-positions theory we developed in Chapter III–VI simplifies considerably in spherical kinematics.

If two positions are given, the second always follows from the first by a rotation. If we take the rotation axis as O_Z and rotation angle as ϕ, the coordinates of the positions P_1, P_2 of two homologous points are related by

$$X_1 = x, \quad Y_1 = y, \quad Z_1 = z \tag{2.1}$$

and

$$X_2 = x\cos\phi - y\sin\phi, \quad Y_2 = x\sin\phi + y\cos\phi, \quad Z_2 = z, \tag{2.2}$$

provided we take the coordinate frames in E and Σ coincident in position 1. If P_1 is not on O_Z, then P_1 and P_2 are different and their join is a line parallel to the O_{XY} plane. Conversely, assuming $\phi \neq \pi$, any line parallel to O_{XY}, not intersecting O_Z, is the join of one point pair P_1P_2. If P_1 is on O_Z, P_2 coincides with P_1; their join is undetermined since any line through P_1 must be considered as the join P_1P_2. The conclusion is that the locus of all joins P_1P_2 consists of two special linear complexes: one consisting of the lines intersecting the line of O_{XY} in V; the other, of all the lines intersecting O_Z. In the general case the locus was seen to be the quadratic complex (Chapter III,

(3.4)); in spherical kinematics the displacement d is zero and the locus degenerates into two linear complexes.

Example 1. Verify that for $d = 0$, the complex degenerates into the two special linear complexes.
Example 2. Consider the special case $\phi = \pi$.
Example 3. Show how the other results of two-positions theory derived in Chapter III are modified for spherical kinematics.

3. Three positions

Three positions theory, as developed in Chapter IV, is of course simplified for the case at hand, but the results are less trivial than for $n = 2$.

In the general case, three positions were completely determined by the three screw axes s_{23}, s_{31}, s_{12} (or s_1, s_2, s_3); we formed this configuration into a spatial triangle by adding the three common normals n_1, n_2, n_3 of the pairs s_2s_3, s_3s_1, s_1s_2 respectively. In spherical kinematics the screw axes are rotation axes, all three passing through O and thus are the edges of a trihedron T with the vertex O. All three normals also pass through O, establishing a second trihedron T′, often called the polar trihedron of T. (Conversely, T is the polar trihedron of T′.) According to Chapter IV, Section 1, three homologous points A_1, A_2, A_3 are the reflections into the edges of T′ of the fundamental (or basic) point A^* of Σ. In the plane V at infinity the situation is the same as in the general case: three homologous points are the reflections into the points N_i (of n_i) at infinity, of a point A^* in V, or, what is actually the same thing, the reflections of A^* into the sides of the polar triangle of N_i with respect to Ω, these sides being the lines at infinity of T.

In Chapter IV, Section 3 we derived the correspondence between the basic point $A^* = (x_1, x_2, x_3)$ and the plane $\alpha(U_1, U_2, U_3, U_4)$ through the three homologous points A_i, the result being a birational cubic relationship between A^* and α. We may do the same thing in spherical kinematics, but we are faced with a difference: If A^* is reflected into the normal n_i, along which a unit vector $(\alpha_i, \beta_i, \gamma_i)$ is chosen, the coordinates of $A_i = (X_i, Y_i, Z_i)$ are seen to be

$$
\begin{aligned}
X_i &= (2\alpha_i^2 - 1)x_1 + 2\alpha_i\beta_i x_2 + 2\alpha_i\gamma_i x_3,\\
Y_i &= 2\beta_i\alpha_i x_1 + (2\beta_i^2 - 1)x_2 + 2\beta_i\gamma_i x_3, \qquad (3.1)\\
Z_i &= 2\gamma_i\alpha_i x_1 + 2\gamma_i\beta_i x_2 + (2\gamma_i^2 - 1)x_3.
\end{aligned}
$$

As in the general case these are linear functions of x_1, x_2, x_3, but they are now also *homogeneous* linear functions. Therefore in the matrix (3.1) of Chapter IV the functions L_{ij} are now homogeneous. This implies that U_1, U_2, U_3 are homogeneous quadratic functions and U_4 is a homogeneous cubic function of

x_1, x_2, x_3. A conclusion is that, for a given sphere (O, R), two basic points on a line through O correspond to parallel planes α, a fact which follows of course directly from the similarity with respect to O of diametrical opposites as was mentioned before. All this has an important influence on the question whether three homologous points can be collinear. In the general case we found two loci for A^*, a twisted cubic c for finite points, derived from the condition that the matrix (3.1) of Chapter IV has a rank of two, and a plane cubic c′ in V, with the equation $x_4 = 0$. In spherical kinematics three homologous points are on the same sphere about O, which implies that in order to be collinear at least two of the points must coincide and therefore A^* must lie on a rotation axis. For instance if A_2 and A_3 coincide they are on s_1, which implies that their base point A^* is on s_1. Hence, the cubic curve c is degenerated into the three edges of T. The curve c′, however, is in general a non-degenerate plane cubic.

A point A has three positions A_i on its sphere, and it seems natural to ask for those points whose three positions are on a great circle. As this means that their plane passes through O the condition reads $U_4 = 0$; hence triplets with this property have their basic point on a cubic cone, with vertex O, which obviously passes through the plane curve c′ at infinity. Clearly then the A_i also lie on cubic cones. The locus of the points A_i ($i = 1$, or, 2, or 3) on a given sphere is therefore the intersection of this sphere and the corresponding cubic cone.

The plane cubic c′ at infinity is the same as for general kinematics and it has the properties mentioned in Chapter IV, Section 4. It passes through the rotation axes at the three points S_{23}, S_{31}, S_{12} and moreover through the intersections with Ω of the polar lines of S_{23}, S_{31}, S_{12} with respect to Ω. Indeed: all these nine points have the property that two of their three positions coincide.

Example 4. Determine the equation of c′ as a function of $\alpha_i, \beta_i, \gamma_i$ by means of (3.1). Show that the cubic cone passes through the edges of T and through the isotropic points in the faces of T′.
Example 5. Show that the locus of the planes through O which contain three homologous points is a cone of class three.

In this section we have dealt with the general three positions theory in spherical kinematics, the positions being defined by three arbitrary axes through O. There are many special cases.

Example 6. Prove the theorem named after Rodrigues and Hamilton: Successive rotations, about three concurrent lines of Σ, through angles equal to twice the angles of the planes formed by the axes, restore E to its original position. (Hence, as we already know, any two successive rotations about a fixed point can be replaced by an equivalent single rotation.)

Example 7. Consider the case where the three rotation axes are mutually orthogonal. Consider the case where the three rotation axes are coplanar.

4. Four and five positions

We consider now, in spherical kinematics, four positions of the moving space. As any one of them follows from any other by a rotation, we obtain six rotation axes. The axis of the displacement from position j to position k will be denoted by s_{jk} $(j, k = 1,2,3,4, j \neq k)$ and its point at infinity by S_{jk}. Two axes with no common index will be called complimentary or opposite; there are three pairs of opposite axes: $(\mathrm{s}_{12}, \mathrm{s}_{34})$, $(\mathrm{s}_{23}, \mathrm{s}_{14})$, $(\mathrm{s}_{31}, \mathrm{s}_{24})$.

As in the general case, the six axes are not independent. We shall study their configuration, which will be seen to be much simpler than in the general case where they must belong to a certain congruence of lines. Two pairs of opposite axes may be chosen arbitrarily, for the two remaining axes we derive a locus, which will be of course a cone with vertex O.

Let $\mathrm{s}_{12}, \mathrm{s}_{34}$ and $\mathrm{s}_{31}, \mathrm{s}_{42}$ be given; the unknown axis s_{23} will be denoted by p. Let $\boldsymbol{s}_{jk}$ be a vector along s_{jk} with Cartesian components $\alpha_{jk}, \beta_{jk}, \gamma_{jk}$; $\boldsymbol{p}$ is a vector along p, and its components are x, y, z. Obviously $\alpha_{jk}, \beta_{jk}, \gamma_{jk}$ are the homogeneous coordinates of S_{jk} in V; x, y, z are those of the point P which is the point of p at infinity.

As we know from general kinematics the condition for p is the following: the angle between the planes $(\mathrm{p}, \mathrm{s}_{12})$ and $(\mathrm{p}, \mathrm{s}_{31})$ must be equal to that between the planes $(\mathrm{p}, \mathrm{s}_{42})$ and $(\mathrm{p}, \mathrm{s}_{34})$.

The normal $\boldsymbol{n}$ to $(\mathrm{p}, \mathrm{s}_{12})$ is $\boldsymbol{p} \times \boldsymbol{s}_{12}$, the normal $\boldsymbol{n}'$ to $(\mathrm{p}, \mathrm{s}_{31})$ is $\boldsymbol{p} \times \boldsymbol{s}_{31}$; the angle between these planes is $\phi_{23}/2$ and we have

$$\tan(\phi_{23}/2) = |\boldsymbol{n} \times \boldsymbol{n}'|/(\boldsymbol{n} \cdot \boldsymbol{n}'). \tag{4.1}$$

If we expand this in terms of the components we obtain after some algebra

$$\tan(\phi_{23}/2) = LW/K, \tag{4.2}$$

with

$$\begin{aligned} W &= (x^2 + y^2 + z^2)^{1/2}, \\ L &= (\beta_{12}\gamma_{31} - \gamma_{12}\beta_{31})x + (\gamma_{12}\alpha_{31} - \alpha_{12}\gamma_{31})y + (\alpha_{12}\beta_{31} - \beta_{12}\alpha_{31})z, \\ K &= (\beta_{12}\beta_{31} + \gamma_{12}\gamma_{31})x^2 + (\gamma_{12}\gamma_{31} + \alpha_{12}\alpha_{31})y^2 \\ &\quad + (\alpha_{12}\alpha_{31} + \beta_{12}\beta_{31})z^2 - (\beta_{12}\gamma_{31} + \gamma_{12}\beta_{31})yz \\ &\quad - (\gamma_{12}\alpha_{31} + \alpha_{12}\gamma_{31})zx - (\alpha_{12}\beta_{31} + \beta_{12}\alpha_{31})xy. \end{aligned} \tag{4.3}$$

The equation $L = 0$ represents the plane (s_{12}, s_{31}). The equation $K = 0$ represents a quadratic cone, passing through s_{12} and s_{31}; it is the locus of the lines p for which $(\phi_{23}/2) = \pi/2$. We have $\phi_{23} = 0$ if $L = 0$ or if p is isotropic (i.e., $W = 0$). The angle ϕ_{23} is undetermined if p coincides with s_{12} or s_{31}, (L and K being both zero) and also if W and K are both zero. This is true for the four intersections of K and the isotropic cone Ω; from this it follows that these four lines are in the tangent planes through s_{12} and s_{31} to Ω.

Example 8. Prove the last statement analytically.
Example 9. Show that the locus of lines for which ϕ_{23} is a constant is a quartic cone.

The same angle, $\phi_{23}/2$, is also the angle between the planes (p, s_{42}) and (p, s_{34}); we obtain a formula analogous to (4.2), L and K being changed into L' and K' (by substituting s_{42} for s_{12}, and s_{34} for s_{31}). The condition $\tan(\phi_{23}/2) = \tan(\phi_{23}/2)$ then implies that the locus of s_{23} (and of s_{14}), if s_{12}, s_{34} and s_{31}, s_{42} are known, is represented by

$$LK' - L'K = 0. \tag{4.4}$$

this locus is therefore a cubic cone Γ, passing through $s_{12}, s_{34}, s_{31}, s_{42}$. If s_{23} is chosen (any generator of Γ), the four positions are completely determined; s_{14}, also on the cone, is determined and so are the six rotation angles associated to the six axes. In our derivation we started from the pairs (s_{12}, s_{31}) and (s_{42}, s_{34}); the result would have been the same if we had chosen (s_{12}, s_{42}) and (s_{31}, s_{34}).

It follows from the definition of Γ that it passes through the intersection of the planes (s_{12}, s_{31}) and (s_{42}, s_{34}), (which is also obvious from (4.4), L and L' being both zero) and through that of (s_{12}, s_{42}) and (s_{31}, s_{34}). Such intersections yield the third generator of Γ lying in the planes containing two axes. As the six axes are equivalent we have in general: the intersection of the planes (s_{jk}, s_{lk}) and (s_{jm}, s_{lm}), (j, k, l, m being mutually different) is a generator of Γ.

Example 10. Show that there are six such special generators.
Example 11. Show that the (four) intersections of K and K' are on Γ.
Example 12. Determine the equation of K if $s_{12} = (a, 0, 1)$ and $s_{31} = (-a, 0, 1)$.

The cubic cone Γ has been derived earlier, it is of course the configuration we called the screw cone in Chapter V, Section 3. In Chapter V our interest was to develop the general case and so we did not dwell on the relationship between the spherical indicatrix of a configuration and spherical kinematics. Clearly though, any aspect of a general displacement which is completely described by its spherical indicatrix must have an equivalence in spherical kinematics. So it is for Sections 3 and 4 of Chapter V: they can be read as describing four position spherical kinematics and they could without any change be inserted in this section.

Example 13. Show that (3.3) of Chapter V yields (4.4).
Example 14. Show that (4.4) is a particular case of (5.14) of Chapter V, and show that the latter gives a more general form of Γ than (4.4).
Example 15. Show that the six π-directions discussed in Chapter V, Section 4 coincide with the six special generators of Example 10.
Example 16. Show that (8.1), (8.2), (8.3) of Chapter V all apply to spherical displacements and that in this case the l_i, l_c correspondence discussed in Chapter V, Section 8, gives those lines l_i which have four homologous positions on a quadratic cone, and l_c is the cone axis.
Example 17. Show that for five positons there are at most six real l_i, l_c pairs.

In Section 3 we have found that the locus of the basic point A^* for which the three positions A_i of A are on a great circle of their sphere is a cubic cone with vertex O, its intersection with V being a cubic c′. Four positions give rise to four such curves, which we denote by $c_{123}, c_{234}, c_{341}$, and c_{412}. An intersection of c_{123} and c_{234}, for example, gives us four homologous points A_1, A_2, A_3, A_4, with the property: $OA_1A_2A_3$ and $OA_2A_3A_4$ are coplanar. This implies that the four points are coplanar provided that A_2 and A_3 are different. The two cubics have nine intersections, but both pass through S_{23} and through the isotropic points of the plane perpendicular to S_{23}. The conclusion is: in spherical kinematics there are on any sphere, about O, in general six quadruples of homologous points (real or imaginary) lying on a great circle.

The rotation (1.1) about O is given by a homogeneous linear transformation of the coordinates x, y, z in E to X, Y, Z in Σ. If four displacements are given such that (x, y, z) is transformed to X_i, Y_i, Z_i $(i = 1, 2, 3, 4)$ the four (finite) homologous points A_i are on one plane if the determinant

$$|X_i \quad Y_i \quad Z_i \quad 1|$$

is equal to zero. Developing this determinant with respect to the last column we can see that: there exists in E a cubic cone, with vertex O, such that the four positions of any point on it are coplanar, and therefore on a circle of their sphere. As four points are coplanar if two of them coincide, the cubic cone passes through the six rotation axes s_{ij} taken as lines in E, these are the axes for the inverse displacements; the so-called image screw axes.

Example 18. Show that the lines from O to the isotropic points perpendicular to s_{ij} do not belong to the cone (a general point on such a line does not have two coinciding positions).

If five positions are given we have ten rotation axes s_{jk} $(j, k = 1, 2, 3, 4, 5; j \neq k)$. We may ask for those points for which the five positions are coplanar and therefore concyclic. Omitting first one position, say the fifth, and then a different position, say the fourth, we have two four position problems. Each (i.e., 1234 and 1235) gives rise to a cubic cone as the locus of coplanar points, and therefore the nine lines of intersection of these two cubic cones must contain the required locus. Since both cones pass through axes s_{12}, s_{23}, and s_{31} (taken as image-screws in E) which do not satisfy the condition, the points we seek are those of the remaining six lines, through O, common to both cones. These lines could perhaps be named after Burmester who solved the analogous problem in plane kinematics. These six lines are not necessarily real: they may be imaginary in pairs.

Example 19. Verify that the l_i, l_c line correspondence, discussed in Chapter V, Section 10, yields exactly this same result. Which phrased in terms of the l_i, l_c correspondence says that for five positions (under spherical displacements) there are at most six lines which each have their homologus positions on right circular cones, and that these cones have the six lines l_c as axes. (ROTH [1967a].)

5. Instantaneous spherical properties

In general instantaneous kinematics, the derivatives of the coordinates of a moving point (for the canonical frame) are given by Chapter II, (6.3). They simplify for spherical kinematics because $\sigma_0 = \lambda = \mu = d_{3_X} = d_{3_Y} = d_{3_Z} = 0$. We obtain

$$(5.1)\quad \begin{array}{llll} X_0 = x, & X_1 = -y, & X_2 = -x + \varepsilon z, & X_3 = (1+\varepsilon^2)y + \gamma_Y z, \\ Y_0 = y, & Y_1 = x, & Y_2 = -y, & Y_3 = -(1+\varepsilon^2)x + ((3/2)\varepsilon - \gamma_X)z, \\ Z_0 = z, & Z_1 = 0, & Z_2 = -\varepsilon x, & Z_3 = -\gamma_Y x + ((3/2)\varepsilon + \gamma_X)y. \end{array}$$

There are no geometric instantaneous invariants of the first order; one, ε, of the second; two, γ_X and γ_Y, of the third.

A moving point (x, y, z) has, up to the third order, at the instant τ the position

$$(5.2)\quad \begin{aligned} X &= x - y\tau - \tfrac{1}{2}(x - \varepsilon z)\tau^2 + (1/6)[(1+\varepsilon^2)y + \gamma_Y z]\tau^3, \\ Y &= y + x\tau - \tfrac{1}{2}y\tau^2 - (1/6)[(1+\varepsilon^2)x - ((3/2)\varepsilon - \gamma_X)z]\tau^3, \\ Z &= z - \tfrac{1}{2}\varepsilon x\tau^2 - (1/6)[\gamma_Y x - ((3/2)\varepsilon + \gamma_X)y]\tau^3, \end{aligned}$$

X, Y, Z are homogeneous linear functions of x, y, z. It is easy to verify that the coefficients (of x, y, z) are, up to the third order, those of an orthogonal matrix. This means that the inverse of (5.2) is obtained as the transpose of its coefficients. Hence, if we solve for x, y, z from (5.2) we obtain:

$$(5.3)\quad \begin{aligned} x &= X + Y\tau - \tfrac{1}{2}(X + \varepsilon Z)\tau^2 - (1/6)[(1+\varepsilon^2)Y + \gamma_Y Z]\tau^3, \\ y &= Y - X\tau - \tfrac{1}{2}Y\tau^2 + (1/6)[(1+\varepsilon^2)X + ((3/2)\varepsilon + \gamma_X)Z]\tau^3, \\ z &= Z + \tfrac{1}{2}\varepsilon X\tau^2 + (1/6)[\gamma_Y X + ((3/2)\varepsilon - \gamma_X)Y]\tau^3, \end{aligned}$$

which gives us the inverse motion, up to the third order, with respect to the canonical frame of the direct motion.

From (5.2) it follows that the points of E which are, for $\tau = 0$, at a cusp of their path are those of the z-axis, different from O.

Example 20. Show that the cuspidal tangent has the direction of the X-axis.

We consider for the time being second order properties; they depend only upon the invariant ε, which we suppose unequal to zero. The osculating plane of the path of (x, y, z) is the plane U through three consecutive points; hence it is the plane through $(x, y, z, 1)$, $(-y, x, 0, 0)$ and $(x - \varepsilon z, y, \varepsilon x, 0)$, which gives us (if x and y are not both zero)

$$(5.4) \qquad \begin{aligned} &U_1 = \varepsilon x^2, \quad U_2 = \varepsilon xy, \quad U_3 = -(x^2 + y^2 - \varepsilon xz), \\ &U_4 = -\varepsilon x(x^2 + y^2 + z^2) + z(x^2 + y^2). \end{aligned}$$

The plane passes through O if $U_4 = 0$. This implies that

$$(5.5) \qquad K \equiv -\varepsilon x(x^2 + y^2 + z^2) + z(x^2 + y^2) = 0$$

is the equation of the cubic cone which is the locus of those points whose paths have three consecutive points in common with a great circle. It is the instantaneous case of the cone considered in Section 3 (which deals with three finitely separated positions).

Example 21. Show that for any point P on O_z, different from O, the osculating plane is not determined; any plane through P parallel to O_X is a possible osculating plane.

Example 22. Show that the cubic cone K has a double line (being then rational) only if $\varepsilon = \pm\frac{1}{2}$: the double line being $x = \pm z$, $y = 0$.

The system of path normals is trivial in this case. For a point $A(x, y, z)$, not on O_z, the direction of the tangent is $(-y, x, 0)$ and the normal plane $-yX + xY = 0$ passes through O_Z. Any normal intersects O_Z; the locus of the normals is the special complex with O_Z as axis; its equation is $p_{12} = 0$, in accordance with Chapter VI, (6.8), the pitch σ_0 being zero in this case.

If P is a point on O_Z, any line l through P is a normal at P, in fact it is even a principal normal. It is a binormal of P if l is in the O_{YZ} plane. The principal normal at a point A is the join of A and the intersection of O_Z and the osculating plane (5.4) of A.

In the general case we have found that a normal is the principal normal of two of its points, real or imaginary. In the case at hand, if l is a line through P on O_Z, one of these points is P; the second is then also real.

Example 23. If l through P is given, determine the second point on l for which l is the principal normal.

Example 24. Show that the binormal at $(x, y, z), x \neq 0$, intersects O_Z at $(0, 0, (x^2+y^2)/\varepsilon x)$.
Example 25. If l is a line intersecting O_Z determine the point on it whose binormal is l.

Curvature theory gives rather simple results for spherical kinematics. The curvature axis is the intersection of two consecutive normal planes. We know, however, that a normal plane also contains the rotation axis; hence the consecutive normal plane contains the consecutive rotation axis. But as all rotation axes pass through O this implies that the curvature axis of every point passes through O. This axis is perpendicular to the osculating plane. The conclusion is: the center of curvature of a point A coincides with the projection M of O on the osculating plane of A. The radius of curvature ρ is $|AM|$. The osculating circle to the path at A is the intersection of the osculating plane of A and the sphere with radius R on which A moves.

From (5.4) it follows that the distance $|OM| = d$ satisfies

$$d^2 = \mathrm{K}^2/N, \tag{5.6}$$

with

$$N = \varepsilon^2 x^2(x^2+y^2) + (x^2+y^2-\varepsilon xz)^2. \tag{5.7}$$

Hence

$$\rho^2 = R^2 - d^2 = (R^2N - \mathrm{K}^2)/N.$$

We obtain

$$\begin{aligned} R^2N - \mathrm{K}^2 &= R^2\varepsilon^2x^2(x^2+y^2) + R^2(x^2+y^2)^2 - 2R^2\varepsilon xz(x^2+y^2) \\ &\quad + R^2\varepsilon^2x^2z^2 - [R^4\varepsilon^2x^2 - 2R^2\varepsilon xz(x^2+y^2) + z^2(x^2+y^2)^2] \\ &= (x^2+y^2)^3. \end{aligned} \tag{5.8}$$

Hence the curvature k of the path of (x, y, z) is given by

$$k = \rho^{-1} = N^{1/2}/(x^2+y^2)^{3/2}. \tag{5.9}$$

Example 26. Show that the equation of the consecutive normal plane is $(-y-(x-\varepsilon z)\tau)X + (x-y\tau)Y - \varepsilon x\tau Z = 0$.
Example 27. Show that the coordinates of M are $X = -\varepsilon x^2u$, $Y = -\varepsilon xyu$, $Z = (x^2+y^2-\varepsilon xz)u$, with $u = \mathrm{K}/N$.
Example 28. Check that M coincides with O if A is a point of the cone (5.5).

In spherical kinematics every path lies on a sphere: they are all spherical curves. In differential geometry a concept has been introduced which is a generalization of elementary curvature. (In plane geometry the curvature of a curve is a measure of its local deviation from a straight line, the tangent at the point.) If A' is a consecutive point of A, the curvature is defined as the limit of

$\Delta\alpha/\Delta s$, $\Delta\alpha$ being the angle between the tangents at A' and A, and Δs the distance AA'.

For a curve p′ on a given surface F there has been introduced the notion of geodesic curvature, which is a measure of the local deviation of p′, at a point A', from the geodesic on F which is tangent to p′ at A. Its definition is similar to ordinary curvature except that $\Delta\alpha$ is now the angle between the geodesics tangent to p′ at A and A' respectively.

In the case at hand F is a sphere and the geodesics are its great circles. We determine the geodesic curvature of an arbitrary circle c′ with radius r on a sphere with radius R, at the point A. Making use of spherical coordinates, chosen so that the Z-axis coincides with the circle axis (Fig. 36), the points of c′ are given by

$$X = R \sin\beta \cos\psi', \quad Y = R \sin\beta \sin\psi', \quad Z = R\cos\beta.$$

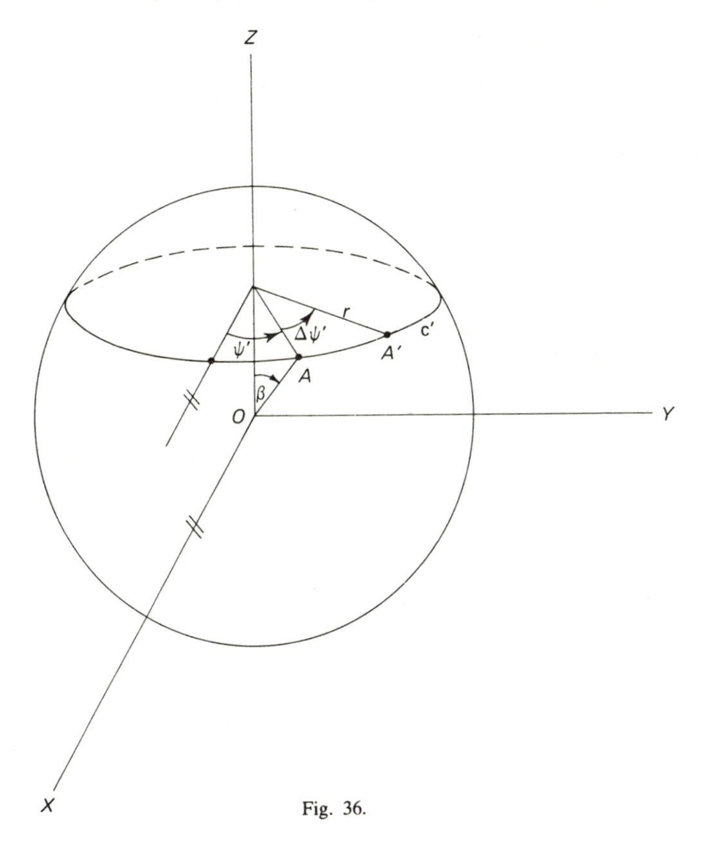

Fig. 36.

For a given circle β is constant and ψ' variable. The direction of the tangent at A to the circle c′ is $(-\sin\psi', \cos\psi', 0)$ and the plane through O and this tangent is $(\cos\beta\cos\psi')X + (\cos\beta\sin\psi')Y - (\sin\beta)Z = 0$. Hence, for the angle $\Delta\alpha$ between the tangential geodesics at A and A' we obtain $\cos(\Delta\alpha) = \cos^2\beta\cos(\Delta\psi') + \sin^2\beta$, which implies $\sin\frac{1}{2}(\Delta\alpha) = \cos\beta\sin\frac{1}{2}(\Delta\psi')$. This means that $\Delta\alpha \to \Delta\psi'\cos\beta$. Furthermore $\Delta s = R(\sin\beta)\Delta\psi'$; hence the geodesic curvature k_g of c′ reads $k_g = R^{-1}\cot\beta$, a well-known formula in the differential geometry on a sphere.

The geodesic curvature of the path is obviously the same as that of its osculating circle. Furthermore we have $\cot\beta = d/\rho$; hence we obtain for the geodesic curvature of a path

$$k_g = d/(R\rho) = K/(R(x^2+y^2)^{3/2}),$$

or explicitly

$$k_g = (-\varepsilon x(x^2+y^2+z^2) + z(x^2+y^2))/((x^2+y^2+z^2)^{1/2}(x^2+y^2)^{3/2}). \tag{5.10}$$

We have $k_g = 0$ if A is on the cubic cone $K = 0$. Therefore this cone is called the inflection cone of the position. We can transform (5.10) into a spherical coordinate system, in which the Z-axis coincides with the screw axis. Using

$$x = R\sin\theta\cos\eta, \quad y = R\sin\theta\sin\eta, \quad z = R\cos\theta,$$

the result is

$$k_g = (-\varepsilon\cos\eta + \sin\theta\cos\theta)/(R\sin^2\theta). \tag{5.11}$$

The center of curvature (i.e., the center of the osculating circle) M is on the line OM with the direction $(-\varepsilon x^2, -\varepsilon xy, x^2+y^2-\varepsilon xz)$ or

$$(-\varepsilon\sin\theta\cos^2\eta, -\varepsilon\sin\theta\cos\eta\sin\eta, \sin\theta - \varepsilon\cos\theta\cos\eta), \tag{5.12}$$

which intersects the sphere at two points, one of them M' being the spherical center of the osculating circle or the "spherical center of curvature" for A. The plane OAM is $yX - xY = 0$; it passes through A and O_Z since it is the normal plane to the path of A. Hence, if we call the point $\theta = 0$ the north pole P on the sphere we have: *the point A, its (spherical) center of curvature M', and P are on a great circle*, the meridian through A. Furthermore, if the angle POM' (that is, the spherical distance PM') is denoted by θ', it follows from (5.12) that

$$\cos\theta' = (\sin\theta - \varepsilon\cos\theta\cos\eta)/[\varepsilon^2\cos^2\eta - 2\varepsilon\sin\theta\cos\theta\cos\eta + \sin^2\theta]^{1/2},$$

or after some algebra

$$\tan\theta' = (-\varepsilon\sin\theta\cos\eta)/(\sin\theta - \varepsilon\cos\theta\cos\eta).$$

Therefore we obtain

(5.13a) $$\varepsilon(\cot\theta - \cot\theta')\cos\eta = 1,$$

which is the relationship between the point $A(\theta, \eta)$ and its spherical center of curvature $M'(\theta', \eta)$.

(5.13) is called the Euler–Savary formula, it is named for the two men who derived the analogous relationship in planar kinematics. If $\eta = \pi/2$, A is in the O_{YZ} plane, we have $\theta' = 0$, hence in this case M' coincides with P for any θ.

The angle η is measured to the plane OAM from the normal plane of the axodes (i.e., plane $O_X O_Z$; as is shown in the next section). If instead we use the angle from the axode tangent plane: we define the angle from the $O_Y O_Z$ plane to OAM as ψ. Hence $\psi = \pi/2 + \eta$ and (5.13a) yields the alternate form

(5.13b) $$\varepsilon(\cot\theta - \cot\theta')\sin\psi = 1.$$

The positive sense of ψ is according to the right-hand-screw rule with O_Z as axis. Obviously ψ is also the angle between axis O_X and the normal to the plane OAM. We have given here an alternative derivation of equation (9.14) of Chapter VI: the angles ψ, θ, θ' in (5.13b) are in fact respectively ψ, α, α_c of Fig. 35 in Chapter VI.

6. Third order properties

In the preceding section we dealt with second order properties, they depend only on the invariant ε. We consider now third order phenomena so that also γ_X and γ_Y have to be taken into account.

The axodes of the spherical motion are cones with vertex at O. Their intersections with a sphere about O will be called the *polhodes* on that sphere.

It follows from (5.2) that, up to the second order, the components of the geometric velocity of a moving point are

(6.1) $$\begin{aligned} X' &= -\tau x + [-1 + \tfrac{1}{2}(1+\varepsilon^2)\tau^2]y + (\varepsilon\tau + \tfrac{1}{2}\gamma_Y\tau^2)z, \\ Y' &= [1 - \tfrac{1}{2}(1+\varepsilon^2)\tau^2]x - \tau y + \tfrac{1}{2}((3/2)\varepsilon - \gamma_X)\tau^2 z, \\ Z' &= -(\varepsilon\tau + \tfrac{1}{2}\gamma_Y\tau^2)x + \tfrac{1}{2}((3/2)\varepsilon + \gamma_X)\tau^2 y, \end{aligned}$$

or, eliminating x, y, z from (6.1) using (5.3),

$$\begin{aligned} X' &= -(1-\tfrac{1}{2}\varepsilon^2\tau^2)Y + (\varepsilon\tau + \tfrac{1}{2}\gamma_Y\tau^2)Z, \\ Y' &= (1-\tfrac{1}{2}\varepsilon^2\tau^2)X + (\tfrac{1}{4}\varepsilon - \tfrac{1}{2}\gamma_X)\tau^2 Z, \\ Z' &= -(\varepsilon\tau + \tfrac{1}{2}\gamma_Y\tau^2)X + (-\tfrac{1}{4}\varepsilon + \tfrac{1}{2}\gamma_X)\tau^2 Y \end{aligned} \tag{6.2}$$

here we recognize the skew matrix of the general case (Chapter II, (5.11) with Chapter II, (6.2) substituted) the coefficients being the components of the angular velocity vector $\boldsymbol{\Omega}$. From (6.2) it follows that, up to the second order, the fixed axode is seen to be

$$X = R(-\tfrac{1}{4}\varepsilon + \tfrac{1}{2}\gamma_X)\tau^2, \quad Y = R(\varepsilon\tau + \tfrac{1}{2}\gamma_Y\tau^2), \quad Z = R(1 - \tfrac{1}{2}\varepsilon^2\tau^2), \tag{6.3}$$

R being a proportionality factor.

Similarly, by differentiating (5.3) and substituting (5.2), it follows that the moving axode, again up to the second order, reads

$$x = R(\tfrac{1}{4}\varepsilon + \tfrac{1}{2}\gamma_X)\tau^2, \quad y = R(\varepsilon\tau + \tfrac{1}{2}\gamma_Y\tau^2), \quad z = R(1 - \tfrac{1}{2}\varepsilon^2\tau^2). \tag{6.4}$$

Example 29. Show by means of (6.1) that for (6.4) we have $X' = Y' = Z' = 0$.
Example 30. Check that both cones pass through the instantaneous rotation axis O_Z and that the common tangent plane through O_Z is the plane $X = 0$; this is in accordance with our definition of the canonic frame.
Example 31. Show that for a constant value of R, (6.3) and (6.4) are the intersections of the axodes and the sphere $(O; R)$.

We shall call the intersections of the axodes and the sphere with $R = 1$ *the polhodes of the spherical motion.* The formulas (6.3) and (6.4) enable us to determine the curvatures of the polhodes in the zero-position.

The plane through three consecutive points of the fixed polhode passes through $(0,0,1,1)$, $(0,1,0,0)$ and $(-\tfrac{1}{2}\varepsilon + \gamma_X, \gamma_Y, -\varepsilon^2, 0)$. Hence this plane is $\varepsilon^2 X - (\tfrac{1}{2}\varepsilon - \gamma_X)Z + (\tfrac{1}{2}\varepsilon - \gamma_X) = 0$. The normal, through O, to this osculating plane intersects the meridian $Y = 0$ (which is the normal plane of the polhode at P) at the point Q', the center of curvature of the polhode. If $\angle POQ' = \xi'$ we obtain

$$\cot \xi' = -(\tfrac{1}{2}\varepsilon - \gamma_X)/\varepsilon^2. \tag{6.5}$$

Similarly, for the moving polhode, the center of curvature at P being Q and $\angle POQ = \xi$, we have

$$\cot \xi = (\tfrac{1}{2}\varepsilon + \gamma_X)/\varepsilon^2. \tag{6.6}$$

From (6.5) and (6.6) follows a relationship between the (spherical) radii of curvature of the fixed and moving polhodes:

$$\varepsilon(\cot \xi - \cot \xi') = 1. \tag{6.7}$$

There is a formal resemblance between (6.7) and (5.13), but they are essentially different. (5.13) expresses a relation between a moving point and the center of curvature of its path, (6.7) gives a correspondence between two centers of curvature. But, still more important: θ and θ' in (5.13) are second order concepts while ξ, ξ' are third order notions. An interesting feature of (6.5) and (6.6) is the fact that ξ and ξ' depend on the second order instantaneous invariant ε and on only one of the third order ones, viz. γ_X, and that γ_Y does not come into the picture.

The formulas (5.13) and (6.7) have given rise to some confusion in theoretical kinematics. To distinguish the two we shall call (6.7) the *second* Euler–Savary formula.

If P, Q and Q' are given, it follows from (6.5) and (6.6) that ε and γ_X are known. This implies that only one of the two third-order invariants is determined by these data. In particular it is obviously not correct to say that third order properties of a motion are the same as those of the motion defined by the rolling of the circle $(Q;\xi)$ on the fixed circle $(Q';\xi')$; ε and γ_X are the same for both, but γ_Y shall in general be different.

Third order properties deal with four consecutive positions of a moving point. We know from Section 5 that there is a locus of lines which have three positions in one plane (which means on any sphere a locus of points for which three positions are on a great circle). We may expect a finite number of lines through O for which four consecutive positions are coplanar. On the sphere this means that four consecutive positions of a point are on a great circle; making use of the terminology of plane geometry, such a point should be called a (spherical) undulation point of its path.

From (5.1) we infer that the condition for such points to exist is that the matrix

$$\left\|\begin{matrix} x & -y & -x+\varepsilon z & (1+\varepsilon^2)y+\gamma_Y z \\ y & x & -y & -(1+\varepsilon^2)x+((3/2)\varepsilon-\gamma_X)z \\ z & 0 & -\varepsilon x & -\gamma_Y x+((3/2)\varepsilon+\gamma_X)y \end{matrix}\right\| \tag{6.8}$$

must have a rank of 2. Hence the determinants derived from (6.8) by omitting the fourth or the third column must be zero. This gives us, respectively, the condition $\mathrm{K}=0$ (5.5) and

$$\mathrm{K}' \equiv (\gamma_X y - \gamma_Y x)(x^2+y^2+z^2) + (3/2)\varepsilon y(x^2+y^2-z^2) = 0. \tag{6.9}$$

Furthermore, the rank of the matrix consisting of the first and the second column must be equal to two, which implies that the three points $x=y=0$, and $x^2+y^2=z=0$ must be excluded. Hence the points we seek are the intersections of the cubics $K = 0$ and $K' = 0$, minus these three points (lying on both) which count as five intersections due to the fact that the curves are tangent at the points $x^2 + y^2 = z = 0$. Hence *there are in general four points on a sphere which are at a given instant at an undulation point of their path*. They are called Ball points, after R. Ball who was the first to derive them for the analogous question in plane kinematics. The number four is in contrast with the result in Section 4 for four finitely separated positions.

Four consecutive points of a spherical path are coplanar and hence lying on a circle, if according to (5.1)

$$C \equiv \begin{vmatrix} x & y & z & 1 \\ -y & x & 0 & 0 \\ -x+\varepsilon z & -y & -\varepsilon x & 0 \\ (1+\varepsilon^2)y + \gamma_Y z & \begin{matrix} -(1+\varepsilon^2)x \\ +((3/2)\varepsilon - \gamma_X)z \end{matrix} & \begin{matrix} -\gamma_Y x \\ +((3/2)\varepsilon + \gamma_X)y \end{matrix} & 0 \end{vmatrix} = 0, \tag{6.10}$$

or

$$(x^2+y^2)[-\gamma_Y x + ((3/2)\varepsilon + \gamma_X)y] - 3\varepsilon^2 xyz = 0. \tag{6.11}$$

The locus of those points is therefore a cubic cone, C, which will be called the circling-point cone. It contains the z-axis and the isotropic lines in the plane $z = 0$. As the z-axis is a double generator, the cone is a rational one.

Example 32. Show that the tangential planes of the double line are $x = 0$ and $y = 0$.
Example 33. Show that

$$x = 3\varepsilon^2 uv, \quad y = 3\varepsilon^2 u^2 v, \quad z = [((3/2)\varepsilon + \gamma_X)u - \gamma_Y](1+u^2)v$$

is a representation of C, by means of the parameters u and v; $u = 0$ and $u = \infty$ denote the double line, $u = \pm i$ the isotropic lines.

7. Fourth order properties

If we want to investigate fourth order instantaneous kinematics, the expressions X_4, Y_4, Z_4 as functions of x, y, z must be available. We did not derive them in the general case, the necessary algebra being rather complicated. We shall determine them here for spherical motion. They will be homogeneous linear functions of x, y, z, say $X_4 = a_1x + a_2y + a_3z$, $Y_4 = b_1x + b_2y + b_3z$, $Z_4 = c_1x + c_2y + c_3z$. Then, in view of (5.2) the matrix transforming x, y, z into X, Y, Z will be, up to the fourth order,

$$
(7.1)\quad \left\| \begin{matrix}
1-\frac{1}{2}\tau^2+(1/24)a_1\tau^4 & -\tau+(1/6)(1+\varepsilon^2)\tau^3 & \frac{1}{2}\varepsilon\tau^2+(1/6)\gamma_Y\tau^3 \\
 & +(1/24)a_2\tau^4 & +(1/24)a_3\tau^4 \\
\tau-(1/6)(1+\varepsilon^2)\tau^3 & 1-\frac{1}{2}\tau^2+(1/24)b_2\tau^4 & (1/6)((3/2)\varepsilon-\gamma_X)\tau^3 \\
+(1/24)b_1\tau^4 & & +(1/24)b_3\tau^4 \\
-\frac{1}{2}\varepsilon\tau^2-(1/6)\gamma_Y\tau^3 & (1/6)((3/2)\varepsilon+\gamma_X)\tau^3 & 1+(1/24)c_3\tau^4 \\
+(1/24)c_1\tau^4 & +(1/24)c_2\tau^4 &
\end{matrix} \right\|
$$

The condition that (7.1) is orthogonal up to the fourth order implies the following six equations for the unknowns a_i, b_i, c_i:

$$
(7.2)\quad \begin{aligned}
&a_1 = 1+\varepsilon^2, \quad b_2 = 1+4\varepsilon^2, \quad c_3 = -3\varepsilon^2, \\
&a_2+b_1 = 0, \quad b_3+c_2 = 4\gamma_Y, \quad c_1+a_3 = 4\gamma_X.
\end{aligned}
$$

To the relations (6.1) for X', Y', Z' we must add now

$$(1/6)(a_1x+a_2y+a_3z)\tau^3, \quad (1/6)(b_1x+b_2y+b_3z)\tau^3, \quad (1/6)(c_1x+c_2y+c_3z)\tau^3$$

respectively. The coordinates x, y, z are obtained as functions of X, Y, Z by the inverse matrix of (7.1), that is by its transpose. Eliminating x, y, z we get X', Y', Z' as functions of X, Y, Z, these being the extension, to the third order, of (6.2); the elements of this skew matrix are the components, in the fixed frame, of the angular rotation vector $\boldsymbol{\Omega}$. After some algebra we obtain

$$
(7.3)\quad \begin{aligned}
\Omega_X &= (-\tfrac{1}{4}\varepsilon+\tfrac{1}{2}\gamma_X)\tau^2+(1/6)(\gamma_Y-b_3)\tau^3, \\
\Omega_Y &= \varepsilon\tau+\tfrac{1}{2}\gamma_Y\tau^2+(\tfrac{1}{4}\varepsilon-(1/6)\gamma_X+(1/6)a_3)\tau^3, \\
\Omega_Z &= 1-\tfrac{1}{2}\varepsilon^2\tau^2-(1/6)a_2\tau^3.
\end{aligned}
$$

There is one more condition to be satisfied: for a canonic system the parameter is normalized so that $\Omega^2=1$. From (7.3) it follows that $\Omega^2=1$, up to the third order, if

$$(7.4)\quad a_2 = 3\varepsilon\gamma_Y.$$

(7.2) and (7.4) give seven equations for the nine unknowns; the remaining two free numbers give us the two fourth order instantaneous invariants say, q_X and q_Y, we expected. We take, for instance, $a_3 = 2\gamma_X+q_X$, $b_3 = 2\gamma_Y-q_Y$; then $c_1 = 2\gamma_X-q_X$, $c_2 = 2\gamma_Y+q_Y$. The result is

$$
\begin{aligned}
X_4 &= (1+\varepsilon^2)x + 3\varepsilon\gamma_Y y + (2\gamma_X + q_X)z,\\
Y_4 &= -3\varepsilon\gamma_Y x + (1+4\varepsilon^2)y + (2\gamma_Y - q_Y)z, \qquad (7.5)\\
Z_4 &= (2\gamma_X - q_X)x + (2\gamma_Y + q_Y)y - 3\varepsilon^2 z.
\end{aligned}
$$

By virtue of (5.1) and (7.5) spherical motion is given up to the fourth order. Any property of this order can be expressed in terms of the instantaneous invariants $\varepsilon, \gamma_X, \gamma_Y, q_X, q_Y$. So, for instance, one may study third order properties of the polhodes (such as their torsion) by means of (7.3) and the analogous formulas (for the components $\omega_x, \omega_y, \omega_z$ of the angular velocity) with respect to the moving frame.

Example 34. Derive the components $\omega_x, \omega_y, \omega_z$ as third order functions of τ by means of (7.3) and the transpose of (7.1).

Example 35. Show that q_X and q_Y are respectively κ_Y and κ_X of Chapter II, (5.11).

As a fourth order problem we can ask the question whether there are points for which five consecutive positions are coplanar and hence on a circle. The condition is that the matrix

$$
\left\|\begin{matrix} -y & x & 0\\ -x+\varepsilon z & -y & -\varepsilon x\\ X_3 & Y_3 & Z_3\\ X_4 & Y_4 & Z_4 \end{matrix}\right\| \qquad (7.6)
$$

has rank two. Such points must lie on the circling-point cone $C \equiv -\varepsilon x(yY_3 + xX_3) + Z_3(x^2+y^2-\varepsilon xz) = 0$ and also on $C' = -\varepsilon x(yY_4 + xX_4) + Z_4(x^2+y^2-\varepsilon xz) = 0$. Their sections with the plane at infinity are the cubics c and c′, having nine common points. We know that c has a double point at $(0,0,1)$, the tangents being $x=0$ and $y=0$; moreover it passes through the isotropic points $(1, \pm i, 0)$. The curve c′ passes through $(0,0,1)$, which is an ordinary point of c′, the tangent being $x=0$; it does not pass through the isotropic points. Hence c and c′ have three intersections, at $x=y=0$, which must be rejected. *The number of points with five consecutive positions on a circle* (real or imaginary) *is therefore six*. They are named after Burmester, who solved the analogous problem for five finitely separated planar positions. If we substitute the coordinates of c, as functions of the parameter u (Example 33) into $c'=0$ we obtain an equation of degree six for the Burmester lines; i.e., the lines from the origin of O_{xyz} through the intersections of c and c′.

8. Time dependent motion

So far we have only considered geometric spherical kinematics. Now we turn to time-dependent motion. We shall restrict ourselves to the velocity and the acceleration distribution.

From Chapter VI, (11.2) we have, in view of $\sigma_0 = \lambda = \mu = 0$,

$$\begin{aligned} &X_0 = x, \quad X_1 = -\omega y, \quad X_2 = -\omega^2 x - \varepsilon_3 y + \varepsilon_2 z, \\ &Y_0 = y, \quad Y_1 = \omega x, \qquad Y_2 = \varepsilon_3 x - \omega^2 y, \\ &Z_0 = z, \quad Z_1 = 0, \qquad Z_2 = -\varepsilon_2 x. \end{aligned} \tag{8.1}$$

ω is the angular velocity, ε_2 and ε_3 are the components of the angular acceleration vector along O_Y and O_Z.

The velocity distribution is that of a rotation about O_Z with angular velocity ω.

X_2, Y_2, Z_2 are the components of the acceleration $\boldsymbol{a}$, of the point $A = (x, y, z)$, along the coordinate axes; the acceleration pole of the general case now coincides with O. As A moves on a sphere we determine the components of $\boldsymbol{a}$ along the path tangent, the meridian, and the radius OA; the direction numbers of these three direction are, if A is not on O_z, $(-y, x, 0)$, $(xz, yz, -(x^2+y^2))$, and (x, y, z) respectively. Hence the tangential component a_t, the normal component a_n, and the radial component a_r are

$$\begin{aligned} a_t &= (\varepsilon_3(x^2+y^2) - \varepsilon_2 yz)/(x^2+y^2)^{1/2} \\ a_n &= (\varepsilon_2 x(x^2+y^2+z^2) - \omega^2 z(x^2+y^2))/((x^2+y^2+z^2)^{1/2}(x^2+y^2)^{1/2}) \\ a_r &= -\omega^2(x^2+y^2)/(x^2+y^2+z^2)^{1/2} \end{aligned} \tag{8.2}$$

a_r is always directed towards O.

The normal component a_n is zero if

$$(\varepsilon_2 x/\omega^2)(x^2+y^2+z^2) - z(x^2+y^2) = 0, \tag{8.3}$$

that is for points on the spherical inflection cone K (5.5); it is a time-independent locus.

That is not the case for the points with $a_t = 0$; their locus has the equation

$$Q \equiv \varepsilon_3(x^2+y^2) - \varepsilon_2 yz = 0, \tag{8.4}$$

which represents a quadratic cone. (Sometimes called the normal cone, because its points have only normal acceleration along the sphere.)

K passes through O_z, and through the isotropic lines $z = 0$, $x^2 + y^2 = 0$, and so does Q. Rejecting these, K and Q have only three generators in common, the acceleration axes we met in Chapter VI, Section 12. Hence, in spherical kinematics, *there are three points* (real or imaginary) *on the sphere which have no acceleration along the sphere*; they could be called the spherical acceleration poles.

Example 36. Show by eliminating z from (8.3) and (8.4) that the acceleration axes satisfy

$$-\omega^2\varepsilon_3 y^3 + (\varepsilon_2^2 + \varepsilon_3^2)xy^2 - \omega^2\varepsilon_3 x^2 y + \varepsilon_3^2 x^3 = 0.$$

CHAPTER VIII

PLANE KINEMATICS

1. Introduction

If $\boldsymbol{P}$ and $\boldsymbol{p}$ are the position vectors of a point in the fixed and the moving space respectively we know that the general spatial motion is represented by

$$\boldsymbol{P} = \mathbf{A}\boldsymbol{p} + \boldsymbol{d}, \tag{1.1}$$

$\mathbf{A}$ being an orthogonal matrix and $\boldsymbol{d}$ the vector displacement of the origin; the number of degrees of freedom is six. If a (finite) point O is fixed during the motion and if this point is taken as the origin, we have $\boldsymbol{d} = 0$. We then no longer have a general motion, but are dealing instead with spherical kinematics, the subject of Chapter VII; as $\boldsymbol{d} = 0$ is equivalent to three conditions the number of degrees of freedom reduces to three.

We consider now the case of a point at infinity being fixed during the motion. If this point is given the homogeneous coordinates $(0, 0, 1, 0)$, we must have in $\mathbf{A} = |a_{ij}|$ the relations $a_{13} = a_{23} = 0$, which imply in view of $\mathbf{A}$ being orthogonal that $a_{31} = a_{32} = 0$, $a_{33} = 1$. The motion is then represented by

$$\begin{aligned} X &= x \cos \phi - y \sin \phi + d_1, \\ Y &= x \sin \phi + y \cos \phi + d_2, \\ Z &= z + d_3; \end{aligned} \tag{1.2}$$

it has, obviously, four degrees of freedom. Indeed, the condition that a point at infinity remains invariant is equivalent to only two relations.

It is easy to analyze a motion of the type (1.2). It is built up of a plane motion, every point remaining in its plane parallel to O_{XY}, followed by a translation in the direction of O_Z; in all planes parallel to O_{XY} the motions are congruent, that is they are essentially the same. Every plane parallel to O_{XY} is transformed into a parallel plane. It is easy to see that the screw axis of the displacement is parallel to O_Z; the rotation angle is ϕ, and the translation along the screw is d_3. The fixed and moving axodes are cylinders with their generators parallel to O_Z; any point of a generator may be taken as its striction point. The motion of the moving axode on the fixed one is a combination of rolling and sliding just as in a general spatial motion, but the axode configuration is much simpler in this case.

Of the two components of the motion (1.2) the planar part is by far the more interesting, and we shall restrict ourselves to it by dealing only with the case $d_3 = 0$. As, moreover, the motions in the parallel planes are all the same the study may be limited to the motion of the plane $z = 0$. The subject of this chapter is therefore plane kinematics. The basic approach in this book has been to start with the general case and to specialize later on, accordingly we have arrived at plane kinematics as a subcase of spatial motion. It is of course possible to develop it autonomously, if we restrict ourselves from the start to two dimensional geometry.

The condition $d_3 = 0$ reduces the parameters of (1.2) to ϕ, d_1 and d_2. Hence the number of degrees of freedom is three.

It is understandable that plane kinematics is historically the most ancient part of our science. Because of its relative simplicity it is by far the most developed branch. Moreover the applications of theoretical kinematics to the analysis and synthesis of mechanisms — a subject not treated in this book — deal mainly with the planar part.

Plane kinematics depends on three degrees of freedom, the same number as for spherical kinematics. The planar case is by far the more simple of the two. In the spherical case the d_i are zero so the three parameters are all enclosed (or hidden) in the orthogonal matrix and we need — as was shown in Chapter VI — Eulerian angles or parameters to exhibit them explicitly. In plane kinematics the orthogonal matrix degenerates so as to depend only on the single parameter ϕ; the two remaining parameters are decoupled and may be taken as translation components. It cannot be denied on the other hand that spherical kinematics is more elegant and more "homogeneous". We have seen that, looking at the plane at infinity, it is isomorphic with a non-Euclidean (i.e., elliptic) geometry, while plane kinematics as a matter of fact is Euclidean. For instance elliptic geometry is completely self-dual, points and lines being equivalent concepts; in the Euclidean plane the line at infinity is a special line without a counterpart among the points. Non-Euclidean geometry, defined in the projective plane, deals with the group of linear transformations leaving the (absolute) conic invariant; in the Euclidean plane the absolute is degenerated into the two isotropic points, and planar displacements are those transformations which leave these points invariant.

However, for plane (Euclidean) kinematics we also require that the determinant of the linear terms in the transformation matrix equals unity. If we drop the latter condition we have equiformal kinematics with four degrees of freedom, on which some remarks will be made in Chapter XII.

2. Two positions theory

It follows from our general theory that if we are given two positions E_1 and E_2 of the moving plane E with respect to the coinciding fixed plane Σ, E_2 follows from E_1 by a rotation about an axis perpendicular to Σ or by a translation. A direct proof follows from Fig. 37. Let A_1, A_2 and B_1, B_2 be two pairs of homologous points, which implies $|A_1B_1| = |A_2B_2|$. The perpendicular bisectors of A_1A_2 and B_1B_2 intersect at P. Then $|PB_1| = |PB_2|$, $|PA_1| = |PA_2|$; hence the triangles PA_1B_1 and PA_2B_2 are congruent, $\angle A_1PB_1 = \angle A_2PB_2$, from which it follows that $\angle A_1PB_1 - \angle A_2PB_1 = \angle A_2PB_2 - \angle A_2PB_1$ which we write as $\angle A_1PA_2 = \angle B_1PB_2 = \phi$. A rotation about P by the angle ϕ transforms A_1 into A_2 and B_1 into B_2 and therefore any point of E_1 into the

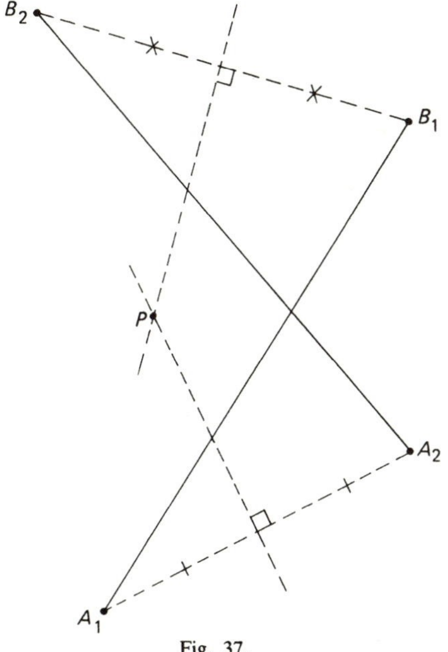

Fig. 37.

homologous point in E_2. Point P is called the pole of the displacement or simply *the pole* (or center; or the roto pole — for rotation pole). If we wish to indicate which two positions are associated with the pole, say i and j, we will use subscripts, i.e. P_{ij}. The perpendicular bisectors are parallel if A_1A_2 and B_1B_2 are parallel; E_2 then follows from E_1 by a translation, and the pole is at infinity. An analytical proof is the following. Let the frames O_{XY} and o_{xy} in Σ and E_1 coincide. The displacement of E_1 into E_2 may be represented by

$$X = x\cos\phi - y\sin\phi + a, \qquad Y = x\sin\phi + y\cos\phi + b. \tag{2.1}$$

(This follows also from (1.2); clearly $d_1 = a$, $d_2 = b$.) An invariant point P follows from (2.1) if we specify the conditions $X = x$, $Y = y$, after some algebra we obtain (if $\phi \neq 0$)

$$X_p = x_p = \tfrac{1}{2}(a - b\cot(\tfrac{1}{2}\phi)), \qquad Y_p = y_p = \tfrac{1}{2}(a\cot(\tfrac{1}{2}\phi) + b). \tag{2.2}$$

Example 1. Prove (2.2) by means of Fig. 38, keeping in mind that the two indicated angles are equal and therefore the triangles $o_2o'M$ and $MP'P$ are similar. (In the figure point o is displaced from position o_1 to o_2, M is the midpoint of o_1o_2, and P' and o' are on a line parallel to the X-axis.)

Example 2. Show that: if two positions of E with a finite rotation center are given, the joins A_1A_2 of homologous points fill up the complete plane Σ. Any finite line of Σ is the join of one point pair.

The formulas (2.1) give us the relationship between the coordinates (x, y) of a point in E and those of the same point measured in Σ. If we solve for x and y we obtain

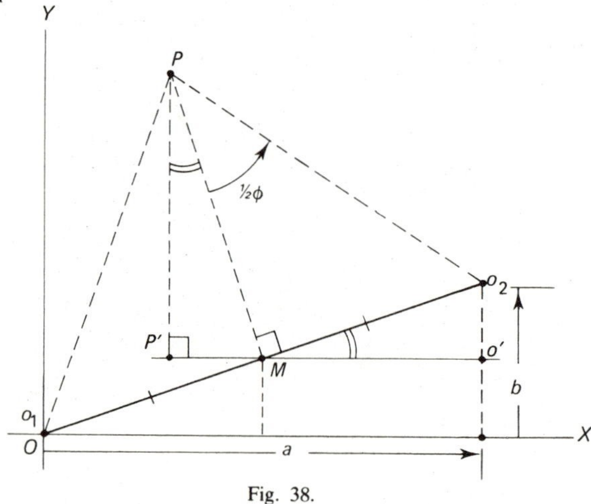

Fig. 38.

(2.3) $x = X\cos\phi + Y\sin\phi + \tilde{a}, \qquad y = -X\sin\phi + Y\cos\phi + \tilde{b},$

with

(2.4) $\tilde{a} = -a\cos\phi - b\sin\phi, \qquad \tilde{b} = a\sin\phi - b\cos\phi,$

(2.3) is a representation of the inverse displacement.

Let $u_1x + u_2y + u_3 = 0$ be a line of E. Its equation in Σ, $U_1X + U_2Y + U_3 = 0$, follows by means of (2.3), and we obtain

$$\begin{aligned} U_1 &= u_1\cos\phi - u_2\sin\phi, \\ U_2 &= u_1\sin\phi + u_2\cos\phi, \qquad (2.5) \\ U_3 &= \tilde{a}u_1 + \tilde{b}u_2 + u_3, \end{aligned}$$

as the transformation formulas for line coordinates.

Example 3. Show from (2.5) that parallel lines are displaced into parallel lines and moreover that the angle between two lines is invariant under planar displacements.

Example 4. Show from (2.4) that $(\tilde{a})^2 + (\tilde{b})^2 = a^2 + b^2$, both sides being the square of the distance between O and o.

3. Three positions theory

A displacement can be represented as a product of two reflections as follows. Through pole P_{12} we take any two lines n_1 and n_2 of Σ which, in addition to intersecting at P_{12}, are at an angle of $\phi_{12}/2$ measured from n_1 to n_2 (Fig. 39). If A^* is an arbitrary point of Σ, and A_i is its reflection into n_i, then A_1 is transformed into A_2 by a rotation about center P_{12} by angle ϕ_{12}. This follows from $|P_{12}A_1| = |P_{12}A^*| = |P_{12}A_2|$ and $\angle A_1P_{12}A_2 = \phi_{12}$. (The statement is the planar version of the theorem given in Chapter I, Section 6.) If P_{12} is at infinity ($\phi_{12} = 0$) we take n_1 and n_2 parallel, with distance $d/2$ (Fig. 40); A_2 then follows from A_1 by a translation, perpendicular to n_i and with distance d. Conversely any displacement of E (a rotation or a translation) may be considered as the product, in the appropriate order, of two reflections. Obviously there are infinitely many such reflections representing any displacement. If the displacement is a rotation about P_{12} with angle ϕ_{12} the line n_1 may be chosen arbitrarily through P_{12}, n_2 is then determined; analogously if the displacement is a translation n_1 may be chosen arbitrarily, perpendicular to the direction of translation. These properties enable us to treat three position theory in plane kinematics using a similar, but of course more simple, development to that used in our study of three position spatial kinematics (Chapter IV).

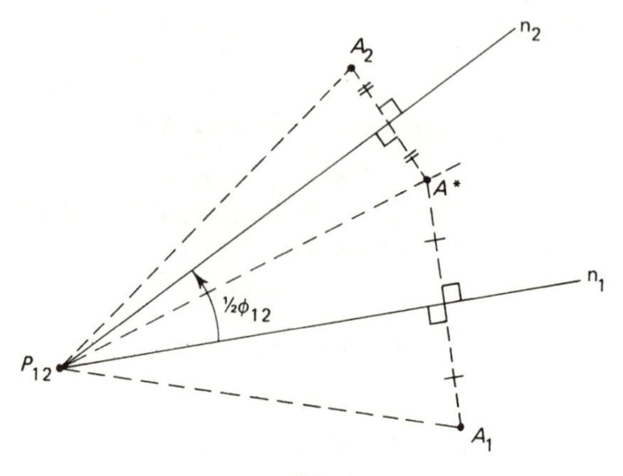

Fig. 39.

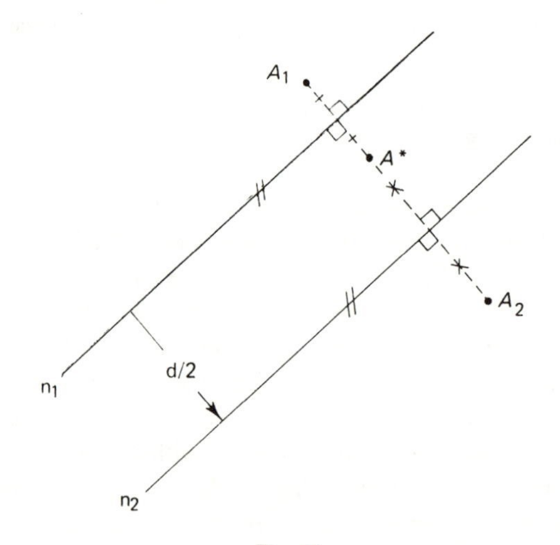

Fig. 40.

Let P_{ij} be the rotation center for the displacement from position i into position j $(i, j = 1, 2, 3;\ i \neq j)$ and ϕ_{ij} the rotation angle. We have $P_{ij} = P_{ji}$, $\phi_{ij} + \phi_{ji} = 0\,(\text{mod.}\,2\pi)$. Furthermore we take all rotation angles as being in the counter-clockwise sense. This in no way limits the displacement, since a rotation of ϕ' in the clockwise direction yields exactly the same position as $2\pi - \phi'$ in the counter-clockwise sense. Three positions are determined if we know P_{12}, ϕ_{12} and P_{31}, ϕ_{31}. We suppose P_{12}, P_{31} to be finite, different points. If

n_1 is taken as the line $P_{12}P_{31}$, and n_2 is the line found by rotating n_1 about P_{12} with rotation angle $\frac{1}{2}\phi_{12}$ (Fig. 41), the displacement $\mathbf{D}_{12}$ is the product of the reflection into n_1 followed by that into n_2; let n_3 be constructed by rotating n_1 about P_{13} with rotation angle $\frac{1}{2}\phi_{13} = -\frac{1}{2}\phi_{31}$. If $\phi_{31} \neq (2\pi - \phi_{12})$ then n_2 and n_3 intersect. If A_1 is an arbitrary point of E_1 and A^* its reflection into n_1, then A_2 and A_3 are the reflections of A^* into n_2 and n_3; this implies that A_3 follows from A_2 by successive reflection into n_2 and n_3. The conclusion is that the intersection of n_2 and n_3 is the center P_{23} and obviously any triple A_1, A_2, A_3 of homologous points is found by reflecting a point A^* into n_1, n_2, n_3 respectively. There are two different cases, given by Figs. 41 and 42, distinguished by $\phi_{12}+\phi_{31}<2\pi$ and $\phi_{12}+\phi_{31}>2\pi$. In the first case the orientation of the triangle $P_{12}P_{23}P_{31}$ is clockwise; in the second case it is anti-clockwise. Our result reads: the three positions of E are determined by the triangle of the three rotation centers $P_{12}P_{23}P_{31}$, the pole-triangle. This figure is comparable to the screw-triangle in spatial kinematics and the trihedron of rotation axes in the spherical case. The three rotation angles are twice the angles of the triangle; it follows easily from the figures that they must be taken with orientation determined so that ϕ_{ij} is measured from the n_i to the n_j side of the triangle; one has always $\phi_{12}+\phi_{23}+\phi_{31} = 0 \,(\text{mod.}\, 2\pi)$. A special case arises if $\phi_{12}+\phi_{31} = 2\pi$; then n_2 and n_3 are parallel (Fig. 43), $\mathbf{D}_{23}$ is a translation, P_{23} is a point at infinity.

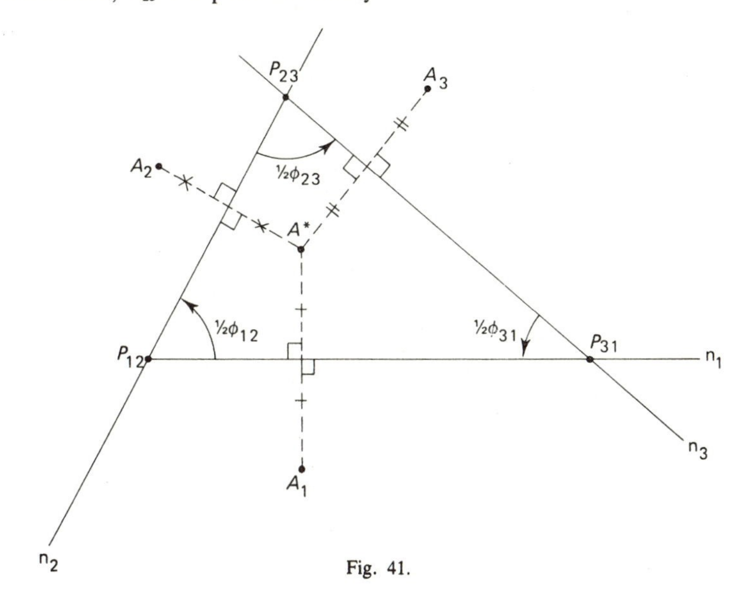

Fig. 41.

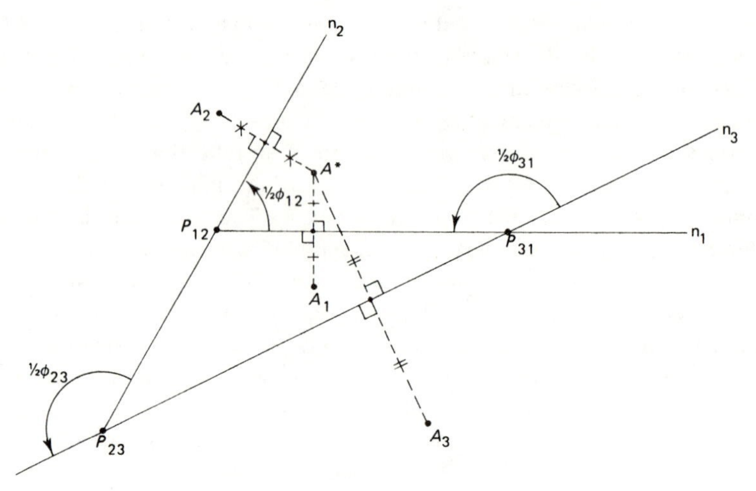
n2
n3
n1
A2
A*
A1
A3
P12
P31
P23
½φ12
½φ31
½φ23

Fig. 42.

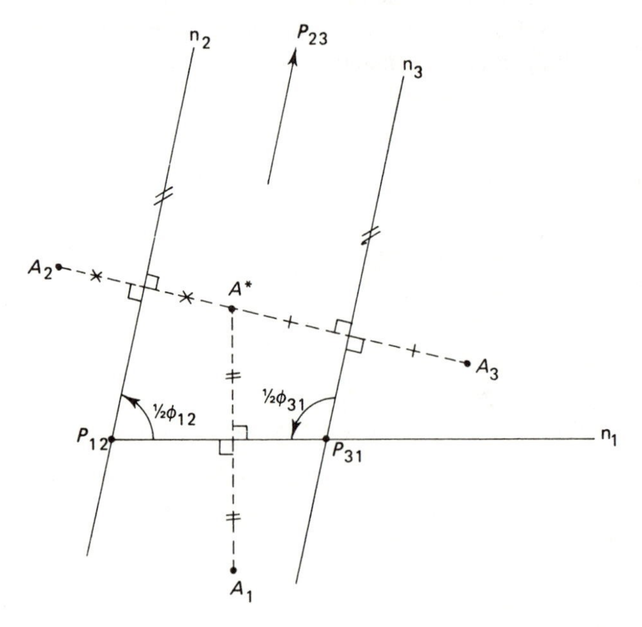
n2
P23
n3
n1
A2
A*
A3
A1
P12
P31
½φ12
½φ31

Fig. 43.

Example 5. Show that if two centers are at infinity, all three are.
Example 6. If n_2 and n_3 are parallel the translation $\mathbf{D}_{23}$ has the distance $2h \sin(\phi_{12}/2) = 2h \sin(\phi_{31}/2)$, h being the side $|P_{12}P_{31}|$.
Example 7. Determine the triad of homologous points if A^* is on n_i or coincides with P_{ij}.

Let A^* be the basic point of a triad $A_1A_2A_3$ (Fig. 44); since $|P_{12}A_1| = |P_{12}A^*| = |P_{12}A_2|$, the triangle $A_1P_{12}A_2$ is isosceles. The perpendicular bisector m_{12} of A_1A_2 coincides therefore with the bisector of the angle $A_1P_{12}A_2$, this implies that $\angle n_1P_{12}m_{12} = \angle A^*P_{12}n_2$; in other words: m_{12} is the reflection of A^*P_{12} into the inner (or outer) bisector of the angle $n_1P_{12}n_2$, or, as the geometers have it, m_{12} is isogonally related to A^*P_{12} with respect to the angle $n_1P_{12}n_2$. As A^* is an arbitrary point, and for any three positions, m_{12}, m_{23}, m_{31} are concurrent (since they pass through the center of the circle defined by A_1, A_2, A_3) we have proved the geometrical theorem: if a point Q is joined to the vertices of a triangle, the isogonally related lines pass through one point Q'; Q and Q' are said to be isogonally conjugate with respect to the triangle (Fig. 45).

Example 8. Show that the relation between Q and Q' is commutative.
Example 9. Show that a vertex of the triangle is isogonally conjugate to any point on the opposite side (and hence every point on a side is isogonally conjugate to the opposite vertex).
Example 10. Determine the (four) points of the plane which are isogonally self-conjugates.

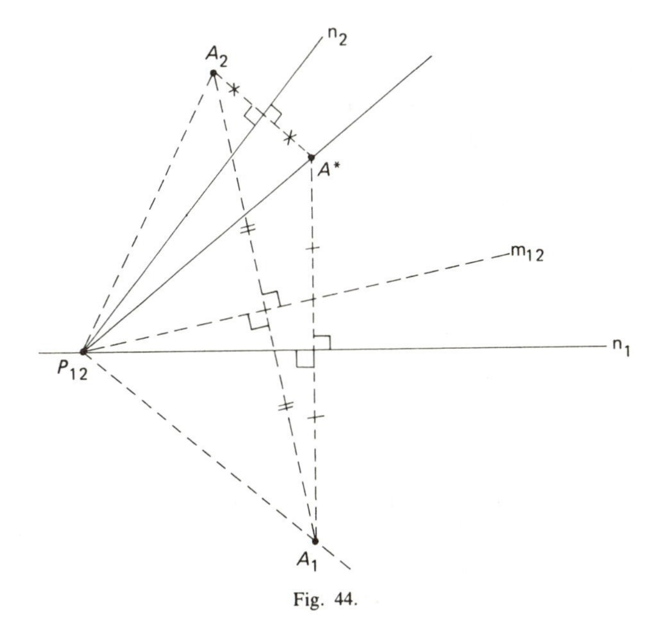

Fig. 44.

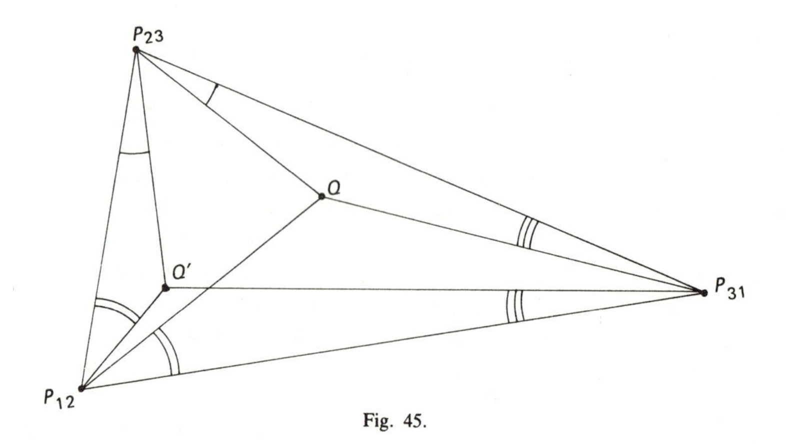

Fig. 45.

Example 11. Prove that in any triangle the orthocenter and the circumcenter are isogonally conjugate points.
Example 12. Prove that the distances of Q to the sides are inversely proportional to those of Q'. (If Q_1, Q_2 and Q_1', Q_2' are respectively the projections of Q and Q' on two sides, the triangles QQ_1Q_2 and $Q'Q_1'Q_2'$ are similar.)
Example 13. Show that the relationship between Q and Q' holds if the triangle has a vertex at infinity; the "bisector" through this vertex is the line half-way between the parallel sides.

From the above theorems we infer: the circumcenter M of three homologous points $A_1A_2A_3$ is isogonally conjugate (with respect to the pole-triangle) to their basic point A^*.

If three homologous points are collinear their circumcircle degenerates and the center M is a point at infinity. Conversely (in view of the commutativity of the isogonal relationship) a triad of collinear points has a basic point isogonally conjugated to a point M at infinity. In Fig. 46 the pole triangle and the direction of M are given; let $\angle MP_{12}P_{31} = \psi$. We construct A^*; the angles indicated are all equal to ψ. If the angles of $P_{23}P_{31}P_{12}$ are α_1, α_2, α_3 then $\angle P_{31}P_{12}A^* = \psi - \alpha_3$, $\angle A^*P_{31}P_{12} = \pi - \psi - \alpha_2$, hence $\angle P_{12}A^*P_{31} = \pi - \alpha_1$. This implies that A^* is on the circumcircle of the pole triangle; conversely any point on this circle has its isogonal conjugate at infinity. The locus of the basic points of three collinear homologous points is the circumcircle c* of the pole triangle.

Example 14. Show that we could have known beforehand (i.e., without recourse to the circumcircle) that the locus of A^* passes through the vertices.
Example 15. Prove that the projections of a point on the circumcircle onto the three sides are collinear (Wallace's theorem).
Example 16. If one vertex of the triangle is at infinity, show that the locus of A^* is the opposite side (which is the circumcircle in this case).

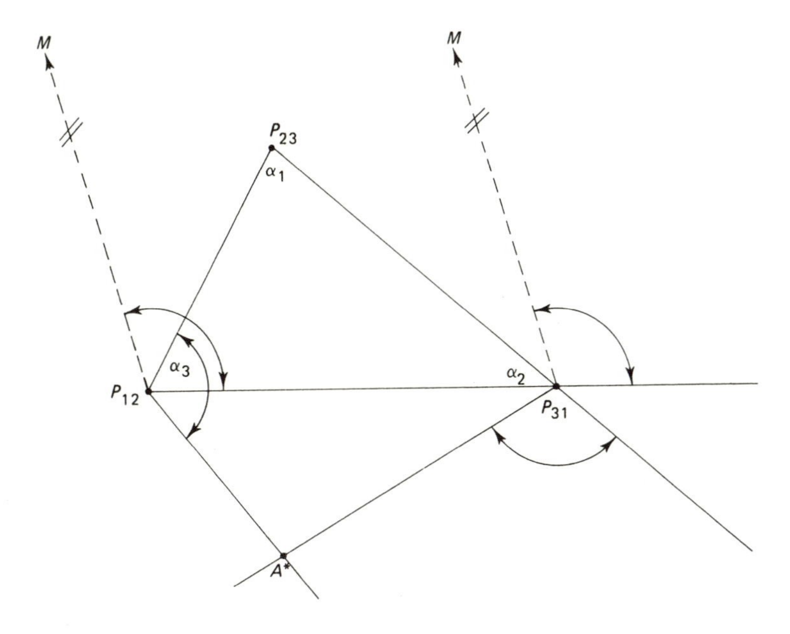

Fig. 46.

A_i is the reflection of the point A^* of Σ into n_i. Hence the locus of the points A of E whose three positions are collinear is a circle c. In the first position, E_1 of E, this circle passes through P_{12}, P_{13}, and P^1_{23} (the reflection of P_{23} into n_1).

In Fig. 47 the orthocenter H of the pole triangle is reflected into n_1; as $\angle P_{12}H_1P_{31} = \angle P_{12}HP_{31} = \alpha_2 + \alpha_3 = \pi - \alpha_1$ the point H_1 is on the circumcircle c^*; in other words: if c^* is reflected, into n_i, to c_i, the circles c_i pass through H $(i = 1, 2, 3)$. In Fig. 48 c_1 and c_2 are drawn, both passing through H; A_1 is an arbitrary point on c_1, S is the (second) intersection of A_1H and c_2. Then we have $\angle A_1P_{12}S = \angle A_1P_{12}P_{31} + \alpha_3 + \angle P_{23}P_{12}S$; the latter angle is equal to $\angle P_{23}HS = \angle A_1HP^1_{23} = \angle A_1P_{12}P^1_{23}$, hence $\angle A_1P_{12}S = \alpha_3 + \angle P_{31}P_{12}P^1_{23} = 2\alpha_3$, which is the rotation angle of $\mathbf{D}_{12}$. The conclusion is that S coincides with A_2. Therefore in general: if A_1, A_2, A_3 are a triad of collinear homologous points, and if the line through them is m_A, then m_A passes through H. The locus of lines which can be drawn through three homologous points is the pencil in Σ with the vertex H.

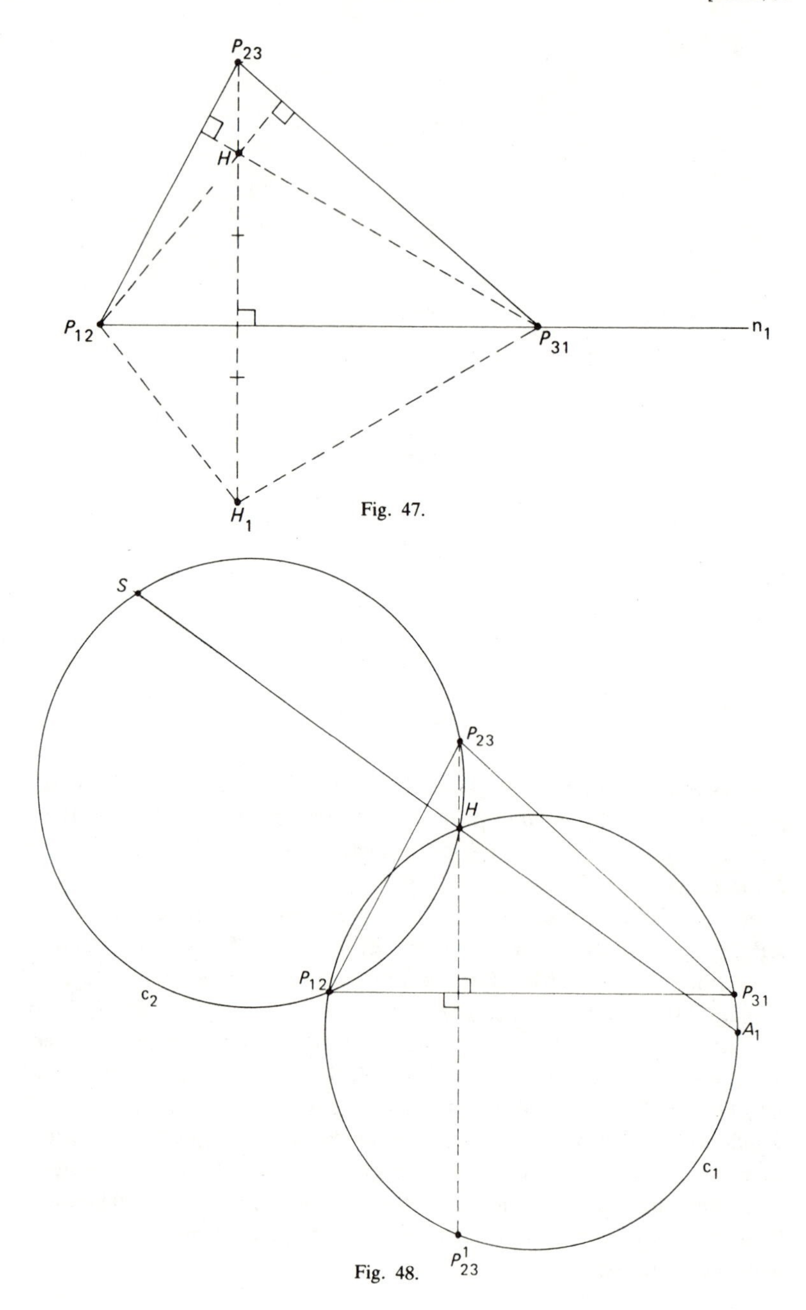

Fig. 47.

Fig. 48.

Example 17. Show that the displacement $\mathbf{D}_{12}$ transforms P_{12}, c_1, P_{31}, P^1_{23}, H, H_1 into P_{12}, c_2, P^2_{31}, P_{23}, H_2 and H respectively.
Example 18. If one vertex of the pole triangle is at infinity the pencil of lines m_A consists of parallel lines, passing through the "orthocenter" of the triangle.

We have derived the theorems associated with three positions by means of geometric arguments. Here follows an analytic treatment: According to (2.1) a position of E with respect to Σ is given by the three numbers (ϕ, a, b). For the three positions determined by (ϕ_i, a_i, b_i), $i = 1, 2, 3$, the three homologous points A_i in Σ, of the point $A = (x, y)$ in E, are collinear if

$$|x \cos \phi_i - y \sin \phi_i + a_i \quad x \sin \phi_i + y \cos \phi_i + b_i \quad 1| = 0, \tag{3.1}$$

or, if we develop this determinant with respect to the last column,

$$\begin{aligned} &4 \sin(\tfrac{1}{2}(\phi_2 - \phi_3)) \sin(\tfrac{1}{2}(\phi_3 - \phi_1)) \sin(\tfrac{1}{2}(\phi_1 - \phi_2))(x^2 + y^2) \\ &\quad + \left[\sum (a_2 \sin \phi_3 - a_3 \sin \phi_2) + \sum (b_3 \cos \phi_2 - b_2 \cos \phi_3)\right] x \\ &\quad + \left[\sum (a_2 \cos \phi_3 - a_3 \cos \phi_2) + \sum (b_2 \sin \phi_3 - b_3 \sin \phi_2)\right] y \\ &\quad + \sum (a_2 b_3 - a_3 b_2) = 0, \end{aligned} \tag{3.2}$$

which confirms that the locus of A is a circle c. Clearly (3.1) is satisfied if two rows are equal, which affirms the fact that c passes through the rotation centers. From (3.2) we see that c is degenerate if two angles ϕ_i are equal, that is if a relative displacement is a translation.

Example 19. Show that when A^* coincides with the orthocenter H of the pole triangle, the circle through the three homologous points A_1, A_2, A_3 is the circumcircle of the pole triangle.
Example 20. Show that when A^* coincides with the incenter of the pole triangle, the circle through the three homologous points A_1, A_2, A_3 has a radius equal to twice the radius of the inscribed circle of the pole triangle.
Example 21. Show that the circle in the previous example is the smallest possible circle through three homologous points.
Example 22. Show that all the points in E which have their three positions on circles of the same radius lie on a tri-circular sextic embedded in E. (ALT [1921], GROENMAN [1950], BOTTEMA [1954].)

The triangle formed by three homologous points is similar to the pedal triangle of the basic point A^* with respect to the pole triangle. Pedal triangles have been extensively studied objects in plane geometry. Of the many theorems we mention only: with respect to any triangle there are always two (real) points — called isodynamic points — whose pedal triangles are equilateral. Transposed to our case we have: if three positions of a plane are given there exist two triads of homologous points which are each the vertices of an equilateral triangle.

So far we have, in three positions theory, studied homologous points. We deal now with homologous lines. Obviously three homologous lines l_1, l_2, l_3 are obtained by reflecting a basic line l^* into the sides of the pole triangle. We ask now which lines are displaced so that their three homologous positions are concurrent. Let S be their common intersection. In Fig. 49 the pole triangle, the line l_3 and the point S on it are given; l_2 follows from l_3 by the rotation $(P_{32}, 2\alpha_1)$, hence there is a point S' on l_3 which is transformed into S, which implies that $\angle Q_1P_{23}S = \alpha_1$, Q_1 being the projection of P_{23} on l_3. Similarly l_1 follows from l_3 by $(P_{31}, 2\alpha_2)$, hence $\angle Q_2P_{31}S = \alpha_2$. Therefore $\angle P_{23}SP_{31} = |\alpha_1| + |\alpha_2| = \pi - |\alpha_3|$. The conclusion is: the locus of the point S, the intersection of three concurrent homologous lines, is the circumcircle of the pole triangle. On the other hand (Fig. 49), if S is a point on the circumcircle, l_3 is found by the condition that $\angle P_{23}Sl_3 = (\pi/2) - \alpha_1 = \angle P_{23}P_{12}H$; hence l_3 passes through H_3, which generalizes to the fact that l_i passes through H_i. In other words: the basic line l^* of three concurrent homologous lines passes through H, and conversely any line of the pencil through H is a basic line l^* of a set of concurrent homologous lines.

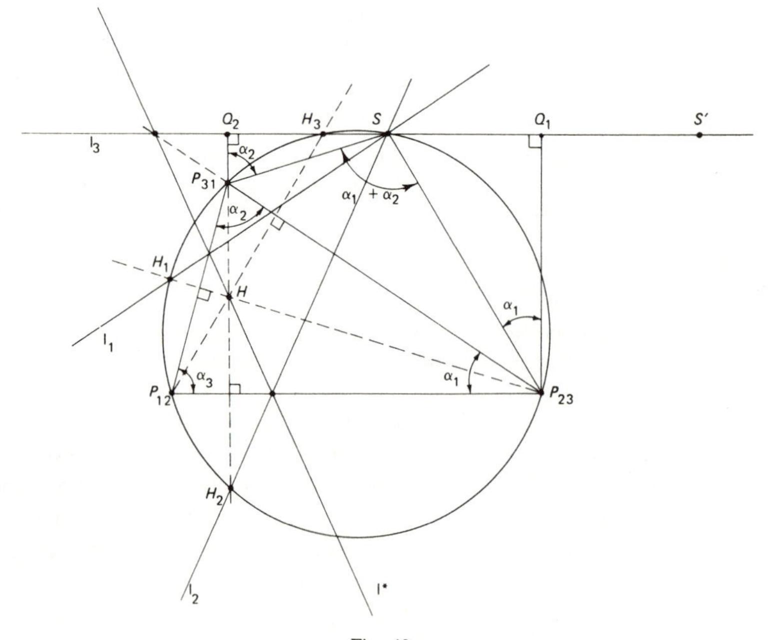

Fig. 49.

Example 23. Determine the triad $\mathbf{l}_i$ passing through a vertex of the pole triangle, and their basic line $\mathbf{l}^*$.
Example 24. Determine the triad $\mathbf{l}_i$ passing through H_1 and their basic line $\mathbf{l}^*$.
Example 25. Determine the sets of three concurrent homologous lines if one vertex of the pole triangle is at infinity.
Example 26. If three homologous lines are not concurrent determine the angles of the triangle they enclose.

In order to treat the problem, of concurrent homologous lines, analytically we make use of the formulas (2.5). If three positions are given by $(\phi_i, \tilde{a}_i, \tilde{b}_i)$, $i = 1, 2, 3$, the condition for concurrency reads:

$$|\, u_1 \cos\phi_i - u_2 \sin\phi_i \quad u_1 \sin\phi_i + u_2 \cos\phi_i \quad \tilde{a}_i u_1 + \tilde{b}_i u_2 + u_3 \,| = 0, \tag{3.3}$$

which gives us an equation of the third degree for u_1, u_2, u_3. Developing this determinant with respect to the third column it is seen to be

$$\sum (\tilde{a}_1 u_1 + \tilde{b}_1 u_2 + u_3) \sin(\phi_3 - \phi_2)(u_1^2 + u_2^2) = 0. \tag{3.4}$$

Hence the locus degenerates. It represents three pencils of lines, two of them having their vertices at the isotropic points I_1, I_2 of the plane; this is as expected because these points are invariant under any displacement. Disregarding the two trivial solutions the only remaining pencil is made up of the lines whose coordinates satisfy

$$\sum (\tilde{a}_1 u_1 + \tilde{b}_1 u_2 + u_3) \sin(\phi_3 - \phi_2) = 0. \tag{3.5}$$

The vertex T of this pencil, whose homogeneous coordinates are

$$\begin{aligned} x &= \sum \tilde{a}_1 \sin(\phi_3 - \phi_2), \qquad y = \sum \tilde{b}_1 \sin(\phi_3 - \phi_2), \\ z &= \sum \sin(\phi_3 - \phi_2) = 4 \sin(\tfrac{1}{2}(\phi_2 - \phi_3)) \sin(\tfrac{1}{2}(\phi_3 - \phi_1)) \sin(\tfrac{1}{2}(\phi_1 - \phi_2)), \end{aligned} \tag{3.6}$$

represents the point in E through which all lines (which have three concurrent homologous positions) must pass. If we invert the motion we must interchange x, y, ϕ_i, a_i, b_i and X, Y, $-\phi_i$, $\tilde{a}_i$, $\tilde{b}_i$ respectively. Under an inversion (3.6) yields the point in Σ with the property that on any line through it lies a triad of homologous points, this is the point H we met before; its homogeneous coordinates are

$$\begin{aligned} X &= \sum a_1 \sin(\phi_3 - \phi_2), \qquad Y = \sum b_1 \sin(\phi_3 - \phi_2), \\ Z &= 4 \prod \sin(\tfrac{1}{2}(\phi_2 - \phi_3)). \end{aligned} \tag{3.7}$$

Example 27. Show by means of (2.2) that (3.7) represents the orthocenter of the pole triangle.

T is a finite point provided no two angles ϕ are equal, which is the case when no displacement $\mathbf{D}_{ij}$ is a translation. From (3.5) it follows that the coordinates of those lines of E whose three positions are concurrent (at a point S) may be written as linear functions of a parameter λ. Then it follows from (3.3) that the coordinates of S (being the minors of two rows) are quadratic functions of λ; hence the locus of S is a conic. But the pencil with vertex T contains the lines TI_1 and TI_2, and for these lines S coincides with I_1 and I_2 respectively. This implies that the locus of S is a circle, which is in accordance with our former results.

Comparing the two methods we have used to treat three positions theory, the geometric and the analytic, it could be said that the former is attractive, giving a visual impression of the situation, but not exact, since it depends on one particular choice of the data; the second is exact, covering all cases, bringing into light imaginary solutions, but formal and not accompanied by an elegant configuration. The advantages and disadvantages are those of the graphical construction method on one hand and the algebraic and computational one on the other. In the study of kinematics, historically the first method has prevailed, in modern times there is a strong tendency in favor of the latter.

Example 28. In the special case of three positions, an analytic treatment can be chosen which makes use of triangle coordinates with respect to the pole triangle. Let (X, Y, Z) be the homogeneous coordinates (the ratios of the distances from the sides) of a point in the fixed plane. If h_i are the sides of the pole triangle, the equation of the line l at infinity reads $h_1X + h_2Y + h_3Z = 0$. If γ_i is the angle between the normals (to h_j and h_k), show that $P_{23}^1 = (-1, 2\cos\gamma_3, 2\cos\gamma_2)$.
Example 29. Verify that the reflection into the line $P_{31}P_{12}$ is represented by $X' = -X$, $Y' = 2X\cos\gamma_3 + Y$, $Z' = 2X\cos\gamma_2 + Z$, by proving that under this displacement the line l and every point on $P_{31}P_{12}$ is invariant and that P_{23} is transformed into P_{23}^1.
Example 30. If the basic point $A^* = (X^*, Y^*, Z^*)$ is given, derive the coordinates (X_i, Y_i, Z_i), $i = 1, 2, 3$, of the homologous points A_i associated with A^*, (in triangular coordinates).
Example 31. Show that the equation $|X_i\ Y_i\ Z_i| = 0$, which expresses that the points A_i are collinear, can be written as $(h_1X^* + h_2Y^* + h_3Z^*)(h_1Y^*Z^* + h_2Z^*X^* + h_3X^*Y^*) = 0$; from this it follows that the locus of A^* consists of the line l and the circumcircle of the pole triangle.
Example 32. If X_i, Y_i, Z_i are the distances of A_i to the sides of the pole triangle (which implies $h_1X_i + h_2Y_i + h_3Z_i = 2F$, F being the area of the triangle), show that the midpoint M_{12} of A_1 and A_2 is given by $(X_1 + X_2,\ Y_1 + Y_2,\ Z_1 + Z_2)$, the point S_{12} at infinity of the line A_1A_2 by $(X_1 - X_2,\ Y_1 - Y_2,\ Z_1 - Z_2)$ and the point at infinity of a normal to A_1A_2 by $(Y^*\sin\gamma_3,\ X^*\sin\gamma_3,\ -X^*\sin\gamma_2 - Y^*\sin\gamma_1)$.
Example 33. Show that the equation of the perpendicular bisector of A_1 and A_2 is $X^*X - Y^*Y = 0$.
Example 34. Show that the circumcenter M of the triangle $A_1A_2A_3$ is (Y^*Z^*, Z^*X^*, X^*Y^*), that is, the point isogonally conjugate to A^* with respect to the pole triangle.

Summing up three positions theory in plane kinematics: we have seen that the pole triangle, its circumcircle and its orthocenter are of primary importance. Three homologous points (lines) are the reflections of a basic point (line) into the sides of the triangle. The three points A_1, A_2, A_3 are collinear on a line m if their basic point A^* is on the circumcircle c in which case m passes through the orthocenter H. On the other hand: three lines l_1, l_2, l_3 are concurrent if their basic line l^* passes through H, the intersection S of the lines lies on c. In both cases there is a correspondence between the points of c and the lines through H, and it is easy to verify that it is the same correspondence for both.

Example 35. Prove the last mentioned statement.

4. Four positions theory

We deal now with four positions E_i $(i = 1, 2, 3, 4)$ of the moving plane E with respect to the fixed plane Σ. Obviously there are six rotation centers P_{ij} $(i, j = 1, 2, 3, 4;\ i \neq j)$ and — as in the corresponding situation in spatial and spherical kinematics — they are not independent. For the case under consideration the reasoning is as follows. In Fig. 50 the pole triangles for positions 123 and 234 are shown; obviously they must have equal angles at their common vertex P_{23}, both being $\frac{1}{2}\phi_{23}$. Hence, if two pairs P_{12}, P_{34} and P_{13}, P_{42} of opposite centers are given (a pair of opposite centers is composed of two centers without a common subscript), P_{23} must have the property that seen from this point the optical distances $P_{12}P_{31}$ and $P_{42}P_{34}$ are equal in size and orientation (or, what is the same thing, $P_{12}P_{42}$ and $P_{31}P_{34}$ have the same

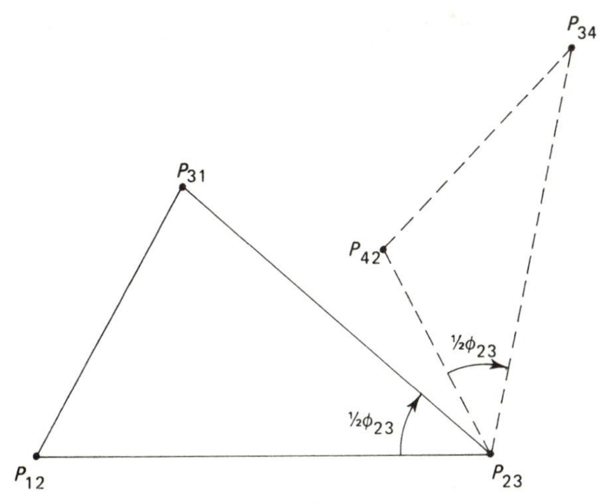

Fig. 50.

optical distance). This gives us the locus for P_{23}, and it is easily seen that P_{14} is also on it. If (p_i, q_i), $i = 1, 2, 3, 4$ are the coordinates of points P_{12}, P_{31}, P_{42}, P_{34} respectively (Fig. 51), $P_{23} = (X, Y)$, α_1 and α_2 the angles respectively of $P_{23}P_{12}$ and $P_{23}P_{31}$ with O_X, we obtain $\tan\alpha = \tan(\alpha_2 - \alpha_1)$, $\tan\alpha_i = (q_i - Y)/(p_i - X)$, hence we have $\tan\alpha = (\tan\alpha_2 - \tan\alpha_1)/(1 + \tan\alpha_2 \tan\alpha_1)$ from which it follows that $\tan\alpha = L_{12}/C_{12}$, with

$$(4.1)\qquad \begin{aligned} L_{12} &\equiv (q_1 - q_2)X - (p_1 - p_2)Y + p_1q_2 - p_2q_1,\\ C_{12} &\equiv X^2 + Y^2 - (p_1 + p_2)X - (q_1 + q_2)Y + p_1p_2 + q_1q_2. \end{aligned}$$

$L_{12} = 0$ is the equation of the line $P_{12}P_{31}$, $C_{12} = 0$ is that of the circle with diameter $P_{12}P_{31}$; these are the loci of P_{23} for $\alpha = 0$ and $\alpha = \pi/2$ respectively. If (p_1, q_1) and (p_2, q_2) are replaced by (p_3, q_3) and (p_4, q_4) in (4.1) we obtain an expression for the angle $P_{42}P_{23}P_{34}$ which must also equal α. Equating our two expressions for $\tan\alpha$ we find that the locus of P_{23} (and P_{14}) has the equation

$$(4.2)\qquad K \equiv L_{12}C_{34} - L_{34}C_{12} = 0,$$

which is comparable to Chapter VII, (4.4) in spherical kinematics. From (4.1) and (4.2) it follows that K is a cubic curve. It is called the pole (or polar) curve (German: Pollagenkurve). It passes through the intersections of the circles $C_{12} = 0$ and $C_{34} = 0$ and therefore through the isotropic points I_1 and I_2: hence K is a circular cubic. It passes through the intersections of L_{12} and C_{12}, which are P_{12} and P_{31}, and also through the intersection of L_{34} and C_{34} (i.e., P_{34} and P_{42}), therefore we know that all six centers P_{ij} are on K. Furthermore K passes through the intersection of L_{12} and L_{34}, that is of $P_{12}P_{13}$ and $P_{42}P_{43}$; this

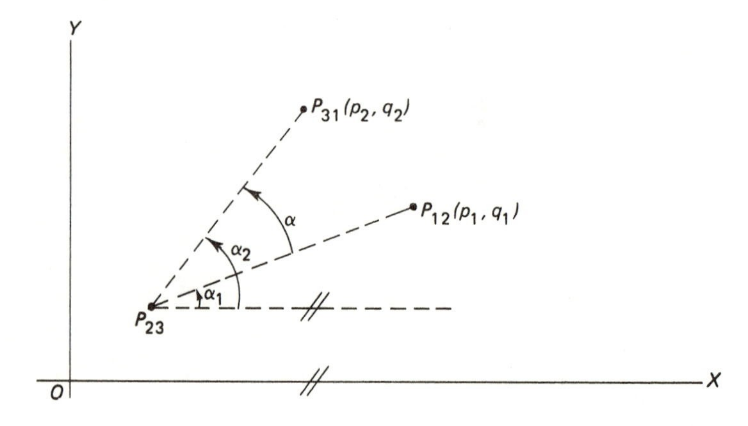

Fig. 51.

generalizes to: it passes through the intersection of $P_{ij}P_{ik}$ and $P_{lj}P_{lk}$ if i, j, k, l are different indices; this point is the third intersection of K and any line joining two non-opposite centers.

Example 36. Prove that there are six such special points on K.

We shall derive the equation of K for a special coordinate system. Suppose first that $P_{12}P_{31}$ and $P_{42}P_{34}$ are not parallel and let their intersection be taken as the origin O (we know it to be a point of K) and a bisector of their angle as the O_X-axis (Fig. 52). Let furthermore $OP_{12} = d_1$, $OP_{31} = d_1'$, $OP_{42} = d_2$, $OP_{34} = d_2'$ and $d_i + d_i' = u_i$, $d_i - d_i' = v_i$ $(i = 1, 2)$. Then, according to (4.1),

$$\text{(4.3)} \qquad \begin{aligned} L_{12} &= v_1(X \sin\alpha - Y\cos\alpha), \\ C_{12} &= X^2 + Y^2 - u_1 X\cos\alpha - u_1 Y \sin\alpha + \tfrac{1}{4}(u_1^2 - v_1^2), \end{aligned}$$

from which L_{34} and C_{34} can be obtained by replacing u_1, v_1, α by u_2, v_2, $-\alpha$ respectively. Hence the equation for K in homogeneous coordinates reads

$$\text{(4.4)} \qquad \begin{aligned} &(X^2 + Y^2)[(v_1 + v_2)X\sin\alpha - (v_1 - v_2)Y\cos\alpha] - \\ &\quad - \sin\alpha\cos\alpha\,(u_1v_2 + u_2v_1)(X^2 + Y^2)Z + \\ &\quad + (u_2v_1 - u_1v_2)XYZ + (k_1X + k_2Y)Z^2 = 0, \end{aligned}$$

k_1 and k_2 being certain functions of the data.

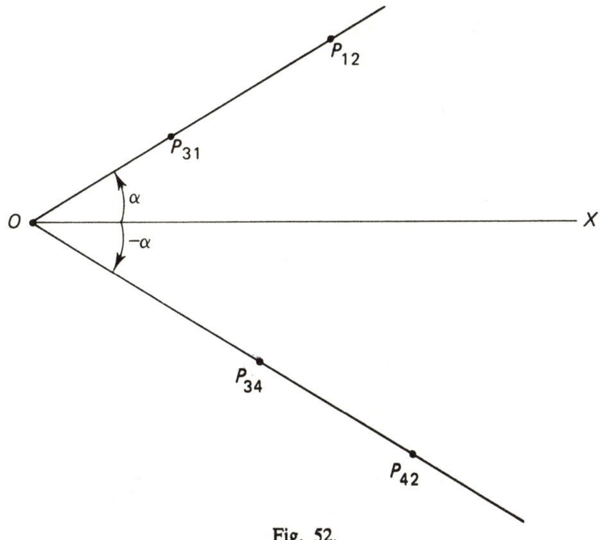

Fig. 52.

A circular cubic has a principal focus, defined as the intersection of its tangents at the isotropic points I_1 and I_2. The tangent to K at its point (X', Y', Z') is given by $\Sigma X(\partial K/\partial X)_{(X=X', Y=Y', Z=Z')} = 0$. Hence the tangent to K at $(1, i, 0)$ has the equation

$$(4.5)\quad 2[(v_1+v_2)\sin\alpha - i(v_1-v_2)\cos\alpha](X+iY) + (u_2v_1-u_1v_2)iZ = 0,$$

and similarly the tangent at I_2 is the complex conjugate of (4.5). From this it follows that the principal focus $F(x_0, y_0, z_0)$ is given by

$$(4.6)\quad \begin{aligned} x_0 &= (u_2v_1-u_1v_2)(v_1-v_2)\cos\alpha,\\ y_0 &= -(u_2v_1-u_1v_2)(v_1+v_2)\sin\alpha,\\ z_0 &= 2(v_1^2+v_2^2-2v_1v_2\cos 2\alpha). \end{aligned}$$

After some algebra we find

$$(4.7)\quad \begin{aligned} &L_{12}(x_0/z_0, y_0/z_0)/C_{12}(x_0/z_0, y_0/z_0) =\\ &= (2(u_2v_1-u_1v_2)\sin 2\alpha)/((u_1^2+u_2^2-v_1^2-v_2^2)-2\cos 2\alpha(u_1u_2-v_1v_2)), \end{aligned}$$

which remains invariant if we interchange u_1 and u_2, v_1 and v_2, α and $-\alpha$. Hence at F $L_{12}/C_{12} = L_{34}/C_{34}$, and the conclusion is: F is a point of K.

In the case, excluded so far, that $P_{12}P_{31}$ and $P_{42}P_{34}$ are parallel, the derivation of the equation for K is slightly different: We take the X-axis halfway between the two parallel lines (Fig. 53). Then $P_{12} = (p_1, h)$, $P_{31} = (p_1', h)$, $P_{42} = (p_2, -h)$, $P_{34} = (p_2', -h)$; putting $p_i - p_i' = v_i$, $p_i + p_i' = u_i$ $(i = 1, 2)$, we obtain

$$(4.8)\quad L_{12} = v_1(-Y+h), \qquad C_{12} = X^2+Y^2-u_1X-2hY+\tfrac{1}{4}(u_1^2-v_1^2)+h^2.$$

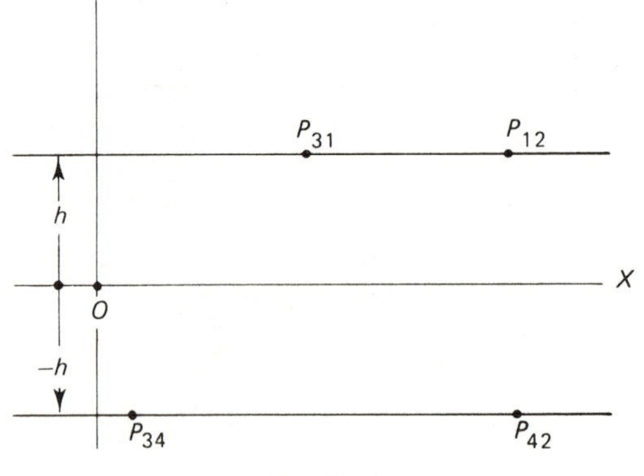

Fig. 53.

L_{34} and C_{34} may be obtained from (4.8) by replacing u_1, v_1 by u_2, v_2 and h by $-h$. The equation for K is now

$$(4.9)\quad \begin{aligned}&(X^2+Y^2)(-v_1+v_2)Y+h(v_1+v_2)(X^2-Y^2)Z+(u_2v_1-u_1v_2)XYZ+\\&+(k_1'X+k_2'Y+k_3'Z)Z^2=0,\end{aligned}$$

where k_1', k_2' and k_3' represent terms containing only u_i, v_i and h. The principal focus $F(x_0, y_0, z_0)$ is now at

$$(4.10)\quad x_0=u_2v_1-u_1v_2,\qquad y_0=-2h(v_1+v_2),\qquad z_0=2(v_1-v_2);$$

after some algebra we obtain

$$(4.11)\quad \begin{aligned}&L_{12}(x_0/z_0, y_0/z_0)/C_{12}(x_0/z_0, y_0/z_0)=\\&=(8h(v_1-v_2))/(16h^2+(u_1-u_2)^2-(v_1-v_2)^2),\end{aligned}$$

which does not change if u_1, u_2, v_1, v_2, h are replaced by u_2, u_1, v_2, v_1, $-h$. Hence F is a point of K.

A circular cubic which passes through its own principal focus is called focal. Focal cubics are well-known in the theory of special plane curves. For instance: the locus of the foci of the conics tangent to four given lines is such a curve (SALMON [1954]). Our conclusion is: the pole curve for four positions of a plane is a focal cubic.

The coordinates of the focus F are given by (4.6) and (4.10) respectively. In the first case $z_0 \neq 0$ (because $|\cos 2\alpha| \neq 1$); in the second case we have $z_0 = 0$ only if $v_1 = v_2 = v$. Hence F is a finite point, unless the given pairs P_{12}, P_{34} and P_{31}, P_{42} (of opposite rotation centers) are opposite vertices of a parallelogram. If they do form a parallelogram, it follows from (4.9) that K is given by

$$(4.12)\quad Z[2hv(X^2-Y^2)+v(u_2-u_1)XY+(k_1'X+k_2'Y+k_3'Z)Z]=0.$$

Hence the theorem, first given by BURMESTER [1888, pp. 620–1]: if P_{12}, P_{34} and P_{31}, P_{42} are opposite vertices of a parallelogram the pole curve degenerates into the line at infinity and an equilateral hyperbola.

In all other cases F is a finite point and it can be chosen as the origin. If we take, moreover, the Y-axis parallel to the (real) asymptote of K the equation for K, (4.4), simplifies to

$$(4.13)\quad X(X^2+Y^2)+\mathfrak{a}(X^2+Y^2)+\mathfrak{b}_1X+\mathfrak{b}_2Y=0.$$

In this standard form of the equation there appear three parameters $\mathfrak{a}$, $\mathfrak{b}_1$, $\mathfrak{b}_2$; these are functions of the coordinates of the four given poles, and they determine the shape of the curve. We derive from (4.13) the following procedure to generate K. The asymptote is seen to be $X + \mathfrak{a} = 0$; we consider the line l with the equation $X = -\frac{1}{2}\mathfrak{a}$, half-way between F and the asymptote (Fig. 54). On any line $Y = \lambda X$ through F there lie, besides F, two points S_1, S_2 of K. Their X-coordinates satisfy $X^2(1+\lambda^2) + X\mathfrak{a}(1+\lambda^2) + \mathfrak{b}_1 + \lambda\mathfrak{b}_2 = 0$, hence their midpoint M is on l. A circle with center $M(-\frac{1}{2}\mathfrak{a}, -\frac{1}{2}\lambda\mathfrak{a})$ has the equation $X^2 + Y^2 + \mathfrak{a}(X + \lambda Y) + \frac{1}{4}\mathfrak{a}^2(1+\lambda^2) - R^2 = 0$; if it passes through S_1, S_2 we have $\frac{1}{4}\mathfrak{a}^2(1+\lambda^2) - R^2 = \mathfrak{b}_1 + \lambda\mathfrak{b}_2$. Hence the circle with S_1S_2 as a diameter reads

$$X^2 + Y^2 + \mathfrak{a}X + \mathfrak{b}_1 + \lambda(\mathfrak{a}Y + \mathfrak{b}_2) = 0, \tag{4.14}$$

which for variable λ represents a pencil of circles. This implies the following point-wise construction of K: intersect the variable line FM and that circle of the pencil whose center is M. This is essentially one of the constructions to generate K given by Burmester in a purely geometrical way (BURMESTER [1888, pp. 614–16], BEYER [1953, pp. 79–80]).

The pencil (4.14) consists of the circles through two basic points Q_1 and Q_2, which are the intersections of the circle $X^2 + Y^2 + \mathfrak{a}X + \mathfrak{b}_1 = 0$ and the line $\mathfrak{a}Y + \mathfrak{b}_2 = 0$. There are two main cases: Q_1, Q_2 can be either real or conjugate imaginary, depending on whether D is positive or negative in

$$D = \mathfrak{a}^4 - 4(\mathfrak{a}^2\mathfrak{b}_1 + \mathfrak{b}_2^2). \tag{4.15}$$

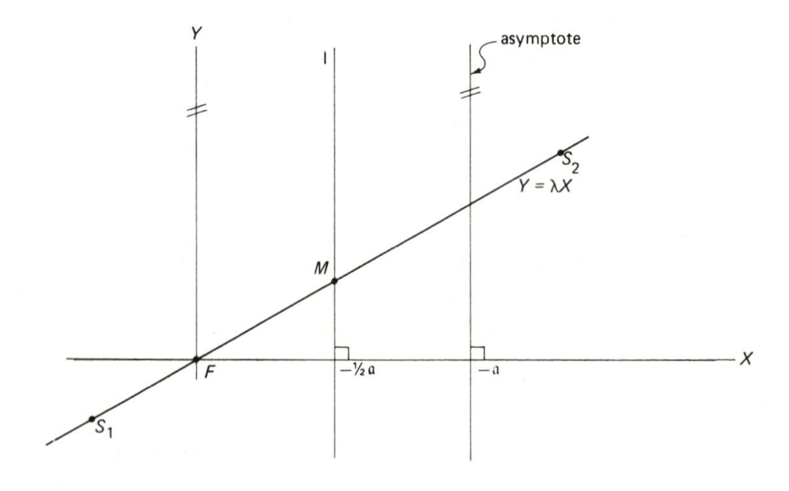

Fig. 54.

It is easy to verify that the two finite intersections of K and l are real (imaginary) if D is negative (positive). In the latter case l separates the two circuits of K, (Fig. 55), in the former there is only one circuit (Fig. 56). If $D = 0$, we have $Q_1 = Q_2 = Q$ and K has a double point at Q (Fig. 57). Indeed, both l and FQ have two coinciding points Q in common with K.

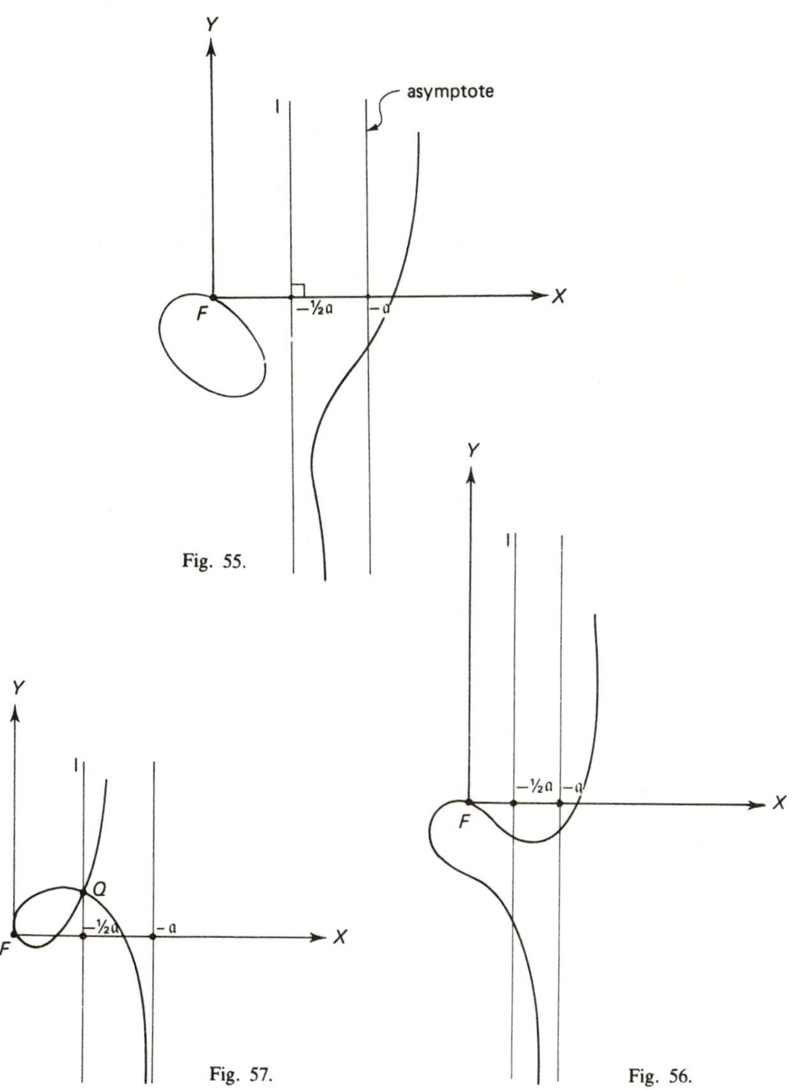

Fig. 55.

Fig. 57.

Fig. 56.

Example 37. Show that Q_1 and Q_2 are points of K.
Example 38. Consider the case $\mathfrak{a} = 0$.
Example 39. Show that K degenerates into a circle and a finite line only if $\mathfrak{a} = \mathfrak{b}_2 = 0$; F is the center of the circle and the line coincides with the Y-axis.
Example 40. Show that K degenerates if the perpendicular bisector of $P_{13}P_{24}$ coincides with that of $P_{14}P_{23}$.
Example 41. Show that K degenerates if P_{14} and P_{23} are both on the perpendicular bisector of $P_{13}P_{24}$.
Example 42. Show that K degenerates if P_{13}, P_{24}, P_{14}, P_{23} are collinear.
Example 43. If $D = 0$, determine the double point of K and derive a rational parametric representation of K in this case.
Example 44. If $D \neq 0$ the curve K has no double point; its genus is one. K is elliptic and could be represented by means of elliptic functions. Its class is six. Through every point of K pass four lines tangent to K at other points. For F, two of these are isotropic; prove that it depends upon the sign of D whether the remaining two tangents are real or imaginary.

The pole curve K is determined if P_{12}, P_{13}, P_{42}, P_{43} are given. On it lies an infinite set of point-pairs P_{14}, P_{23}, which are pairs of opposite rotation centers, we refer to such a pair as a pair of *conjugate points*. From (4.6) it follows that the intersection of $P_{12}P_{13}$ and $P_{42}P_{43}$ (supposed to be a finite point) coincides with the focus F if $u_2v_1 - u_1v_2 = 0$, which implies $d_1d_2' = d_1'd_2$; this means that lines $P_{12}P_{42}$ and $P_{13}P_{43}$ are parallel and therefore (as their intersection is on K) parallel to the asymptote of K (Fig. 58). This gives rise to the following general construction (Fig. 59) for conjugate pairs P, P' (once K is known). If P on K is given, we determine S_1 the third intersection of PF and K, S_2' that of PL and K, then P' is the intersection of LS_1 and FS_2' (L is the point of l at infinity).

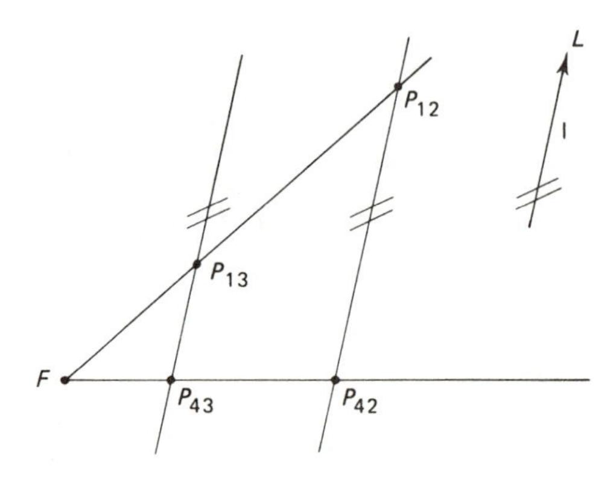

Fig. 58.

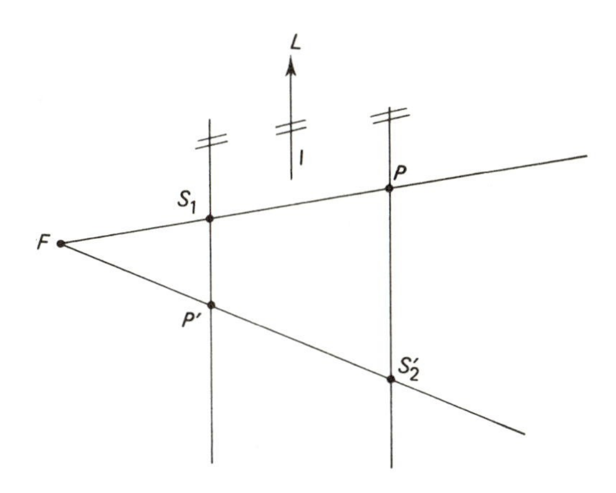

Fig. 59.

Example 45. Show that P and P' are interchangeable.
Example 46. Show that S_1 and S_2' are conjugates.
Example 47. Show that F and L are conjugate points.

If the equation of FP is $Y = \lambda X$ the X-coordinates of both P and S_1 satisfy $\lambda^2(X^2 + \mathfrak{a}X) + \lambda\mathfrak{b}_2 + (X^2 + \mathfrak{a}X + \mathfrak{b}_1) = 0$. Now since P has the same X-coordinate as S_2', and S_1 has the same X-coordinate as P', if the equation of FP' is $Y = \lambda' X$ we obtain $\lambda'^2(X^2 + \mathfrak{a}X) + \lambda'\mathfrak{b}_2 + (X^2 + \mathfrak{a}X + \mathfrak{b}_1) = 0$. Eliminating $(X^2 + \mathfrak{a}X)$ between these two expressions we obtain the following bilinear (and commutative) relation between λ and λ':

$$\mathfrak{b}_2\lambda\lambda' + \mathfrak{b}_1(\lambda + \lambda') - \mathfrak{b}_2 = 0. \tag{4.16}$$

We have considered the curve K in detail because it is an important curve in plane kinematics. It is not only the pole curve; we shall meet it again in a different context when we deal with "center-points" later in this section.

If the six rotaton centers P_{ij} are given (they must of course satisfy the conditions we have set forth in the foregoing), there are four pole triangles, $P_{23}P_{34}P_{42}$, $P_{34}P_{41}P_{13}$, $P_{41}P_{12}P_{24}$, and $P_{12}P_{23}P_{31}$ which we denote by T_1, T_2, T_3, T_4 respectively. Three homologous points A_2, A_3, A_4 are the reflections of a fundamental point, denoted as A_1^*, into the respective sides of T_1. Similar configurations exist for T_2, T_3, T_4. Hence, there appears a configuration (Fig. 60) consisting of the six points P_{ij} and their twelve pertinent joins, four fundamental points A_i^* and four homologous points A_i. It follows from their mutual interdependency that, for instance, $A_2^*A_3^*A_4^*$ are the reflections of A_1

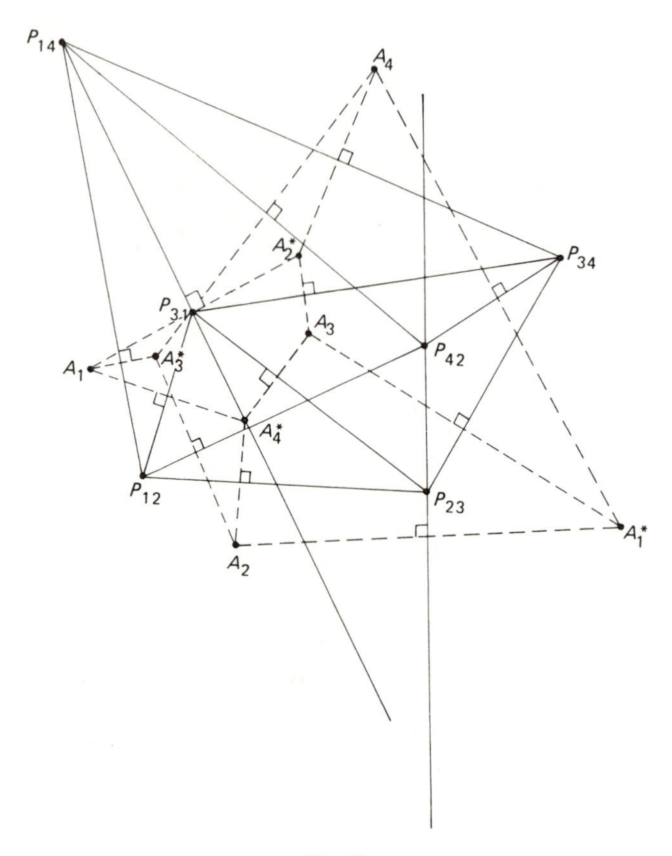

Fig. 60.

into $P_{13}P_{14}$, $P_{14}P_{12}$, $P_{12}P_{13}$. Hence there are four triangles $P_{12}P_{13}P_{14}$, $P_{23}P_{24}P_{21}$, $P_{34}P_{31}P_{32}$ and $P_{41}P_{42}P_{43}$, denoted by $T_1^*, T_2^*, T_3^*, T_4^*$, with the property that A_j^*, A_k^*, A_l^* are the reflections of A_i into the respective sides of T_i^*.

If three positions of E are given, every set of homologous points A_1, A_2, A_3 is on some circle c with center M. We know from Section 3 that any point M of Σ is such a center. If we consider four positions and prescribe A_4 to be on c, M must satisfy one condition. Hence there exists a locus of points M in Σ with this property. In Fig. 61 the circle c and the four points A_i on it are drawn in Σ; let $\angle A_iMA_j = \alpha_{ij}$. The rotation center P_{ij} is on the perpendicular bisector of A_iA_j, that is the bisector of $\angle A_iMA_j$. We have $\angle A_1MP_{12} = \frac{1}{2}\alpha_{12}$,

$\angle A_1MP_{31} = \frac{1}{2}(\alpha_{12} + \alpha_{23})$, hence $\angle P_{12}MP_{31} = \frac{1}{2}\alpha_{23}$; similarly $\angle P_{42}MA_4 = \frac{1}{2}(\alpha_{23} + \alpha_{34})$, $\angle P_{34}MA_4 = \frac{1}{2}\alpha_{34}$ and therefore $\angle P_{42}MP_{34} = \frac{1}{2}\alpha_{23}$. This means that, seen from M, the optical distance $P_{12}P_{31}$ is equal to $P_{42}P_{34}$. The conclusion is that *the locus of M, called the center-point curve of the four positions, coincides with the pole curve* K.

Related to the center-point curve in Σ is the locus in E of the points A whose four positions A_i are concyclic. It is called the *circle-point curve* (German: Kreispunktkurve). It can be determined by considering the curve K for the inverse motion. If A_i are concyclic, with center M, we have $MA_i = R$ for $i = 1, 2, 3, 4$; obviously, for the inverse motion the fixed point A and the positions M_i of M satisfy $AM_i = R$. This implies that the circle-point curve is identical with the center curve of the inverse motion. Hence the pole curve K_P, the center-point curve K_M (both coinciding and lying in Σ) and the circle-point curve K_A (in E) are all focal cubics.

To determine K_A in the position E_k we use the rotation centers P_{ij}^k of the inverse motion: We know that for three positions, if P_{12}, P_{23}, P_{31} are the centers in Σ, we obtain those in E_1 by setting $P_{12}^1 = P_{12}$, $P_{31}^1 = P_{31}$ and P_{23}^1 as

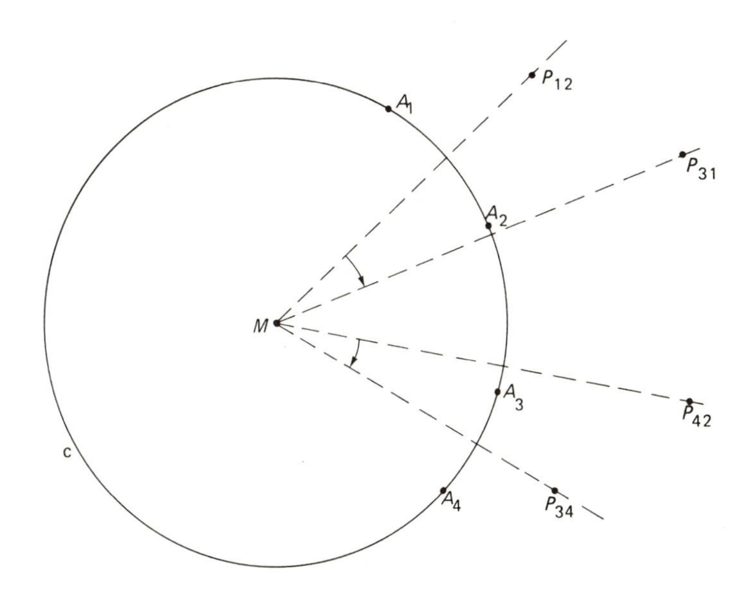

Fig. 61.

the reflection of P_{23} into the line $P_{12}P_{31}$ (Fig. 62). Hence for four positions $P^1_{12} = P_{12}$, $P^1_{13} = P_{13}$, $P^1_{14} = P_{14}$, P^1_{34}, P^1_{42} and P^1_{23} are the reflections of P_{34}, P_{42} and P_{23} into $P_{13}P_{14}$, $P_{14}P_{12}$ and $P_{12}P_{13}$ respectively. K_A is the locus of points with equal optical distance to (for instance) $P_{12}P_{31}$ and $P^1_{42}P^1_{34}$.

As there is a locus of points A for which the four points A_i are concyclic, we may expect a finite number of points A for which A_i are collinear. In Section 3 we found that if A_1, A_2, A_3 are on a line l, this line passes through the orthocenter of the triangle $P_{23}P_{31}P_{12}$, which we denote as H_4; conversely any line through H_4 bears three collinear homologous points. Hence, if H_3 is the orthocenter of the triangle $P_{24}P_{41}P_{12}$ the join $b = H_3H_4$ is such that four homologous points A_i lie on b; if H_3 and H_4 are different points, b is the only line with this property. Hence *there is in general one point B of E such that B_i* ($i = 1, 2, 3, 4$) *are collinear.* It is called the Ball point, being named after R. S. Ball who discovered this point (for the instantaneous case); b is called Ball's line.

Example 48. Prove that the configuration of the centers P_{ij} has the property that the orthocenters H_i of triangles $P_{24}P_{43}P_{32}$, $P_{31}P_{14}P_{43}$, $P_{42}P_{21}P_{14}$, $P_{13}P_{32}P_{21}$ are collinear.
Example 49. If the configuration of the centers P_{ij} is given, determine the four points B_i on Ball's line b.
Example 50. Prove by considering the inverse displacement that there is one line in E such that its four homologous positions in Σ pass through one point.

We have, in planar four-positions theory, so far made use of a method starting from geometrical considerations (such as the P_{ij}-configuration), although we have not excluded algebraic means (such as our equation (4.13)

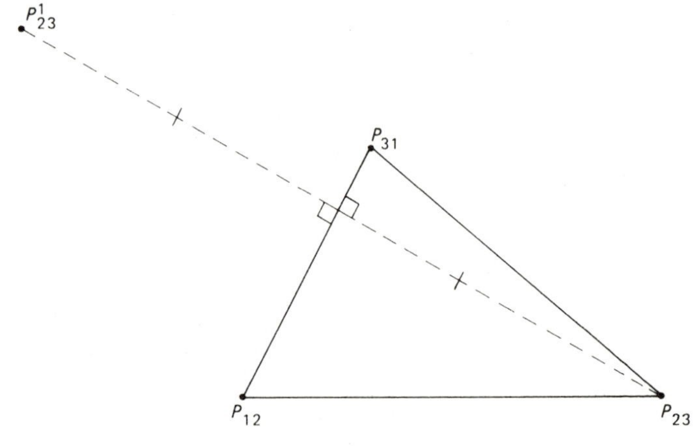

Fig. 62.

for the focal cubic). As we mentioned before, in the history of kinematics there has been a tendency to apply purely geometric reasoning. We could, on the other hand, try to develop a purely analytical theory. It could start by defining four positions of the plane E (with coordinates x, y), with respect to Σ (with coordinates (X, Y)), by means of four displacements each given by three pieces of data: (ϕ_i, a_i, b_i), $i = 1, 2, 3, 4$; then the four positions of (x, y) in Σ would be

$$X_i = x \cos \phi_i - y \sin \phi_i + a_i, \qquad Y_i = x \sin \phi_i + y \cos \phi_i + b_i, \quad i = 1, 2, 3, 4. \tag{4.17}$$

The condition that the homologous points are concyclic is

$$|X_i^2 + Y_i^2 \quad X_i \quad Y_i \quad 1| = 0, \tag{4.18}$$

and this is therefore the equation of the circle-point curve K_A. We have from (4.17)

$$\begin{aligned} X_i^2 + Y_i^2 &= x^2 + y^2 + 2x(a_i \cos \phi_i + b_i \sin \phi_i) \\ &\quad + 2y(-a_i \sin \phi_i + b_i \cos \phi_i) + a_i^2 + b_i^2 \\ &= x^2 + y^2 - 2\tilde{a}_i x - 2\tilde{b}_i y + d_i^2, \end{aligned} \tag{4.19}$$

$\tilde{a}_i$ and $\tilde{b}_i$ being the characteristics of the inverse displacements and $d_i^2 = a_i^2 + b_i^2 = \tilde{a}_i^2 + \tilde{b}_i^2$.

Without any loss of generality we may suppose $\phi_4 = a_4 = b_4 = 0$, which means that the frame in Σ and E_4 coincide. Then subtracting in (4.18) the fourth row from the others it reduces to the 3×3 determinant:

$$\left| \begin{matrix} -2\tilde{a}_i x - 2\tilde{b}_i y + d_i^2 & x(\cos \phi_i - 1) & x \sin \phi_i \\ & -y \sin \phi_i + a_i & +y(\cos \phi_i - 1) + b_i \end{matrix} \right| = 0, \tag{4.20}$$

which shows that K_A is a cubic curve. For the third order terms we obtain, developing (4.20) with respect to the first column,

$$-8 \sum (\tilde{a}_1 x + \tilde{b}_1 y) \sin(\tfrac{1}{2}\phi_2) \sin(\tfrac{1}{2}\phi_3) \sin(\tfrac{1}{2}(\phi_3 - \phi_2))(x^2 + y^2) \tag{4.21}$$

which shows that K_A is a circular cubic. It would cost a great deal of algebra to show that (4.20) is a focal curve and to re-derive its other properties. Therefore we shall not continue this development any further.

If we want to determine Ball's point B, in E, by this method, use can be made of formula (3.2), representing the circle in E which is the locus of the points A whose positions A_1, A_2, A_3 are collinear. Similar circles appear if A_2, A_3, A_4; A_3, A_4, A_1; or A_4, A_1, A_2 are collinear. Ball's point is the common point of the four circles, but the algebra seems cumbersome. Somewhat less complicated is the determination of Ball's line b in Σ, the line on which the four points B_i lie. On it are the orthocenters of the pole triangles T_1, T_2, T_3 and T_4. Making use of (3.7) the line coordinates of b are seen to be the minors of the matrix

$$
(4.22)\qquad \left\Vert \begin{array}{lll}
a_2 \sin(\phi_4 - \phi_3) & b_2 \sin(\phi_4 - \phi_3) & 4\sin(\tfrac{1}{2}(\phi_3 - \phi_4)) \\
\quad + a_3 \sin(\phi_2 - \phi_4) & \quad + b_3 \sin(\phi_2 - \phi_4) & \quad \cdot \sin(\tfrac{1}{2}(\phi_4 - \phi_2)) \\
\quad + a_4 \sin(\phi_3 - \phi_2) & \quad + b_4 \sin(\phi_3 - \phi_2) & \quad \cdot \sin(\tfrac{1}{2}(\phi_2 - \phi_3)) \\
a_3 \sin(\phi_1 - \phi_4) & b_3 \sin(\phi_1 - \phi_4) & 4\sin(\tfrac{1}{2}(\phi_4 - \phi_1)) \\
\quad + a_4 \sin(\phi_3 - \phi_1) & \quad + b_4 \sin(\phi_3 - \phi_1) & \quad \cdot \sin(\tfrac{1}{2}(\phi_1 - \phi_3)) \\
\quad + a_1 \sin(\phi_4 - \phi_3) & \quad + b_1 \sin(\phi_4 - \phi_3) & \quad \cdot \sin(\tfrac{1}{2}(\phi_3 - \phi_4)).
\end{array} \right\Vert
$$

After a great deal of algebra, the three minors are seen to have the common factor $\sin(\phi_4 - \phi_3)$. We introduce the numbers $A_{ij} = -A_{ji} = a_i b_j - a_j b_i$, and furthermore $B_{14} = A_{23}$, $B_{24} = A_{31}$, $B_{34} = A_{12}$, $B_{23} = A_{14}$, $B_{31} = A_{24}$, $B_{12} = A_{34}$. Then the following result is obtained for the coordinates of Ball's line b:

$$
(4.23)\qquad \begin{aligned}
U_1 = -4[&b_1 \sin(\tfrac{1}{2}(\phi_2 - \phi_3)) \sin(\tfrac{1}{2}(\phi_3 - \phi_4)) \sin(\tfrac{1}{2}(\phi_4 - \phi_2)) \\
&- b_2 \sin(\tfrac{1}{2}(\phi_3 - \phi_4)) \sin(\tfrac{1}{2}(\phi_4 - \phi_1)) \sin(\tfrac{1}{2}(\phi_1 - \phi_3)) \\
&+ b_3 \sin(\tfrac{1}{2}(\phi_4 - \phi_1)) \sin(\tfrac{1}{2}(\phi_1 - \phi_2)) \sin(\tfrac{1}{2}(\phi_2 - \phi_4)) \\
&- b_4 \sin(\tfrac{1}{2}(\phi_1 - \phi_2)) \sin(\tfrac{1}{2}(\phi_2 - \phi_3)) \sin(\tfrac{1}{2}(\phi_3 - \phi_1))].
\end{aligned}
$$

U_2 follows from (4.23) by replacing b_i by $-a_i$.

$$
(4.24)\qquad \begin{aligned}
U_3 = {} & B_{14} \sin(\phi_1 - \phi_4) + B_{24} \sin(\phi_2 - \phi_4) + B_{34} \sin(\phi_3 - \phi_4) \\
& + B_{23} \sin(\phi_2 - \phi_3) + B_{31} \sin(\phi_3 - \phi_1) + B_{12} \sin(\phi_1 - \phi_2).
\end{aligned}
$$

Example 51. Show that b is invariant for any permutation of the indices 1, 2, 3, 4; which is of course as it should be.

Until now our discussion of four positions theory has been mainly limited to the displacement of a point of the moving plane E and therefore on the configuration of four homologous points in Σ. Special attention was given to the quadruple with special properties, such as being concyclic or collinear. We now add some remarks about homologous lines.

If l is a line in E and if l_i $(i = 1,2,3,4)$ are its positions in Σ the question whether it is possible to determine l such that l_i are concurrent is already implicitly solved in the foregoing. Indeed l is obviously the Ball line in E for the inverse displacements; the common point B' of l_i is Ball's point in Σ for these displacements.

Example 52. Determine l and B' by means of the formulas (2.5) and (3.5) and compare the results with (4.23) and (4.24).

The problem of determining lines in E such that the homologous lines in Σ are tangent to a circle is not an interesting one; obviously any line in E parallel to the Ball line (of the inverse displacements) has this property; the center of any such circle is the point B'.

In plane kinematics the study of homologous lines is simpler than that of homologous points; this statement will be confirmed when — in the following sections — we consider more than four positions. It stems from the fact that equations (2.5) are simpler than (2.1). The difference between the behavior of points and lines reflects the circumstance that there is no perfect duality in the Euclidean plane.

5. Five positions theory

We consider now five positions of the plane E with respect to Σ.

Any point A of E has five positions A_i $(i = 1,2,3,4,5)$ in Σ. The number of relative rotation centers P_{ij} is obviously equal to *ten*.

We cannot expect that there will be (in general) a point A such that the points A_i are collinear because we have seen that for four positions there is only one point (Ball's point) such that A_1, A_2, A_3, A_4 are collinear. More promising is the problem to determine A such that the five points A_i are concyclic.

We give first a geometric solution. The locus of A_1 such that A_1, A_2, A_3, A_4 are concyclic has been derived in Section 4 and seen to be a cubic curve (which we denoted by K_A) passing through the isotropic points I_1, I_2 and through the centers P_{12}, P_{13}, P_{14}, P^1_{34}, P^1_{42}, P^1_{23}. Analogously the locus of A_1 such that A_1, A_2, A_3, A_5 are concyclic is obtained by replacing position 4 with position 5. The resulting circle-point curve which we denote by $K_{A_{1235}}$ passes through I_1, I_2 and P_{12}, P_{13}, P_{15}, P^1_{35}, P^1_{52}, P^1_{23}. The two circular cubics, K_A and $K_{A_{1235}}$ have nine intersections; among these are I_1, I_2 and P^1_{23}, P_{31}, P_{12}; the remaining four denoted by B^j $(j = 1,2,3,4)$ are the solutions of our problem.

Example 53. Show that the five points I_1, I_2, P^1_{23}, P_{31}, P_{12} must indeed be rejected and that only the B^j satisfy the condition that their five homologous positions are concyclic.

The four points B^j are called the *Burmester points* of the five positions, in honor of the man who was the first to solve the problem (BURMESTER [1876; 1877; 1888, pp. 621–23]). It can be shown by examples that they may be all real, two may be imaginary or all four may be imaginary.

The five positions of a Burmester point B^j_i $(i = 1,\ldots,5)$ are on a circle, the center of which we denote by M^j. There are four such Burmester centers. These are located at the non-trivial intersections of the two center-point curves K_M and $K_{M_{1235}}$. It is easy to see that these four points of Σ are also the Burmester points of the inverse displacements, and that they may also be found as the non-trivial intersections of two circle-point curves, each related to four positions of the inverse displacements. A pair B^j, M^j is called a Burmester pair.

Several special cases arise if the five positions satisfy certain conditions. Apart from being real or imaginary, one or more points or centers may be at infinity.

The determination of the points B^j (or M^j) as intersections of two cubic curves is not an attractive operation. Some easier processes for finding them come from the fact that the Burmester points can be determined as the intersections of two conics. Analytically, this reduces the problem to determining the roots of an equation of degree four (rather than nine).

One such method (HACKMÜLLER [1938a, b], VELDKAMP [1963] makes use of the equation of the focal curve, thus following Burmester's argument. The pole curve (or the center-point curve, identical to it) for the positions 1, 2, 3, 4 is given by (4.2):

$$K \equiv K_M \equiv L_{12}C_{34} - L_{34}C_{12} = 0, \tag{5.1}$$

in which L_{ij} and C_{kl} are defined by (4.1).

$L_{12} = 0$ is the equation of the line $P_{12}P_{31}$, and $C_{12} = 0$ that of the circle with $P_{12}P_{31}$ as diameter; $L_{34} = 0$ is the line $P_{42}P_{34}$ and $C_{34} = 0$ the circle with $P_{42}P_{34}$ as diameter. We write (5.1) as

$$K_M \equiv mZL_{12} - (L_{34} - L_{12})C_{12} = 0, \tag{5.2}$$

with $mZ = C_{34} - C_{12}$. In view of (4.1) we have: $m = 0$ is the equation of the power line (the common chord) of the circles C_{12} and C_{34}; $Z = 0$ is the line at infinity. The center-point curve of the positions 1, 2, 3, 5 is obtained from (5.2) if we replace the subscript 4 with 5:

$$K_{M_{1235}} \equiv m'ZL_{12} - (L_{35} - L_{12})C_{12} = 0, \tag{5.3}$$

where $m'Z = C_{35} - C_{12}$. The curves K_M and $K_{M_{1235}}$ both pass through the isotropic points and through P_{23}, P_{31}, P_{12}. The former lie on $C_{12} = 0$, and on $Z = 0$; furthermore P_{31}, P_{12} lie on $L_{12} = 0$ and on $C_{12} = 0$. Hence if we eliminate ZL_{12} and C_{12} from (5.2) and (5.3) (which are linear homogeneous equations for these expressions) we obtain a relation which is satisfied by P_{23} and the four Burmester centers M^j. The result of the elimination is

$$m(L_{35} - L_{12}) - m'(L_{34} - L_{12}) = 0, \tag{5.4}$$

which represents a conic, passing through these five points. But, obviously, we may derive all together ten such conics, each passing through the four points M^j and one of the rotation centers P_{ij}. Two of them are sufficient to determine the Burmester centers; these are found therefore as the intersections of two conics, the equations of which can be derived if the configuration of the P_{ij} is given.

A more primitive and direct method to determine the Burmester points is the following (BOTTEMA [1964a]) which uses neither the centers P_{ij} nor the properties of focal curves. If we describe the five positions of the moving plane E using the parameters (ϕ_i, a_i, b_i) $(i = 1,2,3,4,5)$, the five homologous points $B_i(X_i, Y_i)$ in Σ of a point $B(x, y)$ in E are given by:

$$\begin{aligned} X_i &= x\cos\phi_i - y\sin\phi_i + a_i, \\ Y_i &= x\sin\phi_i + y\cos\phi_i + b_i, \quad (i = 1,2,3,4,5). \end{aligned} \tag{5.5}$$

A circle C in Σ has the equation

$$c_0(X^2 + Y^2) - 2c_1X - 2c_2Y + c_3 = 0. \tag{5.6}$$

The point B_i is on the circle C if (5.5) satisfy equation (5.6); we substitute (5.5) into (5.6) and obtain the five relations $(i = 1,2,3,4,5)$:

$$\begin{aligned} &c_0(x^2 + y^2) + c_0(a_i^2 + b_i^2) + 2c_0x(a_i\cos\phi_i + b_i\sin\phi_i) \\ &+ 2c_0y(-a_i\sin\phi_i + b_i\cos\phi_i) - 2a_ic_1 - 2b_ic_2 - 2(c_1x + c_2y)\cos\phi_i \\ &+ 2(c_1y - c_2x)\sin\phi_i + c_3 = 0. \end{aligned} \tag{5.7}$$

These are five equations for the five unknowns: x, y and the ratios of c_0, c_1, c_2, c_3.

The coordinate systems in both E and Σ are arbitrary; we specify them such that they coincide in one of the positions, say the first. Then $a_1 = b_1 = \phi_1 = 0$ and the first equation of (5.7) reads

$$c_0(x^2 + y^2) - 2c_1x - 2c_2y + c_3 = 0. \tag{5.8}$$

If we subtract it from any of the four others the unknown c_3 is eliminated. The homogeneous coordinates of the center M of the circle C are (c_1, c_2, c_0); we do not exclude the case $c_0 = 0$. Analogously we introduce homogeneous coordinates (x, y, z) for B in E. Furthermore we set

$$a_i \cos\phi_i + b_i \sin\phi_i = -\tilde{a}_i \quad \text{and} \quad -a_i \sin\phi_i + b_i \cos\phi_i = -\tilde{b}_i. \tag{5.9}$$

The four equations obtained by subtracting (5.8) from (5.7) are $(i = 2, 3, 4, 5)$:

$$\begin{aligned} &\tfrac{1}{2}(a_i^2 + b_i^2)c_0 z - \tilde{a}_i c_0 x - \tilde{b}_i c_0 y - a_i c_1 z - b_i c_2 z \\ &\quad + (1 - \cos\phi_i)(c_1 x + c_2 y) + \sin\phi_i (c_1 y - c_2 x) = 0. \end{aligned} \tag{5.10}$$

These are four equations for the coordinates (x, y, z) of Burmester point B in E and the coordinates (c_1, c_2, c_0) of the corresponding Burmester center M in Σ. The symmetry of the equation is obvious: if E and Σ are interchanged, and thus (x, y, z) and (c_1, c_2, c_0), (a_i, b_i, ϕ_i) and $(\tilde{a}_i, \tilde{b}_i, -\phi_i)$, the equations are invariant. The set (5.10) determines the Burmester pairs of the five positions.

In order to solve the system (5.10) we introduce the seven unknowns

$$\begin{aligned} &u_0 = c_0 z, \quad u_1 = c_1 z, \quad u_2 = c_2 z, \quad u_3 = c_0 x, \\ &u_4 = c_0 y, \quad u_5 = c_1 x + c_2 y, \quad u_6 = c_1 y - c_2 x, \end{aligned} \tag{5.11}$$

satisfying the two quadratic relations

$$u_0 u_5 = u_1 u_3 + u_2 u_4, \qquad u_0 u_6 = u_1 u_4 - u_2 u_3, \tag{5.12}$$

both of which are independent of the displacement parameters. The equations (5.10) are now $(i = 2, 3, 4, 5)$:

$$\begin{aligned} &\tfrac{1}{2}(a_i^2 + b_i^2)u_0 - a_i u_1 - b_i u_2 - \tilde{a}_i u_3 - \tilde{b}_i u_4 \\ &\quad + (1 - \cos\phi_i)u_5 + (\sin\phi_i)u_6 = 0, \end{aligned} \tag{5.13}$$

but these are linear equations for u_i. Hence we have obtained six homogeneous equations for the seven homogeneous unknowns u_i; four of these equations are linear and two are quadratic. The number of solutions is therefore indeed four. If the equations are solved it follows from (5.11) that $B = (u_3, u_4, u_0)$ and $M = (u_1, u_2, u_0)$. Hence by this procedure we obtain the Burmester points and the Burmester centers simultaneously. The method is suitable for numerical computation if the positions are given by the parameters (a_i, b_i, ϕ_i).

We now discuss some aspects of the set of equations (5.12) and (5.13). If the 4×4 determinant

$$\Delta = |\, a_i \quad b_i \quad 1 - \cos\phi_i \quad \sin\phi_i \,| \tag{5.14}$$

is unequal to zero, we can solve (5.13) for u_1, u_2, u_5, u_6 as linear functions of u_0, u_3, u_4. Substituting these into (5.12) gives us two quadratic equations for the coordinates x, y, z of the Burmester points, which are in this way determined as intersections of two conics. The result cannot be that $c_0 = 0$, because this implies $u_0 = u_3 = u_4 = 0$ and therefore* $\Delta = 0$. The conclusion is: if $\Delta \neq 0$ the four Burmester centers are all finite points. Analogously, if

$$(5.15) \qquad \Delta' = | \tilde{a}_i \quad \tilde{b}_i \quad 1 - \cos\phi_i \quad \sin\phi_i |$$

is unequal to zero, we have $z \neq 0$ which implies that the four Burmester points are finite points. Hence $\Delta\Delta' \neq 0$ gives us the condition that all Burmester pairs consist of finite points. If, however, $\Delta = 0$, $\Delta' \neq 0$ the Burmester points are finite points, but for (at least) one of them the corresponding center is at infinity, which means that for that Burmester point the five homologous positions are collinear. This special Burmester point is called in Veldkamp's terminology (VELDKAMP [1963]) a Ball's point with excess one. The relation $\Delta = 0$ has another consequence. It implies that there exist four coefficients λ_i, not all zero, such that

$$(5.16) \qquad \sum \lambda_i a_i = \sum \lambda_i b_i = \sum \lambda_i (1 - \cos\phi_i) = \sum \lambda_i \sin\phi_i = 0.$$

When these conditions apply, multiplying (5.13) by λ_i and summing the resulting equations yields:

$$(5.17) \qquad \tfrac{1}{2} u_0 \sum \lambda_i (a_i^2 + b_i^2) - u_3 \sum \lambda_i \tilde{a}_i - u_4 \sum \lambda_i \tilde{b}_i = 0,$$

but this represents a straight line in E satisfied by the remaining Burmester points (for which c_0 is in general unequal to zero). Hence the theorem: *if one of the four Burmester points B^1 has its corresponding center at infinity, the other three B^2, B^3, B^4 are collinear* (PRIMROSE et al. [1964]). It has also been shown that if B^1 is on the line through the other three, it coincides with one of them. These latter two theorems may be regarded as generalizations of ones due to R. Müller (MÜLLER [1892]) for the instantaneous case.

The Burmester point configuration because of its applicability to the design of four-bar linkages is the most developed aspect of the theory of five homologous points. In the literature some other problems have been investigated: A conic is determined by five of its points, hence any five homologous points determine a conic. One may ask when this conic is a parabola. It has been shown (SANDOR AND FREUDENSTEIN [1967], FREUDENSTEIN et al. [1969]) that the locus, of points in E such that their five positions in Σ are points of a

* This argument relies on the fact that all three homogeneous coordinates cannot simultaneously be zero, i.e., $x = y = z = 0$ and $c_0 = c_1 = c_2 = 0$ are not permitted.

parabola, is a curve of degree twelve, with six-fold points at I_1 and I_2, and passing through the five Ball points related to the five different combinations of four positions. This answer shows that the problem is complicated. We remark that a parabola is not uniquely determined by four points. In fact two parabolas pass through them and these are either real or imaginary.

On the other hand, as a parabola is uniquely determined by four tangents, it seems more promising to ask for the locus of *lines* in E such that their five homologous positions in Σ are tangent to a parabola. We shall show that the answer is relatively simple (BOTTEMA [1970]).

A curve of the second class, in Σ, is given by a quadratic equation of the line coordinates U_1, U_2, U_3. If it is a parabola this equation is satisfied by the line at infinity, that is by $(0, 0, 1)$. Hence the general equation of a parabola reads

$$a_{11}U_1^2 + 2a_{12}U_1U_2 + a_{22}U_2^2 + 2a_{13}U_1U_3 + 2a_{23}U_2U_3 = 0. \tag{5.18}$$

Therefore five lines (U_{1i}, U_{2i}, U_{3i}), $i = 1, 2, 3, 4, 5$, in Σ are tangent to a parabola if the 5×5 determinant

$$|U_{1i}^2 \quad U_{1i}U_{2i} \quad U_{2i}^2 \quad U_{1i}U_{3i} \quad U_{2i}U_{3i}|$$

is equal to zero, which may be written as

$$\mathrm{D} \equiv |U_{1i}^2 + U_{2i}^2 \quad U_{1i}^2 - U_{2i}^2 \quad 2U_{1i}U_{2i} \quad U_{1i}U_{3i} \quad U_{2i}U_{3i}| = 0. \tag{5.19}$$

As U_{1i}, U_{2i}, U_{3i} are linear functions of u_1, u_2, u_3 this equation is of the tenth degree. It can, however, be considerably simplified. It follows from (2.5) that

$$\begin{aligned} U_1^2 + U_2^2 &= u_1^2 + u_2^2, \\ U_1^2 - U_2^2 &= (u_1^2 - u_2^2)\cos 2\phi - 2u_1u_2 \sin 2\phi, \\ 2U_1U_2 &= (u_1^2 - u_2^2)\sin 2\phi + 2u_1u_2 \cos 2\phi. \end{aligned} \tag{5.20}$$

Hence a minor formed from (5.19) using the first three columns yields:

$$\begin{aligned} \mathrm{D}_{ijk} &\equiv \begin{vmatrix} U_{1i}^2 + U_{2i}^2 & U_{1i}^2 - U_{2i}^2 & 2U_{1i}U_{2i} \\ U_{1j}^2 + U_{2j}^2 & U_{1j}^2 - U_{2j}^2 & 2U_{1j}U_{2j} \\ U_{1k}^2 + U_{2k}^2 & U_{1k}^2 - U_{2k}^2 & 2U_{1k}U_{2k} \end{vmatrix} \\ &= (u_1^2 + u_2^2)^3[\sin(2(\phi_k - \phi_j)) + \sin(2(\phi_i - \phi_k)) + \sin(2(\phi_j - \phi_i))] \\ &= 4(u_1^2 + u_2^2)^3 \sin(\phi_j - \phi_k)\sin(\phi_k - \phi_i)\sin(\phi_i - \phi_j). \end{aligned} \tag{5.21}$$

Furthermore we have in view of (2.5)

$$\mathrm{d}_{mn} \equiv \begin{vmatrix} U_{1m}U_{3m} & U_{2m}U_{3m} \\ U_{1n}U_{3n} & U_{2n}U_{3n} \end{vmatrix} = (u_1^2 + u_2^2)\sin(\phi_n - \phi_m)U_{3m}U_{3n}. \tag{5.22}$$

Developing (5.19) by means of the minors of the last two columns shows D has $(u_1^2 + u_2^2)^4$ as a factor. Removing this factor we are left with the equation

$$\sum [\sin(\phi_n - \phi_m)\sin(\phi_j - \phi_k)\sin(\phi_k - \phi_i)\sin(\phi_i - \phi_j) \cdot (\tilde{a}_n u_1 + \tilde{b}_n u_2 + u_3)(\tilde{a}_m u_1 + \tilde{b}_m u_2 + u_3)] = 0, \tag{5.23}$$

Σ being the sum of 10 terms. As (5.23) is of the second degree, the locus of the lines in E such that their five positions in Σ are tangents of a parabola is a conic, which we denote by $\mathfrak{K}$.

We know that there is one line in E such that its four homologous positions in Σ pass through one point. Let l be this Ball line for the four positions 1, 2, 3, 4; hence l_1, l_2, l_3, l_4 pass through one point B, l_5 has an arbitrary position. Any conic with l_i $(i = 1, \ldots, 5)$ as tangents is degenerated into the pencil of lines with vertex B and a second pencil with its vertex on l_5. Among all such conics there is one parabola: obtained if the vertex on l_5 is taken as its point at infinity. Therefore the line l belongs to the locus $\mathfrak{K}$. The conclusion is: the Ball line of any four, out of the five, positions is a tangent of $\mathfrak{K}$. Or, in other words, the locus $\mathfrak{K}$ is that conic determined by the five Ball lines in E, each such line is related to four of the five given positions.

There are of course many special cases.

Example 54. If, for instance, $\phi_1 = \phi_2$ show that $\mathfrak{K}$ is degenerate.
Example 55. If two (and therefore all five) Ball lines coincide, show that $\mathfrak{K}$ is degenerate.

6. Six and more positions

We make some remarks about six and more positions of E with respect to Σ.

In the case of six positions there are obviously 15 relative rotation centers P_{ij} and 15 Ball points B_{ijkl} (one for every group of four positions out of the six).

A natural problem is to determine the locus of points A in E such that A_i $(i = 1, \ldots, 6)$ in Σ are on a conic. This is the case if the 6×6 determinant

$$|X_i^2 \quad X_iY_i \quad Y_i^2 \quad X_i \quad Y_i \quad 1| \tag{6.1}$$

is equal to zero. As X_i, Y_i are linear functions of x, y, the degree of the locus is — at first sight — equal to eight. But (6.1) may be written as

$$|X_i^2 + Y_i^2 \quad X_iY_i \quad Y_i^2 \quad X_i \quad Y_i \quad 1| \tag{6.2}$$

and $X_i^2 + Y_i^2 = x^2 + y^2 +$ linear terms. Hence, if we factor out $(x^2 + y^2)$, the terms of the highest degree are given by a determinant with two equal columns, the first and the last. This implies that the eight order terms vanish. Further analysis (FREUDENSTEIN et al. [1969]) shows that the degree of the locus is in general seven; it has three-fold points at I_1 and I_2 and passes through the 15 centers and the 15 Ball points.

By counting the intersections of this curve with the curve of twelfth order described in Section 5 (i.e., the locus of co-parabolic points for five positions), the number of points in E whose six positions in Σ are on a parabola has been calculated to be 33 (FREUDENSTEIN, et al. [1969]). A much simpler problem is that of determining the lines in E with six homologous positions tangent to a parabola. They are, in view of the results of Section 5, the common tangents of two conics. Each of these conics is in turn tangent to one set of five Ball lines, each set is composed of the Ball lines corresponding to all combinations of four positions out of a given five. Hence, two such sets follow for instance from 1, 2, 3, 4, 5 and 1, 2, 3, 4, 6. In this case then, one of the four common tangents to the two conics is obviously the Ball line associated with the positions 1, 2, 3, 4. This line is spurious since its positions 1, 2, 3, 4 do not define a unique parabola and therefore its 5th and 6th position will generally be tangent to different parabolas. Hence there are generally only three lines with the property we seek (BOTTEMA [1970]).

For six positions, the six homologous lines (U_{1i}, U_{2i}, U_{3i}), $i = 1, \ldots, 6$, of the line (u_1, u_2, u_3) in E, will be tangent to a conic if their coordinates satisfy the condition

$$\mathrm{D}' \equiv |U_{1i}^2 + U_{2i}^2 \quad U_{1i}^2 - U_{2i}^2 \quad 2U_{1i}U_{2i} \quad U_{1i}U_{3i} \quad U_{2i}U_{3i} \quad U_{3i}^2| = 0, \tag{6.3}$$

D′ being a 6×6 determinant. This is at first sight an equation of degree twelve in u_1, u_2, u_3. But we have

$$\begin{aligned} \delta_{ijk} &\equiv \begin{vmatrix} U_{1i}U_{3i} & U_{2i}U_{3i} & U_{3i}^2 \\ U_{1j}U_{3j} & U_{2j}U_{3j} & U_{3j}^2 \\ U_{1k}U_{3k} & U_{2k}U_{3k} & U_{3k}^2 \end{vmatrix} \\ &= -U_{3i}U_{3j}U_{3k}\,[U_{3i}\sin(\phi_j - \phi_k) + U_{3j}\sin(\phi_k - \phi_i) \\ &\qquad + U_{3k}\sin(\phi_i - \phi_j)](u_1^2 + u_2^2), \end{aligned} \tag{6.4}$$

where we have made use of (5.22). Developing (6.3) with respect to the minors of the last three columns and making use of (5.21) it is seen that D′ has the factor $(u_1^2 + u_2^2)^4$. This implies that the locus of the lines of E whose six homologous lines in Σ are tangent to a conic is a curve $\mathfrak{K}_4$ of the fourth class. If l is the Ball line of the positions 1, 2, 3, 4 then l_1, l_2, l_3, l_4 pass through one point B; obviously the six lines l_i are tangents of a conic degenerated into two pencils, one with vertex B, the other with the intersection of l_5 and l_6 as its vertex. This implies that all 15 Ball lines of the six positions are tangent of $\mathfrak{K}_4$.

If we consider seven positions we may use for instance the two curves $\mathfrak{K}_4(1,2,3,4,5,6)$ and $\mathfrak{K}_4(1,2,3,4,5,7)$; they have sixteen common tangents among these are, however, the five Ball lines related to the common positions $1,\ldots,5$. Hence the number of lines in E whose seven homologous lines in Σ are tangents of a conic is equal to eleven (BOTTEMA [1970], where, however, some results must be corrected).

7. Continuous displacements

The fundamental equations of plane kinematics are (2.1):

$$X = x\cos\phi - y\sin\phi + a,$$
$$Y = x\sin\phi + y\cos\phi + b. \tag{7.1}$$

These give the position (X, Y), in the fixed plane Σ, of the point (x, y) of the moving plane E. A displacement is described by the three numbers ϕ, a and b. In the preceding sections we have dealt with sets of discrete displacements. We consider now a continuous set of displacements; it is described if ϕ, a, b are given functions of a parameter t. If t is the time (7.1) defines a motion of E with respect to Σ. Any motion is completely described by the three functions $\phi(t)$, $a(t)$ and $b(t)$. They give us, by means of (7.1), not only the position of any point of E at any time, but also its velocity, its acceleration, and so on. If, for instance, we take constant values for x, y the relations (7.1) are a parametric representation of the path of the point (x, y).

In Chapter II, Section 6 we have distinguished between geometric and time-dependent kinematics. For the latter the parameter t has an essential significance; the velocity and acceleration distribution for instance belong to this subject. Many other properties of the motion are independent of the choice of the parameter and deal with its geometric aspects only: for example the paths, the centrodes and so on. In this and the next section we consider geometric kinematics.

If ϕ is a constant any displacement of the set is a translation. The motion is a simple one. Any line is displaced into a parallel line. The paths of the points of E are congruent curves. The path of one point may be chosen arbitrarily and then the motion is completely known. We shall disregard this special case and suppose that ϕ is variable. Then we may take it as the motion's parameter, a natural one, having a geometric meaning. In this case we have

$$X = x\cos\phi - y\sin\phi + a(\phi),$$
$$Y = x\sin\phi + y\cos\phi + b(\phi), \tag{7.2}$$

and the motion is completely defined by the two functions $a(\phi)$ and $b(\phi)$. All properties depend on these two only. Differentiation with respect to ϕ will be denoted by a prime. The "geometric velocity" of the point x, y is

$$X' = -x\sin\phi - y\cos\phi + a', \qquad Y' = x\cos\phi - y\sin\phi + b'. \tag{7.3}$$

If ϕ depends on the time t we have $\dot{X} = X'\dot{\phi}$, $\dot{Y} = Y'\dot{\phi}$, hence at any moment the ratio of the velocity and the geometric velocity is the same for all points. In particular, if $X' = Y' = 0$ for a certain point, then this point is at rest, independent of the time scale according to which the motion takes place. Hence the instantaneous rotation center (or the pole) of the motion is a geometric concept. It may be defined as the point which at the moment under consideration is at a cusp of its path. $X' = Y' = 0$ are two linear equations for x, y, and the coordinates of the pole P are

$$x_p = a'\sin\phi - b'\cos\phi, \qquad y_p = a'\cos\phi + b'\sin\phi. \tag{7.4}$$

By means of (7.2) we have for its coordinates in Σ:

$$X_p = a - b', \qquad Y_p = a' + b. \tag{7.5}$$

As (7.4) and (7.5) hold for any value of ϕ this implies that (7.4) is a parametric representation of the locus of all points in E which at some position ϕ coincide with the pole P, that is, the moving centrode p_m; (7.5) represents the fixed centrode p_f. At any moment they have P as a common point. The derivative of (7.4) is

$$x_p' = (a' - b'')\cos\phi + (a'' + b')\sin\phi,$$
$$y_p' = -(a' - b'')\sin\phi + (a'' + b')\cos\phi, \tag{7.6}$$

which gives us the components of the geometric velocity along p_m. (This is the geometric rate at which the contact point P between p_m and p_f moves along p_m, and is not the velocity of any one physical point.) This velocity has with respect to O_{XY} the components

$$x_p'\cos\phi - y_p'\sin\phi = a' - b'', \quad x_p'\sin\phi + y_p'\cos\phi = a'' + b', \tag{7.7}$$

but these are, as can be seen by taking the first derivative of (7.5), the components of the change of contact along p_f. Hence the rate of change of contact along p_m is the same (in direction and magnitude) as its change along p_f. This implies two statements: at any moment p_m and p_f are tangent at P, and furthermore the two curves have the same arc element and hence the same arc length between corresponding positions. The arc element s' on both p_m and p_f is given by

$$(s')^2 = (a' - b'')^2 + (a'' + b')^2. \tag{7.8}$$

The relation between p_m and p_f can obviously be described as follows: *the moving centrode rolls without sliding along the fixed centrode.*

We have in particular: p_f is the envelope of the set of congruent curves p_m generated by the motion of E with respect to Σ.

Let γ be an arbitrary curve in E. During the motion it has a set of positions with respect to Σ; this set has in general an envelope in Σ. Let Γ be this envelope and $A(\phi)$ the tangent point of γ and Γ at the position ϕ. This point of contact changes so that as the motion proceeds $A(\phi)$ corresponds to different points along γ and along Γ. Suppose that $X(\phi)$, $Y(\phi)$ is its position in Σ, $x(\phi)$, $y(\phi)$ its position in E. This implies that $X = X(\phi)$, $Y = Y(\phi)$ is a parametric representation of Γ and $x = x(\phi)$, $y = y(\phi)$ is one of γ. For any ϕ the two curves are tangent at A. Obviously we have

$$\begin{aligned} X(\phi) &= x(\phi)\cos\phi - y(\phi)\sin\phi + a, \\ Y(\phi) &= x(\phi)\sin\phi + y(\phi)\cos\phi + b. \end{aligned} \tag{7.9}$$

The geometric velocity $\boldsymbol{v}$ of the change in position of A along γ has in the o_{xy} frame the components

$$x'(\phi), \quad y'(\phi). \tag{7.10}$$

The velocity $\boldsymbol{V}$ of the position of A along Γ has in the frame O_{XY} the components following from (7.9):

$$X' = x'\cos\phi - x\sin\phi - y'\sin\phi - y\cos\phi + a',$$

$$Y' = x'\sin\phi + x\cos\phi + y'\cos\phi - y\sin\phi + b'.$$

The components of this velocity with respect to the o_{xy} frame are

$$\begin{aligned} X'\cos\phi + Y'\sin\phi &= x' - y + a'\cos\phi + b'\sin\phi, \\ -X'\sin\phi + Y'\cos\phi &= x + y' - a'\sin\phi + b'\cos\phi. \end{aligned} \tag{7.11}$$

But, as γ and Γ are tangent at A, the direction of the two velocities $\boldsymbol{v}$ and $\boldsymbol{V}$ must be the same. Let $\boldsymbol{V} = \mathfrak{k}(\phi)\boldsymbol{v}$, $\mathfrak{k}(\phi)$ being the ratio of the velocities of A on Γ and on γ respectively (German: Rollgleitzahl) then from (7.10) and (7.11) it follows

$$x'(1-\mathfrak{k})-y+a'\cos\phi+b'\sin\phi=0$$

(7.12)

$$y'(1-\mathfrak{k})+x-a'\sin\phi+b'\cos\phi=0,$$

which, in general, with a, b and $\mathfrak{k}$ as given functions of ϕ, is a system of differential equations, of order two, for the curve γ. If γ is determined, Γ (being its envelope in Σ) is also known. Hence there is in general, the motion and the function $\mathfrak{k}(\phi)$ being given, a set of ∞^2 curves in E satisfying these conditions. There is, however, one important exception: if $\mathfrak{k}$ is equal to one, in other words if $\boldsymbol{V}=\boldsymbol{v}$, (7.12) is not a system of differential equations; it has only one solution: the moving centrode (7.4). We have then from (7.12) a proof that p_m is the only curve in E which rolls without sliding on its envelope, and have moreover a method to determine pairs of curves γ, Γ with a prescribed $\mathfrak{k}(\phi)$.

Example 56. Show that the system of differential equations for Γ reads $X'(1-\mathfrak{k}^{-1})+Y-a'-b=0$, $Y'(1-\mathfrak{k}^{-1})-X+a-b'=0$.

Example 57. Eliminate y from (7.12) for $\mathfrak{k}\neq 1$ and show that the result is a linear differential equation of the second order.

Example 58. Show that (7.12) can be solved by quadratures if $\mathfrak{k}$ is a constant $\neq 1$. (Theorem due to H. R. MÜLLER [1953].)

Example 59. Let p_f be given by $X(s)$, $Y(s)$, p_m by $x(s)$, $y(s)$, s being the arc length for both. For $s=0$ let $X=Y=x=y=0$. Show that the motion is determined by these data, and that by $\cos\phi = X'x'+Y'y'$, $\sin\phi = Y'x'-X'y'$, $a = X - X'(xx'+yy') - Y'(xy'-yx')$, $b = Y + X'(xy'-yx') - Y'(xx'+yy')$ the motion parameters ϕ, a, b are given as functions of s.

Example 60. When a point on p_m acts as a pole it passes through a cusp of its path; prove that the cuspidal tangent (its direction is determined by $X'':Y''$) is perpendicular to the common tangent of p_m and p_f.

Any motion of E with respect to Σ is represented by (7.2). We now consider the case where it is periodic; this implies that X and Y are periodic functions of ϕ (for any x, y) and therefore it is necessary and sufficient that $a(\phi)$ and $b(\phi)$ are periodic with a period $2n\pi$, n being an integer. Let us suppose $n=1$. The path of any point (x, y) of E is now a closed curved in Σ. We shall determine the enclosed area $\mathfrak{F}(x, y)$. In view of a well-known formula one has

$$2\mathfrak{F}(x,y)=\int_0^{2\pi}(XY'-X'Y)\mathrm{d}\phi$$

(7.13)

$$\begin{aligned}&=\int_0^{2\pi}[(x\cos\phi-y\sin\phi+a)(x\cos\phi-y\sin\phi+b')\\&\quad-(x\sin\phi+y\cos\phi+b)(-x\sin\phi-y\cos\phi+a')]\mathrm{d}\phi\\&=2\pi(x^2+y^2)-2c_1x-2c_2y+c_3,\end{aligned}$$

where the constants c_i are

$$(7.14)\qquad 2c_1=\int_0^{2\pi}[(-a\cos\phi-b\sin\phi)+(a'\sin\phi-b'\cos\phi)]\mathrm{d}\phi,$$

$$(7.15)\qquad 2c_2=\int_0^{2\pi}[(a\sin\phi-b\cos\phi)+(a'\cos\phi+b'\sin\phi)]\mathrm{d}\phi,$$

$$(7.16)\qquad c_3=\int_0^{2\pi}(ab'-a'b)\mathrm{d}\phi.$$

From (7.13) it follows that the locus of the points in E whose paths enclose a given area is a circle. (Theorem due to STEINER [1840].) For a given motion all the circles corresponding to different areas are concentric; their common center being $M(c_1/(2\pi), c_2/(2\pi))$.

(7.14) may be reduced in the following way. Integrating by parts we obtain

$$\int_0^{2\pi}(-a\cos\phi-b\sin\phi)\mathrm{d}\phi=[-a\sin\phi+b\cos\phi]_0^{2\pi}$$
$$+\int_0^{2\pi}(a'\sin\phi-b'\cos\phi)\mathrm{d}\phi,$$

and, as the first term on the right-hand side is zero, we have

$$(7.17)\qquad c_1=\int_0^{2\pi}(a'\sin\phi-b'\cos\phi)\mathrm{d}\phi,$$

and in a similar way

$$(7.18)\qquad c_2=\int_0^{2\pi}(a'\cos\phi+b'\sin\phi)\mathrm{d}\phi.$$

Comparing this with (7.4) we conclude that $(c_1/(2\pi))$, $(c_2/(2\pi))$ are the coordinates of the center of mass of the centrode in E if this curve is uniformly covered with the mass elements $\mathrm{d}\phi$. Hence the center M of the circles coincides with this "mass" center (STEINER [1840]).

Example 61. Consider the analogous problem for a motion with period $2\pi n$.

Now we consider an application of formula (7.13). Let $\mathfrak{C}_{12}$ be a convex curve in Σ (Fig. 63). Two points A_1, A_2 of E, separated by distance $2l$, move along $\mathfrak{C}_{12}$, l being short enough to make this possible. Obviously the motion of E is periodic with period 2π. We introduce a frame in E such that $A_1=(l,0)$, $A_2=(-l,0)$. If $\mathfrak{F}_{12}$ is the area enclosed by $\mathfrak{C}_{12}$, formula (7.13) applied to A_1 and to A_2 gives us

$$2\mathfrak{F}_{12}=2\pi l^2-2c_1l+c_3=2\pi l^2+2c_1l+c_3,$$

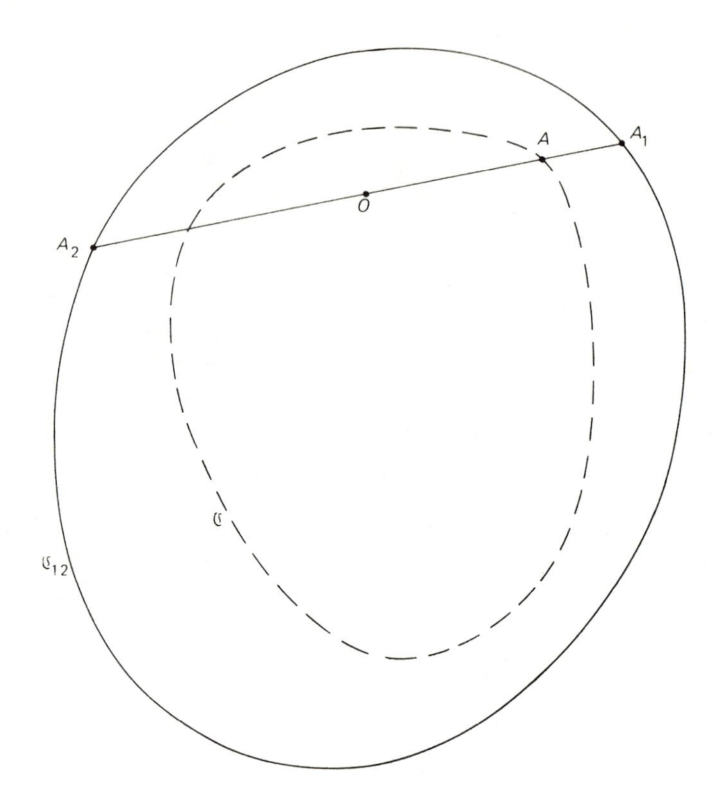

Fig. 63.

which implies $c_1 = 0$. Hence the area $\mathfrak{F}$ enclosed by the path of an arbitrary point $A = (x, y)$ is given by

$$2\mathfrak{F} = 2\pi(x^2 + y^2) - 2c_2 y + c_3.$$

If, in particular, A is the point $(x_1, 0)$ on A_1A_2, with $|x_1| < l$, its path $\mathfrak{C}$ is inside $\mathfrak{C}_{12}$, enclosing the area $\mathfrak{F}$ such that

$$2\mathfrak{F} = 2\pi x_1^2 + c_3.$$

Hence for the ring-shaped region between $\mathfrak{C}_{12}$ and $\mathfrak{C}$ we obtain the area

$$\mathfrak{F}' = \mathfrak{F}_{12} - \mathfrak{F} = \pi(l^2 - x_1^2) = \pi(\overline{A_1A})(\overline{AA_2}), \tag{7.19}$$

a remarkable result (known as Holditch's theorem, HOLDITCH [1858]) because

$\mathfrak{F}'$ depends *only* on the distances from A to A_1 and A_2, being independent of the shape or the size of the convex curve $\mathfrak{C}_{12}$.

The subject of this section has been continuous kinematics; especially the last problem, dealing with a complete motion, belongs to what may be called "kinematics in the large". As we mentioned before this is a rather restricted field of research. Another problem which could be inserted here is that of determining the characteristics (such as order and circularity in the case that these are algebraic curves) of the path of an arbitrary point if the paths of two points of E are given; this would lead to some formulas originally given by S. Roberts. We will put this off to Chapter IX. In the course of its history continuous (plane) kinematics has been developed mainly as the study of instantaneous properties (to be dealt with in the next section) or as the study of special motions, to be treated in Chapter IX.

8. Instantaneous geometric kinematics

This section deals with instantaneous geometric plane kinematics. Which is the study, for a certain instant during a continuous motion, of the differential geometric properties. So if we are interested in, for instance, the path of a point we study its tangent, its curvature, and so on. The best way to deal with this subject analytically is, as we did in the general spatial case, to introduce canonical coordinate systems and to make use of the concept of instantaneous invariants. The circumstances of the motion being restricted to a plane simplifies considerably the general theory. We could specialize the results derived in Chapter II, but we prefer to develop the planar case here autonomously.

One of the most important features of plane motion is the property that the angular velocity has a fixed direction (*viz.* perpendicular to the plane) and thereby essentially loses its vectorial character. In this respect plane kinematics is at the end of a chain: in n-dimensional space the angular velocity could be represented by a skew $n \times n$ matrix; for $n > 3$ this is essentially a tensor, for $n = 3$ it is equivalent to a vector, for $n = 2$ it may be considered as a scalar. This implies that we are able to normalize with respect to angular velocity and introduce as a natural parameter for the motion (provided it is not a translation) the rotation angle ϕ. This is also a reason why planar kinematics is more simple than spherical kinematics where no single parameter describes the angular velocity (although in planar motion we do lose the attractive complete duality that exists for spherical motions).

As we know already, any plane, time-independent, motion may be described by the equations

$$(8.1)\quad X = x\cos\phi - y\sin\phi + a(\phi), \qquad Y = x\sin\phi + y\cos\phi + b(\phi),$$

that is by the two functions $a(\phi)$ and $b(\phi)$. If we consider only one position of the moving plane $E(x, y)$ with respect to $\Sigma(X, Y)$ we may without loss of generality suppose that it is given by $\phi = 0$ and that $a(\phi)$ and $b(\phi)$ are given by the power series

$$(8.2)\qquad a(\phi) = \sum_{n=0}^{\infty} a_n(\phi^n/n!), \qquad b(\phi) = \sum_{n=0}^{\infty} b_n(\phi^n/n!).$$

If the frames are chosen such that they coincide in the "zero-position" we have from (8.1) and (8.2)

$$(8.3)\qquad a_0 = b_0 = 0.$$

In addition we place the common origin of the frames at the pole at the moment $\phi = 0$. In view of (7.5) (or (7.4)) this implies

$$(8.4)\qquad a_1 = b_1 = 0.$$

Furthermore the axes O_X and o_x are chosen along the common tangent (at the pole) of the centrodes. From (7.7) (or (7.6)) it follows

$$(8.5)\qquad a_2 = 0.$$

The two frames are now specified except for the sense of the positive X-axis. If we change this sense then b_2 changes its sign. Assuming that $b_2 \neq 0$ we make a choice by taking the direction in such a way that

$$(8.6)\qquad b_2 > 0.$$

The coinciding frames O_{XY} and o_{xy} defined by (8.3)–(8.6) will be called canonical; they are the systems with respect to which we shall study instantaneous kinematics. Our choice is a natural one: the pole and the pole-tangent have a simple geometrical meaning. That we take the X-axis along the latter (and not the Y-axis) and the inequality (8.6) are mere conventions. The introduction of this canonical frame fails in two exceptional cases: if the pole is at infinity it cannot be taken as the origin, if $b_2 = 0$ the sense of the axes is not determined. We return to these special cases later on and exclude them for the time being.

If we let X_n, Y_n denote the nth derivative of X, Y (with respect to ϕ) evaluated at $\phi = 0$ and measured in the canonical system, from (8.1) it follows that $X_0 = x$, $Y_0 = y$, $X_1 = -y$, $Y_1 = x$, $X_2 = -x$, $Y_2 = -y + b_2$. To determine the higher derivatives we remark that

$$(\cos\phi)_{2n} = (-1)^n, \qquad (\cos\phi)_{2n+1} = 0,$$
$$(\sin\phi)_{2n} = 0, \qquad (\sin\phi)_{2n+1} = (-1)^n$$

and we obtain the following table

$$\begin{aligned} &X_0 = x, \quad X_1 = -y, \quad X_2 = -x, \quad X_3 = y + a_3, \quad X_4 = x + a_4, \ldots \\ &Y_0 = y, \quad Y_1 = x, \quad Y_2 = -y + b_2, \quad Y_3 = -x + b_3, \quad Y_4 = y + b_4, \ldots \end{aligned} \tag{8.7}$$

Example 62. Show that $X_{4n} = x + a_{4n}$, $X_{4n+1} = -y + a_{4n+1}$, $X_{4n+2} = -x + a_{4n+2}$, $X_{4n+3} = y + a_{4n+3}$, and analogously for Y.

Example 63. In Chapter II, (6.3) formulas for the derivatives up to the third have been given in the spatial case. Show that they can be reduced to (8.7) for planar kinematics if we take $\varepsilon = 0$ and $\mu\varepsilon = b_2$. Technically this case is excluded (see statement following Chapter II, (5.2)), but since Chapter II, (5.5) simply states that $\mathbf{d}_2$ is arbitrary in this case, $\mathbf{d}_2$ could be appropriately redefined in Chapter II, (5.6) for this case.

From (8.7) it follows that up to the nth order, the instantaneous properties of the motion depend on the constants $a_3, a_4, \ldots, a_n$ and $b_2, b_3, \ldots, b_n$. These are the instantaneous invariants. Any relation between them reflects a special geometric aspect of a particular motion.

From (8.1) it follows that the inverse motion is given by

$$\begin{aligned} x &= X\cos\phi + Y\sin\phi - a\cos\phi - b\sin\phi, \\ y &= -X\sin\phi + Y\cos\phi + a\sin\phi - b\cos\phi. \end{aligned} \tag{8.8}$$

The rotation angle of this motion is $-\phi$ (with the positive sense of ϕ from O_X towards O_Y) which is measured in the sense of O_X towards $O_{\bar{Y}}$, the opposite direction of O_Y. We introduce new frames such that $\bar{X} = X$, $\bar{Y} = -Y$, $\bar{x} = x$, $\bar{y} = -y$, that means: we reflect the canonical coordinate systems into the pole-tangent. Then we obtain for (8.8)

$$\bar{x} = \bar{X}\cos\phi - \bar{Y}\sin\phi + \bar{a}, \qquad \bar{y} = \bar{X}\sin\phi + \bar{Y}\cos\phi + \bar{b}, \tag{8.9}$$

with

$$\bar{a} = -a\cos\phi - b\sin\phi, \qquad \bar{b} = -a\sin\phi + b\cos\phi. \tag{8.10}$$

Obviously (8.9) is of the same standard form as (8.1). If we write:

$$\bar{a} = \sum_{n=0}^{\infty} \bar{a}_n (\phi^n/n!) \quad \text{and} \quad \bar{b} = \sum_{n=0}^{\infty} \bar{b}_n (\phi_n/n!), \tag{8.11}$$

then from (8.10) we obtain

$$\bar{a}_0 = \bar{b}_0 = 0, \qquad \bar{a}_1 = \bar{b}_1 = 0, \qquad \bar{a}_2 = 0, \qquad \bar{b}_2 = b_2 > 0, \tag{8.12}$$

which are similar to the four conditions (8.2)–(8.6). The conclusion is: the canonical frames for the inverse motion are the reflections into the pole tangent of the canonical frames of the direct motion. The formulas (8.10) enable us to express the instantaneous invariants of the inverse motion in terms of the direct ones. We obtain

$$\tilde{a}_3 = -a_3 - 3b_2, \qquad \tilde{b}_3 = b_3,$$
$$(8.13) \qquad \tilde{a}_4 = -a_4 - 4b_3, \qquad \tilde{b}_4 = -4a_3 + b_4 - 6b_2,$$
$$\tilde{a}_5 = -a_5 + 10a_3 - 5b_4 + 10b_2, \qquad \tilde{b}_5 = -5a_4 + b_5 - 10b_3.$$

Example 64. Derive formulas for $\tilde{a}_n$ and $\tilde{b}_n$.
Example 65. Prove $\tilde{\tilde{a}} = a$, $\tilde{\tilde{b}} = b$ (i.e., the inverse of the inverse yields the original motion).

The formulas in (8.7) give us the derivatives of the coordinates (X, Y) of the moving point (x, y) with respect to the motion parameter. The analogous formulas for the (homogeneous) coordinates U_1, U_2, U_3 of the line (u_1, u_2, u_3) in E follow from (2.5). As the latter may be written

$$(8.14) \qquad U_1 = u_1 \cos\phi - u_2 \sin\phi, \qquad U_2 = u_1 \sin\phi + u_2 \cos\phi,$$
$$U_3 = \tilde{a}u_1 + \tilde{b}u_2 + u_3,$$

we obtain

$$(U_1)_0 = u_1, \quad (U_1)_1 = -u_2, \quad (U_1)_2 = -u_1, \quad (U_1)_3 = u_2, \ldots$$
$$(8.15) \qquad (U_2)_0 = u_2, \quad (U_2)_1 = u_1, \quad (U_2)_2 = -u_2, \quad (U_2)_3 = -u_1, \ldots$$
$$(U_3)_0 = u_3, \quad (U_3)_1 = 0, \quad (U_3)_2 = \tilde{b}_2 u_2, \quad (U_3)_3 = \tilde{a}_3 u_1 + \tilde{b}_3 u_2, \ldots$$

where $(U_i)_n$ denotes the nth derivative of U_i, with respect to ϕ, evaluated at $\phi = 0$ in the canonical coordinate system.

Example 66. Derive the formulas for $(U_1)_n$, $(U_2)_n$, $(U_3)_n$.
Example 67. Determine the counterpart of (8.14) and (8.15) for the inverse motion.

If we want to study instantaneous properties up to the nth order we need the tables (8.7) and (8.15) up to the $(n+1)$th column. This implies that for $n = 1$ there appear no invariants at all, for $n = 2$ there is one, for $n > 2$ the number is $1 + 2(n-2) = 2n - 3$.

Instantaneous kinematics of the nth order is a special case (the "infinitesimal case") of finite $(n+1)$-positions theory. It is more simple because the latter deals with (essentially) $3n$ parameters. Some work has been done on mixed problems, considering n positions of which certain positions are coinciding, i.e., infinitesimally separated (DIZIOĞLU [1967], TESAR [1967,

1968], TESAR AND SPARKS [1968]). Instantaneous kinematics for $n = 1$ is trivial; because no invariant appears, we conclude that up to the first order all planar motions are essentially the same. Since all first order properties relative to the pole are the same, all systems which are placed so that their poles coincide are congruent up to the first order.

Example 68. Prove from (8.7) that the tangent to the path of a point P is perpendicular to the radius vector OP.
Example 69. Prove from (8.15) that the intersection of a moving line l and its consecutive line (that is the point at which l is tangent to its envelope) coincides with the projection of O on l.

In the following sections we consider instantaneous kinematics of the nth order for small values of $n > 1$. There is of course a clear parallelism between these results and those we found for finite positions theory (Sections 3, 4, 5, 6 of this chapter).

9. Second order properties

For $n = 2$ we have from (8.7)

$$X_0 = x, \qquad X_1 = -y, \qquad X_2 = -x,$$
$$Y_0 = y, \qquad Y_1 = x, \qquad Y_2 = -y + b_2, \tag{9.1}$$

which implies that the situation depends on one number, b_2, defined to be positive.

In the general case the moving centrode is given by (7.4) and the fixed centrode by (7.5). From either of these it follows that the (geometric) velocity at which the contact changes along the centrodes has, in our canonical system, the components $(-b_2, 0)$; which gives us a geometric meaning for the invariant b_2. (The centrode contact moves therefore in the direction of the negative X-axis, which is the reason why Veldkamp in his thesis (VELDKAMP [1963]) introduced a canonical frame with the X-axis in the opposite sense to ours.)

The curvature k of the path of a point $A(x, y)$, at the zero-position, is given by the well-known formula from elementary calculus: $k = (X_1Y_2 - X_2Y_1)/(X_1^2 + Y_1^2)^{3/2}$. Substituting from (9.1) it follows that

$$k = (x^2 + y^2 - b_2 y)/(x^2 + y^2)^{3/2}, \tag{9.2}$$

which is valid for any point different from the pole. If $x \to 0$, $y \to 0$ we obtain $k \to \infty$ in accordance with the fact that the moving point coinciding with O

passes through a cusp of its path. From (9.2) it follows that the locus of the points which are at an inflection point of their path is given by

$$x^2 + y^2 - b_2 y = 0. \tag{9.3}$$

The *inflection curve is* therefore *the circle with center* $(0, b_2/2)$ *and diameter* b_2. It is tangent to the X-axis at O (Fig. 64), and is called *the inflection circle.*

If we introduce polar coordinates (r, θ) $(0 \leqslant \theta < \pi, -\infty < r < \infty)$* with O_x as axis, we have $x = r\cos\theta$, $y = r\sin\theta$. For (9.2) we obtain (if $r \neq 0$)

$$k = (r - b_2 \sin\theta)/r^2. \tag{9.4}$$

The radius of curvature ρ is the reciprocal of the curvature, therefore $\rho = k^{-1}$. If M is the center of curvature related to $A(r, \theta)$, then M is on the path normal OA and therefore $M(\bar{r}, \theta)$ with $\bar{r} = r - \rho$. Then we obtain from (9.4):

$$((1/r) - (1/\bar{r}))\sin\theta = 1/b_2, \tag{9.5}$$

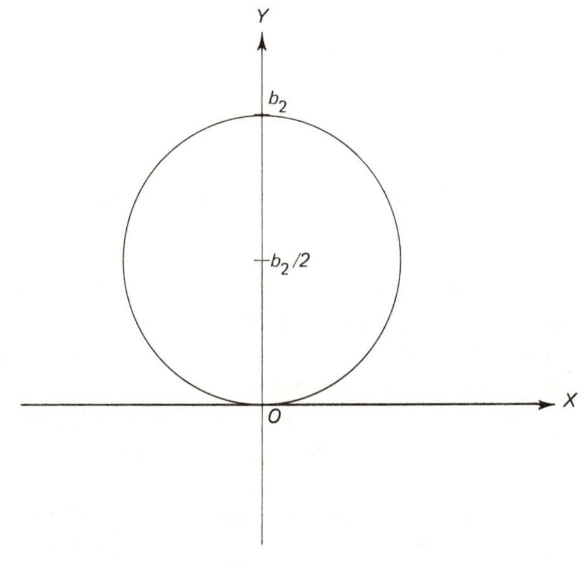

Fig. 64.

*The sign convention here is that θ is measured counterclockwise from the positive O_x axis to the line connecting (x, y) to the origin, extended if necessary into the upper-half plane. r is positive if the point (x, y) is in the upper-half plane and negative if it is in the lower-half.

the (first) Euler–Savary equation. (The same sign convention is used for $\bar{r}$ as for r.)

Example 70. Verify from (9.5) that A is on the inflection circle if $\bar{r} = \infty$.

We can write (9.5) as

$$b_2(\bar{r} - r)\sin\theta = r\bar{r}, \tag{9.6}$$

hence for a fixed value of θ ($\sin\theta \neq 0$) there exists a bi-linear relation between r and $\bar{r}$. Every point M belongs to one point A. If A is on the X-axis, different from O, we have $\sin\theta = 0$, $r \neq 0$ and therefore $\bar{r} = 0$: any point on the pole tangent has its center of curvature at the pole. We can even define the point M if A tends to infinity; for $r \to \infty$ we obtain $\bar{r} = -b_2\sin\theta$ which means that M is on the circle which is the reflection of the inflection circle into O_X. This circle is called the *cuspidal* (or *return*) circle and we shall meet it soon in another context.

Example 71. Show that all points in E with the same magnitude radius of curvature, $|\rho|$, lie on the sixth order tri-circular curve $(x^2+y^2)^3 - \rho^2(x^2+y^2-b_2y)^2 = 0$ and that their corresponding centers lie on $(X^2+Y^2)^3 - \rho^2(X^2+Y^2+b_2Y)^2 = 0$. (These are known as the ρ and ρ_m-curves respectively. They have been studied by ALT [1932a, b].)

Example 72. Use the preceding Example to deduce that the locus of all points which have the same path curvature in two finitely separated positions of their motion is a plane curve, imbedded in E, and that for three separated positions there are only a finite number of points with the same path curvature in all three positions. (Idem.)

If A is on the inflection circle, the inflection tangent of its path is perpendicular to OA and it passes therefore through the point $H = (0, b_2)$, called the inflection pole. For H itself, the path tangent is the tangent to the inflection circle. The cuspidal tangent at O passes also through H (see Example 60).

Our configuration is the limit case of three finitely separated positions: All three vertices P_{ij} of the pole triangle coincide with O and the three sides with O_X. The cuspidal circle is the limit of the circumcircle of the pole triangle, and the inflection pole is the limit of its orthocenter. The inflection circle is the limit of the circumcircles of the image pole triangles. The fundamental point of the three homologous positions A_i, now coinciding at one point A, is the reflection A^* of A into O_X. The isogonal relation of the general three positions theory tends to that between A^* and M.

If the pole O and the pole tangent O_X are known, and if we know moreover the center of curvature $M(\bar{r}, \theta)$ of any point $A(r, \theta)$, $r \neq \bar{r}$ the invariant b_2 is determined by equation (9.6), and therefore the center of curvature of every point of the plane follows if we know the pole, the pole tangent and the center of any one point.

Example 73. Find H if O, O_X, A and M are given.

There is an interesting sequel to the last statement. Suppose that *two* points A_1 and A_2 and their centers of curvature M_1 and M_2 are given, and nothing more. Supposing that the lines M_1A_1 and M_2A_2 are neither coincident nor parallel, their intersection is obviously the pole O. We try to determine the pole tangent O_X. For the time being we introduce an $O_{X'}$-axis along the inner bisector of the angle A_1OA_2 (Fig. 65). Let $\angle A_1O_{X'} = \angle O_{X'}A_2 = \alpha$, and $\angle O_{X'}O_X = \beta$, as yet unknown. Then $A_1 = (-r_1, \theta_1)$, $M_1 = (-\bar{r}_1, \theta_1)$, $A_2 = (r_2, \theta_2)$, $M_2 = (\bar{r}_2, \theta_2)$ with $\theta_1 = \pi - (\beta + \alpha)$, $\theta_2 = \alpha - \beta$. Applying (9.6) to the two pairs A_1, M_1 and A_2, M_2 we obtain after eliminating b_2:

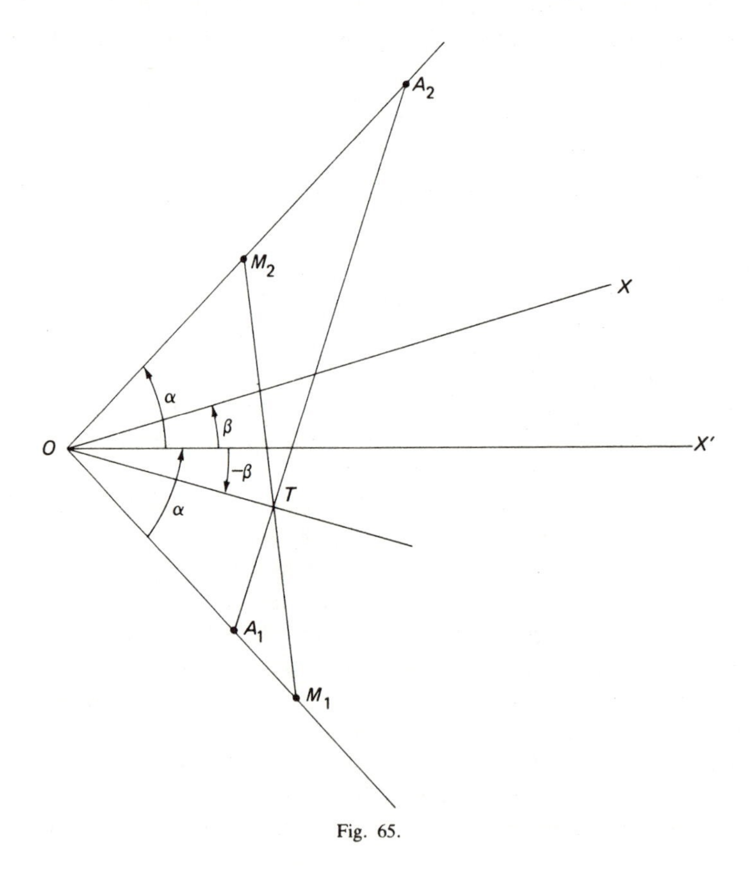

Fig. 65.

(9.7) $$S_2 \sin \theta_1 = S_1 \sin \theta_2,$$

with

(9.8) $$S_2 = r_2\bar{r}_2(\bar{r}_1 - r_1), \qquad S_1 = r_1\bar{r}_1(\bar{r}_2 - r_2).$$

From (9.7) it follows

(9.9) $$\tan \beta = ((S_1 + S_2)/(S_1 - S_2))\tan \alpha,$$

which implies that the pole tangent has been determined. Moreover b_2 may now be calculated from (9.6) and the orientation of O_X chosen in such a way that $b_2 > 0$. The conclusion is: if two points A_i and their centers of curvature M_i are known, the second order configuration is completely determined. This means, in particular, that we now know (at $\phi = 0$) the center of curvature of the path of any point.

Our Fig. 65 has an elegant geometrical property: With respect to the Cartesian frame $O_{X'}$ we have

$$A_1 = (r_1 \cos \alpha, -r_1 \sin \alpha), \qquad M_1 = (\bar{r}_1 \cos \alpha, -\bar{r}_1 \sin \alpha),$$

$$A_2 = (r_2 \cos \alpha, r_2 \sin \alpha), \qquad M_2 = (\bar{r}_2 \cos \alpha, \bar{r}_2 \sin \alpha).$$

Hence the equations of lines A_1A_2 and M_1M_2 are respectively

$$(r_1 + r_2)(\sin \alpha)x + (r_1 - r_2)(\cos \alpha)y - 2r_1r_2 \sin \alpha \cos \alpha = 0,$$

$$(\bar{r}_1 + \bar{r}_2)(\sin \alpha)x + (\bar{r}_1 - \bar{r}_2)(\cos \alpha)y - 2\bar{r}_1\bar{r}_2 \sin \alpha \cos \alpha = 0.$$

For the coordinates of their intersection T we obtain

$$\begin{aligned} x : y &= [r_1r_2(\bar{r}_1 - \bar{r}_2) - \bar{r}_1\bar{r}_2(r_1 - r_2)]\cos \alpha : [\bar{r}_1\bar{r}_2(r_1 + r_2) - r_1r_2(\bar{r}_1 + \bar{r}_2)]\sin \alpha \\ &= [-r_1\bar{r}_1(\bar{r}_2 - r_2) + r_2\bar{r}_2(\bar{r}_1 - r_1)]\cos \alpha : [r_1\bar{r}_1(\bar{r}_2 - r_2) + r_2\bar{r}_2(\bar{r}_1 - r_1)]\sin \alpha \\ &= (-S_1 + S_2)\cos \alpha : (S_1 + S_2)\sin \alpha. \end{aligned}$$

Hence, if $\angle O_{X'}T = \bar{\beta}$, we have, in view of (9.9)

(9.10) $$\tan \bar{\beta} = ((S_1 + S_2)/(-S_1 + S_2))\tan \alpha = -\tan \beta.$$

The pole tangent O_X and the line OT are therefore one another's reflection into the bisector $O_{X'}$; in other words, O_X and OT are isogonally conjugate with respect to the angle A_1OA_2. This is Bobillier's theorem (BOBILLIER [1870]).

Example 74. Consider the case when A_1A_2 and M_1M_2 are parallel.

Example 75. Show that Bobillier's theorem may be used to find M_2 if the pole, the pole tangent, A_1, M_1 and A_2 (not on OA_1) are given.

Example 76. Construct the inflection pole H if O, O_x, A_1, M_1 are given.

If A_1M_1 and A_2M_2 have a point of intersection it coincides with the pole. If they are parallel the pole is at infinity, a situation excluded so far. It remains to determine the pole O in the case when the two lines coincide: Let $\bar{O}$ be an arbitrary point on the line, $\bar{O}A_i = p_i$, $\bar{O}M_i = m_i$. Then it follows, by applying (9.6) twice, eliminating b_2 and putting $O\bar{O} = z$, provided $\sin\theta \neq 0$:

$$(m_1 - p_1)(z + p_2)(z + m_2) - (m_2 - p_2)(z + p_1)(z + m_1) = 0,$$

or

$$\begin{aligned}[(m_1 - m_2) - (p_1 - p_2)]z^2 + 2(m_1p_2 - m_2p_1)z \\ - m_1m_2(p_1 - p_2) + p_1p_2(m_1 - m_2) = 0,\end{aligned} \tag{9.11}$$

a quadratic equation for the unknown distance z. Its discriminant D reads, after some algebra,

$$D = (p_1 - p_2)(m_1 - m_2)(p_1 - m_1)(p_2 - m_2). \tag{9.12}$$

All this leads to two noteworthy conclusions: *If the four points A_i, M_i $(i = 1, 2)$ are not collinear they may be taken arbitrarily and they determine a motion (to the second order) completely and uniquely.* If they are on one line, however, they must be such that $D > 0$, and if so there are two possible poles.

It is clear that when they are collinear the sign of D depends on the order of the four points on their line. There are $4! = 24$ ways to arrange the points, but D does not change its sign for a cyclical transformation of the quadruple, nor if we take the inverse orientation on the line. (The order $(1, 2, 3, 4)$ gives the same sign as $(2, 3, 4, 1)$ and as $(4, 3, 2, 1)$.) Hence there are only three different "projective" orders of the four points, represented for instance by (A_1, M_1, A_2, M_2), (A_1, A_2, M_1, M_2) and (A_1, A_2, M_2, M_1). It is easy to verify that the first and second give $D > 0$ and the third $D < 0$. This implies that the condition *for the collinear points is: the pairs A_1M_2 and A_2M_1 do not separate one another.*

O' was an arbitrary chosen point; if we take it at the midpoint of A_1M_2 we have $p_1 + m_2 = 0$ and (9.11) reads

$$(m_1 + p_2)z^2 + 2(m_1p_2 - m_2p_1)z + (m_1 + p_2)p_1^2 = 0. \tag{9.13}$$

Hence for $m_1 + p_2 \neq 0$ the two roots satisfy $z_1z_2 = p_1^2$, which means that the two possible poles O_1 and O_2 are located so that $(OO_1)(OO_2) = (OA_1)^2 = (OM_2)^2$. The conclusion is: O_1 and O_2 are harmonic with A_1 and M_2 and therefore with A_2 and M_1 as well.

Example 77. Show that O_1 and O_2 are determined by these two harmonic properties.
Example 78. Determine D for the special choice of O at the midpoint of A_1M_2.
Example 79. Consider, in (9.13), the case $m_1 + p_2 = 0$.

We proceed with one more application of the fundamental formula (9.6). It may be written as

$$r : b_2 \sin\theta = (\bar{r} - r) : \bar{r}. \tag{9.14}$$

We know that the geometric velocity of contact along the centrode, $\boldsymbol{u}$, is equal in magnitude to $-b_2$. In Fig. 66 the component $O\bar{O}$ of $\boldsymbol{u}$ along the line perpendicular to OA is drawn, and also $A\bar{A} = OA = r$ which is the geometric velocity of A. As $O\bar{O} = b_2 \sin\theta$, it follows from (9.14) that the center of curvature M related to A is the intersection of OA and $\bar{O}\bar{A}$. This is Hartmann's theorem (HARTMANN [1893]). As $\bar{O}$ is the same for any point on OA it facilitates the construction of centers of curvature.

Example 80. Draw the analog of Fig. 66 for another position of A.
Example 81. Construct A if M is given. (Note that $\angle AO\bar{A} = \pi/4$.)
Example 82. Construct the intersection of OA and the inflection circle.
Example 83. Take two points A_1, A_2 on OA, determine M_1, M_2 and verify that A_1M_2 and A_2M_1 do not separate one another.

Until now our study of second order instantaneous kinematics has focused on the motion of a point and determining the center of curvature of its path. We shall now investigate the motion of a line (u_1, u_2, u_3). Three consecutive lines pass through one point if

$$\begin{vmatrix} (U_1)_0 & (U_2)_0 & (U_3)_0 \\ (U_1)_1 & (U_2)_1 & (U_3)_1 \\ (U_1)_2 & (U_2)_2 & (U_3)_2 \end{vmatrix} = 0, \tag{9.15}$$

or, in view of (8.15) and of $\tilde{b}_2 = b_2$,

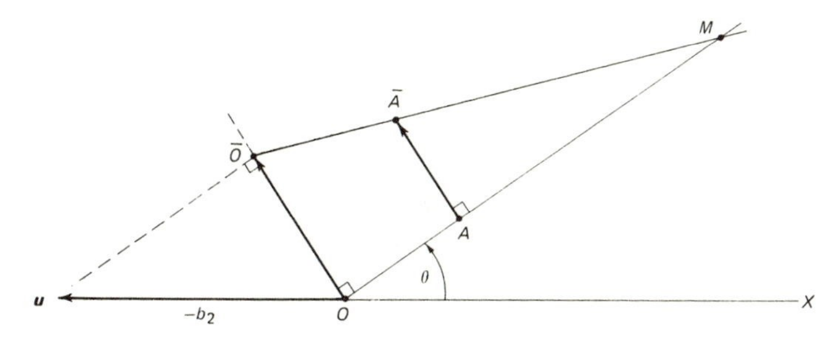

Fig. 66.

$$(9.16)\qquad \begin{vmatrix} u_1 & u_2 & u_3 \\ -u_2 & u_1 & 0 \\ -u_1 & -u_2 & -b_2u_2 \end{vmatrix} = 0,$$

that is

$$(-b_2u_2 + u_3)(u_1^2 + u_2^2) = 0.$$

Hence the locus, of all lines having three concurrent consecutive positions, degenerates into three pencils: the lines through each of the isotropic points I_1 and I_2 (which is obvious because these points are invariant for any motion) and the lines through the point $\bar{H}(0, -b_2)$, the reflection into O_X of the inflection pole H (Fig. 67). The intersection of any line l and its consecutive

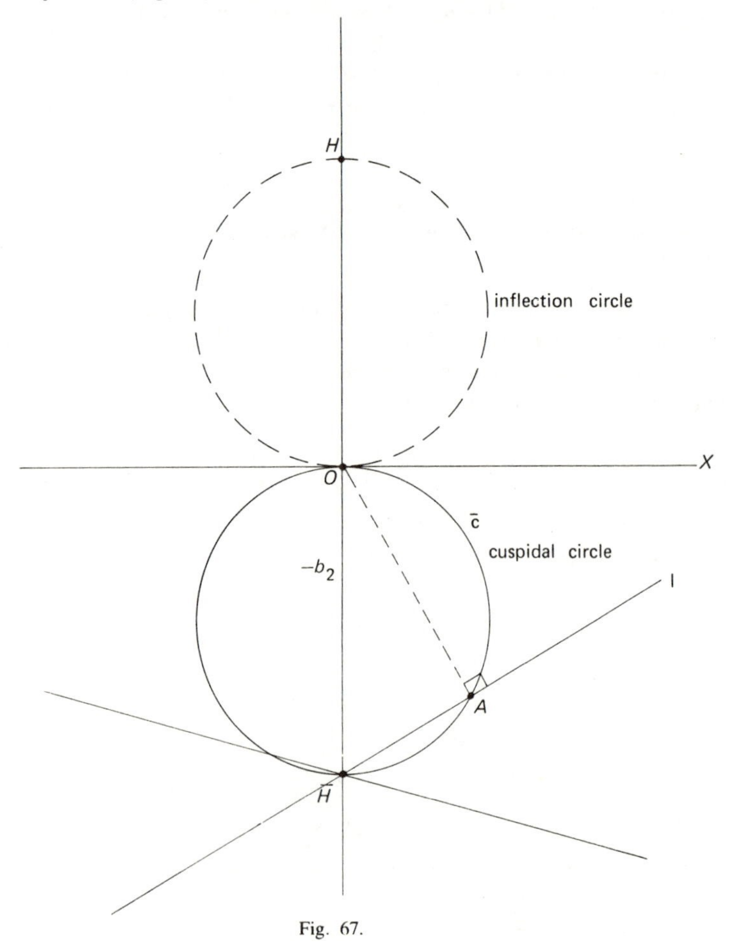

Fig. 67.

line coincides with the projection, A, of O onto l. Hence the locus of A for all lines through $\bar{H}$ is a circle $\bar{c}$, the reflection into O_X of the inflection circle. If three consecutive lines pass through one point A, this point is a cusp of the envelope of the moving line (a cusp being the dual counterpart of an inflection point). Hence $\bar{c}$ is the locus of the cusps of the envelopes of moving lines and it is therefore called the cuspidal circle of the instantaneous position.

Example 84. Show that $\bar{c}$ is the inflection circle of the inverse motion.

If in three *finite* positions theory l is a line for which the homologous positions l_1, l_2, l_3 pass through one point A, and if $\bar{l}$ is parallel to l, then A has equal distances to $\bar{l}_1$, $\bar{l}_2$, $\bar{l}_3$ and it is therefore the center of a circle tangent to these three lines. For the instantaneous case we obtain: if $\bar{l}$ is any line in the moving plane, it is tangent to its envelope $\bar{e}$ at the projection, $\bar{A}$, of O onto $\bar{l}$, and the center $\bar{M}$ of curvature of $\bar{e}$ at $\bar{A}$ is on the cuspidal circle $\bar{c}$; $\bar{M}$ is the intersection of $\bar{c}$ and $O\bar{A}$ (Fig. 68).

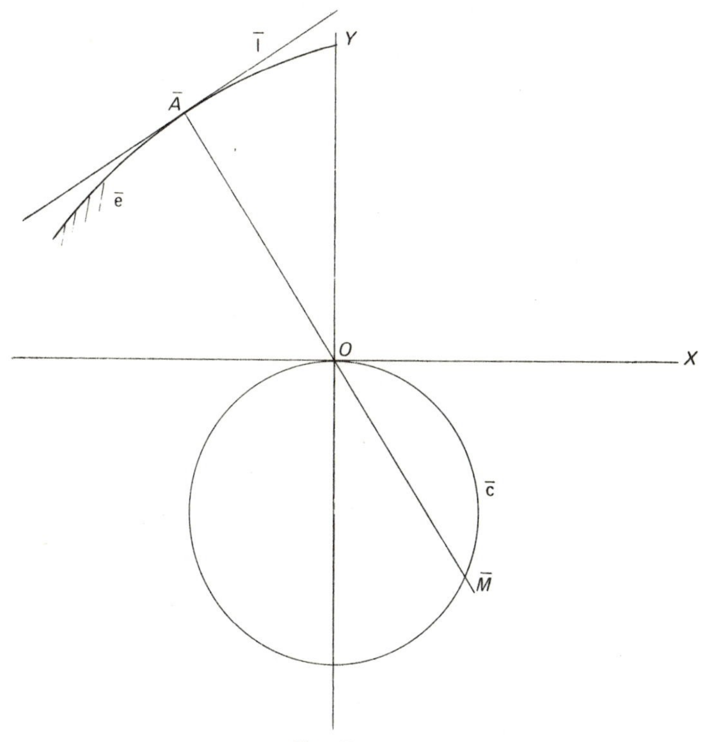

Fig. 68.

Example 85. If $\bar{\mathrm{l}}$ is parallel to O_X the point $\bar{M}$ coincides with $\bar{H}$.
Example 86. If $\bar{\mathrm{l}}$ is perpendicular to O_X the point $\bar{M}$ coincides with O.

This statement about the curvature of the envelope of a straight line is a special case of a general theorem, dealing with the curvature of the envelope of an arbitrary curve in E. It was proved by means of a more-or-less "geometric" argument. We shall give an analytic proof making use of the idea developed in Section 7, dealing with tangency between contacting curves.

Let γ be a curve in E, different from the centrode $\mathrm{p_m}$, and Γ its envelope in Σ. In any position γ is tangent to Γ, A being the tangent point. A is variable on γ and on Γ as well. Its position in E depends on ϕ. If x, y are the coordinates of A in E, $x(\phi)$, $y(\phi)$ is a parametric representation of the curve γ. The coordinates of A in the fixed plane are therefore

$$X = x(\phi)\cos\phi - y(\phi)\sin\phi + a(\phi),$$
$$Y = x(\phi)\sin\phi + y(\phi)\cos\phi + b(\phi),$$

which is a parametric representation of Γ. The components of the (geometric) velocity $\boldsymbol{V}$, of A on Γ, are

$$V_X = X' = x'\cos\phi - y'\sin\phi - x\sin\phi - y\cos\phi + a',$$
$$V_Y = Y' = x'\sin\phi + y'\cos\phi + x\cos\phi - y\sin\phi + b'. \tag{9.17}$$

The components of the velocity $\boldsymbol{v}$, of A along γ, are x', y' in the o_{xy} system and in the O_{XY}-frame they are

$$v_X = x'\cos\phi - y'\sin\phi, \qquad v_Y = x'\sin\phi + y'\cos\phi. \tag{9.18}$$

As γ is tangent to Γ at A, the vectors $\boldsymbol{V}$ and $\boldsymbol{v}$ are linearly dependent. Hence $\boldsymbol{V} = \mathfrak{k}(\phi)\boldsymbol{v}$, $\mathfrak{k}(\phi)$ being a scalar function of ϕ, with $\mathfrak{k} \neq 1$. From (9.17) and (9.18) it follows

$$(x'\cos\phi - y'\sin\phi)(1-\mathfrak{k}) = x\sin\phi + y\cos\phi - a',$$
$$(x'\sin\phi + y'\cos\phi)(1-\mathfrak{k}) = -x\cos\phi + y\sin\phi - b',$$

and therefore, with $g = (1-\mathfrak{k})^{-1}$,

$$x' = g(y - a'\cos\phi - b'\sin\phi),$$
$$y' = g(-x + a'\sin\phi - b'\cos\phi),$$
$$X' = (g-1)(x\sin\phi + y\cos\phi - a'),$$
$$Y' = (g-1)(-x\cos\phi + y\sin\phi - b'). \tag{9.19}$$

At $\phi = 0$ (9.19) becomes

$$x_1 = g_0 y, \qquad y_1 = -g_0 x,$$
$$X_1 = (g_0 - 1)y, \qquad Y_1 = -(g_0 - 1)x. \tag{9.20}$$

The formulas (9.19) express x', y' and X', Y' in terms of the coordinates x, y of A, the functions a and b which determine the motion, and the function g (or $\mathfrak{k}$) which determines, as we know from section 7, the curves γ and Γ. These formulas enable us to derive any property of γ and Γ; we are especially interested in their curvature. We obtain by direct differentiation of (9.19)

$$\begin{aligned}
x'' &= g'(y - a'\cos\phi - b'\sin\phi) \\
&\quad + g(y' - a''\cos\phi + a'\sin\phi - b''\sin\phi - b'\cos\phi), \\
y'' &= g'(-x + a'\sin\phi - b'\cos\phi) \\
&\quad + g(-x' + a''\sin\phi + a'\cos\phi - b''\cos\phi + b'\sin\phi), \\
X'' &= g'(x\sin\phi + y\cos\phi - a') \\
&\quad + (g-1)(x'\sin\phi + x\cos\phi + y'\cos\phi - y\sin\phi - a''), \\
Y'' &= g'(-x\cos\phi + y\sin\phi - b') \\
&\quad + (g-1)(-x'\cos\phi + x\sin\phi + y'\sin\phi + y\cos\phi - b''),
\end{aligned} \tag{9.21}$$

which at $\phi = 0$, and in view of (9.20), reduces to

$$x_2 = g_1 y - g_0^2 x, \quad y_2 = -g_1 x - g_0^2 y - g_0 b_2,$$
$$X_2 = g_1 y - (g_0 - 1)^2 x, \quad Y_2 = -g_1 x - (g_0 - 1)^2 y - (g_0 - 1)b_2. \tag{9.22}$$

If m and M are the centers of curvature at A of γ and Γ respectively, we have, introducing polar coordinates (r, θ) for A:

$$\begin{aligned}
mA &= (x_1^2 + y_1^2)^{3/2}/(x_1 y_2 - y_1 x_2) \\
&= g_0 r^2/(g_0 r + b_2 \sin\theta), \\
MA &= (X_1^2 + Y_1^2)^{3/2}/(X_1 Y_2 - Y_1 X_2) \\
&= ((g_0 - 1)r^2)/((g_0 - 1)r + b_2 \sin\theta).
\end{aligned} \tag{9.23}$$

Hence, if R and R_1 are respectively the directed distances from O to m and M,

$$R = Om = r - mA = (b_2 r \sin\theta)/(g_0 r + b_2 \sin\theta),$$

(9.24)

$$R_1 = OM = r - MA = (b_2 r \sin\theta)/((g_0 - 1)r + b_2 \sin\theta).$$

Eliminating g_0 we obtain

(9.25) $$((1/R) - (1/R_1))\sin\theta = 1/b_2$$

which is called the general Euler–Savary equation, of which the similar looking relation (9.5) is a special case. Comparing the two we formulate the following theorem: if Γ is the envelope in Σ of the curve γ in E, A their tangent point, m the center of curvature of γ at A, M that of Γ at A, then M coincides with the center of curvature (at m) of the curve γ' in Σ described by the point m.

If γ reduces to a point A, m coincides with A and Γ is the path of A; (9.25) reduces to (9.5).

If γ is a straight line then m is at infinity and it follows from (9.25), in view of $R = \infty$, that M is on the cuspidal circle — which verifies a result we have already obtained.

We have excluded the case where γ coincides with the moving centrode p_m, which implies that Γ is different from the fixed centrode p_f. This was necessary since, when we have $\mathfrak{k} = 1$, $g = \infty$; moreover A coincides with O and our derivation is not valid. We shall see later on, however, that (9.25) also holds in this case. But there is an essential difference: As can be seen from (9.23), the center of curvature of the envelope of an arbitrary curve in E is a second order concept, as it depends only on b_2. However, the centers of curvature of p_m and p_f depend upon third order properties, as we shall see in the next section.

Example 87. Show from (9.23) that if γ is a single point A it coincides with m.
Example 88. Apply formula (9.23) to the case where the envelope Γ reduces to a single point (which means that γ continually passes through it during the motion).
Example 89. Show that: If A is an inflection point of γ then the corresponding M is on the cuspidal circle; whereas if A is an inflection point of Γ then M is on the inflection circle.

It is possible to normalize b_2, to say $b_2 = 1$, by a simple linear stretch whereby r is replaced by cr and $\bar{r}$ by $c\bar{r}$, c being a positive number. Hence, all planar motions (excluding translations) are – within a stretch — geometrically identical to the second order. It is in fact possible to make all moving systems completely identical to the second order by: i) bringing their first order poles into coincidence — by a translation; ii) bringing the tangents to their centrodes, at $\phi = 0$, into coincidence — by a rotation; iii) making their b_2's identical — by a stretch.

10. Third order properties

We now consider third order instantaneous kinematics. This topic may be thought of as the limiting case, as the positions approach each other, of four positions theory. All properties depend on the three invariants b_2, a_3, b_3. As we obtained in (8.7):

$$\begin{array}{llll} X_0 = x, & X_1 = -y, & X_2 = -x, & X_3 = y + a_3, \\ Y_0 = y, & Y_1 = x, & Y_2 = -y + b_2, & Y_3 = -x + b_3. \end{array} \tag{10.1}$$

First of all we derive the equation of the *circling-point curve*, the locus of the points in E whose four consecutive positions are on a circle; these points are identically those with stationary curvature. As $k = (X'Y'' - Y'X'')/(X'^2 + Y'^2)^{3/2}$ the condition $k' = 0$ implies

$$(X'^2 + Y'^2)(X'Y''' - X'''Y') - 3(X'X'' + Y'Y'')(X'Y'' - X''Y') = 0, \tag{10.2}$$

which for the $\phi = 0$ positions becomes

$$(X_1^2 + Y_1^2)(X_1Y_3 - X_3Y_1) - 3(X_1X_2 + Y_1Y_2)(X_1Y_2 - X_2Y_1) = 0; \tag{10.3}$$

hence in view of (10.1) we obtain for the circling-point curve

$$F \equiv (x^2 + y^2)(a_3x + b_3y) + 3b_2x(x^2 + y^2 - b_2y) = 0, \tag{10.4}$$

or

$$(x^2 + y^2)[(a_3 + 3b_2)x + b_3y] - 3b_2^2xy = 0. \tag{10.5}$$

Example 90. Show that $(X^2 + Y^2)_0 = x^2 + y^2$, $(X^2 + Y^2)_1 = 2(X_0X_1 + Y_0Y_1) = 0$, $(X^2 + Y^2)_2 = 2b_2y$, $(X^2 + Y^2)_3 = 2[(a_3 + 3b_2)x + b_3y]$.

Example 91. Prove that the locus of the points whose four consecutive positions are concyclic can be derived from

$$|(X^2 + Y^2)_v \quad X_v \quad Y_v \quad 1_v| = 0, \quad v = 0, 1, 2, 3.$$

From (10.5) it follows that, unless $a_3 + 3b_2 = b_3 = 0$ (which is in view of (8.13) equivalent to $\tilde{a}_3 = \tilde{b}_3 = 0$), the circling-point curve is a circular cubic. This is as expected since the circling-point curve is a limiting case of the circle-point curve (4.18) obtained when we studied four finitely separated positions. There is, however, an important simplification here compared with general four positions theory, for (10.5) unlike (4.18) represents a rational curve: It always has a node at O; the tangents at this double point are the pole tangent and the pole normal. The circling-point curve is sometimes also called the "*cubic of stationary curvature*" and by some authors simply the "circle-point curve".

Example 92. Determine the equation of the real asymptote of (10.5).

The *centering-point curve* is the circling-point curve of the inverse motion, its equation follows from (10.5) by replacing x, y, b_2, a_3, b_3 by X, $-Y$, b_2, $-(a_3+3b_2)$, b_3; hence it reads

$$(10.6) \qquad (a_3X+b_3Y)(X^2+Y^2)-3b_2^2XY=0,$$

which is a circular cubic unless $a_3=b_3=0$. It is the locus of the centers of curvature in Σ of all point-paths with stationary curvature when E is in the position $\phi=0$. Clearly it is the limiting case of the center-point curve (5.1) (which is also the pole curve (4.2)).

After some algebra we obtain for the equation of the tangent, to (10.5), at the isotropic point $I_1(1,i,0)$:

$$(10.7) \qquad y=ix+q,$$

with

$$q=(3/2)b_2^2(a_3+3b_2-ib_3)N^{-1},$$

$$N=(a_3+3b_2)^2+b_3^2.$$

The principal focus G of (10.5) is the intersection of (10.7) and its conjugate imaginary line which is the tangent to (10.5) at I_2. We obtain

$$(10.8) \qquad G=(3/2)b_2^2b_3N^{-1},(3/2)b_2^2(a_3+3b_2)N^{-1};$$

by simple algebra we can verify that G satisfies (10.5). Hence the circling-point curve is a focal curve, which is as expected since this is the case for general four positions theory. Although this may be of some interest from a geometric point of view, the point G does not seem to have any special kinematic significance.

The locus of the points for which three consecutive positions are collinear is the inflection circle c with the equation $x^2+y^2-b_2x=0$, from which the pole O must be excluded. Of the six intersections of c and the circling-point curve three coincide with O (this follows easily from (10.4)), and there is one at each of the isotropic points I_1, I_2. The remaining one is the intersection, different from O, of c and the line $a_3x+b_3y=0$. If $a_3\neq 0$, $b_3\neq 0$ this point, which must have four collinear consecutive positions and is therefore Ball's point B, is given by

$$(10.9) \qquad B[-b_2a_3b_3(a_3^2+b_3^2)^{-1},b_2a_3^2(a_3^2+b_3^2)^{-1}].$$

This result still has meaning if $a_3\neq 0$, $b_3=0$; then B coincides with the inflection pole. If, however, $a_3=0$, $b_3\neq 0$, the point (10.9) coincides with O

and the argument is not valid. A very special case arises if $a_3 = b_3 = 0$: then the circling-point curve (10.4) degenerates into the inflection circle and the pole normal, hence every point on c, except O, is a Ball point. In this case the moving plane is said to be in a *Cardan position*, a terminology which will be explained when, in Chapter IX, we deal with a special plane motion.

For a general motion $X'Y'' - X''Y' = 0$ represents the set of inflection circles. Hence, in instantaneous kinematics, the intersections of the circle c and the consecutive inflection circle coincide with the intersections of $X_1Y_2 - X_2Y_1 = 0$ and $X_1Y_3 - X_3Y_1 = 0$, that is of $x^2 + y^2 - b_2y = 0$ and $a_3x + b_3y = 0$. These points are O and B. Hence the inflection circle touches its envelope at the pole and at Ball's point. This envelope consists therefore of two curves: the fixed centrode p_f and the locus of Ball's points, which is called Ball's curve.

Example 93. Show that if $a_3 = b_3 = 0$ the centering-point curve degenerates into the pole tangent, the pole normal and the line at infinity.
Example 94. Show that if $\bar{a}_3 = \bar{b}_3 = 0$, the circling-point curve degenerates into the pole tangent, the pole normal and the line at infinity.

Considering now the instantaneous motion of a line we remark that it is, according to (8.15), described by the formulas

$$
\begin{aligned}
&(U_1)_0 = u_1, \quad (U_1)_1 = -u_2, \quad (U_1)_2 = -u_1, \quad (U_1)_3 = u_2,\\
&(U_2)_0 = u_2, \quad (U_2)_1 = u_1, \quad (U_2)_2 = -u_2, \quad (U_2)_3 = -u_1,\\
&(U_3)_0 = u_3, \quad (U_3)_1 = 0, \quad (U_3)_2 = -\bar{b}_2u_2, \quad (U_3)_3 = \bar{a}_3u_1 - \bar{b}_3u_2.
\end{aligned}
\tag{10.10}
$$

The intersection of the line u_1, u_2, u_3 and its next consecutive position is the point $X = -u_1u_3$, $Y = -u_2u_3$, $Z = u_1^2 + u_2^2$. The third position of this line passes through this point if $u_1^2u_3 + u_2^2u_3 - \bar{b}_2u_2(u_1^2 + u_2^2) = 0$, that is if $-\bar{b}_2u_2 + u_3 = 0$; a result which we already knew (the line should pass through $\bar{H}$, the cuspidal pole). If moreover the fourth position of this line passes through the point, the condition is $\bar{a}_3u_1 - \bar{b}_3u_2 = 0$. There is therefore one line for which four consecutive positions are concurrent, i.e., the line

$$
u_1 : u_2 : u_3 = \bar{b}_3 : \bar{a}_3 : \bar{b}_2\bar{a}_3.
\tag{10.11}
$$

Its four positions pass through the point

$$
\begin{aligned}
(X/Z) &= -u_1u_3/(u_1^2 + u_2^2) = -\bar{b}_2\bar{a}_3\bar{b}_3/(\bar{a}_3^2 + \bar{b}_3^2),\\
(Y/Z) &= -u_2u_3/(u_1^2 + u_2^2) = -\bar{b}_2\bar{a}_3^2/(\bar{a}_3^2 + \bar{b}_3^2);
\end{aligned}
\tag{10.12}
$$

if we compare this with (10.9) and if we keep in mind that the canonic frames of the direct and the inverse motion are one another's reflections into O_X, we see that (10.12) is the Ball point $\tilde{B}$ of the inverse motion — as could be expected.

The coordinates of the instantaneous rotation center are first order quantities of the motion. Hence the curvatures of the centrodes depend on the third order properties of the motion, and are therefore a subject of this section.

For the fixed centrode p_f we have found (7.5)

$$X = a - b', \qquad Y = b + a', \tag{10.13}$$

and for the moving one, p_m, (7.4)

$$x = a' \sin\phi - b' \cos\phi, \qquad y = a' \cos\phi + b' \sin\phi. \tag{10.14}$$

Hence we have for p_f if $\phi = 0$,

$$X_1 = -b_2, \qquad Y_1 = 0, \qquad X_2 = -b_3, \qquad Y_2 = b_2 + a_3,$$

and we obtain for the radius of curvature R_f of p_f at O,

$$-R_f = (X_1^2 + Y_1^2)^{3/2}/(X_1Y_2 - X_2Y_1) = -b_2^2/(a_3 + b_2). \tag{10.15}$$

Analogously, from (10.14),

$$x_1 = -b_2, \qquad y_1 = 0, \qquad x_2 = -b_3, \qquad y_2 = 2b_2 + a_3,$$

and therefore

$$-R_m = (x_1^2 + y_1^2)^{3/2}/(x_1y_2 - x_2y_1) = -b_2^2/(a_3 + 2b_2), \tag{10.16}$$

so that

$$((1/R_m) - (1/R_f)) = 1/b_2. \tag{10.17}$$

This is the *second* Euler–Savary equation. As p_f is the envelope of p_m and the centers of curvature are both on O_Y, which implies $\theta = \pi/2$, the formula (10.17) is in accordance with the general expression (9.25).

Example 95. Show from (10.13) and (10.14): $X = -b_2\phi - \frac{1}{2}b_3\phi^2 + \cdots$, $Y = \frac{1}{2}(a_3 + b_2)\phi^2 + \cdots$, $x = -b_2\phi - \frac{1}{2}b_3\phi^2 + \cdots$, $y = \frac{1}{2}(a_3 + 2b_2)\phi^2 + \cdots$; hence up to the second order the centrodes depend on a_3, b_3, b_2. But to the second order $x - X = 0$, $y - Y = \frac{1}{2}b_2\phi^2$ so that the differences depend only on b_2.

Example 96. Use (10.17) to verify that the convention $b_2 > 0$ implies that the Y-axis is directed from O toward the center of curvature of the *moving centrode*, if the centrodes have their centers (at $\phi = 0$) on opposite sides of their tangent. When their centers are on the same side of the tangent, the Y-axis is directed toward or away from the curvature centers depending upon whether $R_m < R_f$ or $R_m > R_f$ respectively. These three situations representing all possible cases (provided $b_2 > 0$) are shown in Fig. 69.

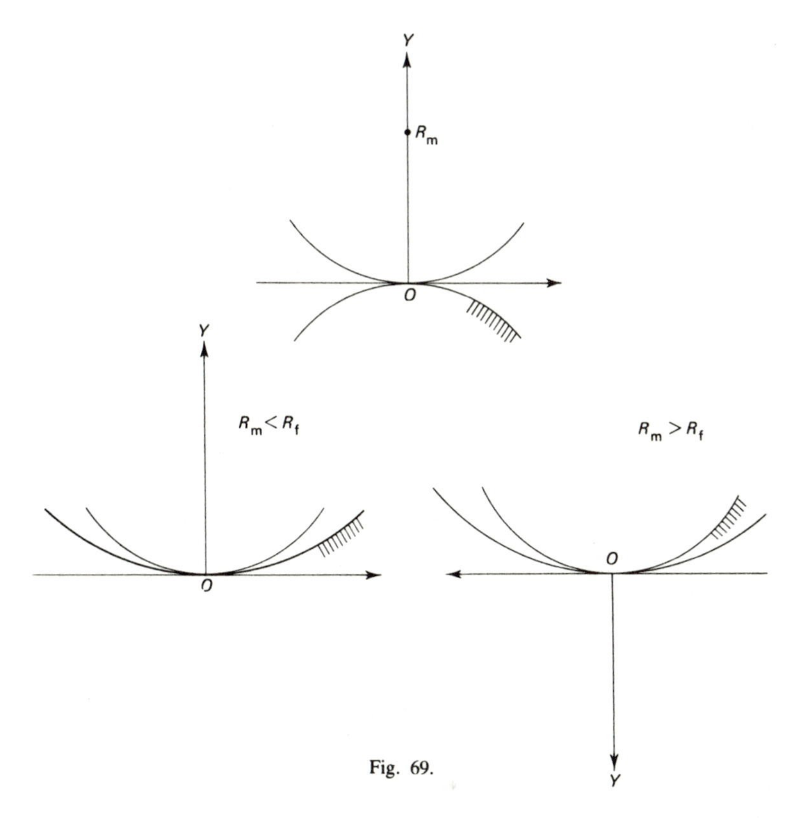

Fig. 69.

We give one more application of third order theory. The radius of curvature of the path p at A is a second order concept, and the same holds for the coordinates of the center of curvature M and hence for its locus, the evolute $\bar{p}$ of the path. This implies that the center of curvature $\bar{M}$ of $\bar{p}$ (at M) is a third order covariant, sometimes called the second center of curvature of p at A. It has been considered by R. Müller (MÜLLER [1891a]) who found a procedure to construct $\bar{M}$; MÜLLER [1891b] also found similar results for evolutes of envelopes of generating curves. We will develop these concepts further at the end of the next section.

Example 97. If $\bar{M} = (\bar{X}, \bar{Y})$, show:

$$\bar{X} = (-b_2xy/N) + (y(x^2 + y^2)F/N^3),$$

$$\bar{Y} = (-b_2y^2/N) - (x(x^2 + y^2)F/N^3),$$

with $N = x^2 + y^2 - b_2y$, while F is given by (10.4).

11. Fourth order properties

In fourth order theory (that is, dealing with five infinitesimal positions) an important problem is to determine those points of E whose paths have in the zero-position five points in common with their osculating circle: the Burmester points for the instantaneous case. The circling-point curve has been derived by means of equation (10.2). The points asked for are on $\mathrm{F} = 0$, given by (10.5), and moreover on the consecutive circling-point curve, which implies that they satisfy the equation found by taking the derivative of (10.2) and putting $\phi = 0$:

$$(11.1)\quad \begin{aligned}(X_1^2+Y_1^2)(X_1Y_4-X_4Y_1+X_2Y_3-X_3Y_2)-(X_1X_2+Y_1Y_2)(X_1Y_3-X_3Y_1)\\ -3(X_1Y_2-X_2Y_1)(X_2^2+Y_2^2+X_1X_3+Y_1Y_3)=0,\end{aligned}$$

which in view of

$$(11.2)\quad \begin{aligned}&X_0=x,\quad X_1=-y,\quad X_2=-x,\quad X_3=y+a_3,\quad X_4=x+a_4,\\ &Y_0=y,\quad Y_1=x,\quad Y_2=-y+b_2,\quad Y_3=-x+b_3,\quad Y_4=y+b_4,\end{aligned}$$

yields the equation

$$(11.3)\quad \begin{aligned}\bar{\mathrm{F}}\equiv[(a_4+4b_3)x+(-4a_3+b_4-6b_2)y+3b_2^2](x^2+y^2-b_2y)\\ +b_2y\,[x^2+y^2+(a_4x+b_4y)]=0.\end{aligned}$$

It represents a circular cubic, passing through O, the tangent at O being $y = 0$. Of the nine intersections of F and $\bar{\mathrm{F}}$, two coincide with I_1, I_2 and three with O. The remaining four are the Burmester points (MÜLLER [1892]). These points were first studied by R. Müller who named them after L. Burmester (who had studied the analogous points for the five-finitely-separated-positions case).

Example 98. Show that $(X^2+Y^2)_4 = 2[(a_4+4b_3)x+(-4a_3+b_4-6b_2)y+3b_2^2]$.
Example 99. Derive (11.3) from the condition $|(X^2+Y^2)_v \quad X_v \quad Y_v \quad 1_v| = 0$, $v = 0, 1, 2, 4$.

In Section 5 we have developed another method to determine the Burmester points, in the general case. We introduced the homogeneous unknowns u_i ($i = 0, 1, 2, 3, 4, 5, 6$), satisfying four linear and two quadratic equations. The Burmester points (and the Burmester centers at the same time) were found in terms of the intersections of two conics. If we apply this procedure to the instantaneous case, the four linear equations are the first four derivatives of:

$$(11.4)\quad c_0(X^2+Y^2)-2c_1X-2c_2Y+c_3=0.$$

Using the same u_i as defined by (5.11) the matrix of the coefficients of the four linear equations becomes

$$(11.5)\qquad \left\|\begin{matrix} 0 & 0 & 0 & 0 & 0 & 0 & 1 \\ 0 & 0 & -b_2 & 0 & b_2 & 1 & 0 \\ 0 & -a_3 & -b_3 & a_3+3b_2 & b_3 & 0 & -1 \\ 3b_2^2 & -a_4 & -b_4 & a_4+4b_3 & -4a_3-6b_2+b_4 & -1 & 0 \end{matrix}\right\|$$

which is much simpler than in the general case. The two quadratic equations remain (5.12), exactly as in the general case. Hence, once (11.5) is substituted for (5.13), the procedure becomes identical to the one discussed in Section 5. The determinants Δ and Δ' (introduced in Section 5 as the determinants of columns 2, 3, 6, 7 and 4, 5, 6, 7 respectively) do simplify in this case. In fact, for the instantaneous case we have

$$(11.6)\qquad \Delta = -a_3b_4 + a_4b_3 - a_3b_2;$$

$\Delta = 0$ is the condition for one Burmester center to be at infinity. This means that the corresponding point is such that its five consecutive positions are collinear; in Veldkamp's terminology it is a Ball point with excess one. Such a point does not exist in general, but will be present if the instantaneous invariants satisfy $\Delta = 0$.

Example 100. Prove the relation $\Delta = 0$ is necessary for there to exist a solution (x, y) of $X_1 : Y_1 = X_2 : Y_2 = X_3 : Y_3 = X_4 : Y_4$.

Example 101. Prove that a Ball point with excess one exists if the plane has a Cardan position (i.e., $a_3 = b_3 = 0$), and determine its coordinates.

Our discussion of point-paths has so far been restricted to those special points which to a given order move on straight lines, circles or other conics. A more general approach would be one which determines those points in E with paths in Σ which, to a given order, describe any specified trajectory. To accomplish this we characterize a plane curve by its curvature k, the curvature of its evolute $\bar{k}$, and the curvature of the evolute of its evolute k^*. If the trajectory p of a moving point A has at $\phi = 0$ a curvature k_0, its center of curvature M is a point on the evolute $\bar{\text{p}}$, of p, which at M has a curvature $\bar{k}_0$ (Fig. 70). Further, the evolute $\bar{\text{p}}$ has an evolute curve p* which contains $\bar{M}$ the center of curvature of M on $\bar{\text{p}}$. The curvature of p* at $\bar{M}$ is denoted by k_0^*. The problem we set ourselves is to determine all points A in E for which, at $\phi = 0$, $k_0 = \lambda_0$, $\bar{k}_0 = \lambda_0\lambda_1$ and $k_0^* = \lambda_0\lambda_2$, where λ_0, λ_1, λ_2 are a specified set of real numbers. It follows from this that

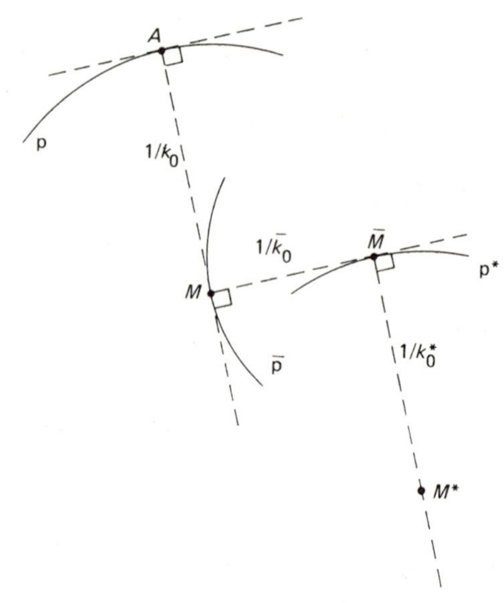

Fig. 70.

(11.6) $$\bar{k}_0 = \lambda_1 k_0.$$

From differential geometry it is known that if $k \neq 0$

(11.7) $$\bar{k}(\mathrm{d}k/\mathrm{d}s) + k^3 = 0$$

where s is the arc length of p. In terms of our parameter ϕ we have $k' = (\mathrm{d}k/\mathrm{d}s)s'$, and hence

(11.8) $$\bar{k}k' + k^3 s' = 0.$$

We know

$$k = (X'Y'' - Y'X'')/(X'^2 + Y'^2)^{3/2}$$

and from this one easily gets k' by differentiating. Also, $s' = (X'^2 + Y'^2)^{1/2}$. Substituting these into (11.8), and then evaluating at $\phi = 0$ by use of (10.1) and (11.6) we find

(11.9) $$(\mathrm{N}^2/\lambda_1) - \mathrm{F} = 0$$

where $\mathrm{F} = 0$ is the circling-point curve given by (10.4), and $\mathrm{N} = x^2 + y^2 - b_2 y = 0$ is the inflection circle. Clearly (11.9) represents a bicircular quartic. It has been named the *quartic of derivative curvature*. It is a rational curve, since it always has a third node. This node is at the pole, and one of its tangents there is the pole tangent O_x. This curve also passes through the Ball point.

Equation (11.9) was first derived by FREUDENSTEIN [1965], and later studied by VELDKAMP [1967b]. It may be regarded as a generalization of the circling-point equation. It is the locus of all points in E which generate paths which, at $\phi = 0$, have curvature ratio $\bar{k}_0/k_0 = \lambda_1$. All such paths are, within a stretch, identical to the third order. Since λ_1 is an intrinsic property of a planar curve which describes it to the third order (within a stretch), (11.9) allows us to determine third order approximations to any curve whatsoever (to obtain the given curve we "stretch" E so that the curvature is λ_0).

Example 102. Show that for a circle $(1/\lambda_1) = 0$, and therefore (11.9) yields the circling point curve.

Example 103. Show that (11.7) may be obtained by taking the derivative of the radius of curvature ρ with respect to the arc lengths s, and substituting the following definitions: $\rho = 1/k$, $d\alpha/d\rho = \bar{k}$, $d\alpha/ds = k$, where α is the angle of the tangent line to curve p at point A.

Example 104. Show that the coordinates of the evolute can be obtained from the original curve by either of the following methods: i) if the curve has coordinates X, Y, the evolute has coordinates X_c, Y_c where $X_c = X - (dY/dX)(W)$, $Y_c = Y + W$ with $W = (1 + (dY/dX)^2)/(d^2Y/dX^2)$; ii) if the equation of the curve is $\mathfrak{f}(X, Y) = 0$ and the normal to it is $Y_c - Y = (-dX/dY)(X_c - X)$, we can eliminate, say, X between these two equations and obtain a new equation $\mathfrak{f}(X_c, Y_c, Y) = 0$. The evolute is the envelope of the normals, i.e., the resultant of eliminating Y between $\mathfrak{f}(X_c, Y_c, Y) = 0$ and $\partial\mathfrak{f}(X_c, Y_c, Y)/\partial Y = 0$.

Provided that $\bar{k} \neq 0$, it is possible to write (11.7) for the evolute curve simply by changing notation:

$$k^*(d\bar{k}/d\bar{s}) + \bar{k}^3 = 0 \tag{11.10}$$

where $\bar{s}$ is the arc length along the evolute $\bar{\mathrm{p}}$. Following a development similar to the foregoing, although somewhat more involved (VELDKAMP [1967b]), one can obtain from (11.10):

$$[(\lambda_1)^2 \mp 3\lambda_2]\mathrm{N}^3 - (\lambda_1)^2\lambda_2(x\mathrm{F} + \mathrm{F}_1) = 0 \tag{11.11}$$

where

$$\mathrm{F}_1 \equiv [(a_4 + 4b_3)x + (b_4 - 4a_3 - 6b_2)y + 3b_2^2]\mathrm{N}\,(x^2 + y^2) - 5\mathrm{F}b_2x + b_2y\,(x^2 + y^2 + a_4x + b_4y)(x^2 + y^2)$$

(11.11) represents a tricircular sextic with a triple point at the pole having O_x as one of its tangents. (11.11) intersects (11.9) at five points other than the pole and the circle points at infinity. These five points were originally derived by FREUDENSTEIN [1965] who named them the generalized Burmester points. The result is that given a motion, up to the fourth order, there are at most five points which approximate any given planar curve characterized by the numbers λ_1, λ_2. Hence, within a stretch, these points move on trajectories for which, at $\phi = 0$, $k_0 = \lambda_0$, $\bar{k}_0 = \lambda_0\lambda_1$, $k_0^* = \lambda_0\lambda_2$. In general λ_0 is a different number for each of these five points. In order to match a given planar curve to the fourth order, λ_1 and λ_2 must be calculated at the desired point on the curve, and then the generalized Burmester points computed using these values. Once a generalized Burmester point is known, E must be "stretched" (i.e., $x \to Cx$, $y \to Cy$, C is a real number) so that the Burmester point's path curvature equals λ_0. FREUDENSTEIN and WOO [1968] have given a further generalization of curve matching concepts using the so-called intrinsic properties of a curve.

Example 105. Consider the Ball point and show that it is in effect a sixth generalized Burmester point (VELDKAMP [1967b]).

In Sections 5 and 6 we have discussed some problems arising when five or more finitely separated positions are given. We leave it to the reader to modify these problems for the instantaneous case. Some results are inserted into the following:

Example 106. The locus of points for which five consecutive positions are on a parabola is a bicircular quartic, passing through the pole O and through Ball's point B; it is tangent at these points to the inflection circle (SANDOR and FREUDENSTEIN [1967]; BOTTEMA [1967]).
Example 107. The locus of points, six consecutive positions of which are on a conic, is a bicircular quintic with double points at O and at B. (Idem.)
Example 108. There are six points in E for which six consecutive positions are on a parabola. (Idem.)
Example 109. There are eleven points for which seven consecutive positions are on a conic. (Idem.)
Example 110. Derive the equation for the locus of all lines (a curve of the second class) which have five consecutive positions tangent to a parabola. Show that the locus is either a hyperbola or degenerate (BOTTEMA [1970]).
Example 111. Derive the equation of the locus of the lines (a curve of the fourth class) which have six consecutive positions tangent to a conic. (Idem.)

In Sections 8–11 of this chapter we have discussed instantaneous properties of the motion by introducing a canonical frame: O at the pole, O_X along the pole tangent. There are some problems where it is advantageous to take other frames. Such an approach has been used in investigating the properties of Ball's curve, that is the locus of Ball's point (either in the moving or in the fixed plane) during a continuous motion. It was found appropriate to take the origin at B, and the X-axis along BP, P being the pole.

Example 112. Prove that in this case, we have instead of (8.7),

$$X_0 = x, \quad X_1 = -y + a_1, \quad X_2 = -x + a_2, \quad X_3 = y + a_3, \quad X_4 = x + a_4, \dots$$

$$Y_0 = y, \quad Y_1 = x, \quad Y_2 = -y, \quad Y_3 = -x, \quad Y_4 = y + b_4, \ldots$$

(BOTTEMA [1966]).

Example 113. If B is a Ball point with excess one, show that it is at a cusp of Ball's curve (R. MÜLLER [1898], [1903]).

In our treatment of instantaneous plane kinematics we have set two restrictions: the pole is a finite point and the invariant $b_2 \neq 0$, which implies that the geometric velocity at which the centrodes change contact points is unequal to zero and that the inflection circle has a positive radius.

If the position has a (finite) pole, this means that the rotation angle ϕ is not stationary, so that it may be chosen as the (geometrical) motion's parameter. If a_v and b_v are zero for every value of v then $a = b = 0$ for all ϕ and the motion is a rotation about a fixed point. If $a_v = b_v = 0$, $v = 0, 1, 2, \ldots, n-1$, but a_n, b_n are not zero, we take the O_X-axis, such that $a_n = 0$. Then we have instead of (8.7)

$$\begin{aligned} &X_0 = x, \quad X_1 = -y, \quad X_2 = -x, \quad X_3 = y, \ldots, \\ &X_n = -y, \quad X_{n+1} = -x + a_{n+1}, \ldots \\ &Y_0 = y, \quad Y_1 = x, \quad Y_2 = -y, \quad Y_3 = -x, \ldots, \\ &Y_n = x + b_n, \quad Y_{n+1} = -y + b_{n+1}, \ldots \end{aligned} \tag{11.12}$$

where we have supposed $n = 4j + 1$. It is clear that the theory simplifies in this case.

Example 114. Consider the case $b_2 = 0$, $b_3 \neq 0$ and modify our results for third and fourth order instantaneous kinematics.

If the invariants are expressed as a power series in terms of the motion parameter, i.e.,

$$a = \sum_{n=3}^{\infty} (a_n/n!)\phi^n, \qquad b = \sum_{n=2}^{\infty} (b_n/n!)\phi^n, \tag{11.13}$$

then for the path of the point o, which is the pole at $\phi = 0$, we have

$$X = \sum_{n=3}^{\infty} (a_n/n!)\phi^n, \qquad Y = \sum_{n=2}^{\infty} (b_n/n!)\phi^n. \tag{11.14}$$

Using (11.14) it is possible to compute the curvature and other properties of the path of o (VELDKAMP [1963]). If $a_3 \neq 0$ the curve described by o has in the interval $-\varepsilon < \phi < \varepsilon$ (where ε is a small quantity) a cusp, and is tangent to the principal normal, $X = 0$; as $\phi \to 0$, the path curvature $k \to \infty$. In general the cusp is an ordinary cusp, in that both branches are on different sides of the tangent. However, if $a_3 = 0$ then as $\phi \to 0$, $k \to -(1/3)(a_4/b_2^2)$ and we have a ramphoid cusp (i.e., both branches are on the same side of the tangent). None of this implies that the trajectory of o taken as a whole has a cusp at the pole, since the curve may retrace itself and pass through $X = 0$, $Y = 0$ several times (VELDKAMP [1963]).

Example 115. If $a = \frac{1}{2}\phi^3 + \frac{1}{2}\phi^4$ amd $b = \frac{1}{2}\phi^2 + \frac{1}{2}\phi^3$, show that the trajectory of o is a curve with a triple point at the origin, with $Y = 0$ as a double tangent and $X - Y = 0$ as a single tangent. (VELDKAMP [1963].)
Example 116. If $a = \frac{1}{4}\phi^4 + \phi^5 + \phi^6$ and $b = \frac{1}{2}\phi^2 + \phi^3$ then the trajectory of o is the parabola $X = Y^2$. Show that the trajectory of o does in fact have a cusp even though the carrier curve apparently does not. (Idem.)
Example 117. Show that if $a_3 = a_4 = 0$ and $a_5 \neq 0$ then, as $\phi \to 0$, o passes through an inflection point of its path. (Idem.)
Example 118. Show that if $a_3 = a_4 = a_5 = 0$ and $a_6 \neq 0$ then, as $\phi \to 0$, o becomes a Ball point. (Idem.)

12. Translations

The preceding discussions on instantaneous kinematics fail if there is no (finite) pole in the zero position. In the case of a finite pole, t being an arbitrary parameter describing the motion, we have $\dot{\phi} \neq 0$ for the zero position $t = 0$. For geometrical problems (those which are independent of the choice of the parameter), it was therefore natural to take $\phi = t$, $\dot{\phi} = 1$.

If there is no finite pole we have $\dot{\phi} = 0$ for $t = 0$. The velocity distribution is then that of a translation (instead of a rotation) of the plane. This is the reason why these positions have been called T-positions by Veldkamp, who discussed them at length (VELDKAMP [1963, p. 127–144]). If, for $t = 0$, $(\mathrm{d}^k\phi/\mathrm{d}t^k) = 0$ $(k = 1, 2, \ldots, n)$ and $(\mathrm{d}^{n+1}\phi/\mathrm{d}t^{n+1}) \neq 0$ the position is called a T-position of order n or simply a T_n-position. In this case it is advisable to introduce a parameter t such that $\phi = (t^{n+1}/(n+1)!)$, and use the notation whereby if $z(t)$ is any function of t, $(\mathrm{d}^n z/\mathrm{d}t^n)_{t=0}$ is denoted by z_n. We shall restrict ourselves to the case $n = 1$, $\phi = \frac{1}{2}t^2$ and therefore: $\phi_0 = \phi_1 = 0$, $\phi_2 = 1$, $\phi_k = 0$ $(k > 2)$.

For the series for $\cos\phi$ we obtain

$$\cos\phi = 1 - (\phi^2/2!) + (\phi^4/4!) - \cdots = 1 - (t^4/(2^2 \cdot 2!)) + (t^8/(2^4 \cdot 4!)) - \cdots$$

This implies

$$(12.1)\qquad \begin{aligned} &(\cos\phi)_0 = 1, \quad (\cos\phi)_1 = (\cos\phi)_2 = (\cos\phi)_3 = 0, \quad (\cos\phi)_4 = -3, \\ &(\cos\phi)_k = 0, \quad (k = 5, 6, 7), \quad (\cos\phi)_8 = 105, \ldots \end{aligned}$$

and analogously

$$(12.2)\qquad \begin{aligned} &(\sin\phi)_0 = 0, \quad (\sin\phi)_1 = 0, \qquad (\sin\phi)_2 = 1,\\ &(\sin\phi)_v = 0 \quad (v = 3,4,5), \qquad (\sin\phi)_6 = -15, \dots . \end{aligned}$$

The motion is given by

$$(12.3)\qquad \begin{aligned} X &= x\cos\phi - y\sin\phi + a(t),\\ Y &= x\sin\phi + y\cos\phi + b(t),\\ \phi &= \tfrac{1}{2}t^2. \end{aligned}$$

Furthermore, let the frames in E and Σ coincide at $t = 0$; then, obviously, $a_0 = b_0 = 0$.

From (12.3), using (12.1) and (12.2), it follows that

$$(12.4)\qquad X_1 = a_1, \qquad Y_1 = b_1.$$

If $a_1 = b_1 = 0$ this implies that at $t = 0$ every point of E is a pole. Excluding this case (sometimes called a T-position of the second kind) we have: every point of E has the same velocity. We choose the positive X-axis parallel to this velocity direction, thereby $a_1 = 0$ and $b_1 > 0$.

From (12.3) we find that $X' = Y' = 0$ for the point in E with the homogeneous coordinates

$$(12.5)\quad x = a'\sin\phi - b'\cos\phi, \quad y = a'\cos\phi + b'\sin\phi, \quad z = t, \quad \phi = \tfrac{1}{2}t^2.$$

The coordinates in Σ of this point are

$$(12.6)\qquad X = at - b', \qquad Y = a' + bt, \qquad Z = t.$$

(12.5) and (12.6) are therefore parametric equations for the centrodes p_m and p_f respectively. For $t = 0$ the pole P is $(1,0,0)$, i.e., the point at infinity of the X-axis. Differentiating (12.5) and (12.6) we find that the tangent at P to both p_m and p_f is the line*

* The tangent to p_m is given by

$$\begin{vmatrix} x & y & z\\ x_0 & y_0 & z_0\\ x_1 & y_1 & z_1 \end{vmatrix} = 0$$

and to p_f by

$$\begin{vmatrix} X & Y & Z\\ X_0 & Y_0 & Z_0\\ X_1 & Y_1 & Z_1 \end{vmatrix} = 0.$$

$$y = a_2.$$

We take the X-axis along this pole tangent, which implies $a_2 = 0$. From (12.3) it then follows that

$$X_2 = -y, \qquad Y_2 = x + b_2.$$

Since the origin is not yet localized on the X-axis, we take O such that for this point $Y_2 = 0$; hence $b_2 = 0$. The two coinciding frames are now completely determined. We obtain the following table for this canonical coordinate system:

$$\begin{array}{llllll} & X_0 = x, & X_1 = 0, & X_2 = -y, & X_3 = a_3, & X_4 = -3x + a_4, \ldots \\ (12.7) & Y_0 = y, & Y_1 = b_1, & Y_2 = x, & Y_3 = b_3, & Y_4 = -3y + b_4, \ldots \end{array}$$

There appears one invariant b_1 (>0) for the first order instantaneous position, no further ones for the second, a_3 and b_3 for the third, etc.

Example 119. Show that the inverse of a T-position is a T-position, and determine $\bar{b}_1$, $\bar{a}_3$, $\bar{b}_3$, $\bar{a}_4$, $\bar{b}_4$.

From (12.7) it follows that the curvature of the path of $A(x, y)$ is given by $(X_1Y_2 - X_2Y_1)/(X_1^2 + Y_1^2)^{3/2} = y/b_1^2$; hence the inflection curve, a circle in the general case, degenerates into $y = 0$, that is, the pole tangent. The center of curvature M of $A(x, y)$ is $[(xy - b_1^2)/y, y]$. The Euler–Savary equation is not recognizable for these special positions and neither is Hartmann's theorem. But the statement that a point A on the pole tangent has its center M at the pole P still holds. So does Bobillier's theorem. Indeed, let $A_1(x_1, y_1)$ and $A_2(x_2, y_2)$, with $y_1 \neq y_2$, be two points and hence $M_i = [(x_iy_i - b_1^2)/y_i, y_i]$, $(i = 1, 2)$, are their associated centers. It is easy to verify that the intersection $T(X, Y)$ of A_1A_2 and M_1M_2 has the ordinate $Y = y_1 + y_2$ (Fig. 71). Hence TP and the pole-tangent are one another's reflections into the "bisector" of A_1P and A_2P. So much for second order properties.

Example 120. Show that in our T-position the origin O plays the part of the inflection pole.

Third order properties depend on the three invariants b_1, a_3, b_3. The condition for the existence of a Ball's point is $X_1 : Y_1 = X_2 : Y_2 = X_3 : Y_3$. Substituting from (12.7) the conclusion is: in the general case, $a_3 \neq 0$, there is no Ball point at all; if, however, $a_3 = 0$, any point of the pole tangent (that is the inflection curve) is a Ball point.

The general equation of the circling-point curve is (10.3)

$$(X_1^2 + Y_1^2)(X_1Y_3 - X_3Y_1) - 3(X_1Y_2 - X_2Y_1)(X_1X_2 + Y_1Y_2) = 0;$$

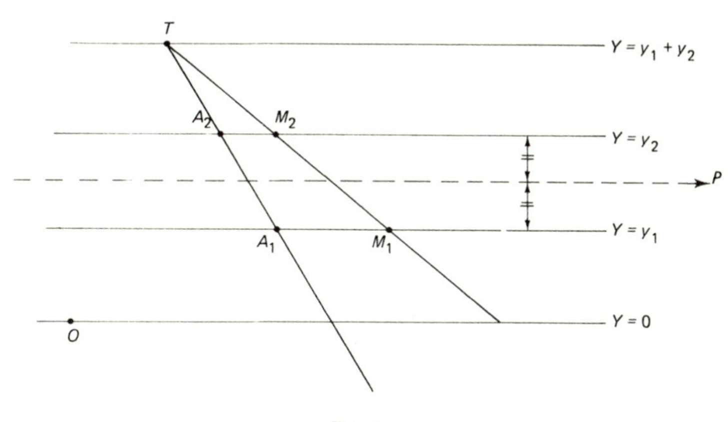

Fig. 71.

hence, in the case at hand, we obtain

$$xy = -(1/3)b_1a_3, \tag{12.8}$$

which represents an orthogonal hyperbola, with O as its center, and O_X and O_Y as its asymptotes (Fig. 72). We find therefore that for a T_1-position the circling-point curve breaks up into a hyperbola and the line at infinity in E. The hyperbola lies in the first and third quadrant if $a_3 < 0$, in the second and fourth if $a_3 > 0$, and degenerates into the coordinate axes if $a_3 = 0$.

In general third order theory we have derived the second Euler–Savary relation (10.17). Obviously there is no direct analog for a T-position.

Example 121. Show from (12.5) and (12.6) that up to the first order we obtain for the centrodes $X = -(b_1/t) - \frac{1}{2}b_3t$, $Y = \frac{1}{2}(a_3 + 2b_1)t$; $x = -(b_1/t) - \frac{1}{2}b_3t$, $y = \frac{1}{2}(a_3 + b_1)t$, which depend on a_3, b_3, b_1. This implies that to this order $X - x = 0$, $Y - y = \frac{1}{2}b_1t$ depend on b_1 only. There is herein a certain analog with the general theory (see Example 95).

Fourth order properties are controlled by b_1, a_3, b_3, a_4, b_4. Burmester points in the general case are the intersections of the circling-point curve and the curve (11.1)

$$(X_1^2 + Y_1^2)(X_2Y_3 - X_3Y_2 + X_1Y_4 - X_4Y_1) - (X_1X_2 + Y_1Y_2)(X_1Y_3 - X_3Y_1)$$
$$- 3(X_1Y_2 - X_2Y_1)(X_2^2 + Y_2^2 + X_1X_3 + Y_1Y_3) = 0,$$

which for our T-position, in view of (12.7), becomes

$$3y(x^2 + y^2) - 3b_1^2x + 4b_1b_3y + a_4b_1^2 = 0. \tag{12.9}$$

This represents a cubic curve; which like the hyperbola passes through

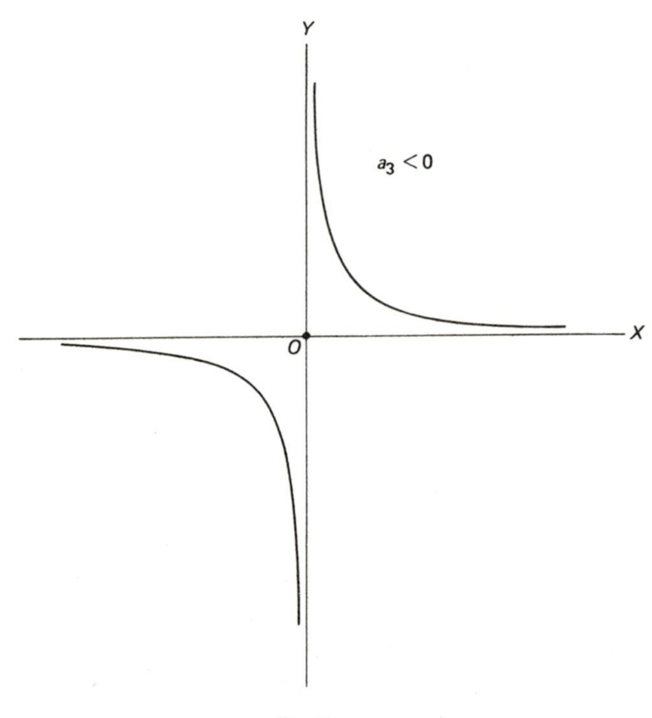

Fig. 72.

$P = (1, 0, 0)$ and has $y = 0$ as its tangent at P. Hence, of the six intersections of (12.8) and (12.9), two coincide with P. Therefore, there are four Burmester points, the same number as in the general case.

Example 122. Show that the y-coordinates of the Burmester points satisfy the equation $3y^4 + 4b_1b_3y^2 + a_4b_1^2y + (1/3)a_3b_1^2(a_3 + 3b_1) = 0$.
Example 123. Show that the center of gravity of the four points is on the pole tangent (VELDKAMP [1963, p. 131]).

For a T-position of the second kind we have $a_1 = b_1 = 0$, which implies that any point of E is a pole or, in other words, the path of every point has in the zero-position a cusp. From (12.1), (12.2) and (12.3) it follows that up to the fourth order terms we have

$$X = x + \tfrac{1}{2}(-y + a_2)t^2 + (1/6)a_3t^3 + (-(1/8)x + (1/24)a_4)t^4,$$
$$Y = y + \tfrac{1}{2}(x + b_2)t^2 + (1/6)b_3t^3 + (-(1/8)y + (1/24)b_4)t^4. \tag{12.10}$$

There is a unique point x, y for which $X_2 = Y_2 = 0$; if we take it as the origin we have $a_2 = b_2 = 0$. Moreover, since $X_3 = a_3$, $Y_3 = b_3$, the vector (X_3, Y_3) is the same for every point. We can take its direction as the Y-axis, if a_3 and b_3 are not both zero, which then makes $a_3 = 0$, $b_3 > 0$. Our table now reads

$$(12.11) \quad \begin{matrix} X_0 = x, & X_1 = 0, & X_2 = -y, & X_3 = 0, & X_4 = -3x + a_4, \dots \\ Y_0 = y, & Y_1 = 0, & Y_2 = x, & Y_3 = b_3, & Y_4 = -3y + b_4, \dots \end{matrix}$$

For a further discussion of this very special case see VELDKAMP ([1963, p. 135–142]).

13. Time as a parameter

Until now we have dealt in this chapter with geometrical, that is time-independent, kinematics. We suppose now that the displacements are such that the three parameters (ϕ, a, b) are functions of the time t. If in the zero-position $\dot{\phi} \neq 0$ we are still able to take ϕ as the geometrical parameter and treat a, b as functions of ϕ. The new element in the discussion is that ϕ is now a function of t (and therefore by implication so are a, b).

The most important properties of the time-dependent motion of a point, of the moving plane, are its velocity and its acceleration, and to some degree its "second acceleration" (the third derivative with respect to t of the position of a point). We introduce, at $t = 0$, the notations $\dot{\phi} = \omega (\neq 0)$, $\ddot{\phi} = \varepsilon$, $\dddot{\phi} = \delta$.

The velocity distribution is very simple. We have

$$(13.1) \qquad \dot{X} = -\omega y, \qquad \dot{Y} = \omega x;$$

which is a rotation about O with angular velocity ω. Indeed, (13.1) follows from $\dot{X} = x(\cos\phi)^{\cdot} - y(\sin\phi)^{\cdot} + \dot{a}$, $\dot{Y} = x(\sin\phi)^{\cdot} + y(\cos\phi)^{\cdot} + \dot{b}$ since for $t = 0$: $(\cos\phi)^{\cdot} = 0$, $(\sin\phi)^{\cdot} = \omega$, $\dot{a} = a_1\omega = 0$, $\dot{b} = b_1\omega = 0$.

Similarly for the acceleration we note that $(\cos\phi)^{\cdot\cdot} = -\omega^2$, $(\sin\phi)^{\cdot\cdot} = \varepsilon$, $\ddot{a} = a_2\omega^2 + a_1\varepsilon = 0$, $\ddot{b} = b_2\omega^2 + b_1\varepsilon = b_2\omega^2$; it follows

$$(13.2) \qquad \ddot{X} = -\omega^2 x - \varepsilon y, \qquad \ddot{Y} = \varepsilon x - \omega^2 y + \omega^2 b_2.$$

From these expressions it follows that the acceleration of the velocity pole of E is always $\omega^2 b_2$, directed toward the inflection pole (i.e., along the positive Y-axis). Clearly the velocity pole cannot also be an acceleration pole unless $\omega = 0$ or $b_2 = 0$.

As $\omega \neq 0$, we have $\omega^4 + \varepsilon^2 \neq 0$. Hence there is one point, the acceleration pole Γ, with zero acceleration, viz

(13.3) $$x_\Gamma = -\omega^2 \varepsilon b_2/(\omega^4 + \varepsilon^2), \quad y_\Gamma = \omega^4 b_2/(\omega^4 + \varepsilon^2).$$

If we introduce polar coordinates r, θ, with $x = r\cos\theta$, $y = r\sin\theta$, where v_r, v_θ, are the components of the velocity vector $\boldsymbol{v}$ in the r- and θ-direction and α_r, α_θ those of the acceleration $\boldsymbol{\alpha}$, we can project the Cartesian components of (13.1) and (13.2) onto the polar directions and obtain

(13.4) $$v_r = 0, \quad v_\theta = \omega r, \quad \alpha_r = \omega^2(-r + b_2\sin\theta), \quad \alpha_\theta = \varepsilon r + \omega^2 b_2 \cos\theta.$$

Obviously α_r and α_θ are, in our canonical system, the components of $\boldsymbol{\alpha}$ normal and tangential to the path respectively. We have $\alpha_r = 0$ if $r = b_2\sin\theta$, which implies that the locus of the points for which the normal component of acceleration vanishes is the inflection circle; this could have been expected since $\alpha_r = (v^2/\rho)$, ρ being the radius of curvature of the path. The locus of the points with $\alpha_\theta = 0$ has the equation $\varepsilon r + \omega^2 b_2\cos\theta = 0$, which represents for $\varepsilon \neq 0$ the circle n (sometimes called *the normal circle*), tangent at O to O_Y, with diameter $\omega^2 b_2/|\varepsilon|$ (Fig. 73); for $\varepsilon = 0$ the equation represents the O_Y-axis.

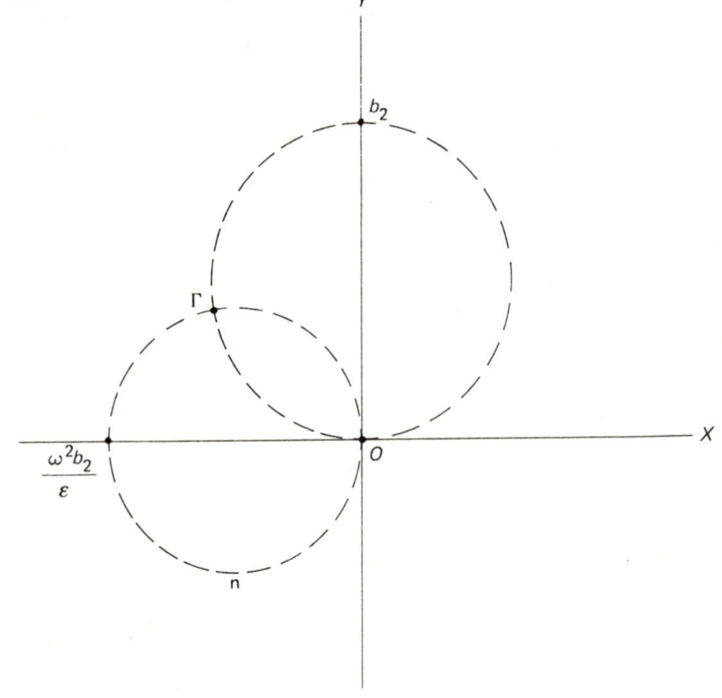

Fig. 73.

The acceleration pole Γ is therefore the intersection, different from O, of the inflection circle and the circle n. (Also associated with these circles are the names of Bresse and De La Hire.)

Example 124. Show that the point (13.3) is on the inflection circle $x^2+y^2-b_2y=0$.
Example 125. Show that Γ coincides with the inflection pole H if $\varepsilon=0$.
Example 126. Derive the (first) Euler–Savary equation (9.6) from (13.4) using $\alpha_r=(v^2/\rho)$ and $\bar{r}=r+\rho$.

From the definition of Γ it follows from (13.2) that

(13.5) $$0=-\omega^2x_\Gamma-\varepsilon y_\Gamma, \qquad 0=\varepsilon x_\Gamma-\omega^2y_\Gamma+\omega^2b_2;$$

subtracting this from (13.2) yields

(13.6) $$\ddot{X}=-\omega^2(x-x_\Gamma)-\varepsilon(y-y_\Gamma), \quad \ddot{Y}=\varepsilon(x-x_\Gamma)-\omega^2(y-y_\Gamma).$$

Hence, if we translate the two frames so that the origin O is displaced to Γ, the new coordinates of a point being $\xi=x-x_\Gamma$, $\eta=y-y_\Gamma$, the acceleration distribution is described by

(13.7) $$\ddot{X}=-\omega^2\xi-\varepsilon\eta, \qquad \ddot{Y}=\varepsilon\xi-\omega^2\eta.$$

If the polar coordinates of a point A are r, θ (Fig. 74) then the components α_1, α_2 of the acceleration vector along ΓA and orthogonal to ΓA are

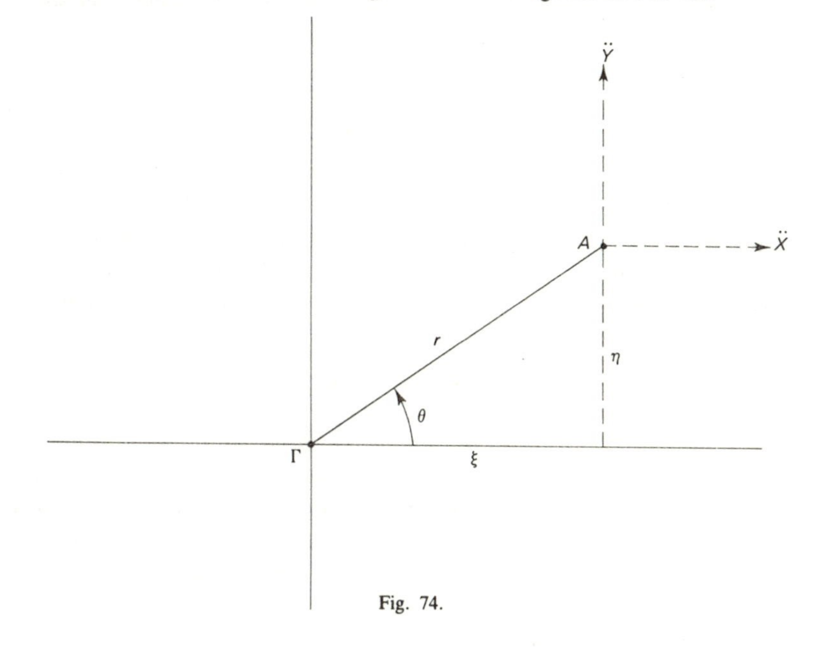

Fig. 74.

$$\alpha_1 = \ddot{X}\cos\theta + \ddot{Y}\sin\theta = -\omega^2 r,$$
$$(13.8) \qquad \alpha_2 = -\ddot{X}\sin\theta + \ddot{Y}\cos\theta = \varepsilon r,$$

which gives us a simple description of the vector $\boldsymbol{\alpha}$ with respect to the acceleration pole Γ.

It follows that the velocity distribution is the same as that of a plane rotating about O with angular velocity ω; the acceleration distribution is that of a plane rotating about Γ with angular velocity ω and angular acceleration ε.

Example 127. Show that the angle β between the vector $\boldsymbol{\alpha}$ and $A\Gamma$ is the same for any point and that $\tan\beta = (-\varepsilon/\omega^2)$.
Example 128. Show that $\angle\Gamma OO_Y$ is equal to β.
Example 129. Verify that if $\varepsilon = 0$ the acceleration of every point is directed towards the inflection pole H.

The components of the second acceleration (the "pulse" or "jerk"; *German*, Rück) of the point x, y are

$$\dddot{X} = x(\cos\phi)^{\dddot{}} - y(\sin\phi)^{\dddot{}} + \dddot{a},$$
$$(13.9) \qquad \dddot{Y} = x(\sin\phi)^{\dddot{}} + y(\cos\phi)^{\dddot{}} + \dddot{b},$$

which (instantaneously at $t = 0$, for the canonical frames) yield after some algebra:

$$\dddot{X} = -3\omega\varepsilon x - (\delta - \omega^3)y + a_3\omega^3,$$
$$(13.10) \qquad \dddot{Y} = (\delta - \omega^3)x - 3\omega\varepsilon y + (3b_2\varepsilon + b_3\omega^2)\omega.$$

Example 130. Show that there is only one point with zero second acceleration, unless $\varepsilon = \delta - \omega^3 = 0$; in the latter case the second acceleration of every point is equal.

Time-dependent instantaneous kinematics for a T-position is very simple. It is easy to derive by direct differentiation of (12.3), if we let the parameter t represent the time and set $\phi = \frac{1}{2}\varepsilon t^2 + (1/6)\delta t^3$. We find, at $t = 0$, using the same canonical frame as for (12.7),

$$\dot{X} = 0, \qquad \ddot{X} = -\varepsilon y, \qquad \dddot{X} = -\delta y + a_3,$$
$$(13.11) \qquad \dot{Y} = b_1, \qquad \ddot{Y} = \varepsilon x, \qquad \dddot{Y} = \delta x + b_3.$$

From which it may be concluded that: the velocity is $(0, b_1)$ for every point. For $\varepsilon \neq 0$ the acceleration pole coincides with the origin; the acceleration of a point A is orthogonal to OA and equal to $\varepsilon(OA)$; if $\varepsilon = 0$ the acceleration of every point is zero. If $\delta \neq 0$ the pole of the second acceleration is $(-b_3/\delta, a_3/\delta)$.

CHAPTER IX

SPECIAL MOTIONS

1. General remarks

In all preceding chapters we dealt with motions in the general sense, deriving theorems valid for any motion whatsoever. We consider now specified motions, thus specializing the general theory. Our examples, besides being often interesting in themselves, also serve as specific illustrations of our general results. The present chapter borders more than any other upon what may be called applied kinematics, a subject not explicitly inserted in this book. Kinematics has been applied for engineering purposes in two ways: there are machines producing motions and it may be useful to investigate or even to predict the kinematic properties of these motions; in doing so one deals with kinematic analysis. A more modern application is kinematic synthesis: if one wants to construct a machine such that it generates a prescribed motion (say, the path of a certain point or its velocity) kinematics may be used as a tool. In this respect the word machine has a rather restricted sense: its stands for an abstract model, an approach to reality, consisting of a finite number of rigid bodies linked together in some way, thus restricting the movability of the components so that the system has, in general, one, and sometimes more, degrees-of-freedom. Such a set is called a mechanism. As the shape of the bodies is not of primary importance, each of them may be considered as a moving space and is therefore subject to the kinematic theory as developed in the preceding chapters.

A motion can be defined in various ways. It may be done in a purely analytical manner by giving (in the general displacement formula $\boldsymbol{P} = \mathbf{A}\boldsymbol{p} + \boldsymbol{d}$) the matrix $\mathbf{A}$ and the vector $\boldsymbol{d}$ explicitly as specified functions of a parameter. Examples will follow in Section 6. Usually, however, the motion is described in geometrical terms. We know that a 3-dimensional space E moving with respect to a fixed space Σ has six degrees of freedom. Hence *its motion is specified if the displacements of E satisfy five conditions.* The condition that a specific point A of E remains on a given surface of Σ (a plane, a sphere, etc.) is a single one; if the path of A is prescribed it satisfies two conditions; if it is permanently at rest three conditions must be satisfied. If a line l of E is constrained to intersect permanently a given line of Σ, or to be tangent to a given surface, or more general, to belong to a given complex of lines in Σ, it must satisfy one condition; if l must obey two such requirements (or more general, if it must belong to a given congruence of lines in Σ) the condition is equivalent to two single ones; the number is three if l is compelled to be a generator of a ruled surface (for instance a ruled quadric surface or a pencil of lines); the number is four if l, as a whole, must be invariant. Analogously for a moving plane: passing through a fixed point or being tangent to a given surface imposes one conditon, belonging to a given developable of planes (for instance if it must pass through a given line of Σ) is equivalent to two, having an invariant position is equivalent to three single conditions.

The constraints may be on more complicated elements than the foregoing elementary examples of points, lines or planes: if a sphere must remain tangent to a given plane we have one condition, if the sphere rolls (without slipping) on it the condition is threefold. All kinds of combinations may exist. If the total number of (equivalent) single conditions is equal to five, there is a reasonable possibility that a motion satisfying the constraints exists, but this is by no means a certainty. The set of conditions may be dependent (which would mean that the degree of freedom for E is more than one) or contradictory (and no motion possible), or it may not have any *real* solution. A trivial example of the latter circumstance: if two points at a distance a are compelled to remain on two skew lines respectively, with distance b, obviously no motion is possible if $a < b$.

We may add that a motion may be defined also by its kinematic properties, for instance by its two axodes. A combination of analytical and geometrical data is also possible; the motion of Section 2 is an example of this kind.

Example 1. Show that in the following examples the primitive counting of constraints gives the (decisive) number five: a) five points each remain in a given plane; b) two points remain on a circle, a third on a sphere; c) a point remains on a line, two lines each intersect a given line, a plane passes through a point; d) two planes each pass through a line, a third through a point.

There are some generally important motions called *algebraic motions*: these are characterized by the properties that they generate algebraic loci such that the path of every point is an algebraic curve, every plane describes an algebraic developable, every line an algebraic ruled surface. In Section 11 we shall see that there exist in plane kinematics certain formulas giving the order and the circularity of any path if these characteristic numbers are known for two paths. No analogously general relations seem to exist in spatial kinematics, although there are some for rather simple cases, see for example DARBOUX [1897] and SOMMERVILLE [1934, p. 382]. However, there do exist simple equalities between characteristic numbers for a motion and the inverse motion, given by CHASLES [1837]:

Let the paths of the direct motion be of order n. This means: a fixed plane α in Σ has n intersections with every path; in other words, a moving point P coincides in n positions with a point of α. For the inverse motion P is a fixed point and α a moving plane, and the conclusion is: in n positions of α it passes through P or, in other words, the class of the developable described by α is equal to n. Conversely the class of a developable described by a moving plane under the direct motion is equal to the order of the path of a point under the inverse motion.

If the order of the ruled surface S generated by a moving line l under the direct motion is m this means that S has m points in common with a line l′ of Σ, that is: l intersects l′ in m positions of the moving space E. It follows immediately that, for the inverse motion, any line l′ of Σ generates a ruled surface of order m.

We will refer to the foregoing two paragraphs as Chasles' theorem.

2. The Frenet–Serret motion

The first special motion we study furnishes an illustration of the general theory wherein the axodes and their striction curves can be determined explicitly and yet are not trivial. This motion is defined in terms of a space curve Γ fixed in Σ:

The motion is such that the moving frame o_{xyz} moves with o along Γ while rotating so that the x and y axes always coincide with, respectively, the tangent and principal normal of Γ. This means that as o coincides with a point Q of Γ the o_{xyz} frame coincides with the Frenet–Serret trihedron at Q: $Q_{\xi\eta\rho}$. This trihedron consists of the tangent Q_ξ, the principal normal Q_η, and the binormal Q_ρ, which are three mutually orthogonal axes. Obviously, the geometry of this motion is completely defined by Γ.

If $\boldsymbol{P}$ and $\boldsymbol{p}$ are the position vectors, of a moving point, measured in Σ and E respectively, we know that for any motion

$$P = \mathbf{A}p + d \tag{2.1}$$

$\mathbf{A}$ is an orthogonal matrix and $\boldsymbol{d}$ is the displacement vector of the origin, both depend upon the motion parameter.

For our special motion, because o moves along Γ, $\boldsymbol{P} = \boldsymbol{d}$ represents the space curve Γ. We will use the arc length, s, of Γ as the motion parameter, and use primes to denote derivatives with respect to s.

If $\boldsymbol{t} = (t_1, t_2, t_3)$, $\boldsymbol{n} = (n_1, n_2, n_3)$, $\boldsymbol{b} = (b_1, b_2, b_3)$ are the unit vectors along Q_ξ, Q_η and Q_ρ, differential geometry gives us

$$\boldsymbol{t} = \boldsymbol{d}', \quad \boldsymbol{n} = r\boldsymbol{d}'', \quad \boldsymbol{b} = \boldsymbol{t} \times \boldsymbol{n}, \tag{2.2}$$

$r = k^{-1}$ being the radius of curvature of Γ. Then we have

$$\mathbf{A} = \begin{Vmatrix} t_1 & n_1 & b_1 \\ t_2 & n_2 & b_2 \\ t_3 & n_3 & b_3 \end{Vmatrix}. \tag{2.3}$$

From (2.1) it follows

$$\boldsymbol{P}' = \mathbf{A}'\boldsymbol{p} + \boldsymbol{d}'. \tag{2.4}$$

The Frenet–Serret formulas read

$$\boldsymbol{t}' = k\boldsymbol{n}, \quad \boldsymbol{b}' = -\tau\boldsymbol{n}, \quad \boldsymbol{n}' = -k\boldsymbol{t} + \tau\boldsymbol{b}, \tag{2.5}$$

k being the curvature and τ the torsion of the curve Γ. From (2.1) we obtain

$$\boldsymbol{p} = \mathbf{A}^{-1}(\boldsymbol{P} - \boldsymbol{d}), \tag{2.6}$$

and as $\mathbf{A}^{-1} = \mathbf{A}^{\mathrm{T}}$, we have from (2.4) and (2.6), eliminating $\boldsymbol{p}$:

$$\boldsymbol{P}' = \mathbf{\Omega}(\boldsymbol{P} - \boldsymbol{d}) + \boldsymbol{d}', \tag{2.7}$$

with $\mathbf{\Omega} = \mathbf{A}'\mathbf{A}^{\mathrm{T}}$, or explicitly, by means of (2.5),

$$\mathbf{\Omega} = \begin{Vmatrix} 0 & -(b_3k + t_3\tau) & (b_2k + t_2\tau) \\ (b_3k + t_3\tau) & 0 & -(b_1k + t_1\tau) \\ -(b_2k + t_2\tau) & (b_1k + t_1\tau) & 0 \end{Vmatrix}, \tag{2.8}$$

a skew matrix, as could be expected.

The components, in Σ, of the (geometric) instantaneous angular velocity vector $\boldsymbol{\Omega}$ are therefore

$$\Omega_X = b_1k + t_1\tau, \quad \Omega_Y = b_2k + t_2\tau, \quad \Omega_Z = b_3k + t_3\tau. \tag{2.9}$$

Its components with respect to the moving frame follow from $\boldsymbol{\omega} = \mathbf{A}^{\mathrm{T}}\mathbf{\Omega}$, and we obtain for the vector $\boldsymbol{\omega}$:

$$\omega_x = \tau, \quad \omega_y = 0, \quad \omega_z = k. \tag{2.10}$$

So much for the direction of the screw axis. To determine its position in the o_{xyz}-frame, suppose that it intersects the plane o_{xy} at $(x_0, y_0, 0)$ and that the translation parallel to the screw axis is $(\tau\sigma_0, 0, k\sigma_0)$. The velocity vector of the origin would be $(ky_0 + \tau\sigma_0, -kx_0, -\tau y_0 + k\sigma_0)$ and as (by virtue of $\boldsymbol{P}_0' = \boldsymbol{d}' = \boldsymbol{t}$) it must be $(1,0,0)$ we have $x_0 = 0$, $y_0 = k/(k^2 + \tau^2)$, $\sigma_0 = \tau/(k^2 + \tau^2)$. The conclusion is (Fig. 75): the instantaneous screw axis intersects o_y, the principal normal, at the point S, such that $oS = k/(k^2 + \tau^2)$; it is parallel to the plane o_{xz}; the components of $\boldsymbol{\omega}$ are $(\tau, 0, k)$ and that of the translation vector $[\tau^2/(k^2 + \tau^2), 0, k\tau/(k^2 + \tau^2)]$. Note that, as developed here, the angular and the linear velocities are geometrical concepts with the arc length of Γ acting as "time".

The case $\tau = 0$ (Γ is then a plane curve) will be dealt with in Section 11.

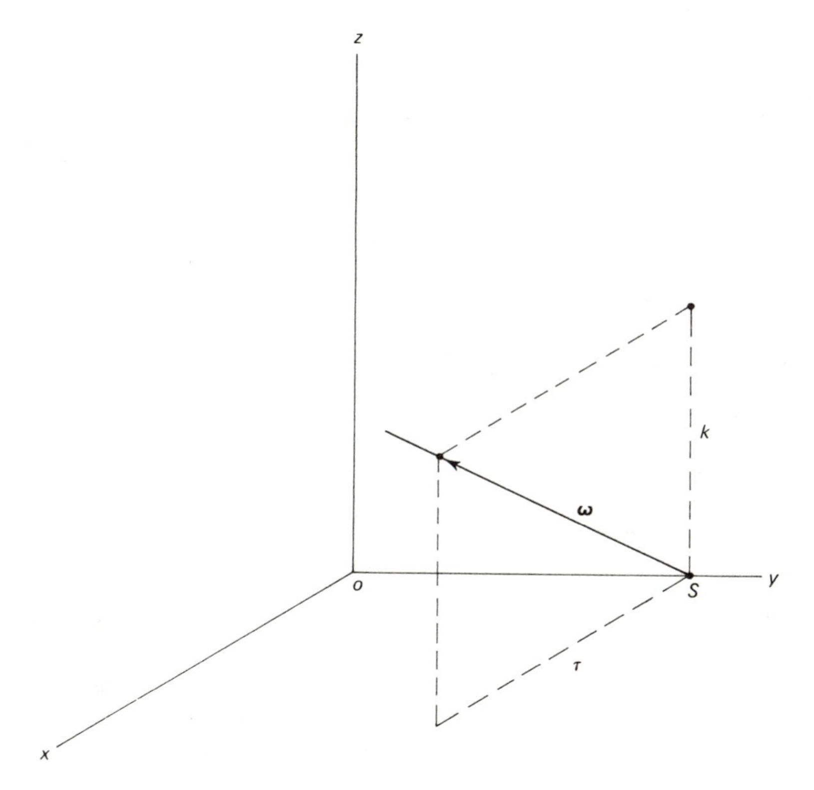

Fig. 75.

As the instantaneous screw always intersects the principal normal at a right angle, its locus—the moving axode—is a special type of ruled surface called a conoid. The axode in the fixed space Σ follows by means of (2.1).

Example 2. Determine the pitch of the instantaneous screw.

Example 3. Show that the Plücker coordinates p_{ij} of the generators of the moving axode are

$$p_{14} = \tau, \quad p_{24} = 0, \quad p_{34} = k,$$

$$p_{23} = -k^2/(k^2+\tau^2), \quad p_{31} = 0, \quad p_{12} = k\tau/(k^2+\tau^2).$$

Example 4. If $\boldsymbol{d} = (a, b, c)$, show that the coordinates P_{ij} of the fixed axode are

$$P_{14} = b_1 k + t_1 \tau, \quad \text{and cyclically for } P_{24} \text{ and } P_{34},$$

$$P_{23} = (cb_2 - bb_3)k + (ct_2 - bt_3)\tau + (b_1\tau - t_1 k)k/(k^2+\tau^2),$$

and cyclically for P_{31} and P_{12}.

Verify that $\Sigma P_{14} P_{23} = 0$.

Example 5. Determine the screw axis and the axodes if Γ is a helical curve on a general (i.e., not necessarily right-angular) cylinder (k/τ being a constant).

Example 6. Determine the screw axis and the axodes if Γ is a common helix (k and τ being both constants).

The principal normal is the common perpendicular of two consecutive screw axes; hence S is the striction point. The conclusion is that the striction curve in the moving space is the straight line $x = z = 0$.

Example 7. Show that a parametric representation of the striction curve on the fixed axode is

$$X = a + n_1 q, \quad Y = b + n_2 q, \quad Z = c + n_3 q, \quad q = k/(k^2+\tau^2).$$

This motion has been studied by DARBOUX [1887] who introduced it in order to derive the Frenet–Serret formulas by kinematic means. The geometric angular-velocity vector $\boldsymbol{\Omega}$ is named, after him, the Darboux vector.

3. Darboux's motion

In a classical investigation DARBOUX [1881] has completely solved the problem of determining all possible motions of E with respect to Σ such that the path of every point of E is a plane curve. There are of course trivial examples, for instance the case in which the planes of the paths are all parallel. We suppose that every plane of Σ containes the path of at least one point. Following Darboux we consider first the inverse motion, which has also been studied by MANNHEIM [1889]; it has some interest in itself and we deal with it in Section 4. Obviously it is a motion of Σ with respect to E such that any plane α_1 of Σ passes permanently through a fixed point P_1 of E. Let α_2 be a plane of Σ parallel to α_1 at the distance d; α_2 passes through P_2. Hence α_1 describes a developable of planes in E, all passing through P_1 and such that the distance of P_2 to any plane of the developable equals d. This implies that the developable consists of the tangent planes of a circular cone with vertex P_1. As α_1 was arbitrarily chosen, any plane of Σ generates such a cone. In view of Chasles' theorem of Section 1, we conclude that (if the motion under consideration is possible) the paths of the direct motion are conics.

Without any loss of generality we may suppose that α_1 has the equation $Z + e_1 = 0$ and therefore the coordinates $U_1 = 0$, $U_2 = 0$, $U_3 = 1$, $U_4 = e_1$; α_2 is therefore the plane $(0, 0, 1, e_2)$; let α_1 pass through $P_1 = (0, 0, z_1)$, α_2 through $(0, 0, z_2)$. The formulas for the transformation of plane coordinates give us then that a plane $(0, 0, 1, e_i)$ passes during the motion through a point $(0, 0, z_i)$. In other words: if the planes of a parallel set of planes each pass through a fixed point of E these points are all on a line $\mathrm{l}(\alpha)$, associated with the set. The angle $\theta(\alpha)$ between $\mathrm{l}(\alpha)$ and any plane of the set is constant during the motion and so is therefore the angle between $\mathrm{l}(\alpha)$ and its projection $\mathrm{l}'(\alpha)$ on such a plane. So far we have only derived properties of the motion that deal with the direction of lines, that is with the motion of the points in the plane at infinity, or (if we take an arbitrary origin O and speak in analytic terms) with the matrix $\mathbf{A}$ of the motion equation. Isolating the rotational part $\mathbf{A}^T$ for this inverse motion we obtain that this corresponds to a motion about O such that any line l' through O describes, during the motion, a circular cone with axis l. The question whether such a motion is possible is equivalent to one in spherical kinematics: can a sphere move about its center in such a way that every point on it describes a circle? The answer is obviously positive: this is of course the case when the sphere rotates about a fixed line through O. We shall prove that this is the only possibility. This may be done by means of formula (6.11) of Chapter VII which gives us the locus of points on the sphere with (instantaneously) four coplanar consecutive positions. In our case every point must satisfy this equation, which implies that $\varepsilon = (\gamma_X = \gamma_Y) = 0$. Hence during the motion we must have $\varepsilon = 0$ and therefore $\boldsymbol{\omega}$ is a constant. The conclusion is that the lines $\mathrm{l}(\alpha)$ for various sets of parallel planes are all parallel.

If $\mathbf{A}^T$ represents the set of rotations about a fixed axis a the same holds for $\mathbf{A}$; a is a fixed line in both Σ and E. We introduce frames such that a coincides with both O_Z and o_z; we take the rotation angle ϕ as the motion parameter. Considering now the direct motion and adding the translation part, we conclude that the motion is necessarily of the following type

$$
(3.1)\qquad
\begin{aligned}
X &= x\cos\phi - y\sin\phi + d_1(\phi),\\
Y &= x\sin\phi + y\cos\phi + d_2(\phi),\\
Z &= z + d_3(\phi).
\end{aligned}
$$

We remark that O_Z and o_z are chosen parallel to the axis a but the origins and the other coordinate axes are still arbitrary. The functions $d_i(\phi)$ will now be determined. Let (x_i, y_i, z_i) be three points of E and $(U_{1i}, U_{2i}, U_{3i}, U_{4i})$ the planes of their paths in Σ $(i = 1,2,3)$. This implies that

$$
(3.2)\qquad
\begin{aligned}
&U_{1i}[x_i\cos\phi - y_i\sin\phi + d_1(\phi)] + U_{2i}[x_i\sin\phi + y_i\cos\phi + d_3(\phi)]\\
&\quad + U_{3i}[z_i + d_3(\phi)] + U_{4i} = 0,
\end{aligned}
$$

must be satisfied for $i = 1,2,3$ and for all values of ϕ. This gives us three linear equations for d_1, d_2, d_3. Furthermore we may suppose that the determinant $|U_{1i}\ U_{2i}\ U_{3i}|$ is unequal to zero; in the other case the three planes would be parallel to one line and we started from the assumption that all planes of Σ contain at least one path. Then by solving (3.2), we would obtain d_i as linear functions of $\cos\phi$ and $\sin\phi$. Hence (3.1) becomes

$$
(3.3)\qquad
\begin{aligned}
X &= x\cos\phi - y\sin\phi + c_{11}\cos\phi + c_{12}\sin\phi + c_{13},\\
Y &= x\sin\phi + y\cos\phi + c_{21}\cos\phi + c_{22}\sin\phi + c_{23},\\
Z &= z + c_{31}\cos\phi + c_{32}\sin\phi + c_{33},
\end{aligned}
$$

with constant c_{ij}.

We make use of the following coordinate transformations

$$
\begin{aligned}
&X = X' + c_{13}, \quad Y = Y' + c_{23}, \quad Z = Z';\\
&x = x' - c_{11}, \quad y = y' - c_{21}, \quad z = z' - c_{31} - c_{33},
\end{aligned}
$$

with the result

$$
\begin{aligned}
X' &= x'\cos\phi - y'\sin\phi + e_1\sin\phi,\\
Y' &= x'\sin\phi + y'\cos\phi + e_2\sin\phi,\\
Z' &= z' + c_{32}\sin\phi - c_{31}(1-\cos\phi),\\
e_1 &= c_{12} + c_{21}, \quad e_2 = -c_{11} + c_{22}.
\end{aligned}
$$

Furthermore we rotate the coordinate systems so that

$$
\begin{aligned}
X'' &= X'\cos\beta + Y'\sin\beta, \quad Y'' = -X'\sin\beta + Y'\cos\beta,\\
x'' &= x'\cos\beta + y'\sin\beta, \quad y'' = -x'\sin\beta + y'\cos\beta,
\end{aligned}
$$

with $\tan\beta = -e_1/e_2$, then suppressing the double-primes the motion is expressed by

$$(3.4)\qquad \begin{aligned} X &= x\cos\phi - y\sin\phi,\\ Y &= x\sin\phi + y\cos\phi + a\sin\phi,\\ Z &= z + b\sin\phi + c(1-\cos\phi), \end{aligned}$$

where a, b, c are constants. Note that the two frames coincide for $\phi = 0$.

It is easy to show that all paths of (3.4) are plane curves. Indeed, using $\tan(\phi/2) = t$, $\sin\phi = 2t/(1+t^2)$, $\cos\phi = (1-t^2)/(1+t^2)$, the homogeneous coordinates X, Y, Z, W (in Σ) can be expressed as quadratic functions of the parameter t; hence all paths are conics and therefore plane curves. Moreover, as their intersections with $W = 0$ follow from $t = \pm i$ they are ellipses.

Example 8. Show that two points on a parallel to o_z describe congruent paths.
Example 9. Show that the points of the line $x = -(a^2+b^2-c^2)/(2a)$, $y = bc/a$, describe circles.
Example 10. Show that the path of a point in $w = 0$ is a conic in $W = 0$; when does it degenerate?

If (U_1, U_2, U_3, U_4) is the plane in Σ of the path of the point (x, y, z, w) in E, we must have $\Sigma U_iX = 0$ for all values of ϕ. That gives us three equations, linear in U_i and linear in x, y, z, w as well. If we solve with respect to U_i the solution is in general

$$(3.5)\qquad \begin{aligned} &U_1 = (cx + by + acw)w^2, \quad U_2 = (-bx + cy)w^2,\\ &U_3 = (x^2+y^2+axw)w, \quad U_4 = -(x^2+y^2+axw)(z+cw), \end{aligned}$$

and if we solve with respect to x, y, z, w we obtain

$$(3.6)\qquad \begin{aligned} &x = -(aU_2^2 + bU_2U_3 - cU_1U_3)U_3, \quad y = (aU_1U_2 + bU_1U_3 + cU_2U_3)U_3,\\ &z = -(cU_3 + U_4)(U_1^2 + U_2^2), \qquad w = (U_1^2 + U_2^2)U_3. \end{aligned}$$

Hence there exists a birational cubic relationship between the points of the moving space and the planes of their paths. There are singularities in both directions. If in (3.5) we have $U_i = 0$ $(i = 1,2,3,4)$ the plane is undetermined, an infinity of planes passes through the path, which is therefore a straight line.

Example 11. Show that the locus of these exceptional points is the line m, parallel to o_z, given by $x = -ac^2/(b^2+c^2)$, $y = -abc/(b^2+c^2)$; prove that the path of a point A on m is given by $X = acu/(b^2+c^2)$, $Y = abu/(b^2+c^2)$, $Z = z + c + u$, with $u = b\sin\phi - c\cos\phi$; show that the path is a line segment and determine its end points; show that the straight paths are parallel and all in one plane; show that all line segments have the same length.
Example 12. Show that two points of E on a line parallel to o_z have paths in parallel planes.
Example 13. Show that circular paths lie in the set of parallel planes given by

$$U_1 = 2ac, \quad U_2 = 2ab, \quad U_3 = -a^2+b^2+c^2, \quad U_4 = -(z+c)(-a^2+b^2+c^2).$$

From (3.6) it follows that a plane in Σ contains in general the path of *one* point. Those planes for which the moving point is not uniquely determined are exceptional, these are the planes $U_3 = U_4 = 0$ and the planes through O_Z.

Example 14. Show that the exceptional plane $(U_1, U_2, 0, 0)$ contains the path of all points on the line $x = -aU_2^2/(U_1^2 + U_2^2)$, $y = aU_1U_2/(U_1^2 + U_2^2)$; consider in particular the plane $(b, -c, 0, 0)$.

It follows at once from (3.4) that during the motion the coordinate axis o_z remains parallel to O_Z and that the angular velocity vector, in the fixed and in the moving space, is parallel to these axes. This implies that the fixed axode and the moving axode are both cylinders, with their generators perpendicular to O_{XY} and o_{xy} respectively. Hence the instantaneous screw axis is the locus of the points with velocity parallel to O_Z. It follows from (3.4) that the geometric velocity is

$$\begin{aligned} \dot{X} &= -x \sin\phi - y\cos\phi, \\ \dot{Y} &= x\cos\phi - y\sin\phi + a\cos\phi, \\ \dot{Z} &= b\cos\phi + c\sin\phi. \end{aligned} \tag{3.7}$$

The equations $\dot{X} = \dot{Y} = 0$ are satisfied by

$$x = -a\cos^2\phi = -\tfrac{1}{2}a(1+\cos 2\phi), \quad y = a\sin\phi\cos\phi = \tfrac{1}{2}a\sin 2\phi. \tag{3.8}$$

The equation of the moving axode C_m is therefore

$$x^2 + y^2 + ax = 0, \tag{3.9}$$

which represents a circular cylinder with diameter a, tangent to the o_{yz}-plane along o_z.

From (3.4) and (3.8) it follows that the position of the instantaneous screw axis in the fixed space is given by

$$X = -a\cos\phi, \quad Y = a\sin\phi, \tag{3.10}$$

and the fixed axode C_f is

$$X^2 + Y^2 = a^2, \tag{3.11}$$

again a circular cylinder, but with diameter $2a$. The moving axode is internally tangent to the fixed axode. The Darboux motion is generated if C_m rolls along the inside of C_f while it translates parallel to the axes of the cylinders; the magnitude of this translation depends on the rotation angle ϕ and is equal to $b\sin\phi + c(1-\cos\phi)$.

Example 15. Prove that the path of a point on C_m is in a plane through O_Z; all points on a generator of C_m have their paths in the same plane. (We knew already that planes through O_Z are exceptional, containing an infinity of paths.)

Example 16. Show that the locus m (mentioned in Example 11), of the points with a linear path, is a generator of C_m.

Example 17. Consider the path of a point on the axis of the cylinder C_m; show that its projection on O_{XY} is a circle.

Example 18. If O_Z is a vertical, prove that all moving points have their highest (lowest) position at the same time, i.e., if $\cos\phi = \pm c/N$, $\sin\phi = \mp b/N$, $N = (b^2+c^2)^{1/2}$; show that at these moments the velocity distribution is that of a rotation.

Example 19. Determine the center of the ellipse described by (x, y, z).

From (3.4) we obtain the transformation from the coordinates of a line p_{ij} of E to its coordinates P_{ij} in Σ:

$$
\begin{aligned}
P_{14} &= p_{14}\cos\phi - p_{24}\sin\phi,\\
P_{24} &= p_{14}\sin\phi + p_{24}\cos\phi,\\
P_{34} &= p_{34},\\
P_{23} &= -p_{31}\sin\phi + p_{23}\cos\phi - ap_{34}\sin\phi + k(\phi)(p_{14}\sin\phi + p_{24}\cos\phi),\\
P_{31} &= p_{31}\cos\phi + p_{23}\sin\phi + k(\phi)(-p_{14}\cos\phi + p_{24}\sin\phi),\\
P_{12} &= p_{12} + ap_{14}\sin\phi\cos\phi - ap_{24}\sin^2\phi,
\end{aligned}
\tag{3.12}
$$

with $k(\phi) = b\sin\phi + c(1-\cos\phi)$.

If we rationalize $\sin\phi$ and $\cos\phi$ by means of $\tan(\phi/2) = t$, the P_{ij} are seen to be functions of t of the fourth degree; (P_{14}, P_{24}, and P_{34} contain the factor $(1+t^2)$). Hence a line l of E describes in Σ a (rational) ruled surface R_4 of order four. As any point A on l describes an ellipse, the surface contains a system of ∞^1 ellipses. It may be generated by the joins of corresponding points of two projective ellipses (corresponding points having the same value of t) and is therefore a well-known surface in line geometry.

Let A and B be two points of the moving line l and α and β the planes of their paths. The path of B has two intersections B_1 and B_2 (real or imaginary) with α; if A_1, A_2 are the corresponding positions of A, the lines A_1B_1 and A_2B_2 are the generators of R_4. Hence the intersection of α and R_4 (which must be of order four) consists of the ellipse described by A and two generators. In other words: during its motions the line l is twice in α, and each time α is a tangent plane since it contains l and the velocity vector of A; hence α is a double tangent plane of R_4. The system of planes α, for variable A, is therefore the double tangent developable of R_4. As the coordinates (x, y, z, w) of A are linear functions of a parameter it follows from (3.5) that the developable is of the third class. The plane $W = 0$ at infinity belongs to the developable; its intersection with R_4 consists of the conic described by the point at infinity of l and two conjugate imaginary lines: if $1+t^2=0$ we have $P_{14}=P_{24}=P_{34}=0$.

Any plane of the developable contains two generators; their intersection S is a double point of R_4, i.e., an arbitrary line through S has four points in common with R_4, of which two coincide with S. The locus of S is the double curve of R_4; it can be shown to be a twisted cubic.

We have, in the foregoing, considered the path of a point and the surface generated by a line when E is constrained to a Darboux motion. The apparatus we have developed enabled us to investigate the developable generated in Σ by a plane of E; if we obtain the formulas for R_4 and use (3.4), R_4 is seen to be of the fourth class. We return to this problem in the next section.

The Darboux motion is a much studied example of a spatial motion. We shall meet it in another context in Section 4; a generalization will be given in Section 5. The motion, as given by (3.4) contains three parameters a, b, c;* hence there are a variety of cases. An important special one, defined by $a=0$, will be dealt with in Section 7; another with $a=b=c=0$ in Section 10. For certain special cases the general properties given above must be modified.

A derivation of the Darboux motion making use of spatial instantaneous invariants has been given by VELDKAMP [1967a].

4. Mannheim's motion

In this section we make some remarks about the inverse of Darboux's motion. We already know some of its properties because it was the starting point of the preceding section. In this so-called Mannheim motion every plane passes through a fixed point; apart from trivial cases it is the only motion with this property. Moreover every moving plane envelopes a circular cone.

Solving for x, y, z from (3.4) and interchanging x, y, z, ϕ with $X, Y, Z, -\phi$, the Mannheim motion reads

$$\begin{aligned} X &= x\cos\phi - y\sin\phi - a\sin^2\phi, \\ Y &= x\sin\phi + y\cos\phi + a\sin\phi\cos\phi, \\ Z &= z + b\sin\phi - c(1-\cos\phi). \end{aligned} \tag{4.1}$$

* A. GRÜNWALD [1906] claims that by a suitable transformation of the frames the constant c can be removed, but this statement does not seem to be correct.

The axodes C_m and C_f of Section 3 change their roles. Hence for the case at hand the fixed axode is a circular cylinder of diameter a, and is on the interior of the moving one which has the diameter $2a$.

The path of a moving point follows immediately from (4.1). With homogeneous coordinates and the rational parameter t we obtain

$$
\begin{aligned}
X &= x(1-t^4) - 2yt(1+t^2) - 4awt^2, \\
Y &= 2xt(1+t^2) + y(1-t^4) + 2awt(1-t^2), \\
Z &= z(1+t^2)^2 + 2bwt(1+t^2) - cwt^2(1+t^2), \\
W &= w(1+t^2)^2.
\end{aligned} \tag{4.2}
$$

Hence the path of an arbitrary point is a rational space curve of the fourth order, a so-called quartic of the second kind. Any path is for $t = \pm i$ tangent to the plane at infinity; the tangent points are the same for all paths, they are seen to be the isotropic points I_1 and I_2 of the plane $Z = 0$.

Example 20. Show that the tangent to the path at I_1 (or I_2) is the same for all paths.

A quartic of the second kind has the property that it lies on *one* quadric. To determine the equation of the latter we may proceed as follows. From (4.1) we deduce that $X^2 + Y^2, X, Y, Z$ and Z^2 can be written as linear functions of $\sin\phi$, $\cos\phi$, $\sin^2\phi$ and $\sin\phi\cos\phi$. By eliminating these four terms from the five equations, we obtain after some algebra

$$
\begin{aligned}
&(b^2+c^2)(X^2+Y^2) - a^2Z^2 + 2ac(cX+bY) \\
&\quad - 2aZ(cx+by-az+ac) - \mathrm{P} = 0,
\end{aligned} \tag{4.3}
$$

with

$$
\begin{aligned}
\mathrm{P} = {}&(b^2+c^2)(x^2+y^2) - a^2z^2 + 2ac(cx+by) \\
&- 2az(cx+by-az+ac).
\end{aligned} \tag{4.4}
$$

Example 21. Verify that (4.1) satisfies (4.3) for all values of ϕ.

From (4.3) it follows that the unique quadric Q through the path of $A(x, y, z)$ is a quadric of revolution; its axis is parallel to O_Z.

The center M of (4.3) is seen to be

$$
\begin{aligned}
X_M &= -ac^2/(b^2+c^2), \quad Y_M = -abc/(b^2+c^2), \\
Z_M &= -(cx+by-az+ac)/a.
\end{aligned} \tag{4.5}
$$

Hence all quadrics associated with a given motion have the same axis of revolution: $X = X_M$, $Y = Y_M$. Furthermore, for all points of E satisfying

$cx + by - az = d_1$, $\mathrm{P} = d_2$, in which d_1 and d_2 are constants, the quadric (4.3) is the same. Therefore, although there are ∞^3 paths, the system (4.3) represents only ∞^2 quadrics. In other words, there are sets of ∞^1 points of E moving on the same quadric. Such a set consists of the points of a conic k, in E, determined by d_1 and d_2; there are ∞^2 such conics. The plane of k is a moving plane of E. Characteristic of the Mannheim motion is the property that any moving plane passes through a fixed point S of Σ; for the plane $cx + by - az = d_1$ this point S is seen to be $X = -ac^2/(b^2 + c^2)$, $Y = -abc/(b^2 + c^2)$, $Z = -c - d_1/a$, hence in view of (4.5) it coincides with the center M of the quadric.

Example 22. Verify the coordinates of S.

Any point $A(x, y, z)$ of E determines a pair of constants d_1 and d_2. If A is real the plane $cx + by - az = d_1$ and the quadric $\mathrm{P} = d_2$ have a real intersection. This is only the case if $g = d_1^2 + 2acd_1 - d_2 \leqq 0$.

Example 23. Show that the quadric $\mathrm{P} = d_2$ intersects the plane $cx + by - az = d_1$ in the curve given by $(bx - cy)^2 + g = 0$ and the plane; hence we have the condition $g \leqq 0$ for a real intersection.

The quadric Q (4.3) has a real intersection with the plane at infinity and it is therefore hyperbolic. Its discriminant D is equal to $a^2(b^2 + c^2)^2(d_2 - d_1^2 - 2acd_1) = -a^2(b^2 + c^2)^2 g$; this implies that for real moving points we have $D \geqq 0$. If $D > 0$ the signature of Q is zero, hence Q is a hyperboloid of one sheet, with real generators, and the same holds of course for $\mathrm{P} = 0$. If $g = 0$ the quadric Q is a cone, and the conic k degenerates into a double line; this line moves so that it passes permanently through the fixed point M. The lines with this property are the counterpart of those points which under Darboux motion describe a straight line instead of an ellipse.

During a Mannheim motion a moving line generates, in view of Chasles theorem of Section 1, a quartic ruled surface. A moving plane envelopes a circular cone.

There are some interesting special cases, if $a = 0$, or $b = c = 0$, to which we shall return in later sections.

5. Schoenflies' motion

The following motion has been studied at some length by SCHOENFLIES [1892]: a plane α of E moves in itself and is also subjected to a translation. If α is taken as the o_{xy}-plane (in E) and a plane parallel to it in Σ as O_{XY}, the motion is given by

$$X = x\cos\phi - y\sin\phi + d(\phi),$$
$$(5.1)\qquad Y = x\sin\phi + y\cos\phi + e(\phi),$$
$$Z = z + f(\phi),$$

with three arbitrary functions d, e and f. Any plane parallel to α remains parallel to itself. Two points on a line parallel to o_z describe congruent paths. The instantaneous screw axis is parallel to O_Z and to o_z. The axodes are cylinders with their generators parallel to O_Z. They intersect O_{XY} along two curves, the centrodes of the planar part of the motion. Both Darboux's and Mannheim's motion are special cases and it was in fact Schoenflies' intention to generalize them.

The motion is too special and on the other hand not special enough to be very interesting. We shall restrict ourselves to instantaneous kinematics and determine how the properties of a general motion are modified for this case. We introduce canonical systems for the planar component of the motion and suppose that the origins of the spatial frames coincide for $\phi = 0$. Then we obtain the following scheme

$$X_0 = x,\quad X_1 = -y,\quad X_2 = -x,\quad X_3 = y + d_3, \ldots,$$
$$(5.2)\qquad Y_0 = y,\quad Y_1 = x,\quad Y_2 = -y + e_2,\quad Y_3 = -x + e_3, \ldots,$$
$$Z_0 = z,\quad Z_1 = f_1,\quad Z_2 = f_2,\quad Z_3 = f_3, \ldots .$$

If in the expressions (3.4) for the Darboux motion we transform the frames by means of

$$x = y' - a,\quad y = -x',\quad X = Y' - a,\quad Y = -X',\quad z = z',\quad Z = Z',$$

then omitting the primes we obtain

$$X = x\cos\phi - y\sin\phi,$$
$$(5.3)\qquad Y = x\sin\phi + y\cos\phi + a(1 - \cos\phi),$$
$$Z = z + b\sin\phi + c(1 - \cos\phi);$$

it is easy to verify that we have now introduced canonical frames. We have $d(\phi) = 0$, $e(\phi) = a(1 - \cos\phi)$, $f(\phi) = b\sin\phi + c(1 - \cos\phi)$ hence:

$$d_n = 0 \text{ (for every } n\text{)}:$$
$$(5.4)\qquad e_0 = e_1 = 0,\quad e_2 = a,\quad e_3 = 0,\quad e_4 = -a, \ldots,$$
$$f_0 = 0,\quad f_1 = b,\quad f_2 = c,\quad f_3 = -b,\quad f_4 = -c, \ldots .$$

By inversion of (5.3), i.e., interchanging X, Y, ϕ with $x, y, -\phi$, we obtain for Mannheim's motion

$$X = x\cos\phi - y\sin\phi + a\sin\phi(1-\cos\phi),$$
(5.5)
$$Y = x\sin\phi + y\cos\phi - a\cos\phi(1-\cos\phi),$$
$$Z = z + b\sin\phi - c(1-\cos\phi),$$

with the instantaneous invariants

$$d_0 = 0,\quad d_1 = 0,\quad d_2 = 0,\quad d_3 = 3a,\quad d_4 = 0,\dots,$$
(5.6)
$$e_0 = 0,\quad e_1 = 0,\quad e_2 = -a,\quad e_3 = 0,\quad e_4 = 7a,\dots,$$
$$f_0 = 0,\quad f_1 = b,\quad f_2 = -c,\quad f_3 = -b,\quad f_4 = c,\dots.$$

We now derive some instantaneous properties of Schoenflies' motion, by means of (5.2).

The tangent at (x, y, z) to its path is seen to be the line with the coordinates

$$p_{14} = y,\quad p_{24} = -x,\quad p_{34} = -f_1,$$
(5.7)
$$p_{23} = f_1 y - xz,\quad p_{31} = -f_1 x - yz,\quad p_{12} = x^2 + y^2.$$

Hence the locus of the path tangents reads

(5.8) $$f_1(p_{14}^2 + p_{24}^2) + p_{12}p_{34} = 0$$

which represents, for $f_1 \neq 0$, a tetrahedral complex; which is also the case for the general motion in space. This complex is also the locus of tangents for the Darboux and the Mannheim motions if $b \neq 0$.

The osculating plane at (x, y, z), that is the plane through three consecutive positions of the point, is seen to be

$$U_1 = f_2 x + f_1 y - f_1 e_2,\quad U_2 = -f_1 x + f_2 y,\quad U_3 = x^2 + y^2 - e_2 y,$$
(5.9)
$$U_4 = -(x^2 + y^2)(z + f_2) + e_2 yz + e_2 f_1 x.$$

As in the general case this gives us a cubic relationship between the plane and the corresponding point.

Example 24. Show that the osculating planes of all points of a plane $z = z_0$ pass through one point, i.e., through $(0, e_2, z_0 + f_2)$.
Example 25. Apply (5.9) to the Darboux motion and show that the osculating plane is the plane of the path.
Example 26. Show that the normal plane of the path at (x, y, z) is $yX - xY - f_1 Z + f_1 z = 0$.
Example 27. Show that the locus of all path normals is a linear complex and determine its equation.

Three consecutive points of the path of (x, y, z, w) are collinear if the rank of the matrix

$$\left\| \begin{matrix} x & y & z & w \\ -y & x & f_1 w & 0 \\ -x & -y+e_2 w & f_2 w & 0 \end{matrix} \right\| \tag{5.10}$$

is two. This is the case if either $x = f_1 f_2 e_2/(f_1^2 + f_2^2)$, $y = f_1^2 e_2/(f_1^2 + f_2^2)$ or $w = 0$, $x^2 + y^2 = 0$. Hence the inflection curve, a twisted cubic in general, degenerates for the Schoenflies motion into a finite line, "the inflection line" parallel to O_Z and two imaginary lines at infinity, all three passing through $(0, 0, 1, 0)$. At least for the Darboux motion, this degeneration could have been expected: an ellipse has an inflection point only if it is a straight line.

Example 28. Show that for the Darboux motion the points on the inflection line are those for which the path is a line segment.

For general spatial motion the locus in E of the points with a stationary osculating plane (that is, points for which four consecutive positions are coplanar) is a cubic surface. In our case it follows from (5.2) that its equation is

$$w \begin{vmatrix} -y & x & f_1 \\ -x & -y+e_2 & f_2 \\ y+d_3 & -x+e_3 & f_3 \end{vmatrix} = 0,$$

which implies that it degenerates into the plane at infinity and the circular cylinder

$$\begin{aligned} &(f_1+f_3)(x^2+y^2)+(d_3 f_2 - e_3 f_1)x \\ &\quad + (d_3 f_1 - e_2 f_1 + e_3 f_2 - e_2 f_3)y - d_3 e_2 f_1 = 0. \end{aligned} \tag{5.11}$$

Example 29. Show that the inflection line is on the cylinder, as could be expected.
Example 30. Determine the cylinder (5.11) for the Darboux and for the Mannheim motion.

For more properties of this motion we refer the reader to Schoenflies' paper.

6. Analytically determined motions

The special motions considered in Sections 2, 3, 4, 5 have been defined in various ways by means of some of their geometric properties. There is another, more formal method which gives us a special motion in a purely analytic manner. The general motion in space is expressed by $\boldsymbol{P} = \mathbf{A}\boldsymbol{p} + \boldsymbol{d}$; therefore a special motion is defined any time we specify the matrix $\mathbf{A}$ and the vector $\boldsymbol{d}$ as functions of a parameter t. To develop this idea we represent the orthogonal matrix $\mathbf{A}$ by means of the Euler parameters, as was done in Chapter VI, (2.1). If $N = c_0^2 + c_1^2 + c_2^2 + c_3^2$, $\boldsymbol{d} = (d_1, d_2, d_3)$ and if we put $d_i = g_i/(Ng_0)$, then any spatial displacement is given, in terms of the homogeneous coordinates X, Y, Z, W in Σ and x, y, z, w in E, as

$$
\begin{aligned}
X &= [(c_0^2 + c_1^2 - c_2^2 - c_3^2)x + 2(c_1c_2 - c_0c_3)y + 2(c_1c_3 + c_0c_2)z]g_0 + g_1w, \\
Y &= [2(c_2c_1 + c_0c_3)x + (c_0^2 - c_1^2 + c_2^2 - c_3^2)y + 2(c_2c_3 - c_0c_1)z]g_0 + g_2w, \\
Z &= [2(c_3c_1 - c_0c_2)x + 2(c_3c_2 + c_0c_1)y + (c_0^2 - c_1^2 - c_2^2 + c_3^2)z]g_0 + g_3w, \\
W &= [(c_0^2 + c_1^2 + c_2^2 + c_3^2)w]g_0.
\end{aligned} \tag{6.1}
$$

Hence a spatial motion is defined if we specify c_i and g_i as functions of t $(i = 0, 1, 2, 3)$; g_0 is introduced for the sake of uniformity. The sets c_i and g_i are both homogeneous quadruples of numbers, where the c_i are not all zero and $g_0 \neq 0$.

From (6.1) it follows: if c_i and g_i are algebraic functions of t, the motion is algebraic; which means that the path of a point, the ruled surface described by a moving line, and the developable generated by a moving plane, are all algebraic.

Furthermore, if c_i and g_i are rational functions of t, all these varieties are rational. If $g_0 = 1$, g_i of degree m_1 $(i = 1, 2, 3)$, c_i of degree m_2, a path is of order $\max(m_1, 2m_2)$. Hence, if $m_1 = 2$, $m_2 = 1$ any path is a conic, which implies that (6.1) represents a Darboux motion.

Example 31. Let $c_1 = c_2 = 0$, $c_0 = 1$, $c_3 = t$, $g_0 = 1$, $g_1 = 0$, $g_2 = 2at$, $g_3 = 2bt + c(1 - t^2)$; show that (6.1) represents the motion (3.4).

If $m_1 = 4$, $m_2 = 2$ any path is a rational quartic; this is also the case if c_i are linear, g_0 quadratic and g_i $(i = 1, 2, 3)$ quartic functions.

Example 32. If $c_1 = c_2 = 0$, $c_0 = 1$, $c_3 = t$, $g_0 = (1 + t^2)$, $g_1 = -4at^2$, $g_2 = 2at(1 - t^2)$, $g_3 = 2bt(1 + t^2) - 2ct^2(1 + t^2)$, show that (6.1) represents Mannheim's motion (4.2).

We make use of (6.1) to define a cubic motion, all paths being rational cubic curves. Such a motion appears obviously if c_i and g_0 are linear and g_i cubic functions of t, $(i = 1, 2, 3)$. To give an example we put $c_0 = c_3 = 0$, $c_1 = 1$, $c_2 = t$, $g_0 = 1$, $g_1 = t^3$, $g_2 = t^2$, $g_3 = 1$, which gives us the following cubic motion

$$X = (1 - t^2)x + 2ty + t^3 w,$$
$$Y = 2tx - (1 - t^2)y + t^2 w,$$
$$(6.2) \qquad Z = -(1 + t^2)z + w,$$
$$W = (1 + t^2)w.$$

Example 33. Show that (6.2) has the following properties: it is a Schoenflies motion; the fixed and the moving axode are cylinders of order seven and six respectively; the path of (x, y, z) is a twisted cubic unless $x = 0$; if $x = 0$, $w \neq 0$, the path is a cubic in the plane $wY + (2y + w)Z - (y - z + w)W = 0$; the inverse motion is a quintic motion.

7. Line-symmetric motions

Any displacement of the space E with respect to the coinciding space Σ is, as we know, either a screw displacement, with axis l, rotation angle ϕ, and a translation d parallel to l, or as a special case a pure translation. In the first case the inverse displacement is the screw displacement with axis l, rotation angle $-\phi$, and a translation $-d$ parallel to l. Hence the two displacements are the same if and only if $\phi = -\phi(\text{mod. } 2\pi)$ and $d = 0$, which (apart from the trivial case $\phi = 0$) implies $\phi = \pi$; if the direct displacement is a translation it can never be identical with its inverse. The conclusion is: the only displacement which is identical with its inverse is a half-turn about a line l, or in other words the reflection into the line l. Such a displacement will be called symmetric; it is the starting point for a class of special motions to be studied in this section.

Analytically a displacement is given by $\boldsymbol{P} = \mathbf{A}\boldsymbol{p} + \boldsymbol{d}$; the inverse is therefore $\boldsymbol{p} = \mathbf{A}^{-1}(\boldsymbol{P} - \boldsymbol{d})$. Hence a symmetric displacement satisfies the equations

$$(7.1) \qquad \mathbf{A}^{-1} = \mathbf{A}, \qquad \mathbf{A}^{-1}\boldsymbol{d} + \boldsymbol{d} = 0.$$

$\mathbf{A}$ being an orthogonal matrix satisfies $\mathbf{A}^{-1} = \mathbf{A}^{\mathrm{T}}$; hence the first equation (7.1) implies that $\mathbf{A}$ is a symmetric matrix. Making use again of Chapter VI, (2.1), which expresses an orthogonal matrix by means of the parameters c_i $(i = 0, 1, 2, 3)$ it follows immediately that it is only symmetric if $c_0 = 0$; this is as expected in view of the geometric meaning of this parameter: $c_0 = \cos\frac{1}{2}\phi$; hence $c_0 = 0$ characterizes the half-turns. Symmetric orthogonal matrices depend therefore on c_1, c_2, c_3; these parameters have also a geometric meaning: they are the direction numbers of the axis l_0 of the rotation represented by $\mathbf{A}$. As they are homogeneous parameters, not all zero, we may suppose $\boldsymbol{c} = (c_1, c_2, c_3)$ to be a unit vector: $c_1^2 + c_2^2 + c_3^2 = 1$. This implies that

$c_1^2 - c_2^2 - c_3^2 = 2c_1^2 - 1$, etc. Hence, in view of Chapter VI, (2.1) the half-turn about the axis $\boldsymbol{c}$ through O is given by $\boldsymbol{P} = 2(\boldsymbol{c} \cdot \boldsymbol{p})\boldsymbol{c} - \boldsymbol{p}$.

We know that the product of a rotation about an axis l_0 and the translation $\boldsymbol{d}$ is a pure rotation (about the axis l, parallel to l_0) if and only if $\boldsymbol{d}$ is orthogonal to l_0. A vector $\boldsymbol{d}$ orthogonal to $\boldsymbol{c}$ can always be written as $\boldsymbol{c} \times \boldsymbol{b}$, $\boldsymbol{b}$ being an arbitrary vector. But as $\boldsymbol{c} \times (\boldsymbol{b} + k\boldsymbol{c}) = \boldsymbol{c} \times \boldsymbol{b}$, we may without loss of generality suppose that $\boldsymbol{c} \cdot \boldsymbol{b} = 0$. Moreover, as the rotation about l_0 is a half-turn, the line l passes through $\frac{1}{2}\boldsymbol{d}$ (Fig. 76); the first Plücker vector of l is $\boldsymbol{c}$, if the second is $\boldsymbol{b}$ we have $\boldsymbol{b} = \boldsymbol{c} \times \frac{1}{2}\boldsymbol{d}$ and therefore $\boldsymbol{d} = 2(\boldsymbol{b} \times \boldsymbol{c})$. Summing up we obtain: the half-turn about the line l, determined by its Plücker vectors $(\boldsymbol{c}, \boldsymbol{b})$ is represented by

$$\boldsymbol{P} = 2(\boldsymbol{c} \cdot \boldsymbol{p})\boldsymbol{c} - \boldsymbol{p} + 2(\boldsymbol{b} \times \boldsymbol{c}). \tag{7.2}$$

Example 34. Verify that (7.2) satisfies the conditions (7.1).

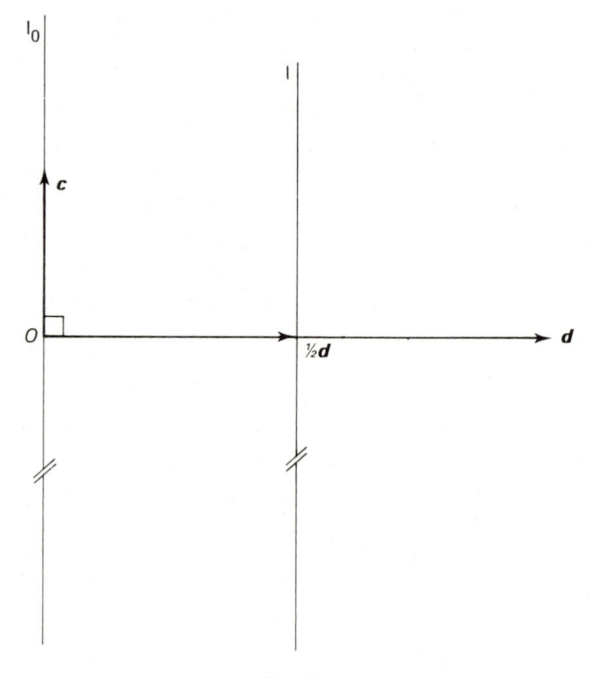

Fig. 76.

If we expand in terms of coordinates, (7.2) is seen to be

$$\begin{aligned} X &= (2c_1^2-1)x + 2c_1c_2y + 2c_1c_3z + 2(b_2c_3 - b_3c_2), \\ Y &= 2c_2c_1x + (2c_2^2-1)y + 2c_2c_3z + 2(b_3c_1 - b_1c_3), \\ Z &= 2c_3c_1x + 2c_3c_2y + (2c_3^2-1)z + 2(b_1c_2 - b_2c_1), \end{aligned} \tag{7.3}$$

with $c_1^2+c_2^2+c_3^2=1$, $c_1b_1+c_2b_2+c_3b_3=0$; (7.3) is the general expression of a symmetric displacement.

Example 35. Show that there are ∞^4 symmetric displacements.

We have preferred to derive (7.2) and (7.3) in a way that exhibits the connection with the Euler parameters. A more elementary method would be to start from a line l with Plücker vectors $(\boldsymbol{c}, \boldsymbol{b})$ and to calculate the reflection (X, Y, Z) into l of a point (x, y, z).

Example 36. Derive (7.3) by this latter method.

It is seen from (7.3) that all coefficients of the transformation are quadratic functions of the parameters c_i and b_i. From (7.3), which gives the transformation of the point coordinates, those of the line and the plane coordinates can be derived with the result that the coefficients of both are also quadratic functions of c_i and b_i.

Example 37. Verify this statement by calculating the corresponding two matrices, making use of the two relations for c_i and b_i.

Let a frame O_{XYZ} be given in the fixed space Σ and let E_0 be that position of the moving space E, with the frame o_{xyz}, at which the two frames coincide. Let also be given, in Σ, a continuous set of lines $\mathrm{l}(t)$, determined by their vectors $\boldsymbol{c}(t), \boldsymbol{b}(t)$; the locus of these lines is a certain ruled surface R′. Reflecting E_0 into the line $\mathrm{l}(t)$ yields $E(t)$. If t varies, a continuous set of positions of E with respect to Σ appears and we have defined a certain motion of E. Such a motion, completely determined by the ruled surface R′ will be called a (line-) symmetric motion or a Krames-motion; R′ is its basic surface.

It has a remarkable property: $E(t)$ is obtained by reflecting E_0 into $\mathrm{l}(t)$, or what is the same thing, by reflecting Σ into $\mathrm{l}(t)$. Hence at any moment the positions of E and Σ are one anothers' reflections into a line. But this means that the two spaces are interchangable and that the inverse motion is characterized by a series of positions of Σ each determined by reflecting Σ into a line of E. The locus of these lines is obtained as follows: reflect each line of R′ into the line $l(t_0)$ of R′ then we get a ruled surface R in E, congruent to R′. Under the direct motion R, fixed in E, moves in Σ. Obviously R and R′ have at any moment a line in common, $l(t_0)$ at the moment t_0. The tangent plane α to R′ at a point B (on the common line) has as its reflection the tangent plane at B to R, and since the plane α passes through $l(t_0)$ it is invariant under the reflection, which implies that, at any moment, R and R′ are raccording along their common line.

The relation between E and Σ has the consequence that any property of the direct motion has its counterpart for the inverse motion. So, for instance, the set of all paths in Σ is the reflection of the set of paths in E. Especially, the fixed and moving axodes are, at any moment, one anothers' reflection; they are congruent surfaces.

These axodes may be determined as follows. The instantaneous screw axis for the moment t is the limit, as $\Delta t \to 0$, of the screw axis for the displacement $E(t) \to E(t+\Delta t)$. This displacement takes place by the reflection into $l(t)$ being followed by that into $l(t+\Delta t)$. The construction for the resultant of two displacements has been dealt with in Chapter IV. If it is applied to two half-turns it is easily seen that the axis of the product displacement coincides with the common perpendicular of the two half-turn axes. Hence the instantaneous screw axis s′ at the instant t is the common perpendicular of $l(t)$ and the consecutive generator of R′, a line called the central line of $l(t)$ on the ruled surface; its intersection with $l(t)$ is the central point (or striction point) on $l(t)$. The result is: the fixed axode A′ is the ruled surface composed of the central lines of R′; the moving axode A is the analog figure of R. A moves with respect to A′ in the well-known rolling-sliding manner described in Chapter VI, Section 5; any point of s (the generator of A, coinciding with s′) has its velocity along s′. A and A′ are reflections with respect to l; s and s′ coincide because they are perpendicular to l. The motion of R with respect to R′ is not a rolling-sliding motion.

A Krames motion is determined by its basic surface R′ hence by the functions $\boldsymbol{c}(t)$, $\boldsymbol{b}(t)$. From (7.3) it follows that the motion is algebraic if R′ is algebraic. Moreover: if the order of R′ is n, then (in general) the paths are curves of order $2n$, a moving line generates a ruled surface of order $2n$, a moving plane describes a developable of class $2n$.

The motions under consideration have been studied at length, in a series of papers by Krames, who has given many applications of the general theory. His examples are of two kinds. In some cases he recognized that certain known motions belong to the class of line-symmetric ones and he determined the basic surfaces R′ by which they can be generated. In other cases he started from a chosen basic surface and in this way he found some interesting special motions.

Here, we will give a brief discussion of some special Krames motions, many more particulars can be found in the original papers (KRAMES [1937 a–f, 1940]).

The general Darboux motion is given by (3.4); its axodes are circular cylinders with diameters a and $2a$. Since a Krames motion has congruent axodes, a Darboux motion can obviously only be line-symmetric if the parameter a is zero. Then it follows from (3.4) that the motion is the result of a rotation about the Z-axis and a harmonic translation, with the same period, in the O_Z-direction. This special case will be called a vertical Darboux motion (German: aufrechte Ellipsenbewegung, or volkommen steiler Darbouxscher Umschwung). The remaining parameters are b and c, but in this special case by a suitable rotation of the frame o_{xyz} about o_z and a certain translation in the o_z direction we can have c vanish. The motion is then represented by

$$\begin{aligned} X &= x \cos \phi - y \sin \phi, \\ Y &= x \sin \phi + y \cos \phi, \\ Z &= z + h \sin \phi. \end{aligned} \tag{7.4}$$

Example 38. Show that the foregoing transformation of coordinates yields (7.4) if the rotation angle is $\arctan(-b/c)$, the translation distance c, and $h = (b^2 + c^2)^{1/2}$.
Example 39. Show that a vertical Darboux motion is a (special) Mannheim motion; any moving plane envelopes a circular cone.
Example 40. Show that the points whose paths are straight lines are those of o_z.
Example 41. Show that the ellipses described by the points of E all have the same (linear) eccentricity h.

The formulas (7.3) give us the position of the point (x, y, z) of E_0 when this space is reflected into the line $(\boldsymbol{c}, \boldsymbol{b})$. If $\boldsymbol{c}$ and $\boldsymbol{b}$ are functions of a parameter t these lines are the generators of the basic surface R′. Suppose that for $t = 0$ the line coincides with O_X; then for $t = 0$ we have $X = x$, $Y = -y$, $Z = -z$. Hence if we reflect the y and z axes about o_x the O_{XYZ} and o_{xyz} frames coincide at $t = 0$, and the Krames motion may be written as

$$\begin{aligned} X &= (2c_1^2 - 1)x - 2c_1c_2y - 2c_1c_3z + 2(b_2c_3 - b_3c_2), \\ Y &= 2c_1c_2x + (1 - 2c_2^2)y - 2c_2c_3z + 2(b_3c_1 - b_1c_3), \\ Z &= 2c_3c_1x - 2c_3c_2y + (1 - 2c_3^2)z + 2(b_1c_2 - b_2c_1). \end{aligned} \tag{7.5}$$

To demonstrate that the vertical Darboux motion is a Krames motion we must show that $\boldsymbol{c}$ and $\boldsymbol{b}$ in (7.5) can be chosen as functions of a parameter in such a way that the formulas (7.5) are identical with (7.4). This implies

$c_1 = \cos(\phi/2)$, $c_2 = \sin(\phi/2)$, $c_3 = 0$, $b_3 = 0$, $2(b_1c_2 - b_2c_1) = h \sin\phi$. Keeping in mind that $b_1c_1 + b_2c_2 = 0$ we obtain $b_1 = h\sin^2(\phi/2)\cos(\phi/2)$, $b_2 = -h\sin(\phi/2)\cos^2(\phi/2)$. Hence, putting $\tan(\phi/2) = t$ and factoring the term $\cos^3(\phi/2)$ from each component, the surface R′ is given by

$$(7.6)\qquad \begin{aligned} &c_1 = 1 + t^2, \quad c_2 = t(1+t^2), \quad c_3 = 0, \\ &b_1 = ht^2, \quad b_2 = -ht, \quad b_3 = 0, \end{aligned}$$

from which it follows that it is a cubic surface. To determine its equation we note that $c_3 = 0$, $b_3 = 0$ implies that the generators of R′ intersect O_Z orthogonally. For the generator through point X, Y, Z we have $c_1 = X$, $c_2 = Y$, $c_3 = 0$, $b_1 = YZ$, $b_2 = -XZ$, $b_3 = 0$. Hence, using either the parameter t or ϕ, the surface is known. In terms of ϕ we have

$$(7.7)\qquad X = k\cos(\phi/2), \qquad Y = k\sin(\phi/2), \quad Z = h\sin(\phi/2)\cos(\phi/2),$$

where k is an arbitrary constant, from which it follows

$$(7.8)\qquad (X^2 + Y^2)Z = hXY,$$

which is the standard equation for a well-known surface: Cayley's cylindroid, also called Plücker's conoid. The conclusion is: the vertical Darboux motion is the Krames motion with a cylindroid as its basic surface.

Example 42. If an arbitrary point is reflected into the generators of R′ we obtain the path of its motion. Show that the pedal curve of any point with respect to a cylindroid is an ellipse. (A theorem due to Appell [1900].)

The axodes of a Schoenflies motion are two cylinders with parallel generators. Such a motion can only be line-symmetric if the cylinders are congruent. As an example we consider the case when they are both circular cylinders with radius ρ, externally tangent during the motion. Then for suitably chosen frames the motion is represented by

$$(7.9)\qquad \begin{aligned} X &= x\cos\phi - y\sin\phi - 2\rho\sin(\phi/2), \\ Y &= x\sin\phi + y\cos\phi + 2\rho\cos(\phi/2), \\ Z &= z + 2g(\phi), \end{aligned}$$

$g(\phi)$ being an arbitrary function.

If we compare (7.9) with (7.5) we obtain for the generators $\boldsymbol{c}, \boldsymbol{b}$ of the basic surface R′

$$(7.10)\qquad \begin{aligned} &c_1 = \cos(\phi/2), \quad c_2 = \sin(\phi/2), \quad c_3 = 0 \\ &b_1 = g(\phi)\sin(\phi/2), \quad b_2 = -g(\phi)\cos(\phi/2), \quad b_3 = \rho. \end{aligned}$$

Example 43. Show that the generators of R′ are parallel to the O_{XY}-plane.
Example 44. Consider the case $g(\phi) = a \sin(\phi/2)$. Show that: the equation of R′ reads $(X^2 + Y^2)Z^2 + 2\rho aXZ - a^2Y^2 + \rho^2a^2 = 0$; the path of a point is in general a rational quartic; the paths of points on o_z are circles; the intersection of R′ and a plane $Z = h$ consists of the line $Z = W = 0$ at infinity and of two generators of R′ which intersect at a point of the hyperbola $Y = 0$, $XZ + \rho a = 0$ which is the double curve of R′. Consider the case $\rho = 0$.

We define now a (line-symmetric) motion starting from a given basic surface R′. As an example we take R′ to be a hyperbolic paraboloid with the standard equation $(X^2/a^2) - (Y^2/b^2) = 2Z/c$. One of the two systems of generators on R′ is determined by $(X/a) + (Y/b) = 2t^{-1}Z/c$, $(X/a) - (Y/b) = t$; all these generators are parallel to the plane $bX - aY = 0$. Their Plücker coordinates are seen to be

$$
\begin{aligned}
&p_{14} = 2a, \quad &&p_{24} = 2b, \quad &&p_{34} = 2ct,\\
&p_{23} = bct^2, \quad &&p_{31} = act^2, \quad &&p_{12} = -2abt.
\end{aligned}
\tag{7.11}
$$

If we substitute this in (7.3), keeping in mind that c_1, c_2, c_3; b_1, b_2, b_3 are proportional to (7.11), and that $\Sigma c_i^2 = 1$, we obtain for the motion, in homogeneous coordinates,

$$
\begin{aligned}
X &= (a^2 - b^2 - c^2t^2)x + 2aby + 2actz + at(2b^2 + c^2t^2)w,\\
Y &= 2abx + (-a^2 + b^2 - c^2t^2)y + 2bctz - bt(2a^2 + c^2t^2)w,\\
Z &= 2actx + 2bcty + (-a^2 - b^2 + c^2t^2)z + c(b^2 - a^2)t^2w,\\
W &= (a^2 + b^2 + c^2t^2)w.
\end{aligned}
\tag{7.12}
$$

The formulas (7.12) represent a cubic motion. The paths are in general twisted cubics. The points at infinity of a path follow from $W = 0$, which gives us either $t = \infty$ or $c^2t^2 = -(a^2 + b^2)$. For any x, y, z, w we have: for $t = \infty$ the real points $X = a$, $Y = -b$, $Z = W = 0$; for $c^2t^2 = -(a^2 + b^2)$ two imaginary points at infinity, these lie at the intersection of $bX - aY = 0$ and $X^2 + Y^2 + Z^2 = 0$.

Example 45. Verify the latter statements by direct substitution of t into (7.12).

This implies that the two imaginary points at infinity of a path are isotropic points. Hence the paths are, metrically, special twisted cubics, often called cubical circles. Moreover the three points at infinity of a path are the same for all paths.

Example 46. Determine the path of a point in the plane at infinity.
Example 47. Discuss the formulas (7.12) for the special case $a = b$.

The fixed axode of the motion is in general the locus of the central tangents of the basic surface. The lines (7.11) are all parallel to the plane $bX - aY = 0$. Hence the central tangents $(\boldsymbol{n}, \boldsymbol{m})$, being common perpendiculars of consecutive generators, are orthogonal to this plane and their locus is therefore a cylinder. We obtain

$$\begin{aligned} &n_1 = -2b(a^2+b^2), \quad n_2 = 2a(a^2+b^2), \qquad n_3 = 0, \\ &m_1 = ac(a^2-b^2)t^2, \quad m_2 = bc(a^2-b^2)t^2, \quad m_3 = -2(a^4-b^4)t, \end{aligned} \tag{7.13}$$

which implies that the cylinder is quadratic. Moreover there is only one line (namely $t = \infty$) in the plane at infinity. The conclusion is: the axodes of the Krames motion with a paraboloid as basic surface are parabolic cylinders.

Example 48. Show from (7.13) that the equation of the fixed axode reads $2(a^4 - b^4)Z = c(bY + aX)^2$.

Example 49. Show that the formulas (7.12) simplify by the following transformation of the fixed frame:

$$X_1 = (bX - aY)d^{-1}, \quad Y_1 = (aX + bY)d^{-1}, \quad Z_1 = Z, \quad d = (a^2 + b^2)^{1/2}.$$

Example 50. Show that there is a line in the moving space the paths of whose points are straight lines (Krames [1937d, p. 150]).

Example 51. Show that the ruled surface generated by a line moving according to the motion (7.12) is of the third order; the direction cone is a circular cone (Krames [1937d, p. 151]).

Bricard [1926] has given a spatial motion such that all point-paths are spherical curves; Krames [1937b] has shown that this motion is line-symmetric, the basic surface being a spherical conoid. Such a surface is defined as the locus of lines which intersect a fixed line orthogonally and which are tangent to a fixed sphere.

Let the fixed line be O_Z and the equation of the fixed sphere $(X - d)^2 + Y^2 + Z^2 = \rho^2$. A line with the equations $X : Y = \cos\beta : \sin\beta$, $Z = h$ is seen to be tangent to the sphere if $h^2 = \rho^2 - d^2 \sin^2\beta$, which is the fundamental relation between the variables h and β. The generators $(\boldsymbol{c}, \boldsymbol{b})$ of the conoid R′ are

$$\begin{aligned} &c_1 = \cos\beta, \qquad c_2 = \sin\beta, \qquad\quad c_3 = 0, \\ &b_1 = h\sin\beta, \quad b_2 = -h\cos\beta, \quad b_3 = 0, \end{aligned} \tag{7.14}$$

Substituting this into (7.5) we obtain the following motion

$$\begin{aligned} X &= x\cos 2\beta - y\sin 2\beta, \\ Y &= x\sin 2\beta + y\cos 2\beta, \\ Z &= z + 2h, \end{aligned} \tag{7.15}$$

which is a rotation about O_Z combined with a periodic translation parallel to O_Z, defined by $h^2 = \rho^2 - d^2 \sin^2 \beta$. From (7.15) it follows that

$$X^2 + Y^2 + Z^2 = x^2 + y^2 + z^2 + 4hz + 4h^2.$$

We have $2h = Z - z$; moreover $h^2 = \rho^2 - d^2(1 - \cos 2\beta)/2$, and $Xx + Yy = (x^2 + y^2)\cos 2\beta$. Hence we obtain

$$\begin{aligned} &X^2 + Y^2 + Z^2 - (2d^2/(x^2+y^2))(Xx + Yy) \\ &\quad - 2zZ - x^2 - y^2 + z^2 - 2(2\rho^2 - d^2) = 0, \end{aligned} \tag{7.16}$$

which shows that any point (x, y, z) remains, during the motion, on a sphere: every path is a spherical curve.

The center M of the sphere is given by

$$X_M = d^2 x/(x^2 + y^2), \quad Y_M = d^2 y/(x^2 + y^2), \quad Z_M = z. \tag{7.17}$$

Example 52. Show that M follows from (x, y, z) by an inversion with respect to o_z.
Example 53. Determine the radius of the sphere; it is independent of z.
Example 54. Show that there is a cubic relationship between the moving point and the center M.

From (7.15) we have moreover $X^2 + Y^2 = x^2 + y^2$; this implies that the path of a moving point is the intersection of the sphere and a circular cylinder. The paths are therefore in general biquadratic curves of the first kind. In contrast to the preceding examples, the Bricard motion is not rational: it is of genus *one*.

Example 55. The motion (7.15) can be described by elliptic functions. If $d < \rho$, $k = d/\rho$, then by means of Jacobi functions with modulus k we obtain

$$X = x(1 - 2\mathrm{sn}^2 t) - 2y\, \mathrm{sn}\, t\, \mathrm{cn}\, t, \quad Y = 2x\, \mathrm{sn}\, t\, \mathrm{cn}\, t + y(1 - 2\mathrm{sn}^2 t),$$

$$Z = z + 2\rho\, \mathrm{dn}\, t.$$

In the special case $d = \rho$ (that is if the sphere of the conoid is tangent to O_Z) the motion is rational; we have $h = \rho \cos \beta$; the paths are rational biquadratic curves of the first kind.

Example 56. For this case, write (7.5) in a rational form. Determine the double point of the general path; show that all the points pass through the node of their path simultaneously.

The axodes of the Bricard motion are trivial; they consist of the O_Z and the o_z axis respectively.

Example 57. Determine the order of the surface generated by a moving line and the class of the developable described by a moving plane.

As our last example of a Krames motion we consider the case where the basic surface R′ consists of one of the systems of generators on a hyperboloid H. It will be seen to be the so-called Bennett motion. Let the equation of H be

$$(7.18)\qquad (X^2/a^2)+(Y^2/b^2)-(Z^2/c^2)=W^2,\quad a^2-b^2\geqq 0.$$

One set of generators is given by

$$b(cX-aZ)-tac(bW-Y)=0,\quad tb(cX+aZ)-ac(bW+Y)=0.$$

From this it follows that the Plücker coordinates of a generator are

$$(7.19)\qquad \begin{aligned} p_{14}&=a(t^2-1), & p_{24}&=-2bt, & p_{34}&=-c(t^2+1),\\ p_{23}&=bc(t^2-1), & p_{31}&=-2act, & p_{12}&=ab(t^2+1). \end{aligned}$$

Example 58. Show that the set of lines is the intersection of the three linear complexes $bcp_{14}-ap_{23}=0$, $acp_{24}-bp_{31}=0$, $abp_{34}+cp_{12}=0$.

Example 59. Show that for $t=\tan(\phi/2)$ the set (7.19) may be written $p_{14}=a\cos\phi$, $p_{24}=-b\sin\phi$, $p_{34}=-c$, $p_{23}=bc\cos\phi$, $p_{31}=-ac\sin\phi$, $p_{12}=ab$.

If we substitute (7.19) into (7.3) we obtain the Bennett motion

$$(7.20)\qquad \begin{aligned} X&=[(a^2-c^2)t^4-2(a^2+2b^2+c^2)t^2+(a^2-c^2)]x\\ &\quad -4abt(t^2-1)y-2ac(t^4-1)z+4a(b^2+c^2)t(t^2+1)w,\\ Y&=-4abt(t^2-1)x+[-(a^2+c^2)t^4+2(a^2+2b^2-c^2)t^2-(a^2+c^2)]y\\ &\quad +4bct(t^2+1)z+2b(a^2+c^2)(t^4-1)w,\\ Z&=-2ac(t^4-1)x+4bct(t^2+1)y\\ &\quad +[-(a^2-c^2)t^4+2(a^2-2b^2+c^2)t^2-(a^2-c^2)]z\\ &\quad +4c(a^2-b^2)t(t^2-1)w,\\ W&=[(a^2+c^2)t^4+2(-a^2+2b^2+c^2)t^2+(a^2+c^2)]w. \end{aligned}$$

A first conclusion is: the paths of the motion are in general twisted quartics of the second kind. Further properties follow from the circumstance that the set (7.19) contains four isotropic lines. Indeed the conic at infinity of H (7.18) has four intersections with the isotropic conic Ω. They are seen to be

$$(7.21)\qquad \begin{aligned} &J_1(iaq_2, bq_1, cq_3), && J_1'(-iaq_2, bq_1, cq_3),\\ &J_2(iaq_2, -bq_1, cq_3), && J_2'(-iaq_2, -bq_1, cq_3), \end{aligned}$$

with

$$(7.22)\qquad q_1=(a^2+c^2)^{1/2},\quad q_2=(b^2+c^2)^{1/2},\quad q_3=(a^2-b^2)^{1/2}.$$

J_1 and J_1' are conjugate imaginary points and so are J_2, J_2'. Through each point J there passes one generator of the set (7.19); we shall denote them by l_1, l_1', l_2, l_2' respectively. We know that the reflection into an isotropic line l (with its point at infinity J on Ω) is a singular displacement: it transforms an arbitrary point into J; furthermore a point P in the isotropic plane through l is transformed into all points of the line PJ. Applying this to the motion under consideration we obtain: all paths pass through the four points (7.21); the paths are circular curves and their isotropic points are the same for all paths.

Let $\alpha_1, \alpha_1', \alpha_2, \alpha_2'$ be the isotropic planes through l_1, l_1', l_2, l_2' respectively. Then the path of a point P in α_1, say, contains the line PJ_1; it degenerates into this line and a twisted cubic (passing through J_1', J_2, J_2'). More interesting still is the path of a point on the intersection of two planes α; obviously it degenerates into two isotropic lines and a conic. The latter passes through the two remaining isotropic points and it is therefore a circle. The four planes α give rise to six intersections, the edges of the tetrahedron of which the α's are the faces. As two conjugate imaginary planes have a real intersection we obtain: there are two skew lines in E, the intersection m_1 of α_1, α_1' and the intersection m_2 of α_2, α_2', with the property that the path of any point on m_1 or m_2 is a circle.

The line l_1 passes through J_1; its first Plücker vector is therefore (iaq_2, bq_1, cq_3) and hence according to (7.19) the second reads $(ibcq_2, acq_1, -abq_3)$; it follows that l_1 passes through $(aq_1, -ibq_2, 0, q_3)$. The tangent at J_1 to Ω is $iaq_2X + bq_1Y + cq_3Z = W = 0$; this implies that the plane α_1 through l_1 and this tangent has the equation

$$iaq_2X + bq_1Y + cq_3Z - iq_1q_2q_3W = 0.$$

For m_1' in Σ we obtain therefore

$$aX - q_1q_3W = 0, \quad bq_1Y + cq_3Z = 0, \tag{7.23}$$

and analogously for m_2':

$$aX + q_1q_3W = 0, \quad bq_1Y - cq_3Z = 0. \tag{7.24}$$

In the original position of E its frame coincides with that of Σ. Hence the two special lines m_1, m_2 in E follow from (7.23) and (7.24) by replacing X, Y, Z, W by x, y, z, w.

m_1 and m_2 intersect o_x orthogonally at the points $(q_1q_3, 0, 0, a)$ and $(-q_1q_3, 0, 0, a)$; their distance is $2q_1q_3/a$; if γ is their angle then $\tan(\gamma/2) = bq_1/cq_3$.

An arbitrary point on m_1 describes a circle through J_2 and J_2'. Hence the circles described by the points of m_1 are in parallel planes, passing through the line J_2J_2'. The pole of this line with respect to Ω is, however, the point at infinity of the line m_2'. Hence the parallel planes are perpendicular to m_2'.

Let M be the center of the circle described by the point P_1 on m_1. Then the distance MP_1 is constant during the motion. Hence, during the inverse motion M describes a circle with center P_1. Since the motion is line-symmetric the inverse motion is congruent with the direct motion. The conclusion is that M is on m_2', for during the inverse motion a point of m_2' describes a circle in a plane perpendicular to m_1. Summing up, the Bennett motion has the following property. In E there exist a pair of skew lines m_1, m_2 and in Σ a pair m_1', m_2' such that the figure m_1', m_2' is congruent with m_1, m_2. During the motion any point on m_1 describes a circle, in a plane perpendicular to m_2', with center on m_2'; a point on m_2 describes a circle, in a plane perpendicular to m_1', and with its center on m_1'.

It follows from (7.21) that the equation of J_2, J_2' reads $cq_3Y + bq_1Z = W = 0$. Hence the planes of the paths of points on m_1 are parallel to the plane $cq_3Y + bq_1Z = 0$, which is indeed perpendicular to m_2', given by (7.24). An arbitrary point P_1 on m_1 is obtained from (7.23): $x = q_1q_3$, $y = \lambda cq_3$, $z = -\lambda bq_1$, $w = a$. Its positions during the motion follow from (7.20). For $t = 0$ it is seen to be

$$X = (a^2 - c^2)q_1q_3 - 2abc\lambda q_1, \quad Y = -(a^2 + c^2)\lambda cq_3 - 2ab(a^2 + c^2),$$

$$Z = 2acq_1q_3 + (a^2 - c^2)\lambda bq_1, \quad W = (a^2 + c^2)a,$$

which is a point on the path of P_1. Hence the plane of this path is $cq_3Y + bq_1Z + \lambda a(c^2 - b^2)W = 0$. This plane's intersection with m_2', that is the center of the path of P_1, reads $X = -q_1q_3(b^2 + c^2)$, $Y = \lambda cq_3(b^2 - c^2)$, $Z = \lambda bq_1(b^2 - c^2)$, $W = a(b^2 + c^2)$.

Example 60. Show that the line m_1 (m_2) generates a hyperboloid of revolution with m_2' (m_1') as axis.

The fixed axode of the motion is the locus of the common perpendiculars of two consecutive generators of the hyperboloid H, given by (7.19). It is a rational ruled surface of the sixth order.

Example 61. Prove the latter statement.

We have considered the Bennett motion as a special case of a line-symmetric motion. It was derived by BENNETT [1913] in terms of a certain mechanism for which two moving points describe circles. Although the motion has been studied by several geometers, it was KRAMES [1937e] who recognized it as a symmetric motion with a hyperboloid as basic surface. GROENEVELD [1954] has written a monograph on this motion.

8. Plane-symmetric motions

In the preceding section we defined a class of special motions by considering the series of positions of the space E_0 when it is reflected into a set of lines in Σ. We deal now with an analogous procedure. Let a continuous set of ∞^1 planes $\mathrm{U}(t)$ be given in Σ; E_0 coincides with Σ (provided we let frame o_{xyz} coincide with O_{XYZ}). If E_0 is reflected into the planes $\mathrm{U}(t)$ of the developable we obtain a series of positions $E(t)$. There is, however, a fundamental difference from the former construction. If we reflect a space into a line we obtain a congruent space, if we reflect into a plane the transformed space is symmetric with the original but it has the opposite orientation. However, if we change the orientation of $E(t)$, by putting $x = -\bar{x}$, $y = -\bar{y}$, $z = -\bar{z}$, the set of positions do define a motion of E with respect to Σ. We shall call it a plane-symmetric motion; $\mathrm{U}(t)$ is its basic developable.

If U_i are the coordinates of a plane U, with $U_1^2 + U_2^2 + U_3^2 = 1$, it is easy to derive that the reflection of a point x, y, z into U has the coordinates $x' = (-U_1^2 + U_2^2 + U_3^2)x - 2U_1U_2y - 2U_1U_3z - 2U_1U_4$, and so on for y' and z' by cyclic substitution. Hence, omitting the bars, we obtain for the general plane-symmetric motion

$$
\begin{aligned}
X &= (2U_1^2 - 1)x + 2U_1U_2y + 2U_1U_3z - 2U_1U_4, \\
Y &= 2U_2U_1x + (2U_2^2 - 1)y + 2U_2U_3z - 2U_2U_4, \qquad (8.1) \\
Z &= 2U_3U_1x + 2U_3U_2y + (2U_3^2 - 1)z - 2U_3U_4.
\end{aligned}
$$

If we compare this with Chapter III, (12.12) we see that the motion which transforms E_0 into $E(t)$ is a screw motion with its rotation equal to a half-turn about the axis (U_1, U_2, U_3) through O, and its translation, parallel to this axis, equal to $-2U_4$. We are interested, however, in the displacement which carries $E(t_1)$ into $E(t_2)$. If we reflect $E(t_1)$ into $\mathrm{U}(t_1)$ we obtain E_0 which when reflected into $\mathrm{U}(t_2)$ yields $E(t_2)$. Hence, this displacement is the product of the reflection into $\mathrm{U}(t_1)$ followed by that into $\mathrm{U}(t_2)$. The product of two plane reflections is obviously a rotation about the intersection of the two planes (or a translation if the planes are parallel). The conclusion is: if E has a plane-symmetric motion any position follows from any other by a rotation (and not a screw-displacement as in the general case). This is especially the case for two consecutive positions; hence the instantaneous screw is at any instant a pure rotation. Its axis is the intersection of two consecutive planes of the developable generated by $\mathrm{U}(t)$. Therefore, the fixed axode is the surface of the tangents to the curve associated with the developable. Hence the axode is a developable ruled surface, which also follows from the fact that the pitch of the instantaneous screw, and hence the distribution parameter of the axode, is zero. Furthermore the moving axode is obvious obtained by reflecting the fixed axode into a plane of the developable U(t).

Example 62. Show that the inverse motion of (8.1) is obtained by interchanging $X, Y, Z, U_1, U_2, U_3, U_4, x, y, z$ with $x, y, z, U_1, U_2, U_3, -U_4, X, Y, Z$.

If the basic developable consists of parallel planes the motion is a translation; if the planes pass through a line l the motion consists of the rotation about l. A quadratic developable is a quadratic cone (this is the dual theorem of: a quadratic curve is a plane curve); if the developable is (any) cone the motion has a fixed point and it is therefore spherical.

To discuss a non-trivial example of a plane-symmetric motion we choose as basic developable the set of osculating planes of the cubic parabola $X = t^3$, $Y = t^2$, $Z = t$, $W = 1$. Then we have $U_1 = 1$, $U_2 = -3t$, $U_3 = 3t^2$, $U_4 = -t^3$. For (8.1) we obtain, since $U_1^2 + U_2^2 + U_3^2 = 1 + 9t^2 + 9t^4$,

$$
\begin{aligned}
X &= (1 - 9t^2 - 9t^4)x - 6ty + 6t^2z + 2t^3w,\\
Y &= -6tx + (-1 + 9t^2 - 9t^4)y - 18t^3z - 6t^4w,\\
Z &= 6t^2x - 18t^3y + (-1 - 9t^2 + 9t^4)z + 6t^5w,\\
W &= (1 + 9t^2 + 9t^4)w.
\end{aligned}
\tag{8.2}
$$

The paths of this motion are rational quintic curves, all passing (at $t = \infty$) through the point $(0, 0, 1, 0)$ at infinity; the other intersections with $W = 0$ are imaginary. The fixed axode is the surface of the tangents to the cubic parabola, which is a (developable) ruled surface of the fourth order.

Example 63. Show that the motion is in general of order $2n$ if the basic developable is of class n. In (8.2) it is diminished by one because the plane at infinity belongs to the developable.

Example 64. Show that the fixed axode for (8.1) is given by the following line coordinates of its generators

$$
\begin{aligned}
p_{14} &= U_2\dot{U}_3 - U_3\dot{U}_2, & p_{24} &= U_3\dot{U}_1 - U_1\dot{U}_3, & p_{34} &= U_1\dot{U}_2 - U_2\dot{U}_1,\\
p_{23} &= U_1\dot{U}_4 - U_4\dot{U}_1, & p_{31} &= U_2\dot{U}_4 - U_4\dot{U}_2, & p_{12} &= U_3\dot{U}_4 - U_4\dot{U}_3.
\end{aligned}
$$

9. Motions defined by geometrical conditions

In the preceding sections examples of spatial motions have been given by various methods. Yet another procedure would be the following one. As the degree of freedom of a moving space is equal to six a motion can be defined if we impose five simple conditions. Such a condition arises for instance if a moving point must remain on a given surface of the fixed space. One could ask, for instance, for a formula giving the order of a motion if five points of E must each remain on an algebraic surface, each having a given order, in Σ. In contrast to the much simpler case of plane kinematics, such a formula does not seem to be known. Even for the case where all five surfaces are planes the problem is a complicated one since so many cases have to be considered: Some planes may be parallel, or parallel to the same line; of the moving points three may be on a line or four of them on a plane. DARBOUX [1897] has derived some results for this problem and we shall make some remarks about it in Chapter X on n-parameter motion.

Another interesting possibility arises if one prescribes the path of a moving point; it is equivalent to two simple conditions. For instance we could study the motion at which two points of E are compelled to remain on given curves and a third point on a given surface. It seems that no systematic account has been developed for such problems. Much work has been done on the motion for which two points A_1, A_2 describe given circles, but obviously these conditions do not define a one-parameter motion; the motion of the line A_1A_2, however, is determined in this case and it has been shown that the path of a point on it is a curve of order eight. This problem is strongly related to the theory of mechanisms (in fact it is a spatial four-bar motion).

10. Special spherical motions

We consider now some special *spherical* motions. If we specialize the examples of spatial motions, dealt with in the preceding sections, to spherical motion the results are often trivial. So, for instance, the Darboux motion (Section 3) is only spherical if $a = b = c = 0$, which reduces it to a rotation about a fixed axis. The same holds for the Bricard motion (Section 7), which is spherical for $h = 0$.

A general Krames motion is only spherical if the basic surface R′ is a cone. If its vertex coincides with the origin we have in (7.3) $b_1 = b_2 = b_3 = 0$. Hence we obtain

$$
\begin{aligned}
X &= (c_1^2 - c_2^2 - c_3^2)x + 2c_1c_2y + 2c_1c_3z,\\
Y &= 2c_2c_1x + (-c_1^2 + c_2^2 - c_3^2)y + 2c_2c_3z,\\
Z &= 2c_3c_1x + 2c_3c_2y + (-c_1^2 - c_2^2 + c_3^2)z,\\
W &= (c_1^2 + c_2^2 + c_3^2)w.
\end{aligned}
\tag{10.1}
$$

c_i are functions of a parameter t. As an example we choose as basic surface the quadratic cone $(X^2/a^2)+(Y^2/b^2)-(Z^2/c^2)=0$, with $a \geqq b$. Then we have

$$c_1 : c_2 : c_3 = a(1-t^2) : 2bt : c(1+t^2). \tag{10.2}$$

The motion is given by (10.1) with (10.2) substituted into it. The paths are quartics of the second kind; the unique quadric defined by each path is the sphere on which the moving point remains. The four points at infinity of a path are isotropic; they are the same for all paths: J_1, $J_1' = (\pm iaq_2, bq_1, cq_3)$ and J_2, $J_2' = (\pm iaq_2, -bq_1, cq_3)$, with $q_1=(a^2+c^2)^{1/2}$, $q_2=(b^2+c^2)^{1/2}$, $q_3=(a^2-b^2)^{1/2}$. For a point P of E in an isotropic plane, through O and a point J, the path degenerates into the line PJ and a twisted cubic; for a point in two such planes the path consists of two lines and a circle. The isotropic planes through J_1 and J_1' intersect in the real line m_1 with $x:y:z=0:cq_3:-bq_1$; it has the property that all points on it describe circles, in planes parallel to $cq_3Y+bq_1Z=0$; a second line m_2 with $x:y:z=0:cq_3:bq_1$ has analogous properties. The circles described by points on m_1 (and, respectively m_2) have their centers on m_2 (or respectively m_1). This motion is a limit case of the Bennett motion.

Example 65. Verify these properties analytically by means of (10.1) and (10.2).
Example 66. Determine, for a point on m_1, the plane and the center of its path.
Example 67. Show that the lines m_1 and m_2 describe circular cones.

The fixed axode of the motion is the locus of the central tangents of the basic cone. An instantaneous axis passes through O, it is normal to two consecutive generators and thus to the tangent plane of the cone. Hence the fixed axode is determined by $X:Y:Z=bc(t^2-1):-2act:ab(t^2+1)$; its equation is $a^2X^2+b^2Y^2-c^2Z^2=0$. The moving axode is congruent with it; its positions in Σ are obtained if the fixed axode is reflected into a generator of the basic cone. The motion is generated by the rolling of a quadratic cone on a congruent one, such that at any moment corresponding generators coincide.

Example 68. Show that the path of a point on the moving axode is a twisted quartic of the second kind with a cusp.

A plane-symmetric spherical motion arises if the basic developable consists of planes through the center of the sphere. The fixed axode is the cone enveloped by the developable; the moving axode is a cone, it can be obtained from the fixed axode by reflecting it into a plane of the developable.

Example 69. Determine the motion if the basic developable consists of the tangent planes of a quadratic cone.

Any displacement with point O fixed is given by

$$
(10.3)\qquad
\begin{aligned}
X &= (c_0^2 + c_1^2 - c_2^2 - c_3^2)x + 2(c_1c_2 - c_0c_3)y + 2(c_1c_3 + c_0c_2)z,\\
Y &= 2(c_2c_1 + c_0c_3)x + (c_0^2 - c_1^2 + c_2^2 - c_3^2)y + 2(c_2c_3 - c_0c_1)z,\\
Z &= 2(c_3c_1 - c_0c_2)x + 2(c_3c_2 + c_0c_1)y + (c_0^2 - c_1^2 - c_2^2 + c_3^2)z,\\
W &= (c_0^2 + c_1^2 + c_2^2 + c_3^2)w.
\end{aligned}
$$

A specific spherical motion is represented by (10.3) if the parameters c_i are specified functions of t. As there are three degrees of freedom such a motion is in general defined by two simple conditions. Hence we can obtain a motion by prescribing the paths of two points of the moving space. If A is a point of E its path is similar to that of every point on the line OA. We will therefore restrict ourselves to points on the same sphere about O.

As an example we suppose that two points A_1, A_2 of E describe circles Γ_1, Γ_2 on their sphere, with spherical center M_i and spherical radius ρ_i ($i = 1, 2$). Such a motion is called a spherical coupler motion or a spherical four-bar motion, the four "bars" being the arcs M_1M_2, M_1A_1, M_2A_2, A_1A_2 which define the motion; A_1A_2 is called the coupler of the spherical quadrilateral $M_1M_2A_2A_1$.

The condition that, under spherical motion, A_i remains on Γ_i is equivalent to the condition that A_i remains on the plane U^i of Γ_i. Let the equation of U^i be $U_1^iX + U_2^iY + U_3^iZ + U_4^iW = 0$ and let furthermore $A_i = (x_i, y_i, z_i)$. Hence, eliminating X, Y, Z, W by means of (10.3) we obtain two quadratic equations $\mathrm{Q}_i = 0$, $i = 1, 2$ for the homogeneous unknowns c_j; the coefficients depend on the constants U_k^i, x_i, y_i, z_i. In the three dimensional c_j-space the two conditions represent two quadrics. Their intersection, a quartic space curve γ of the first kind, is an image of the set of positions of E during the motion.

γ is in general a curve of genus one; c_j may be written as elliptic functions of a parameter. Such functions represent a fourth order curve and as, in view of (10.3), the path of an arbitrary point is represented by quadratic functions of c_j we draw the conclusion: the paths of the spherical coupler motion are curves of order eight and genus one.

The motion under consideration is essentially defined by the four bars of the quadrilateral; hence there are ∞^4 spherical coupler motions. There is a multitude of special cases. First of all there is no motion possible if any bar is more than the sum of the remaining three; then the two quadrics have no real common point, γ is an imaginary curve. Furthermore one (or both) of the circles may be great circles; two (or more) bars (adjacent or opposite) may be equal; a bar may be equal to $\pi/2$; the planes of the circles may be perpendicular; the sum of two bars may be equal to that of the other two (which implies that there are positions at which the four points M_i, A_i are on a great circle). All these special cases have their counterpart in the behavior of the curve γ; it may have a double point or it may be degenerate. All this has its influence on the set of paths; we already know that even in the general case there are points, namely A_1, A_2, the paths of which are not of the eighth but of the second order; in the special cases the order of the paths of a set of points or even of all points may be less than eight.

We restrict ourselves here to these general remarks on the spherical coupler motion, a complete discussion of all cases would obviously require an elaborate amount of detail.

Example 70. Show that the line-symmetric motion with a quadratic cone as its basic surface (treated above) is a special, spherical, coupler motion; the paths are rational quartic curves.
Example 71. We choose the frame O_{XYZ} such that $M_1, M_2 = (\cos\frac{1}{2}g, \mp\sin\frac{1}{2}g, 0)$ and the frame O_{xyz} such that $A_1, A_2 = (\cos\frac{1}{2}k, \mp\sin\frac{1}{2}k, 0)$. Derive the equations of the quadrics Q_1, Q_2 and discuss their intersection γ.

11. Special plane motions

There is an extensive literature on *plane* motions, both from purely geometric considerations and from the view point of application to the theory of plane mechanisms. Within the scope of this book we restrict ourselves to some important examples, for more details one may refer to numerous text-books and special papers on the subject.

The Frenet–Serret motion, considered in Section 2, simplifies considerably for the planar case. A curve Γ being given in the fixed plane Σ, the motion is defined as follows. The origin of the moving frame o_{xy} has Γ as its path and the o_x-axis rotates so that it always coincides with the tangent to Γ. In the formulas of Section 2 we have now $\tau = 0$.

Example 72. Show that the pole P of the motion coincides with the corresponding center of curvature of Γ; and that the Darboux screw is a rotation about the perpendicular to the plane, through P.
Example 73. Show that any plane motion is a Frenet–Serret motion; for Γ we may take any evolvent of the fixed centrode (the latter being the evolute of Γ).

A line-symmetric plane motion may be defined in the same way as in the spatial and the spherical cases, starting from a basic set of lines. In the plane case these lines are the set of tangents to a curve in the fixed plane Σ. But, there is an essential difference in this case: Confining ourselves to the plane, a reflection into a line, lying in the plane, gives us a figure which is symmetric with the original one and not congruent as it is in space. (Of course, congruent figures result from reflecting about lines normal to the plane, this yields what we could consider as planar reflections about a point.) Moreover a reflection into a line l of Σ is the same procedure as the reflection into the plane through l perpendicular to Σ. Hence line-symmetry and plane-symmetry are identical concepts in the planar case.

If any line l of the basic set is represented by its normal equation such a set may be given by

$$X \sin \psi + Y \cos \psi = h(\psi), \tag{11.1}$$

$h(\psi)$ being an arbitrary function. If the frames in Σ and in the original position E_0 of the moving plane coincide, the reflection of (x, y) into the line (11.1) is seen to be

$$\begin{aligned} \bar{X} &= x \cos 2\psi - y \sin 2\psi + 2h(\psi) \sin \psi, \\ \bar{Y} &= -x \sin 2\psi - y \cos 2\psi + 2h(\psi) \cos \psi, \end{aligned} \tag{11.2}$$

which transforms E_0 into a plane $\bar{E}(\psi)$ with the opposite orientation. To generate a continuous set of positions with the same orientation as Σ we reflect $\bar{E}(\psi)$ into a fixed line l_0 of Σ; the new set of positions defines a motion of E with respect to Σ which is called a symmetric motion.

To derive a standard representation of such a motion we take O_{XY} such that for $\psi = 0$ the line (11.1) coincides with O_X; moreover, let the tangent point (of the line (11.1)) with its envelope be the origin. This implies $h_0 = 0$, $h_1 = 0$. Furthermore let the fixed line l_0 be O_X. The motion is now given by

$$\begin{aligned} X &= x \cos 2\psi - y \sin 2\psi + 2h(\psi) \sin \psi, \\ Y &= x \sin 2\psi + y \cos 2\psi - 2h(\psi) \cos \psi. \end{aligned} \tag{11.3}$$

The most characteristic property of a symmetric plane motion is that *the fixed and the moving centrode are symmetric curves with respect to the instantaneous common pole tangent.*

Example 74. From (11.3) derive that the moving centrode is given by

$$x = h(\psi) \sin \psi + h'(\psi) \cos \psi, \quad y = h(\psi) \cos \psi - h'(\psi) \sin \psi,$$

and the fixed centrode by

$$X = h(\psi) \sin \psi + h'(\psi) \cos \psi, \quad Y = -h(\psi) \cos \psi + h'(\psi) \sin \psi.$$

If we compare (11.3) with the general plane motion $X =$

$x \cos\phi - y \sin\phi + a(\phi)$, $Y = x \sin\phi + y \cos\phi + b(\phi)$, we get $\phi = 2\psi$, $a = 2h(\psi)\sin\psi$, $b = -2h(\psi)\cos\psi$. The general motion is defined by two arbitrary functions, a symmetric motion depends on the function $h(\psi)$ only.

From (11.3) it follows that the frames of E and Σ are canonic for the zero-position $\psi = 0$. Hence the instantaneous invariants a_n, b_n can be calculated in terms of the numbers $h_k = d^k h(0)/d\psi^k$. We obtain

$$\begin{aligned} &a_0 = 0, \quad a_1 = 0, \quad a_2 = 0, \quad a_3 = (3/4)h_2, \\ &a_4 = \tfrac{1}{2}h_3, \quad a_5 = (5/16)(h_4 - 2h_2), \ldots, \\ &b_0 = 0, \quad b_1 = 0, \quad b_2 = -\tfrac{1}{2}h_2, \quad b_3 = -\tfrac{1}{4}h_3, \\ &b_4 = -(1/8)(h_4 - 6h_2), \quad b_5 = -(1/16)(h_5 - 10h_3), \ldots . \end{aligned} \tag{11.4}$$

Example 75. Show that any motion is "symmetric up to the second order"; it is symmetric up to the third if $2a_3 + 3b_2 = 0$, to the fourth if moreover $a_4 + 2b_3 = 0$, to the fifth if moreover $2a_5 + 5b_4 - 5b_2 = 0$. (VELDKAMP [1963].)

Example 76. Discuss the symmetric motion if the centrodes are ellipses, each with semi-axes p and q. Show that the path of a moving point is in general a rational quartic and determine its three double points. Show that the foci of the moving centrode describe circles. Consider the case $p = q$.

A moving plane has three degrees of freedom. Hence a plane motion is in general determined if the paths of two points are prescribed. The classical example is the four-bar motion, for this case the two given points describe circles.

Let the points $M_1 = (-M, 0)$, $M_2 = (M, 0)$ in the fixed plane Σ be the centers of the circles, and R_1 and R_2 their radii; let $A_1 = (-m, 0)$ and $A_2 = (m, 0)$ in E have these circles as their respective paths. The motion will only be possible if none of the four bars $2M, 2m, R_1, R_2$ is larger than the sum of the remaining three. We shall show how the equation of the path of an arbitrary point (x, y) of E can be derived.

The motion is given by

$$X = x \cos\phi - y \sin\phi + a, \quad Y = x \sin\phi + y \cos\phi + b. \tag{11.5}$$

We have $X(A_1) = -m \cos\phi + a$, $Y(A_1) = -m \sin\phi + b$, upon eliminating a and b by means of (11.5):

$$X(A_1) = X - (x + m)\cos\phi + y \sin\phi,$$

$$Y(A_1) = Y - y \cos\phi - (x + m)\sin\phi.$$

The condition that A_1 remains on $(M_1; R_1)$ reads $[X(A_1) + M]^2 + [Y(A_1)]^2 = R_1^2$ which gives rise to a *linear* equation for $\cos\phi$ and $\sin\phi$:

$$(11.6)\quad \begin{aligned} &2[(X+M)(x+m)+Yy]\cos\phi-2[(X+M)y-(x+m)Y]\sin\phi=\\ &=(X+M)^2+(x+m)^2+Y^2+y^2-R_1^2. \end{aligned}$$

Analogously we obtain from A_2

$$(11.7)\quad \begin{aligned} &2[(X-M)(x-m)+Yy]\cos\phi-2[(X-M)y-(x-m)Y]\sin\phi=\\ &=(X-M)^2+(x-m)^2+Y^2+y^2-R_2^2. \end{aligned}$$

Adding and subtracting (11.6) and (11.7) we get the somewhat simpler equations

$$(11.8)\quad \begin{aligned} &4[Xx+Yy+Mm]\cos\phi-4[Xy-Yx]\sin\phi=\\ &=2[X^2+Y^2+x^2+y^2+M^2+m^2]-(R_1^2+R_2^2),\\ &4[Xm+xM]\cos\phi-4[yM-Ym]\sin\phi=4[XM+xm]-(R_1^2-R_2^2), \end{aligned}$$

from which $\cos\phi$ and $\sin\phi$ can be determined. Then using the condition $\cos^2\phi+\sin^2\phi=1$ we obtain a relation, between (X, Y) and (x, y), which is the equation of the path of (x, y), usually called a *coupler curve*. From (11.8) it is easily seen that the equation remains the same if we interchange X, Y, M and x, y, m; this could be expected because the inverse motion is that of the four-bar with fixed centers A_1, A_2 and the moving bar M_1M_2.

The equation of the path is too complicated to be of much use but we can at least derive from (11.8) that the curve is of order six and, after some algebra, that the sixth order terms are $(X^2+Y^2)^3$ and that those of the fifth and the fourth order have the factors $(X^2+Y^2)^2$ and (X^2+Y^2) respectively. The conclusion is: the coupler curve is a tri-circular sextic. It may be derived moreover that the line M_jI_i is tangent to the curve at I_i, with $i=1,2$, $j=1,2$. This means that M_1 and M_2 are foci of the coupler curve. (They are in fact singular foci since the isotropic points I_1, I_2 are on the coupler curve.)

Another analytic approach to the curve is the following one. Using the same coordinate systems, let $\angle A_1M_1O=\theta_1$, $\angle A_2M_2O=\theta_2$, then $A_1=(-M+R_1\cos\theta_1, R_1\sin\theta_1)$, $A_2=(M-R_2\cos\theta_2, R_2\sin\theta_2)$. The condition $A_1A_2=2m$ gives us, with $2M=g$, $2m=h$,

$$(11.9)\quad \begin{aligned} &(R_1^2+R_2^2+g^2-h^2)-2R_1g\cos\theta_1-2R_2g\cos\theta_2\\ &\quad+2R_1R_2(\cos\theta_1\cos\theta_2-\sin\theta_1\sin\theta_2)=0, \end{aligned}$$

or if $\tan(\frac{1}{2}\theta_1)=u/w$, $\tan(\frac{1}{2}\theta_2)=v/w$ and provided $(w^2+u^2)(w^2+v^2)\neq 0$, $w\neq 0$:

$$(11.10)\quad S_3T_4u^2v^2-S_2T_1u^2w^2-S_1T_2v^2w^2-8R_1R_2uvw^2+S_4T_3w^4=0,$$

with

$$
(11.11)\qquad
\begin{aligned}
S_1 &= -R_1 + R_2 + h + g, & T_1 &= -R_1 + R_2 + h - g,\\
S_2 &= R_1 - R_2 + h + g, & T_2 &= R_1 - R_2 + h - g,\\
S_3 &= R_1 + R_2 - h + g, & T_3 &= R_1 + R_2 - h - g,\\
S_4 &= R_1 + R_2 + h - g, & T_4 &= R_1 + R_2 + h + g.
\end{aligned}
$$

A position of the coupler plane E undergoing a four-bar motion is determined by a pair (θ_1, θ_2) satisfying (11.9); hence a position is mapped as a point of the (image) curve k, in the (u, v, w)-plane, given by equation (11.10). Obviously k is a quartic curve with the double points $B_1 = (1, 0, 0)$ and $B_2 = (0, 1, 0)$. It has in general no other double points, which means that k is a curve of genus one, an elliptic curve, the coordinates of whose points may be expressed by elliptic functions of a parameter t.

This shows that the general four-bar motion is an elliptic motion and that its paths are in general curves of genus one. The double points B_1 and B_2 of k do not correspond to positions of E. Other special points of k are those for which $(u^2 + w^2)(v^2 + w^2) = 0$, this equation represents four lines, two through B_1 and two through B_2. They also intersect k at six other points, which are conjugate imaginary in pairs, namely $C_1, C_1' = (1, -1, \pm i)$, $C_2, C_2' = [R_1 - g, -(R_1 + g), \pm i(R_1 + g)]$, $C_3, C_3' = [R_2 - g, -(R_2 + g), \pm i(R_2 - g)]$.

For the angle ϕ between o_x and O_X we have (Fig. 77):

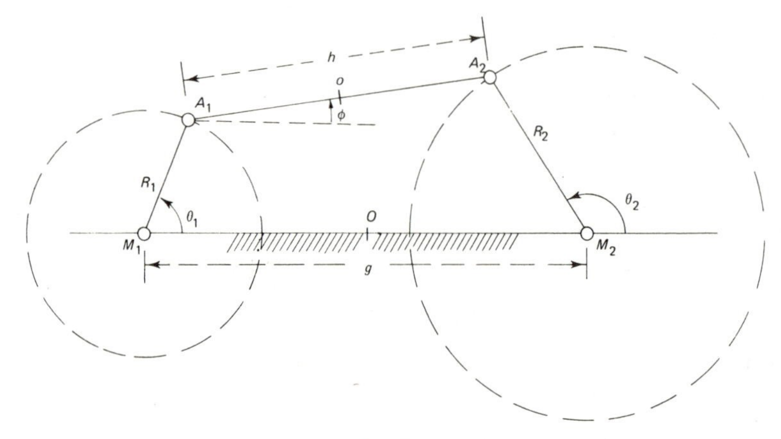

Fig. 77.

$$\cos\phi = [X(A_2) - X(A_1)]/h, \quad \sin\phi = [Y(A_2) - Y(A_1)]/h;$$

furthermore $X(o) = [X(A_1) + X(A_2)]/2$, $Y(o) = [Y(A_1) + Y(A_2)]/2$. Hence in terms of θ_1 and θ_2 the four-bar motion is given by

$$\begin{aligned} X &= h^{-1}[g - R_2\cos\theta_2 - R_1\cos\theta_1]x - h^{-1}[R_2\sin\theta_2 - R_1\sin\theta_1]y \\ &\quad + [R_1\cos\theta_1 - R_2\cos\theta_2]/2, \\ Y &= h^{-1}[R_2\sin\theta_2 - R_1\sin\theta_1]x + h^{-1}[g - R_2\cos\theta_2 - R_1\cos\theta_1]y \\ &\quad + [R_1\sin\theta_1 + R_2\sin\theta_2]/2, \end{aligned} \tag{11.12}$$

provided θ_1, θ_2 satisfy (11.9). Expressed in terms of the parameters u, v, w the motion reads in homogeneous coordinates

$$\begin{aligned} X &= h^{-1}[g(u^2+w^2)(v^2+w^2) - R_2(w^2-v^2)(u^2+w^2) - R_1(w^2-u^2)(v^2+w^2)]x \\ &\quad - h^{-1}[2R_2vw(u^2+w^2) - 2R_1uw(v^2+w^2)]y \\ &\quad + \tfrac{1}{2}[R_1(w^2-u^2)(v^2+w^2) - R_2(w^2-v^2)(u^2+w^2)]z, \end{aligned} \tag{11.13}$$

analogously for Y, and furthermore

$$Z = (u^2+w^2)(v^2+w^2)z,$$

with the condition that u, v, w satisfy (11.10).

The intersection of a path with the line at infinity is given by $(u^2+w^2)(v^2+w^2) = 0$; hence a moving point is at infinity at the six positions corresponding to the points C_j, C'_j $(j = 1, 2, 3)$ on the image curve k. For these singular positions all points of E are transformed into the isotropic points I_1, I_2 of Σ. We have by this once more shown that the path is a tricircular sextic, and moreover that its genus is *one*.

Example 77. Show that for C_1, C'_1 "the" position of a moving point consists of all points on an isotropic line. Determine the position of a moving point corresponding to the images C_2, C'_2 and C_3, C'_3.

A remarkable, although somewhat isolated, theorem on the coupler curve was given by S. ROBERTS [1875]. Let $M_1A_1A_2M_2$ be a four-bar and A a specified point in the moving plane (Fig. 78). We construct the parallelograms $M_1A_1AA'_1$ and $AA_2M_2A'_2$, the triangles A'_1AG_1 and AA'_2G_2 both similar to A_1A_2A, and then the parallelogram $G_1AG_2M_3$. We make use of planar vector-algebra, or what is the same thing, complex numbers. If we let $M_1M_2 = \boldsymbol{p}_0$, $M_1A_1 = \boldsymbol{p}_2$, $A_1A_2 = \boldsymbol{p}_1$, $A_2M_2 = \boldsymbol{p}_3$, $A_1A = \boldsymbol{p}$, $\angle A_2A_1A = \alpha$, then $\boldsymbol{p} = (p/p_1)e^{i\alpha}\boldsymbol{p}_1$, $A'_1G_1 = (p/p_1)e^{i\alpha}\boldsymbol{p}_2$, $AG_2 = (p/p_1)e^{i\alpha}\boldsymbol{p}_3$. From this it follows that $M_1M_3 = M_1A'_1 + A'_1G_1 + G_1M_3 = (p/p_1)e^{i\alpha}(\boldsymbol{p}_1 + \boldsymbol{p}_2 + \boldsymbol{p}_3) = (p/p_1)e^{i\alpha}\boldsymbol{p}_0$.

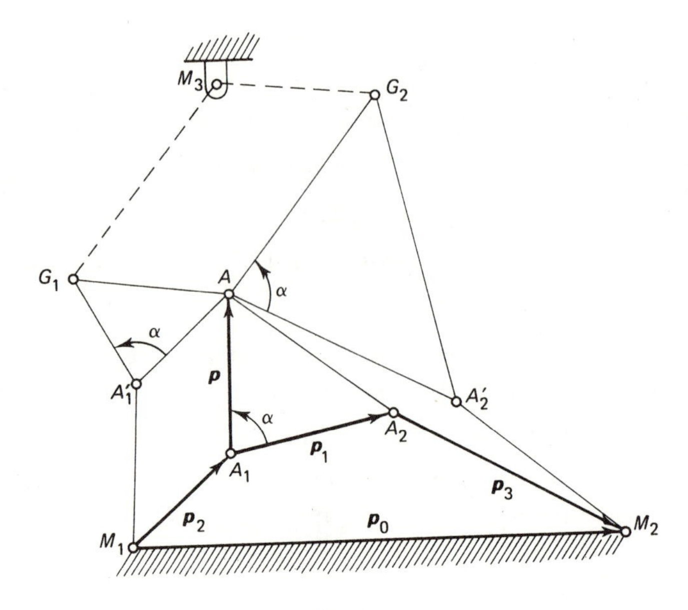

Fig. 78.

Hence M_3 is a fixed point, independent of the position of the four-bar; it follows furthermore that the vertex M_3 is located so that triangle $M_1M_2M_3$ is similar to the "coupler-triangle" A_1A_2A. A consequence is that the path of A for the four-bar motion of $M_1A_1A_2M_2$ coincides with its path under two other four-bar motions, namely $M_2A_2'G_2M_3$ and $M_3G_1A_1'M_1$, the three motions being equivalent only with respect to the coupler curve described by A. We knew that M_1, M_2 are foci of the curve, M_3 is now seen to be the third one.

Two (or more) motions of the plane E such that for a certain point of E the paths coincide are called *path-cognate* motions.

A non-degenerate curve of order n has at most $\frac{1}{2}(n-1)(n-2)$ double points; for $n = 6$ this number is 10. The coupler curve has genus one, hence it has 9 double points. I_1 and I_2 being three-fold points count for three double points each. This implies that the curve has three finite double points. Let D be a double point of the path of A. Point A is taken as the third vertex of the coupler triangle A_1A_2A, and $\angle A_1AA_2 = \gamma$ is its vertex angle. Since D is a double point there are two positions of this triangle with the same vertex D, say $A_1'A_2'D$ and $A_1''A_2''D$ (Fig. 79). It is always possible to think of the second following from the first position by a rotation about D, with angle 2η. As $M_1A_1' = M_1A_1''$ and $DA_1' = DA_1''$, the line M_1D is the bisector of $\angle A_1'DA_1''$;

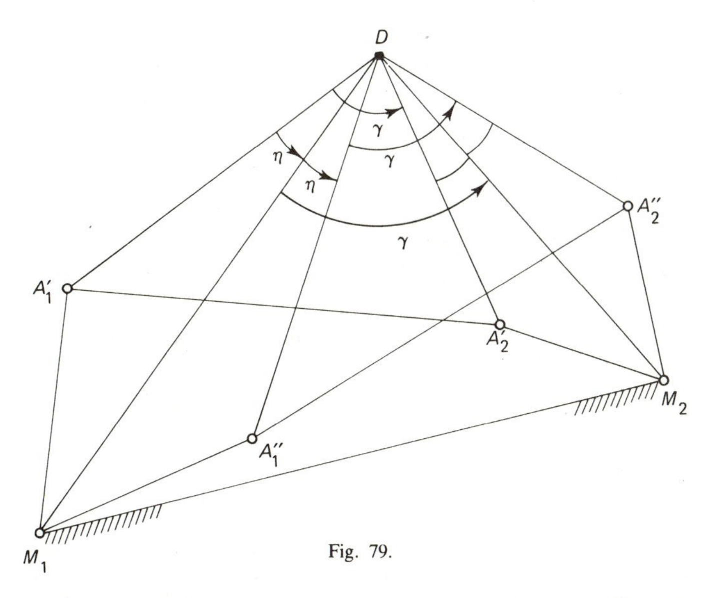

Fig. 79.

analogously for M_2D. Hence the lines DM_1, DM_2 follow from DA_1', DA_2' by a rotation η about D, and therefore $\angle M_1DM_2 = \gamma$ which is the vertex angle of the coupler triangle (or $\pi - \gamma$ if the figure is slightly different). As $\angle M_1DM_2 = \angle M_1M_3M_2$ the conclusion is: D is on the circumcircle of $M_1M_2M_3$.

This circle has 12 intersections with the coupler curve, of which six are such that three each coincide with I_1, I_2. The six finite intersections are therefore the three double points, each counted twice.

The coupler curve has been the subject of many geometric investigations. The configuration of the points M_1, M_2, M_3 and the three double points D_1, D_2, D_3, all on the same circle, has remarkable properties: any D is isogonally conjugate, with respect to the triangle $M_1M_2M_3$, to the point at infinity of the join of the other two (Cayley's theorem, Cayley [1875]); the same statement holds if M_i and D_i are interchanged. The points D can be all real or two of them are imaginary. The points in E whose paths belong to the first category are separated from the others by a *transition* curve of the tenth order (Müller [1889]); the points on this curve have a path with a "self-contact point". If a double point is real it can be a node or a cusp or an isolated point. The locus of the points describing a path with a cusp is the moving centrode, a curve of order eight, too complicated to clarify the motion. There are curves, found by Alt ([1921], Bottema [1954]) for which the double points coincide with the foci. Mayer [1937] has investigated coupler curves with three cusps. Other constructions to generate a coupler curve have been given by HIPPISLEY ([1920]) and by WEISS ([1942]). The curve has been treated by means of so-called isotropic coordinates by HAARBLEICHER ([1933]) and by GROENMAN ([1950]).

Example 78. Consider the path of a point A on A_1A_2 and show that it has three collinear double points.

The positions of the moving plane of the four-bar motion have been mapped on the image curve k (11.10), a quartic of genus one. Its equation depends on S_i and T_i. For a real four-bar we have $S_i > 0$ $(i = 1, 2, 3, 4)$ and $T_4 > 0$, but T_1, T_2, T_3 can have different signs, from which follow various shapes of k and therefore of the paths of the motion. Special cases arise if one T is zero, and still more special cases if two of them are zero. If one T is zero the four-bar has the property that the sum of two bars is equal to that of the other two, which implies that there are positions at which M_1, M_2, A_1, A_2 are collinear: the four-bar is said to fold (or be stretchable).

Example 79. Show that for $T_1 = 0$ the image curve k (11.10) has a self-contact at B_1 (which is the equivalent of two coinciding double points); B_1 represents a position of the four-bar in this case. For $T_2 = 0$ we have analogous properties of B_2. For $T_3 = 0$ the point $(0, 0, 1)$ is a double point. In all three cases the image curve, the motion and therefore all paths are rational.

Example 80. Consider the cases $T_2 = T_3 = 0$, $T_3 = T_1 = 0$, $T_1 = T_2 = 0$; in all three the image curve degenerates and so does every path. The four-bar is doubly stretchable.

Other special cases of the four-bar motion arise if, for instance, $R_2 = \infty$; one point of E describes a circle, another a straight line. The motion is the slider-crank motion (Fig. 80).

Let $(M_1; R_1)$ be the circle $X^2 + Y^2 - R^2 = 0$, and the line $X - d = 0$. If $A_1 = (0, 0)$ and $A_2 = (k, 0)$ we obtain the relations $X^2(A_1) + Y^2(A_1) - R^2 = 0$, $X(A_2) = d$. Hence by means of $X = x \cos\phi - y \sin\phi + a$, $Y = x \sin\phi + y \cos\phi + b$, and eliminating a and b, using the same procedure as we did for (11.6), we arrive at

$$\begin{aligned} &-2(xX + yY)\cos\phi + 2(yX - xY)\sin\phi + (X^2 + Y^2 + x^2 + y^2 - R^2) = 0, \\ &(k - x)\cos\phi + y \sin\phi + (X - d) = 0, \end{aligned} \tag{11.14}$$

with the solution

$$\begin{aligned} N\cos\phi &= yX^2 - 2xXY - yY^2 - 2dyX + 2dxY - y(x^2 + y^2 - R^2), \\ N\sin\phi &= (k + x)X^2 + 2yXY + (k - x)Y^2 \\ &\qquad - 2dxX - 2dyY + (k - x)(x^2 + y^2 - R^2), \\ N &= -2kyX - 2(x^2 + y^2 - kx)Y. \end{aligned} \tag{11.15}$$

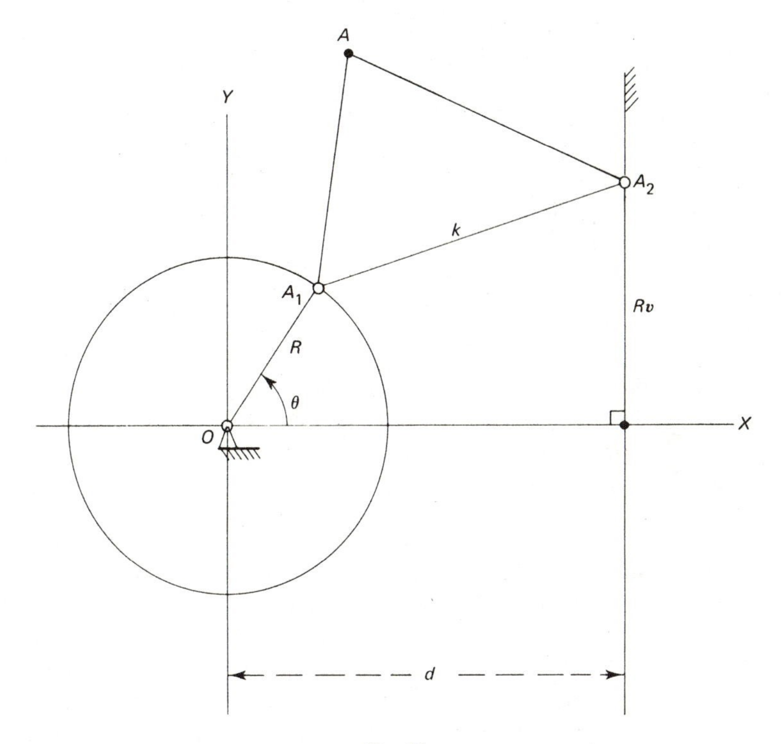

Fig. 80.

From $\cos^2\phi + \sin^2\phi = 1$ follows the equation of the path in Σ of the point $A(x, y)$ of E. It is a quartic; the fourth order terms are

$$(11.16)\quad (X^2 + Y^2)[((k + x)^2 + y^2)X^2 + 4kyXY + ((k - x)^2 + y^2)Y^2].$$

Hence the path is unicircular. It intersects the line at infinity at the isotropic points I_1, I_2 and two other points.

Example 81. Derive the complete equation of the path.

Example 82. Show that the non-isotropic points at infinity are in general conjugate imaginary, but if $x^2 + y^2 - k^2 = 0$ they are real and coinciding. Determine the equation of the path in this latter case.

Example 83. Determine the path if $x^2 + y^2 - R^2 = 0$.

Example 84. Show that O is the focus of every path.

The inverse of the slider-crank motion is such that a point of the moving plane remains on a circle, while a given line in E passes permanently through a fixed point of Σ. The general relation between X, Y and x, y (which we obtain from (11.15)) gives us the point-path equations for this inverse motion, sometimes called a turning-block-crank motion.

Example 85. Show that the point-paths of this inverse motion are in general tri-circular sextics.

Another analytic approach to the slider-crank motion makes use of an image-curve. Let (Fig. 80) $A_1 = (R\cos\theta, R\sin\theta)$, $A_2 = (d, Rv)$. Then if $\tan\frac{1}{2}\theta = u$, the condition that $(A_1A_2)^2 = k^2$ yields the following relation (between u and v):

$$\begin{aligned} &R^2u^2v^2 + (R+d+k)(R+d-k)u^2 + R^2v^2 \\ &\quad - 4R^2uv + (R-d+k)(R-d-k) = 0. \end{aligned} \tag{11.17}$$

It represents the general image curve (11.10) for the special case at hand. (11.17) is a quartic of genus one, with the double points $(1,0,0)$ and $(0,1,0)$. Hence the slider-crank motion is elliptic; its paths are in general quartics of genus one, and therefore they have two double points.

Example 86. Consider the special cases of the folding slider-crank ($R + d - k = 0$, etc.).

Roberts' configuration (Fig. 78) specializes for the slider-crank motion. M_2 is now the point at infinity of O_X. As the triangles $M_1M_2M_3$ and A_1A_2A are similar, M_3 is also at infinity and so are A_2' and G_2; but A_1' and G_1 are still finite points (Fig. 81). The path of A as a coupler point of the original slider-crank is the same as its path under the slider-crank motion defined by $M_1A_1'G_1M_3$, (A being the vertex of the coupler triangle $A_1'G_1A$); A_1' describes the circle with center M_1, and G_1 the straight line perpendicular to G_1M_3.

Example 87. Show that the "circumcircle" of $M_1M_2M_3$ is a straight line through M_1; determine its direction and prove that the (two) double points of the path of A are on it (this can be done either analytically or by means of a construction analogous to Fig. 78).

For the inverse motion a line l of E passes through a fixed point A' of Σ and a point A_1 of E remains on a circle in Σ. A special case appears if this circle is a straight line l′. This motion is obviously self-inverse (and it can still be considered as a limit case of the four-bar motion). It is called a conchoid motion because of the shape of its path.

Let A_1 be $x = y = 0$, $A' : X = Y = 0$, $\mathrm{l} : x - d = 0$, $\mathrm{l}' : X - D = 0$. Introducing the conditions that l passes through A' and A_1 remains on l′; then eliminating a and b as we did before, we obtain two linear equations for $\cos\phi$ and $\sin\phi$:

$$\begin{aligned} X\cos\phi + Y\sin\phi + (d - x) &= 0, \\ x\cos\phi - y\sin\phi + (D - X) &= 0. \end{aligned}$$

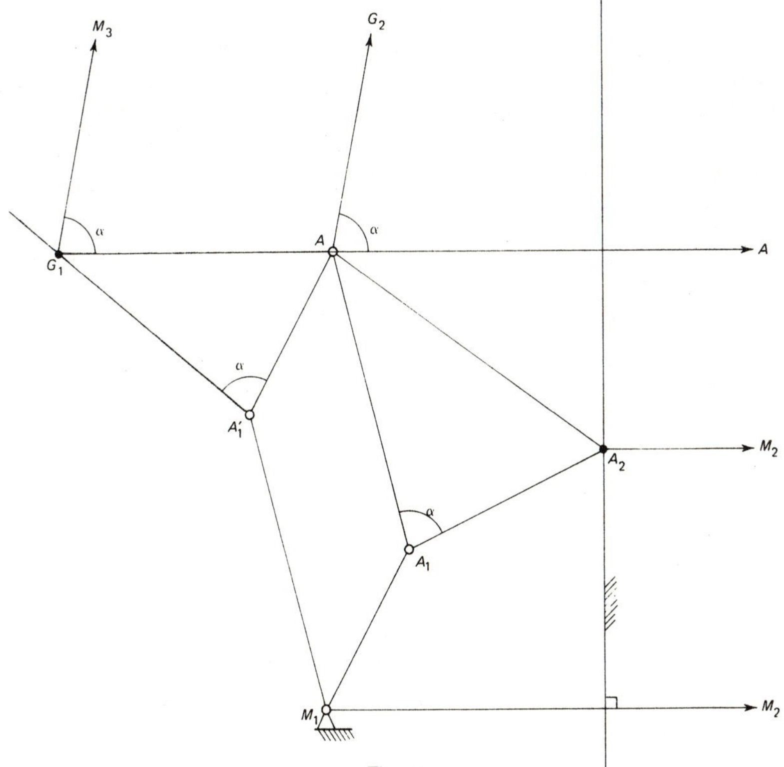

Fig. 81.

After eliminating ϕ this gives as the equation of the path of x, y:

$$(X^2 + Y^2)(X - D)^2 + 2(yY - xX)(X - D)(x - d)$$
$$+ (x^2 + y^2)(x - d)^2 - (yX + xY)^2 = 0,$$

which represents a circular quartic, called a conchoid.

Example 88. Show analytically that the motion is self-inverse in the sense that the equation remains the same if X, Y, D are interchanged with x, y, d.
Example 89. Consider the case $D = d$; the motion is symmetric.
Example 90. Show that with $a = D$, $b = -(d + D\cos\phi)/\sin\phi$ and $\tan\frac{1}{2}\phi = t$, the homogeneous coordinates of the path are:

$$X = xt(1 - t^2) - 2yt^2 + Dt(1 + t^2)z,$$
$$Y = 2xt^2 + yt(1 - t^2) - \tfrac{1}{2}(1 + t^2)[D + d - (D - d)t^2]z,$$
$$Z = t(1 + t^2)z.$$

A still more special four-bar motion, called Cardan or Cardanic motion after Geronimo Cardano, is that for which two points of E are compelled to remain on two straight lines of Σ. It is trivial if these lines are parallel. We take them perpendicular and will show that this covers the general case as well.

Let $A_1 = (-R, 0)$ remain on $X = 0$ and $A_2 = (R, 0)$ on $Y = 0$ (Fig. 82). For $A(x, y)$ we have

$$X = (R + x)\cos\phi - y\sin\phi,$$
$$Y = -(R - x)\sin\phi + y\cos\phi, \tag{11.18}$$

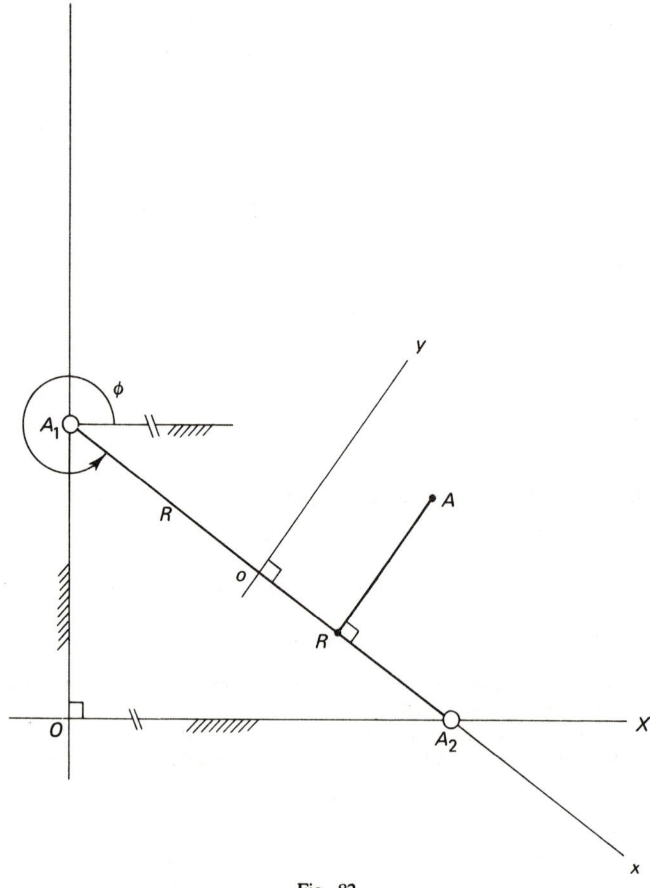

Fig. 82.

which proves that the path of A is in general an ellipse. Eliminating ϕ from (11.18) we obtain

$$[(x-R)^2+y^2]X^2-4RyXY+[(x+R)^2+y^2]Y^2-(x^2+y^2-R^2)^2=0, \tag{11.19}$$

for the equation of the path of (x, y). This represents in general an ellipse with center at O. For $x^2+y^2=R^2$ it degenerates; if $x = R\cos\theta$, $y = R\sin\theta$, the equation reads $(X\sin\frac{1}{2}\theta - Y\cos\frac{1}{2}\theta)^2 = 0$, hence every point on the circle with A_1A_2 as diameter remains on a line through O. This justifies our having taken two perpendicular lines to describe the motion. The only point of E whose path is a circle is the origin o (the midpoint of A_1, A_2); its radius is R. The Cardan motion is often called the elliptic motion.

The formulas (11.18) may be written:

$$\begin{aligned} X &= x\cos\phi - y\sin\phi + R\cos\phi, \\ Y &= x\sin\phi + y\cos\phi - R\sin\phi. \end{aligned} \tag{11.20}$$

Hence $\dot{X} = \dot{Y} = 0$ implies for the pole P

$$x_p = R\cos 2\phi, \qquad y_p = -R\sin 2\phi, \tag{11.21}$$

and

$$X_p = 2R\cos\phi, \quad Y_p = -2R\sin\phi. \tag{11.22}$$

The moving centrode is the circle $(o; R)$, the fixed centrode is $(O; 2R)$. A Cardan motion is generated by the rolling of any circle within a circle with twice its radius. The path of a point on the moving centrode is a line-segment of length $4R$ (Fig. 83).

If we transform (11.20) using $X' = Y$, $Y' = -X+2R$, $x' = y$, $y' = -x+R$, then, omitting the primes, the elliptic motion is represented by

$$X = x\cos\phi - y\sin\phi, \quad Y = x\sin\phi + y\cos\phi + 2R(1-\cos\phi). \tag{11.23}$$

Hence, $a = 0$, $b = 2R(1-\cos\phi)$, which implies $a_0 = a_1 = a_2 = 0$, $b_0 = b_1 = 0$; this means that for $\phi = 0$ we have canonical frames. The Cardan motion is therefore characterized by the following instantaneous invariants

$$a_k = 0 \text{ (for all } k), \quad b_0 = 0, \quad b_{2k+1} = 0, \quad b_{2k} = (-1)^{k+1}2R. \tag{11.24}$$

Example 91. Show that the inflection circle (which has diameter b_2) coincides with the moving centrode.

Example 92. Show that every point on the inflection circle is a Ball-point.

Example 93. Determine the circling-point curve for the Cardan motion.

Example 94. Show that the Cardan motion is a (folding) slider-crank motion.

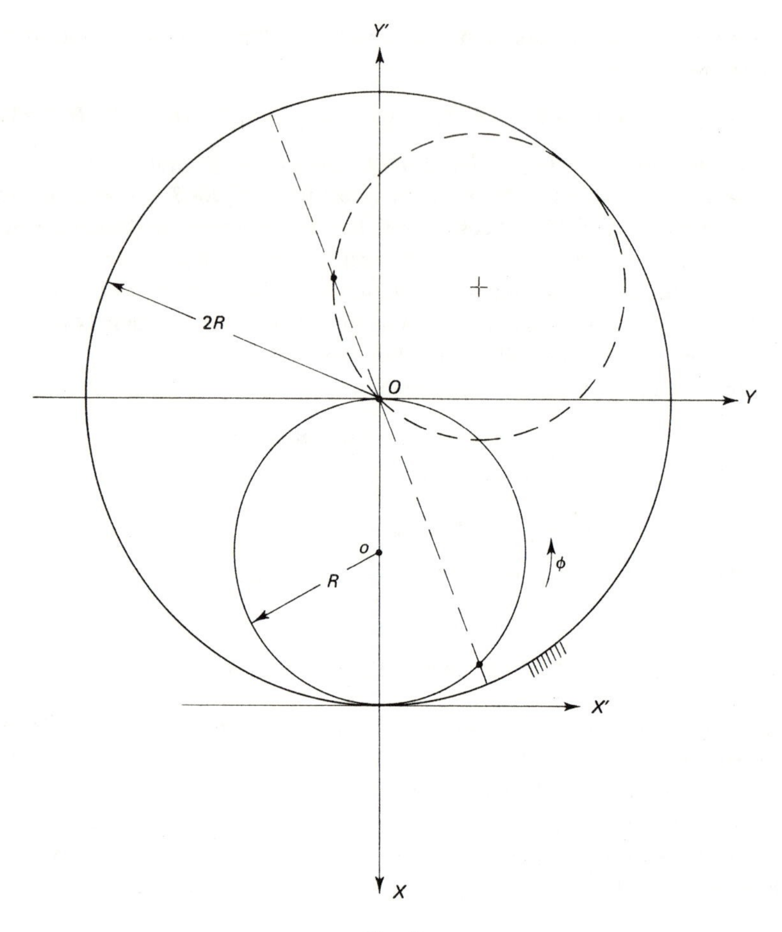

Fig. 83.

A moving plane is said to be at a Cardan position if there is a Cardan motion such that instantaneously the paths of the two motions hyper-osculate. There has been some controversy about this concept which was cleared up recently (BOTTEMA [1949], FREUDENSTEIN [1960]). Obviously any motion (with $b_2 \neq 0$) is up to the second order a Cardan motion (with $R = b_2/2$), but to the third order only if $a_3 = b_3 = 0$.

The inverse of a Cardan motion is called a cardioid motion. It may be defined as a motion such that two lines of E each pass permanently through a different fixed point of Σ. It is a rational motion because the Cardan motion is rational. The paths are given by (11.19) if X, Y are constant and x, y variable. Hence the path is in general a rational bicircular quartic (a curve called a limacon). The motion is generated if a circle with radius $2R$ rolls with internal contact on a circle with radius R.

Example 95. The path has one finite double point; determine the condition (on X, Y) to be satisfied if it is a node, a cusp or an isolated point. (If the limacon has a cusp it is known as a cardioid, which has given the name to the motion.)
Example 96. Show by means of (11.23) that the cardioid motion is represented by

$$X = x \cos \phi - y \sin \phi - 2R \sin \phi(1 - \cos \phi),$$

$$Y = x \sin \phi + y \cos \phi + 2R \cos \phi(1 - \cos \phi).$$

Prove that it is a rational motion of the fourth order. Determine the instantaneous invariants a_k, b_k $(k = 0, 1, 2, 3, 4;$ $a_0 = a_1 = a_2 = 0,$ $a_3 = -6R,$ $a_4 = 0;$ $b_0 = b_1 = 0,$ $b_2 = 2R,$ $b_3 = 0,$ $b_4 = -14R)$. Determine the conditions for the invariants of a motion so that it is, up to the second, the third, or the fourth order, a cardioid motion.

We have completed our discussion of the planar four-bar motion and its various sub-cases. We shall return to some of its properties in Chapter XI.

Now, we deal with some general theorems of plane kinematics which may be applied to some special motions discussed earlier.

From Chasles' theorems (in space) about the relations between the order of paths etc. for the direct and for the inverse motion (Section 1), we deduce for the plane, using analogous arguments, the following statement: if for a direct (algebraic) motion the order of the paths is, in general, equal to n, then *for the inverse motion the class of the envelope generated by a moving line is, in general, equal to n*, and of course the converse is also true. Keeping in mind that the inverse of the general four-bar motion is a general four-bar motion, that of a slider-crank motion a turning-block-crank motion, and that of a Cardan motion a cardioid motion, we may derive some new properties of these special plane motions.

Example 97. Show that the class of the envelope of a moving line for the four-bar motion is six, for the slider-crank motion again six, for the turning-block-crank motion four, for the Cardan motion also four and for the cardioid motion two (Bottema [1976]).

Another general problem has been raised (and solved) by Roberts [1870, 1876]: A plane motion is determined if two moving points A_1, A_2 are compelled to describe given paths. If the orders of these paths are n_1 and n_2 respectively can the order of the path of a general point be derived? It appears that this is not possible. The problem is more refined, we must know not only the orders of the paths but moreover their circularity: the number of times they pass through each isotropic point.

The pertinent formulas are derived by Roberts by means of the so-called "principle of preservation of number" (German: Prinzip der Erhaltung der Anzahl), a much discussed and somewhat dubious argument no longer used by modern algebraic geometers. Roughly speaking it comes to this: if a geometric problem has in general n solutions then it has n solutions (or an infinite number) in a special case. The converse of this is the argument used by Roberts. To give an example: two curves with orders m_1 and m_2 have m_1m_2 intersections (real, imaginary, coinciding) because this is the number if the curves degenerate into m_1 and m_2 lines respectively.

Let the path K_i of A_i be of order n_i and circularity c_i $(i = 1, 2)$; a degenerate specimen of K_i consists of c_i circles and $n_i - 2c_i$ straight lines. The motion under consideration degenerates into c_1c_2 four-bar motions, plus $c_1(n_2 - 2c_2) + c_2(n_1 - 2c_1)$ slider-crank motions, plus $(n_1 - 2c_1)(n_2 - 2c_2)$ Cardan motions. The orders of the paths of these special motions are 6, 4 and 2 respectively. Hence the order n and the circularity c in the general, non-degenerate, case are given by

$$n = 6c_1c_2 + 4c_1(n_2 - 2c_2) + 4c_2(n_1 - 2c_1) + 2(n_1 - 2c_1)(n_2 - 2c_2),$$

or

$$n = 2n_1n_2 - 2c_1c_2, \tag{11.25}$$

and $c = 3c_1c_2 + c_1(n_2 - 2c_2) + c_2(n_1 - 2c_1)$ or

$$c = n_1c_2 + n_2c_1 - c_1c_2. \tag{11.26}$$

(11.25) and (11.26) are Roberts' formulas.

Example 98. If $c_1 = c_2 = 0$ we obtain $n = 2n_1n_2$, $c = 0$.
Example 99. If $c_2 = 0$ we obtain $n = 2n_1n_2$, $c = n_2c_1$.
Example 100. Determine n and c if the two given paths are ellipses, or an ellipse and a circle.
Example 101. Determine n and c if one path is an ellipse and the other a straight line. The answer is $n = 4$, $c = 0$. But, for the Cardan motion ($n = 2$, $c = 0$) these conditions are also satisfied. Clarify the apparent contradiction. *This example contains a warning with respect to the use of Roberts' formulas.*

A plane motion may also be defined by its centrodes. The most simple example is the case where both centrodes are circles; these motions are called *cycloidal motions*. Let (Fig. 84) the fixed centrode be given by $X^2 + Y^2 = R^2$ and the moving centrode by $x^2 + y^2 - r^2 = 0$. In the initial position their tangent point A_0 is $X = 0$, $Y = R$; $x = 0$, $y = -r$. We take $R > 0$; if the circles are externally (internally) tangent we take $r > 0$ $(r < 0)$. In an arbitrary position B_1 is the tangent point. Let $\angle A_0MB_1 = \psi$. Then arc $A_0B_1 = R\psi$, hence arc $A_0B_0 =$ arc $A_1B_1 = R\psi$ and therefore $r\theta = R\psi$. From this it follows

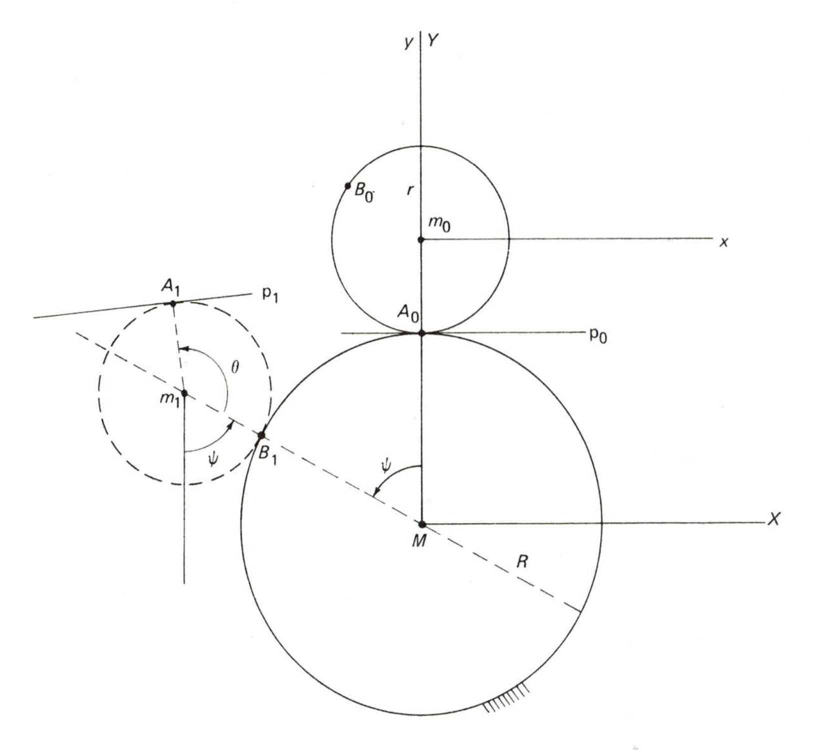

Fig. 84.

that the circle $(m;r)$ has been rotated about the angle $\phi = \psi + \theta = ((R + r)/r)\psi$, which implies $\psi = k\phi$, with $k = r/(R + r)$.

The coordinates of m_1 are $X = -(R + r)\sin\psi$, $Y = (R + r)\cos\psi$. It follows from all this that the motion is represented by

$$
\begin{aligned}
X &= x\cos\phi - y\sin\phi - (R + r)\sin k\phi, \\
Y &= x\sin\phi + y\cos\phi + (R + r)\cos k\phi.
\end{aligned}
\tag{11.27}
$$

The path of a point is called a cycloid. It shows a variety of shapes, depending on x, y as well as k. The point m always describes a circle. If k is an irrational number no other point ever returns to its initial position and a generic point-path has infinitely many intersections with a suitably chosen circle about M. Hence the motion is in general transcendental. It is algebraic, however, if k is rational.

Let k be equal to p/q, p and q being (positive or negative) integers without a common factor. Let $\phi = q\phi_1$. Then we have for (11.27)

$$\begin{aligned} X &= x \cos q\phi_1 - y \sin q\phi_1 - (R + r)\sin p\phi_1, \\ Y &= x \sin q\phi_1 + y \cos q\phi_1 + (R + r)\cos p\phi_1. \end{aligned} \tag{11.28}$$

It is well-known that $\cos N\phi_1$ and $\sin N\phi_1$ may be written as polynomials of degree N in $\cos\phi_1$ and $\sin\phi_1$. If moreover we introduce $\tan(\phi_1/2) = t$ it follows that the general path is a rational curve of order $2n$, n being the larger of the two numbers $|p|$ and $|q|$.

Example 102. Show that the rational cycloid is circular; it is tangent to the line at infinity at the isotropic points.
Example 103. Show from the foregoing that the Cardan motion is the cycloidal motion with $k = -1$ and check the degree of the equation for its paths.
Example 104. Show that we have the cardioid motion for $k = 2$; check the degree of the equation of its paths.
Example 105. If k is the characteristic number of a cycloid motion and $\bar{k}$ that of its inverse motion show that $k + \bar{k} = 1$.
Example 106. Consider the case $R = -3r$; show that the order of the paths is four. If a moving point is on the centrode its path has three cusps; the point generates Steiner's hypocycloid (or deltoid).
Example 107. Show that every cycloid can be generated by two different cycloid motions (which are therefore cognate).

In order to determine the instantaneous invariants of a cycloidal motion we transform (11.27) letting $X' = X$, $Y' = Y - R$, $x' = x$, $y' = y + r$. Omitting the primes, the motion reads (for frames which are obviously canonical):

$$X = x\cos\phi - y\sin\phi + a, \quad Y = x\sin\phi + y\cos\phi + b,$$

with

$$a = r\sin\phi - (R + r)\sin k\phi, \quad b = -R - r\cos\phi + (R + r)\cos k\phi. \tag{11.29}$$

From which the invariants follow

$$\begin{aligned} &a_0 = a_1 = a_2 = 0, \quad a_3 = (k^2 - 1)r, \quad a_4 = 0, \\ &a_5 = -(k^4 - 1)r, \quad a_6 = 0, \quad a_7 = (k^6 - 1)r, \ldots, \\ &b_0 = b_1 = 0, \quad b_2 = -(k - 1)r, \quad b_3 = 0, \\ &b_4 = (k^3 - 1)r, \quad b_5 = 0, \quad b_6 = -(k^5 - 1)r, \ldots . \end{aligned} \tag{11.30}$$

Example 108. Any motion is instantaneously cycloidal up to the second order; derive the conditions for it being so up to the third and the fourth order.
Example 109. Determine the inflection circle diameter for a cycloidal motion.
Example 110. Determine the circle-point curve for a cycloidal motion.

Example 111. Investigate the Ball-point and the Burmester points for a cycloidal motion.

Special cycloidal motions are those for which either R or r is infinite. In the first case a circle rolls on a straight line; in the second we have its inverse motion.

Example 112. Derive the equations for these motions; show that they are always transcendental. Determine the instantaneous invariants. [For the first case we obtain, for the canonical frames, $a = -r(\phi - \sin\phi)$, $b = r(1 - \cos\phi)$.]

CHAPTER X

n-PARAMETER MOTIONS

1. Two-parameter spatial motions

A space E has *six* degrees of freedom when it moves without constraints with respect to a coinciding fixed space Σ. If E has to satisfy $m \leq 5$ independent conditions the degree of freedom diminishes to $n = 6 - m$. Kinematics in the usual sense deals with the case $n = 1$: the positions of E depend on *one* parameter t, which is generally taken to be *time*. When $n = 1$, the locus generated by a moving point, plane or line is a curve, a developable, or a ruled surface respectively.

In this chapter we consider the case $n > 1$, which means we are dealing with "n-parameter motions" which we denote by M_n. First, we suppose $n = 2$. The positions of E depend then on two parameters, λ and μ say. The locus of a point is now (in general) a surface, which is called its trajectory surface (French: surface trajectoire; German: Bahnfläche); that of a plane is the set of tangent planes of a surface, enveloped by the plane; the locus of a moving line is its trajectory congruence (German: Bahnkongruenz). The connection with time, a one-dimensional phenomenon, has vanished, but in cases where we can take $\lambda(t)$, $\mu(t)$ as arbitrary functions of one parameter t we have defined a motion in the ordinary sense, which "belongs" to M_2, and to which the results of our preceding chapters may be applied.

We deal now with *instantaneous* two-parameter motion. At a certain zero-position we may study the differential-geometrical properties of a trajectory surface, its tangent plane, its normal, its curvature, etc. We may also study the analogous notions for a moving plane or line.

For the zero-position we suppose $\lambda = \mu = 0$. If $F(\lambda, \mu)$ is any function of λ and μ we denote $(\partial F/\partial\lambda)(0,0)$ by F_λ, $(\partial^2 F/\partial\lambda\partial\mu)(0,0)$ by $F_{\lambda\mu}$, etc.

The general two-parameter motion is given by

$$P(\lambda, \mu) = A(\lambda, \mu)p + d(\lambda, \mu), \tag{1.1}$$

$\boldsymbol{p}$ being the position vector in E of a point P of E, $\boldsymbol{P}$ that of P in Σ, $\mathbf{A}$ an orthogonal matrix, and $\boldsymbol{d}$ the position vector in Σ of the origin of coordinates in E.

If the frames in E and Σ coincide in the zero-position, we have $\mathbf{A}_0 = \mathbf{I}$, $\boldsymbol{d}_0 = 0$, $\boldsymbol{P}_0 = \boldsymbol{p}$.

From (1.1) it follows that

$$\boldsymbol{P}_\lambda = \mathbf{A}_\lambda \boldsymbol{p} + \boldsymbol{d}_\lambda, \qquad \boldsymbol{P}_\mu = \mathbf{A}_\mu \boldsymbol{p} + \boldsymbol{d}_\mu. \tag{1.2}$$

As $\mathbf{A}$ is orthogonal we have

$$\mathbf{A}\mathbf{A}^{\mathrm{T}} = \mathbf{I}.$$

which implies

$$\mathbf{A}_\lambda + \mathbf{A}_\lambda^{\mathrm{T}} = 0, \qquad \mathbf{A}_\mu + \mathbf{A}_\mu^{\mathrm{T}} = 0. \tag{1.3}$$

Hence $\mathbf{A}_\lambda$ and $\mathbf{A}_\mu$ are skew matrices and we may put

$$\mathbf{A}_\lambda = \mathbf{J} = \begin{Vmatrix} 0 & -j_3 & j_2 \\ j_3 & 0 & -j_1 \\ -j_2 & j_1 & 0 \end{Vmatrix}, \qquad \mathbf{A}_\mu = \mathbf{L} = \begin{Vmatrix} 0 & -l_3 & l_2 \\ l_3 & 0 & -l_1 \\ -l_2 & l_1 & 0 \end{Vmatrix}. \tag{1.4}$$

Furthermore, let $\boldsymbol{d} = (a, b, c)$, then $\boldsymbol{d}_\lambda = (a_\lambda, b_\lambda, c_\lambda)$, $\boldsymbol{d}_\mu = (a_\mu, b_\mu, c_\mu)$.

Consider now an arbitrary M_1 belonging to our M_2; then $\lambda(t)$, $\mu(t)$ are arbitrary functions of t. Let $\lambda_0 = \mu_0 = 0$. For M_1 we obtain the first derivative with respect to t:

$$\boldsymbol{P}_1 = \boldsymbol{P}_\lambda \lambda_1 + \boldsymbol{P}_\mu \mu_1, \tag{1.5}$$

or, introducing Cartesian frames, O_{XYZ} and o_{xyz}, coinciding in the zero-position but for the time being otherwise arbitrary,

$$\begin{aligned} X_1 &= -(j_3\lambda_1 + l_3\mu_1)y + (j_2\lambda_1 + l_2\mu_1)z + a_\lambda\lambda_1 + a_\mu\mu_1, \\ Y_1 &= (j_3\lambda_1 + l_3\mu_1)x - (j_1\lambda_1 + l_1\mu_1)z + b_\lambda\lambda_1 + b_\mu\mu_1, \\ Z_1 &= -(j_2\lambda_1 + l_2\mu_1)x + (j_1\lambda_1 + l_1\mu_1)y + c_\lambda\lambda_1 + c_\mu\mu_1. \end{aligned} \tag{1.6}$$

In these formulas j_i, l_i, a_λ, a_μ, b_λ, b_μ, c_λ, c_μ are data from M_2, while λ_1, μ_1 depend on M_1 (all these quantities are evaluated at the zero-position). Our aim is to determine the instantaneous screw axis of M_1.

The components of the angular velocity vector $\boldsymbol{\omega}$ are

$$\omega_1 = j_1\lambda_1 + l_1\mu_1, \qquad \omega_2 = j_2\lambda_1 + l_2\mu_1, \qquad \omega_3 = j_3\lambda_1 + l_3\mu_1. \tag{1.7}$$

Eliminating λ_1, μ_1 we see that ω_1, ω_2, ω_3 are linearly dependent:

$$\begin{vmatrix} j_2 & l_2 \\ j_3 & l_3 \end{vmatrix} \omega_1 + \begin{vmatrix} j_3 & l_3 \\ j_1 & l_1 \end{vmatrix} \omega_2 + \begin{vmatrix} j_1 & l_1 \\ j_2 & l_2 \end{vmatrix} \omega_3 = 0. \tag{1.8}$$

Since the coefficients of ω_i in (1.8) depend only on data from M_2, the conclusion is: the angular velocity vectors for all M_1's of M_2 (and therefore their screw axes) are parallel to a certain plane. We take the frames such that O_{XY} is parallel to this plane. This implies $\omega_3 = 0$ and hence $j_3 = l_3 = 0$.

Our parameters λ, μ of M_2 may be transformed into λ^*, μ^* by $\lambda^* = F(\lambda, \mu)$, $\mu^* = G(\lambda, \mu)$, with $F(0,0) = G(0,0) = 0$. This means we have homogeneous linear transformations of λ_1, μ_1. We normalize the parameters so that $\lambda_1^* = j_1\lambda_1 + l_1\mu_1$, $\mu_1^* = j_2\lambda_1 + l_2\mu_1$. Then $a_\lambda\lambda_1 + a_\mu\mu_1$ is transformed into a new linear homogeneous function in λ_1^*, μ_1^* which we write as $a_\lambda^*\lambda_1^* + a_\mu^*\mu_1^*$, etc. Omitting the stars we obtain for (1.6):

$$\begin{aligned} X_1 &= \mu_1 z + a_\lambda\lambda_1 + a_\mu\mu_1, \\ Y_1 &= -\lambda_1 z + b_\lambda\lambda_1 + b_\mu\mu_1, \\ Z_1 &= -\mu_1 x + \lambda_1 y + c_\mu\lambda_1 + c_\mu\mu_1. \end{aligned} \tag{1.9}$$

The normalized parameters are such that $\lambda_1 = \omega_1$, $\mu_1 = \omega_2$. Note that the frames have a Z-axis of a determined direction, but that they are otherwise still arbitrary.

We now determine the screw axis of M_1. It is the locus of the points whose velocities are parallel to $\boldsymbol{\omega}$. These points satisfy

$$\begin{aligned} \mu_1 z + a_\lambda\lambda_1 + a_\mu\mu_1 &= \sigma_0\lambda_1, \\ -\lambda_1 z + b_\lambda\lambda_1 + b_\mu\mu_1 &= \sigma_0\mu_1, \\ -\mu_1 x + \lambda_1 y + c_\lambda\lambda_1 + c_\mu\mu_1 &= 0, \end{aligned} \tag{1.10}$$

σ_0 being the pitch of the instantaneous screw. From the first two equations we obtain

$$\begin{aligned} \sigma_0 &= [a_\lambda\lambda_1^2 + (a_\mu + b_\lambda)\lambda_1\mu_1 + b_\mu\mu_1^2]/(\lambda_1^2 + \mu_1^2), \\ z &= [b_\lambda\lambda_1^2 + (-a_\lambda + b_\mu)\lambda_1\mu_1 - a_\mu\mu_1^2]/(\lambda_1^2 + \mu_1^2). \end{aligned} \tag{1.11}$$

The screw axis is the intersection of the plane given by the second equation of (1.11) and the plane represented by the third equation of (1.10). Hence for its Plücker coordinates we have $p_{14} : p_{24} : p_{12} = \lambda_1 : \mu_1 : -(c_\lambda\lambda_1 + c_\mu\mu_1)$, which implies $p_{12} + c_\lambda p_{14} + c_\mu p_{24} = 0$. This means that all screw axes intersect a line r with coordinates r_{ik} given by $r_{34} = 1$, $r_{23} = c_\lambda$, $r_{31} = c_\mu$, $r_{12} = r_{14} = r_{24} = 0$, which is a line parallel to the Z-axis. We take the Z-axis along r, which gives the further reduction $c_\lambda = c_\mu = 0$. We have

$$X_1 = \mu_1 z + a_\lambda \lambda_1 + a_\mu \mu_1,$$

$$(1.12) \qquad Y_1 = -\lambda_1 z + b_\lambda \lambda_1 + b_\mu \mu_1,$$

$$Z_1 = -\mu_1 x + \lambda_1 y.$$

The Z-axis is now fixed, but the origin O and the O_X- and O_Y-axis are still arbitrary.

We note that the screw axis (and the pitch) depend only on the ratio $\lambda_1 : \mu_1$; hence (although there are ∞^2 motions M_1) in the neighborhood of the zero-position there are ∞^1 screw axes, their locus therefore being a ruled surface R. If (x, y, z) is a point of R it follows from (1.10) that $-\mu_1 x + \lambda_1 y = 0$, while z is given by (1.11). Hence, eliminating μ_1, λ_1, the equation of R reads

$$(1.13) \qquad (x^2 + y^2)z - b_\lambda x^2 + (a_\lambda - b_\mu)xy + a_\mu y^2 = 0,$$

and it is seen to be a cubic ruled surface. By a suitable transformation $x = x'\cos\psi - y'\sin\psi$, $y = x'\sin\psi + y'\cos\psi$, $z = z' + e$ the coefficients a_μ and b_λ vanish. For the new frame we let $a_\lambda = k_1$, $b_\mu = k_2$ and omitting the primes we obtain for (1.12)

$$X_1 = \mu_1 z + k_1 \lambda_1,$$

$$(1.14) \qquad Y_1 = -\lambda_1 z + k_2 \mu_1,$$

$$Z_1 = -\mu_1 x + \lambda_1 y,$$

and for (1.13)

$$(1.15) \qquad (x^2 + y^2)z + (k_1 - k_2)xy = 0.$$

The coordinate systems are now completely determined; we call them canonical.

(1.14) shows that the M_2 has two instantaneous invariants of the first order, k_1 and k_2. The surface R depends only on their difference. The screw axis of a M_1 belonging to M_2 reads

$$(1.16) \qquad x : y = \lambda_1 : \mu_1, \qquad z = (k_2 - k_1)\lambda_1 \mu_1 / (\lambda_1^2 + \mu_1^2),$$

and its pitch

$$(1.17) \qquad \sigma_0 = (k_1 \lambda_1^2 + k_2 \mu_1^2)/(\lambda_1^2 + \mu_1^2).$$

The cubic ruled surface R, given by its standard equation (1.15), is well-known in algebraic geometry. It is called Cayley's cylindroid or Plücker's conoid and it has been introduced into kinematics (although in another context) by BALL [1876, 1900].

Example 1. Determine the ψ and e by which (1.13) transforms into (1.15).
Example 2. Show that R has two directrices, intersected by all generators: the z-axis and the line at infinity of o_{xy}. The first is a double line of R so two screw axes pass through each of its points; the other is single.
Example 3. Show that R lies between two parallel planes.
Example 4. Determine the points on o_z through which pass two orthogonal generators.
Example 5. Determine the two generators in the plane at infinity.
Example 6. Show that R contains ∞^2 ellipses.

Real generators pass through the point $(0, 0, z)$ only if $4z^2 \leq (k_1 - k_2)^2$; for the two limit points $z = \pm(k_1 - k_2)/2$ they coincide with the so-called dorsal lines of R. These satisfy $\lambda_1 = \pm\mu_1$, and the pitch is $(k_1 + k_2)/2$ for both; they are parallel to the planes $x = \pm y$.

The axes o_x and o_y are generators of R, the pitches being k_1 and k_2 respectively; this gives an interpretation of the two instantaneous invariants. They are moreover the extreme values of the pitch σ_0.

According to (1.17) there are two generators, s_1 and s_2, for which $\sigma_0 = 0$, so that they are axes of pure rotations. s_1, s_2 are, however, only real if $k_1 k_2 \leq 0$. We shall meet s_i once more in the next section. They are represented by

$$x : y = \sqrt{k_2} : \sqrt{-k_1}, \qquad z = \sqrt{-k_1 k_2}. \tag{1.18}$$

If $k_1 = k_2$ the locus R degenerates into the pencil of lines through o in the plane $z = 0$ (and the two isotropic planes through o_z). All screws then have equal pitches. If $k_1 + k_2 = 0$ the lines s_i are real and they coincide with the dorsal lines.

If $k_1 = 0$, $k_2 \neq 0$ the lines s_1, s_2 both coincide with o_x; for $k_1 \neq 0$, $k_2 = 0$ they coincide with o_y.

If $k_1 = k_2 = 0$ all instantaneous motions are pure rotations; the origin of the moving system is at rest. M_2 is instantaneously a spherical motion; all rotation axes are in the plane $z = 0$.

Example 7. If $\lambda_1/\mu_1 = \cot\phi$ $(-(\pi/2) < \phi < (\pi/2))$ the locus of the endpoints of the screw vector (if its starting point is on o_z and its length equals σ_0) is given by

$$x = (k_1 \cos^2\phi + k_2 \sin^2\phi)\cos\phi, \qquad y = (k_1 \cos^2\phi + k_2 \sin^2\phi)\sin\phi,$$

$$z = (k_2 - k_1)\sin\phi\cos\phi.$$

Example 8. By introducing $\tan\frac{1}{2}\phi = \tau$ $(-1 < \tau < 1)$ show that the locus of the endpoints is an arc of a rational sextic space curve.

Summarizing the main result of this section we have: up to the first order any M_2 is represented by

$$X = x + \mu z + k_1\lambda, \qquad Y = y - \lambda z + k_2\mu, \qquad Z = z - \mu x + \lambda y, \tag{1.19}$$

λ, μ being the parameters of the motion and k_1, k_2 the instantaneous invariants of the zero-position.

2. First order properties

All first order properties of a two-parameter motion follow from the canonical formulas (1.19). If (x, y, z) is any moving point, the tangent plane of its trajectory surface, at the zero-position, passes through the directions $X_\lambda : Y_\lambda : Z_\lambda = k_1 : -z : y$ and $X_\mu : Y_\mu : Z_\mu = z : k_2 : -x$. For this plane U we obtain

$$\begin{aligned} &U_1 = xz - k_2 y, \qquad U_2 = yz + k_1 x, \qquad U_3 = z^2 + k_1 k_2, \\ &U_4 = -(x^2 + y^2 + z^2)z - (k_1 - k_2)xy - k_1 k_2 z. \end{aligned} \tag{2.1}$$

which expresses the (cubic) relationship between a moving point and the tangent plane of its trajectory surface.

Example 9. Determine the plane for a point on the z-axis and for a point at infinity.

From (2.1) it follows that the Plücker coordinates n_{ij} of the normal to the trajectory plane are the minors of the matrix

$$\begin{Vmatrix} xz - k_2 y & yz + k_1 x & z^2 + k_1 k_2 & 0 \\ x & y & z & 1 \end{Vmatrix}, \tag{2.2}$$

which gives us

$$\begin{aligned} &n_{14} = xz - k_2 y, \qquad n_{24} = yz + k_1 x, \qquad n_{34} = z^2 + k_1 k_2, \\ &n_{23} = k_1(xz - k_2 y), \qquad n_{31} = k_2(yz + k_1 x), \qquad n_{12} = -(k_1 x^2 + k_2 y^2). \end{aligned} \tag{2.3}$$

Hence all normals satisfy

$$p_{23} = k_1 p_{14}, \qquad p_{31} = k_2 p_{24}, \tag{2.4}$$

which implies that the *locus of the normals is* the intersection of two linear complexes and therefore *a linear congruence*. This theorem was first derived by SCHÖNEMANN [1855] and independently by MANNHEIM [1875].

From (2.4) it follows that the two lines r_{ij} intersecting all normals are

$$(2.5)\qquad \begin{aligned} r_{14} &= k_2, & r_{24} &= \pm\sqrt{-k_1k_2}, & r_{34} &= 0, \\ r_{23} &= -k_1k_2, & r_{31} &= \mp k_2\sqrt{-k_1k_2}, & r_{12} &= 0, \end{aligned}$$

and we recognize the lines s_i given by (1.18). Hence all normals intersect the axes of the two instantaneous (pure) rotation axes.

The congruence of normals is hyperbolic, parabolic or elliptic if $k_1k_2 < 0$, $k_1k_2 = 0$ (but k_i not both zero) or $k_1k_2 > 0$ respectively. Through each of the ∞^3 points of E passes a normal, there are, however, only ∞^2 normals. This is as expected: if P and Q are two moving points their velocity components along PQ are equal; hence if Q is a point on the normal n of P then n must be the normal of Q. Any normal is the normal for all its points.

Mannheim has drawn attention to a noteworthy analog between spatial M_2-motion and planar M_1-motion. In the latter there are ∞^2 points, each describing a path and therefore each having an instantaneous normal; there are, however, only ∞^1 normals. Their locus is the pencil of lines through the pole; there is in general only one normal through a given point, the pole being the only exception. In the spatial M_2-motion the lines s_1 and s_2 (real or imaginary) are analogous to the pole, there is in general one normal through a given point, points on s_i are the only exceptions. We shall return to this analog in the next section.

Through any point P on s_1 pass ∞^1 lines of the congruence *viz.* all lines through P in the plane (P, s_2); all these lines are normal to P's velocity. The conclusion is: the tangent plane at P is undetermined, any plane containing the line q through P perpendicular to the plane (P, s_2) acts as a tangent plane. In other words: although we deal with an M_2-motion P can instantaneously only move in one direction, that of q. Similar conditions hold for all points on s_2. The movability of a point on s_i is reduced from two to one degree of freedom. The analog in the plane is the fact that the pole is instantaneously at rest.

Example 10. Consider the case $k_1k_2 = 0$, which means that the congruence of normals is parabolic. Determine the (only possible) velocity direction of a point on s $(= s_1 = s_2)$.
Example 11. Show that every plane of Σ is the tangent plane of one of its points.

We consider now the first order motion of a plane (u_1, u_2, u_3, u_4) of E. From (1.19) it follows that, up to the first order in λ and μ,

$$(2.6)\qquad x = X - \mu Z - k_1\lambda, \quad y = Y + \lambda Z - k_2\mu, \quad z = \mu X - \lambda Y + Z,$$

which implies

$$
\begin{aligned}
U_1 &= u_1 + \mu u_3,\\
U_2 &= u_2 - \lambda u_3,\\
U_3 &= -\mu u_1 + \lambda u_2 + u_3,\\
U_4 &= -\lambda k_1 u_1 - \mu k_2 u_2 + u_4.
\end{aligned} \tag{2.7}
$$

Every M_1 out of M_2 gives us a characteristic line in the moving plane; these lines belong to a pencil. Its vertex (the "characteristic point" of the plane, which is the tangent point with its envelope) has as its homogeneous coordinates the minors of the matrix

$$
\left\|\begin{matrix} u_1 & u_2 & u_3 & u_4 \\ 0 & -u_3 & u_2 & -k_1u_1 \\ u_3 & 0 & -u_1 & -k_2u_2 \end{matrix}\right\|. \tag{2.8}
$$

Hence this point is

$$
\begin{aligned}
X &= [k_1u_1^2 + k_2(u_2^2 + u_3^2)]u_2 - u_1u_3u_4,\\
Y &= -[k_1(u_1^2 + u_3^2) + k_2u_2^2]u_1 - u_2u_3u_4,\\
Z &= (k_1 - k_2)u_1u_2u_3 - u_3^2u_4,\\
W &= u_3(u_1^2 + u_2^2 + u_3^2).
\end{aligned} \tag{2.9}
$$

The coordinates of the characteristic point of a moving plane are therefore cubic functions of its coordinates u_i.

We know that a moving point remains, for the linear approximation, in the tangent plane of its trajectory surface; we have now the dual: a plane moves about its characteristic point.

Example 12. Show that the formulas (2.9) are the inverse of (2.1), if we interchange X and x, U_1 and u_1, etc. This can be done by proving that the line through (2.9) perpendicular to u belongs to the congruence of normals (2.4). The relation between a point and its tangent plane is birational.

Example 13. The only (real) planes whose characteristic point is at infinity are those parallel to O_z. Determine the characteristic point of a plane through O_z.

Example 14. Show that the exceptional planes of (2.9) are those perpendicular to s_1 or to s_2. If a plane u is perpendicular to s_1 all points of the projection of s_2 on u are characteristic points of u; the motion of u is restricted to a rotation about this projection.

Example 15. The point where a moving plane u is tangent to its envelope is the intersection with u and that line of the s_1, s_2 congruence which is perpendicular to u; formulate the analogous property of a moving line in planar M_1-motion.

The motion of a line is more complicated than the first order motion of a point or a plane. Its motion follows from (1.19) (or from (2.7)), p_{ij} and P_{ij} being its coordinates in E and Σ respectively,

$$
\begin{aligned}
&P_{14} = p_{14} + \mu p_{34}, \qquad P_{24} = p_{24} - \lambda p_{34}, \qquad P_{34} = p_{34} - \mu p_{14} + \lambda p_{24},\\
&P_{23} = p_{23} + \mu(p_{12} - k_2 p_{34}) + \lambda^2 p_{23} + \lambda\mu(p_{31} - k_2 p_{24}) + \mu^2 k_2 p_{14},\\
&P_{31} = p_{31} - \lambda(p_{12} - k_1 p_{34}) + \lambda^2 k_1 p_{24} + \lambda\mu(p_{23} - k_1 p_{14}) + \mu^2 p_{31},\\
&P_{12} = p_{12} + \lambda(p_{31} - k_1 p_{24}) - \mu(p_{23} - k_2 p_{14}) + (\lambda^2 k_1 + \mu^2 k_2) p_{34}.
\end{aligned} \tag{2.10}
$$

In the differential geometry of lines there is a well-known theorem that any line l of a congruence is intersected by *two* of its neighbors; the points of intersection are called the foci of l, the planes through l and these neighbors are the focal planes of l. If we substitute (2.10) into $\Sigma P_{14} p_{23} = 0$, the terms linear in λ, μ vanish and we obtain

$$
\begin{aligned}
&[p_{23} p_{14} + k_1(p_{24}^2 + p_{34}^2)]\lambda^2 + [p_{31} p_{14} + p_{23} p_{24} - (k_1 + k_2) p_{14} p_{24}]\lambda\mu\\
&\qquad + [p_{31} p_{24} + k_2(p_{14}^2 + p_{34}^2)]\mu^2 = 0,
\end{aligned} \tag{2.11}
$$

a quadratic equation for $\lambda : \mu$, whose roots enable us in principle to determine the foci and the focal planes of the line $\mathrm{l} = (p_{ij})$. This seems to give rise, however, to complicated algebra.

Example 16. Show by means of (2.4) and (2.11) that the foci of a line l which coincides with a normal are its intersections with s_1 and s_2.
Example 17. Show that the locus of the lines with coinciding foci is a complex of the fourth order.

Another method to study the instantaneous behavior of an arbitrary line l is the following. Not making use of the canonical frames but normalizing the parameters as we did before, we start from (1.6) and obtain (1.9) except that $l_3 \neq 0$, $j_3 \neq 0$; the motion up to linear terms is given by

$$
\begin{aligned}
X &= x - (j_3\lambda + l_3\mu)y + \mu z + a_\lambda\lambda + a_\mu\mu,\\
Y &= (j_3\lambda + l_3\mu)x + y - \lambda z + b_\lambda\lambda + b_\mu\mu,\\
Z &= -\mu x + \lambda y + z + c_\lambda\lambda + c_\mu\mu.
\end{aligned} \tag{2.12}
$$

We take the frames such that l coincides with the z-axis; hence $p_{34} = 1$ and all other p_{ij} are zero. Then, from (2.12),

$$
\begin{aligned}
&P_{14} = \mu, \qquad P_{24} = -\lambda, \qquad P_{34} = 1,\\
&P_{23} = -\lambda(c_\lambda\lambda + c_\mu\mu) - (b_\lambda\lambda + b_\mu\mu),\\
&P_{31} = (a_\lambda\lambda + a_\mu\mu) - \mu(c_\lambda\lambda + c_\mu\mu),\\
&P_{12} = \mu(b_\lambda\lambda + b_\mu\mu) + \lambda(a_\lambda\lambda + a_\mu\mu).
\end{aligned} \tag{2.13}
$$

The neighbors that intersect l follow from $P_{12}=0$, i.e.,

$$a_\lambda\lambda^2+(a_\mu+b_\lambda)\lambda\mu+b_\mu\mu^2=0. \tag{2.14}$$

Let the roots be $\lambda':\mu'$ and $\lambda'':\mu''$, then the foci (real or conjugate imaginary) are at $z'=-(a_\lambda\lambda'+a_\mu\mu')/\mu'=(b_\lambda\lambda'+b_\mu\mu')/\lambda'$ and analogously for z''. We specify the origin (as yet arbitrary) by taking it at the midpoint of the foci, which is always a real point. Hence $z'+z''=0$, which by means of (2.14) gives us

$$a_\mu=b_\lambda(=d). \tag{2.15}$$

Let $\mathrm{n}(\lambda,\mu)$ be the common perpendicular between l and the neighboring position of l given by (2.13). If n passes through the point (x,y,z) we have $n_{14}=x$, $n_{24}=y$, $n_{34}=0$, $n_{23}=yz$, $n_{31}=-xz$, $n_{12}=0$ and moreover $\lambda:\mu=x:y$. To the first order, n intersects P_{ij} if

$$-x(b_\lambda x+b_\mu y)+y(a_\lambda x+a_\mu y)+(x^2+y^2)z=0, \tag{2.16}$$

which implies that the lines n are the generators of a cylindroid. Through any point on l $(0,0,z)$ there pass two generators, determined by

$$(z-b_\lambda)x^2+(a_\lambda-b_\mu)xy+(z+a_\mu)y^2=0. \tag{2.17}$$

They are real if (in view of (2.15))

$$4z^2\leqslant(a_\lambda-b_\mu)^2+4d^2, \tag{2.18}$$

from which the two (real) limit points on l follow; they are at equal distances from O. Our main result is: up to the first order, the motion of a line l is such that the common perpendiculars (which depend only on the ratio $\lambda:\mu$) between l and its consecutive positions are the generators of a cylindroid (Blaschke [1960]). By rotating the frame about O_z equation (2.16) may be reduced to the standard form $(x^2+y^2)z+hxy=0$.

Example 18. Show that the rotation angle, θ, is given by $\tan 2\theta=d/(2(a_\lambda-b_\mu))$ and determine h.

3. Second order properties

In this section we deal with second order properties of instantaneous two-parameter kinematics in space. From the fundamental relation for an orthogonal matrix $\mathbf{A}(\lambda,\mu)$:

$$\mathbf{A}\mathbf{A}^{\mathrm{T}}=\mathbf{I}, \tag{3.1}$$

we have seen it follows that $\mathbf{A}_\lambda = \mathbf{J}$, $\mathbf{A}_\mu = \mathbf{L}$ (from (1.4)). Differentiating (3.1) twice and then putting $\lambda = \mu = 0$ so that $\mathbf{A} \to \mathbf{A}_0 = \mathbf{I}$, we obtain

$$\begin{aligned} \mathbf{A}_{\lambda\lambda} + \mathbf{A}^{\mathrm{T}}_{\lambda\lambda} &= 2\mathbf{J}^2, \\ \mathbf{A}_{\lambda\mu} + \mathbf{A}^{\mathrm{T}}_{\lambda\mu} &= (\mathbf{JL} + \mathbf{LJ}), \\ \mathbf{A}_{\mu\mu} + \mathbf{A}^{\mathrm{T}}_{\mu\mu} &= 2\mathbf{L}^2. \end{aligned} \tag{3.2}$$

As the right-hand-sides of (3.2) are symmetric matrices, we have

$$\mathbf{A}_{\lambda\lambda} = \mathbf{J}^2 + \mathbf{F}, \qquad \mathbf{A}_{\lambda\mu} = \tfrac{1}{2}(\mathbf{JL} + \mathbf{LJ}) + \mathbf{G}, \qquad \mathbf{A}_{\mu\mu} = \mathbf{L}^2 + \mathbf{H}, \tag{3.3}$$

where $\mathbf{F}$, $\mathbf{G}$, $\mathbf{H}$ are skew matrices; we shall denote their elements by f_i, g_i, h_i $(i = 1,2,3)$ using the convention implied by (1.4). Hence, up to the second order terms, we have

$$\begin{aligned} \boldsymbol{P} = {} & [\mathbf{I} + \mathbf{J}\lambda + \mathbf{L}\mu + \tfrac{1}{2}((\mathbf{J}^2 + \mathbf{F})\lambda^2 + (\mathbf{JL} + \mathbf{LJ} + 2\mathbf{G})\lambda\mu + (\mathbf{L}^2 + \mathbf{H})\mu^2)]\boldsymbol{p} \\ & + \boldsymbol{d}_\lambda\lambda + \boldsymbol{d}_\mu\mu + \tfrac{1}{2}(\boldsymbol{d}_{\lambda\lambda}\lambda^2 + 2\boldsymbol{d}_{\lambda\mu}\lambda\mu + \boldsymbol{d}_{\mu\mu}\mu^2). \end{aligned} \tag{3.4}$$

If we use the canonical system, which was fully determined by first order notions, we have (as seen from (1.19)) in terms of the two invariants k_1 and k_2,

$$\boldsymbol{J} = \begin{Vmatrix} 0 & 0 & 0 \\ 0 & 0 & -1 \\ 0 & 1 & 0 \end{Vmatrix}, \qquad \boldsymbol{L} = \begin{Vmatrix} 0 & 0 & 1 \\ 0 & 0 & 0 \\ -1 & 0 & 0 \end{Vmatrix}, \tag{3.5}$$
$$\boldsymbol{d}_\lambda = (k_1, 0, 0), \qquad \boldsymbol{d}_\mu = (0, k_2, 0);$$

whence

$$\mathbf{J}^2 = \begin{Vmatrix} 0 & 0 & 0 \\ 0 & -1 & 0 \\ 0 & 0 & -1 \end{Vmatrix}, \quad \mathbf{L}^2 = \begin{Vmatrix} -1 & 0 & 0 \\ 0 & 0 & 0 \\ 0 & 0 & -1 \end{Vmatrix}, \quad \mathbf{JL} + \mathbf{LJ} = \begin{Vmatrix} 0 & 1 & 0 \\ 1 & 0 & 0 \\ 0 & 0 & 0 \end{Vmatrix}. \tag{3.6}$$

Expanding (3.4) and recalling that the components of $\boldsymbol{d}$ are (a, b, c), the second order representation of the M_2-motion is:

$$\begin{aligned} X = {} & x + z\mu + k_1\lambda + \tfrac{1}{2}(-f_3y + f_2z)\lambda^2 + (\tfrac{1}{2}y - g_3y + g_2z)\lambda\mu \\ & + \tfrac{1}{2}(-x - h_3y + h_2z)\mu^2 + \tfrac{1}{2}a_{\lambda\lambda}\lambda^2 + a_{\lambda\mu}\lambda\mu + \tfrac{1}{2}a_{\mu\mu}\mu^2, \\ Y = {} & y - z\lambda + k_2\mu + \tfrac{1}{2}(-y + f_3x - f_1z)\lambda^2 + (\tfrac{1}{2}x + g_3x - g_1z)\lambda\mu \\ & + \tfrac{1}{2}(h_3x - h_1z)\mu^2 + \tfrac{1}{2}b_{\lambda\lambda}\lambda^2 + b_{\lambda\mu}\lambda\mu + \tfrac{1}{2}b_{\mu\mu}\mu^2, \\ Z = {} & z + y\lambda - x\mu + \tfrac{1}{2}(-z - f_2x + f_1y)\lambda^2 + (-g_2x + g_1y)\lambda\mu \\ & + \tfrac{1}{2}(-z - h_2x + h_1y)\mu^2 + \tfrac{1}{2}c_{\lambda\lambda}\lambda^2 + c_{\lambda\mu}\lambda\mu + \tfrac{1}{2}c_{\mu\mu}\mu^2. \end{aligned} \tag{3.7}$$

This shows that although there are only two first order invariants, there are eighteen for the second order: f_i, g_i, h_i $(i = 1,2,3)$ and the nine second order partial derivatives of a, b, c. This implies that second order spatial M_2 theory is a complicated matter (BLASCHKE [1960, p. 81]). On the other hand the kinematicians of the second half of the 19th century, especially MANNHEIM [1894, pp. 127–160], obtained some interesting results using their ingeneous geometrical methods. Restricting ourselves to the motion of a point we shall derive by analytical means some theorems on the curvature of its trajectory surface (BOTTEMA [1971]).

We have seen that there exists a certain analog between spatial M_2- and planar M_1-motion. In the latter one considers a normal, which is determined by its direction, and a point A on it given by its distance r from the pole (thus introducing polar coordinates for A). Of interest is the relationship between r and the center of curvature of A, which is also on the normal; this relation is the Euler–Savary equation.

Analogously we shall consider a fixed normal n, in space, by giving its direction cosines α, β, γ; if we assume $\gamma \neq 0$, excluding thereby horizontal normals, n (which intersects s_1 and s_2) is completely determined by its direction. Moreover we fix the orientation of n with the condition $\gamma > 0$. If A_0 is the intersection of n and the plane $Z = 0$, a point A on n is determined by the distance $A_0A = r$ $(-\infty < r < \infty)$. We have now introduced "polar coordinates" $(r; \alpha, \beta, \gamma)$ for A.

Any normal satisfies (2.4); hence $p_{23} = k_1\alpha$, $p_{31} = k_2\beta$ and therefore $A_0 = (k_2\beta/\gamma, -k_1\alpha/\gamma, 0)$. For A we obtain

$$x = (k_2\beta/\gamma) + \alpha r, \quad y = (-k_1\alpha/\gamma) + \beta r, \quad z = \gamma r. \tag{3.8}$$

To study the curvature of the trajectory surface we derive first some first order expressions. The arc-element on the surface is defined by the first differential form

$$\mathrm{d}s^2 = E_1\mathrm{d}\lambda^2 + 2E_2\mathrm{d}\lambda\,\mathrm{d}\mu + E_3\mathrm{d}\mu^2,$$

with

$$E_1 = \sum X_\lambda^2, \qquad E_2 = \sum X_\lambda X_\mu, \qquad E_3 = \sum X_\mu^2. \tag{3.9}$$

From (1.14) and (3.8) it follows

$$\begin{aligned} X_\lambda &= k_1, \quad Y_\lambda = -\gamma r, \quad Z_\lambda = (-k_1\alpha/\gamma) + \beta r, \\ X_\mu &= \gamma r, \quad Y_\mu = k_2, \quad Z_\mu = (-k_2\beta/\gamma) - \alpha r, \end{aligned} \tag{3.10}$$

and we obtain

$$
\begin{aligned}
\gamma^2 E_1 &= \gamma^2(\beta^2+\gamma^2)r^2 - 2k_1\alpha\beta\gamma r + k_1^2(\alpha^2+\gamma^2),\\
\gamma^2 E_2 &= -\gamma^2\alpha\beta r^2 + [k_1(\alpha^2+\gamma^2) - k_2(\beta^2+\gamma^2)]\gamma r + k_1k_2\alpha\beta,\\
\gamma^2 E_3 &= \gamma^2(\alpha^2+\gamma^2)r^2 + 2k_2\alpha\beta\gamma r + k_2^2(\beta^2+\gamma^2).
\end{aligned}
\tag{3.11}
$$

From this it follows

$$\Delta_1 = E_1E_3 - E_2^2 = (\gamma^2r^2 + k_1k_2)^2/\gamma^2. \tag{3.12}$$

The arc-element is singular ($\Delta_1 = 0$), as could be expected, if A is the intersection of n and s_1 or s_2.

Furthermore we have, from (3.10),

$$\boldsymbol{N} = \boldsymbol{P}_\lambda \times \boldsymbol{P}_\mu = [(\gamma^2r^2 + k_1k_2)/\gamma]\boldsymbol{n}, \tag{3.13}$$

where $\boldsymbol{n}$ is the unit vector along the directed normal.

To determine the curvature of the trajectory surface we need its second differential form

$$D_1 \mathrm{d}\lambda^2 + 2D_2 \mathrm{d}\lambda\, \mathrm{d}\mu + D_3 \mathrm{d}\mu^2, \tag{3.14}$$

with

$$\Delta_1^{\frac{1}{2}} D_1 = \boldsymbol{P}_{\lambda\lambda}\cdot\boldsymbol{N}, \qquad \Delta_1^{\frac{1}{2}} D_2 = \boldsymbol{P}_{\lambda\mu}\cdot\boldsymbol{N}, \qquad \Delta_1^{\frac{1}{2}} D_3 = \boldsymbol{P}_{\mu\mu}\cdot\boldsymbol{N}, \tag{3.15}$$

or, in view of (3.12) and (3.13), if $\Delta_1 \neq 0$,

$$D_1 = \boldsymbol{P}_{\lambda\lambda}\cdot\boldsymbol{n}, \qquad D_2 = \boldsymbol{P}_{\lambda\mu}\cdot\boldsymbol{n}, \qquad D_3 = \boldsymbol{P}_{\mu\mu}\cdot\boldsymbol{n},$$

which gives us by means of (3.7) and (3.8)

$$
\begin{aligned}
D_1 &= -(\beta^2+\gamma^2)r + \gamma^{-1}[k_1\alpha\beta + f_3(k_1\alpha^2 + k_2\beta^2) - \gamma(f_1k_1\alpha + f_2k_2\beta)]\\
&\quad + (a_{\lambda\lambda}\alpha + b_{\lambda\lambda}\beta + c_{\lambda\lambda}\gamma),\\
D_2 &= \alpha\beta r + \gamma^{-1}[-\tfrac{1}{2}(k_1\alpha^2 - k_2\beta^2) + g_3(k_1\alpha^2 + k_2\beta^2) - \gamma(g_1k_1\alpha + g_2k_2\beta)]\\
&\quad + (a_{\lambda\mu}\alpha + b_{\lambda\mu}\beta + c_{\lambda\mu}\gamma),\\
D_3 &= -(\alpha^2+\gamma^2)r + \gamma^{-1}[-k_2\alpha\beta + h_3(k_1\alpha^2 + k_2\beta^2) - \gamma(h_1k_1\alpha + h_2k_2\beta)]\\
&\quad + (a_{\mu\mu}\alpha + b_{\mu\mu}\beta + c_{\mu\mu}\gamma).
\end{aligned}
\tag{3.16}
$$

From this it follows

$$\Delta_2 = D_1D_3 - D_2^2 = \gamma^2r^2 + \gamma^{-1}rQ_1 + \gamma^{-2}Q_2, \tag{3.17}$$

Q_1 and Q_2 are homogeneous quartic functions of α, β, γ, the coefficients of which depend on the twenty instantaneous invariants of the motion.

Differential geometry of surfaces has it that all second order properties depend on the fundamental expressions N, E_i, D_i $(i = 1, 2, 3)$, which are given here in terms of the polar coordinates of the moving point. We apply our results to the determination of the Gaussian curvature K (also called the measure of curvature, or specific curvature; German: Krümmungsmass), which is the product of the two principal curvatures, of the trajectory surface. It is given by $K = \Delta_2/\Delta_1$, and therefore

$$K = (r^2 + \gamma^{-3} r Q_1 + \gamma^{-4} Q_2)/(r^2 + (k_1 k_2/\gamma^2))^2 \tag{3.18}$$

which gives us the Gaussian curvature of the trajectory surface of a variable point on a fixed normal. The denominator is zero if A is one of the two intersections (real or imaginary) of the normal and the lines s_1, s_2. The numerator has two zeros, r_1 and r_2, real or imaginary, and (3.18) may be written

$$K = (r - r_1)(r - r_2)/(r^2 + q)^2, \tag{3.19}$$

with $q = k_1 k_2/\gamma^2$. We have $K = 0$ if $r = r_1$ or $r = r_2$; hence on any normal there are two points (of E) which while in the zero-position pass through a parabolic point of their trajectory surface. If these points are imaginary all real points of n are at elliptic points of their surface; if they are real and distinct there are on n both elliptic and hyperbolic points.

(3.19) is the analog of the Euler–Savary equation in M_1-planar motion, which may be written

$$K = (r - r_0)/r^2, \tag{3.20}$$

r_0 being the distance from the pole to the intersection of the normal and the inflection circle.

Until now we have excluded the special normals parallel to the plane $z = 0$. There are two sets: those intersecting s_1 and parallel to s_2 and those intersecting s_2 and parallel to s_1. If s_1 intersects the z-axis at S_1 a normal n_1 of the first set may be determined by the distance $S_1 A_0 = d$, A_0 being the intersection of n_1 and s_1; a point A on n_1 is then given by $r' = AA_0$.

Example 19. Show that the Gaussian curvature of A is given by $K = (A^* r' + B^*)/(r')^2$; A^*, B^* are functions of the invariants and the normal [BOTTEMA [1971]).

We have derived an expression for K in terms of spatial polar coordinates because we wished to establish an analog with the planar M_1-motion. We shall now deal with the curvature of the trajectory surface in terms of Cartesian coordinates.

From (1.19) and (3.13) it follows that

$$N = [(xz - k_2y), (yz + k_1x), (z^2 + k_1k_2)], \tag{3.21}$$

and furthermore

$$E_1 = y^2 + z^2 + k_1^2, \quad E_2 = (k_1 - k_2)z - xy, \quad E_3 = x^2 + z^2 + k_2^2, \tag{3.22}$$

hence

$$\Delta_1 = (xz - k_2y)^2 + (yz + k_1x)^2 + (z^2 + k_1k_2)^2, \tag{3.23}$$

which means

$$\Delta_1 = N \cdot N.$$

Example 20. Show that $\Delta_1 = 0$ represents a quartic surface whose only real points are those of the line $z = 0$, $w = 0$ and those of s_1, s_2 if these lines are real. The surface is a ruled one, its generators intersect s_1, s_2 and the isotropic conic.

From (3.7) and (3.21) it follows

$$\begin{aligned}
\Delta_1^{1/2}D_1 = P_{\lambda\lambda} \cdot N = F_1 &= (-f_3y + f_2z + a_{\lambda\lambda})(xz - k_2y) \\
&\quad + (-y + f_3x - f_1z + b_{\lambda\lambda})(yz + k_1x) \\
&\quad + (-z - f_2x + f_1y + c_{\lambda\lambda})(z^2 + k_1k_2), \\
\Delta_1^{1/2}D_2 = F_2 &= (\tfrac{1}{2}y - g_3y + g_2z + a_{\lambda\mu})(xz - k_2y) \\
&\quad + (\tfrac{1}{2}x + g_3x - g_1z + b_{\lambda\mu})(yz + k_1x) \\
&\quad + (-g_2x + g_1y + c_{\lambda\mu})(z^2 + k_1k_2), \\
\Delta_1^{1/2}D_3 = F_3 &= (-x - h_3y + h_2z + a_{\mu\mu})(xz - k_2y) \\
&\quad + (h_3x - h_1z + b_{\mu\mu})(yz + k_1x) \\
&\quad + (-z - h_2x + h_1y + c_{\mu\mu})(z^2 + k_1k_2).
\end{aligned} \tag{3.24}$$

F_i is a cubic polynomial in x, y, z. Obviously $F_i = 0$ represents a cubic surface, passing through s_1 and s_2. The third order terms of F_1, F_2, F_3 are $-z(y^2 + z^2)$, xyz, $-z(x^2 + z^2)$ respectively. Moreover $\Delta_2 = (F_1F_3 - F_2^2)/\Delta_1$.

Also of importance in curvature theory is the factor

$$\Delta_3 = E_1D_3 - 2E_2D_2 + E_3D_1 = (1/\Delta_1^{1/2})G(x, y, z). \tag{3.25}$$

G is a quintic polynomial, the fifth order terms being $-z^3(x^2 + y^2 + z^2)$.

As is well-known the principal curvatures κ_1, κ_2 at a point of a surface are the two roots of the equation

$$\Delta_1\kappa^2 - \Delta_3\kappa + \Delta_2 = 0, \tag{3.26}$$

that is

$$\Delta_1^2\kappa^2 - G\Delta_1^{1/2}\kappa + (F_1F_3 - F_2^2) = 0, \tag{3.27}$$

from which κ_1 and κ_2 may be found in terms of x, y, z and the twenty instantaneous invariants.

The Gaussian curvature $K = \kappa_1\kappa_2$ is given by

$$K = (F_1F_3 - F_2^2)/\Delta_1^2. \tag{3.28}$$

The locus of the points with $K = 0$ is represented by $F_1F_3 - F_2^2 = 0$, if we exclude the singular points on s_1 and on s_2. This locus is therefore a sextic surface S_6, a theorem already proved by MANNHEIM [1875]. It passes twice through s_1 and s_2. The sixth order terms are seen to be $z^4(x^2 + y^2 + z^2)$; hence S_6 passes through the isotropic conic and it intersects the plane at infinity in the line $l(z = w = 0)$ counted four times. For a fixed value of z the polynomials F_i are quadratic functions of x, y. Hence a plane parallel to O_{xy} intersects S_6 in a quartic curve, which shows that l is a double line of S_6. An arbitrary line in E has six intersections with S_6, representing six points passing through a parabolic point of their trajectory surface. If, however, the line is a normal, four (two times two) intersections (with s_1 and s_2) do not correspond to such points. Hence on a normal there are only two parabolic points, as we found before. On a special normal there is only one.

This sextic surface is the analog of the inflection circle of the planar case. The pole must be excluded from the locus; on an arbitrary line there are two points (real or imaginary) passing instantaneously through an inflection point of their path; if the line is a normal there is only one.

Another locus follows from (3.26). The mean curvature is defined by $(\kappa_1 + \kappa_2)/2$, it is zero if $G = 0$. Hence the locus of these points is a quintic surface S_5, passing through s_1 and s_2 which must be excluded from the locus. The surface intersects the plane at infinity in the isotropic conic and the line l counted three times.

Example 21. Show that l is a double line of S_5.

If the two principal curvatures at a point of a surface are equal the point is called an umbilic; it is defined by the conditions $\kappa_1 = \kappa_2 \neq 0$. The point is elliptic; its Dupin's indicatrix, an ellipse in general, is a circle.

Example 22. Show that the locus of the moving points which are instantaneously at an umbilic of their trajectory surface is a surface S_{10} of order ten. Determine its intersection with the plane at infinity.

An interesting M_2 occurs if we take the motion of E as the *rotation* about two screw axes: S_A in Σ and S_B in E. We assume furthermore that S_A and S_B always maintain a fixed relative distance, $2h$, and a fixed relative angle, 2α. As parameters λ and μ we take the rotation angles about S_A and S_B respectively. The loci of points in E which have several finitely separated positions on spheres, planes, circles, cylinders or lines in Σ were derived by ROTH [1967c], who considered such displacements as special *similarity transformations*, and showed that the loci degenerate for these special displacements. The instantaneous case has recently been studied by TSAI [1977] who proceeded as follows: If in the zero position we take the coordinate system o_{xyz} in E coinciding with the frame O_{XYZ} in Σ and we take the origin O as the midpoint of the common normal between S_A and S_B, the Z-axis along the common normal (pointing from S_A toward S_B), and the X-axis along the internal bisector of S_A and S_B; we have the configuration shown in Fig. 85.

We can express this M_2-motion as the ordered product of two M_1-motions: M_B about S_B followed by M_A about S_A. From Chapter III, (12.11)–(12.13) it follows that for M_B we have:

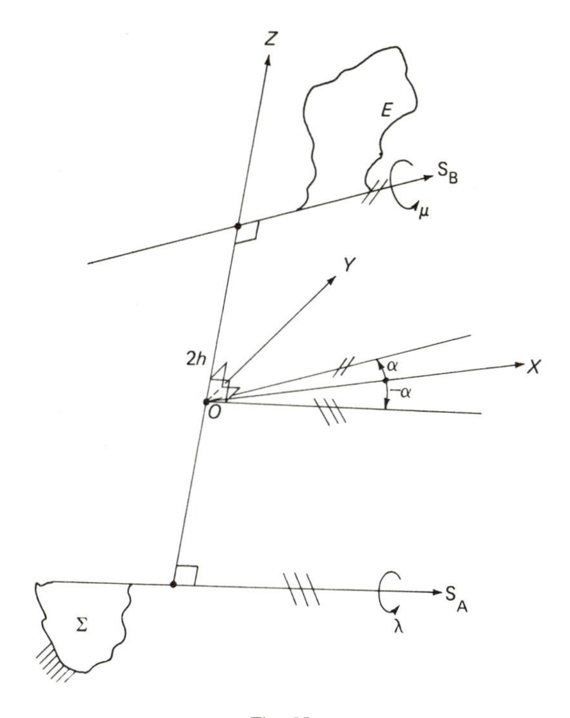

Fig. 85.

$\mathbf{M}_B =$

$$\left\|\begin{matrix} \sin^2\alpha(\cos\mu - 1) + 1 & \cos\alpha\sin\alpha(1 - \cos\mu) & \sin\alpha\sin\mu & -h\sin\alpha\sin\mu \\ \cos\alpha\sin\alpha(1 - \cos\mu) & \cos^2\alpha(\cos\mu - 1) + 1 & -\cos\alpha\sin\mu & h\cos\alpha\sin\mu \\ -\sin\alpha\sin\mu & \cos\alpha\sin\mu & \cos\mu & h(1 - \cos\mu) \\ 0 & 0 & 0 & 1 \end{matrix}\right\|$$

where we have written the result in the 4×4 form (Chapter I, (8.5)). Similarly, for $\mathbf{M}_A$ we have:

$\mathbf{M}_A =$

$$\left\|\begin{matrix} \sin^2\alpha(\cos\lambda - 1) + 1 & \cos\alpha\sin\alpha(\cos\lambda - 1) & -\sin\alpha\sin\lambda & -h\sin\alpha\sin\lambda \\ \cos\alpha\sin\alpha(\cos\lambda - 1) & \cos^2\alpha(\cos\lambda - 1) + 1 & -\cos\alpha\sin\lambda & -h\cos\alpha\sin\lambda \\ \sin\alpha\sin\lambda & \cos\alpha\sin\lambda & \cos\lambda & h(\cos\lambda - 1) \\ 0 & 0 & 0 & 1 \end{matrix}\right\|$$

The resulting M_2-motion is determined by the matrix product $\mathbf{M}_A\mathbf{M}_B$. Differentiating this product it follows that in the zero position

$$\begin{aligned} X_\lambda &= (-\sin\alpha)z - h(\sin\alpha); & X_\mu &= (\sin\alpha)z + h(-\sin\alpha) \\ Y_\lambda &= (-\cos\alpha)z - h(\cos\alpha); & Y_\mu &= (-\cos\alpha)z + h(\cos\alpha) \\ Z_\lambda &= (\sin\alpha)x + (\cos\alpha)y; & Z_\mu &= (-\sin\alpha)x + (\cos\alpha)y \end{aligned} \tag{3.29}$$

Hence the first order properties depend upon the two numbers: h and α.

Example 23. Show that the tangent plane of the trajectory surface of (x, y, z) has the coordinates $2\cos\alpha(-xz\sin\alpha + yh\cos\alpha)$, $2\sin\alpha(-yz\cos\alpha + xh\sin\alpha)$, $-2\sin\alpha\cos\alpha(z^2 - h^2)$, $2\sin\alpha\cos\alpha(x^2z + y^2z + z^3 - 2xyh - zh^2)$ (TSAI [1977]).

Example 24. For the M_1-motions in which μ is a function of λ, show that the geometric velocity of (x, y, z) is

$$X' = (\mathrm{r} - 1)(\sin\alpha)z - h(\mathrm{r} + 1)\sin\alpha, \quad Y' = -(\mathrm{r} + 1)(\cos\alpha)z + h(\mathrm{r} - 1)\cos\alpha,$$
$$Z' = (-(\mathrm{r} - 1)\sin\alpha)x + (\mathrm{r} + 1)(\cos\alpha)y, \quad \text{where} \quad \mathrm{r} = d\mu/d\lambda \text{ (Idem.)}$$

Example 25. Show that in this case the cylindroid (1.15) has the equation $(x^2 + y^2)z\sin\alpha\cos\alpha - hxy = 0$ (Idem.)

Taking the second derivatives of the product $\mathbf{M}_A\mathbf{M}_B$, we obtain:

$$\begin{aligned} X_{\lambda\lambda} &= (-\sin^2\alpha)x + (-\sin\alpha\cos\alpha)y \\ Y_{\lambda\lambda} &= (-\sin\alpha\cos\alpha)x + (-\cos^2\alpha)y \\ Z_{\lambda\lambda} &= -z - h \end{aligned} \tag{3.30}$$

$$\begin{aligned} X_{\mu\mu} &= (-\sin^2\alpha)x + (\sin\alpha\cos\alpha)y \\ Y_{\mu\mu} &= (\sin\alpha\cos\alpha)x + (-\cos^2\alpha)y \\ Z_{\mu\mu} &= -z + h \end{aligned} \tag{3.31}$$

$$\begin{aligned} X_{\lambda\mu} &= (\sin^2\alpha)x + (-\sin\alpha\cos\alpha)y \\ Y_{\lambda\mu} &= (\sin\alpha\cos\alpha)x + (-\cos^2\alpha)y \\ Z_{\lambda\mu} &= (-\cos 2\alpha)z + h(\cos 2\alpha) \end{aligned} \tag{3.32}$$

From which the terms for the Gaussian curvature expression (3.28) follow:

$$F_1 = \sin 2\alpha(z+h)[(x\sin\alpha - y\cos\alpha)^2 + (z-h)^2]$$

$$F_2 = \sin 2\alpha(z-h)[-x^2\sin^2\alpha + y^2\cos^2\alpha + (z^2-h^2)\cos 2\alpha]$$

$$F_3 = \sin 2\alpha(z-h)[(x\sin\alpha + y\cos\alpha)^2 + (z+h)^2]$$

$$\begin{aligned} \Delta_1 = {} & z^2(x^2+y^2+z^2)\sin^2 2\alpha - 4xyzh\sin 2\alpha \\ & + 4h^2(x^2\sin^4\alpha + y^2\cos^4\alpha - 2z^2\sin^2\alpha\cos^2\alpha) + h^4\sin^2(2\alpha). \end{aligned}$$

Substituting into (3.28) yields the Gaussian curvature:

$$K = \frac{\sin^2(2\alpha)(z-h)[\mathrm{F}(x,y,z)]}{(\Delta_1)^2} \tag{3.33}$$

where $\mathrm{F}(x, y, z)$ is a fifth degree polynomial in x, y, z. It is easily seen from (3.33) that the locus of all parabolic points is the plane $z - h = 0$ and the fifth order surface $\mathrm{F}(x, y, z)$. The fifth order terms of $\mathrm{F}(x, y, z)$ are $z^3(x^2+y^2+z^2)\sin^2 2\alpha$, hence the surface passes through the isotropic conic, and the line $w = z = 0$ three times. This surface has been studied by TSAI [1977].

Example 26. Show that S_A and S_B are singular double lines of $\mathrm{F}(x, y, z)$.

4. Continuous two-parameter spatial motions

As was the case for a M_1, a specific two-parameter motion may be defined in various ways. The most formal is the analytical procedure of considering the general displacement formula (Chapter IX, (6.1)) with c_i and g_i taken as arbitrary functions of λ and μ.

Apart from trivial cases, the most simple cases are those for which c_i

($i = 0, 1, 2, 3$) are linear functions, g_0 is a constant, and g_i ($i = 1, 2, 3$) are functions of order two at most. Then for a fixed (x, y, z) the homogeneous coordinates (X, Y, Z, W) are quadratic functions of λ and μ. They represent a surface, well-known in classical algebraic geometry, usually called Steiner's surface or the Roman surface (see, for example, SOMMERVILLE [1934]). For the sake of brevity we shall denote such an M_2 as a Steiner-motion. The surface has some interesting properties: it is a quartic surface with three double lines; there are four (tangent) planes whose intersections with the surface are conics counted twice; any tangent plane intersects the surface in two conics; the surface contains a set of ∞^2 conics.

Example 27. Show the surface is of the fourth order by determining the number of intersections with an arbitrary line.
Example 28. Show that the surface contains ∞^2 conics or, in kinematical terms, show that the Steiner M_2-motion contains ∞^2 Darboux M_1-motions.

The Steiner motion has been studied at length by DARBOUX [1897]. Following him we remark that, as c_i are linear functions of λ, μ there exists a linear relation between the four. Hence by a suitable linear transformation of the parameters we may arrive at $c_0 = 0$ and two new independent parameters written as c_1/c_3, c_2/c_3, or more elegantly as the three homogeneous parameters c_1, c_2, c_3. The Steiner motion is then represented by

$$
\begin{aligned}
X &= (c_1^2 - c_2^2 - c_3^2)x + 2c_1c_2y + 2c_1c_3z + g_1(c_1, c_2, c_3)w, \\
Y &= 2c_2c_1x + (-c_1^2 + c_2^2 - c_3^2)y + 2c_2c_3z + g_2(c_1, c_2, c_3)w, \\
Z &= 2c_3c_1x + 2c_3c_2y + (-c_1^2 - c_2^2 + c_3^2)z + g_3(c_1, c_2, c_3)w, \\
W &= (c_1^2 + c_2^2 + c_3^2)w,
\end{aligned}
\tag{4.1}
$$

g_i being arbitrary homogeneous quadratic polynomials of c_i. The parameters of the motion have now a geometrical meaning: they are the direction numbers of the screw-motion (4.1). Its rotational part is a half-turn.

Example 29. Show that the Plücker coordinates of the screw axis of (4.1) are

$$p_{14} = c_1s, \qquad p_{24} = c_2s, \qquad p_{34} = c_3s,$$
$$p_{23} = \tfrac{1}{2}(c_2g_3 - c_3g_2), \qquad p_{31} = \tfrac{1}{2}(c_3g_1 - c_1g_3), \qquad p_{12} = \tfrac{1}{2}(c_1g_2 - c_2g_1),$$

with $s^2 = c_1^2 + c_2^2 + c_3^2$. These formulas represent the congruence of the screw axes.
Example 30. Show that the translation distance of (4.1) reads $(g_1c_1 + g_2c_2 + g_3c_3)/s^3$; there are ∞^1 pure rotations in the set.

DARBOUX [1897] has shown that for a Steiner motion there are in general ten points whose trajectory surface is a plane.

Another method to define an M_2 is an extension of Krames' procedure for obtaining an M_1. In the latter case one started with an arbitrary ruled surface R in Σ and reflected E into its generators, thus constructing ∞^1 positions of E with respect to Σ. To obtain an M_2 we can reflect E into the lines of a congruence C in Σ. If C consists of the lines of a plane or of those through a point the results are trivial, the trajectory surface of any moving point being a plane or a sphere respectively.

For a non-trivial example we assume that C is a linear congruence, consisting of the transversals of two real, skew and perpendicular lines l_1 and l_2. Let l_1 be $y = 0$, $z = h$ and l_2 be $x = 0$, $z = -h$. From this it follows that the lines of C can be taken as

$$p_{14} = \lambda, \qquad p_{24} = \mu, \qquad p_{34} = 2h, \qquad p_{23} = \mu h, \qquad p_{31} = \lambda h, \qquad p_{12} = -\lambda\mu. \tag{4.2}$$

Making use of Chapter IX, (7.3) the M_2 motion is seen to be

$$\begin{aligned} X &= (\lambda^2 - \mu^2 - 4h^2)x + 2\lambda\mu y + 4\lambda h z + 2\lambda(\mu^2 + 2h^2)w, \\ Y &= 2\lambda\mu x + (-\lambda^2 + \mu^2 - 4h^2)y + 4\mu h z - 2\mu(\lambda^2 + 2h^2)w, \\ Z &= 4\lambda h x + 4\mu h y + (-\lambda^2 - \mu^2 + 4h^2)z - 2h(\lambda^2 - \mu^2)w, \\ W &= (\lambda^2 + \mu^2 + 4h^2)w. \end{aligned} \tag{4.3}$$

This is obviously not a Steiner motion because the translation terms are of degree three. To give an idea of the motion described by (4.3), we determine the trajectory surface of the origin. It is given by

$$X = 2\lambda(\mu^2 + 2h^2), \qquad Y = -2\mu(\lambda^2 + 2h^2), \qquad Z = -2h(\lambda^2 - \mu^2), \qquad W = (\lambda^2 + \mu^2 + 4h^2), \tag{4.4}$$

which implies

$$2hX = \lambda(Z + 2hW), \qquad 2hY = \mu(Z - 2hW);$$

eliminating λ and μ, we obtain the equation of the surface explicitly:

$$(X^2 + Y^2 + Z^2)Z - 2h(X^2 - Y^2 + 2hZW)W = 0, \tag{4.5}$$

which shows that it is a cubic surface.

Example 31. Determine the intersection of the surface and the plane at infinity; prove that the (finite) intersection with a plane $Z = k$ is an ellipse and determine its semi-axes; show that the surface lies inside the parallel planes $Z = \pm 2h$.

A third method of constructing an M_2 is to impose constraints on E equivalent to four simple conditions. We shall discuss an example: We obtain a two-fold condition if a curve of E is compelled to pass permanently through a fixed point of Σ. Hence we obtain an M_2 if two lines l_1, l_2 of E pass through the points A_1, A_2 of Σ respectively. We take l_1, l_2 to be orthogonal skew lines with distance $2h$; we introduce the frame in E such that

$$l_1\colon y = 0, \quad z = h; \qquad l_2\colon x = 0, \quad z = -h.$$

Furthermore, in Σ, let

$$A_1 = (0, 0, m), \qquad A_2 = (0, 0, -m).$$

It is clear that the motion is only possible if $h \leqslant m$. If E is in a position such that l_i passes through A_i ($i = 1, 2$) and it is then rotated about O_Z, it obviously still is in a possible position. The conclusion is: all trajectory surfaces are surfaces of revolution with axis O_Z.

The general motion is given by

$$\boldsymbol{P} = \mathbf{A}\boldsymbol{p} + \boldsymbol{d},$$

$\mathbf{A}$ is an orthogonal matrix $\|a_{ij}\|$ and $\boldsymbol{d} = (a, b, c)$. If the point $(x_0, 0, h)$ of l_1 coincides with A_1 and the point $(0, y_0, -h)$ of l_2 coincides with A_2 the following conditions must be satisfied

$$\begin{aligned} a_{11}x_0 + a_{13}h + a &= 0, & a_{12}y_0 - a_{13}h + a &= 0, \\ a_{21}x_0 + a_{23}h + b &= 0, & a_{22}y_0 - a_{23}h + b &= 0, \\ a_{31}x_0 + a_{33}h + c &= m, & a_{32}y_0 - a_{33}h + c &= -m. \end{aligned} \tag{4.6}$$

Eliminating a, b, c we obtain

$$\begin{aligned} a_{11}x_0 - a_{12}y_0 + 2a_{13}h &= 0, \\ a_{21}x_0 - a_{22}y_0 + 2a_{23}h &= 0, \\ a_{31}x_0 - a_{32}y_0 + 2a_{33}h &= 2m. \end{aligned} \tag{4.7}$$

Multiplying these equations, respectively, by a_{11}, a_{21}, a_{31}, then adding them, and then repeating this process using a_{12}, a_{22}, a_{32} and a_{13}, a_{23},. a_{33}, we get

$$x_0 = 2ma_{31}, \quad y_0 = -2ma_{32}, \quad h = ma_{33}. \tag{4.8}$$

Hence if $h/m = \cos\theta$ (which is possible in view of $h \leqslant m$), the result is

$$a_{33} = \cos\theta, \tag{4.9}$$

which means that the angle between o_z and O_Z has the constant value θ. This is, however, one of the Eulerian angles (Chapter VI, Section 3) and it seems appropriate to choose the other two as the motion's parameters. The components a, b, c follow from (4.6) and (4.8) and are seen to be

$$a = m(-a_{11}a_{31} + a_{12}a_{32}), \qquad b = m(-a_{21}a_{31} + a_{22}a_{32}),$$
$$c = m(-a_{31}^2 + a_{32}^2). \tag{4.10}$$

Therefore, making use of the formulas (3.1) of Chapter VI we obtain the following explicit representation of the M_2:

$$\begin{aligned} X &= (\cos\psi\cos\xi - \sin\psi\sin\xi\cos\theta)x \\ &\quad + (-\sin\psi\cos\xi - \cos\psi\sin\xi\cos\theta)y \\ &\quad + (\sin\xi\sin\theta)z - m\sin\theta(\cos 2\psi\sin\xi\cos\theta + \sin 2\psi\cos\xi), \\ Y &= (\cos\psi\sin\xi + \sin\psi\cos\xi\cos\theta)x \\ &\quad + (-\sin\psi\sin\xi + \cos\psi\cos\xi\cos\theta)y \\ &\quad + (-\cos\xi\sin\theta)z + m\sin\theta(\cos 2\psi\cos\xi\cos\theta - \sin 2\psi\sin\xi), \\ Z &= (\sin\psi\sin\theta)x + (\cos\psi\sin\theta)y + (\cos\theta)z \\ &\quad + m\cos 2\psi\sin^2\theta, \end{aligned} \tag{4.11}$$

with ξ and ψ as parameters. Putting $\tan\frac{1}{2}\xi = \lambda$, $\tan\frac{1}{2}\psi = \mu$ we see that the motion is rational.

Example 32. Show that for ψ = constant (4.11) represents a rotation about the Z-axis; indeed $X^2 + Y^2$ and Z as well do not depend on ξ. This confirms analytically that any trajectory surface is a surface of revolution. Also show that $\cos\theta = h/m$ follows directly from the fact that the projection of A_1A_2 on o_z has a constant length.

Example 33. Consider the border case $h = m$, that is $\theta = 0$; (4.11) is now $X = x\cos(\xi+\psi) - y\sin(\xi+\psi)$, $Y = x\sin(\xi+\psi) + y\cos(\xi+\psi)$, $Z = z$ and M_2 is degenerated into the M_1 of rotations about O_Z, as could have been expected.

Example 34. Consider the case $\xi = 0$; show that the relation (4.11) yields $Y\sin\theta - Z\cos\theta + z = 0$; prove that the trajectory surface of any point is generated by rotating a rational bicircular quartic plane curve about O_Z.

Equation (4.11) is a parametric representation of the trajectory surface of (x, y, z). To obtain an explicit equation we must eliminate the parameters ψ and ξ from the three equations. We know that the result must be an equation of the type $F((X^2 + Y^2), Z) = 0$. From (4.11) it follows after some algebra,

$$\begin{aligned} X^2 + Y^2 + Z^2 &= x^2 + y^2 + z^2 + m^2\sin^2\theta \\ &\quad - 2m\sin\theta(x\sin\psi - y\cos\psi) \end{aligned} \tag{4.12}$$

and therefore

$$X^2+Y^2+Z^2+2mZ = x^2+y^2+z^2+2mz\cos\theta + m^2\sin^2\theta(4\cos^2\psi-1)+4m(\cos\psi\sin\theta)y, \tag{4.13}$$

$$X^2+Y^2+Z^2-2mZ = x^2+y^2+z^2-2mz\cos\theta + m^2\sin^2\theta(4\sin^2\psi-1)-4m(\sin\psi\sin\theta)x, \tag{4.14}$$

which are quadratic equations for $\cos\psi$ and $\sin\psi$ respectively. If we solve these, make use of $\sin^2\psi+\cos^2\psi=1$ and rationalize by squaring twice, we obtain

$$\begin{aligned}&[(S-m^2\sin^2\theta)^2-(x^2+y^2)S+2m(x^2-y^2)(Z-z\cos\theta)\\&-m^2\sin^2\theta(x^2+y^2)+2x^2y^2]^2-4x^2y^2(S+m^2\sin^2\theta)^2\\&+4x^2y^2(x^2+y^2)(S+m^2\sin^2\theta)+16m^2x^2y^2(Z-z\cos\theta)^2\\&-8mx^2y^2(x^2-y^2)(Z-z\cos\theta)-4x^4y^4=0,\end{aligned} \tag{4.15}$$

where $S=X^2+Y^2+Z^2-z^2$.

Hence for this M_2, the trajectory surface (of revolution) of an arbitrary point $A(x, y, z)$ of E is a surface of order eight. Its intersection with the plane at infinity is the isotropic conic counted four times.

There are some special cases: If A is on the common perpendicular of the moving lines l_1, l_2, we have $x=y=0$ and the trajectory is the sphere

$$X^2+Y^2+Z^2=z^2+m^2\sin^2\theta, \tag{4.16}$$

counted four times.

If A is on l_1 we have $y=0$, $z=h=m\cos\theta$; its trajectory surface is seen to be the quartic surface (counted twice) with the equation

$$(X^2+Y^2+Z^2-m^2)^2-x^2[X^2+Y^2+(Z-m)^2]=0, \tag{4.17}$$

it has the isotropic conic as a double curve and it is therefore a *cyclide*.

It should be noted that although a moving point A remains, during the M_2, on its surface (4.15), there may be points on the surface which A may not reach in any real position of E; this is a consequence of the process used to eliminate the motion parameters.

To investigate this we may determine from (4.11) the maximum and the minimum values Z_M and Z_m of Z, considered as a function of ψ. The locus of the moving point is then that part of the surface of revolution lying between or on the parallel planes $Z=Z_M$ and $Z=Z_m$.

The determination of these extreme values of Z depends in general on the solution of an equation of degree four.

Example 35. Show that for $x = y = 0$ we obtain $Z_M = z \cos\theta + m \sin^2\theta$, $Z_m = z \cos\theta - m \sin^2\theta$. Prove that the locus of $A(0, 0, z)$ is the zone on the sphere (4.16) between the two parallel circles of radii $\sin\theta\,|z \pm m \cos\theta|$.
Example 36. Determine Z_M and Z_m for a point $(x, 0, m \cos\theta)$ on l_1 and the zone on the cyclide (4.17) which is its locus for the motion.
Example 37. The surface (4.17) passes through $A_1(0, 0, m)$ for any point A on l_1; show from (4.13) that A passes through A_1 for a real position only if $x^2 \leq 4m^2 \sin^2\theta$.
Example 38. Determine the trajectory surface for a point $(0, y, -h)$ on l_2.
Example 39. Intersect the cyclide (4.17) with a plane $Z = k$; show that the intersection consists of two circles and investigate whether they are real or imaginary.
Example 40. We have determined M_2 by means of the Eulerian angles; if we make use of the Euler parameters c_i $(i = 1, 2, 3, 4)$, the fundamental relation $a_{33} = \cos\theta$ reads

$$(c_0^2 - c_1^2 - c_2^2 + c_3^2) = \cos\theta(c_0^2 + c_1^2 + c_2^2 + c_3^2).$$

Show that the c_i may be expressed as $c_0 = (1 - \lambda\mu)\cos\frac{1}{2}\theta$, $c_1 = (1 + \lambda\mu)\sin\frac{1}{2}\theta$, $c_2 = (\lambda - \mu)\sin\frac{1}{2}\theta$, $c_3 = (\lambda + \mu)\cos\frac{1}{2}\theta$; derive the equations for the motion in terms of the rational parameters λ and μ.

The inverse motion of our M_2 is the following. In E, now the fixed space, two orthogonal skew lines l_1, l_2, with distance $2h$ are given; two points A_1, A_2 (with distance $2m$) of the moving space Σ move along l_1, l_2 respectively. Obviously the motion of the line $l = A_1A_2$ is an M_1-motion, while Σ may rotate about l in any position of the latter.

Example 41. Derive the equation of the locus of l and prove that it is a quartic ruled surface; determine its intersection with the plane at infinity; show that all generators make the constant angle θ with o_z.

In plane kinematics any M_1 may be defined by the rolling of a moving centrode on a fixed centrode. It is obvious from the preceding developments that there does not exist an analog of this for an M_2 in space. If a surface F is compelled to roll on a fixed surface F′ they must be tangent at a point A which is instantaneously at rest, which implies that the motion is instantaneously spherical. It is well-known from dynamics that the rolling conditions are in general not integrable (the conditions are called non-holonomic). Integrability takes place if F and F′ are isometric and thus applicable one on the other.

5. Two-parameter spherical motions

We deal now with two-parameter motions in spherical kinematics. The pertinent formulas follow immediately from the general spatial M_2 theory, if we specialize the motion by assuming that the translation part is permanently equal to zero. Hence, making use of canonical frames, the first and the second order instantaneous properties for spherical M_2-motion are in view of (1.14) and (3.7) given by

$$X_\lambda = 0, \quad Y_\lambda = -z, \quad Z_\lambda = y, \qquad X_\mu = z, \quad Y_\mu = 0, \quad Z_\mu = -x, \tag{5.1}$$

and

$$\begin{aligned} X_{\lambda\lambda} &= -f_3 y + f_2 z, & Y_{\lambda\lambda} &= -y + f_3 x - f_1 z, & Z_{\lambda\lambda} &= -z - f_2 x + f_1 y, \\ X_{\lambda\mu} &= \tfrac{1}{2} y - g_3 y + g_2 z, & Y_{\lambda\mu} &= \tfrac{1}{2} x + g_3 y - g_1 z, & Z_{\lambda\mu} &= -g_2 x + g_1 y, \\ X_{\mu\mu} &= -x - h_3 y + h_2 z, & Y_{\mu\mu} &= h_3 x - h_1 z, & Z_{\mu\mu} &= -z - h_2 x + h_1 y. \end{aligned} \tag{5.2}$$

This implies that there are no instantaneous invariants of the first order and nine of the second. The rotation axes in the zero-positions are the lines through O in the plane $z = 0$; the points of this plane are singular, they are only able to move parallel to O_Z. The curvature of trajectory surfaces is not interesting, because any point of E moves on a sphere with center O. The set of normals are the lines through O.

Example 42. Check the formula (3.18) (for the Gaussian curvature of the trajectory surface) for the case of spherical motion.

A continuous spherical M_2 may be defined by giving the Eulerian angles (or Euler parameters) as functions of two parameters λ, μ. As a spherical motion has three degrees of freedom it could also be done by introducing *one* constraint.

Example 43. Consider the spherical M_2 for which the point $x = y = 0$, $z = h > 0$ is compelled to remain in the plane $Z = H > 0$, with $h > H$. Show that any point on o_z has an M_1-motion, its path being a circle. Determine the spherical locus covered by an arbitrary point of the moving space E. Consider the inverse motion.

6. Two-parameter plane motions

Plane two-parameter motions have been treated in different ways (Boulad [1916], Darboux [1916], Koenigs [1917], van der Woude [1926], Blaschke and Müller [1956]). We could do it by specializing the spatial M_2 theory, but we prefer an independent development.

A general plane motion is given by

$$X = x\cos\phi - y\sin\phi + a, \quad Y = x\sin\phi + y\cos\phi + b;$$

an M_2 is defined if ϕ, a, b are given as functions of two parameters λ, μ. Accepting a certain asymmetry we take ϕ and a as the parameters. Hence

$$X = x\cos\lambda - y\sin\lambda + l\mu, \quad Y = x\sin\lambda + y\cos\lambda + b(\lambda,\mu), \tag{6.1}$$

where the constant length l (which may be chosen as the unit of length) has been introduced for the sake of elegance: both λ and μ are now dimensionless variables. We choose the two frames to coincide at $\lambda = \mu = 0$; hence $b(0,0) = 0$.

If we consider time as the motion parameter then $\lambda(t)$, $\mu(t)$ define an M_1 belonging to (6.1). Its pole P for the zero position is seen to be

$$x_p = -(b_\lambda \lambda_1 + b_\mu \mu_1)/\lambda_1, \quad y_p = l\mu_1/\lambda_1, \tag{6.2}$$

which depends only on the ratio λ_1/μ_1. Hence the poles for all possible M_1 satisfy

$$lx_p + b_\mu y_p + lb_\lambda = 0; \tag{6.3}$$

their locus is a *straight line* p, called the polar line for the zero-position of the M_2.

The (coinciding) frames are as yet arbitrary. We take the y-axis along p. This implies $b_\lambda = b_\mu = 0$, and we obtain

$$X_\lambda = -y, \quad X_\mu = l, \quad Y_\lambda = x, \quad Y_\mu = 0, \tag{6.4}$$

which shows that M_2 has no instantaneous invariants of the first order.

For an M_1 belonging to M_2 we have

$$\dot{X} = -y\lambda_1 + l\mu_1, \quad \dot{Y} = x\lambda_1, \tag{6.5}$$

the velocity of $A(x,y)$ being orthogonal to AP with $x_p = 0$, $y_p = l\mu_1/\lambda_1$ (Fig. 86). There is one M_1 which is instantaneously a translation: $\lambda_1 = 0$, its direction being parallel to O_X. Any point not on p is able to move in any direction; the exceptional points are those of the y-axis, they can only move parallel to O_X. If we take another time-scale t, the ratio λ_1/μ_1 does not change and as we are only interested in time-independent properties we may without loss of generality normalize the time-scale by supposing $\lambda_1^2 + \mu_1^2 = 1$. This implies that for the point $A_0 = (l, 0)$ we have $\dot{X} = l\mu_1$, $\dot{Y} = l\lambda_1$ and therefore $\dot{X}^2 + \dot{Y}^2 = l^2$; for all M_1 of M_2, the scalar value of the velocity of A_0 has the constant value l. In other words: the hodograph of A_0 is a circle.

If we put $\lambda_1 = \cos\psi$, $\mu_1 = \sin\psi$, it follows from (6.5)

$$\dot{X} = -y\cos\psi + l\sin\psi, \quad \dot{Y} = x\cos\psi, \tag{6.6}$$

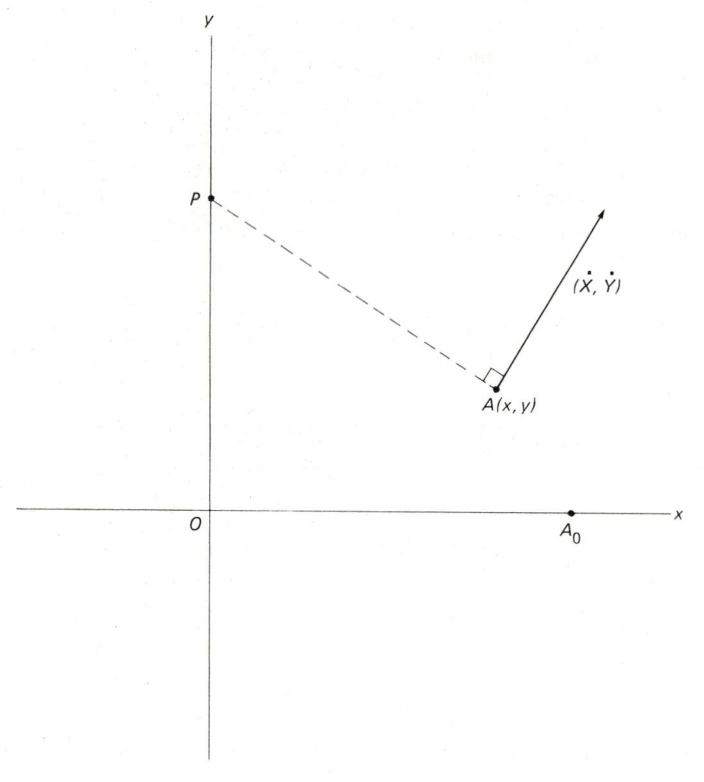

Fig. 86.

which means that the hodograph of $A(x, y)$ is an ellipse $e(x, y)$ whose explicit equation is seen to be

$$x^2\dot{X}^2 + 2xy\dot{X}\dot{Y} + (y^2 + l^2)\dot{Y}^2 = l^2x^2. \tag{6.7}$$

Elementary analytical geometry shows that the area F of e is $\pi l|x|$, this being a measure of some kind for the movability of A. It depends only on x; for a point on the polar line we have $F = 0$.

In terms of ψ the pole P is seen to be $(0, l \tan \psi)$ and the corresponding angular velocity $\omega = \cos \psi$.

Example 44. Determine the semi-axes of the ellipse $e(x, y)$.

We consider now second order properties. Differentiating (6.1) twice and then putting $\lambda = \mu = 0$, we obtain

$$X_{\lambda\lambda} = -x, \qquad X_{\lambda\mu} = 0, \qquad X_{\mu\mu} = 0,$$
(6.8)
$$Y_{\lambda\lambda} = -y + b_{\lambda\lambda}, \qquad Y_{\lambda\mu} = b_{\lambda\mu}, \qquad Y_{\mu\mu} = b_{\mu\mu}.$$

Our Y-axis has been fixed, but the X-axis is still arbitrary. If we chose it along the line $Y = b_{\lambda\lambda}$ we have $b_{\lambda\lambda} = 0$. The canonical frame is now completely determined if we orient O_Y so that $b_{\mu\mu} > 0$, and we see that there are two second order instantaneous invariants $b_{\lambda\mu} = b'$ and $b_{\mu\mu} = b'' \geqslant 0$. Hence any M_2 may, up to the second order, be represented in a suitably chosen frame by

$$X = x - y\lambda + l\mu - \tfrac{1}{2}x\lambda^2,$$
(6.9)
$$Y = y + x\lambda + b'\lambda\mu + \tfrac{1}{2}b''\mu^2.$$

This enables us to determine the polar line up to the first order. Considering the M_1 given by $\lambda(t)$, $\mu(t)$ and putting $\dot{X} = \dot{Y} = 0$, we obtain (restricting the results to contain, at highest, linear terms)

(6.10) $$x_p = -b'\lambda \tan\psi - \mu(b' + b''\tan\psi), \quad y_p = l\tan\psi;$$

the (∞^2) polar lines in the neighborhood of the zero-position are obtained by eliminating ψ:

(6.11) $$lx + (b'\lambda + b''\mu)y + lb'\mu = 0.$$

Differentiating (6.10) with respect to t and then putting $t = 0$ yields the components of the pole velocity in the zero-position:

(6.12) $$(x_p)_1 = -2b'\sin\psi - b''\sin^2\psi/\cos\psi, \quad (y_p)_1 = l\dot{\psi}/\cos^2\psi.$$

This implies that for a pole $(0, l\tan\psi)$ the (∞^1) velocities have the property that the x-component is constant while the y-component varies. In other words: the hodograph of the pole velocities is a line parallel to the y-axis. (The pole velocity is the rate at which different points in E become poles; it is not the velocity of any point.)

We know from plane kinematics that the diameter of the inflection circle is equal to the pole velocity divided by the angular velocity ω and moreover that its direction is found by rotating the pole velocity about P by angle $-\pi/2$. Hence the locus of the inflection poles B corresponding to the M_1's with pole P is a straight line m, parallel to O_X, intersecting O_Y at the point B_0 such that (Fig. 87)

(6.13) $$PB_0 = 2b'\tan\psi + b''\tan^2\psi,$$

while $OP = l\tan\psi$.

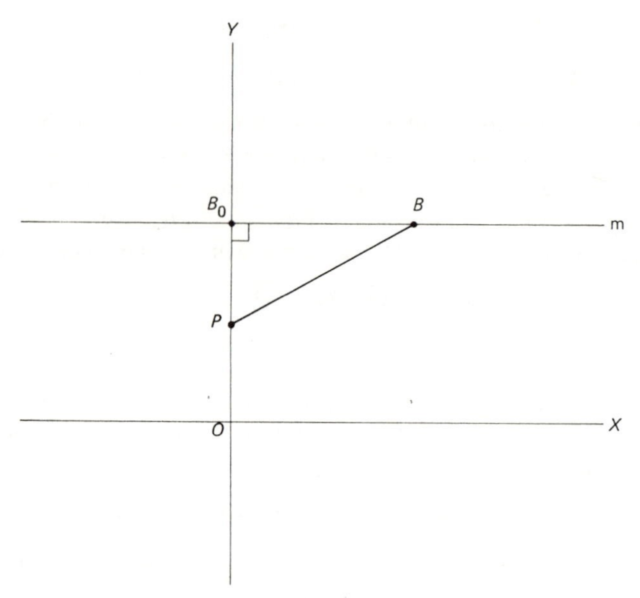

Fig. 87.

If $OP = y$, $OB_0 = y^*$, we have

$$PB_0 = (2b'/l)y + (b''/l^2)y^2 \tag{6.14}$$

$$y^* = ((l + 2b')/l)y + (b''/l^2)y^2, \tag{6.15}$$

which shows that y^* is a quadratic function of y.

Example 45. Show that there are two points on O_Y where P and B_0 coincide. Determine their kinematic meaning.

To any point P there corresponds a B_0, but to B_0 there corresponds a P only if $y^* \geqslant -(l + 2b')^2/4b''$. Hence the locus of B_0 is a half-line and that of the inflection poles B is a half-plane.

Example 46. Consider the case $b'' = 0$.

If P is the pole and B the inflection pole, than PB is a diameter of the inflection circle, which passes therefore through B_0. The set of inflection circles corresponding to P is the pencil of circles through P and B_0.

The first order properties of a one-parameter motion M_1 belonging to M_2 are known if P on the polar line p is given. Those of the second order follow if, moreover, B on the line m corresponding to P, (6.15), is given. By means of the Euler–Savary equation or the Bobillier construction the center of curvature for the point-path of any point of the moving plane can be determined.

Example 47. Derive the equation, with two parameters, of all inflection-circles of M_2.
Example 48. If A and α are two different given points not on p, show that there is in general an M_1 belonging to M_2 such that α is the center of curvature of A.
Example 49. Determine the possible centers of curvature for the paths of a point A on the polar line p.

7. Examples of three-parameter motions

It is understandable that n-parameter motions for $n>2$ have not been given much attention by kinematicians. In plane and in spherical kinematics the most general motions have a degree-of-freedom equal to three and an M_3 for these cases could not be defined by an equality condition (at most it might be restricted by an inequality); no M_n for $n>3$ can exist.

In space some 3-parameter motions have been considered. One is mentioned, but not discussed, by KOENIGS [1897, p. 241].

As an example we consider the case where the space E moves such that three points B_1, B_2, B_3 on a given line l remain in three given planes V_1, V_2, V_3 of Σ. The motion is in a certain aspect a generalization of the Cardan motion in a plane. It has obviously three degrees of freedom, but it is of a special kind. In any position of l the space E can freely rotate about l. Hence l itself and each of its points have essentially a two-degrees-of-freedom motion.

We consider the case where V_i are three mutually orthogonal planes and we take them as the Cartesian frame O_{XYZ} in Σ. In E we take o_{xyz} such that o_x coincides with l; let $B_i=(b_i,0,0)$. As always the motion will be represented by $\boldsymbol{P}=\mathbf{A}\boldsymbol{p}+\boldsymbol{d}$, $\mathbf{A}$ being an orthogonal matrix $\|a_{ij}\|$. Then for B_1 to be permanently in $X=0$, etc., it follows that $d_i=-a_{i1}b_i$. Our motion is therefore

$$X=a_{11}(x-b_1)+a_{12}y+a_{13}z,\qquad Y=a_{21}(x-b_2)+a_{22}y+a_{23}z,$$
$$Z=a_{31}(x-b_3)+a_{32}y+a_{33}z, \tag{7.1}$$

a_{i1} satisfy $a_{11}^2+a_{21}^2+a_{31}^2=1$ but they are otherwise arbitrary.

From this it follows that the trajectory surface of a point $B=(b,0,0)$ on l, different from the points B_i, has the equation

$$X^2/(b-b_1)^2+Y^2/(b-b_2)^2+Z^2/(b-b_3)^2=1, \tag{7.2}$$

which shows that the surface is *an ellipsoid*, with O_X, O_Y, O_Z as its axes of symmetry. If B coincides with B_1, we have since $a_{21}^2+a_{31}^2\leq 1$:

$$X = 0, \qquad Y^2/(b_1 - b_2)^2 + Z^2/(b_1 - b_3)^2 \leq 1, \tag{7.3}$$

hence the locus of the positions of B_1 consists of *the points on or inside an ellipse*, and similar for B_2 and B_3.

Example 50. Show that (in general) there are three points on l for which the trajectory surface is an ellipsoid of revolution.
Example 51. When is the ellipse given by (7.3) a circle?

The line l has ∞^2 positions in Σ and its locus is therefore a congruence, which we denote by C. Since l joins B_1 and B_2, which according to (7.1) have coordinates $B_1 = [0, a_{21}(b_1 - b_2), a_{31}(b_1 - b_3), 1]$ and $B_2 = [a_{11}(b_2 - b_1), 0, a_{31}(b_2 - b_3), 1]$, its Plücker vectors $(\boldsymbol{P}, \boldsymbol{Q})$ have coordinates

$$\begin{aligned} &P_1 = a_{11}, \qquad P_2 = a_{21}, \qquad P_3 = a_{31}, \\ &Q_1 = (b_2 - b_3)a_{21}a_{31}, \qquad Q_2 = (b_3 - b_1)a_{31}a_{11}, \\ &Q_3 = (b_1 - b_2)a_{11}a_{21}. \end{aligned} \tag{7.4}$$

These formulas give a representation of C by means of the parameters a_{11}, a_{21}, a_{31} satisfying $a_{11}^2 + a_{21}^2 + a_{31}^2 = 1$. From (7.4) it follows that C *belongs to the tetrahedral complex* with the equation

$$b_1P_1Q_1 + b_2P_2Q_2 + b_3P_3Q_3 = 0. \tag{7.5}$$

The planes V_i and the plane at infinity are the faces of the associated tetrahedron. This result could be expected because l intersects the four planes at four points with a constant cross-ratio.

The planes V_i ($i = 1, 2, 3$) are singular planes of C. Indeed, if l is in V_1, say, it is sufficient that B_2 is on O_Z and B_3 on O_Y. Hence l has in V_1 a Cardan motion. There are ∞^1 lines l in V_1, the tangents of an astroid, a curve of class four.

The degree of C is the number of lines l passing through a (non-singular) point P. Let P be in V_1. Through P pass two kinds of lines l. If B_1 does not coincide with P then l is in V_1; four such lines pass through P. If B_1 coincides with P then l is a common generator of two circular (half-) cones, with vertex B_1 and with their axes perpendicular to V_2 and V_3 respectively; there are two such lines, real or imaginary. The conclusion is: *the degree of* C *is six.*

Example 52. Show that the six lines l through an arbitrary point are generators of a quadratic cone.
Example 53. Determine the lines of C through O.

The class of C is the number of lines in a (non-singular) plane U. Let U intersect O_X, O_Y, O_Z in the distinct points S_1, S_2, S_3. If B_1 is on S_2S_3, B_2 on

S_3S_1 then B_3 describes (again in view of Cardan's motion) an ellipse. This has two points, real or imaginary, on S_1S_2. Hence two lines l are in U; the class of C is two. C is seen to be a (6,2) congruence (STURM [1893]).

Example 54. If $a_{i1} = \Delta c_i$ then $\Delta^2(c_1^2 + c_2^2 + c_3^2) = 1$. If $\Delta \to \infty$ and therefore $c_1^2 + c_2^2 + c_3^2 \to 0$, show that $P_1 = P_2 = P_3 = 0$ and $\Sigma\, Q_2^2 Q_3^2 (b_2 - b_3)^{-2} = 0$. This implies that the plane V at infinity is a singular plane of C; the lines l in V are the tangents of a curve of class four; three tangents are real, all others are imaginary.

Example 55. Determine the lines of C through a point at infinity. Determine the equations of C from (7.4), one of these is (7.5).

We now give a second example of a motion with three parameters. DARBOUX [1897] has considered the case where the vertices A_1, A_2, A_3 of a triangle in the moving space are compelled to remain in the planes V_1, V_2, V_3 of a trihedron in Σ respectively. The only property he proved is a negative one: There is no fourth point whose motion is planar.

We deal with the special case of the three planes V_i being mutually orthogonal and we take them as the coordinate planes of the Cartesian frame O_{XYZ} in Σ. Furthermore we suppose the triangle $A_1A_2A_3$ to be acute. Then there are two (real) points such that their joins to A_i are mutually orthogonal; we take one of them as the origin o in the moving space and oA_1, oA_2, oA_3 as o_x, o_y, o_z. Hence $A_1 = (p, 0, 0)$, $A_2 = (0, q, 0)$, $A_3 = (0, 0, r)$; we suppose that A_1 remains in the plane $X = 0$, A_2 in $Y = 0$, A_3 in $Z = 0$. Then the M_3 is represented by

$$
\begin{aligned}
X &= a_{11}(x - p) + a_{12}y + a_{13}z,\\
Y &= a_{21}x + a_{22}(y - q) + a_{23}z, \qquad (7.6)\\
Z &= a_{31}x + a_{32}y + a_{33}(z - r),
\end{aligned}
$$

where $\| a_{ij} \|$ is an arbitrary orthogonal matrix; it may be expressed by means of the Euler parameters (Chapter VI, (2.1)), and the motion is then given in terms of the four homogeneous numbers c_i.

During the motion a point A of E occupies ∞^3 positions in Σ and we may expect that A covers in general a three-dimensional region $G(A)$ of Σ. Our aim could be to determine $G(A)$ for each point A of E.

We take first some special points. Let A_0 coincide with the origin of E. Then its positions are according to (7.6)

$$
\begin{aligned}
X &= -p(c_0^2 + c_1^2 - c_2^2 - c_3^2)/N,\\
Y &= -q(c_0^2 - c_1^2 + c_2^2 - c_3^2)/N,\\
Z &= -r(c_0^2 - c_1^2 - c_2^2 + c_3^2)/N, \qquad (7.7)\\
N &= c_0^2 + c_1^2 + c_2^2 + c_3^2.
\end{aligned}
$$

The coordinates X, Y, Z depend on c_i^2, four homogeneous positive numbers. Hence if A_0 occupies a certain position in Σ it does so for all positions $(c_0, \pm c_1, \pm c_2, \pm c_3)$ of E. This implies that if no c_i is equal to zero, A_0 is at the same point of Σ for eight positions of E; this number reduces to four if one parameter c_i is equal to zero, to two if two parameters are zero, and to a single one if three numbers c_i are zero. There are four possibilities for the latter case: $c_0 = 1,\ c_1 = c_2 = c_3 = 0$; $c_1 = 1,\ c_0 = c_2 = c_3 = 0$, etc. The corresponding positions of A_0 are:

$$B_0 = (-p, -q, -r), \qquad B_1 = (-p, q, r),$$
$$B_2 = (p, -q, r), \qquad B_3 = (p, q, -r).$$

B_i are the vertices of a tetrahedron B; the planes of its faces are given by

$$\begin{aligned} &(X/p)+(Y/q)+(Z/r)-1=0, \quad -(X/p)+(Y/q)+(Z/r)+1=0,\\ &(X/p)-(Y/q)+(Z/r)+1=0, \quad (X/p)+(Y/q)-(Z/r)+1=0, \end{aligned} \tag{7.8}$$

respectively.

If X_0, Y_0, Z_0 is an attainable point of A_0 then (7.7) considered as three linear equations for the unknowns c_i^2 must have solutions with non-negative ratios. We obtain

$$\begin{aligned} c_0^2 : c_1^2 : c_2^2 : c_3^2 = -&((X_0/p)+(Y_0/q)+(Z_0/r)-1):\\ &(-(X_0/p)+(Y_0/q)+(Z_0/r)+1):\\ &((X_0/p)-(Y_0/q)+(Z_0/r)+1):\\ &((X_0/p)+(Y_0/q)-(Z_0/r)+1), \end{aligned} \tag{7.9}$$

and this implies that X_0, Y_0, Z_0 is a point of the closed region B. If for instance $c_2 = c_3 = 0$, we have $X = -p$,

$$Y = -q(c_0^2 - c_1^2)/(c_0^2 + c_1^2), \qquad Z = -r(c_0^2 - c_1^2)/(c_0^2 + c_1^2),$$

which means that the position of A_0 is on the edge B_0B_1, covering the closed interval between the two vertices. If $c_3 = 0$ the locus of A_0 is the closed triangle $B_0B_1B_2$, if no c_i is zero A_0 is at an internal point of B. Summing up: *the region* G *covered by the origin of* o_{xyz} *consists of the vertices, the edges, the faces and the internal points of the tetrahedron* B.

Example 56. Show that opposite edges of B have equal length; B is an equifacial tetrahedron.
Example 57. Show that there is another point of E whose region is a tetrahedron: the reflection of o into the plane $A_1A_2A_3$.

We consider now the region $G(A)$ for the arbitrary point $A(x, y, z)$ of E. For the time being we suppose

$$(x-p)(y-q)(z-r) \neq 0. \tag{7.10}$$

If A_0 is at B_i the point A is at D_i, where

$$\begin{aligned} D_0 &= [(x-p),(y-q),(z-r)], \\ D_1 &= [(x-p), -(y-q), -(z-r)], \\ D_2 &= [-(x-p),(y-q), -(z-r)], \\ D_3 &= [-(x-p),-(y-q),(z-r)]. \end{aligned} \tag{7.11}$$

In view of (7.10) the points D_i are not coplanar and they are therefore the vertices of a tetrahedron D. If A_0 moves along an edge of B, B_2B_3 say, then $c_0 = c_1 = 0$ and the path of A is given by

$$\begin{aligned} X &= -(x-p), \\ Y &= ((c_2^2 - c_3^2)/(c_2^2 + c_3^2))(y-q) + (2c_2c_3/(c_2^2+c_3^2))z, \\ Z &= (2c_2c_3/(c_2^2+c_3^2))y - ((c_2^2-c_3^2)/(c_2^2+c_3^2))(z-r), \end{aligned} \tag{7.12}$$

or if $c_3/c_2 = \tan(\phi/2)$:

$$\begin{aligned} X = -(x-p), \qquad & Y = (y-q)\cos\phi + z\sin\phi, \\ & Z = y\sin\phi - (z-r)\cos\phi, \end{aligned} \tag{7.13}$$

which represents a (complete) ellipse, denoted by k_{23} and passing through D_2 and D_3, its plane being $X = -(x-p)$. In this way the vertices of D are joined by six ellipses k_{ij}, corresponding to the edges B_iB_j of B. Any point on B_iB_j between B_i and B_j corresponds to two points of k_{ij}. We determine now the locus of A if A_0 is on a face of the tetrahedron B, say the face $B_1B_2B_3$. Then $c_0 = 0$ and we obtain, in homogeneous coordinates

$$\begin{aligned} X &= (c_1^2 - c_2^2 - c_3^2)(x-p) + 2c_1c_2y + 2c_1c_3z, \\ Y &= 2c_2c_1x + (-c_1^2 + c_2^2 - c_3^2)(y-q) + 2c_2c_3z, \\ Z &= 2c_3c_1x + 2c_2c_3y + (-c_1^2 - c_2^2 + c_3^2)(z-r), \\ W &= c_1^2 + c_2^2 + c_3^2, \end{aligned} \tag{7.14}$$

which represents a Steiner quartic surface F_0. It passes through D_1, D_2, D_3 and the ellipses k_{23}, k_{31}, k_{12} are on it. Hence the points of the triangle $B_1B_2B_3$ correspond to those of the region on F_0 bordered by the three ellipses. Any point inside $B_1B_2B_3$ corresponds to four points on F_0. (It is well-known that a

Steiner surface contains ∞^2 conics and that through two given points passes a unique conic.)

All this leads to the following conclusion. The region $G(A)$ covered by an arbitrary point of E has in general the following shape; it could be called a pseudo-tetrahedron T. *It has four vertices, any three of them are joined by a part of a Steiner surface* F_i $(i = 0, 1, 2, 3)$; F_i *and* F_j *have the ellipse* k_{ij} *in common.*

The ellipse k_{23}, corresponding to $c_0 = c_1 = 0$, is in the plane $X = -(x-p)$; analogously k_{01}, with $c_2 = c_3 = 0$, is in the plane $X = x - p$. Moreover it is easy to prove that k_{01} follows from k_{23} by reflection into the plane $X = 0$. Hence two "opposite" ellipses are congruent. The "face" $D_1D_2D_3$ of T is the part of F_0 cut off by the planes $X = -(x-p)$, $Y = -(y-q)$, $Z = -(z-r)$. If A tends to A_0 the vertices D_i tend to B_i, the faces of T become more flat and their limits are the planes of B.

Example 58. Show that the tetrahedron D is equifacial.
Example 59. Determine the explicit equations of k_{23} and k_{01}.

We consider now the special moving points for which (7.10) does not hold. Among these are the points A_i. The positions of A_1 for instance are given by

$$
\begin{aligned}
&X = 0, \qquad Y = (2p(c_0c_3 + c_2c_1)/N) - (q(c_0^2 - c_1^2 + c_2^2 - c_3^2)/N),\\
(7.15)\quad &Z = (2p(-c_0c_2 + c_3c_1)/N) - (r(c_0^2 - c_1^2 - c_2^2 + c_3^2)/N).
\end{aligned}
$$

If A_0 is at B_i the point A_1 is at $\mathfrak{E}_i$:

$$\mathfrak{E}_0 = (0, -q, -r), \qquad \mathfrak{E}_1 = (0, q, r), \qquad \mathfrak{E}_2 = (0, -q, r), \qquad \mathfrak{E}_3 = (0, q, -r),$$

the vertices of a rectangle. Furthermore it is easy to verify that if A_0 moves along B_iB_j then A_1 moves along $\mathfrak{E}_i\mathfrak{E}_j$. But to conjecture that $G(A_1)$ coincides with the rectangle is wrong: if $c_0 = c_3 = 0$, $c_1 = c_2$ the point A_1 is at $(0, p, r)$ which is outside the rectangle if $p > q$. From (7.15) it follows

$$
\begin{aligned}
&(Y+q)c_0^2 + (Y-q)c_1^2 + (Y+q)c_2^2 + (Y-q)c_3^2 - 2pc_0c_3 - 2pc_1c_2 = 0,\\
(7.16)\quad &(Z+r)c_0^2 + (Z-r)c_1^2 + (Z-r)c_2^2 + (Z+r)c_3^2 + 2pc_0c_2 - 2pc_1c_3 = 0,
\end{aligned}
$$

which for given values of Y and Z represent two quadrics in the projective (c_0, c_1, c_2, c_3)-space. The region $G(A_1)$ is now determined by means of an inequality in Y and Z which expresses the condition that these two quadrics have at least one real point of intersection. We shall not derive this condition because it requires complicated algebra.

Another special moving point, not satisfying (7.10), is $H = (p, q, r)$. Its positions are given by

$$X = (2(-c_0c_3 + c_1c_2)q/N) + (2(c_0c_2 + c_1c_3)r/N),$$

$$(7.17) \qquad Y = (2(c_0c_3 + c_2c_1)p/N) + (2(-c_0c_1 + c_2c_3)r/N),$$

$$Z = (2(-c_0c_2 + c_3c_1)p/N) + (2(c_0c_1 + c_3c_2)q/N).$$

If A_0 is at any of the vertices of B then H is always at O. If A_0 moves along B_2B_3 or along B_0B_1 then H moves along the line l_1 with equations $X = 0$. $Y = r \sin \phi_1$, $Z = q \sin \phi_1$, its endpoints being $(0, r, q)$ and $(0, -r, -q)$. Analogously the edges B_3B_1 and B_0B_2 both correspond to l_2: $Y = 0$, $Z = p \sin \phi_2$, $X = r \sin \phi_2$ and the edges B_1B_2 and B_0B_3 to l_3: $Z = 0$, $X = q \sin \phi_3$, $Y = p \sin \phi_3$. The lines l_i pass through O. Here ϕ_1, ϕ_2, ϕ_3 are motion parameters.

The faces of B correspond to Steiner surfaces F_i. It is easy to verify that all four pass through l_1, l_2 and l_3. (It is well-known that every Steiner surface has three double lines, passing through a triple-point.)

Example 60. Consider the region covered by the point $(x, 0, 0)$, $x \neq p$.
Example 61. Consider the region covered by $(p, q, 0)$.
Example 62. Consider the motion if $A_1A_2A_3$ is equilateral: $p = q = r$.
Example 63. Consider the motion for the limiting case $p \to 0$; the triangle is rectangular.

Our first example of this section covered the special case when A_1, A_2, A_3 are collinear. We saw that the motion becomes essentially an M_2. This completes our discussion of the second example.

If a motion is the result of the product of two screw motions (about two different axes) then instantaneously, to the first order, it is irrelevant in which sequence the two component motions occur. The same is true for the product of any number of screw motions. If each component screw is taken about a specified axis and has a known pitch, then any such system of n-screws describes a certain n-parameter motion. Such systems were originally studied by RODRIGUES [1840], BALL [1876] and some of the other classical geometers. Recently interest has been renewed in these motions especially in regard to determining the instantaneous mobility of linkwork (SHARIKOV [1961], VOINEA and ATANASIU [1962], HUNT [1968], WALDRON [1969], etc.). It is mainly necessary to consider the case $n = 1, 2, 3$, since results for $n = 4, 5$ follow from $n = 2, 1$ by considering the so-called reciprocal screw system.

Example 64. Show that for the $n = 3$ given by three screws of pitch σ_{0X}, σ_{0Y}, σ_{0Z}, each located along the respective coordinate axis, the resultant screws have pitch σ_0 and pass through X, Y, Z as given by

$$(\sigma_{0Z} - \sigma_0)X^2 + (\sigma_{0Y} - \sigma_0)Y^2 + (\sigma_{0Z} - \sigma_0)Z^2 + (\sigma_{0X} - \sigma_0)(\sigma_{0Y} - \sigma_0)(\sigma_{0Z} - \sigma_0) = 0$$

(BALL [1900], HUNT [1970]).

Example 65. Show that three of the resultant screws pass through an arbitrary point (X, Y, Z); at least one of these screws is always real, and the sum of their pitches is always the same; i.e., $(\sigma_{0X} + \sigma_{0Y} + \sigma_{0Z})$ (HUNT [1970]).

CHAPTER XI

A MAPPING OF PLANE KINEMATICS

1. The mapping

A displacement $\mathbf{D}(a, b, \phi)$ in the plane is determined by three numbers. This has led to the idea of mapping the set of displacements onto the points of a three-dimensional space, and also to the idea of using this map to derive theorems of plane kinematics. This was done first by GRÜNWALD [1911] and soon after by BLASCHKE [1911]. In their monograph on plane kinematics, BLASCHKE and MÜLLER [1956] have given a treatment of the method making use of quaternions. Valuable contributions to the subject are also due to BETH [1937, 1938, 1949], who preferred to apply geometric reasoning. In this chapter we deal with the mapping in a more elementary way.

Let O_{XY} be a Cartesian frame in the fixed plane Σ and o_{xy} one in the moving plane E. The displacement $\mathbf{D}(a, b, \phi)$ is given by

$$\begin{aligned} X &= x\cos\phi - y\sin\phi + a, \\ Y &= x\sin\phi + y\cos\phi + b, \end{aligned} \tag{1.1}$$

in which a, b are the coordinates of o and ϕ $(0 \leq \phi < 2\pi)$ is the rotation angle. If $a = b = \phi = 0$ the two frames coincide and E is in its "zero-position". If $\mathbf{D}$ is not a translation, that is if $\phi \neq 0$, we know that there exists a rotation, with its center at $X = x = x_p$, $Y = y = y_p$ and with rotation angle ϕ, which transforms E from its zero-position to (a, b, ϕ). We introduce a three-dimensional projective space Σ' with homogeneous coordinates X_i $(i = 1, 2, 3, 4)$ and we define the mapping by

$$X_1 : X_2 : X_3 : X_4 = x_p u : y_p u : u : 1, \tag{1.2}$$

in which $u = \tan\frac{1}{2}\phi$, and we exclude for the time being the case $\phi = \pi$.

In order to determine the image point (1.2) in terms of the parameters a, b, ϕ we must solve for x_p and y_p. Setting $X = x$, $Y = y$, in (1.1) and using half-angles we have:

$$x \sin \tfrac{1}{2}\phi + y \cos \tfrac{1}{2}\phi = a/(2\sin \tfrac{1}{2}\phi),$$
(1.3)
$$-x \cos \tfrac{1}{2}\phi + y \sin \tfrac{1}{2}\phi = b/(2\sin \tfrac{1}{2}\phi),$$

which gives us the values of x, y for the center P:

(1.4) $$x_P = \tfrac{1}{2}(a - b \cot \tfrac{1}{2}\phi), \qquad y_P = \tfrac{1}{2}(a \cot \tfrac{1}{2}\phi + b),$$

and the image point is therefore

(1.5) $$X_1 : X_2 : X_3 : X_4 = (au - b) : (a + bu) : 2u : 2,$$

or, multiplying by $\cos \frac{1}{2}\phi$ (which we supposed to be $\neq 0$),

(1.6) $$X_1 : X_2 : X_3 : X_4 = (a \sin \tfrac{1}{2}\phi - b \cos \tfrac{1}{2}\phi) : (a \cos \tfrac{1}{2}\phi + b \sin \tfrac{1}{2}\phi) : 2\sin \tfrac{1}{2}\phi : 2\cos \tfrac{1}{2}\phi.$$

We also define (1.6) as the image point for the cases $\phi = 0$ and $\phi = \pi$ which have been excluded so far; obviously it has sense for both of these cases whereas (1.2) has none when $\phi = \pi$. Hence all displacements have now a unique image; the zero-position is the point $O = (0, 0, 0, 1)$, translations ($\phi = 0$) and half-turns ($\phi = \pi$) are mapped onto the points of the planes $X_3 = 0$ and $X_4 = 0$ respectively.

We ask now if each point of Σ' is the image of a displacement. Accordingly, we try to determine a, b, ϕ if the X_i are given. If $X_3 = 0$, $X_4 \neq 0$ we have at once $a = 2X_2/X_4$, $b = -2X_1/X_4$, $\phi = 0$; if $X_4 = 0$, $X_3 \neq 0$ we obtain $a = 2X_1/X_3$, $b = 2X_2/X_3$, $\phi = \pi$. If $X_3 \neq 0$, $X_4 \neq 0$ then

$$u = \tan \tfrac{1}{2}\phi = X_3/X_4,$$
(1.7)
$$\cos \phi = (1 - u^2)/(1 + u^2) = (X_4^2 - X_3^2)/(X_3^2 + X_4^2),$$
$$\sin \phi = 2u/(1 + u^2) = 2X_3X_4/(X_3^2 + X_4^2).$$

Furthermore

$$a - b \cot \tfrac{1}{2}\phi = 2X_1/X_3,$$
$$a \cot \tfrac{1}{2}\phi + b = 2X_2/X_3,$$

which gives us

(1.8) $$a = 2(X_1X_3 + X_2X_4)/(X_3^2 + X_4^2), \quad b = 2(X_2X_3 - X_1X_4)/(X_3^2 + X_4^2).$$

These formulas also yield the correct result if either X_3 or X_4 is zero.

From (1.7) and (1.8) we see that any *real* image point is the map of a displacement provided X_3 and X_4 are not both zero. Hence in Σ' the only points which do *not* correspond to a displacement are the points on the real line l given by the equations $X_3 = X_4 = 0$.

If we deal with *imaginary* points of Σ' we see from (1.7) and (1.8) that they are maps of (imaginary) displacements provided $X_3^2 + X_4^2 \neq 0$. Hence there are two conjugate imaginary planes in Σ', V_1 with equation $X_3 - iX_4 = 0$ and V_2 with $X_3 + iX_4 = 0$, the points of which are not the image of a displacement of the type (1.1). The (real) singular line l is the intersection of V_1 and V_2.

The displacement (1.1) expressed in terms of the image point coordinates X_i is now seen to be:

$$
\begin{aligned}
X &= (X_4^2 - X_3^2)x - 2X_3X_4y + 2(X_1X_3 + X_2X_4)z,\\
Y &= 2X_3X_4x + (X_4^2 - X_3^2)y + 2(X_2X_3 - X_1X_4)z, \qquad (1.9)\\
Z &= (X_3^2 + X_4^2)z.
\end{aligned}
$$

The determinant Δ of this linear transformation is equal to $(X_3^2 + X_4^2)^3$, and we have a non-singular transformation if $\Delta \neq 0$.

Example 1. Show that if $\Delta \neq 0$, the inverse transformation of (1.9) reads

$$
\begin{aligned}
x &= (X_4^2 - X_3^2)X + 2X_3X_4Y + 2(X_1X_3 - X_2X_4)Z,\\
y &= -2X_3X_4X + (X_4^2 - X_3^2)Y + 2(X_2X_3 + X_1X_4)Z,\\
z &= (X_3^2 + X_4^2)Z.
\end{aligned}
$$

If $X_3 = X_4 = 0$ all coefficients of (1.9) are zero; the transformation has no meaning and we indeed exclude the image points on l. If however $\Delta = 0$, X_3 and X_4 not both equal to zero, (1.9) is singular but not senseless. We take for instance $X_3 = iX_4 \neq 0$ (which means that the image point is in V_1 but not on l), this yields

$$
\begin{aligned}
X &= 2X_4^2x - 2iX_4^2y + 2(iX_1 + X_2)X_4z,\\
Y &= 2iX_4^2x + 2X_4^2y + 2(iX_2 - X_1)X_4z, \qquad (1.10)\\
Z &= 0,
\end{aligned}
$$

and we have $Y = iX$ for any X_1, X_2 and for any x, y, z. Denoting the isotropic points $(1, i, 0)$ and $(1, -i, 0)$ of E and Σ by I_1 and I_2, we must accept that *any point for which X, Y are not both equal to zero is, after the singular displacement, situated at* I_1. An exception must be made for those points for which $X = Y = 0$, the transforms would then lead to $(0,0,0)$ which is undetermined. From (1.10) these points satisfy

$$X_4x - iX_4y + (iX_1 + X_2)z = 0,$$

or, in view of $X_4 = -iX_3$, $x_p = X_1/X_3$, $y_p = X_2/X_3$,

$$i(x - x_p z) + (y - y_p z) = 0 \tag{1.11}$$

which is the line PI_2. The formulas would allow us to consider any point of the plane as the transform of any point Q of PI_2, but we want to define the displacement of Q on arguments of continuity. As (1.11) is linear, the transform of PI_2 must be a straight line. P must be on it because for a non-singular displacement P is conjugate to itself. But I_1 must also be on it, for all points of a line intersecting PI_2 and not on it are displaced to I_1. Hence our definition: *any point on PI_2 has as its transform all points of PI_1*. This also seems reasonable from the following argument. We consider all rotations about P; the path of any point at a distance ρ from P (real or imaginary, but unequal to zero) is a non-degenerate circle. For all non-singular rotation angles, P is at rest, a point Q_1 on PI_1 travels along PI_1, a point Q_2 on PI_2 goes along PI_2.

If $\tan\frac{1}{2}\phi = i$ (or, what is the same thing, $\tan\phi = i$) we have a singular displacement, called *isotropic* and denoted by $\mathbf{D}^+$: all points not on PI_2 move to I_1, which we knew to be a point of their paths (seeing now moreover that they arrive there all at the same instant), any point on PI_2, however, must be considered to be situated at all points of PI_1 simultaneously. For the conjugate isotropic rotation $\mathbf{D}^-$, with $\tan\frac{1}{2}\phi = -i$, we have the same phenomenon with I_1 and I_2 interchanged. The consequence is that not only a point with $\rho \neq 0$ describes a complete circle (including the points I_1, I_2) but that the same holds for the points with $\rho = 0$. These are the points on PI_1 and PI_2 and especially P itself, clearly such points have irregular motions. Any point Q_2 on PI_2, different from P, moves in the traditional way along PI_2 and then by $\mathbf{D}^+$ occupies the complete line PI_1; a point Q_1 on PI_1 travels along PI_1 but by $\mathbf{D}^-$ fills the line PI_2; the center P is at rest for any non-isotropic displacement, but occupies PI_1 by $\mathbf{D}^+$ and PI_2 by $\mathbf{D}^-$. Any point on the "zero" circle has this circle as its complete path.

All circles with center P have a common tangent at I_1 which coincides with PI_1, because such a line is perpendicular to itself; and similar for I_2. Isotropic displacements play an essential part in our mapping. The ∞^2 displacements $\mathbf{D}^+$ (one for any finite center P) are mapped on the points of V_1, not on l; the $\mathbf{D}^-$ are mapped on those of V_2, not on l. A $\mathbf{D}^+$ is determined by its center P and its image point is $(x_p, y_p, 1, -i)$ in V_1, and analogously for $\mathbf{D}^-$. Conversely a point $(X_1, X_2, 1, -i)$ of V_1 corresponds to the displacement for which any point of PI_2 with $P = (X_1, X_2)$ moves to all points of PI_1, and all other points go to I_1.

2. The geometry of the image space

The image of a displacement, as defined by (1.2), depends on the arbitrarily chosen zero position of E as well as the frame in Σ. This means that there are ∞^6 mappings. We now study the relationships between these mappings. Obviously we obtain another mapping if we take another frame in Σ and, independently, another frame in E. Then the new zero position is implicitly determined as that position of E for which its frame coincides with that of Σ.

We take a new frame in Σ, with coordinates X', Y' given by the relations

$$X' = X\cos\alpha - Y\sin\alpha + p, \qquad Y' = X\sin\alpha + Y\cos\alpha + q, \tag{2.1}$$

and a new one in E, with coordinates x', y':

$$x = x'\cos\beta - y'\sin\beta + r, \qquad y = x'\sin\beta + y'\cos\beta + s. \tag{2.2}$$

Eliminating x, y by means of (2.1), (2.2) and

$$X = x\cos\phi - y\sin\phi + a, \qquad Y = x\sin\phi + y\cos\phi + b, \tag{2.3}$$

we obtain

$$X' = x'\cos\phi' - y'\sin\phi' + a', \quad Y' = x'\sin\phi' + y'\cos\phi' + b', \tag{2.4}$$

with

$$\begin{aligned} \phi' &= \phi + \alpha + \beta, \\ a' &= r\cos(\phi+\alpha) - s\sin(\phi+\alpha) + a\cos\alpha - b\sin\alpha + p, \\ b' &= r\sin(\phi+\alpha) + s\cos(\phi+\alpha) + a\sin\alpha + b\cos\alpha + q. \end{aligned} \tag{2.5}$$

Substituting (2.4) into (1.6) allows us to obtain a relation between the old and the new image points:

$$\begin{aligned} X_1' &= (\cos\gamma_1)X_1 - (\sin\gamma_1)X_2 + a_{13}X_3 + a_{14}X_4, \\ X_2' &= (\sin\gamma_1)X_1 + (\cos\gamma_1)X_2 + a_{23}X_3 + a_{24}X_4, \\ X_3' &= \qquad\qquad\qquad\qquad (\cos\gamma_2)X_3 - (\sin\gamma_2)X_4, \\ X_4' &= \qquad\qquad\qquad\qquad (\sin\gamma_2)X_3 + (\cos\gamma_2)X_4, \end{aligned} \tag{2.6}$$

the six parameters $\gamma_1, \gamma_2, a_{ij}$ are functions of α, β, p, q, r, s.

Example 2. Show that $2\gamma_1 = \alpha - \beta$, $2\gamma_2 = -(\alpha+\beta)$ and determine a_{13}, a_{14}, a_{23}, a_{24}.

Obviously a *group* G of ∞^6 linear transformations in Σ' is defined by (2.6), which determines the Σ'-geometry. In the study of planar kinematics only those notions and relations, in Σ', are relevant which are independent of the two frames, or in other words which are invariant for the group (2.6).

From (2.6), it is easy to verify that the line l ($X_3 = X_4 = 0$) is invariant for $\mathbf{G}_6$, and also the points J_1 $(1, i, 0, 0)$ and J_2 $(1, -i, 0, 0)$ on it and the planes V_1 ($X_3 = iX_4$) and V_2 ($X_3 = -iX_4$) through it. Thus, as could have been expected, the singular elements of the mapping are independent of the frames.

The group is not completely defined by its fixed elements. If a linear transformation

$$X' = \| a_{ij} \| X \tag{2.7}$$

leaves l, J_1, J_2, V_1, V_2 invariant, we have $a_{11} - a_{22} = 0$, $a_{12} + a_{21} = 0$, $a_{33} - a_{44} = 0$, $a_{34} + a_{43} = 0$, $a_{31} = a_{32} = a_{41} = a_{42} = 0$, which are eight relations for essentially fifteen parameters, and from this a group $\mathbf{G}_7$ would follow. It is seen from (2.6) that we have moreover $a_{11}a_{22} - a_{12}a_{21} = a_{33}a_{44} - a_{34}a_{43}$.

The group $\mathbf{G}_6$ (2.6) has sub-groups with kinematical interpretations. If we do not alter the zero-position of E, but change the frames in E and Σ in the same way, it follows from (2.1) and (2.2) that $\beta = -\alpha$ and therefore $\phi' = \phi$, which is obvious. Moreover the rotation centers P are the same as before, but they have other coordinates, i.e., $x'_p = x_p \cos\alpha - y_p \sin\alpha + p$, $y'_p = x_p \sin\alpha + y_p \cos\alpha + q$. The corresponding transformations in Σ' read

$$\begin{aligned} X'_1 &= X_1 \cos\alpha - X_2 \sin\alpha + pX_3, \\ X'_2 &= X_1 \sin\alpha + X_2 \cos\alpha + qX_3, \\ X'_3 &= X_3, \\ X'_4 &= X_4, \end{aligned} \tag{2.8}$$

which is the general one leaving the origin O invariant, a "rotation" of Σ' about O. These rotations constitute a sub-group $\mathbf{G}_3$.

Example 3. Show that for (2.8) any point on the line $X_1 : X_2 : X_3 = (p \sin\frac{1}{2}\alpha - q \cos\frac{1}{2}\alpha) : (p \cos\frac{1}{2}\alpha + q \sin\frac{1}{2}\alpha) : 2\sin\frac{1}{2}\alpha$ is invariant. (Hence, this line is the rotation axis.)

If we maintain the frame in Σ but choose another zero-position in E, we have in (2.1) $\alpha = p = q = 0$. Hence $\phi' = \phi + \beta$, $\gamma_1 = \gamma_2 = -\frac{1}{2}\beta$, $a' = r\cos\phi - s\sin\phi + a$, $b' = r\sin\phi + s\cos\phi + b$ and the new image point is given by

$$\begin{aligned} X'_1 = X_1 \cos\tfrac{1}{2}\beta + X_2 \sin\tfrac{1}{2}\beta - \tfrac{1}{2}X_3(r\cos\tfrac{1}{2}\beta + s\sin\tfrac{1}{2}\beta) \\ + \tfrac{1}{2}X_4(r\sin\tfrac{1}{2}\beta - s\cos\tfrac{1}{2}\beta), \end{aligned}$$

$$
\begin{aligned}
X_2' &= -X_1 \sin\tfrac{1}{2}\beta + X_2 \cos\tfrac{1}{2}\beta + \tfrac{1}{2}X_3(r \sin\tfrac{1}{2}\beta - s\cos\tfrac{1}{2}\beta) \\
&\quad + \tfrac{1}{2}X_4(r\cos\tfrac{1}{2}\beta + s \sin\tfrac{1}{2}\beta), \\
X_3' &= \qquad\qquad X_3 \cos\tfrac{1}{2}\beta + X_4 \sin\tfrac{1}{2}\beta, \\
X_4' &= \qquad\qquad -X_3 \sin\tfrac{1}{2}\beta + X_4 \cos\tfrac{1}{2}\beta.
\end{aligned} \tag{2.9}
$$

Example 4. Show that the fixed points of (2.9) belong to two lines, one in V_1 through J_2, the other in V_2 through J_1.

A geometric interpretation of the transformation (2.9) will be given later.

In Euclidean three dimensional geometry the configuration of the fixed elements (the "absolutum") consists of a plane and an imaginary conic in it; for non-Euclidean geometry it is a general quadric, real for hyperbolic, imaginary for elliptic geometry. In Σ'-geometry it consists of two imaginary planes. Hence the geometry in the image space of our mapping is not identical with one of the classical metric geometries. It may be considered as a borderline case of elliptic geometry and is therefore often called quasi-elliptic.

Σ'-geometry is a rather primitive one. Moreover it interests us only insofar as it is an image of plane kinematics. We introduce in Σ' the following concepts: the distance of two points, the angle of two planes, the parallelism of two lines.

We define the (angular) distance between two image points as follows: if A_i is the image of the displacement $\mathbf{D}_i$ $(a_i, b_i; \phi_i)$, $i = 1, 2$, the distance A_1A_2 is defined by $(A_1A_2) = \phi_2 - \phi_1$. In view of its kinematic meaning this is obviously a concept invariant for Σ' transformations (2.6).

Example 5. Show that the distance between the image points with coordinates X_i, Y_i reads $\phi = 2 \arctan [(X_4Y_3 - X_3Y_4)/(X_3Y_3 + X_4Y_4)]$. Prove that it is invariant under transformation (2.6).

Example 6. Prove the following properties. If A_1, A_2, A_3 are collinear then $(A_1A_2) + (A_2A_3) + (A_3A_1) = 0$; $(A_1A_2) = -(A_2A_1)$.

If the line A_1A_2 intersects l the distance $(A_1A_2) = 0$; the line is a *null-line*. The distance between any two points of a plane through l is zero; the plane is a *null-plane*. If U and W are two null-planes the distance from any point of U to any of W is always the same. (All this implies that spheres and circles are trivial configurations in Σ'-geometry.)

Example 7. If the line A_1A_2 intersects V_1 and V_2 at S_1 and S_2 write (A_1A_2) as a function of the cross-ratio $(A_1A_2S_1S_2)$.

The absolutum of Σ', consisting of the line l, the two points J_1, J_2 on it and the two planes V_1, V_2 through it, is a self-dual configuration. Hence there is complete duality in Σ'-geometry (as in elliptic geometry, but not in Euclidean geometry). Hence we may define the angle between two planes U and W in the same manner as the distance between two points. If U_i and W_i are the coordinates of the planes we define their angle ψ by

$$\psi = 2\arctan\,[(U_2W_1 - U_1W_2)/(U_1W_1 + U_2W_2)]. \tag{2.10}$$

Example 8. Two planes have angle zero (they are "Σ'-parallel") if they intersect l at the same point.

Example 9. If the planes U and W intersect l at T_1 and T_2 write ψ(UW) as a function of the cross-ratio $(J_1J_2T_1T_2)$.

We consider now two pencils of lines p_1 and p_2. p_1 consists of the lines m_1 in V_1 through J_2, p_2 of the lines m_2 in V_2 through J_1. They have l in common, but all other lines are imaginary. If m_1 belongs to p_1 the conjugate imaginary line belongs to p_2.

Any (real) line q in Σ' has an intersection S_1 with V_1, S_2 with V_2, hence it intersects a line m_1 and the conjugate line m_2. Conversely: the join of a point of V_1 and the conjugate point of V_2 is a real line.

We define two lines q and q′, which are not null-lines, to be Σ'-parallel if they intersect the same lines m_1 and m_2 of the pencils p_1 and p_2; two null-lines are Σ'-parallel if they have the same intersection with l. Contrary to the situation in Euclidean space *two parallel lines are in general skew*. But they have some properties which justify the term parallel. From the definition it follows immediately that if q_1, q_2 are parallel and q_2, q_3 are parallel then q_1, q_3 are parallel; hence parallelism is a transitive property. Moreover: if q is a given line and Q a given point there is always one line q′ through Q parallel with q. Indeed, if q is not a null line it determines a line m_1 and the conjugate m_2; q′ is seen to be the unique line through Q intersecting both m_1 and m_2, it is the (real) intersection of the planes (Q, m_1) and (Q, m_2). If q is a null-line, q′ is the join of Q and the intersection of q and l. Hence the parallel axiom is satisfied in Σ'-geometry. All lines parallel to a given line, and therefore mutually parallel, constitute the ∞^2 lines of a linear congruence (an elliptic one, because the two common transversals are imaginary). These lines all have the same "Σ'-direction".

Example 10. If q_{ij} and q'_{ij} are the Plücker coordinates of two lines which are not null-lines, prove that they are Σ'-parallel if the two conditions

$$q_{34}(q'_{41} + q'_{32}) = q'_{34}(q_{41} + q_{32}), \quad q_{34}(q'_{31} + q'_{24}) = q'_{34}(q_{31} + q_{24})$$

are satisfied. Determine the conditions for the parallelism of two null-lines ($q_{34} = 0$, $q'_{34} = 0$).

Having now completed the introduction of some necessary metric concepts in Σ' we can return to the set of transformations (2.8) and (2.9). For (2.8), the origin O and a line through it, mentioned in Example 3, are fixed.

Example 11. Verify that the phrase "rotation about the axis $\mathfrak{a}$" is justified by proving that any plane through $\mathfrak{a}$ is transformed into another through $\mathfrak{a}$; all transformed planes undergo the same change of angle from their former positions. Determine this "rotation angle".
Example 12. Show that any plane through l is invariant for (2.8).

We consider now the Σ'-transformation (2.9), which resulted when we took another zero-position, leaving the frame in Σ unaltered. By (2.9) the origin O has gone to the point $(r \sin \frac{1}{2}\beta - s \cos \frac{1}{2}\beta,\ r \cos \frac{1}{2}\beta + s \sin \frac{1}{2}\beta,\ 2\sin \frac{1}{2}\beta,\ 2\cos \frac{1}{2}\beta)$, this is the image point of the displacement $\mathbf{D}(r, s; \beta)$. But this implies that all points of Σ' have moved the distance β from their original position. Moreover, it is easy to verify (for instance by means of the criterion of Example 10) that all lines through the points (X_1, X_2, X_3, X_4) and (X_1', X_2', X_3', X_4') are Σ'-parallel. Hence by a transformation (2.9) all points of Σ' are moved the same distance along mutually parallel lines. Such a transformation may be called a "Σ'-translation".

We have shown that two sets of transformations in Σ', both with a clear kinematical meaning are comparable to rotations and translations of Euclidean space. We add that the general Σ'-transformation (2.6) is the product of a transformation from one set by one from the other set (as is obvious from their kinematical meaning); it is a Σ'-screw-motion. Such a motion has an axis $\mathfrak{a}$ and is built up from a rotation about $\mathfrak{a}$ and a translation, Σ'-parallel with $\mathfrak{a}$. It has four fixed points: J_1, J_2 and the intersections S_1, S_2 of $\mathfrak{a}$ and V_1, V_2.

Example 13. Show that the general Σ'-transformation has six fixed lines and four fixed planes.
Example 14. Show that the eigenvalues of (2.6) are $e^{\pm i\gamma_1}$, $e^{\pm i\gamma_2}$. What are the conditions γ_1, γ_2 must satisfy if the Σ'-motion is to be a general screw, a rotation, a translation? Consider the case $\gamma_1 = \gamma_2 = 0$.
Example 15. If a Σ'-translation transforms the arbitrary plane U into U′, for all such planes, the intersections (U, U′) are all parallel and the angle between U′ and U is constant.

3. Straight lines in the image space

We have obtained some insight into the metric properties of Σ', and now we return to our starting point: the study of the image space as a map of plane kinematics. Our program will be twofold: on one hand we ask how certain phenomena in kinematics are represented by the mapping as properties in Σ', on the other hand we investigate the kinematic meaning of certain configurations in the image space. We start with the latter approach.

Obviously a set of n points of Σ' corresponds to that of n finitely separated positions of E with respect to Σ; a *curve* k in Σ' is the image of a continuous motion of the moving plane; a *surface* F in Σ' corresponds to a motion with two independent parameters. It is clear that not only projective properties of these configurations in Σ' are of interest but also their metric ones, that is their behavior with respect to $\mathbf{G}_6$ or, what is the same thing, their positions with regard to the absolutum consisting of l, J_1, J_2, V_1, V_2.

The displacements of E are defined with respect to a certain arbitrarily chosen zero-position which corresponds to the origin O in the image space. If $Q(X_1, X_2, X_3, X_4)$ is an image point and $X_3 \neq 0$, in view of (1.2), it corresponds to a displacement with center $x_p = X_1/X_3$, $y_p = X_2/X_3$ and the rotation angle ϕ is given by $\tan \frac{1}{2}\phi = X_3/X_4$. Obviously the line OQ intersects the null-plane $X_3 = X_4$ at the point $(x_p, y_p, 1, 1)$. All points of a null-plane $X_3 = \lambda X_4$ correspond to displacements with the same rotation angle. If $X_3 = 0$ the point Q corresponds to a translation of E, for which according to (1.8) we have $a = 2X_2/X_4$, $b = -2X_1/X_4$, the rotation center is at infinity $x_p : y_p : z_p = X_1 : X_2 : 0$.

A line p *in* Σ', *through* O, not in $X_3 = 0$, given by $X_1 : X_2 : X_3 = p_1 : p_2 : p_3$, $p_3 \neq 0$, *corresponds to the set of all rotations about the same center* $(x_p = p_1/p_3, y_p = p_2/p_3)$. If p through O lies in $X_3 = 0$ it corresponds to a set of mutually parallel translations of E, with variable displacements.

A line p not through O, not a null-line, would be the image of the set of rotations of E with center A_2, say, if the zero-position O' had been chosen on p. By the displacement of E corresponding to the change of zero-position from O' to O let A_2 be transformed into A_1. Hence a point of p is the image of a displacement of E which transforms A_1 into A_2 followed by an arbitrary rotation about A_2. The conclusion is: any point of p corresponds to a displacement of E for which A_1 is transformed into A_2. The rotation centers are on the perpendicular bisector of A_1, A_2. That they are on a line is also seen directly in the map: the intersection of plane (O, p) and the plane $X_3 = X_4$ is a straight line. The same holds if p, not through O, is a null-line; moreover the rotation angle is now the same for all rotations. Such a line corresponds to a set of rotations the centers of which are on a line and they transform a point A_1 at infinity into a point A_2 again at infinity.

For any line p in the image space Σ' we have now determined the corresponding meaning in plane kinematics: *it is associated with an ordered pair of points* A_0, A_1 *of* Σ, *and each point of* p *corresponds to a displacement which transforms the point of* E *with zero-position* A_0 *to the position* A_1.

Our mapping considered in the other direction is one which represents the lines of three dimensional space by oriented point pairs of a plane. This leads to the study of line geometry by means of a plane map and has been extensively treated by MÜLLER and KRUPPA [1923].

Example 16. If the coordinates of p are p_{ij} show that the points A_0, A_1 in Σ are $(p_{14}+p_{23}, p_{24}+p_{31}, p_{34})$ and $(p_{14}-p_{23}, p_{24}-p_{31}, p_{34})$ respectively. Verify that A_0 and A_1 coincide if p passes through O and that they are at infinity if p is a null-line. If p does not pass through O the line of corresponding rotation centers in Σ is $p_{23}X + p_{31}Y + p_{12}Z = 0$.
Example 17. Let p and q be two lines of Σ' and (A_0, A_1), (B_0, B_1) the oriented points pairs of Σ corresponding to them. Determine the configuration of the four points if p and q intersect.
Example 18. If p and q are Σ'-parallel show that A_0 and B_0 coincide; also show that the converse is true.

4. Curves in the image space

Any curve k in Σ' is the image of a motion **M** of E with respect to Σ; it determines the motion completely and all the motion's properties may be derived from the properties of this single curve k. If the points of k are given as functions of the time t the motion is also given with its time schedule. For now we restrict ourselves to geometric kinematics, not taking the time dependence into account.

If the coordinates of the points of k are given as functions of a parameter, the path of any point (x, y) of E for the motion **M** follows immediately from the fundamental equations (1.9). If k is an algebraic curve all paths are algebraic; if k is a rational curve all paths are rational. We shall call the *motion* algebraic and rational respectively. In this section we intend to derive the order and the circularity of the plane paths if the order n of the algebraic image curve k has been given; k is of course in general a space curve. The most simple case occurs if k does not intersect the singular line l. Then all points of k represent displacements of E. From (1.9) it follows immediately that all paths have order $2n$. Moreover k has n intersections with V_1; any point with $X_3 = iX_4$ however, as seen by direct substitution into (1.9), gives us $X:Y:Z = 1:i:0$. The analogous result holds for V_2. Hence the theorem: *if the image curve* k *is an algebraic curve of order n which does not intersect* l, *the paths of all points of E for the corresponding motion* **M** *are curves of order* $2n$ *and circularity* n.

If S is an intersection of k and V_1 its coordinates are $(s_1, s_2, 1, -i)$, s_1 and s_2 being, in general, complex numbers. One of the intersections of k and V_2 is S', the conjugate imaginary of S. The point S is the image of an isotropic displacement $\mathbf{D}^+$ (see Section 1), the center P of which in Σ is (s_1, s_2). The path of any point of E has therefore a tangent at I_1 coinciding with PI_1 and at I_2 a tangent $P'I_2$ (P' is the conjugate of P). In view of the properties of isotropic displacements any point of PI_2 has the line PI_1 as part of its path, which is therefore degenerate. Analogously any point of $P'I_1$ has $P'I_2$ as a part of its path. It follows from this that the (real) intersection G of PI_2 and $P'I_1$ has a path to which the lines PI_1 and $P'I_2$ belong. This means that the path of G is of order $2n-2$ and circularity $n-1$. Obviously *there are n points of E with this property*. If two or more coincide the order of the path of such a point is further diminished.

For every path the (real) intersection F of PI_1 and $P'I_2$ is the intersection of conjugate imaginary tangents to the path at I_1 and I_2. It is therefore a special focus of every path.* *There are thus n special foci, they are the same for every path* generated by the motion represented by k. Two corresponding points such as F and G are called the Laguerre-pair of the pair (P, P'); they are the reflection of one another with respect to the (real) line PP'. It is possible that P and P' coincide; then F and G coincide at the same point. Summarizing we have: *if the image curve* k *of a motion* **M** *has order n and does not intersect* l, *the path of a point of E is in general a n-circular curve of order* $2n$. *There are n points of E for which the path is essentially a* $(n-1)$*-circular curve of order* $(2n-2)$. *All paths have n special foci which are at the same points for all paths.*

Exceptional situations appear if k intersects l. Such an intersection S is not the image of a displacement, but we shall investigate the behavior of a moving point of E in the limit as the point on k approaches S.

Let $S=(s_1, s_2, 0, 0)$ be a real point of intersection of k and l. If k has been given by $X_i(t)$, t being a parameter, and if S corresponds to $t=0$, for small values of t we have

$$X_1 = s_1+\alpha_1 t, \quad X_2 = s_2+\alpha_2 t, \quad X_3=\alpha_3 t, \quad X_4=\alpha_4 t \tag{4.1}$$

(s_1, s_2 not both zero, α_3, α_4 not both zero). Substituting this in (1.9) we observe that X, Y, Z have the factor t; dividing by t and restricting ourselves to linear terms we obtain

$$\begin{aligned} X &= (\alpha_3^2-\alpha_4^2)tx - 2\alpha_3\alpha_4 ty + 2[(s_1\alpha_3+s_2\alpha_4)+(\alpha_1\alpha_3+\alpha_2\alpha_4)t]z, \\ Y &= 2\alpha_3\alpha_4 tx + (\alpha_3^2-\alpha_4^2)ty + 2[(s_2\alpha_3-s_1\alpha_4)+(\alpha_2\alpha_3-\alpha_1\alpha_4)t]z, \\ Z &= (\alpha_3^2+\alpha_4^2)tz, \end{aligned} \tag{4.2}$$

which gives us for $t \to 0$

$$X:Y:Z = (s_1\alpha_3+s_2\alpha_4):(s_2\alpha_3-s_1\alpha_4):0 \tag{4.3}$$

for any finite point of E. As $s_1\alpha_3+s_2\alpha_4 = s_2\alpha_3-s_1\alpha_4=0$ would imply either $s_1^2+s_2^2=0$ or $\alpha_3^2+\alpha_4^2=0$, both excluded, the conclusion is: all paths pass

* Any tangent to a curve from I_1 intersects the corresponding tangent from I_2 in a point called a focus of the curve. If the tangency occurs at I_1 and I_2 respectively the intersection point is called a special (or singular, or principal, or Laguerre) focus.

through the same real point B at infinity. As seen from (4.2) the tangent at B depends on (x, y, z), hence the paths have in general different asymptotes at B.

If one intersection of k and V_1 is on l, there remain $n-1$ intersections, not on l, corresponding to isotropic $\mathbf{D}^+$ displacements, and similarly for V_2. The conclusion is: if k has one real intersection with l, the path of any point of E is a $(n-1)$-circular curve of order $2n-1$; with one real point at infinity which is the same for all paths generated by the motion k.

Example 19. Check the results so far for $n = 1$, the image curve being a line which does not or does intersect l.

The phenomenon of k intersecting l at a real point may occur more than once. If there are m real intersections ($m \leqslant n-1$), the result is that the path of any point of E is a $(n-m)$-circular curve of order $2n-m$. The m infinite, non-isotropic, points of the paths are the same for all paths. There are several sub-cases: two or more intersections of k and l may coincide (either because an intersection is a multiple point of k, or because l is a tangent of k to a certain order or a tangent at a multiple point of k). In all such cases the preceding results must be modified accordingly.

If a (real) curve k has imaginary intersections with l they appear in conjugate pairs. Suppose that k intersects l in two conjugate imaginary points S and S'. We suppose first that S, S' do not coincide with J_1, J_2 and also that the tangents at S, S' are not in V_1 or V_2. This implies that, making use of the notations of (4.1), we still have $s_1^2+s_2^2 \neq 0$, $\alpha_3^2+\alpha_4^2 \neq 0$. Hence the above argument is valid and the result is: if k, of order n, has two conjugate imaginary intersections S, S' with l, not coinciding with J_1, J_2 and such that the tangents of k at S, S' are not in V_1 or V_2, the paths of all points of E are $(n-2)$-circular curves of order $(2n-2)$. Every path has the same two conjugate imaginary, non-isotropic, points at infinity. It is now easy to extend this to the case where k has more than one pair of imaginary intersections with l.

In all cases that follow we again take k to have two conjugate imaginary intersections with l, denoted by S and S', while the tangents of k at S and S' are r and r′ respectively.

First, let S and S' not coincide with J_1, J_2, but let r be in V_1, and thus r′ in V_2. This implies that in (4.1), if $t = 0$ is the value of the parameter at S, we have $\alpha_3 = i\alpha_4$. Substituting this in (4.3) we obtain for B the point $(s_1 i + s_2):(s_2 i - s_1):0 = (1:i:0)$, that is I_1. Moreover, as seen from (4.2) the tangent at I_1 is the line at infinity ($Z = 0$). The plane V_1 has n intersections with k, but three are on l (S counted twice, S' counted once) and therefore $n-3$ intersections, not on l, correspond to displacements $\mathbf{D}^+$. But since B coincides with J_1 the path is circular for two different reasons: one, as always, because k has points corresponding to isotropic positions, and the second because of its behavior with respect to l. If the respective circularities are denoted by c_1 and c_2 respectively, we have in our case $c_1 = n-3$, $c_2 = 2$, which gives as the complete circularity c of the path $c = c_1 + c_2 = n-1$. The degree of the path is $2n-2$, for the path has no points at infinity outside I_1, I_2. If at the case at hand (r in V_1, r′ in V_2) we have moreover $S = J_1$, $S' = J_2$ we have $s_1 : s_2 = 1 : i$, but that has no influence on the argument: the point B again coincides with I_1. The result is different if $S = J_2$, $S' = J_1$, then B is undetermined; this case must be dealt with specially.

Before doing this we consider the situation where $S = J_1$, $S' = J_2$ while r and r′ are not in V_1 or V_2. In (4.1) we have now $s_1 : s_2 = 1 : i$ (if $t = 0$ is the parameter value of J_1), but $\alpha_3^2 + \alpha_4^2 \neq 0$. For B we obtain $X : Y : Z = (\alpha_3 + i\alpha_4) : (i\alpha_3 - \alpha_4) : 0 = 1 : i : 0$, that is I_1 once more, but the tangent at I_1 does not coincide with the line at infinity. The conclusion is $c_1 = n-2$, $c_2 = 1$, $c = n-1$.

The last case we must investigate is: $S = J_1$, $S' = J_2$, r lies in V_2, r′ in V_1. Or (if $t = 0$ gives J_1): $s_1 : s_2 = 1 : i$, $\alpha_3 : \alpha_4 = -i : 1$, which means that in (4.2) all terms without t are zero and the limit for $t \to 0$ is senseless. It is clear that the representation of k in the neighborhood of l must be given up to second order terms in t. It we put $s_1 = s$, $s_2 = is$, $\alpha_3 = -i\alpha$, $\alpha_4 = \alpha$, we have

$$X_1 = s + \alpha_1 t + \beta_1 t^2, \qquad X_2 = is + \alpha_2 t + \beta_2 t^2,$$
$$X_3 = -i\alpha t + \beta_3 t^2, \qquad X_4 = \alpha t + \beta_4 t^2. \tag{4.4}$$

Substituting this in the equations (1.9) of the paths we obtain, after dividing by t^2, and then letting $t \to 0$

$$\begin{aligned} X &= \alpha^2 x + i\alpha^2 y + [s(\beta_3 + i\beta_4) + \alpha(\alpha_2 - i\alpha_1)]z, \\ Y &= -i\alpha^2 x + \alpha^2 y + [is(\beta_3 + i\beta_4) - i\alpha(\alpha_2 - i\alpha_1)]z, \\ Z &= 0. \end{aligned} \tag{4.5}$$

Hence if the point on k approaches J_1 the path in Σ tends to a point at infinity. The point at infinity depends on (x, y, z) and is therefore, in general, different for different paths. (Those points of E lying on a line through I_1 go to the same point.) For J_2 we obtain the analogous result. The number of intersections of k and V_1, not on l, is $n-3$ (because two coincide with J_2 and one with J_1). Hence $c_1 = n-3$, $c_2 = 0$, $c = n-3$; the order of a path is $2(n-3)+2 = 2n-4$.

If we are interested in the order N of the paths of a motion and their circularity c it follows that we must distinguish five cases, summarized in the following table; d is the number of non-isotropic points at infinity (of a path).

(4.6)

Type	Intersections of k and l	N	c	d
I	No intersections	$2n$	n	0
II	One real intersection	$2n-1$	$n-1$	1
III	A pair of conjugate imaginary intersections; S and S' not J_1 or J_2; r, r′ not in V_1 or V_2.	$2n-2$	$n-2$	2
IV	A pair S, S'; all cases different from III and V.	$2n-2$	$n-1$	0
V	$S=J_1$, $S'=J_2$; r in V_2, r′ in V_1.	$2n-4$	$n-3$	2

If k has m_2 intersections of type II and m_j pairs of type j ($j=3,4,5$) we have

$$
\begin{aligned}
N &= 2n - m_2 - 2m_3 - 2m_4 - 4m_5,\\
c &= n - m_2 - 2m_3 - m_4 - 3m_5, \qquad (4.7)\\
d &= m_2 + 2m_3 + 2m_5.
\end{aligned}
$$

Example 20. In case V the paths have two non-isotropic points at infinity, B_1 and B_2, which are in general conjugate imaginary. Show that there are points of E for which B_1 and B_2 coincide, (which means that the path has a real isolated point at infinity, derived for an imaginary displacement) and prove from (4.5) that their locus is the circle $\alpha^2[(\alpha x+\alpha_2 z)^2+(\alpha y-\alpha_1 z)^2]-s^2(\beta_3^2+\beta_4^2)=0$.

5. Plane image curves

In this section we determine those motions **M** of E which correspond to image curves k which are plane curves.

It should be noted that the origin O of the image space Σ' is an arbitrarily chosen point. If we investigate the motion corresponding to a given image curve k we may always take O at a point of k. This means that one of the positions of E during its motion is singled out as the zero-position.

Let U be the plane of k. Then U passes through O. If U is the plane $X_3 = 0$, all points correspond to translations; a curve k in it represents a general translatory motion, wherein all the paths are congruent. For a general plane U containing k and not passing through l, the plane $X_3 = X_4$ intersects U in a straight line s′. Each point of U lies on a line u′ connecting O and one point of s′. The line pencil (O, u') completely covers U. We know, from Section 3, that all the points of any one line u′ have only one rotation center X_p, Y_p. Since the points of s′ have the coordinates $(X_p, Y_p, 1, 1)$, see Section 3, it follows that there is a linear relationship between X_p and Y_p, hence all the rotation centers (in Σ) lie on one line s in Σ. The conclusion is then that the points of U correspond to the set **D** of ∞^2 displacements of E for which the locus of the rotation center P is a straight line, s of Σ, the rotation angles being arbitrary.

The points of k correspond to a continuous subset of **D**. Hence k corresponds to a motion **M** of the following type: any position of E is obtained by rotating the zero-position about a center P on s, the rotation angle ϕ being a continuous (and, in general, many valued) function of P.

We know from Chapter VIII that the rotation about P can always be considered as the product of the reflections into two lines through P (the angle between them being $\frac{1}{2}\phi$). As the first of these lines we take s, which passes through P, and for the second line we take a line through P making an angle of $\phi/2$ with s, we denote this line as s(ϕ). There will be a line s(ϕ) corresponding to each point of k. We denote the envelope of the lines s(ϕ) as the plane curve K, in Σ, hence s(ϕ) are the tangents of K; s is one of them: s(0) = s. The displacements of E may be found as follows: reflect the zero position into the fixed tangent s of K, and reflect the result into an arbitrary tangent of K. If K′ is the reflection of K into s, a set of positions of K′ are obtained by reflecting K into the set of its own tangents. However, each such position of K is a position of E (since it can also be obtained if we consider K′ a curve of E and reflect it twice from the zero-position, first about s and then about s(ϕ)). We recognize the general symmetrical motion of E (Chapter IX, Section 8), K being the fixed and K′ the moving centrode. Hence, *a plane image curve corresponds to a symmetrical motion of E, and conversely a symmetrical motion has a plane image curve.*

If the image curve k is of order n, any line through O, in U, has n points in common with it (O being one of them); hence to any rotation center P, in Σ, there belong n values of ϕ ($\phi = 0$ being one of them); through P pass therefore n lines s(ϕ) (s itself being one of them). The conclusion is: *the centrodes* K *and* K′ *are of class n.*

The plane U of k has one (real) point of intersection S with l. In general S is not on k; if k passes m times through S ($0 \leq m \leq n-1$) the path of a point of E for the corresponding (symmetrical) motion is, according to (4.7), a $(n-m)$-circular curve of order $2n-m$.

6. Quadratic motions

A motion is quadratic if the image curve is a (non-degenerate) conic, and hence a plane curve in a plane U. In view of Section 5, all quadratic motions are symmetrical, the centrodes K and K′ are curves of the second class and therefore conics. In general k has no intersection with l; then the paths are in view of (4.7) rational bicircular quartic curves. If k has a point S in common with l (S being the intersection of l and U), we have $m = 1$: the paths are rational circular cubic curves with one real point at infinity, the same for all paths. If Q is any point on k the intersection of OQ and the plane $X_3 = X_4$ gives us the point (x_p, y_p, z_p, z_p) and thus the coordinates of the rotation center P of the displacement represented by Q. Therefore the intersection of $X_3 = X_4$ and the tangent of K at O gives us the instantaneous rotation center of the zero-position, which is a point of the centrode K. This point is at infinity if the tangent represents a translation; which is the case if it passes through S (which gives $z_p = 0$). As O can be chosen arbitrarily on k the conclusion is that K has a point at infinity if some tangent of k passes through S. Hence K is an ellipse if S is inside k, a hyperbola if S is outside k and a parabola if S is on k.

If k does not pass through S it has two intersections with V_1 (and V_2) not on l. The corresponding motion has two isotropic positions. In view of Section 4 there are two points of E with paths which (each) contain two isotropic lines; hence there are two points which each have a path which is a circle. The analogous argument shows that in case k passes through S there is one point of E with a straight line path.

We have now derived by means of our mapping the most important properties of the motion defined by the rolling of a conic on a symmetrical conic. In Chapter IX we used the order of the point paths to name the motion. In this chapter we often use the image curve's order instead.

7. Displacements for which one point stays on a circle

In the preceding sections we considered some configurations (a straight line, a plane curve, a conic) in the image space Σ' and asked for the motions of

E to which they correspond. We now study the mapping in the other direction: we start from some given motions and derive their images in Σ'. In classical plane motions, such as we dealt with in Chapter VIII, one or more points of the moving plane describe circles. Therefore we investigate first of all those displacements of E for which a given point A is permanently on a given circle c of Σ. As this means one condition for the motion, we may expect the image to be a certain surface in Σ'. If (X_1, X_2, X_3, X_4) is an image point the displacement of E is given by the fundamental relations (1.9) from which

$$X^2+Y^2=(X_3^2+X_4^2)^2(x^2+y^2)-4(X_3^2+X_4^2)(X_1X_3-X_2X_4)xz$$
$$-4(X_3^2+X_4^2)(X_2X_3+X_1X_4)yz+4(X_3^2+X_4^2)(X_1^2+X_2^2)z^2. \tag{7.1}$$

So that if A is (x, y, z), and the equation of the circle c is

$$C_0(X^2+Y^2)+2C_1XZ+2C_2YZ+C_3Z^2=0, \tag{7.2}$$

the displacements for which A is always on c satisfy $X_3^2+X_4^2=0$ (which is trivial) or

$$\mathrm{H}\equiv C_0z^2(X_1^2+X_2^2)+(-C_0x+C_1z)zX_1X_3+(-C_0y+C_2z)zX_2X_3$$
$$+(-C_0y-C_2z)zX_1X_4+(C_0x+C_1z)zX_2X_4+(-C_1y+C_2x)zX_3X_4$$
$$+\tfrac{1}{4}[C_0(x^2+y^2)-2C_1xz-2C_2yz+C_3z^2]X_3^2 \tag{7.3}$$
$$+\tfrac{1}{4}[C_0(x^2+y^2)+2C_1xz+2C_2yz+C_3z^2]X_4^2=0.$$

This represents a quadric surface H; the coefficients depend on the ratios of C_0, C_1, C_2, C_3 and on those of x, y, z. Hence (7.3) is a system of ∞^5 quadrics. It obviously is not a linear system.

To study the surface H we choose the origin O on it (then the last term of (7.3) vanishes) or, what is the same condition, A is on c at the zero-position. Furthermore we take the frame in E and Σ such that $C_1=C_2=0$, $C_0C_3<0$, which may be done without any loss of generality if c is a real, non-degenerate, circle. Then for $z\neq 0$ (which means that A is a finite point of E) we obtain for (7.3) (provided $C_0(x^2+y^2)+C_3z^2=0$),

$$z(X_1^2+X_2^2)-x(X_1X_3-X_2X_4)-y(X_2X_3+X_1X_4)=0, \tag{7.4}$$

which is therefore the equation of H for a suitably chosen zero-position and a suitably chosen frame in E and Σ. The discriminant of (7.4) is $(x^2+y^2)^2$; hence H is a general quadric. Furthermore, if $X_3=X_4=0$ we obtain $X_1^2+X_2^2=0$ which implies that H passes through J_1 and J_2. If $X_3=iX_4$ we have

$$z(X_1^2+X_2^2)+[(-ix-y)X_1+(x-iy)X_2]X_4=0,$$

or

(7.5) $$(-iX_1+X_2)[izX_1+zX_2+(x-iy)X_4]=0.$$

This implies that the intersection of H and V_1 degenerates into two lines, intersecting at the point $T=(ix+y,-x+iy,2iz,2z)$, the two lines being TJ_1 and TJ_2. In the same manner we may prove that the intersection of H and V_2 consists of the two lines $T'J_2$ and $T'J_1$, T' being the conjugate imaginary to T.

If a quadric passes through J_1 and J_2, having V_1 and V_2 as tangent planes it must satisfy four conditions. There are ∞^9 quadrics in space, therefore there are ∞^5 quadrics with these properties.

If c is a degenerate real circle it may be given the equation $Y=0$, which implies $y=0$ if A is on c for the zero-position. Hence $C_0=C_1=C_3=0$, $C_2\neq 0$ and the equation (7.3) is seen to be

(7.6) $$zX_2X_3-(zX_1-xX_3)X_4=0,$$

which shows that the locus which we denote for this case by H* is again a non-singular quadric. It contains the line l. The intersection with V_1 consist of l and the line $X_3-iX_4=zX_1-i(zX_2-xX_4)=0$, which passes through J_2; the intersection of H* and V_2 consists of l and a line through J_1. Summing up we have: the locus of the image points of those displacements of E for which a given point A of E is on a given circle c of Σ, is a quadric H, passing through J_1 and J_2 and tangent to V_1 and V_2 at the points T and T'. If c degenerates into a finite straight line, T coincides with J_2 and T' with J_1, and l is a generator of H.

Example 21. Determine the system of ∞^4 special quadrics H*, corresponding to displacements for which A stays on a straight line (take $C_0=0$ in (7.3)).

The quadric H with equation (7.4) contains the (real) line $X_1=X_2=0$; it is therefore a hyperboloid. The same holds for (7.6) because l is on it. The conclusion is that H in the general and in the special case contains two sets of real generators.

The question arises then as to the meaning of these sets on the kinematical side of the mapping. We know that a line of Σ' corresponds to the set of all displacements of E for which a point A_1 is displaced to A_2, A_1 and A_2 being two fixed points.

Let c be the circle in Σ, A_0 the zero-position of the moving point and A another position. If A_0 and A are on c obviously any rotation which displaces A_0 into A has its image on H. The set, **G**, of these rotations, however, corresponds to a line g of Σ' and thus one of the sets of generators of the hyperboloid is accounted for. A generator g corresponds to a set of rotations the centers of which are on a line through the center M of c (Fig. 88). Two such sets of rotations $\mathbf{G}_1$ and $\mathbf{G}_2$ have no common element: two generators g_1, g_2 are mutually skew.

The meaning of the second set may be found as follows. Let m be any line through A_0 (Fig. 89), M' the reflection of M into m. Consider all rotations which displace M' into M. If P on m is the center of such a rotation, the rotation angle must be $M'PM$. It displaces A_0 into its reflection into the radius PM, and therefore into a point A on the circle. Hence the rotations $\mathbf{G}'$ displacing M' into M correspond to a generator g′ on H. Any line m through A_0 gives rise to such a generator. Two sets $\mathbf{G}'_1$ and $\mathbf{G}'_2$ have no rotation in common, which implies that the corresponding generators g'_1 and g'_2 are skew. All rotations of a set $\mathbf{G}$ displace A_0 into the same point, those of a set $\mathbf{G}'$ displace A_0 into a variable point of c. Hence a set $\mathbf{G}$ and a set $\mathbf{G}'$ have one displacement in common: two generators g and g′ intersect.

Example 22. Determine the set $\mathbf{G}$ and the set $\mathbf{G}'$ to which the identical displacement belongs; they correspond to the two generators of H through the origin O.
Example 23. Determine the two sets of displacements of E corresponding with the generators of the special hyperboloid H*.

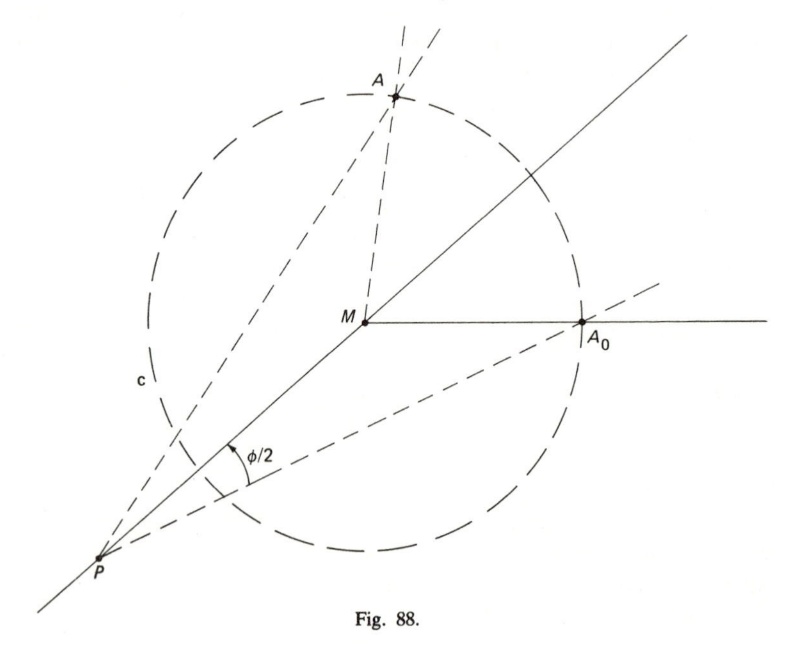

Fig. 88.

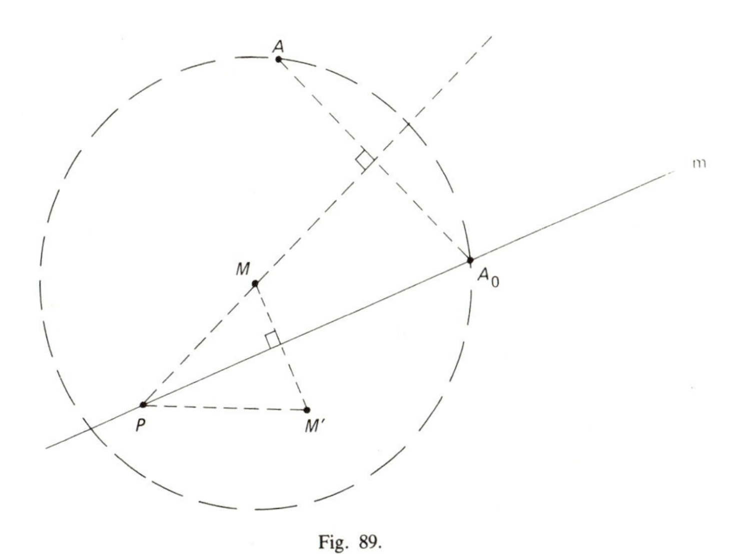

Fig. 89.

Example 24. Discuss the use of equation (7.3) in solving the problem named after Burmester: if five positions of E with respect to Σ are given, find the points of E which have their five homologous positions on a circle in Σ.

Example 25. Verify that the solution to this problem is exactly the same as was given in Chapter VIII.

8. The general four-bar motion

We obtain a four-bar motion of E if a point A_1 of E is compelled to stay on a circle $(M_1; R_1)$ of Σ, and a second point A_2 on a second circle $(M_2; R_2)$. In view of the results of the preceding section we obtain: a four-bar motion of E is mapped on the intersection k_4 of two quadrics H_1 and H_2; k_4 is the well-known quartic space curve of the first kind.

H_1 and H_2 both pass through J_1 and J_2; hence k_4 passes through J_1 and J_2. The tangents r_1 and r_2 at these points, however, are not in V_1 or V_2.

Example 26. Show that r_1 lying in V_1 or V_2 would imply that M_1 and M_2 are not both either real or coinciding.

Therefore the behavior of k_4 with respect to the absolutum is such that it belongs to type IV of (4.6). As $n = 4$ we have $N = 6$, $m_4 = 1$, $c_1 = 2$, $c_2 = 1$, $c = 3$: the paths of the moving points are tricircular curves of order 6. The general k_4 is a curve of genus 1, i.e., an elliptic curve; any path is therefore, according to a theorem of algebraic geometry, in general also elliptic. As $c_1 = 2$ there are two points of E with paths which degenerate into two isotropic lines and a curve of order 4. These points are seen to be A_1 and A_2, the paths of which are the circles with centers at M_1 and M_2 respectively, each counted twice.

We shall derive a parametric representation of the image curve and (by means of the mapping) one for the paths, that is of the coupler curve of any point of E. We know beforehand that we need elliptic functions. Let the centers of the circles be $M_1 = (-\frac{1}{2}g, 0)$, $M_2 = (\frac{1}{2}g, 0)$, $(g > 0)$; the radii are R_1 and R_2. We take the frame in E such that for $A_1 : x = -\frac{1}{2}h, y = 0$, and for A_2: $x = \frac{1}{2}h$, $y = 0$, $(h > 0)$. The lengths of the four bars are $M_1M_2 = g$, $M_1A_1 = R_1$, $M_2A_2 = R_2$, $A_1A_2 = h$. For the coefficients of the circle with center M_1 we have $C_0 = 1$, $C_1 = \frac{1}{2}g$, $C_2 = 0$, $C_3 = \frac{1}{4}g^2 - R_1^2$, and analogously for the circle centered at M_2. From (7.3) it follows that the corresponding quadrics in Σ' are

$$(8.1)\quad \begin{aligned} \mathrm{H}_1 &\equiv 4(X_1^2 + X_2^2) + 2(g+h)X_1X_3 + 2(g-h)X_2X_4 + [\tfrac{1}{4}(g+h)^2 - R_1^2]X_3^2 \\ &\quad + [\tfrac{1}{4}(g-h)^2 - R_1^2]X_4^2 = 0 \\ \mathrm{H}_2 &\equiv 4(X_1^2 + X_2^2) - 2(g+h)X_1X_3 - 2(g-h)X_2X_4 + [\tfrac{1}{4}(g+h)^2 - R_2^2]X_3^2 \\ &\quad + [\tfrac{1}{4}(g-h)^2 - R_2^2]X_4^2 = 0. \end{aligned}$$

The intersection k_4 of H_1 and H_2 is also the intersection of $\mathrm{H}_1 \pm \mathrm{H}_2 = 0$, that is of

$$(8.2)\quad \begin{aligned} \mathrm{H}_1' &\equiv 8(X_1^2 + X_2^2) + F_1X_3^2 + F_2X_4^2 = 0, \\ \mathrm{H}_2' &\equiv 4(g+h)X_1X_3 + 4(g-h)X_2X_4 + FX_3^2 + FX_4^2 = 0, \end{aligned}$$

with

$$(8.3)\quad \begin{aligned} F_1 &= \tfrac{1}{2}(g+h)^2 - R_1^2 - R_2^2, \\ F_2 &= \tfrac{1}{2}(g-h)^2 - R_1^2 - R_2^2, \\ F &= -R_1^2 + R_2^2. \end{aligned}$$

The eigenvalues of the pencil $\mathrm{H}_1' + \lambda \mathrm{H}_2'$ are the roots of

$$(8.4)\quad \begin{vmatrix} 8 & 0 & 2(g+h)\lambda & 0 \\ 0 & 8 & 0 & 2(g-h)\lambda \\ 2(g+h)\lambda & 0 & F_1 + \lambda F & 0 \\ 0 & 2(g-h)\lambda & 0 & F_2 + \lambda F \end{vmatrix} = 0,$$

that is of

$$[(g+h)^2\lambda^2 - 2F\lambda - 2F_1][(g-h)^2\lambda^2 - 2F\lambda - 2F_2] = 0. \tag{8.5}$$

We introduce the following abbreviations

$$\begin{aligned} S_1 &= -g+h+R_1+R_2, & T_1 &= g+h+R_1+R_2,\\ S_2 &= g-h+R_1+R_2, & T_2 &= g+h-R_1-R_2,\\ S_3 &= g+h-R_1+R_2, & T_3 &= g-h+R_1-R_2,\\ S_4 &= g+h+R_1-R_2, & T_4 &= -g+h+R_1-R_2, \end{aligned} \tag{8.6}$$

$$S = S_1S_2S_3S_4, \qquad T = T_1T_2T_3T_4, \tag{8.7}$$

$$\begin{aligned} G = &-g^4-h^4-R_1^4-R_2^4+2g^2h^2+2g^2R_1^2+2g^2R_2^2\\ &+2h^2R_1^2+2h^2R_2^2+2R_1^2R_2^2. \end{aligned} \tag{8.8}$$

They satisfy the following relations

$$\begin{aligned} S &= G+8ghR_1R_2, & T &= G-8ghR_1R_2,\\ S+T &= 2G, & S-T &= 16ghR_1R_2>0. \end{aligned} \tag{8.9}$$

The discriminants of the two quadratic factors of (8.5) are, in view of (8.3),

$$\begin{aligned} D_1 &= F^2+2(g+h)^2F_1 = T_1T_2S_3S_4,\\ D_2 &= F^2+2(g-h)^2F_2 = S_1S_2T_3T_4. \end{aligned} \tag{8.10}$$

Hence the four roots λ_i of (8.4) are, provided $g \neq h$,

$$\lambda_{1,2} = (F \pm D_1^{1/2})(g+h)^{-2}, \qquad \lambda_{3,4} = (F \mp D_2^{1/2})(g-h)^{-2}. \tag{8.11}$$

For the special case $g = h$ we have (if $R_1 \neq R_2$) $\lambda_3 = (R_1^2+R_2^2)(-R_1^2+R_2^2)^{-1}$, $\lambda_4 = \infty$.

Example 27. Show that $D_1 - D_2 = 8gh\,(g^2+h^2-R_1^2-R_2^2)$.

A four-bar with sides g, h, R_1, R_2 is only physically realizable if the length of every side is less than the sum of the remaining three. Hence we have always $S_i > 0$ $(i = 1,2,3,4)$ and moreover $T_1 > 0$.

The twisted quartic curve is of the general type if the four roots λ_i are mutually different. It is easy to verify that λ_1, λ_2 are both unequal to λ_3 or λ_4. Therefore, two roots are only equal if $D_1D_2 = 0$, which implies $T_2T_3T_4 = 0$; in this case the sum of the lengths of two bars is equal to the sum of the other two.

From (8.10) it follows: if $T_2 > 0$, $T_3T_4 > 0$, all roots are real; if $T_2T_3T_4 < 0$, two are real and two imaginary; if $T_2 < 0$, $T_3T_4 < 0$, all roots are imaginary.

We consider first the general case of four different roots λ_i. For their cross-ratio we obtain

$$\delta = (\lambda_1\lambda_2\lambda_3\lambda_4) = (\lambda_4 - \lambda_2)(\lambda_3 - \lambda_1)(\lambda_3 - \lambda_2)^{-1}(\lambda_4 - \lambda_1)^{-1},$$

and after some algebra

$$\delta = [G - (ST)^{1/2}][G + (ST)^{1/2}]^{-1}, \tag{8.12}$$

or in view of (8.9),

$$\delta = [(S^{1/2} - T^{1/2})(S^{1/2} + T^{1/2})^{-1}]^2, \tag{8.13}$$

by which formula the characteristic cross-ratio of the quartic curve k_4 has been expressed in terms of the lengths of the four bars. If $T > 0$ (that is if the roots are either all real or all imaginary) δ is real; if $T < 0$, δ is a number of the unit circle of the complex plane.

For $\lambda = \lambda_i$ equation (8.4) is satisfied, and $H_1' + \lambda_i H_2' = 0$ $(i = 1, 2, 3, 4)$ are therefore the four quadratic cones through k_4. For the time being we suppose $g \neq h$. Then the four vertices are $[\lambda_1(g + h), 0, -4, 0]$, $[\lambda_2(g + h), 0, -4, 0]$, $[0, \lambda_3(g - h), 0, -4]$, $[0, \lambda_4(g - h), 0, -4]$; they are mutually conjugate points with respect to any quadric through k_4, or, in other words, the vertices of their common polar tetrahedron.

In the following, the equation for the pencil of quadrics through k_4 is studied for the case of four real roots λ_i, the conditions being $T_2 > 0$, $T_3T_4 > 0$. For this case it follows from their definition (8.11) that $\lambda_1 > \lambda_2$, $\lambda_4 > \lambda_3$. Furthermore (8.13) implies that $0 < \delta < 1$, from which it follows that on the λ-axis the intervals $\lambda_2 < \lambda < \lambda_1$ and $\lambda_3 < \lambda < \lambda_4$ have no common points. Further study (for instance, by considering $g \rightarrow h$) shows that the first interval is to the left (right) of the second if $R_1 < R_2$ ($R_1 > R_2$). The two cranks M_1A_1 and M_2A_2 play equivalent roles in the configuration, so we may without any loss of generality suppose $R_1 < R_2$. This implies that λ_i satisfy the following inequalities

$$\lambda_2 < \lambda_1 < \lambda_3 < \lambda_4. \tag{8.14}$$

Example 28. Show that the roots λ_i cannot be real and different if $R_1 = R_2$.

As $T_3 + T_4 = 2(R_1 - R_2)$ the inequality $R_1 < R_2$ implies $T_3 < 0$, $T_4 < 0$ and therefore, by Grashof's well-known criteria, we have a crank-rocker motion.

As λ_i are real, the vertices of the four cones are real. We take them as the vertices of a coordinate tetrahedron. If the new coordinates are Y_i, the (real) transformation reads

$$X_1 = \lambda_1(g+h)Y_1 + \lambda_2(g+h)Y_2,$$

$$X_2 = \lambda_3(g-h)Y_3 + \lambda_4(g-h)Y_4,$$

$$X_3 = -4Y_1 - 4Y_2,$$

$$X_4 = -4Y_3 - 4Y_4.$$

We obtain the following equations for the quadrics (8.2),

$$\text{(8.16)}\qquad \begin{aligned} H_1' &\equiv \lambda_1 D_1^{1/2}Y_1^2 - \lambda_2 D_1^{1/2}Y_2^2 - \lambda_3 D_2^{1/2}Y_3^2 + \lambda_4 D_2^{1/2}Y_4^2 = 0,\\ H_2' &\equiv -D_1^{1/2}Y_1^2 + D_1^{1/2}Y_2^2 + D_2^{1/2}Y_3^2 - D_2^{1/2}Y_4^2 = 0, \end{aligned}$$

in which all mixed terms have vanished. A second transformation

$$\text{(8.17)}\qquad D_1^{1/4}Y_1 = Z_1,\quad D_1^{1/4}Y_2 = Z_2,\quad D_2^{1/4}Y_3 = Z_3,\quad D_2^{1/4}Y_4 = Z_4,$$

reduces (8.16) to the simpler form

$$\text{(8.18)}\qquad \begin{aligned} H_1' &\equiv \lambda_1 Z_1^2 - \lambda_2 Z_2^2 - \lambda_3 Z_3^2 + \lambda_4 Z_4^2 = 0,\\ H_2' &\equiv -Z_1^2 + Z_2^2 + Z_3^2 - Z_4^2 = 0. \end{aligned}$$

k_4 is the intersection of $H_1' + \lambda_3 H_2' = 0$ and $H_1' + \lambda_4 H_2' = 0$, that is of the two cones

$$\text{(8.19)}\qquad \begin{aligned} -Z_1^2 + \alpha^{-2}Z_2^2 + \gamma^{-2}Z_4^2 &= 0,\\ -Z_1^2 + k^2\alpha^{-2}Z_2^2 + \beta^{-2}Z_3^2 &= 0, \end{aligned}$$

where, since $0<\delta<1$ we have replaced δ by k^2, and

$$\alpha^2 = (\lambda_3-\lambda_1)/(\lambda_3-\lambda_2) > 0,\qquad \beta^2 = (\lambda_4-\lambda_1)/(\lambda_4-\lambda_3) > 0,$$

$$\gamma^2 = (\lambda_3-\lambda_1)/(\lambda_4-\lambda_3) > 0,\qquad k^2 = (\lambda_1\lambda_2\lambda_3\lambda_4),\quad 0<k^2<1,$$

in view of (8.14).

The Jacobian functions $\operatorname{sn} t$, $\operatorname{cn} t$ and $\operatorname{dn} t$ satisfy the relations

$$\text{(8.20)}\qquad -1 + \operatorname{sn}^2 t + \operatorname{cn}^2 t = 0,\qquad -1 + k^2\operatorname{sn}^2 t + \operatorname{dn}^2 t = 0.$$

By comparing (8.19) and (8.2) the following parametric representation of k_4 by means of elliptic functions (with modulus k) can be obtained.

$$\text{(8.21)}\qquad Z_1 = 1,\qquad Z_2 = \alpha \operatorname{sn} t,\qquad Z_3 = \beta \operatorname{dn} t,\qquad Z_4 = \gamma \operatorname{cn} t$$

or, by means of the original coordinates,

$$
(8.22)\qquad
\begin{aligned}
X_1 &= (g+h)D_1^{-1/4}(\lambda_1 + \lambda_2\alpha \operatorname{sn} t),\\
X_2 &= (g-h)D_2^{-1/4}(\lambda_3\beta \operatorname{dn} t + \lambda_4\gamma \operatorname{cn} t),\\
X_3 &= -4D_1^{-1/4}(1+\alpha \operatorname{sn} t),\\
X_4 &= -4D_2^{-1/4}(\beta \operatorname{dn} t + \gamma \operatorname{cn} t),
\end{aligned}
$$

which is valid for $D_1 > 0$, $D_2 > 0$, $R_1 < R_2$, $g \neq h$.

If (8.22) is substituted into (1.9) we obtain a parametric representation for the coupler curve generated by any point of the moving plane.

Example 29. Show that (8.22) passes through J_1 and J_2; and that the corresponding values of $\operatorname{sn} t$, $\operatorname{cn} t$, $\operatorname{dn} t$ are $\operatorname{sn} t = -(\lambda_3-\lambda_2)^{1/2}(\lambda_3-\lambda_1)^{-1/2}$; $\operatorname{cn} t = \pm i(\lambda_1-\lambda_2)^{1/2}(\lambda_3-\lambda_1)^{-1/2}$; $\operatorname{dn} t = \mp i(\lambda_1-\lambda_2)^{1/2}(\lambda_4-\lambda_1)^{-1/2}$.

Example 30. Determine the two intersections of (8.22) with V_1, not lying on I; find the centers of the corresponding isotropic rotations.

Example 31. Consider the case $g = h$, excluded so far. Show that the vertices of the four cones are $[2g\lambda_1, 0, -4, 0]$, $[2g\lambda_2, 0, -4, 0]$, $[0, 0, 0, 1]$ and $[0, 1, 0, 0]$. Determine the parametric equations of k_4 for this case.

From the analytical geometry of quadrics, it follows that if the four eigenvalues of a pencil are all real the quartic curve consists of two circuits. This implies that all coupler curves of our crank-rocker motion have the same property.

There are two other cases of four distinct eigenvalues which arise when there are two imaginary or four imaginary roots. These cases can be worked out by means of elliptic functions in a similar manner. The curve k_4 has then only one circuit and so have the corresponding coupler curves.

Example 32. If two roots are imaginary we have the following possibilities: $T_2 > 0$, $T_3 < 0$, $T_4 > 0$; $T_2 > 0$, $T_3 > 0$, $T_4 < 0$; $T_2 < 0$, $T_3 < 0$, $T_4 < 0$. Determine the corresponding types of motion of the four-bars for each of these three sub-cases.

Example 33. If all roots are imaginary there are two sub-cases: $T_2 < 0$, $T_3 < 0$, $T_4 > 0$; $T_2 < 0$, $T_3 > 0$, $T_4 < 0$. Determine the corresponding types of motion.

9. Folding four-bars

In the preceding section we supposed the four roots λ_i to be different. We deal now with the special four-bars for which $T = T_1T_2T_3T_4 = 0$, which implies that the sum of two sides is equal to the sum of the other two; hence the four-bar has a folding position: this is the position at which the points M_1, M_2, A_1, A_2 are collinear. For the time being we suppose that only one of the expressions T_2, T_3, T_4 is equal to zero; then two roots λ_i are equal, but different from both of the other two. In this case the quadrics H_1 and H_2 have a tangent point and their intersection is a quartic curve of the first kind, with a double point and hence a rational curve. The motion is rational, all point-paths are rational tricircular curves of order six.

There are several sub-cases: any of the three expressions T_2, T_3, T_4 can be zero, and further discussion depends on the signs of the other two. All cases may be dealt with in a similar manner. We restrict ourselves to one example: let $T_4 = 0$; then in view of $R_1 < R_2$ we have $T_3 < 0$; let us suppose furthermore $T_2 > 0$. Hence

$$g + R_2 = h + R_1 \qquad (9.1)$$
$$g + h > R_1 + R_2,$$

and therefore $h > R_2$, $g > R_1$, $g - h < 0$.

From this it follows that in the folding position the four vertices are in the order $M_1A_1M_2A_2$ (Fig. 90). We choose in Σ the same frame as before, such that $M_1 = (-\frac{1}{2}g, 0)$, $M_2 = (\frac{1}{2}g, 0)$. If we take the collinear position as the zero-position, we have furthermore $A_1 = (-\frac{1}{2}g + R_1, 0)$, $A_2 = (\frac{1}{2}g + R_2, 0)$ which implies $A_1A_2 = h$. Using these coordinates for A_i, the equations of the quadrics in the image space which express that A_i is on circle $(M_i; R_i)$ are in analogy to (8.1):

$$(9.2)\qquad \begin{aligned} H_1 &\equiv 4(X_1^2 + X_2^2) + 4(g - R_1)X_1X_3 + 4R_1X_2X_4 + g(g - 2R_1)X_3^2 = 0. \\ H_2 &\equiv 4(X_1^2 + X_2^2) - 4(g + R_2)X_1X_3 + 4R_2X_2X_4 + g(g + 2R_2)X_3^2 = 0. \end{aligned}$$

Both quadrics pass through the origin O, and their common tangent plane at this point is $X_2 = 0$. Of the four eigenvalues of the pencil $H_1 + \lambda H_2$, two are equal to $-R_1/R_2$, the others are the roots of

$$R_2^2\lambda^2 - 2(2gh - R_1R_2)\lambda + R_1^2 = 0, \qquad (9.3)$$

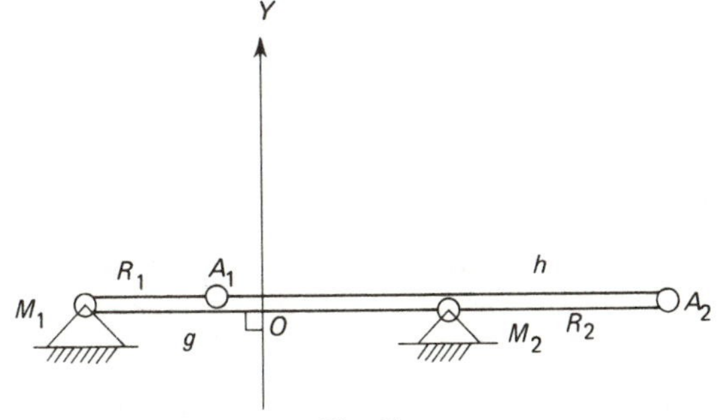

Fig. 90.

the discriminant of which is $4gh(g-R_1)(g+R_2)>0$; hence the two other cones are real. The cone (counted twice) with vertex O has the equation

$$(9.4)\qquad \begin{aligned} R_2\mathrm{H}_1 - R_1\mathrm{H}_2 \equiv 4(R_2-R_1)(X_1^2+X_2^2)+4g(R_1+R_2)X_1X_3 \\ +g(gR_2-gR_1-4R_1R_2)X_3^2=0. \end{aligned}$$

Its generators in the plane $X_2=0$ are

$$(9.5)\qquad X_2=0,\qquad 2(R_2-R_1)X_1+e_{1,2}X_3=0,$$

with

$$(9.6)\qquad e_{1,2}=g(R_1+R_2)\pm 2(ghR_1R_2)^{1/2}.$$

They are the two (real) tangents ρ_1, ρ_2 of k_4 at O; clearly O is therefore a node of k_4. To derive a parametric representation of k_4 we intersect it with the pencil of planes through ρ_1:

$$(9.7)\qquad 2(R_2-R_1)X_2=t[2(R_2-R_1)X_1+e_1X_3],$$

every plane of this pencil intersects k_4 in the point O, counted three times. If (9.7) is substituted into (9.4) the result is

$$(9.8)\qquad [2(R_2-R_1)X_1+e_1X_3][2(R_2-R_1)(1+t^2)X_1+(e_2+e_1t^2)X_3]=0.$$

The second factor gives us, with proportionality factor σ,

$$\sigma X_1=e_2+e_1t^2,\qquad \sigma X_3=-2(R_2-R_1)(1+t^2),$$

while from (9.7) it follows that $\sigma X_2=(e_2-e_1)t$; X_4 may be found after substitution of X_1, X_2, X_3 into the equation of any quadric through k_4, different from (9.4), for example $\mathrm{H}_1-\mathrm{H}_2=0$. This gives us the following representation of the (rational) curve k_4,

$$(9.9)\qquad \begin{aligned} X_1 &= (e_2-e_1)t(e_2+e_1t^2), \\ X_2 &= (e_2-e_1)^2t^2, \\ X_3 &= -2(R_2-R_1)(e_2-e_1)t(1+t^2), \\ X_4 &= -2(1+t^2)[(g+h)(e_2+e_1t^2)+g(R_2^2-R_1^2)(1+t^2)]. \end{aligned}$$

Example 34. Show that (9.9) has a double point at O and passes through J_1, J_2.

Substituting (9.9) into (1.9), we obtain a rational parametric representation of all coupler curves of the motion. As X_3 and X_4 both have the factor $(1+t^2)$ all terms of (1.9) will have this factor, and hence these curves are indeed of order six.

Example 35. Derive the equations for the coupler curves and determine the three values of t at which they all pass through I_1 (I_2).

Example 36. Show that all coupler curves have a node at this zero-position, and determine the tangents at the node for the point (x, y).
Example 37. Show that ρ_1 and ρ_2 correspond to the two instantaneous centers of the motion for the zero-position and determine the coordinates of these centers.

The motion we considered was that of a simple folding four-bar $T_4 = 0$, $T_2 \neq 0$, $T_3 \neq 0$. The cases for which either $T_2 = 0$ or $T_3 = 0$, the other being unequal to zero, can be dealt with in a similar manner.

A more special motion arises if two of the expressions T_i are zero: the four-bar may fold in two different ways. As seen from (8.10), T_3 and T_4 play equivalent roles, different from that of T_2. Therefore two main cases must be considered: $T_2 = 0$ and T_3 or T_4 is zero, and $T_3 = T_4 = 0$.

In the first case we have from (8.10) $D_1 = D_2 = 0$; the λ-equation of H_1 and H_2 has two pairs of equal roots. Hence k_4 has two double points and is therefore degenerate. We suppose $T_2 = 0$, $T_4 = 0$, $T_3 \neq 0$ which implies $R_1 = g$, $R_2 = h$, $h \neq g$; two pairs of adjacent sides of the four-bar are equal.

Making use of the coordinate systems introduced above for the simple folding four-bar, the quadrics (9.2) become

$$\text{(9.10)} \quad \begin{aligned} H_1 &\equiv 4(X_1^2 + X_2^2) + 4gX_2X_4 - g^2X_3^2 = 0, \\ H_2 &\equiv 4(X_1^2 + X_2^2) - 4(g+h)X_1X_3 + 4hX_2X_4 + g(g+2h)X_3^2 = 0, \end{aligned}$$

and the eigenvalues are $\lambda_1 = \lambda_2 = -g/h$, $\lambda_3 = \lambda_4 = g/h$. The two cones of the pencil, each counted twice, are

$$\text{(9.11)} \quad \begin{aligned} hH_1 - gH_2 &\equiv 4(h-g)(X_1^2 + X_2^2) + 4g(h+g)X_1X_3 \\ &\qquad - g^2(g+3h)X_3^2 = 0, \\ hH_1 + gH_2 &\equiv 4(h+g)(X_1^2 + X_2^2) - 4g(h+g)X_1X_3 + 8ghX_2X_4 \\ &\qquad + g^2(h+g)X_3^2 = 0. \end{aligned}$$

Their vertices are respectively O and $O' = (g, 0, 2, 0)$. It is easy to verify that their join $X_2 = 0$, $2X_1 - gX_3 = 0$ is on both cones. Hence k_4 degenerates into the line OO' and a twisted cubic k_3 through O and O'. Any plane $2X_2 = t(2X_1 - gX_3)$ through this chord has one more point in common with k_3, which is seen to be

$$\text{(9.12)} \quad \begin{aligned} X_1 &= g[(h-g)(1+t^2) - 4h]t, \\ X_2 &= -4ght^2, \\ X_3 &= 2(h-g)(1+t^2)t, \\ X_4 &= 2(h+g)(1+t^2), \end{aligned}$$

and thus a parametric representation for k_3 has been found.

Our conclusion is that the motion (of the type with two folding positions being considered) degenerates into a trivial rotation, the image of which is the line OO', and the essential part, a cubic motion whose image is k_3.

Example 38. Verify that the "trivial" rotation is about the center M_2.

For $t = \pm i$, k_3 passes through J_1 and J_2; for $t = 0$ it passes through O, and for $t = \infty$ through O', these being the collinear positions. The tangents of k_3 at J_1 and J_2 are not in V_1 or V_2; hence, in view of the formulas (4.6), the paths of the cubic motion are rational bicircular curves of order four. The coupler curve for any point of E may be found by substituting (9.12) into (1.9), keeping in mind that both X_3 and X_4 have $(1+t^2)$ as a factor.

Example 39. Derive from (9.12) the relation between the parameter t and the rotation angle ϕ of the corresponding displacement. Show that $\phi = 0$ and $\phi = \pi$ correspond to O and O'.

The preceding four-bar with two folding positions was characterized by $T_2 = T_4 = 0$; as we mentioned before the case $T_2 = T_3 = 0$ is completely similar. Quite another case arises if $T_3 = T_4 = 0$, $T_2 \neq 0$, which implies $g = h$, $R_1 = R_2$ $(= R)$; opposite sides of the four-bar are equal. We have $D_1 \neq 0$, $D_2 = 0$, which means $\lambda_3 = \lambda_4$ but λ_1 and λ_2 are not equal.

We now choose the coordinate system as we did in Section 8, for the case of the general four-bar, which implies $M_1 = (-\frac{1}{2}g, 0)$, $M_2 = (\frac{1}{2}g, 0)$, $A_1 = (-\frac{1}{2}g, 0)$, $A_2 = (\frac{1}{2}g, 0)$. The two quadrics (8.1) are now

$$\text{(9.13)} \qquad \begin{aligned} H_1 &\equiv 4(X_1^2 + X_2^2) + 4gX_1X_3 + (g^2 - R^2)X_3^2 - R^2X_4^2 = 0, \\ H_2 &\equiv 4(X_1^2 + X_2^2) - 4gX_1X_3 + (g^2 - R^2)X_3^2 - R^2X_4^2 = 0. \end{aligned}$$

We have $\lambda_3 = \lambda_4 = -1$, and the corresponding quadric $H_1 - H_2 = 0$ is seen to be $X_1X_3 = 0$, which means that this (doubly counted) cone degenerates into the planes $X_1 = 0$ and $X_3 = 0$. Hence k_4 degenerates into two conics k_2 and k_2' (with two common points), and the corresponding motion degenerates into two quadratic motions (such as considered in Section 6). k_2', in $X_3 = 0$, corresponds to a set of translations. k_2 is the image of a symmetric motion, the centrodes of which are conics. There are two cases, depending upon whether $g > R$ or $g < R$. In the first case k_2 is an ellipse while the centrodes are hyperbolas, whereas if $g < R$ k_2 is a hyperbola and the centrodes are ellipses (which is in accordance with the criterion given in Section 6). We recognize in the two aggregrates of k_4 the images of a parallelogram and an antiparallelogram motion of E. The complete path of a point consists of a circle and a rational bicircular curve of order four.

The most special four-bar is that for which $T_2 = T_3 = T_4 = 0$, which means

$g = h = R_1 = R_2$. In (9.13) the term X_3^2 vanishes; k_1 degenerates into two straight lines, corresponding to rotations. Every path consists of three circles.

Example 40. The reader who is familiar with the projective classification of quartic space curves by means of elementary divisors can verify that the general four-bar corresponds to the Segre symbol [1 1 1 1]; the singly folding corresponds to [2 1 1], and the doubly folding of the first kind to [2 2], that of the second kind to [(1 1) 1 1], and the last one to [(1 1) 2].

10. The slider-crank motion

Until now we have dealt with four-bars where two points of E describe non-degenerate circles in Σ. We consider now the case in which one circle is a straight line.

In Section 7 we proved that the image surface of those positions of E for which one of its points is constrained to remain on a line of Σ is a hyperboloid H^*, not only passing through J_1 and J_2 but, moreover, having V_2 and V_1 as tangent planes at these points; l is a generator of H^*. If the point A_1 of E remains on a non-degenerate circle $c_1 = (M; R)$ and the point A_2 on the straight line m, the corresponding motion is a slider-crank motion. It is that special case of a four-bar in which M_2 is at infinity (in the direction perpendicular to m) and $R_2 = \infty$. The image curve k of such a motion is the intersection of a general hyperboloid of type H and a special one H^*. Hence k is a quartic space curve of the first kind, elliptic in general, passing through J_1 and J_2, such that its tangents r and r′ at these points are in V_2 and V_1 respectively. In view of Table (4.6), the curve k is one of type V. Hence $N = 4$, $c = 1$, $d = 2$; the paths for the slider-crank motion are elliptic curves of order 4, with simple points at I_1, I_2, and with two other points at infinity which are in general different for different paths. In order to describe the motion analytically we take the frames in Σ and E such that c_1 is the circle $(0,0; R)$, m the line $Y - h' = 0$ $(h' \geq 0)$, $A_1 = (0,0)$, $A_2 = (h, 0)$. Then the quadrics on which the image point is situated are (in view of (7.3)),

$$\text{(10.1)} \qquad H \equiv X_1^2 + X_2^2 - \tfrac{1}{4}R^2(X_3^2 + X_4^2) = 0,$$

$$\text{(10.2)} \qquad H^* \equiv -2X_2X_3 + 2X_1X_4 - 2hX_3X_4 + h'(X_3^2 + X_4^2) = 0.$$

The eigenvalues λ_i of the pencil $RH^* + 2\lambda H = 0$ are the roots of $|RH^* + 2\lambda H| = 0$, or

$$\text{(10.3)} \quad \lambda^4 - 4(h'/R)\lambda^3 + (2/R^2)(R^2 - 2h^2 + 2h'^2)\lambda^2 - 4(h'/R)\lambda + 1 = 0,$$

which is a reciprocal equation for λ. Introducing the expressions

$$
U_1 = h + h' + R, \qquad U_2 = h + h' - R,
$$
$$
U_3 = h - h' + R, \qquad U_4 = h - h' - R, \tag{10.4}
$$

we obtain

$$
R\lambda_1 = h' + h + (U_1U_2)^{1/2}, \qquad R\lambda_3 = h' - h + (U_3U_4)^{1/2},
$$
$$
R\lambda_2 = h' + h - (U_1U_2)^{1/2}, \qquad R\lambda_4 = h' - h - (U_3U_4)^{1/2}. \tag{10.5}
$$

By suitable transformation of the coordinates X_i, the equations of H and H* can now be reduced to sums of squares, and a parametric representation of the image curve can be derived by means of elliptic functions in a manner similar to that used for the general four-bar. From this would follow the parametric equations for any path.

It is easy to see that the slider-crank mechanism is real if and only if $h \geqslant |R - h'|$. From this it follows $U_1 > 0$, $U_2 \geqslant 0$, $U_3 \geqslant 0$. Hence there are two main cases: $U_4 > 0$, $U_4 < 0$. In the first case all λ_i are real, in the second λ_1, λ_2 are real, λ_3, λ_4 are conjugate imaginary. In the first case k_4 consists of two circuits and the same holds for any path; in the latter there is only one circuit.

Example 41. Show that the cross-ratio of the image curve is

$$
[-h'^2 + h^2 + R^2 - (U_1U_2U_3U_4)^{1/2}][-h'^2 + h^2 + R^2 + (U_1U_2U_3U_4)^{1/2}]^{-1}.
$$

Example 42. Show that λ_i are a harmonic quadruple if $h'^2 = h^2 + R^2$.
Example 43. Show that for $U_4 < 0$ there exists a triangle with sides h', h, R.
Example 44. The path of a point $A(x, y)$ of E has two non-isotropic points Q_1, Q_2 at infinity. Show that they are determined by $(x^2 + y^2 - 2hx + h^2)X^2 + 4hyXY + (x^2 + y^2 + 2hx + h^2)Y^2 = 0$. Determine the points Q_1, Q_2 for A_1, and for A_2. Prove that the points Q_1, Q_2 are, in general, conjugate imaginary, and that the locus of the points of E for which Q_1, Q_2 coincide is the circle $(A_1; h)$ (compare Example 20). The path of such a point has an isolated double point at infinity; determine this point.

A special slider-crank is one in which m passes through the center of the circle M; it is the so-called in-line type. Then $h' = 0$ and the equation of H and H*, and the roots λ_i are simplified. It belongs to the first case if $R > h$, and to the second if $R < h$. The image curve and the paths are still elliptic.

Example 45. Determine the roots λ_i, the reduced equations of H, H*, and the parametric representation of k and of all paths, in the case $h' = 0$.
Example 46. Show that the cross-ratios of k in the case $h' = 0$ read h^2/R^2, R^2/h^2, $-u^2/R^2$, $-R^2/u^2$, u^2/h^2, h^2/u^2, with $u^2 = h^2 - R^2$.

The slider-crank has a folding position if $U_4 = h - h' - R = 0$. Then there exists a position at which M_1, A_1, A_2 and the point M_2 at infinity are collinear. We have now in view of (10.4): $\lambda_3 = \lambda_4$. The image curve has a double point (corresponding to the folding position); k and all paths are rational.

Example 47. Derive a rational parametric representation of k and of any point-path, for a folding slider-crank.
Example 48. Discuss the cases in which one or more of the expressions U_2, U_3, U_4 are zero.

11. Elliptic motion

A still more special case appears if both circles of the four-bar configuration are straight lines. If A_1 is constrained to move on the line m_1 in Σ, and A_2 on the line m_2, the image curve of the motion is the intersection of two special hyperboloids H_1^* and H_2^*. Since both hyperboloids have l as a generator (the points of l do not correspond to positions of E) this image curve is in general a twisted cubic k_3.

Furthermore, since both H_1^* and H_2^* have at J_1 the tangent plane V_2 and at J_2 the tangent plane V_1, their intersection k_3 passes through J_1 and J_2, its tangents r, r′ at these points lie in V_2 and V_1. Hence k_3 is a curve of type V of Table (4.6); and as $n = 3$ we obtain $N = 2$, $c = 0$, $d = 2$. The paths of the motion are conics and their points at infinity Q_1, Q_2 are in general different for different paths. All quadrics of the pencil $H_1^* + \lambda H_2^*$ are special hyperboloids H^*, and as k_3 lies on all of these, there are ∞^1 points of E which remain on a line of Σ. Furthermore the equation (7.3) is linear in C_i and therefore these lines are those of a pencil of lines in Σ.

Omitting the trivial case where m_1, m_2 are parallel, we may without loss of generality suppose them to be the lines $Y = 0$ and $X = 0$. If $A_1 = (p, 0)$, $A_2 = (-p, 0)$ we have from (7.3)

$$H_1^* \equiv -X_1X_4 + X_2X_3 + pX_3X_4 = 0,$$
(11.1)
$$H_2^* \equiv X_1X_3 + X_2X_4 + \tfrac{1}{2}p(X_3^2 - X_4^2) = 0.$$

Example 49. Show by means of (11.1) and (7.3) that the locus of points A in E the paths of which are straight lines is the circle $x^2 + y^2 = p^2$.
Example 50. Intersect H_1^*, H_2^* by the pencil of planes $X_3 = tX_4$ and show that

$$X_1 = pt(3 - t^2), \qquad X_2 = p(1 - 3t^2),$$
$$X_3 = 2t(1 + t^2), \qquad X_4 = 2(1 + t^2)$$

is a parametric representation of k_3.
Example 51. Prove that apart from the special hyperboloids H_1^*, H_2^* the only hyperboloid through k_3 tangent to V_1, V_2 reads

$$4(X_1^2 + X_2^2) - p^2(X_3^2 + X_4^2) = 0;$$

show from (7.3) that the path of $A_0(0, 0)$ is a circle.
Example 52. Determine the points Q_1, Q_2 at infinity of the path of the point $A(x, y)$ of E;

prove that they are in general conjugate imaginary; they coincide only if the path of A is a straight line.

Example 53. Show that the twisted cubic

$$X_1 = -Rt^2, \qquad X_2 = Rt^3, \qquad X_3 = t(1+t^2), \qquad X_4 = (1+t^2),$$

is the image curve of an elliptic motion. Prove that it passes through O, its tangent there is $X_1 = X_2 = 0$ and its osculating plane $X_2 = 0$.

The following table summarizes the most important results of Sections 8–11.

	Image curve k	Coupler curve in Σ order	genus	circularity	
g, h, R_1, R_2 are finite	$T = T_2T_3T_4 \neq 0$, k is a quartic of genus *one*.	6	1	3	general four-bar
g, h, R_1, R_2 are finite	*One* $T_i = 0$, k is a rational quartic.	6	0	3	simple foldable four-bar
g, h, R_1, R_2 are finite	$T_2 = 0$ and either $T_3 = 0$ or $T_4 = 0$, k degenerates into a twisted cubic and a chord.	4 2	0 0	2 1	twice foldable four- bar; two pairs of adjacent sides are equal
g, h, R_1, R_2 are finite	$T_2 \neq 0$, $T_3 = T_4 = 0$, k degenerates into two conics.	4 2	0 0	2 1	twice foldable four- bar; opposite sides are equal
g, h, R_1, R_2 are finite	$T_2 = T_3 = T_4 = 0$, k degenerates into a conic and two lines.	2 2 2	0 0 0	1 1 1	the four sides are equal
	$g = \infty$, *one* $R_i = \infty$, $h \neq R + h'$, k is a quartic curve of genus one; the tangents at J_1, J_2 are in V_2 and V_1.	4	1	1	general slider-crank
	$g = \infty$, *one* $R_i = \infty$, $h = R + h'$, k is a rational quartic; the tangents at J_1, J_2 are in V_2 and V_1.	4	0	1	foldable slider-crank
	$g = \infty$, $R_1 = R_2 = \infty$, k is a twisted cubic, the tangents at J_1, J_2 are in V_2 and V_1.	2	0	0	elliptic motion

12. The inverse of a four-bar motion

The general relationship between a motion and its inverse, as it presents itself in the image space, will be dealt with in Section 14. We consider here the inverse of the four-bar motion. If the point A of E is constrained to move on the circle $c(M; R)$ in Σ, the relative motion of E and Σ is such that the point A of E and the point M of Σ remain at the constant distance R. Hence the inverse motion is such that M of Σ moves on the circle $(A; R)$ in E. Its image surface is therefore a hyperboloid H of the set (7.3).

We have seen in Section 10 that if c is a straight line this surface is a special one: it is a hyperboloid H* tangent to V_1 and V_2 at J_2 and J_1 respectively.

The inverse of this motion, however, is different from the direct motion. It is defined by the condition that a line m of E passes permanently through a fixed point P of Σ. Let m be the line $ux + vy + wz = 0$, P the point a_1, b_1, c_1. Making use of the formulas of the inverse motion (Section 1; Example 1) the condition is seen to be

$$\begin{aligned} \mathrm{H}^{**} \equiv 2c_1u(X_1X_3 - X_2X_4) + 2c_1v(X_2X_3 + X_1X_4) + 2(b_1u - a_1v)X_3X_4 \\ -(a_1u + b_1v - c_1w)X_3^2 + (a_1u + b_1v + c_1w)X_4^2 = 0, \end{aligned} \tag{12.1}$$

which represents a hyperboloid passing through J_1, J_2 such that the tangent planes at these points are V_1 and V_2 respectively. Hence, in comparison with H*, the planes are interchanged.

This result enables us to derive the image curves of some classical plane motions.

If the point A of E is constrained to move on a circle in Σ and a line m of E to pass through a point P of Σ, the image curve of the motion is the intersection of a general hyperboloid H and a special one H**. The image curve is then a quartic curve k_4, which is elliptic in general and passes through J_1, J_2, the tangents at these points lie in V_1, V_2 respectively. Hence k_4 is of type IV in Table (4.6). We have $n = 4$, $N = 6$, $c = 3$, $d = 0$. This motion is the inverse of the slider-crank motion. Its paths are elliptic, tricircular curves of order six.

If a point A of E moves on a line s_1 of Σ, and a line s of E passes through a point P of Σ, the image curve is the intersection of a special hyperboloid H* and a special hyperboloid H**. As both have l as a generator, the intersection is a twisted cubic k_3. To discuss its position with respect to l we introduce coordinates systems such that $A = (0, a', 1)$, $P = (0, p, 1)$, $s = y = 0$, $s_1 = Y = 0$. Then from (7.3)

$$\mathrm{H}^* \equiv 2X_2X_3 - 2X_1X_4 - a'(X_3^2 - X_4^2) = 0,$$

and from (12.1)

$$\text{(12.2)} \qquad \mathrm{H}^{**} \equiv 2X_2X_3 + 2X_1X_4 - p(X_3^2 - X_4^2) = 0.$$

$\mathrm{H}^* + \mathrm{H}^{**} = 0$, $\mathrm{H}^* - \mathrm{H}^{**} = 0$ are two quadratic cones through k_3, with vertices $S_1 = (1,0,0,0)$ and $S_2 = (0,1,0,0)$. Hence k_3 intersects l at these points and is seen to be of type II in Table (4.6), with $m_1 = 2$. It follows from this that $n = 3$, $N = 4$, $c = 1$, $d = 2$. The paths are rational circular curves of order four; the two non-isotropic points at infinity are the same for all paths. The inverse of the motion is of the same type.

Example 54. Show that a parametric representation of the k_3 under consideration is

$$X_1 = (a' - p)t(1 - t^2),$$
$$X_2 = -(a' + p)(1 - t^2),$$
$$X_3 = 4t^2,$$
$$X_4 = 4t.$$

Example 55. Derive the parametric equations for the path of any point of E, and show that the common non-isotropic points at infinity coincide at a double point.
Example 56. Discuss the special cases $a' = 0$, $p = 0$, $a' = \pm p$.

Our final application deals with the following situation: two lines of E are constrained to pass through two given points of Σ. The motion is the inverse of the elliptic motion. The image curve is the intersection of two special hyperboloids H_1^{**}, H_2^{**}; it is a twisted cubic k_3: passing through J_1, J_2 with tangents lying in V_1, V_2. It is of type III of Table (4.6). We have $n = 3$, $N = 4$, $c = 2$, $d = 0$. The paths are rational, bicircular curves of order four.

Example 57. Introduce suitable frames in E and Σ, derive the equations of H_1^{**} and H_2^{**}, a parametric representation of the image curve k_3 and one for an arbitrary path.

13. Cycloidal motions

A cycloidal motion is defined as a motion for which both the fixed and moving centrodes are circles. Let the fixed circle, in Σ, be $(M; R)$ and the moving one, in E, be the circle $(m; r)$. The zero-position is taken as the position at which the m_y-axis and the M_Y-axis coincide. We take the two circles to be externally tangent (Fig. 84, Chapter IX, Section 11, P. 351). $\angle A_0MB_1 = \psi$, arc $A_0B_1 =$ arc $A_1B_1 = R\psi$. In the new position E has been rotated from the zero-position by the angle $\phi = \psi + (R/r)\psi$; hence $\psi = (r/(R+r))\phi$. The coordinates of m are $X = -(R+r)\sin\psi$, $Y = (R+r)\cos\psi$. It follows from this that the motion is given by

$$
\begin{aligned}
X &= x\cos\phi - y\sin\phi + a,\\
Y &= x\sin\phi + y\cos\phi + b,\\
a &= -(R+r)\sin(r\phi/(R+r)),\\
b &= (R+r)\cos(r\phi/(R+r)).
\end{aligned}
\tag{13.1}
$$

So in view of (1.6) the image point of the position is

$$
\begin{aligned}
X_1 &= -(R+r)\cos(p\phi/2),\\
X_2 &= (R+r)\sin(p\phi/2),\\
X_3 &= 2\sin(\phi/2),\\
X_4 &= 2\cos(\phi/2)
\end{aligned}
\tag{13.2}
$$

with

$$p = (R-r)/(R+r). \tag{13.3}$$

For variable values of ϕ these equations are a parametric representation of the image curve k. The same expressions hold if the moving circle is internally tangent to the fixed one, provided that we take r negative. We remark that p is always a finite number, since $R + r = 0$ would mean that the circles coincide. The center m of the moving circle describes the circle $(M; R + r)$. Hence, as follows from Section 7, *the image curve of a cycloidal motion is on a quadric of the set* H. Indeed, (13.2) satisfies (7.3) which for this case becomes

$$\mathrm{H} \equiv X_1^2 + X_2^2 - \tfrac{1}{4}(R+r)^2(X_3^2 + X_4^2) = 0. \tag{13.4}$$

We have supposed for the time being that R and r are finite numbers. There are two limiting cases: $R = \infty$ and $r = \infty$. In the first case a circle of radius r rolls on a straight line ("simple cycloidal motion"), in the second, the inverse of the first, a straight line rolls on a fixed circle of radius R.

Example 58. If the "circle" in Σ is the line $Y = 0$, and that in E the circle $(m; r)$, show that if the zero-position is suitably chosen the image curve k is given by

$$
\begin{aligned}
X_1 &= -r\phi\sin(\tfrac{1}{2}\phi),\\
X_2 &= -r\phi\cos(\tfrac{1}{2}\phi),\\
X_3 &= 2\sin(\tfrac{1}{2}\phi),\\
X_4 &= 2\cos(\tfrac{1}{2}\phi),
\end{aligned}
$$

a curve on

$$\mathrm{H}^* \equiv X_1X_4 - X_2X_3 = 0.$$

Its intersections with V_1 are on the generator of H^* in V_1, passing through J_2; for $\phi \to \infty$ the intersection approaches J_2.

Example 59. Show that the image curve of the inverse motion reads

$$X_1 = r\phi \sin(\tfrac{1}{2}\phi),$$

$$X_2 = -r\phi \cos(\tfrac{1}{2}\phi),$$

$$X_3 = 2\sin(\tfrac{1}{2}\phi),$$

$$X_4 = 2\cos(\tfrac{1}{2}\phi).$$

Its intersections with V_1 are on a line through J_1.

The cycloidal motion is in general transcendental, for its image curve has in general infinitely many intersections with a plane. We obtain an algebraic motion if and only if, in (13.2), the number p is rational. The image curve is then an algebraic curve on H.

There are on H two systems of generators, G_1 and G_2. It is easily seen that an algebraic curve on H has the same number of intersections, s_i say, with any line of $\mathbf{G}_i$ $(i = 1, 2)$, the degree of the curve being $s_1 + s_2$.

The meaning of the generators for the E, Σ configuration has been investigated in Section 3. Let g_1, g_2 be the generators through the image point of the zero-position. Then the points of g_1 correspond to the displacements of E for which m_0 is the rotation center; those of g_2 are the displacements with rotation center M (Example 22). Hence s_1 is the number of times, during the complete cycloidal motion, that m passes through its initial position m_0; s_2 is the number of times a position of E follows from the initial one by a rotation about M. Therefore, if $|R|:|r| = q_1 : q_2$, q_1 and q_2 being positive, relatively prime integers, we have $s_1 = q_2$, $s_2 = q_1$. *The image curve of an algebraic cycloidal motion is an algebraic curve on the hyperboloid* H, *of order* $q_1 + q_2$, *intersecting the generators of one system* q_1 *times and those of the other system* q_2 *times.*

Depending on the signs of R, r, $R + r$ and $R - r$, different types of cycloidal motion may be considered.

If $R = r$ we have $p = 0$ and in (13.2) $X_2 = 0$; the image curve is a plane section of H, and consequently a conic; we recognize the quadratic motion dealt with in Section 6. The corresponding cycloidal motion appears if a circle rolls externally on a congruent circle. Obviously we have $q_1 = q_2 = 1$.

The case $R + r = 0$ is not a possible one, because the two circles coincide. As seen from (13.4) H would be a pair of conjugate imaginary planes.

We may from now on suppose $R - r$ and $R + r$ to be unequal to zero, whence $q_1 \neq q_2$.

Consider first the case $R > 0$, $r > 0$, $R - r > 0$, and consequently $q_1 > q_2$.

We introduce in (13.2) the parameter η defined by $\phi = (q_1 + q_2)\eta$ and we obtain

$$(13.5)\qquad \begin{aligned} X_1 &= -(R+r)\cos(n_1\eta/2),\\ X_2 &= (R+r)\sin(n_1\eta/2),\\ X_3 &= 2\sin(n_2\eta/2),\\ X_4 &= 2\cos(n_2\eta/2), \end{aligned}$$

in which $n_1 = q_1 - q_2$, $n_2 = q_1 + q_2$. Consider first the case that q_1, q_2 have the same parity, that is that both are odd. Then n_1 and n_2 are even numbers, and we can use the trigonometric identities

$$(13.6)\qquad \begin{aligned} \cos \upsilon\eta &= ((1+it)^{2\upsilon} + (1-it)^{2\upsilon})/(2(1+t^2)^{\upsilon}),\\ \sin \upsilon\eta &= ((1+it)^{2\upsilon} - (1-it)^{2\upsilon})/(2i(1+t^2)^{\upsilon}), \end{aligned}$$

where $t = \tan(\eta/2)$. Substituting into (13.5) and multiplying X_i by $(1+t^2)^{n_2/2}$ to remove the denominators, the result is

$$(13.7)\qquad \begin{aligned} X_1 &= -(R+r)(1+t^2)^{q_2}\mathrm{P}_1(t),\\ X_2 &= (R+r)(1+t^2)^{q_2}\mathrm{P}_2(t),\\ X_3 &= \mathrm{P}_3(t),\\ X_4 &= \mathrm{P}_4(t), \end{aligned}$$

in which P_1, P_2, P_3, P_4 stand for polynomials in t of degrees $q_1 - q_2$, $q_1 - q_2 - 1$, $q_1 + q_2 - 1$ and $q_1 + q_2$ respectively. We have obtained a rational parametric representation of k; the curve is indeed seen to be of order $q_1 + q_2$. It passes through the points $K_1 = (0,0,1,-i)$ and $K_2 = (0,0,1,i)$, the tangent points of H and the planes V_1 and V_2 respectively. Moreover any intersection of V_1 and k satisfies $\tan(n_2\eta/2) = i$, the n_2 roots of which are all $t = i$. The conclusion is that the $q_1 + q_2$ common points of k and V_1 coincide with K_1; K_1J_1 has q_1, and K_1J_2 has q_2 points in common with k.

If q_1 and q_2 have different parity, one being even and the other odd, n_1 and n_2 are both odd. The argument must be modified somewhat but leads to the same result. The case $R > 0$, $r > 0$, $R - r < 0$ may be treated along the same lines if we make use now of $n_1 = q_1 - q_2$, $n_2 = q_1 + q_2$, putting a minus sign before X_2.

Summarizing we have: *if the centrodes of an algebraic cycloidal motion are externally tangent its image curve* k *is a rational* (q_1, q_2)*-curve on* H, *having* $q_1 + q_2$ *points in common with* V_i $(i = 1, 2)$ all of which coincide at K_i.

Example 60. Show that the paths in Σ are rational (q_1+q_2)-circular curves of order $2(q_1+q_2)$; they all have the same set of tangents at $I_1(I_2)$.

The situation is different if R and r have opposite signs, that is if the centrodes are internally tangent. If $R>0$, $r<0$, $R+r>0$, the numbers n_1, n_2 are positive, but we have now $n_1>n_2$. The roles of the coordinates, X_1, X_2 on one hand and X_3, X_4 on the other hand, are interchanged. The plane W_1 with equation $X_2-iX_1=0$ has q_1+q_2 points in common with k, all coinciding at J_1; the intersections of k and $\mathrm{W}_2 \equiv X_2+iX_1=0$ coincide with J_2. Of the tangents of k at J_1, q_1-q_2 coincide with J_1K_1 and none with J_1K_2. Similar results follow if $R>0$, $r<0$, $R+r<0$. *If the centrodes are internally tangent and* $|R|:|r|=q_1:q_2$, *the image curve is a rational* (q_1,q_2)*-curve on* H; if $q_1>q_2$ it passes q_2 times through J_1 and J_2, q_1-q_2 tangents at J_1 (J_2) are in V_1 (V_2); if $q_1<q_2$ k passes q_1 times through J_1, J_2 with q_2-q_1 tangents in V_2, V_1 respectively.

Example 61. Determine the type of k (in Table (4.6)) for the two cases $q_1>q_2$ and $q_1<q_2$.
Example 62. Determine the order and the circularity of the paths in both cases.
Example 63. Show that if $R:r=2:1$, we have $q_1=2$, $q_2=1$, $n_1=1$, $n_2=3$, $\cos 3\eta = \cos\eta\,(1-4\sin^2\eta)$, $\sin 3\eta=\sin\eta\,(3-4\sin^2\eta)$, and the image curve is the twisted cubic

$$X_1=-3r(1+t^2),$$

$$X_2=3rt(1+t^2),$$

$$X_3=2t(3-t^2),$$

$$X_4=2(1-3t^2).$$

Hence, the paths are rational, tricircular curves of order six.
Example 64. Show that if $R:r=2:-1$, k is the twisted cubic

$$X_1=r(1-3t^2),$$

$$X_2=-rt(3-t^2),$$

$$X_3=2t(1+t^2),$$

$$X_4=2(1+t^2),$$

and the paths are conics; the motion is the elliptic motion.
Example 65. Consider the case $R:r=1:-2$.
Example 66. Investigate the relation between the image curve of a cycloidal motion and that of its inverse motion.

14. The inverse motion

In our mapping of a motion we considered in Section 1 the displacements of the plane E with respect to the fixed plane Σ. We may interchange the roles of the two planes and consider the motion of Σ with respect to E: the inverse motion, as distinguished from the original, the direct motion. One motion determines the other one; the inverse of the inverse motion is the direct motion.

The formulas (1.2) define the image of the direct displacement. The inverse displacement has the same rotation center but the opposite rotation angle. Therefore, if $Q(X_1, X_2, X_3, X_4)$ is the image of the direct motion and $\tilde{Q}(\tilde{X}_1, \tilde{X}_2, \tilde{X}_3, \tilde{X}_4)$ that of its inverse we have

$$\tilde{X}_1 = X_1, \qquad \tilde{X}_2 = X_2, \qquad \tilde{X}_3 = X_3, \qquad \tilde{X}_4 = -X_4. \tag{14.1}$$

This is a transformation of the image space which may be called the reflection into the origin O. It is however not a Σ'-displacement: the points J_1, J_2 are each invariant for (14.1) but *the planes* V_1 *and* V_2 *are interchanged.* If k is the image curve of a motion, that of the inverse motion, denoted by $\tilde{\mathrm{k}}$, follows from k by (14.1). This implies that the two motions have much in common; k and $\tilde{\mathrm{k}}$ are projectively equivalent. They have the same order, the same genus, the same number and types of special points, and so on. But we know that the properties of the motion depend not only on these characteristics, but also on the behavior of k with respect to J_1, J_2, V_1, V_2. It follows immediately from (14.1) that k and $\tilde{\mathrm{k}}$ have identical positions with respect to J_1 and J_2, but the position of $\tilde{\mathrm{k}}$ with respect to V_1 (V_2) is the same as that of k with respect to V_2 (V_1). Consulting the fundamental Table (4.6) we see at once that if k is either of the types I, II or III, $\tilde{\mathrm{k}}$ will be of the same type. But this does not hold for types IV and V. First of all we have to split type IV into two sub-cases: the case IV′, $S = J_1$, $S' = J_2$, r in V_1, r′ in V_2, and the case IV″, covering all cases of IV different from IV′. Then we have obviously: if k is of type IV″, $\tilde{\mathrm{k}}$ is of that same type, and inversely too. But if k is of type IV′, $\tilde{\mathrm{k}}$ is of type V and if k is of type V, $\tilde{\mathrm{k}}$ is of type IV′.

Hence if k has m_j intersections with l of type j ($j = 2, 3, 5$), m'_4 of type IV′, m''_4 of type IV″, we have

$$\begin{aligned} \tilde{m}_j &= m_j \quad (j = 2, 3), \\ \tilde{m}''_4 &= m''_4, \\ \tilde{m}'_4 &= m_5, \\ \tilde{m}_5 &= m'_4, \end{aligned} \tag{14.2}$$

and accordingly, in view of (4.7), we obtain the following formulas relating the order and the circularity of the paths of the direct and the inverse motion

$$\tilde{N} = N + 2(m_5 - m'_4), \quad \tilde{c} = c + 2(m_5 - m'_4), \quad \tilde{d} = d - 2(m_5 - m'_4); \tag{14.3}$$

which shows that the paths of the direct and the inverse motions have different orders and different circularities in all cases for which $m'_4 \neq m_5$.

Example 67. Show that $\tilde{\tilde{N}} = N$, $\tilde{\tilde{c}} = c$, $\tilde{\tilde{d}} = d$.
Example 68. Check the formulas (14.3) when the direct motion is the elliptic motion.
Example 69. Check (14.3) for the case when the direct motion is the general algebraic cycloidal motion.
Example 70. Check (14.3) for the slider-crank motion.

15. The centrodes

If two positions of E with respect to Σ are represented by the image points $Q_1(X_1, X_2, X_3, X_4)$ and $Q_2(Y_1, Y_2, Y_3, Y_4)$ we can find using (1.9), that the center, in Σ, of the rotation which transforms one position into the other is given by

$$
\begin{aligned}
X_p &= (X_1 Y_4 - X_4 Y_1) - (X_2 Y_3 - X_3 Y_2),\\
Y_p &= (X_1 Y_3 - X_3 Y_1) - (X_4 Y_2 - X_2 Y_4),\\
Z_p &= X_3 Y_4 - X_4 Y_3,
\end{aligned}
\tag{15.1}
$$

or, if p_{ij} are the Plücker coordinates of the line Q_1Q_2,

$$
X_p = p_{14} - p_{23}, \qquad Y_p = p_{13} - p_{42}, \qquad Z_p = p_{34}.
\tag{15.2}
$$

Let Q(t) be the image curve k of a motion, t being a parameter. If Q_2 is the point $Q(t+\Delta t)$, the center of the rotation which transforms Q_1 into Q_2 approaches the instantaneous center (or centro, or first order pole) for the position Q_1 as $\Delta t \to 0$. This implies that the *fixed centrode* of the motion is given by

$$
\begin{aligned}
X_f &= (X_1 X'_4 - X_4 X'_1) - (X_2 X'_3 - X_3 X'_2),\\
Y_f &= (X_1 X'_3 - X_3 X'_1) - (X_4 X'_2 - X_2 X'_4),\\
Z_f &= X_3 X'_4 - X_4 X'_3,
\end{aligned}
\tag{15.3}
$$

the prime denoting differentiation with respect to the parameter t. As *the moving centrode* is the fixed centrode of the inverse motion it is according to (14.1),

$$
\begin{aligned}
x_m &= (X_1 X'_4 - X_4 X'_1) + (X_2 X'_3 - X_3 X'_2),\\
y_m &= -(X_1 X'_3 - X_3 X'_1) - (X_4 X'_2 - X_2 X'_4),\\
z_m &= X_3 X'_4 - X_4 X'_3.
\end{aligned}
\tag{15.4}
$$

Making use of (15.2) the fixed centrode is given by

$$X_f = p_{14} - p_{23}, \qquad Y_f = p_{13} - p_{42}, \qquad Z_f = p_{34}, \tag{15.5}$$

and the moving centrode by

$$x_m = p_{14} + p_{23}, \qquad y_m = -p_{13} - p_{42}, \qquad z_m = p_{34}, \tag{15.6}$$

where $p_{ij}(t)$ are the Plücker coordinates of the tangent to k at the point $Q(t)$. If the tangent intersects the line l ($p_{34} = 0$) the motion is instantaneously a translation and the centro is at infinity.

The centro would be undetermined if in (15.5) $X_f = Y_f = Z_f = 0$. Hence $p_{14} = p_{23}$, $p_{13} = p_{42}$, $p_{34} = 0$, which implies in view of the fundamental relation, $p_{14}p_{23} + p_{13}p_{42} + p_{12}p_{34} = 0$, either

$$p_{14} = p_{23} = 1, \qquad p_{13} = p_{42} = i, \qquad p_{34} = 0, \tag{15.7}$$

or

$$p_{14} = p_{23} = 1, \qquad p_{13} = p_{42} = -i, \qquad p_{34} = 0, \tag{15.8}$$

both times with arbitrary p_{12}. The lines (15.7) are those of the pencil w_1 in V_1 with vertex J_2; those of (15.8) are the conjugate imaginary lines w_2 in V_2 through J_1.

The set of tangents of the image curve k constitute the tangent surface T of k; it is algebraic if k is algebraic. If T does not contain any line of w_1, w_2 all tangents correspond to a well-determined centro. The order of the fixed centrode is equal to the number of solutions of a linear equation between the X_f, Y_f, Z_f of (15.5); that is to say it is equal to the number of tangents of k intersecting a straight line. This number r, the *rank* of k, is also the order of T. If, however, for $t = t_0$ the tangent $p_{ij}(t)$ belongs to the pencil w_1 or w_2, the coordinates X_f, Y_f, Z_f of (15.5) have the factor $(t - t_0)$ which diminishes the order of the centrode by one. Similar arguments apply to the moving centrode. Summarizing we have: if r is the rank of the image curve k of a motion, and if s_1 tangents of k belong to the pencils $(V_1; J_2)$ and to $(V_2; J_1)$, s_2 tangents to $(V_1; J_1)$ and to $(V_2; J_2)$, the order of the fixed centrode is $r - 2s_1$; that of the moving centrode is $r - 2s_2$.

Example 71. It is well-known that the rank of the intersection of two quadrics is equal to eight; determine the order of the centrodes of a general four-bar motion.

Example 72. Check the order of the centrodes of the elliptic motion and of its inverse motion.

Example 73. Show that for the general rational motion of order n, the order of both centrodes is $2n - 2$. If the image of a symmetrical motion is a plane curve k, verify that the centrode's order equals the class of k.

The fixed centrode (15.5) passes through I_1 if $p_{34} = 0$ and $p_{13} - p_{42} =$

$i(p_{14}-p_{23})$. This implies that the line p_{ij} either passes through J_1 or is lying in V_1. Analogously, if it passes through I_2, the tangent passes through J_2 or is lying in V_2. Similar conditions hold for the moving centrode.

Example 74. Determine the circularities of the centrodes of the general four-bar motion.
Example 75. Check the circularities of the centrodes of the elliptic motion.

16. Instantaneous kinematics

A motion of the plane E with respect to Σ is represented by a curve k in the image space Σ'. A position of E corresponds to a point of k. Obviously the instantaneous properties of the position correspond to the local characteristics of k at the point under consideration, such as the tangent, the osculating plane, the Σ'-curvature, and so on. In view of (2.7) the Σ'-arc-element of the curve k is $d\phi$.

The image point of the displacement $\mathbf{D}(a, b; \phi)$ has been defined by (1.6),

$$X_1 : X_2 : X_3 : X_4 = \tag{16.1}$$
$$= (a \sin(\tfrac{1}{2}\phi) - b\cos(\tfrac{1}{2}\phi)) : (a\cos(\tfrac{1}{2}\phi) + b\sin(\tfrac{1}{2}\phi)) : 2\sin(\tfrac{1}{2}\phi) : 2\cos(\tfrac{1}{2}\phi).$$

The image of the zero-position $a = b = \phi = 0$ is the origin, $O = (0,0,0,1)$. Furthermore, if we introduce canonical frames in E and Σ, which implies $a_1 = b_1 = a_2 = 0$, we obtain for the differential quotients, with respect to arc lengths s, of X_i at the origin, up to the third order,

$$\begin{array}{llll} X_1' = 0, & X_2' = 0, & X_3' = 1, & X_4' = 0, \\ X_1'' = -b_2, & X_2'' = 0, & X_3'' = 0, & X_4'' = -\tfrac{1}{2}, \\ X_1''' = -b_3, & X_2''' = a_3 + (3/2)b_2, & X_3''' = -\tfrac{1}{4}, & X_4''' = 0. \end{array} \tag{16.2}$$

Hence the use of canonical frames gives us the results that in the image space the tangent of k at O has the equations $X_1 = X_2 = 0$ and the osculating plane is $X_2 = 0$.

To study higher order instantaneous properties we should develop the differential geometry of Σ' and introduce such notions as Σ'-curvature and Σ'-torsion. We shall not do this, instead we restrict ourselves to some examples.

Up to the second order the curve k coincides with its osculating plane, which means that any motion is up to the second order a symmetrical motion. If the osculating plane is stationary the curve is a plane one up to the third order. The condition for this is $X_1''X_2''' - X_2''X_1''' = 0$. From (16.2) this yields $X_2''' = 0$ or $2a_3 + 3b_2 = 0$, in accordance with Chapter VIII.

Example 76. Determine X_2^{IV} and derive the condition that the osculating plane has five consecutive points in common with k. Check that this is in accordance with the condition that the motion is instantaneously symmetrical to the fourth order (Chapter VIII).

This example illustrates the notion of two motions having instantaneously a contact of some order (which implies that all the paths of both motions have the corresponding order of contact at each point of the plane), the motions being identical up to a certain order. This is of importance if we want a given motion to be instantaneously compared with or approximated by a simpler motion. In our example the latter was a symmetrical motion.

We give a second example, asking whether a given motion is to a certain extent instantaneously identical with another simple motion, the elliptic motion. If the image curve of the latter passes through O, with the tangent $X_1 = X_2 = 0$ and the osculating plane $X_2 = 0$, its equations are given in Section 11, Example 53; with the arc length as parameter (i.e., $t = \tan(\phi/2)$) they read

$$
\begin{aligned}
X_1 &= -R\sin^2(\tfrac{1}{2}\phi)\cos(\tfrac{1}{2}\phi),\\
X_2 &= R\sin^3(\tfrac{1}{2}\phi),\\
X_3 &= \sin(\tfrac{1}{2}\phi),\\
X_4 &= \cos(\tfrac{1}{2}\phi),
\end{aligned}
\tag{16.3}
$$

which implies for $\phi = 0$:

$$
\begin{array}{llll}
X_1 = 0, & X_2 = 0, & X_3 = 0, & X_4 = 1,\\
X_1' = 0, & X_2' = 0, & X_3' = \tfrac{1}{2}, & X_4' = 0,\\
X_1'' = -\tfrac{1}{2}R, & X_2'' = 0, & X_3'' = 0, & X_4'' = -\tfrac{1}{4},\\
X_1''' = 0, & X_2''' = (3/4)R, & X_3''' = -(1/8), & X_4''' = 0.
\end{array}
\tag{16.4}
$$

If we compare (16.2) and (16.4) we see that the ratios of the coordinates are identical up to the second order provided $R = b_2$, which means that any motion is up to that order identical with one well-defined elliptic motion. Furthermore the third derivatives are only the same if $a_3 = b_3 = 0$, which are therefore the conditions for a Cardan position of the moving plane; this is in accordance with Chapter VIII.

17. Motions with two degrees of freedom

If a set of displacements depends on two parameters, the locus of their image points is a surface. Hence the study of plane motions with two degrees of freedom, denoted by $\mathbf{M}_2$, corresponds to the study of surfaces in the Σ'-geometry.

Example 77. Consider an $\mathbf{M}_2$ corresponding to a plane in the image space. Prove that any $\mathbf{M}_1$ belonging to $\mathbf{M}_2$ is a symmetric motion.

Important examples are those for which a given point A of the moving plane E is compelled to stay on a given curve p of Σ. In Section 10 we considered the case of p being a circle or a straight line.

If p is an arbitrary curve of Σ, A_0 the position of A on p at the zero-position, A_1 any other position of A on p, then all displacements $A_0 \to A_1$ belong to the set and therefore the images of these displacements lie on the image surface F. But we know that the locus of these images is a straight line. Hence F is a ruled surface. Furthermore: the lines corresponding to the displacements $A_0 \to A_1$ and $A_0 \to A_2$ are Σ'-parallel (Example 18). This implies that the generators of F are mutually Σ'-parallel, they intersect the same lines of the isotropic pencils m_1 and m_2, through J_1 in V_2 and J_2 in V_1 respectively. The conclusion is: F is a ruled surface with two directrices.

Let p be an algebraic curve of order N and suppose that it is not circular. Its equation is $P[X, Y, Z] = 0$. If we let A be the point (x, y, z) of E, in view of (1.9) the equation of the image surface F under consideration reads

$$(17.1)\quad \begin{aligned} P[(X_4^2 - X_3^2)x - 2X_3X_4y + 2(X_1X_3 + X_2X_4)z, 2X_3X_4x + (X_4^2 - X_3^2)y \\ + 2(X_2X_3 - X_1X_4)z, (X_3^2 + X_4^2)z] = 0, \end{aligned}$$

which implies that F is of order $2N$. If we substitute $X_3 = \lambda X_4$ in (17.1) the left-hand-side has the factor X_4^N and the conclusion is: l is an N-fold line on F.

If p is of order N and c-circular its equation is

$$(17.2)\quad \begin{aligned} (X^2 + Y^2)^c R_{N-2c}(X, Y) + (X^2 + Y^2)^{c-1} ZR_{N-2c+1}(X, Y) + \cdots + \\ + (X^2 + Y^2)Z^{c-1} R_{N-c-1}(X, Y) + Z^c R_{N-c}(X, Y, Z) = 0, \end{aligned}$$

in which R_v stands for a polynomial of degree v.

Without any loss of generality we may suppose that A is the origin in E. Then (1.9) simplifies to

$$(17.3)\quad X = 2(X_1X_3 + X_2X_4), \quad Y = 2(X_2X_3 - X_1X_4), \quad Z = X_3^2 + X_4^2,$$

which implies

$$(17.4)\qquad X^2 + Y^2 = 4(X_1^2 + X_2^2)(X_3^2 + X_4^2).$$

If we substitute (17.3) into (17.2) the left-hand-side has the factor $(X_3^2 + X_4^2)^c$, which means that F degenerates into the pair of planes V_1, V_2 counted c times and the locus proper of order $2N-2c$, the line l being $(N-2c)$-fold on it. Combining the foregoing, our result is: the displacements of E such that a given point of E is compelled to remain on a curve p of Σ, of order N and circularity c, correspond to a ruled surface F of order $2N-2c$, of which l is a $(N-2c)$-fold generator; F has two directrices m_1 and m_2, both $(N-c)$-fold lines, m_1 in V_1 through J_2, m_2 in V_2 through J_1.

Through any point of m_1 or m_2 pass $(N-c)$ generators of F. The intersection of F and V_1 consists of the $(N-c)$-fold line m_1, the $(N-2c)$-fold generator l, and c generators through J_1.

Example 78. Analytically verify the nature of the intersection of F and V_1, by means of (17.2), (17.3) and (17.4).
Example 79. Check the result if p is a straight line ($N=1$, $c=0$).
Example 80. Check the result if p is a circle ($N=2$, $c=1$).

We investigate the case where p is an ellipse with semi-axes $\mathfrak{a}$ and $\mathfrak{b}$ ($\mathfrak{a}\neq\mathfrak{b}$). For a suitable chosen frame (Fig. 91) its equation is $\mathfrak{b}^2X^2-2\mathfrak{a}\mathfrak{b}^2XZ+\mathfrak{a}^2Y^2=0$. Hence the image surface F is

$$\begin{aligned}\mathfrak{b}^2(X_1X_3+X_2X_4)^2+\mathfrak{a}^2(X_2X_3-X_1X_4)^2 \\ -\mathfrak{a}\mathfrak{b}^2(X_1X_3+X_2X_4)(X_3^2+X_4^2)=0.\end{aligned} \tag{17.5}$$

F is a quartic ruled surface with three double lines: the line l and the directrices m_1, m_2.

The ellipse can be represented by

$$X=2\mathfrak{a},\qquad Y=2\mathfrak{b}t,\qquad Z=1+t^2. \tag{17.6}$$

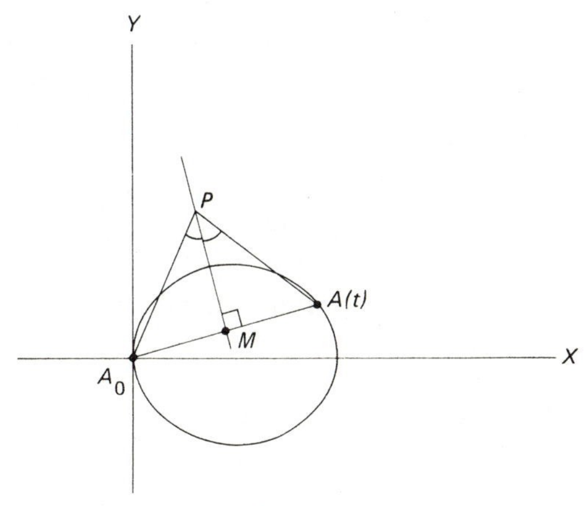

Fig. 91.

In the zero-position A is at A_0. All rotation centers for displacements $A_0 \to A$ lie on the perpendicular bisector MP of A_0A; P has the Cartesian coordinates

$$(\mathfrak{a} - \lambda\, \mathfrak{b}t)/(1+t^2), \qquad (\mathfrak{b}t + \lambda\, \mathfrak{a})/(1+t^2),$$

and since $A_0M = (\mathfrak{a}^2 + \mathfrak{b}^2t^2)^{1/2}(1+t^2)^{-1}$ and $MP = \lambda A_0M$ we have $\tan(\frac{1}{2}\phi) = \lambda^{-1}$. Hence the image point of the displacement $A_0 \to A(t)$ with center P is from (1.2)

$$(17.7) \qquad X_1 = \mathfrak{a} - \lambda\, \mathfrak{b}t, \quad X_2 = \mathfrak{b}t + \lambda\, \mathfrak{a}, \quad X_3 = 1 + t^2, \quad X_4 = \lambda(1+t^2),$$

and this is therefore the representation of the image surface F by means of the two parameters t and λ; F is a rational surface. A constant value of t corresponds to a generator.

Example 81. Show that the Plücker coordinates of a generator are

$$p_{14} = \mathfrak{a}(1+t^2), \qquad p_{24} = \mathfrak{b}t(1+t^2), \qquad p_{34} = (1+t^2)^2,$$

$$p_{23} = \mathfrak{a}(1+t^2), \qquad p_{31} = \mathfrak{b}t(1+t^2), \qquad p_{12} = -(\mathfrak{a}^2 + \mathfrak{b}^2t^2).$$

l corresponds to $t = \pm i$ and it is therefore an (isolated) double generator. Show that all generators belong to the linear complexes $p_{14} - p_{23} = 0$ and $p_{24} - p_{31} = 0$. Show that the directrices with coordinate q_{ij} are

$$q_{14} = -1, \qquad q_{24} = \pm i, \qquad q_{34} = 0,$$

$$q_{23} = 1, \qquad q_{31} = \mp i, \qquad q_{12} = 0.$$

Example 82. Investigate the image surface of those displacements for which a point of E remains on a parabola or on a hyperbola of Σ.

So far we have considered certain sets $\mathbf{M}_2$ and determined their image surfaces F in Σ'. We now make some remarks about the opposite mapping: given a surface F in Σ' we ask for the corresponding set $\mathbf{M}_2$ of positions of E with respect to Σ.

Let F be an algebraic surface of order n. We know that the set of displacements for which a point A of E moves from its zero-position A_0 to the point A of Σ is mapped on a line $\rho(A_0A)$ of Σ'. This line has in general n points in common with F. Hence the theorem: the surface F in Σ', of order n, corresponds to a $\mathbf{M}_2$ with the property that any point of E coincides n times with any point of Σ. But this is of course only true if we also accept imaginary positions, since the intersections of $\rho(A_0A)$ and F *need not* be real. If n is odd, however, at least one intersection is real, and we may conclude: the $\mathbf{M}_2$ corresponding to a surface F of odd order has the property that the positions of any point A of E fill the complete Σ, that is to say, A is able to move to any point of Σ. If on the other hand all intersections of $\rho(A_0A)$ and F are imaginary the position A is inaccessible to A_0, which is therefore compelled to remain within a certain region of Σ.

Example 83. Consider the case when F is a plane and therefore $n = 1$.

The lines $\rho(A_0A)$ for a fixed A_0 and variable A constitute a linear line congruence $L(A_0)$, built up of all lines intersecting two conjugate lines of the pencils m_1 and m_2 in V_1 and V_2 respectively. Hence the position of F with respect to the lines of $L(A_0)$ determines topologically the domain of A_0; Σ will be divided in a finite number of regions $G_0, G_1, \ldots, G_n$, such that A_0 passes s times through any point of G_s; some regions may be empty. A_0 is on a borderline of a region in those positions of E which correspond to points on F where the line $L(A_0)$ through it is a tangent of F. We note that the regions depend on A_0; their number and their shape may be different for different points A_0.

Example 84. The image surface of an $\mathbf{M}_2$ is the cone $X_1^2 + X_2^2 - \lambda^2 X_3^2 = 0$. Determine the accessible region of the origin O and of an arbitrary point of E.

Example 85. The image surface is the hyperboloid $X_1X_2 = \lambda^2 X_3X_4$. Determine the accessible region of any point of E. If the region is not the complete plane Σ, show that it is bordered by a bicircular quartic curve.

Example 86. Determine the characteristic property of an $\mathbf{M}_2$, the image surface of which is a ruled surface. Apply this to the two preceding examples.

Example 87. Determine the characteristic property of an $\mathbf{M}_2$, the image surface of which is a ruled surface with generators belonging to a linear congruence $L(A_0)$. Show that the accessible "region" of A_0 is a curve.

Instantaneous properties of an $\mathbf{M}_2$ could be studied by means of our mapping. We would need, however, the differential-geometrical theory of surfaces in Σ'-geometry, which will not be developed here. We remark only that, up to the first order, F may be replaced by its tangent plane at the origin. From this follows immediately the well-known theorem: *the locus of the instantaneous centers of the M_1's belonging to M_2 is a straight line* (Chapter X, Section 6).

18. Algebraic motion with two given paths

If two points of E describe algebraic paths, the image curve of the motion of E is the intersection of two algebraic surfaces and therefore algebraic; this implies that the path of any point is algebraic.

Example 88. Show that if one path is transcendental, there is at most *one* algebraic path.

If the points A_1, A_2 of E are compelled to remain on the curves p_1, p_2 of Σ respectively, the motion of E is completely defined. In the Σ'-space this corresponds to the intersection k of the surfaces F_1, F_2, as investigated in Section 17. It is of course possible that k has no real points: then the motion is impossible. The condition, necessary and sufficient, to obtain a real motion is obviously $d_m < A_1A_2 < d'_m$, d_m (d'_m) being the minimum (maximum) distance between a point of p_1 and one of p_2.

Let p_i be a c_i-circular curve of order N_i ($i = 1,2$). Then the motion thus defined has as its image curve k the intersection of two image surfaces F_i ($i = 1,2$) which are ruled surfaces of order $2N_i - 2c_i$, with l as a $(N_i - 2c_i)$-fold generator. The intersection degenerates into the line l, counted $(N_1 - 2c_1)(N_2 - 2c_2)$ times. This implies that the order of the proper intersection k is

$$
\begin{aligned}
n &= (2N_1 - 2c_1)(2N_2 - 2c_2) - (N_1 - 2c_1)(N_2 - 2c_2) \\
&= 3N_1N_2 - 2N_1c_2 - 2N_2c_1.
\end{aligned} \tag{18.1}
$$

Example 89. Show that J_1 and J_2 are $[(N_1 - c_1)(N_2 - c_2)]$-fold points of k.

We know that not only the order of k is of importance but also its position with respect to J_1, J_2, V_1, V_2. The intersection of V_1 and F_i consists of the directrix in V_1 (a $(N_i - c_i)$-fold line of F_i) and c_i lines through J_1. Hence k has $c_1(N_2 - c_2) + c_2(N_1 - c_1)$ points in V_1, not on l; furthermore in view of the multiplicity of J_1, J_2, on k, and of the order of k, we determine the characteristic numbers of k in Table (4.6) to be

$$
m_2 = m_3 = 0, \qquad m_4 = c_1c_2, \qquad m_5 = N_1N_2 - N_1c_2 - N_2c_1. \tag{18.2}
$$

From (4.7), (18.1) and (18.2) it follows that (for the order N and the circularity c of the path of an arbitrary point of E)

$$
N = 2N_1N_2 - 2c_1c_2, \qquad c = N_1c_2 + N_2c_1 - c_1c_2; \tag{18.3}
$$

these are the formulas of S. Roberts which we derived in Chapter IX.

19. The mapping of indirect displacements

Besides displacements $\mathbf{D}$ we have considered (in Chapter I, Section 7) orthogonal transformations $\mathbf{D}'$ of a plane, called indirect displacements (or symmetries), which change its orientation to the opposite one. They interchange the isotropic points I_1, I_2. The general expression for a $\mathbf{D}'(a, b, \phi)$ is:

$$(19.1) \quad X = -x\cos\phi - y\sin\phi + a, \qquad Y = -x\sin\phi + y\cos\phi + b.$$

We can easily show that (19.1) has an invariant line s, its axis, given by the equation

$$(19.2) \qquad x\cos(\tfrac{1}{2}\phi) + y\sin(\tfrac{1}{2}\phi) - \tfrac{1}{2}(a\cos(\tfrac{1}{2}\phi) + b\sin(\tfrac{1}{2}\phi))z = 0,$$

and that it may be considered as the product (in any order) of the reflection into s and the translation in the direction of s, with distance

$$(19.3) \qquad d = \tfrac{1}{2}(-a\sin(\tfrac{1}{2}\phi) + b\cos(\tfrac{1}{2}\phi)).$$

We shall represent the indirect displacement $\mathbf{D}'(a, b, \phi)$ (19.1) by a *plane* in the image space Σ', with the equation $U_1X_1 + U_2X_2 + U_3X_3 + U_4X_4 = 0$, the coefficients U_i being given by

$$(19.4) \qquad \begin{aligned} U_1 &= 2\cos(\tfrac{1}{2}\phi), \\ U_2 &= 2\sin(\tfrac{1}{2}\phi), \\ U_3 &= -(a\cos(\tfrac{1}{2}\phi) + b\sin(\tfrac{1}{2}\phi)), \\ U_4 &= -a\sin(\tfrac{1}{2}\phi) + b\cos(\tfrac{1}{2}\phi). \end{aligned}$$

To answer the question whether any plane is the image of an indirect displacement we determine a, b, ϕ from (19.4), if the U_i are given. After some algebra we obtain for the indirect displacement, in homogeneous coordinates:

$$(19.5) \qquad \begin{aligned} X &= -(U_1^2 - U_2^2)x - 2U_1U_2y - 2(U_1U_3 + U_2U_4)z, \\ Y &= -2U_1U_2x + (U_1^2 - U_2^2)y + 2(U_1U_4 - U_2U_3)z, \\ Z &= (U_1^2 + U_2^2)z, \end{aligned}$$

from which it follows that any plane U with $U_1^2 + U_2^2 \neq 0$ corresponds to an indirect displacement. Moreover if $U_1^2 + U_2^2 = 0$, without $U_1 = 0$, $U_2 = 0$, formula (19.5) represents a singular transformation: if $U_1 = iU_2$ all points go to I_2, if $U_1 = -iU_2$ they are transformed to I_1. If, however, $U_1 = U_2 = 0$, the transformation is meaningless. Hence: the only planes not corresponding to an indirect displacement are the planes through l; planes through J_1 or J_2, but not through both, correspond to singular symmetries; all other planes correspond to non-singular symmetries.

Special symmetries are the reflections, characterized by $d = 0$. Hence reflections correspond to planes for which $U_4 = 0$, they are the planes through the origin. There is obviously a strong similarity between the image points of displacements $\mathbf{D}(a, b, \phi)$ and the image planes of indirect displacements $\mathbf{D}'(a, b, \phi)$, as is seen by comparing (1.6) and (19.4). Excluded for the former are points on l, while for the latter we exclude planes through l. Singular displacements correspond to points of the planes V_1, V_2, singular indirect displacements to planes through J_1, J_2. Special displacements, half turns, correspond to points of a plane $(X_4 = 0)$; special indirect displacements, reflections, to planes through a point $(U_4 = 0)$.

Example 90. Show that the inverse transformation of (U_1, U_2, U_3, U_4) is given by $(U_1, U_2, U_3, -U_4)$.

Example 91. If the indirect displacement $\mathbf{D}'_1$ has the image plane U, and $\mathbf{D}'_2$ the image plane V, there is a displacement $\mathbf{D}$ with image point X transforming the first position of E into its second position. Prove that it is given by

$$
\begin{aligned}
X_4 &= U_1V_1 + U_2V_2,\\
X_3 &= U_1V_2 - U_2V_1,\\
X_2 &= -U_1V_3 - U_2V_4 + U_3V_1 + U_4V_2,\\
X_1 &= -U_1V_4 + U_2V_3 - U_3V_2 + U_4V_1.
\end{aligned}
$$

Example 92. If $\mathbf{D}'_1$ and $\mathbf{D}'_2$ are both reflections, show that the center of $\mathbf{D}$ is the intersection of the axes of $\mathbf{D}'_1$ and $\mathbf{D}'_2$. Check whether the converse theorem is true. Show that $\mathbf{D}$ is a translation if and only if the axes of $\mathbf{D}'_1$ and $\mathbf{D}'_2$ are parallel. Show that $\mathbf{D}$ is a half turn if and only if the axes are perpendicular.

Let X be the image point of a displacement $\mathbf{D}$ which gives E the position E_1, and U the image plane of an indirect displacement $\mathbf{D}'$ transforming E into the position E_2. Then E_1 is transformed into E_2 by $\mathbf{D}'\mathbf{D}^{-1}$ which is an indirect displacement, with the image plane U′ say. $\mathbf{D}^{-1}$ follows from (1.9), and $\mathbf{D}'$ is given by (19.5). Multiplying the respective transformation matrices we obtain

$$
\begin{aligned}
U'_1 &= U_1X_4 - U_2X_3,\\
U'_2 &= U_1X_3 + U_2X_4,\\
U'_3 &= -U_1X_2 + U_2X_1 + U_3X_4 - U_4X_3,\\
U'_4 &= U_1X_1 + U_2X_2 + U_3X_3 + U_4X_4.
\end{aligned}
\tag{19.6}
$$

U′ is a reflection if $U'_4 = 0$. Hence the theorem: *if the image point of the displacement* $\mathbf{D}$ *lies in the image plane of the indirect displacement* $\mathbf{D}'$, *the two corresponding positions of E are transformed one into the other by a reflection, and the converse is also true.*

This gives us a new interpretation of a plane in the image space, considered as the locus of its points: it corresponds to the set of motions which yield positions of E generated as if a fixed, symmetrical (i.e., indirect) position of E is reflected in all lines of Σ.

Example 93. Show that any plane curve is the image of a symmetrical motion (Section 5).
Example 94. Prove that *three* given positions of E are always the reflections, into appropriate axes, of one symmetrical (i.e., indirect) position of E. (This theorem has been fundamental in our study of plane three positions theory, Chapter VIII.)
Example 95. A plane and a line in Σ' have a point of intersection; determine the corresponding property in plane kinematics.
Example 96. Two planes in Σ' have a line of intersection. Give an interpretation of this property.

CHAPTER XII

KINEMATICS IN OTHER GEOMETRIES

1. Introduction

Until now we have mainly studied kinematics in two and three-dimensional Euclidean spaces. In this chapter we treat the kinematics associated with geometries of a different structure. According to the well-known Klein principle, a geometry may be defined as the study of the invariants of a group of transformations. Extending this concept to kinematics we define a transformation of the group as a displacement in the geometry under consideration; a continuous set of transformations which depend on one or more parameters is called a motion. Concepts such as trajectories and instantaneous kinematics have, in the new geometries, a meaning analogous to those of the classic theory. In this chapter we consider: kinematics in n-dimensional Euclidean geometry, in particular the cases $n = 4$, and $n = 5$; the kinematics of equiform geometry, an extension of the ordinary theory because it accepts also similarity transformations; the kinematics of two types of non-Euclidean geometry; the kinematics of affine geometry, and some other examples.

In all these cases we try to follow the main themes of this book: finite positions theory, instantaneous kinematics (with emphasis on canonical frames and instantaneous invariants), and special motions. We have confined ourselves to a concise treatment of all examples and restricted ourselves to the most important features, the choice of subjects being determined primarily by the attractiveness of the kinematics concerned, the simplicity (or complexity) of the matter, and the present state of knowledge. Some topics that are almost completely omitted are the motions of lines and planes, the inverse motion, and motions with more than one parameter.

A survey of kinematical theory in various geometries has been given by H. R. Müller [1970] along with an extensive list of references.

2. Kinematics in *n*-dimensional Euclidean space

We know from Chapter I that Euclidean n-dimensional spaces exhibit different behavior, as far as kinematical properties are concerned, depending upon if n is either an even or an odd number. As an example of both possibilities we now consider the cases $n = 4$ and $n = 5$.

A (one-parameter) motion in four-dimensional Euclidean space is analytically represented by

$$P = \mathbf{A}p + d, \tag{2.1}$$

P and p being the position vectors (or their column matrices) of a point in the fixed and the moving spaces Σ and E respectively; $\mathbf{A}$ is an orthogonal 4×4-matrix and d is the position vector of the origin of the moving system. Both $\mathbf{A}$ and d depend on the parameter t. Introducing notations which are somewhat different from those in the preceding chapters we denote the Cartesian coordinates of the points P and p by X_1, X_2, X_3, X_4 and x_1, x_2, x_3, x_4 respectively. For any function $\mathrm{f}(t)$ we denote $\mathrm{df}/\mathrm{d}t$ by $\dot{\mathrm{f}}$ and its value at $t = 0$ by f'.

We know from Chapter I, Section 4 that a displacement in E^4 has in general one invariant point and moreover two mutually orthogonal invariant planes through it. Hence for the instantaneous case there is one point with velocity zero, the pole, which we take as the origin O of the canonical frames. Moreover there are two orthogonal planes through O and they rotate about O with a certain angular velocity. We take these planes to be $x_3 = x_4 = 0$ and $x_1 = x_2 = 0$ respectively. It follows from this that the first order derivates for a moving point in the zero position are expressed by

$$\begin{aligned} X_1' &= -\omega_1 x_2, \\ X_2' &= \omega_1 x_1, \\ X_3' &= -\omega_2 x_4, \\ X_4' &= \omega_2 x_3, \end{aligned} \tag{2.2}$$

ω_1 and ω_2 being the two angular velocities; we exclude the singular cases $\omega_1\omega_2 = 0$. (Note that the axes Ox_1, Ox_2 in the one invariant plane and Ox_3, Ox_4 in the other are not yet determined.) If we consider time-dependent properties (2.2) contains two instantaneous invariants ω_1, ω_2; there is one geometric invariant: the ratio $\omega_1 : \omega_2$. (2.2) gives the velocity vector of any point $A(x_i)$; it is perpendicular to the radius vector OA.

Example 1. Show that all points on a line OA have parallel velocity vectors.

Example 2. Show that the plane through OA and the path tangents of its points is given, for instance, by the two equations

$$(-\omega_2x_2x_4-\omega_1x_1x_3)X_1+(-\omega_1x_2x_3+\omega_2x_1x_4)X_2+\omega_1(x_1^2+x_2^2)X_3=0,$$
$$(\omega_2x_2x_3-\omega_1x_1x_4)X_1+(-\omega_1x_2x_4-\omega_2x_1x_3)X_2+\omega_1(x_1^2+x_2^2)X_4=0.$$

Example 3. Show that the set of ∞^4 path tangents is built up of ∞^3 plane *parallel* pencils, one through every point at infinity.

Example 4. If U_i $(i=1,\dots,5)$ are 3-space coordinates show that the normal-space of the path is given by

$$U_1=-\omega_1x_2,\quad U_2=\omega_1x_1,\quad U_3=-\omega_2x_4,\quad U_4=\omega_2x_3,\quad U_5=0;$$

it passes through O and it is the same for all points on OA. Any 3-space through O is the normal space of all points on a line through O (perpendicular to the 3-space).

Example 5. Consider in (2.2) the special case $\omega_1=0$, $\omega_2\neq 0$ and prove that every point of the plane $x_3=x_4=0$ is a pole.

Example 6. A linear transformation for $n+1$ *homogeneous* coordinates has in general $n+1$ invariant points. Verify that, for the "infinitesimal transformation" (2.2) with $n+1=5$, these points are the origin and the isotropic points of the invariant planes $x_3=x_4=0$ and $x_1=x_2=0$.

Example 7. Derive the first order expressions, analogous to (2.2), for $2m$-dimensional Euclidean kinematics; show that they contain m time-dependent and $m-1$ geometrical invariants. Determine the $2m+1$ invariant points.

Let the frames in the moving and the fixed space coincide in the zero-position, but be otherwise arbitrary. Then from (2.1) it follows that

$$(2.3)\qquad \begin{aligned}X_1'&=\omega_{12}X_2+\omega_{13}X_3+\omega_{14}X_4+a_1',\\X_2'&=\omega_{21}X_1+\omega_{23}X_3+\omega_{24}X_4+a_2',\\X_3'&=\omega_{31}X_1+\omega_{32}X_2+\omega_{34}X_4+a_3',\\X_4'&=\omega_{41}X_1+\omega_{42}X_2+\omega_{43}X_3+a_4',\end{aligned}$$

where $\omega_{ij}=-\omega_{ji}$, and a_i' are the components of $\boldsymbol{d}'$. These expressions give the velocity vector of a point in terms of its coordinates in the fixed space.

The pole is found from (2.3) using the conditions $X_i'=0$ $(i=1,2,3,4)$; this gives us four linear equations for its coordinates. The determinant D of the set is $D=(\omega_{14}\omega_{23}+\omega_{24}\omega_{31}+\omega_{34}\omega_{12})^2$. If $D=0$ the system has either no (finite) solution or there are infinitely many. We exclude this singular case. Making use of the properties of an orthogonal matrix, for $D\neq 0$ the solution in terms of homogeneous coordinates is seen to be:

$$(2.4)\qquad \begin{aligned}X_1&=a_2'\omega_{34}+a_3'\omega_{42}+a_4'\omega_{23},\\X_2&=a_1'\omega_{43}+a_3'\omega_{14}+a_4'\omega_{31},\\X_3&=a_1'\omega_{24}+a_2'\omega_{41}+a_4'\omega_{12},\\X_4&=a_1'\omega_{32}+a_2'\omega_{13}+a_3'\omega_{21},\\X_5&=\omega_{14}\omega_{23}+\omega_{24}\omega_{31}+\omega_{34}\omega_{12}.\end{aligned}$$

To study second order properties we make use of formulas (2.5), (2.7) of Chapter II. For our case elements of the matrix $\mathbf{B}_1$ are given by the coefficients in (2.2); this implies that $\mathbf{C}_2$ is a diagonal matrix with the terms $-\omega_1^2$, $-\omega_1^2$, $-\omega_2^2$, $-\omega_2^2$. The components of the position vector $\boldsymbol{d}$ (i.e., a_1, a_2, a_3, a_4) in (2.1) are zero for $t=0$ and also satisfy $a_i'=0$ if we choose the origin of coordinates at the pole. We obtain for the velocity components $\dot{X}_i$, up to the first order

$$(2.5)\qquad \begin{aligned} \dot{X}_1 &= -t\omega_1^2x_1+(-\omega_1+t\varepsilon_{12})x_2+t\varepsilon_{13}x_3+t\varepsilon_{14}x_4+a_1''t,\\ \dot{X}_2 &= (\omega_1+t\varepsilon_{21})x_1-t\omega_1^2x_2+t\varepsilon_{23}x_3+t\varepsilon_{24}x_4+a_2''t,\\ \dot{X}_3 &= t\varepsilon_{31}x_1+t\varepsilon_{32}x_2-t\omega_2^2x_3+(-\omega_2+t\varepsilon_{34})x_4+a_3''t,\\ \dot{X}_4 &= t\varepsilon_{41}x_1+t\varepsilon_{42}x_2+(\omega_2+t\varepsilon_{43})x_3-t\omega_2^2x_4+a_4''t, \end{aligned}$$

where $\varepsilon_{ij}=-\varepsilon_{ji}$ are the elements of the skew matrix $\mathbf{B}_2$. If we solve $\dot{X}_i=0$ $(i=1,2,3,4)$ for x_i, restricting ourselves to first order terms, we get

$$(2.6)\qquad x_1=-ta_2''/\omega_1,\quad x_2=ta_1''/\omega_1,\quad x_3=-ta_4''/\omega_2,\quad x_4=ta_3''/\omega_2,$$

which gives, to the first order, the position of the pole in the moving space, or, in other words, the pole velocity for the zero-position. Our coordinate system is not completely determined, because the x_1 and x_2 axes are as yet arbitrary in their plane, and the same holds for the x_3 and x_4 axes. Therefore, we fix the frame with the condition that the projections of the polar velocity on the invariant planes are directed along the x_1- and the x_3-axis respectively (or in other words that the polar velocity vector is in the plane $x_2=x_4=0$). This implies $a_1''=a_3''=0$. The canonical system is now completely defined, and we obtain from (2.5) the second order relations

$$(2.7)\qquad \begin{aligned} X_1'' &= -\omega_1^2x_1+\varepsilon_{12}x_2+\varepsilon_{13}x_3+\varepsilon_{14}x_4,\\ X_2'' &= \varepsilon_{21}x_1-\omega_1^2x_2+\varepsilon_{23}x_3+\varepsilon_{24}x_4+a_2'',\\ X_3'' &= \varepsilon_{31}x_1+\varepsilon_{32}x_2-\omega_2^2x_3+\varepsilon_{34}x_4,\\ X_4'' &= \varepsilon_{41}x_1+\varepsilon_{42}x_2+\varepsilon_{43}x_3-\omega_2^2x_4+a_4'', \end{aligned}$$

which, together with (2.2), in principle equip us to answer all questions that could be asked about second order instantaneous properties of the motion. It contains the first order invariants ω_1^2, ω_2^2 and moreover the eight second order invariants consisting of the six ε_{ij} as well as a_2'' and a_4''. All these are time-dependent. Obviously (2.7) gives us the acceleration vector of any point of the moving space. Solving $X_i''=0$ $(i=1,2,3,4)$ for x_i we conclude that there exists in general one point in the moving space whose acceleration is equal to zero: the acceleration pole.

The polar velocity vector for the canonical systems reads $(-a_2''/\omega_1, 0, -a_4''/\omega_2, 0)$; it is time-dependent.

The curvature theory of the paths is contained in the sets (2.2) and (2.7). We shall not fully develop it. Here we restrict ourselves to determining the locus of the points which in the zero-position are at an inflection of their paths. These points satisfy the equations $X_i'' = \lambda X_i'$, which are four linear equations for x_k $(k = 1,2,3,4)$ all containing the linear parameter λ. If we introduce homogeneous coordinates x_1, x_2, x_3, x_4, x_5 and then solve for the x_i, we see that x_1 and x_3 are cubic polynomials in λ, x_2 and x_4 are quadratic and x_5 is a quartic polynomial. Hence the inflection curve is a rational quartic curve (a so-called norm curve) in the moving space. It passes through the origin (when $\lambda = \infty$, which must be excluded from the locus) and through the acceleration pole (when $\lambda = 0$). It is the analog of the inflection circle in planar kinematics.

Example 8. Prove that the tangent of the inflection curve at O coincides with the pole-velocity vector.

We could also have obtained (2.7) using the method considered in Chapter II, Section 3: Restricting ourselves to the rotational part of the motion (given in arbitrary frames), differentiating with respect to t, and then eliminating x_i we obtain

$$\dot{X}_i = \omega_{12}X_2 + \omega_{13}X_3 + \omega_{14}X_4,$$

and so forth by cyclic substitution.

Differentiating once more and eliminating $\dot{X}_i$ yields

$$\begin{aligned}\ddot{X}_1 = &-(\omega_{12}^2 + \omega_{13}^2 + \omega_{14}^2)X_1 + (\dot{\omega}_{12} + \omega_{13}\omega_{32} + \omega_{14}\omega_{42})X_2 \\ &+ (\dot{\omega}_{13} + \omega_{12}\omega_{23} + \omega_{14}\omega_{43})X_3 + (\dot{\omega}_{14} + \omega_{12}\omega_{24} + \omega_{13}\omega_{34})X_4,\end{aligned} \tag{2.8}$$

and so forth by cyclic substitution.

Then introducing the canonical frames we can substitute $X_i = x_i$, $\omega_{12} = -\omega_1$, $\omega_{34} = -\omega_2$; and zero for the remaining ω_{ij}. Hence from (2.8)

$$\ddot{X}_1 = -\omega_1^2 x_1 + \dot{\omega}_{12}x_2 + \dot{\omega}_{13}x_3 + \dot{\omega}_{14}x_4, \tag{2.9}$$

and so forth by cyclic substitution,

which, if compared to (2.7) gives us $\varepsilon_{ij} = \dot{\omega}_{ij}$. Hence the skew matrix $\|\varepsilon_{ij}\|$ is the derivative of $\|\omega_{ij}\|$, the numbers ε_{ij} are therefore the angular accelerations for the zero-position.

What was done here for first and second order instantaneous kinematics in four-dimensional space may easily be expanded to $2m$-dimensional space. There is in general one pole and moreover there are m invariant planes through it. The generalization of (2.2) contains m invariants ω_i, and that of (2.7) an additional $m(2m-1)$ invariants $\varepsilon_{ij} = \dot{\omega}_{ij}$ and also m invariants from the part of the motion described by the displacement vector $\boldsymbol{d}$.

Example 9. Consider the foregoing for the case $m = 1$, i.e., when we are dealing with planar kinematics.

Example 10. Show that the inflection curve in $2m$-dimensional kinematics is a rational $2m$-ic, passing through the pole and through the acceleration pole.

Example 11. Express the pole velocity vector in terms of the instantaneous invariants for a $2m$-dimensional space.

Kinematics in $(2m+1)$-dimensional space is less elegant and more difficult to handle than in $2m$-dimensional space. Because the determinant of a skew $(2m+1)\times(2m+1)$ matrix is always zero it follows that there is in general no (finite) pole. There is, however, a real invariant point at infinity, which gives rise to a screw-axis l; every point of l has its velocity in the direction of l. Any $2m$-space, orthogonal to l, is translated (instantaneously, to the first order) as a whole in this same direction, and within such a space the rules of $2m$-dimensional kinematics are valid up to the first order. So, for $m = 2$, for instance, if we take X_5 parallel to l, first-order properties are expressed by virtue of (2.2) as

$$(2.10) \qquad X_1' = -\omega_1 x_2, \quad X_2' = \omega_1 x_1, \quad X_3' = -\omega_2 x_4, \quad X_4' = \omega_2 x_3, \quad X_5' = v,$$

with three time-dependent invariants ω_1, ω_2, v. The origin of the canonical frame is not yet determined. This may be done using second-order considerations just as we did for the 3-space, i.e., we make use of the striction point concept; we remark that two neighboring screw axes always lie in a 3-space, and have therefore a common perpendicular. Having established the basic concepts we stop here leaving any detailed development to the reader.

Special continuous motions in n-dimensional space E^n can be defined by imposing $(k-1)$ simple conditions on the general displacement, where $k = \frac{1}{2}n(n+1)$ is the degree-of-freedom of a free moving E^n. Another, analytical, technique is to express the general displacement in terms of k parameters: $\frac{1}{2}n(n-1)$ of them for the rotational part (as given by Cayley's representation of an orthogonal matrix) and n for the motion of the origin. (If the origin is taken on l these n are the so-called translation terms.) A motion is defined if all parameters are specified as functions of a parameter t.

Most often it is convenient to use a homogeneous set of parameters in which case there are $k+1$ of them and the rotation is given in terms of $\frac{1}{2}n(n-1)+1$ *homogeneous* parameters.

As an example we mention the case $n=4$. Cayley's representation is found as follows (BALTZER [1881]). The starting point is the 4×4 matrix

$$\mathbf{C} = \left\| \begin{array}{cccc} c_0 & c_{12} & c_{13} & c_{14} \\ c_{21} & c_0 & c_{23} & c_{24} \\ c_{31} & c_{32} & c_0 & c_{34} \\ c_{41} & c_{42} & c_{43} & c_0 \end{array} \right\|, \tag{2.11}$$

with $c_{ij}+c_{ji}=0 \quad (i\neq j)$. It is easy to verify that $N=|\mathbf{C}|= c_0^4+(c_{12}^2+c_{13}^2+c_{14}^2+c_{23}^2+c_{24}^2+c_{34}^2)c_0^2+\Delta^2$, with $\Delta = c_{14}c_{23}+c_{24}c_{31}+c_{34}c_{12}$. The rotational part of the motion may be given in terms of an orthogonal matrix $\|a_{ij}\|$ which is derived from $\mathbf{C}$ by the following formulas

$$a_{ii} = (2c_0C_{ii}/N)-1, \qquad a_{ij}=(2c_0C_{ij}/N) \quad (i\neq j) \tag{2.12}$$

where C_{ij} is the minor of the ij^{th} element of the matrix $\mathbf{C}$. It may be shown that (nearly) all orthogonal matrices can be found by this procedure; (there are some exceptional ones which are limit cases of (2.12)). The elements of $\|a_{ij}\|$ depend on six parameters: the ratios of c_0, c_{ij}, which may be normalized by, say, the condition $N=1$.

We find from (2.12):

$$\begin{aligned} N\,a_{11} &= c_0^4+c_0^2(c_{23}^2+c_{34}^2+c_{42}^2-c_{12}^2-c_{13}^2-c_{14}^2)-\Delta^2, \\ N\,a_{12} &= 2c_0[c_0^2c_{12}-c_0(c_{13}c_{23}+c_{14}c_{24})+c_{34}\Delta], \quad \text{etc.} \end{aligned} \tag{2.13}$$

The elements of $\|a_{ij}\|$ are quartic functions of the parameters; hence if we take each parameter to be a linear function of t, and the four coordinates for the moving origin to be functions of t of degree four (at most) multiplied by N^{-1}, we have defined a special motion in 4-space for which all paths are (rational) quartic curves. This may be considered as an analog of the Cardan motion in the plane.

Cayley's procedure is valid for arbitrary n. In order to derive an orthogonal matrix in n-space we must start from an $n\times n$-matrix $\mathbf{C}$ similar to (2.11). If n is even the elements of a_{ij} given by (2.12) are functions of degree n of the parameters c_0, c_{ij}; if n is odd, N has the factor c_0 and the a_{ij} are thus of degree $n-1$.

Example 12. Derive by Cayley's method the orthogonal matrices for $n=2$.

Example 13. Derive the orthogonal matrices for $n=3$ and show that they are identical with those represented by the Euler parameters.

In 5-space the elements of an orthogonal matrix are quartic functions of the (eleven) homogeneous parameters c_0, c_{ij}. If they are all chosen as linear functions of t and the translation components are taken as functions of degree four (at most), we have defined in 5-space a motion for which all paths are quartic curves. A quartic curve, however, always lies in (at most) a 4-space. The motion is therefore an analog of the Darboux motion in 3-space.

Some remarks will now be made about 2- and 3-parameter motions in four-dimensional space. An M_2 is represented by

$$P = \mathbf{A}(t_1, t_2)\boldsymbol{p} + \boldsymbol{d}(t_1, t_2) \tag{2.14}$$

where t_1 and t_2 are the parameters. An M_1 belonging to (2.14) is defined if t_1 and t_2 are given as (arbitrarily chosen) functions of t. We take the zero-position at $t_1 = t_2 = t = 0$. From (2.14) we have

$$P' = ((\partial\mathbf{A}/\partial t_1)\lambda_1 + (\partial\mathbf{A}/\partial t_2)\lambda_2)\boldsymbol{p} + (\partial \boldsymbol{d}/\partial t_1)\lambda_1 + (\partial \boldsymbol{d}/\partial t_2)\lambda_2, \tag{2.15}$$

with $\lambda_i = t'_i$ $(i = 1, 2)$. If we write (2.15) in terms of the coordinates and if we eliminate x_i we obtain in view of (2.3):

$$X'_1 = \omega_{12}X_2 + \omega_{13}X_3 + \omega_{14}X_4 + a'_1, \quad \text{and so forth in a cyclic manner.} \tag{2.16}$$

Here $\omega_{ij} = r_{ij}\lambda_1 + s_{ij}\lambda_2$, $a'_i = r_i\lambda_1 + s_i\lambda_2$, where r_{ij}, s_{ij}, r_i and s_i follow from the given M_2, while λ_1, λ_2 depend on the M_1 we have chosen. The pole of this M_1 follows from the equations $X'_i = 0$; because these are homogeneous linear expressions in terms of λ_1 and λ_2, the pole depends only on the ratio $\lambda_1 : \lambda_2$. Hence, there is at any moment a locus of ∞^1 poles for an M_2. The system of four equations for two homogeneous parameters λ_1, λ_2 has a rank which is at most two. This implies that in the zero-position all poles are in one plane. Their coordinates follow from (2.4) and are therefore homogeneous quadratic functions of λ_1, λ_2. The conclusion is: the locus of the poles of a 2-parameter motion in four-dimensional space, at a certain moment, is a conic.

For a 3-parameter motion M_3 we have a similar property. In (2.16) we would have ω_{ij} and a'_i given as homogeneous linear functions of three parameters. Since the rank of the equations for $X'_i = 0$ would be at most three, it follows that all instantaneous poles lie in a 3-space. Moreover, their coordinates are quadratic functions of three homogeneous parameters: hence the locus of the poles at each moment is a Steiner surface.

Example 14. Show that the ∞^2 conics on this surface correspond to the loci of the poles of the M_2's belonging to the M_3.

Example 15. Show that the instantaneous locus of the poles of an M_4 is a three-dimensional manifold; the coordinates of its points are quadric functions of four homogeneous parameters.

3. Equiform kinematics

A displacement in Euclidean space is a transformation which leaves the distance between any pair of points invariant; we deal now with transformations such that every distance is multiplied by the same number, called the scale factor, denoted by s. *The displaced configuration is similar but not congruent to the original.* The kinematics corresponding to such transformations will be called *equiform*.

In three-dimensional space an equiform displacement is represented by

$$P = s\mathbf{A}p + d; \tag{3.1}$$

for $s = 1$ we have congruent configurations, for $s = 0$ the transformation is singular; if $s > 0$ the transformed configuration is directly similar with the original, for $s < 0$ they are symmetrically similar. It is a matter of convention whether to accept or to reject the case $s < 0$.

For a displacement we have now seven degrees of freedom; to the classical six we add the parameter s. For Cartesian frames the displacement is given by

$$\begin{aligned} X &= s(a_{11}x + a_{12}y + a_{13}z) + a, \\ Y &= s(a_{21}x + a_{22}y + a_{23}z) + b, \\ Z &= s(a_{31}x + a_{32}y + a_{33}z) + c, \end{aligned} \tag{3.2}$$

where $\| a_{ij} \|$ is an orthogonal matrix. If we introduce homogeneous coordinates we see that the transformation for points in the plane at infinity is independent of s. Hence there are in general three invariant points in this plane, one real point A_1 and the two isotropic points in the plane orthogonal to the direction of A_1. For a possible finite invariant point we must solve the equations $X = x$, $Y = y$, $Z = z$ which gives us three linear equations with the determinant

$$\Delta = \begin{vmatrix} a_{11} - s^{-1} & a_{12} & a_{13} \\ a_{21} & a_{22} - s^{-1} & a_{23} \\ a_{31} & a_{32} & a_{33} - s^{-1} \end{vmatrix}. \tag{3.3}$$

We know however (Chapter I, Section 4) that an orthogonal matrix in three dimensions has three eigenvalues, one equal to 1 and two imaginary ones. Hence, unless $s^{-1} = s = 1$, we have $\Delta \neq 0$. This implies that an equiform displacement has in general one finite invariant point A_0. If we take A_0 as origin and A_0A_1 as o_z, the displacement equations (3.2) take on the standard form

$$X = s(x \cos \phi - y \sin \phi),$$
$$(3.4) \qquad Y = s(x \sin \phi + y \cos \phi),$$
$$Z = sz.$$

This shows that the displacement may built up by a rotation about the axis O_Z, followed by a dilation (or a homothety) with center O and scale factor s.

The fact that an equiform displacement has an invariant point was already known to CHASLES [1831]. It makes equiform kinematics in many respects easier to handle than classical three-dimensional kinematics, which appears as the singular case where A_0 coincides with A_1 and their join becomes, in the limit, the screw-axis. On the other hand it should be kept in mind that equiform geometry does not deal directly with distance and length; two points have no invariant, any pair of points may be displaced into any other pair; three points, however, have (two) invariants: the angles of their triangle or the ratios of their mutual distances. A general linear transformation of four homogeneous variables has in general four different invariant points; equiform displacements belong to this general class, ordinary displacements do not.

A motion in equiform geometry is defined if the parameters of (3.2), including s, are given as functions of t. We remark that an actual angular velocity exists and so does the line along which the velocity vector of a point is directed, but the scalar value of this velocity is not defined.

We consider the instantaneous kinematics of equiform three-dimensional geometry. If we differentiate the expressions (3.2) with respect to t and eliminate x, y, z we obtain (provided $a_0 = b_0 = c_0 = 0$; i.e., at $t = 0$ the origins of the coordinate frames coincide):

$$\dot{X} = (\dot{s}/s)X - \omega_3 Y + \omega_2 Z + \dot{a},$$
$$(3.5) \qquad \dot{Y} = \omega_3 X + (\dot{s}/s)Y - \omega_1 Z + \dot{b},$$
$$\dot{Z} = -\omega_2 X + \omega_1 Y + (\dot{s}/s)Z + \dot{c},$$

where the ω_i are functions of t, their dependence on a_{ij} is known from ordinary kinematics. Since the frames coincide, $s = 1$ at $t = 0$ (the zero-position), and if $\mathrm{d}f/\mathrm{d}t$ for $t = 0$ is denoted by f_1, we have

$$X_1 = s_1 X - \omega_3 Y + \omega_2 Z + a_1,$$
$$(3.6) \qquad Y_1 = \omega_3 X + s_1 Y - \omega_1 Z + b_1,$$
$$Z_1 = -\omega_2 X + \omega_1 Y + s_1 Z + c_1,$$

where ω_i is now the value of this function at $t = 0$ and similar for X, Y, Z. A point is at rest if $X_1 = Y_1 = Z_1 = 0$, which gives rise to three linear equations for X, Y, Z the determinant being

$$\Delta = s_1(s_1^2 + \omega_1^2 + \omega_2^2 + \omega_3^2), \tag{3.7}$$

which is $\neq 0$ if $s_1 \neq 0$. Hence there is a finite invariant point, the pole of the motion. If we take it as the origin and moreover the rotation axis along O_Z; with $\omega_1 = \omega_2 = 0$, $\omega_3 = \omega$, we obtain

$$\begin{aligned} X_1 &= s_1 X - \omega Y, \\ Y_1 &= \omega X + s_1 Y, \\ Z_1 &= s_1 Z, \end{aligned} \tag{3.8}$$

which gives us the motion, up to the first order, with respect to a frame for which the axes O_X and O_Y are not yet determined. There are two first order instantaneous invariants, ω and s_1, both time-dependent; their ratio is a geometric invariant. (Note that s is a dimensionless number.)

From (3.8) it follows that the path tangent at (X, Y, Z) has the Plücker coordinates

$$\begin{aligned} &Q_{14} = s_1 X - \omega Y, \qquad Q_{24} = \omega X + s_1 Y, \qquad Q_{34} = s_1 Z \\ &Q_{23} = \omega XZ, \qquad Q_{31} = \omega YZ, \qquad Q_{12} = -\omega(X^2 + Y^2). \end{aligned} \tag{3.9}$$

Eliminating X, Y, Z the complex of the path tangents is seen to be

$$s_1(Q_{14}Q_{31} - Q_{24}Q_{23}) - \omega Q_{12}Q_{34} = 0, \tag{3.10}$$

which represents a quadratic complex.

Example 16. Prove that (3.10) is a tetrahedral complex; the vertices of the tetrahedron are: the origin, the point at infinity of O_Z and the isotropic points of the plane $Z = 0$.

Example 17. Show that the coordinates of the normal plane to the path are

$$U_1 = s_1 X - \omega Y, \quad U_2 = \omega X + s_1 Y, \quad U_3 = s_1 Z, \quad U_4 = -s_1(X^2 + Y^2 + Z^2).$$

Setting $\dot{X} = \dot{Y} = \dot{Z} = 0$ in (3.5) we find the coordinates (X_p, Y_p, Z_p) of the pole $P(t)$ for any set of coinciding frames:

$$NX_p = \dot{a}((\dot{s}/s)^2 + \omega_1^2) + \dot{b}((\dot{s}/s)\omega_3 + \omega_1\omega_2) + \dot{c}(-(\dot{s}/s)\omega_2 + \omega_1\omega_3), \tag{3.11}$$

the other two follow by cyclic substitution. Here

$$N = -(\dot{s}/s)((\dot{s}/s)^2 + \omega_1^2 + \omega_2^2 + \omega_3^2).$$

Differentiating (3.11) and then substituting $t = 0$, we obtain the pole velocity vector for the zero-position taken in accordance with (3.8):

(3.12)
$$X_p' = -(a_2 s_1 + b_2\omega)/(s_1^2+\omega^2), \qquad Y_p' = -(b_2 s_1 - a_2\omega)/(s_1^2+\omega^2),$$
$$Z_p' = -c_2/s_1.$$

$\ddot{X}$, $\ddot{Y}$, $\ddot{Z}$ follow from (3.5); if we eliminate $\dot{X}$, $\dot{Y}$, $\dot{Z}$, the result is, after the substitution $t = 0$,

(3.13)
$$X_2 = (s_2 + s_1^2 - \omega^2)X - (\varepsilon_3 + 2s_1\omega)Y + \varepsilon_2 Z + a_2,$$
$$Y_2 = (\varepsilon_3 + 2s_1\omega)X + (s_2 + s_1^2 - \omega^2)Y - \varepsilon_1 Z + b_2,$$
$$Z_2 = -\varepsilon_2 X + \varepsilon_1 Y + (s_2 + s_1^2)Z + c_2,$$

where ε_i are the values of $\dot{\omega}_i$ for $t = 0$. O_X and O_Y are still arbitrary. We choose them in such a way that the point coinciding with the pole moves, to the second order, in the O_{XZ}-plane. Hence $b_2 = 0$, and the canonical system is completely determined. The formulas (3.8) and (3.13) (the latter with $b_2 = 0$) characterize the motion of any point up to the second order. The set contains the instantaneous invariants ω, s_1, ε_1, ε_2, ε_3, s_2, a_2, c_2.

Example 18. Determine (from (3.12)) the direction of the pole-tangent in the canonical frame.
Example 19. Show that there is at any moment, in general, one point with zero-acceleration.
Example 20. Prove that the locus of the points which are at an inflection point of their path is a twisted cubic curve, passing through the pole and through the acceleration pole.
Example 21. In order to make b_2 vanish in (3.13) the frame must be rotated about O_Z through the angle θ. Determine θ; show that s_2, s_1, ω, ε_3, c_2 do not change; determine the new values of a_2, ε_1, ε_2.

From (3.13) we may derive the osculating plane of the path, the curvature axis as the intersection of two consective normal planes, and the center of curvature. Obviously the concept of radius of curvature does not exist in a geometry which does not deal with length.

It seems that not much attention has been given in the literature to spatial equiform kinematics. The plane case, however, has been developed at length, mainly be R. MÜLLER [1910a, b, c] and by KRAUSE [1910, 1920]. It will be treated here using, for the instantaneous case, methods adopted in former chapters.

An equiform displacement in the plane is represented by

(3.14) $$X = s(x\cos\phi - y\sin\phi) + a, \quad Y = s(x\sin\phi + y\cos\phi) + b.$$

Example 22. Show that the displacement has in general one finite invariant point and determine its standard equations.

A motion is defined if in (3.14) one gives ϕ, a, b, s as functions of t. If we restrict ourselves to geometrical properties, the angle ϕ can be taken as the independent variable. Then it follows from (3.14)

$$\dot{X} = \dot{s}(x \cos\phi - y \sin\phi) + s(-x \sin\phi - y \cos\phi) + \dot{a},$$

(3.15)

$$\dot{Y} = \dot{s}(x \sin\phi + y \cos\phi) + s(x \cos\phi - y \sin\phi) + \dot{b},$$

and the conditions $\dot{X} = \dot{Y} = 0$ give us the pole P with the coordinates x_p, y_p, such that

$$(\dot{s}^2 + s^2)x_p = -\dot{s}(\dot{a} \cos\phi + \dot{b} \sin\phi) + s(\dot{a} \sin\phi - \dot{b} \cos\phi),$$

(3.16)

$$(\dot{s}^2 + s^2)y_p = \dot{s}(\dot{a} \sin\phi - \dot{b} \cos\phi) + s(\dot{a} \cos\phi + \dot{b} \sin\phi),$$

which for variable ϕ represents the moving centrode. The centrode in the fixed plane follows if one substitutes (3.16) into (3.14):

$$(\dot{s}^2 + s^2)X_P = -s(\dot{s}\dot{a} + s\dot{b}) + (\dot{s}^2 + s^2)a,$$

(3.17)

$$(\dot{s}^2 + s^2)Y_P = s(s\dot{a} - \dot{s}\dot{b}) + (\dot{s}^2 + s^2)b,$$

We shall deal now with finitely separated positions of a plane E with respect to a coinciding fixed plane Σ. If we have two positions and the frames are chosen to be coinciding in the first, then the second is defined by

(3.18) $X = s(x \cos\phi - y \sin\phi) + a, \qquad Y = s(x \sin\phi + y \cos\phi) + b.$

It is easily seen that there, in general, is one finite invariant point,

$$X = x = [-s(b \sin\phi + a \cos\phi) + a]/N,$$

(3.19)

$$Y = y = [s(a \sin\phi - b \cos\phi) + b]/N,$$

$$N = (s \cos\phi - 1)^2 + s^2 \sin^2\phi,$$

the only exception occurs if $\phi = 0$, $s = 1$ which would mean we have a Euclidean translation. If we take the invariant point as the common origin we obtain

(3.20) $X = s(x \cos\phi - y \sin\phi), \qquad Y = s(x \sin\phi + y \cos\phi).$

For $\phi = 0$, $s \neq 1$ we have a dilation with scale factor s and center O. Any line through O is an invariant line. When $\phi \neq 0$, we find that there is no (real) finite invariant line.

We consider now three positions of E with respect to Σ. The displacement from position i to position j will be a rotation with center P_{ij} and angle ϕ_{ij}, and a dilation with this same center and the scale factor s_{ij}. As the product of displacements $1 \to 2$, $2 \to 3$, $3 \to 1$ is unity we have

(3.21) $\phi_{12} + \phi_{23} + \phi_{31} = 0 \ (\text{mod. } 2\pi); \quad s_{12}s_{23}s_{31} = 1.$

We denote the ratios, of the lengths of a line segment in the three positions, by the homogeneous numbers $k_1 : k_2 : k_3$; then $s_{12} = k_2/k_1$, $s_{23} = k_3/k_2$, $s_{31} = k_1/k_3$.

Example 23. Show that two positions of plane E with respect to plane Σ are given by four pieces of data, and that three positions are given by eight data. Show that this last number is in accordance with specifying the positions by giving the four points P_{23}, P_{31}, P_{12} and P_{23}^1.

Consider the first position E_1 of E (Fig. 92). It has three special points: P_{12}, P_{31} and the point P_{23}^1 which is transformed into P_{23} by the displacement D_{12} and by D_{13} as well. From this it follows that $P_{12}P_{23}^1 = a_2k_1/k_2$, $P_{31}P_{23}^1 = a_3k_1/k_3$, where a_i $(i = 1, 2, 3)$ are the sides of the pole triangle $P_{12}P_{23}P_{31}$. Hence if the pole triangle is given and also the point P_{23}^1, the three positions E_i are known. Indeed: s_{12} and s_{31} (and therefore s_{23} by (3.21)) are determined and so are ϕ_{12} and ϕ_{31} (and therefore ϕ_{23}): as can be seen from the figure $\phi_{12} = \gamma + \gamma'$, $\phi_{31} = \beta + \beta'$, $\phi_{23} = \alpha + \alpha'$. In particular we obtain the second and the third position of the triangle $P_{12}P_{31}P_{23}^1$ which are seen to be $P_{12}P_{31}^2P_{23}$ and $P_{12}^3P_{31}P_{23}$ respectively, both similar to the former (Fig. 93).

If we denote the ratios of the distances of a point to $P_{12}P_{31}$, $P_{23}P_{12}$, $P_{31}P_{23}$ by

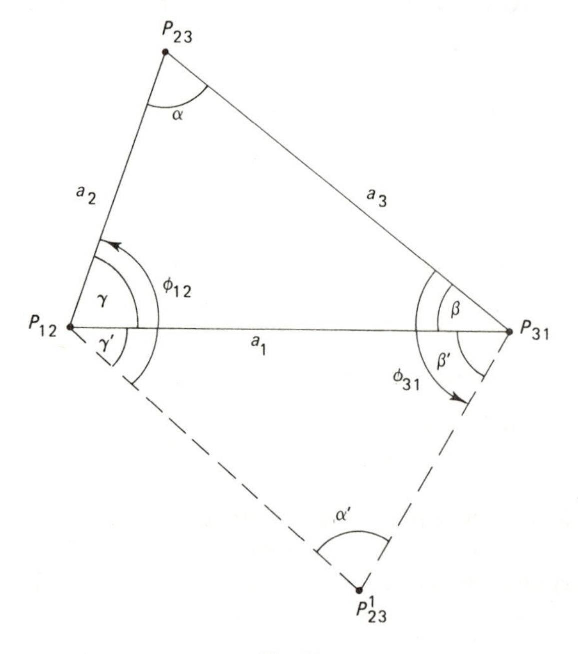

Fig. 92.

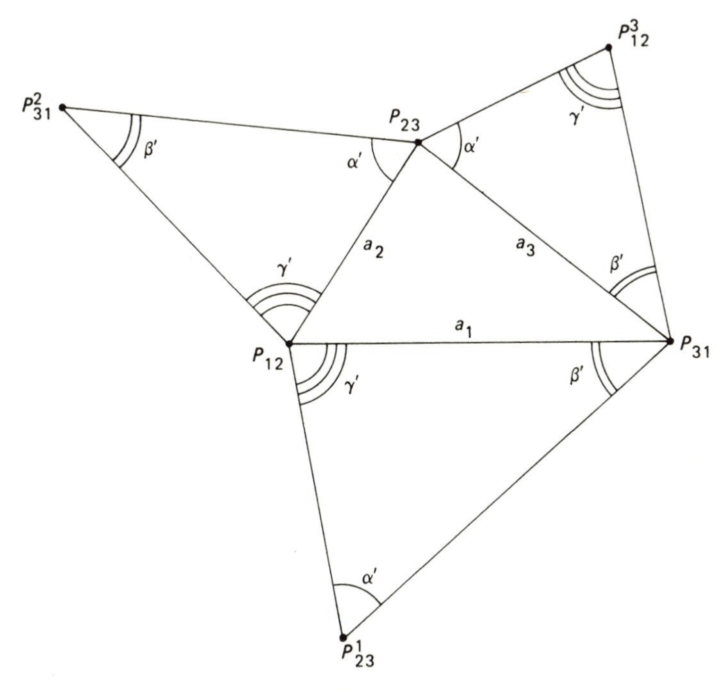

Fig. 93.

$z_1 : z_2 : z_3$ then we have for P^1_{23} $z_2 : z_3 = a_2(k_1/k_2)\sin\phi_{12} : a_3(k_1/k_3)\sin\phi_{31} = a_2k_3\sin\phi_{12} : a_3k_2\sin\phi_{31}$, which is therefore at the same time the equation of the line $P_{23}P^1_{23}$.

In the same way we obtain for $P_{31}P^2_{31}$ $z_3 : z_1 = a_3k_1\sin\phi_{23} : a_1k_3\sin\phi_{12}$, and for $P_{12}P^3_{12}$ $z_1 : z_2 = a_1k_2\sin\phi_{31} : a_2k_1\sin\phi_{23}$. Hence the three lines are Cevians of the pole triangle; they pass through one point H (Fig. 94), determined by

$$z_1 : z_2 : z_3 = a_1k_2k_3\sin\phi_{31}\sin\phi_{12} : a_2k_3k_1\sin\phi_{12}\sin\phi_{23} : a_3k_1k_2\sin\phi_{23}\sin\phi_{31}. \tag{3.22}$$

Example 24. Show that for $k_1 = k_2 = k_3$ the configuration becomes the one for ordinary kinematics: $\alpha' = \alpha$, $\beta' = \beta$, $\gamma' = \gamma$; $\phi_{23} = 2\alpha$, $\phi_{31} = 2\beta$, $\phi_{12} = 2\gamma$. Verify that (3.22) are in this case the coordinates of the orthocenter of the pole triangle.

We have (Fig. 94):

$$P_{12}P_{23} : P_{12}P^1_{23} = P_{12}P^2_{31} : P_{12}P_{31}(= s_{12}).$$

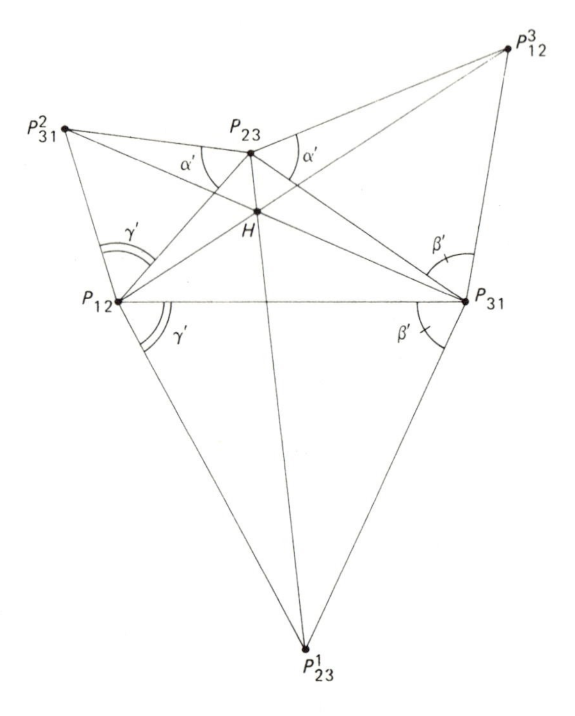

Fig. 94.

Hence the triangles $P^1_{23}P_{12}P_{23}$ and $P_{31}P_{12}P^2_{31}$ are similar. This implies that the first triangle may be transformed into the other by a rotation through angle γ', about center P_{12}, followed by a dilation. Since the latter does not change the direction of any line we conclude that $\angle P^1_{23}HP_{31}$ is equal to γ' and hence $\angle P_{31}HP_{23} = \pi - \gamma'$. The point H has therefore the property that, from the viewpoint H, the sides of the pole triangle appear under the angles $\pi - \alpha'$, $\pi - \beta'$ and $\pi - \gamma'$ respectively. This implies that the four points H, P_{12}, P^1_{23}, P_{31} are on a circle, and analogously H, P_{23}, P^2_{31}, P_{12} and H, P_{31}, P^3_{12}, P_{23}, are on circles. An important conclusion is the following: three positions of a plane E with respect to Σ in equiform kinematics are completely determined by the pole triangle $P_{23}P_{31}P_{12}$ and the point H. Indeed: C_i are the circles through H and two vertices, P^1_{23} is the intersection of $P_{23}H$ and C_1, etc.

The configuration of Fig. 94 is, independent of its kinematical meaning, well-known in geometry. It is there built up from a triangle $A_1A_2A_3$ on the sides of which are constructed outwardly the triangles $A'_1A_2A_3$, $A_1A'_2A_3$ and

$A_1A_2A_3'$, all three similar to a given second triangle $B_1B_2B_3$. Then A_1A_1', A_2A_2', A_3A_3' pass through a point M, called the meta-pole of $A_1A_2A_3$ with respect to $B_1B_2B_3$; it is also the common point of the circumcircles of the three associated triangles (JOHNSON [1960]).

Example 25. In Fig. 94 we have $\alpha + \alpha' < \pi$, $\beta + \beta' < \pi$, $\gamma + \gamma' < \pi$. Consider the case when these inequalities are not all satisfied, which implies that H is outside the polar triangle or on its perimeter.

For arbitrary frames, the three positions (X_i, Y_i) in Σ of the point (x, y) of E are given by

$$
\begin{aligned}
X_i &= s_i(x \cos \phi_i - y \sin \phi_i) + a_i, \\
Y_i &= s_i(x \sin \phi_i + y \cos \phi_i) + b_i, \quad i = 1, 2, 3.
\end{aligned}
\tag{3.23}
$$

Three homologous points are collinear if $|X_i\ Y_i\ 1| = 0$, from which, after developing the determinant, we see that the locus of the points of E for which the three positions in Σ are collinear is a circle C. If C_i is C in the ith position, the points P_{12}, P_{31} and P_{23}^1 are on C_1; this follows since for P_{12} of E_1 the first and the second position coincide, for P_{31} the first and third, for P_{23}^1 the second and the third (at P_{23}). The conclusion is: C_1 is the circumcircle of P_{12}, P_{31}, P_{23}^1; C_2 is that of P_{23}, P_{12}, P_{31}^2; C_3 that of P_{31}, P_{23}, P_{12}^3; we know that the three circles C_i pass through H. The radii of C_1, C_2, C_3 are in the ratio $k_1 : k_2 : k_3$.

We now turn to the displacement of lines. From (3.18) it follows that the line $ux + vy + w = 0$ is transformed into $UX + VY + W = 0$, such that

$$
\begin{aligned}
&U = u \cos \phi - v \sin \phi, \qquad V = u \sin \phi + v \cos \phi, \\
&W = -(a \cos \phi + b \sin \phi)u + (a \sin \phi - b \cos \phi)v + sw.
\end{aligned}
\tag{3.24}
$$

For three positions defined by s_i, ϕ_i, a_i, b_i $(i = 1, 2, 3)$ three homologous lines are concurrent if $|U_i\ V_i\ W_i| = 0$. If we develop this determinant it is immediately seen that it has $u^2 + v^2$ as a factor; indeed: the isotropic lines do satisfy the condition. Excluding these trivial cases the condition is seen to be linear in u, v, w. The conclusion is: those lines of E for which three positions are concurrent belong to a pencil: its vertex is of course a point of E.

We apply this result to the inverse displacements, by interchanging E and Σ. Keeping in mind that if for the direct displacement a point of E moves along a line of Σ then in the inverse displacement a line of Σ passes permanently through a point of E, we obtain: the lines of Σ which each pass through three homologous points of E, all pass through one point. But $P_{23}P_{23}^1$, $P_{31}P_{31}^2$, $P_{12}P_{12}^3$ are three such lines; we already know that they are concurrent (and we have now even given a new proof). Summing up we have: every point of C (in E) has three positions on one line, all these lines pass through H. If A_1 is an arbitrary point on C_1 then A_2 and A_3 are found as the second intersections of A_1H with C_2 and C_3 respectively (Fig. 95).

Example 26. Determine A_2 and A_3 if A_1 coincides with H.

We have seen that the lines l of E such that l_1, l_2, l_3 pass through one point are those of a pencil. Let the vertex of the pencil be denoted by D (in E) and the intersection point of l_i by S (in Σ). Again making use of the inverse motion we conclude: the locus of all S, a figure in Σ, is the circumcircle c_0 of the polar triangle. The point D_1 is the point in E_1 which acts as H for the inverse motion; this implies that D_1 is the (second) intersection of $P_{23}P_{23}^1$ with c_0, D_2 and D_3 are those of $P_{31}P_{31}^2$ and $P_{12}P_{12}^3$ with this circle. If l_1 passes through D_1 then l_2 and l_3 must pass through D_2 and D_3 respectively; S is the other intersection of each l_i and c_0 (Fig. 96).

Three positions theory in plane equiform kinematics as developed here

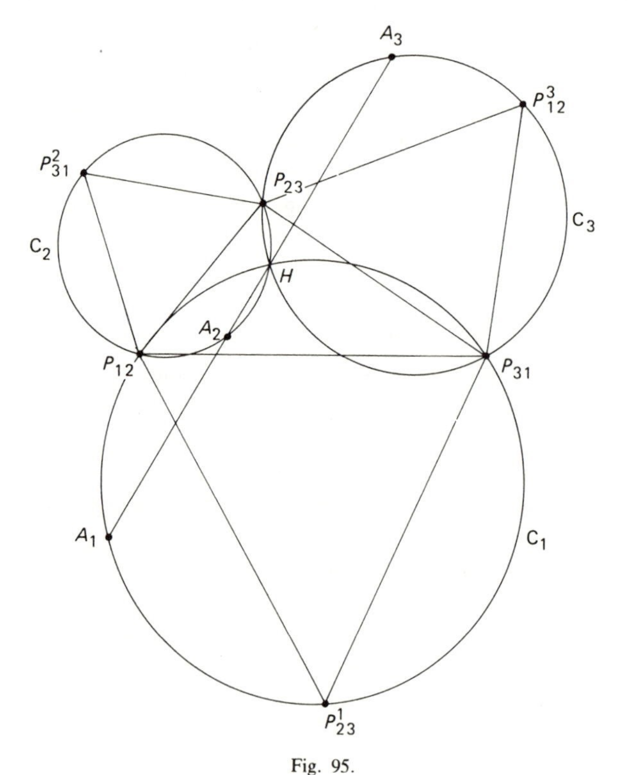

Fig. 95.

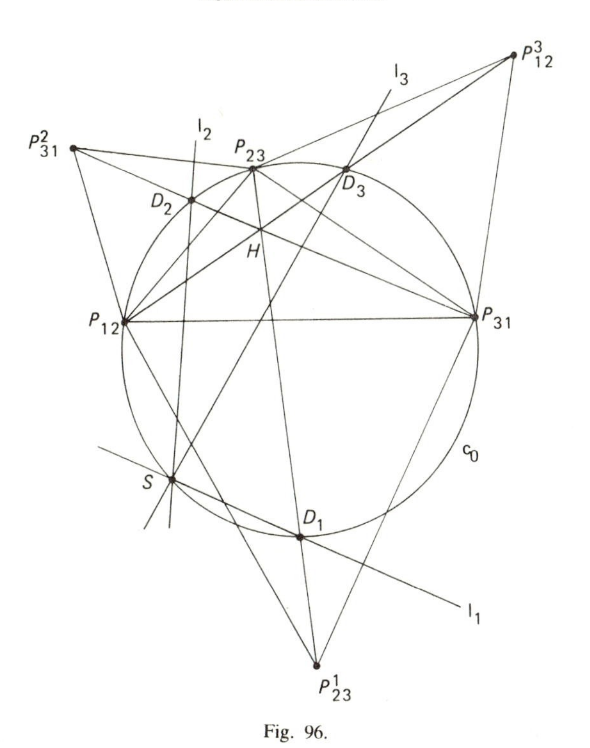

Fig. 96.

appears as a generalization of the analogous theory in ordinary kinematics; in the latter the three positions of each point are determined if the pole triangle and hence also its orthocenter is given; in our more general case the orthocenter is replaced by an arbitrarily chosen point of Σ. In ordinary kinematics we have introduced the concept of the *basic point* A^* of a triad of homologous points; the latter are the reflections of A^* into the sides of the pole triangle. In order to extend this notion to equiform kinematics we make use of a transformation T, called an axial affinity, defined as follows. A line $\mathfrak{a}$, the axis, is given and also a line $\mathfrak{l}$ (not parallel to $\mathfrak{a}$) and a number k, unequal to zero or unity (Fig. 97). To transform the arbitrary point A, the line $\mathfrak{l}'$ is drawn through A parallel to $\mathfrak{l}$, intersecting $\mathfrak{a}$ at S. A is transformed by T into A', on $\mathfrak{l}'$, so that $SA' = k(SA)$.

Example 27. Show that the transformation is linear and that the line at infinity is invariant, which implies that T is a (special) affinity.

Example 28. Determine the invariant points of T.

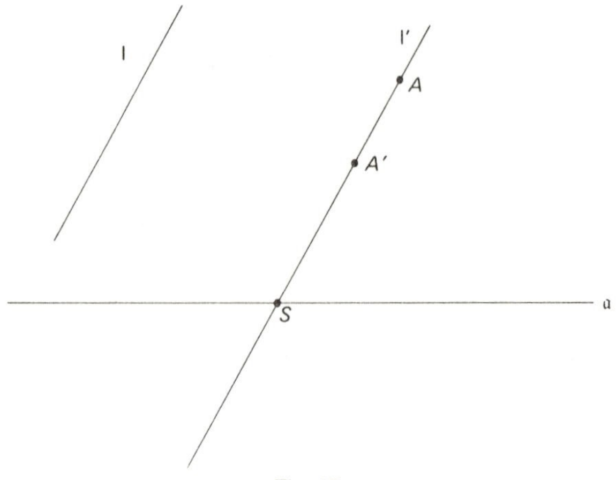

Fig. 97.

Example 29. Show that T is completely determined by the axis a and a pair A, A'; determine B' if B is given.
Example 30. Show that the reflection into a is an axial affinity.

If the pole triangle and the points P^1_{23}, P^2_{31}, P^3_{12} are given we define three axial affinities T_i as follows: T_1 has the axis $P_{12}P_{31}$ and transforms P_{23} into P^1_{23}, and analogously for T_2 and T_3. Let A^* be an arbitrary point in the plane Σ and let T_i transform A^* into A_i. We assert that A_1, A_2, A_3 are three homologous points. Obviously $T_2T_1^{-1}$, being the product of two affine transformations is an affinity; it transforms A_1 into A_2. By T_1^{-1} the points P_{12}, P_{31}, P^1_{23} are transformed into P_{12}, P_{31}, P_{23}, and these points are transformed by T_2 in P_{12}, P^2_{31}, P_{23}; hence $T_2T_1^{-1}$ transforms P_{12}, P_{31}, P^1_{23} into P_{12}, P^2_{31}, P_{23}, and since an affine transformation is determined by three pairs of conjugate points, $T_2T_1^{-1}$ transforms a triangle into a similar triangle and it is therefore the equiform transformation from position 1 into 2; this concludes the proof (Fig. 98).

Example 31. Show that T_i are the reflections into the sides of the pole triangle if H coincides with the orthocenter; the scale factor k is equal to -1 in this case.
Example 32. Determine A_i $(i = 1, 2, 3)$ if the basic point A^* is on a side of the pole triangle.
Example 33. Determine A_i if A^* coincides with one of the vertices of the pole triangle.
Example 34. Determine A_i if A^* coincides with H.
Example 35. Show that the locus of A^* such that A_i are collinear is an ellipse, passing through the vertices of the pole triangle.
Example 36. Show that the scale factor k_1 of T_1 is equal to

$$-(\sin\alpha \sin\beta' \sin\gamma')/(\sin\alpha' \sin\beta \sin\gamma),$$

and analogously for k_2 and k_3.

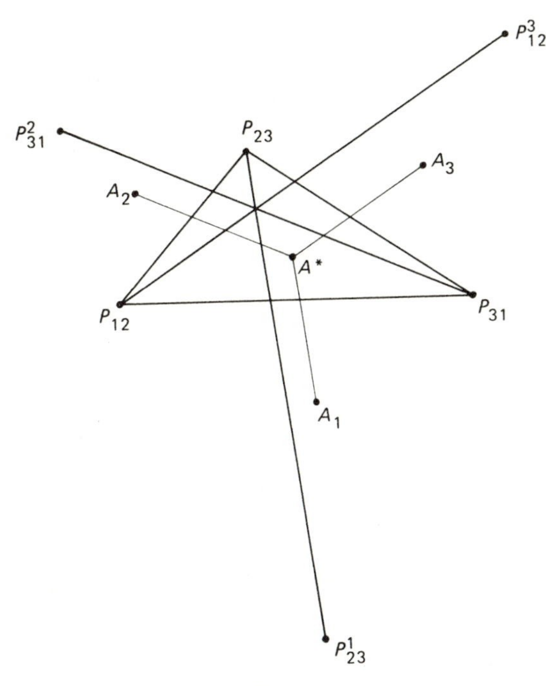

Fig. 98.

In the three-positions theory of ordinary kinematics there exists a birational relationship between the basic point A^* and the circumcenter M of the three homologous points A_i associated with A^*; (A^* and M are isogonal conjugates with respect to the pole triangle). This theorem cannot be extended to equiform kinematics; indeed we shall see further on that such a birational relationship does not even exist in the instantaneous case (Müller's theorem).

We shall derive the relationship between the point $A(x, y)$ of the moving plane and the circumcenter M of the homologous points A_i $(i = 1, 2, 3)$. The coordinates of A_i are

$$X_i = s_i(x \cos \phi_i - y \sin \phi_i) + a_i,$$
$$Y_i = s_i(x \sin \phi_i + y \cos \phi_i) + b_i, \tag{3.25}$$

where ϕ_i, s_i, a_i, b_i are the constants defining the three positions. The circumcircle of A_1, A_2, A_3 has the equation

$$F_0(X^2+Y^2)+F_1X+F_2Y+F_3=0, \tag{3.26}$$

its coefficients being the determinants of the 4×3-matrix $\|X_i^2+Y_i^2\ X_i\ Y_i\ 1\|$. From (3.25) it follows that

$$\begin{aligned} X_i^2+Y_i^2&=s_i^2(x^2+y^2)+\mathrm{l}_i(x,y),\\ X_2Y_3-X_3Y_2&=s_2s_3(x^2+y^2)\sin(\phi_3-\phi_2)+\mathrm{l}_i'(x,y), \end{aligned} \tag{3.27}$$

and cyclically for $X_3Y_1-X_1Y_3$ and $X_1Y_2-X_2Y_1$; l_i and l_i' are linear functions of x, y. We obtain

$$F_0:F_1:F_2=[c_0(x^2+y^2)+\mathrm{L}_0(x,y)]:[c_1(x^2+y^2)\mathrm{L}_1(x,y)+\mathrm{Q}_1(x,y)]:$$
$$[c_2(x^2+y^2)\mathrm{L}_2(x,y)+\mathrm{Q}_2(x,y)],$$

where c_0, c_1, c_2 are constants, L_0, L_1, L_2 linear functions and Q_1, Q_2 quadratic functions. The center M of the circle (3.26) has the coordinates $-F_1/2F_0$ and $-F_2/2F_0$. Hence we obtain the following relationship between the homogeneous coordinates of M and those of A:

$$\begin{aligned} X_M&=c_1(x^2+y^2)\mathrm{L}_1(x,y,z)+z\,\mathrm{Q}_1(x,y,z),\\ Y_M&=c_2(x^2+y^2)\mathrm{L}_2(x,y,z)+z\,\mathrm{Q}_2(x,y,z),\\ Z_M&=-2z[c_0(x^2+y^2)+z\,\mathrm{L}_0(x,y,z)], \end{aligned} \tag{3.28}$$

which is obviously a cubic relationship. It has five singular points; indeed, if A coincides with one of the three poles in the moving plane (that is if for instance $X_2=X_3$, $Y_2=Y_3$) we have $X_M=Y_M=Z_M=0$ and from (3.27) it follows that this is also the case when A coincides with an isotropic point ($x^2+y^2=0$, $z=0$). This implies that the points of a line $p_1X_M+p_2Y_M+p_3Z_M=0$ correspond to those of a cubic curve in the (x,y,z)-plane, every cubic passes through the same five singular points. Hence a point M corresponds to the intersections of two such curves, the number of essential intersections being $3\times 3-5=4$. We have derived the theorem: *a point of the fixed plane is the circumcenter of four triads of homologous points (real or imaginary).*

Example 37. Show that the line $Z_M=0$ contains the centers corresponding to a degenerate cubic consisting of the combination of the line at infinity and a circle, which is the locus of A if A_i are collinear.

Example 38. Show that in the case of ordinary kinematics ($s_1=s_2=s_3=1$) we have $c_1=c_2=0$; the relationship is quadratic, with three singular points; it is birational (because $2\times 2-3=1$).

We illustrate the theorem by an example. Let the three positions be given by

$$
(3.29)\qquad \begin{aligned} &X_1 = x, \qquad Y_1 = y; \qquad X_2 = -sy - p, \qquad Y_2 = sx + sp;\\ &X_3 = sy + p, \qquad Y_3 = -sx + sp. \end{aligned}
$$

From this it follows that (Fig. 99)

$$
\begin{aligned} &P_{12} = (-p, 0), \qquad P_{23} = (0, sp), \qquad P_{31} = (p, 0);\\ &P_{12}^3 = (p, 2sp), \qquad P_{23}^1 = (0, -s^{-1}p), \qquad P_{31}^2 = (-p, 2sp). \end{aligned}
$$

The pole triangle is isosceles with two angles equal to ψ. We have

$$
\begin{aligned} &\phi_{12} = \pi/2, \qquad \phi_{23} = \pi, \qquad \phi_{31} = \pi/2,\\ &s_{12} = s = \tan\psi, \qquad s_{23} = 1, \qquad s_{31} = s^{-1} = \cot\psi. \end{aligned}
$$

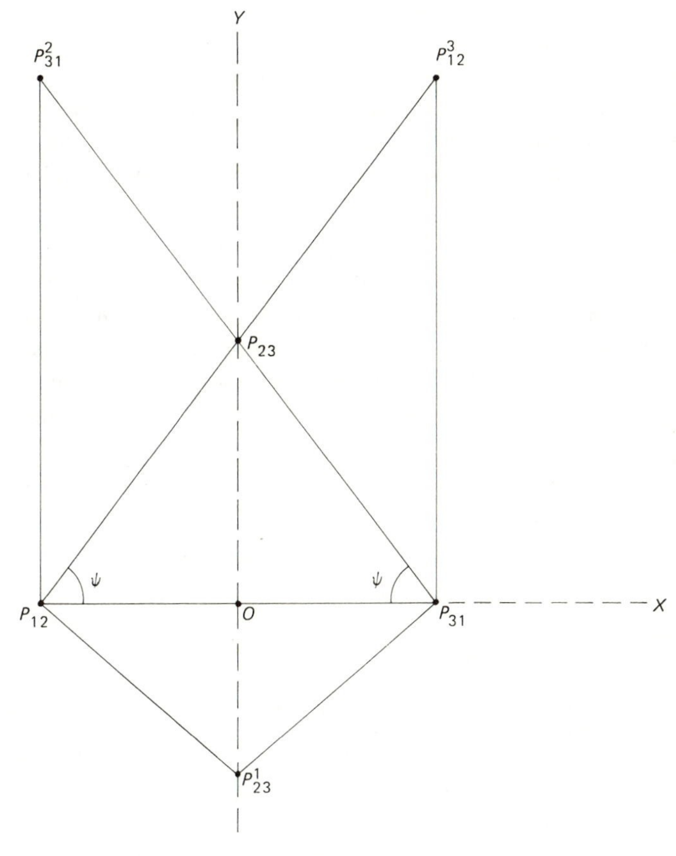

Fig. 99

The point H coincides with P_{23}.

Substituting (3.29) into the matrix $\|X_i^2 + Y_i^2\ X_i\ Y_i\ 1\|$ we obtain after some algebra the following relationship between the displaced point $A(x, y)$ and the circumcenter $M(X_M/Z_M, Y_M/Z_M)$ of its three positions

$$
\begin{aligned}
X_M &= s(s^2-1)(x^2+y^2)x - spx[-4sy + p(s^2-1)],\\
Y_M &= s(s^2-1)(x^2+y^2)y - (s^2+1)px^2 + (3s^2-1)py^2\\
&\qquad + s(s^2+3)p^2y + (s^2+1)p^3,\\
Z_M &= -2[s(x^2+y^2) - (s^2-1)py - sp^2].
\end{aligned} \tag{3.30}
$$

We shall determine A if M coincides with O; hence we must solve the set $X_M = 0$, $Y_M = 0$. The first equation represents a degenerate cubic, consisting of O_Y and a circle C_2 passing through P_{12} and P_{31}; $Y_M = 0$ represents a nondegenerate circular cubic C_3, passing through P_{12}, P_{23}, P_{31}. The points asked for are: the two intersections S_1, S_2 (different from P_{23}^1) of O_Y and C_3, and the two intersections S_3, S_4 (different from P_{12}, P_{31} and the isotropic points) of C_2 and C_3. The four points can be determined by elementary algebra.

Example 39. Determine the (x, y) coordinates of S_i $(i = 1, 2, 3, 4)$. Show that S_3S_4 is parallel to O_X. Show that S_1, S_2 can be real or imaginary. Show that S_3, S_4 can be real or imaginary.

Equiform three positions theory could be treated by means of triangular coordinates with respect to the pole triangle in the same way as was done in Chapter VIII (Example 28) for ordinary kinematics. This leads, however, to complicated algebra. Some results are given in the following examples.

Example 40. If (X, Y, Z) are the coordinates of a point in the fixed plane, show that the axial affinity with respect to $P_{31}P_{12}$ is represented by

$$
\begin{aligned}
X' &= -(\sin\alpha \sin\beta' \sin\gamma')X,\\
Y' &= (\sin\alpha \sin\beta' \sin(\gamma+\gamma'))X + (\sin\alpha' \sin\beta \sin\gamma)Y,\\
Z' &= (\sin\alpha \sin\gamma' \sin(\beta+\beta'))X + (\sin\alpha' \sin\beta \sin\gamma)Z.
\end{aligned}
$$

Show that for $\alpha' = \alpha$, $\beta' = \beta$, $\gamma' = \gamma$ these formulas are identical with those for ordinary kinematics (Chapter VIII, Example 29).

Example 41. Consider the special case $\alpha' = (\pi/2) - (\alpha/2)$, $\beta' = (\pi/2) - (\beta/2)$, $\gamma' = (\pi/2) - (\gamma/2)$. Show that H coincides with the incenter of the pole triangle and P_{23}^1, P_{31}^2, P_{12}^3 with its excenters. Determine the coordinates (x_i, y_i, z_i) of three homologous points A_i associated with the basic point $A^*(x, y, z)$. Show that the locus of A^* in the case of three collinear points A_i is given by

$$(x \sin\alpha + y \sin\beta + z \sin\gamma)(yz \cos(\alpha/2) + zx \cos(\beta/2) + xy \cos(\gamma/2)) = 0,$$

representing the line at infinity and a circumscribed ellipse of the pole triangle.

We consider a zero-position with coinciding frames which implies $\phi = 0$,

$s_0 = 1$, $a_0 = b_0 = 0$. Furthermore we take the origin at the pole P; hence $a_1 = b_1 = 0$. Then from (3.15) it follows that

$$X_1 = s_1 x - y, \qquad Y_1 = x + s_1 y, \tag{3.31}$$

which gives us the tangent to the path of the moving point $A = (x, y)$ in the zero-position. For $s_1 = 0$ we have the Euclidean case: we assume $s_1 \neq 0$.

For the angle γ between the tangent and the line OA we have $\gamma = \beta - \alpha$ (Fig. 100), with $\tan \beta = (x + s_1 y)/(s_1 x - y)$, $\tan \alpha = y/x$; hence $\tan \gamma = s_1^{-1}$. The path tangent at A has a constant angle ($\neq \pi/2$) with the line OA. All points on OA have parallel path tangents.

Example 42. In plane equiform kinematics three points A_i and their path tangents t_i are known ($i = 1, 2, 3$). Determine the pole. [Hint: if S_{ij} is the intersection of t_i and t_j the pole P is on the circles $A_1A_2S_{12}$, $A_2A_3S_{23}$ (and $A_3A_1S_{31}$). This construction fails if t_1, t_2, t_3 pass through one point on the circle $A_1A_2A_3$ (R. MÜLLER [1907]).]

Differentiating (3.16) with respect to ϕ and substituting $\phi = 0$, we obtain

$$(s_1^2 + 1)(x_p)_1 = -s_1 a_2 - b_2, \qquad (s_1^2 + 1)(y_p)_1 = a_2 - s_1 b_2, \tag{3.32}$$

and the same procedure for (3.17) gives

$$(s_1^2 + 1)(X_p)_1 = -s_1 a_2 - b_2, \qquad (s_1^2 + 1)(Y_p)_1 = a_2 - s_1 b_2, \tag{3.33}$$

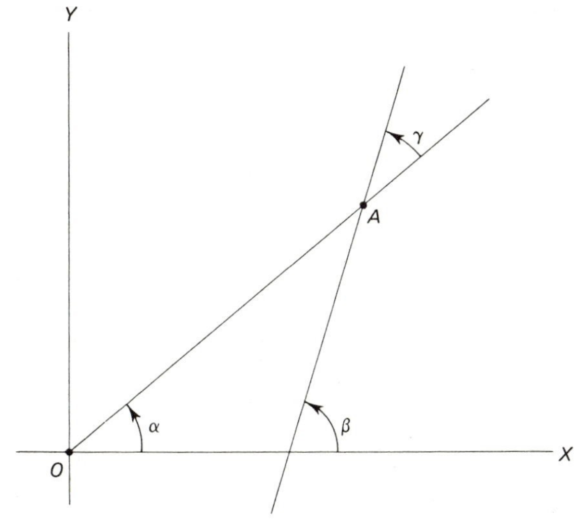

Fig. 100.

which shows that the centrodes p_m and p_f are tangent to one-another at the pole. We take the common tangent to be O_x, which implies $a_2 = s_1 b_2$; we choose the orientation of the axes so that $s_1 a_2 + b_2 > 0$ (we exclude the singular case $(x_p)_1 = (y_p)_1 = 0$). The canonical frame is now completely determined. By differentiating (3.15), and then setting $\phi = 0$, we obtain the following scheme:

$$
\begin{aligned}
&X_0 = x, && Y_0 = y,\\
&X_1 = s_1 x - y, && Y_1 = x + s_1 y,\\
&X_2 = (s_2 - 1)x - 2s_1 y + s_1 b_2, && Y_2 = 2s_1 x + (s_2 - 1)y + b_2,\\
&X_3 = (s_3 - 3s_1)x + (-3s_2 + 1)y + a_3, && Y_3 = (3s_2 - 1)x + (s_3 - 3s_1)y + b_3,
\end{aligned}
\tag{3.34}
$$

and so on.

All instantaneous properties up to the nth order depend on the instantaneous invariants s_k $(k = 1, \ldots, n)$, a_k $(k = 3, 4, \ldots, n)$, b_k $(k = 2, 3, \ldots, n)$.

Example 43. Derive the expressions for X_n, Y_n.
Example 44. Show that for the inverse motion (3.14) yields $\tilde{\phi} = -\phi$, $\tilde{s} = s^{-1}$, $\tilde{a} = -s^{-1}(a \cos \phi + b \sin \phi)$, $\tilde{b} = s^{-1}(a \sin \phi - b \cos \phi)$. Derive the canonical frames for the inverse motion and determine its instantaneous invariants up to the third order.

It follows from (3.14) that a line (u, v, w) of the moving plane is transformed into (U, V, W) such that

$$
\begin{aligned}
&U = u \cos \phi - v \sin \phi, \qquad V = u \sin \phi + v \cos \phi,\\
&W = -(a \cos \phi + b \sin \phi)u + (a \sin \phi - b \cos \phi)v + sw,
\end{aligned}
\tag{3.35}
$$

from which we obtain for the canonical frames

$$
\begin{aligned}
&U_0 = u, && V_0 = v, && W_0 = w,\\
&U_1 = -v, && V_1 = u, && W_1 = s_1 w,\\
&U_2 = -u, && V_2 = -v, && W_2 = -s_1 b_2 u - b_2 v + s_2 w,\\
&U_3 = v, && V_3 = -u, && W_3 = -(a_3 + 3b_2)u + (3s_1 b_2 - b_3)v + s_3 w, \text{ etc.}
\end{aligned}
\tag{3.36}
$$

We shall derive some second order instantaneous properties by means of (3.34) and (3.36).

A moving point passes through an inflection point of its path if $X_1 Y_2 - X_2 Y_1 = 0$; the locus of these points, the inflection curve is therefore

$$
C_1 \equiv (1 + 2s_1^2 - s_2)(x^2 + y^2) - b_2(1 + s_1^2)y = 0,
\tag{3.37}
$$

which represents a circle c tangent to O_x at the pole (Fig. 101). If A is a point on c the tangent to its path intersects c again at B which is the same for all points A because $\angle OAB = \psi$, with $\tan\psi = s_1^{-1}$. Hence there exists an inflection pole B, on c, but not lying on the pole normal O_Y.

Example 45. Show that $B = (b_2 s_1/N, b_2/N)$, with $N = 1 + 2s_1^2 - s_2$.
Example 46. Show that for $1 + 2s_1^2 - s_2 = 0$ the inflection circle degenerates into O_x and the line at infinity; determine B in this case.

A moving line $l = (u, v, w)$, according to (3.26), intersects its consecutive line at $S = [(s_1 v - u)w/(u^2+v^2), (-s_1 u - v)w/(u^2+v^2)]$; S is the point where l is tangent to its envelope.

Example 47. Show that the angle between l and OS is equal to ψ (Fig. 102).

The lines which in three consecutive positions pass through one point satisfy $|U_i\ V_i\ W_i| = 0$, $i = 0, 1, 2$. Their locus is, in view of (3.36) given by

$$(u^2+v^2)[-s_1 b_2 u - b_2 v + (s_2+1)w] = 0, \tag{3.38}$$

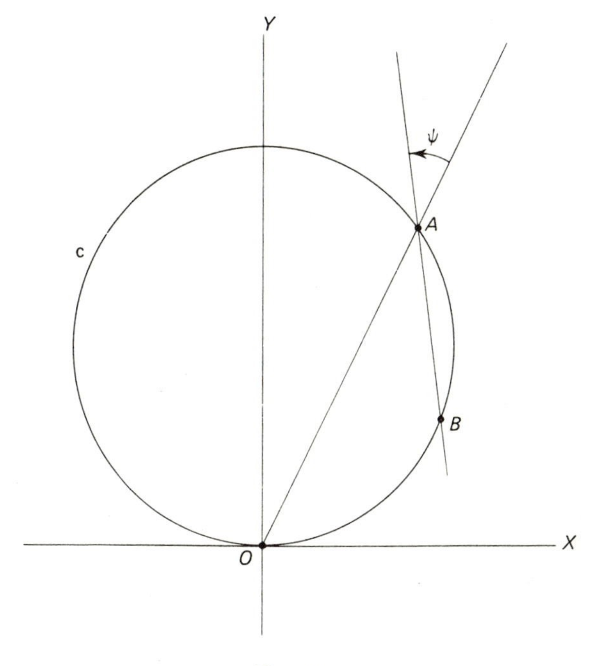

Fig. 101.

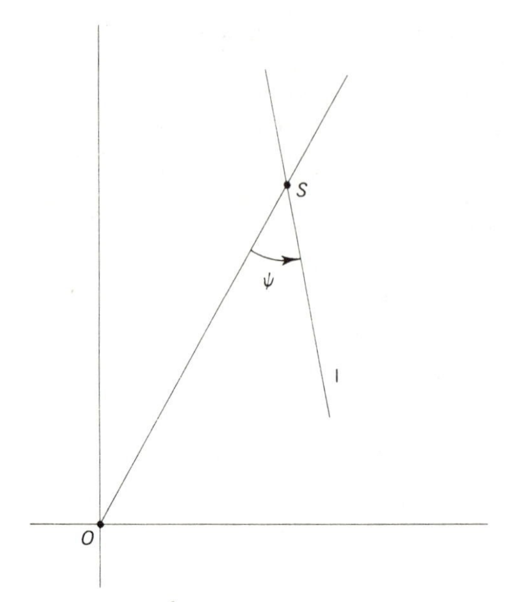

Fig. 102.

which implies that it consists of the isotropic lines and also of those passing through $K = [-s_1b_2/(1+s_2), -b_2/(1+s_2)]$, we call this point the cuspidal pole of the zero-position.

Example 48. Show that BK passes through O, that its angle with O_x is ψ, and determine the ratio $KO : OB$ (Fig. 103).

The points S corresponding to the lines through K are seen to lie on the circle, called the cuspidal circle, $\tilde{c}$, with the equation

$$(3.39) \qquad (1+s_2)(x^2+y^2)+b_2(1+s_1^2)y=0,$$

passing through K and tangent to O_x at O (Fig. 103).

Example 49. Determine in which ways Fig. 103 and the corresponding one in the Euclidean case agree and disagree.

Although the concept of "radius of curvature" does not exist in equiform geometry (since there is no such concept as length or distance) one may still deal with the center of curvature of a curve. It may be defined as the intersection of two consecutive normals or as the center of the circle through three consecutive points. From this it follows that the center M corresponding to the moving point A has the coordinates

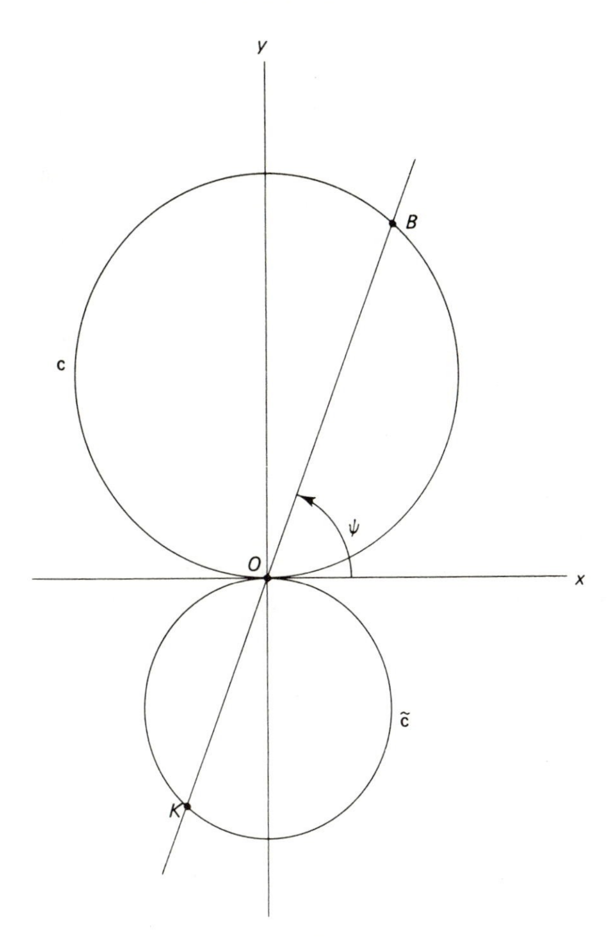

Fig. 103.

$$X_M = X_0 - Y_1(X_1^2 + Y_1^2)/(X_1 Y_2 - X_2 Y_1),$$

$$Y_M = Y_0 + X_1(X_1^2 + Y_1^2)/(X_1 Y_2 - X_2 Y_1),$$

which gives us by means of (3.34), making use of homogeneous coordinates

$$\begin{aligned} X_M &= (x^2 + y^2)[(s_1^2 - s_2)x - s_1(1 + s_1^2)y] - b_2(1 + s_1^2)xyz, \\ Y_M &= (x^2 + y^2)[s_1(1 + s_1^2)x + (s_1^2 - s_2)y] - b_2(1 + s_1^2)y^2 z, \\ Z_M &= [(1 + 2s_1^2 - s_2)(x^2 + y^2) - b_2(1 + s_1^2)yz]z, \end{aligned} \tag{3.40}$$

which expresses the coordinates of M in terms of those of A and the instantaneous invariants up to the second order. The right-hand sides are cubic polynomials. The formulas (3.40) are of course more complicated than those in Euclidean kinematics because they depend on s_1 and s_2 (which are zero in the usual theory). But, as R. MÜLLER [1910a] pointed out, there is also another important difference: In the elementary case (3.40) can be solved for x, y, z, giving (in general) *one* solution; A determines M, conversely M determines A: the relationship between a point and the corresponding center of curvature is birational. This is no longer the case in equiform kinematics. We shall prove Müller's theorem: to each M there correspond *two* points A (real or imaginary).

It follows from (3.40) that

$$\begin{aligned}&[s_1(1+s_1^2)x+(s_1^2-s_2)y-b_2s_1^2z]X_M-[(s_1^2-s_2)x-s_1(1+s_1^2)y+b_2s_1z]Y_M\\ &\quad+b_2s_1(s_1x+y)Z_M=0,\end{aligned}\tag{3.41}$$

which can be verified by direct substitution from (3.40). It is a bilinear relation between X_M, Y_M, Z_M and x, y, z which yields the equation of a straight line for the locus of A if M is given. Furthermore M lies on the path normal of A (Fig. 104). Hence, $\angle MAO=(\pi/2)-\psi$, which implies that A is on a specific

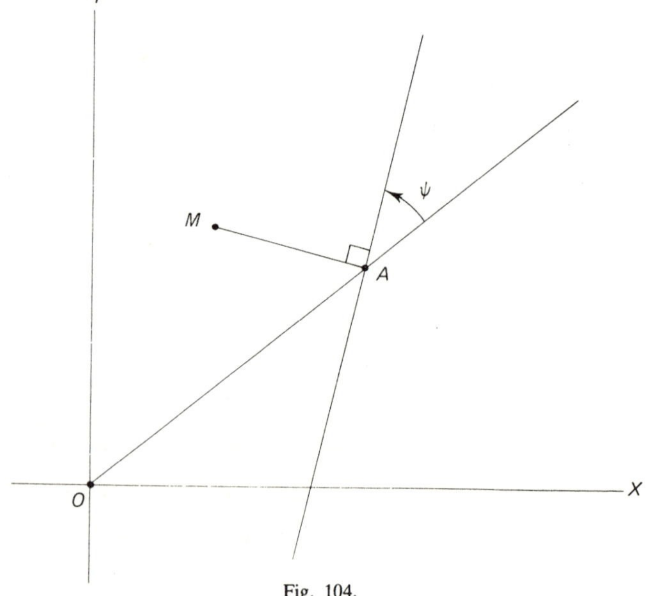

Fig. 104.

circle through O and M, with the equation (using Cartesian, not homogeneous, coordinates)

$$s_1(x^2+y^2)-(s_1X_M+Y_M)x-(s_1Y_M-X_M)y=0. \tag{3.42}$$

Example 50. Verify (3.42) by means of (3.40).

If $M(X_M, Y_M)$ is given, $A(x, y)$ follows from (3.41) and (3.42), that is from the intersection of the circle and the straight line

$$\begin{aligned}&[s_1(1+s_1^2)X_M-(s_1^2-s_2)Y_M+b_2s_1^2]x\\&+[(s_1^2-s_2)X_M+s_1(1+s_1^2)Y_M+b_2s_1]y\\&-b_2s_1(s_1X_M+Y_M)=0,\end{aligned} \tag{3.43}$$

and there are indeed in general two solutions, real or imaginary.

Example 51. Show that there are two different real points A if M is outside a certain (rational) bicircular quartic with a cusp at O (R. MÜLLER [1910a], KRAUSE [1920]).

We consider some third order properties: A moving point A will pass through an undulation point of its path (the tangent having four coinciding points in common with the curve) if $X_1:Y_1=X_2:Y_2=X_3:Y_3$. The equality of the first two ratios implies that A is on the inflection circle. The condition $X_1Y_3-X_3Y_1=0$ gives us by means of (3.34):

$$C_2\equiv(2s_1+3s_1s_2-s_3)(x^2+y^2)+(s_1b_3-a_3)x-(s_1a_3+b_3)y=0, \tag{3.44}$$

which represents a circle through O. It has in general one real intersection with the inflection circle (in addition to the trivial one at O). The conclusion is: there is, as a rule, one point passing through an undulation point of its path, the point of Ball for equiform kinematics.

Example 52. Determine the coordinates of Ball's point. Consider the special cases: 1) there is no proper Ball point because the second intersection of the two circles coincides with O; 2) every point of the inflection circle is a Ball point; 3) the inflection circle is degenerate.

Example 53. Show that the intersections of two consecutive inflection circles are the pole and Ball's point.

The centers of curvature of the centrodes are also third order concepts. The equations of the centrodes are given by (3.16) and (3.17). Let m and f be the centers of curvature at O of the moving and the fixed centrode respectively. They are on the pole normal. Since we have $(x_p)_0=(y_p)_0=0$, $(x_p)_1=-b_2$, $(y_p)_1=0$ it follows that the center $m(x_m, y_m)$ is at

$$x_m=0, \qquad y_m=b_2^2/(y_p)_2$$

which gives after differentiating y_p, in (3.16), two times

(3.45) $\quad y_m = b_2^2(1+s_1^2)^2/[2b_2(1-s_2+s_1^2(3+s_2))+(a_3-s_1b_3)(1+s_1^2)],$

and by the same procedure

(3.46) $\quad Y_f = b_2^2(1+s_1^2)^2/[b_2(1-2s_2+s_1^2(4+2s_2-s_1^2))+(a_3-s_1b_3)(1+s_1^2)],$

from which it follows

(3.47) $$(1/y_m)-(1/Y_f)=1/b_2,$$

which shows that the (second) Euler-Savary relation holds in equiform kinematics.

Four consecutive points of a path are on a circle if $|(X^2+Y^2)_n, X_n, Y_n|=0$ $(n=1,2,3)$, or

$$(X_1Y_3-X_3Y_1)(X_1^2+Y_1^2)-3(X_1X_2+Y_1Y_2)(X_1Y_2-X_2Y_1)=0.$$

By means of (3.34) we obtain

(3.48) $$(1+s_1^2)(x^2+y^2)C_2-3C_3C_1=0,$$

where C_1 and C_2 are given by (3.37) and (3.44) and furthermore

(3.49) $$C_3 \equiv s_1(1+s_2)(x^2+y^2)+b_2(1+s_1^2)x.$$

Hence the locus of the points with stationary curvature, the circling point curve, represented by (3.48) is a bicircular quartic K_4; it also has a node at O, with O_x and O_y as tangents. Since K_4 has three double points, it is a rational curve.

Example 54. Derive a rational parametric representation of K_4 by intersecting (3.48) with the set of circles $x^2+y^2=\lambda x$; any circle has one finite intersection with K_4, different from O.
Example 55. Show that K_4 passes through Ball's point.

We shall give some examples of special continuous motions in plane equiform kinematics. The moving plane has four degrees of freedom, which implies that in general a (one-degree-of-freedom) motion with be defined if the paths of three moving points are given. A simple case is a motion for which three points $A_i(x_i, y_i)$ remain on three lines $L_i \equiv l_iX+m_iY+n_i=0$ respectively $(i=1,2,3)$. Then the following equations must be satisfied

(3.50) $\quad l_i[s(x_i\cos\phi-y_i\sin\phi)+a]+m_i[s(x_i\sin\phi+y_i\cos\phi)+b]+n_i=0,$

which are three linear equations for s, a, b. The solutions are

(3.51)
$$\begin{aligned} s &= C\Delta^{-1},\\ a &= (A_1\sin\phi+A_2\cos\phi)\Delta^{-1},\\ b &= (B_1\sin\phi+B_2\cos\phi)\Delta^{-1},\\ \Delta &= A\sin\phi+B\cos\phi, \end{aligned}$$

where A, B, A_1, B_1, A_2, B_2, C are constants depending on the given data; $C \neq 0$ if the three lines L_i are not concurrent. Without any loss of generality we may suppose that the moving and the fixed frame coincide for $\phi = 0$. In which case $C = B$, $A_2 = 0$, $B_2 = 0$ and the motion is

$$\begin{aligned} X &= (1/(A \sin \phi + B \cos \phi))[B(x \cos \phi - y \sin \phi) + A_1 \sin \phi] \\ &= ((A_1 - By)u + Bx)/(Au + B), \\ Y &= (1/(A \sin \phi + B \cos \phi))[B(x \sin \phi + y \cos \phi) + B_1 \sin \phi] \\ &= ((B_1 + Bx)u + By)/(Au + B), \end{aligned} \tag{3.52}$$

with $u = \tan \phi$.

As X, Y are linear functions of the parameter u we have: if the paths of three points are straight lines the path of every point is a straight line.

From (3.52) we see that X, Y are independent of u if $(A_1 - By) : A = x$ and $(B_1 + Bx) : A = y$; hence $x = (A_1A - B_1B)/(A^2 + B^2)$, $y = (A_1B + AB_1)/(A^2 + B^2)$ is a fixed point during the motion. If we choose it as the common origin of the frames we have $A_1 = B_1 = 0$ and the motion reads, with $A/B = k$:

$$\begin{aligned} X &= (k \sin \phi + \cos \phi)^{-1}(x \cos \phi - y \sin \phi), \\ Y &= (k \sin \phi + \cos \phi)^{-1}(x \sin \phi + y \cos \phi). \end{aligned} \tag{3.53}$$

For $\tan \phi = -k^{-1}$ we have a singular position: all points are at infinity. As the X- and Y-axes are not yet determined we may take them in such a way that the singular position occurs when ϕ is equal to $\pi/2$; hence $k = 0$ and we have the following standard representation of the motion

$$\begin{aligned} X &= (\cos \phi)^{-1}(x \cos \phi - y \sin \phi), \\ Y &= (\cos \phi)^{-1}(x \sin \phi + y \cos \phi). \end{aligned} \tag{3.54}$$

From this it follows, by eliminating ϕ, that the path of (x, y) reads

$$xX + yY - (x^2 + y^2) = 0. \tag{3.55}$$

Conversely, the line $UX + VY + W = 0$ is the path of the point $x = -UW/(U^2 + V^2)$, $y = -VW/(U^2 + V^2)$, which shows that any line is in general the path of some one point. There is therefore a birational quadratic relationship between a moving point and its path.

Example 56. Determine the singular elements of this relationship.

The inverse motion of (3.54) is represented by

$$x = \cos \phi(X \cos \phi + Y \sin \phi), \qquad y = \cos \phi(-X \sin \phi + Y \cos \phi), \tag{3.56}$$

which shows that the path of any point is a circle.

Example 57. Show that the equation of the circle described by (X, Y) is $x^2 + y^2 - xX - yY = 0$.

Every path passes through the origin; all the moving points arrive simultaneously at this point when the motion parameter reaches the singular value $\phi = \pi/2$. Any circle through O is a possible path. Every point of the fixed plane is the center of one path.

Example 58. Determine the envolope of a moving line for the motion (3.54) and for the inverse motion (3.56).

Our example of a continuous motion in the equiform plane is related to a well-known problem in elementary geometry. If a triangle ABC is given one may study the set of triangles PQR, inscribed into ABC (P on BC, etc.), such that PQR is similar to a given triangle $P_0Q_0R_0$. It may be shown that there exists one point O, such that its pedal triangle $P_1Q_1R_1$ with respect to ABC is similar to $P_0Q_0R_0$. The other inscribed triangles follow from $P_1Q_1R_1$ by a rotation of angle ϕ about O, and a dilation with center O and the scale factor $s = (\cos\phi)^{-1}$. Obviously there appears an equiform motion, the paths of P_1, Q_1, R_1 being the lines BC, CA, AB respectively.

If an equiform motion has a fixed point it may be represented by $X = s(x\cos\phi - y\sin\phi)$, $Y = s(x\sin\phi + y\cos\phi)$. It is then defined if the path of one other point is known.

Example 59. Let a point, $(m, 0)$ say, describe the circle $X^2 + Y^2 - 2MX + D = 0$. Determine s; show that any point describes a circle. For $D = 0$ the motion is the one given by (3.56).

Several special motions in equiform kinematics have been investigated by BURMESTER [1874a,b, 1878], MÜLLER [1910c], and KRAUSE [1910]. We mention the following results: if the paths of two points are straight lines and that of a third point a circle, the path of a general point is a conic; if three points describe circles the general path is a tri-circular sextic, the same curve as the four-bar coupler curve of ordinary kinematics. Similarly, KRAUSE [1910] has pointed out that every planar trajectory generated by an equiform motion is also the trajectory of an ordinary (Euclidean) motion. Such results lead to the direct extension of Roberts' formulas (ROBERTS [1870, 1876]) to equiform planar kinematics.

4. Non-Euclidean kinematics

Non-Euclidean geometry in n-space may be defined as the study of geometric properties which are invariant for the group of those linear transformations in projective n-space which leave a given quadratic variety Ω

invariant. There are two types: for elliptic geometry Ω is of the type represented by the equation $X_1^2 + X_2^2 + \cdots + X_{n+1}^2 = 0$. X_i being homogeneous projective coordinates; for hyperbolic geometry Ω is of the type $X_1^2 + X_2^2 + \cdots + X_n^2 - X_{n+1}^2 = 0$. Euclidean geometry is more or less a singular border-line case between the two; it is in some respect more simple but it lacks their elegance. In view of their definitions there exists much similarity between elliptic and hyperbolic geometry but there is an essential difference with respect to the reality of the quadratic in both cases; in the first Ω has no real points, in the second there are ∞^{n-1}. A projective transformation which leaves Ω invariant is a non-Euclidean *displacement*; a continuous set of displacements depending on one parameter t defines a non-Euclidean *motion*. The degree of freedom for both geometries in n-space is equal to that for the Euclidean case: $\frac{1}{2}n(n+1)$.

For $n \leq 3$, especially for the instantaneous case, our knowledge of non-Euclidean kinematics is fairly well advanced; there even exists a text-book on the subject, the third volume of Garnier's *Cours* (GARNIER [1951]).

In what follows we give a survey of the matter, restricting ourselves to the main topics without dealing with the many special cases which arise.

There is a relationship between elliptic geometry in n-space and Euclidean geometry in $(n+1)$-space: Elliptic geometry is identical with the Euclidean geometry of the n-space *at infinity*, Ω playing the role of the isotropic variety. This implies a strong resemblance between elliptic n-space geometry and spherical Euclidean geometry in $(n+1)$-space; there is no identity, however, because one point at infinity corresponds to two (diametrically opposite) points on the (unit) sphere. Elliptic geometry is therefore somewhat simpler than the latter. In both a displacement may be represented by

$$P = \mathbf{A}p, \tag{4.1}$$

$\mathbf{A}$ being an orthogonal $(n+1)\times(n+1)$ matrix; P and p are column matrices whose elements are Cartesian coordinates (in E^{n+1}) in the Euclidean case and homogeneous projective ones (in E^n) for the elliptic case. For any n the matrix $\mathbf{A}$ may be expressed in terms of Cayley's $\frac{1}{2}n(n+1)$ parameters c_i, as discussed in Section 2 of this chapter. The example given there for a 4×4-matrix gives us the general expression for a displacement in elliptic 3-space.

In view of the foregoing and our discussion about eigenvalues of orthogonal matrices (Chapter I, Section 4), a displacement in elliptic 3-space has in general four different invariant points on Ω composed of two pairs of conjugate imaginary points. This implies that it has two real invariant lines l and l* which are moreover conjugate with respect to Ω and therefore orthogonal in elliptic geometry. Hence, for a suitably chosen coordinate system (the vertices being those of a polar tetrahedron of Ω) with homogeneous coordinates X, Y, Z, W and x, y, z, w, such a displacement is represented by

$$X = x\cos\phi - y\sin\phi, \qquad Y = x\sin\phi + y\cos\phi,$$
$$Z = z\cos\psi - w\sin\psi, \qquad W = z\sin\psi + w\cos\psi, \tag{4.2}$$

this may be called a screw motion with the two (equivalent) axes l: $(x = y = 0)$ and l*: $(z = w = 0)$. The invariant points are $S_1, S_2 = (1, \pm i, 0, 0)$ and $S_3, S_4 = (0, 0, 1, \pm i)$. For $\phi = 0, \psi \neq 0$ and for $\phi \neq 0, \psi = 0$ we obtain pure rotations. If we consider two positions, the first being the reference position $\phi = \psi = 0$, the coordinates of the lines joining two homologous points are seen to be

$$p_{14} = xz\sin\psi + xw(\cos\psi - \cos\phi) + yw\sin\phi, \text{ etc.}$$

Example 60. Determine the six expressions p_{ij} and show that they satisfy $(\sin\phi\sin\psi)\cdot(p_{14}^2 + p_{23}^2 + p_{24}^2 + p_{31}^2) = 2(1 - \cos\phi\cos\psi)p_{12}p_{34}$, which shows that the locus of the joins of homologous points is a tetrahedral complex, the associated tetrahedron being $S_1S_2S_3S_4$.

If in (4.2) we have $\phi = \pi$, $\psi = 0$ then $X = -x$, $Y = -y$, $Z = z$, $W = w$, which represents a reflection into l. Similarly $\phi = 0$, $\psi = \pi$ gives a reflection into l*. The product of these two reflections is a half-turn about both axes. The product of reflections into the arbitrary lines l_1 and l_2 can be shown to be a screw motion, the axes of which are the two common perpendiculars of l_1 and l_2 (or: the two transversals of the four lines l_1, l_1^*, l_2, l_2^*). On the other hand any screw motion may be written as the product of two reflections and in ∞^1 many ways. By analogous arguments as were used for ordinary kinematics (Chapter IV, Section 1) we are then able to treat three positions theory in elliptic space by studying the configuration of three arbitrary axes l_{12}, l_{23}, l_{31}: from which follow $l_{12}^*, l_{23}^*, l_{31}^*$ as their polar lines with respect to Ω, and furthermore n_1, n_1^* as the transversals of $l_{12}, l_{12}^*, l_{31}, l_{31}^*$, and analogously n_2, n_2^* and n_3, n_3^*. The three homologous positions A_i of a point of the moving space are then seen to be the reflections into n_i of a "fundamental" point A, representing the triplet.

Example 61. Verify that the degree of freedom for the definition of three positions is the same as for specifying three arbitrary lines.

We set ourselves the problem of determining, in the theory of three finitely separated positions, the locus of the points in the moving space whose three homologous positions are collinear. If these are X_i, Y_i, Z_i, W_i, $(i = 1, 2, 3)$ the condition reads: the 4×3-matrix $\| X_i \; Y_i \; Z_i \; W_i \|$ has rank two. Keeping in mind that its elements are homogeneous linear functions of x, y, z, w, and in view of general theorems on *determinental loci* (see ROOM [1938]), we conclude that the locus in the moving space is a curve Γ_{123} of order $(3^2 - 2^2 + 1 =)$ six. It must have twelve intersections with Ω, in accordance with the fact that it passes through the four invariant points of each of the displacements D_{12}, D_{23} and D_{31}, which obviously belong to the locus.

It is easily seen that if four finitely separated positions are given, the locus, in the moving space, of those points A for which the homologous points A_i $(i = 1, 2, 3, 4)$ are coplanar is a quartic surface, passing through Γ_{234}, Γ_{341}, Γ_{412}, Γ_{123}.

The study of four and more finitely separated positions can follow the method of screw triangles presented for the Euclidean case (Chapter V) by virtue of the fact that analogous configurations of axes exist for displacements in elliptic spaces.

We consider now instantaneous kinematics in elliptic space. Obviously there exist two instantaneous screw axes l and l*, conjugate with respect to Ω. If the frames coincide in the zero-position and if we take l to be $x = y = 0$ and l* therefore $z = w = 0$, the first order formulas are

$$X_1 = -\omega_1 y, \quad Y_1 = \omega_1 x, \quad Z_1 = -\omega_2 w, \quad W_1 = \omega_2 z, \tag{4.3}$$

where ω_1, ω_2 are (time-dependent) instantaneous invariants. In the general case $\omega_1\omega_2 \neq 0$ and $\omega_1 \neq \omega_2$.

Example 62. Determine the coordinates p_{ij} of the path tangent m at $A(x, y, z, w)$ and show that the locus of these tangents is the tetrahedral complex $\omega_1\omega_2(p_{14}^2 + p_{23}^2 + p_{24}^2 + p_{31}^2) = (\omega_1^2 + \omega_2^2)p_{12}p_{34}$.
Example 63. Show that $A^* = (-\omega_1 y, \omega_1 x, -\omega_2 w, \omega_2 z)$ is the point on m conjugate to A with respect to Ω, that the normal plane at A is given by $U_1 = -\omega_1 y$, $U_2 = \omega_1 x$, $U_3 = -\omega_2 w$, $U_4 = \omega_2 z$ and that the locus of path normals is the linear complex $\omega_1 p_{12} + \omega_2 p_{34} = 0$.

To derive the formulas for higher orders we make use of the general expressions for the development of an orthogonal matrix, given in Chapter II, (2.5), (2.7). For the second order we obtain

$$\left\|\begin{matrix} X_2 \\ Y_2 \\ Z_2 \\ W_2 \end{matrix}\right\| = \left\|\begin{matrix} -\omega_1^2 & \beta_{12} & \beta_{13} & \beta_{14} \\ \beta_{21} & -\omega_1^2 & \beta_{23} & \beta_{24} \\ \beta_{31} & \beta_{32} & -\omega_2^2 & \beta_{34} \\ \beta_{41} & \beta_{42} & \beta_{43} & -\omega_2^2 \end{matrix}\right\| \cdot \left\|\begin{matrix} x \\ y \\ z \\ w \end{matrix}\right\|, \tag{4.4}$$

with $\beta_{ij} + \beta_{ji} = 0$.

Our coordinate systems are not yet completely determined. They may be transformed by the set of ∞^2 transformations

$$x = x'\cos\phi - y'\sin\phi, \qquad y = x'\sin\phi + y'\cos\phi,$$

$$z = z'\cos\psi - w'\sin\psi, \qquad w = z'\sin\psi + w'\cos\psi,$$

and similarly for X', Y', Z', W'. In which case in (4.4) the β_{ij} are also transformed, we obtain in particular

$$\begin{aligned} \beta'_{14} &= \beta_{14}\cos\phi\cos\psi - \beta_{23}\sin\phi\sin\psi \\ &\quad - \beta_{13}\cos\phi\sin\psi + \beta_{24}\sin\phi\cos\psi, \\ \beta'_{23} &= \beta_{23}\cos\phi\cos\psi - \beta_{14}\sin\phi\sin\psi \\ &\quad + \beta_{24}\cos\phi\sin\psi - \beta_{13}\sin\phi\cos\psi. \end{aligned} \tag{4.5}$$

We determine the orientation of the axes so that in the initial position ϕ and ψ are such that $\beta'_{14} = \beta'_{23} = 0$.

Example 64. Show that these equations are bilinear quadratics in $\tan\phi$ and $\tan\psi$; prove that we obtain real solutions for these unknowns.

Our canonical frames are now completely determined and we have

$$\begin{aligned} X_2 &= -\omega_1^2 x - \beta_{21} y + \beta_{13} z, \\ Y_2 &= \beta_{21} x - \omega_1^2 y - \beta_{42} w, \\ Z_2 &= -\beta_{13} x - \omega_2^2 z + \beta_{34} w, \\ W_2 &= \beta_{42} y - \beta_{34} z - \omega_2^2 w, \end{aligned} \tag{4.6}$$

with four instantaneous invariants β_{21}, β_{13}, β_{42}, β_{34} for the second order. The formulas (4.3) and (4.6) describe completely the properties of the motion up to this order.

Example 65. Show that the coordinates of the osculating plane of the path of $A(x, y, z, w)$ are polynomials of the third degree in x, y, z, w.
Example 66. Determine the osculating plane of a point on the screw axis l given by $x = y = 0$.
Example 67. Show that the osculating planes at $(1,0,0,0)$, $(0,1,0,0)$, $(0,0,1,0)$ and $(0,0,0,1)$ are $W = 0$, $Z = 0$, $Y = 0$ and $X = 0$ respectively, from which follows a geometrical characterization of the canonical frames.
Example 68. Show that the inflection curve is of order six, having three coinciding intersections with Ω at each of the invariant points $(1, \pm i, 0, 0)$ and $(0, 0, 1, \pm i)$.
Example 69. Show that the determinant of (4.6) is $\omega_1^4\omega_2^4 + \omega_1^4\beta_{34}^2 + \omega_2^4\beta_{21}^2 + \omega_1^2\omega_2^2(\beta_{13}^2 + \beta_{42}^2) + (\beta_{21}\beta_{34} - \beta_{13}\beta_{42})^2$, from which it follows that an acceleration pole does, in general, not exist.
Example 70. Show by means of Chapter II, (2.5), (2.7) how the expressions for higher derivatives can be obtained and that there are six instantaneous invariants of order n if $n \geq 3$.

We mention some special continuous motions in elliptic space. If the general displacement is written by means of Cayley's formula for a 4×4-

orthogonal matrix (Chapter I, (5.7)), and if the six parameters c_i are chosen as linear functions of the parameter t, we obtain a motion such that the paths are in general rational quartic space curves.

Another motion is defined when in (4.2) we take ϕ and ψ as functions of t; then we have a continuous screw-motion with fixed axes. A noteworthy special case appears if $\phi = \psi$; for $\tan\phi = \tan\psi = t$ we obtain

$$X = x - yt, \quad Y = xt + y, \quad Z = z - wt, \quad W = zt + w, \tag{4.7}$$

which implies that every path is a straight line. The motion could be called a (pseudo-) translation. Every path has ∞^1 common perpendiculars with both axes, and all these have equal lengths; a path and an axis are so-called Clifford parallels. For $t = \infty$ every point arrives at the conjugate point to its initial position.

Example 71. Show that the set of paths is a tetrahedral complex. All paths are different.
Example 72. Consider the elliptic motion

$$X = ax - by - cz,$$
$$Y = bx + ay + cw,$$
$$Z = cx + az - bw,$$
$$W = -cy + bz + aw,$$

where a, b, c are arbitrary functions of t. Show that all paths are plane curves (the point (x, y, z, w) moves in the plane $wX + zY - yZ - xW = 0$) and furthermore that all paths are congruent. Every plane contains one path. (BOTTEMA [1944a].)
Example 73. Show that in elliptic space, motions exist analogous to the Darboux motions in ordinary kinematics: any path is a conic (BOTTEMA [1944b]).

In hyperbolic space the absolute quadric Ω may be given the equation $X^2 + Y^2 + Z^2 - W^2 = 0$. Although the corresponding geometry has much resemblance to the elliptic one there are, in view of questions about reality, striking differences. Ω now has real points; the points "at infinity" are those on and outside Ω, the region inside Ω is the proper part of the space. Any hyperbolic displacement has in general four invariant points on Ω; two (S_1, S_2) are real and two (S_3, S_4) conjugate imaginary. There are two real invariant lines l and l*, which are conjugate with respect to Ω; l has points inside Ω, l* is completely outside. If we choose them as $x = y = 0$ and $z = w = 0$ respectively the displacement is seen to be

$$\begin{aligned} X &= x\cos\phi - y\sin\phi, & Y &= x\sin\phi + y\cos\phi, \\ Z &= z\cosh\psi + w\sinh\psi, & W &= z\sinh\psi + w\cosh\psi. \end{aligned} \tag{4.8}$$

It is a screw displacement of which the Z, W part could be called the "translation" component.

Example 74. Derive the line coordinates of the join of two homologous points, in two separated positions, and show that they satisfy $(\sin\phi \sinh\psi)(p_{31}^2 + p_{23}^2 - p_{14}^2 - p_{24}^2) = 2(1-\cos\phi\cosh\psi)p_{12}p_{34}$. It represents a tetrahedral complex, the corresponding tetrahedron being $S_1S_2S_3S_4$.

The possibility of considering any displacement as the product of two line-reflections holds in hyperbolic kinematics and therefore three, four and five positions theory may be dealt with in an analogous manner as in the Euclidean (and the elliptic) case.

From (4.8) it follows that instantaneous kinematics of the first order is expressed by

$$X_1 = -\omega_1 y, \quad Y_1 = \omega_1 x, \quad Z_1 = \omega_2 w, \quad W_1 = \omega_2 z. \tag{4.9}$$

Example 75. Show that the locus of path tangents is the tetrahedral complex $\omega_1\omega_2(p_{31}^2 + p_{23}^2 - p_{14}^2 - p_{24}^2) = (\omega_1^2 - \omega_2^2)p_{12}p_{34}$.

Example 76. Show that the normal plane of the moving point $A(x, y, z, w)$ has the coordinates $(-\omega_1 y, \omega_1 x, \omega_2 w, \omega_2 z)$*; the locus of normals is the linear complex $\omega_1 p_{12} - \omega_2 p_{34} = 0$.

It is easy to verify that in the hyperbolic case the second order formulas are, instead of (4.4),

$$\begin{aligned} X_2 &= -\omega_1^2 x + \beta_{12} y + \beta_{13} z + \beta_{14} w, \\ Y_2 &= \beta_{21} x - \omega_1^2 y + \beta_{23} z + \beta_{24} w, \\ Z_2 &= \beta_{31} x + \beta_{32} y + \omega_2^2 z + \beta_{34} w, \\ W_2 &= \beta_{41} x + \beta_{42} y + \beta_{43} z + \omega_2^2 w, \end{aligned} \tag{4.10}$$

with $\beta_{12} + \beta_{21} = 0$, $\beta_{23} + \beta_{32} = 0$, $\beta_{31} + \beta_{13} = 0$, $\beta_{14} - \beta_{41} = 0$, $\beta_{24} - \beta_{42} = 0$, $\beta_{34} - \beta_{43} = 0$. In a similar way as in elliptic kinematics we can determine a canonical system such that $\beta_{14} = \beta_{23} = 0$.

Example 77. Apply the çoordinate transformation $X' = X\cos\phi - Y\sin\phi$, $Y' = X\sin\phi + Y\cos\phi$, $Z' = Z\cosh\psi + W\sinh\psi$, $W' = Z\sinh\psi + W\cosh\psi$, and similar for x', y', z', w' and determine β'_{14} and β'_{23}. Show that $\beta'_{14} = \beta'_{23} = 0$ gives rise to two quadratic equations in $\tan\phi$ and $\tanh\psi$ with real solutions; one solution yields $|\tanh\psi| < 1$, which guarantees real solutions for ϕ and ψ.

Summarizing, we have in hyperbolic space, up to the second order, the following scheme

* The reader is reminded that in hyperbolic geometry the scalar product of two vectors $\boldsymbol{a}_i(x_i, y_i, z_i, w_i)$, $i = 1, 2$, is $\boldsymbol{a}_1\boldsymbol{a}_2 = x_1x_2 + y_1y_2 + z_1z_2 - w_1w_2$.

$$\begin{aligned}
&X_0 = x, && X_1 = -\omega_1 y, && X_2 = -\omega_1^2 x - \beta_{21} y + \beta_{13} z,\\
&Y_0 = y, && Y_1 = \omega_1 x, && Y_2 = \beta_{21} x - \omega_1^2 y + \beta_{42} w,\\
&Z_0 = z, && Z_1 = \omega_2 w, && Z_2 = -\beta_{13} x + \omega_2^2 z + \beta_{34} w,\\
&W_0 = w, && W_1 = \omega_2 z, && W_2 = \beta_{42} y + \beta_{34} z + \omega_2^2 w,
\end{aligned} \tag{4.11}$$

with two instantaneous invariants of the first and four of the second order, all time-dependent. All second order properties may be derived from (4.11).

Example 78. Derive the formulas for third order properties; they contain six new instantaneous invariants.

We mention only two special motions in hyperbolic space. If in (4.8) we take $\psi = 0$, $\phi = t$ we have rotations about l, all paths being circles; if $\phi = 0$, $\psi = t$ we obtain "translations", the paths are conics, called "distance lines" or circles with their center in the infinite region. The counterpart of (4.7), all paths being straight lines, does not exist; there are no (real) Clifford parallels in the hyperbolic case.

We finish this section with some remarks about plane non-Euclidean kinematics.

In the elliptic plane, with $\Omega \equiv X^2 + Y^2 + Z^2 = 0$, any displacement may be represented by

$$X = x\cos\phi - y\sin\phi, \quad Y = x\sin\phi + y\cos\phi, \quad Z = z, \tag{4.12}$$

the rotation center is $(0,0,1)$ and the invariant points on Ω are $(1, \pm i, 0)$.

Three positions are completely defined by the rotation centers P_{23}, P_{31}, P_{12}; the rotation angles ϕ_{ij} are twice the angles of the pole triangle, satisfying the inequality $\phi_{23} + \phi_{31} + \phi_{12} > 2\pi$. Three homologous points A_1, A_2, A_3 are the reflections of a *basic* point A^* into the sides of the pole triangle. The circumcenter of A_1, A_2, A_3 is the isogonal conjugate (with respect to the pole triangle) of A^*. The locus of points A^* such that A_1, A_2, A_3 are collinear is a cubic curve passing through the three rotation centers and through six points of Ω. All these properties follow by specializing the spatial results.

Instantaneous elliptic plane kinematics up to the third order is given by the scheme (derived from the general theory of Chapter II, (2.5), (2.7) analogously to Chapter II, (5.10))

$$X_0 = x, \quad X_1 = -\omega y, \quad X_2 = -\omega^2 x - \beta_3 y + \beta_2 z,$$

$$X_3 = -3\beta_3\omega x - \gamma_3 y + \gamma_2 z,$$

$$Y_0 = y, \quad Y_1 = \omega x, \quad Y_2 = \beta_3 x - \omega^2 y,$$

(4.13)
$$Y_3 = \gamma_3 x - 3\beta_3\omega y + ((3/2)\beta_2\omega - \gamma_1)z,$$
$$Z_0 = z, \qquad Z_1 = 0, \qquad Z_2 = -\beta_2 x,$$
$$Z_3 = -\gamma_2 x + ((3/2)\omega\beta_2 + \gamma_1)y,$$

with the invariants ω, β_2, β_3, γ_1, γ_2, γ_3 all time-dependent.*

The inflection curve** is given by $|X_i\, Y_i\, Z_i| = 0$, $i = 0, 1, 2$ and is seen to be

$$-\beta_2 x(x^2 + y^2 + z^2) + \omega^2(x^2 + y^2)z = 0, \tag{4.14}$$

which represents a cubic curve, passing through O (with tangent $x = 0$) and intersecting (actually, touching) Ω three times at each of the points $(1, \pm i, 0)$.

If $X(t)$, $Y(t)$, $Z(t)$ is a curve in the fixed plane and if we normalize the homogeneous coordinates by the condition $X^2 + Y^2 + Z^2 = 1$, then the line coordinates of the normal are $(\dot{X}, \dot{Y}, \dot{Z})$; the center of curvature M of the curve, the intersection of two consecutive normals, is therefore

$$X_M = \dot{Y}\ddot{Z} - \dot{Z}\ddot{Y}, \qquad Y_M = \dot{Z}\ddot{X} - \dot{X}\ddot{Z}, \qquad Z_M = \dot{X}\ddot{Y} - \dot{Y}\ddot{X}. \tag{4.15}$$

Hence instantaneously the normal at $A(x, y, z)$ coincides with OA and the center M is, in view of (4.13):

$$X_M = \beta_2 x^2, \qquad Y_M = \beta_2 xy, \qquad Z_M = \beta_2 xz - \omega^2(x^2 + y^2), \tag{4.16a}$$

and conversely

$$x = \beta_2 X_M^2, \qquad y = \beta_2 X_M Y_M, \qquad z = \beta_2 X_M Z_M + \omega^2(X_M^2 + Y_M^2). \tag{4.16b}$$

The formulas (4.16) express the birational quadratic relationship between a point A and the corresponding center of curvature M. They are in terms of projective coordinates. If we introduce elliptic metrics, the distances OA and OM being ρ and ρ_1 respectively and $\angle XOA = \angle XOM = \theta$, then we have from the well-known formulas

$$\tan\rho = (x^2 + y^2)^{\frac{1}{2}}/z, \quad \tan\rho_1 = (X_M^2 + Y_M^2)^{\frac{1}{2}}/Z_M, \quad \tan\theta = y/x = Y_M/X_M,$$

which together with (4.16) yields

$$(\cot\rho - \cot\rho_1)\cos\theta = \omega^2/\beta_2, \tag{4.17a}$$

the elliptic version of the (first) Euler-Savary formula.

* We use different symbols for the invariants here to emphasize we have an elliptic rather than a Euclidean motion. Clearly however β here is analogous to ε in Chapter II, (5.10) and the subscripts 2 and 3 have simply replaced Y and Z respectively.

** Which is the locus of all points with zero geodesic curvature.

The reader should note that here the angle θ is measured to the ray (i.e., OA) from the common normal of the centrodes. If, as is customary, we measure the ray angle from the pole tangent then we use the angle η $(0 \leq \eta < \pi)$, measured counterclockwise from the pole tangent to the ray, such that $\theta = \eta - \pi/2$ and therefore

$$(\cot \rho - \cot \rho_1)\sin \eta = \omega^2/\beta_2. \tag{4.17b}$$

Example 79. Show that for any point on O_y, different from O, the corresponding center M coincides with O.
Example 80. Show that there are at any moment three acceleration poles.
Example 81. Verify that O_Y is the pole tangent.

We mention some third order properties of the motion. A Ball point is on the inflection curve and also satisfies the equation $|X_i\ Y_i\ Z_i| = 0$, $i = 0, 1, 3$. From this determinant and (4.14) we obtain $yzY_3 + xzX_3 - (x^2 + y^2)Z_3 = 0$, which in view of (4.13) represents a cubic curve, passing through O (but with its tangent different than O_Y) and through $(1, \pm i, 0)$. There are therefore in general $(3 \times 3 - 3 =)$ six Ball points. The circling point curve is the locus of points with the property that three consecutive normals are concurrent. Hence its equation reads $|X_i\ Y_i\ Z_i| = 0$, $i = 1, 2, 3$, which follows explicitly from (4.13). It is a cubic curve.

Example 82. Determine from (4.13) the centers of curvature of the fixed and the moving centrodes at O.

We mention some special continuous motions in the elliptic plane. If the general displacement is written in terms of the Euler parameters c_i $(i = 0, 1, 2, 3)$ and if we take these to be linear functions of t we have defined a motion whose paths are all conics.

It has been shown that the motion (comparable with Cardan motion in the ordinary plane) such that two points move on perpendicular lines has the following properties: the paths are quartic curves of genus one, with three centers and are tangent to Ω four times; the fixed centrode is a quartic, the moving one a conic; the paths of the inverse motion are curves of order eight and genus one. (Bottema [1975a]).

Kinematics in the hyperbolic plane is more complicated than that in the elliptic plane because several cases must be distinguished. It has been treated by Garnier [1951], Frank [1971], Tölke [1974] and others.

A displacement has two invariant points on the absolute conic $\Omega \equiv x^2 + y^2 - z^2 = 0$. If they are imaginary their join is real and the third invariant point lies inside Ω; if they are real the latter is outside Ω; there is a special case, all three invariant points coinciding at one point of Ω, the tangent there being the only invariant line.

In the first case the displacement is given, using a suitably chosen coordinate system, by

$$X = x\cos\phi - y\sin\phi,\quad Y = x\sin\phi + y\cos\phi,\quad Z = z, \tag{4.18}$$

the center being $(0,0,1)$. In the second case we obtain

$$X = x,\quad Y = y\cosh\psi + z\sinh\psi,\quad Z = y\sinh\psi + z\cosh\psi, \tag{4.19}$$

with the center $(1,0,0)$ outside Ω. For the sake of completeness we mention the borderline case, which in rational form may be written

$$\begin{aligned} X &= (1-\tfrac{1}{2}u^2)x + uy + \tfrac{1}{2}u^2 z,\quad Y = -ux + y + uz,\\ Z &= -\tfrac{1}{2}u^2 x + uy + (1+\tfrac{1}{2}u^2)z. \end{aligned} \tag{4.20}$$

Example 83. Show that (4.20) leaves Ω invariant; the only invariant point is $(1,0,1)$ and the invariant line is $x - z = 0$. Prove that the eigenvalues of the transformation matrix are all equal to *one.*

Example 84. Write (4.18) and (4.19) in rational form.

Hyperbolic three positions theory may be dealt with in the same way as the Euclidean and the elliptic case. Three positions are again determined by three rotation centers P_{23}, P_{31}, P_{12}, but as each of them may be either inside or outside Ω (or, in the bordeline case, on Ω) for a complete treatment several possibilities must be taken into account. The concept of the fundamental (or basic) point representing a triple of homologous points still holds but the "isogonal relationship" must be modified.

Obviously for instantaneous kinematics three different cases must be considered. Restricting ourselves to the two main cases we have for the first (up to the second order)

$$\begin{aligned} X_0 &= x, & X_1 &= -\omega y, & X_2 &= -\omega^2 x - \beta_1 y,\\ Y_0 &= y, & Y_1 &= \omega x, & Y_2 &= \beta_1 x - \omega^2 y + \beta_2 z,\\ Z_0 &= z, & Z_1 &= 0, & Z_2 &= \beta_2 y, \end{aligned} \tag{4.21}$$

and for the second

$$\begin{aligned} X_0 &= x, & X_1 &= 0, & X_2 &= \beta_1 z,\\ Y_0 &= y, & Y_1 &= \omega z, & Y_2 &= \omega^2 y + \beta_2 z,\\ Z_0 &= z, & Z_1 &= \omega y, & Z_2 &= \beta_1 x + \beta_2 y + \omega^2 z. \end{aligned} \tag{4.22}$$

Both (4.21) and (4.22) may be obtained by a suitable specialization of (4.11).

Example 85. Determine, in both cases, the inflection curve.
Example 86. Determine, in both cases, the birational quadratic relationship between a moving point A and the corresponding center M of curvature.
Example 87. Show that the locus of A such that M is on Ω is a quartic curve; determine the locus of M if A is on Ω.
Example 88. Show that there are in both cases three acceleration poles, lying on the inflection curve.
Example 89. Add to the schemes (4.21) and (4.22) the expressions for X_3, Y_3, Z_3, similar to (4.13); show that there are in general six Ball points; derive the equation of the circling-point curve.

We mention some special motions in plane hyperbolic kinematics: If in (4.18) ϕ is variable, any path is a circle. If in (4.19) ψ is variable, all paths are ("concentric") distance lines. If u is variable in (4.20) any path is a so-called *horicycle*, a conic which has four coinciding points in common with Ω.

Example 90. Show, by modifying the Euler parameters for the hyperbolic plane, that there are motions whose paths are conics.

5. Affine kinematics

Affine geometry in 3-space deals with those properties which are invariant for general linear transformations of three non-homogeneous coordinates. An affine displacement is represented by

$$\boldsymbol{P} = \mathbf{A}\boldsymbol{p} + \boldsymbol{d}, \tag{5.1}$$

where $\boldsymbol{P}$ and $\boldsymbol{p}$ are column vectors whose elements are the three coordinates in the fixed and moving space, $\mathbf{A}$ is an arbitrary 3×3-matrix and $\boldsymbol{d}$ a column vector giving the displacement of the origin. An affine space has therefore 12 degrees of freedom. If $\Delta = |\mathbf{A}|$, we suppose $\Delta \neq 0$ thus excluding singular affinities. Euclidean and equiform displacements are sub-groups of the affine group. Another sub-group is that for which $\Delta = 1$. It is called the equiaffine group and the corresponding differential geometry has been developed by BLASCHKE [1923]. Affine geometry is less specialized and therefore less rich in concepts than Euclidean and non-Euclidean geometry. It is itself a sub-geometry of projective geometry and may be defined as dealing with those projective transformations which leave a special plane, the plane at infinity, invariant. Affine geometry admits parallel lines, and also the ratio of distances on the same or on parallel lines as well as the ratio of volumes (volumes according to (5.1) are multiplied by Δ). The concept of angle or of orthogonality is unknown in affine geometry.

If $\mathbf{A}$ is a unit matrix and $\boldsymbol{d} \neq 0$ (5.1) represents a pure translation; any point at infinity is invariant and there are no finite invariant points. Excluding this special case and some others for which the eigenvalues of **A** are not all distinct, we have in general one finite invariant point and three invariant points B_1, B_2, B_3 at infinity, either all three are real or one is real and two are conjugate imaginary. Hence by a suitable choice of frames an affine displacement may be represented by

$$X = a_1 x, \qquad Y = a_2 y, \qquad Z = a_3 z, \tag{5.2}$$

or by

$$X = a_1(x \cos \phi - y \sin \phi), \quad Y = a_1(x \sin \phi + y \cos \phi), \quad Z = a_3 z, \tag{5.3}$$

with $a_i \neq 0$, $a_i \neq 1$.

Example 91. Show that there are three finite invariant lines and three finite invariant planes, real or imaginary.

Example 92. Show that (5.2) is an equiaffine displacement if $a_1 a_2 a_3 = 1$; (5.3) is one if $a_1^2 a_3 = 1$.

Either formulas (5.2) or (5.3) give the conditions for a two-positions theory in affine space. We consider one application: In (5.2) the line coordinates of the join of two homologous points are seen to be

$$\begin{aligned} &p_{14} = l_1 x, \qquad p_{24} = l_2 y, \qquad p_{34} = l_3 z, \\ &p_{23} = (l_2 - l_3) yz, \qquad p_{31} = (l_3 - l_1) zx, \qquad p_{12} = (l_1 - l_2) xy, \end{aligned} \tag{5.4}$$

with $l_i = a_i - 1$.

Hence the locus of these joins is given by

$$l_2 l_3 p_{14} p_{23} + l_3 l_1 p_{24} p_{31} + l_1 l_2 p_{34} p_{12} = 0, \tag{5.5}$$

which represents a tetrahedral complex, the corresponding tetrahedron being that of the three coordinate planes and the plane at infinity.

Example 93. Show that the (constant) cross-ratio of the four intersections of a join and the faces of the tetrahedron is (for a certain order of the points) equal to $l_3(l_1 - l_2)/(l_2(l_1 - l_3))$.

Example 94. In the case given by (5.3), determine the coordinates of the join of two homologous points and the equation of the locus (of joins).

If in (5.1) the matrix **A** and the vector ***d*** are functions of t, it defines an affine motion. In the instantaneous case we obtain for the first order

$$X_1 = b_1 x, \qquad Y_1 = b_2 y, \qquad Z_1 = b_3 z, \tag{5.6}$$

for the case (5.2), and

$$X_1 = bx - cy, \qquad Y_1 = cx + by, \qquad Z_1 = b_3 z, \tag{5.7}$$

for the case (5.3); both have three instantaneous (time-dependent) invariants.

Example 95. Determine the complex of path tangents in both cases.
Example 96. Show that for equiaffine kinematics we have $b_1 + b_2 + b_3 = 0$ in (5.6), and $2b + b_3 = 0$ in (5.7).

By (5.2) and (5.3) the canonical frames are completely determined. From this it follows that expressions for the second order quantities X_2, Y_2, Z_2 are arbitrary non-homogeneous linear functions of x, y, z, which implies that we have 12 instantaneous invariants for the second order.

Example 97. If U_i $(i = 1, 2, 3, 4)$ are the coordinates of the osculating plane of the path, show that U_i $(i = 1, 2, 3)$ are quadratic functions of x, y, z and U_4 is a cubic function.
Example 98. Determine the pole tangent, in terms of the first and the second order instantaneous invariants, both in the fixed and the moving space and show that they coincide.

We consider now plane affine kinematics. The general displacement is represented by

$$X = \alpha x + \beta y + a, \qquad Y = \gamma x + \delta y + b. \tag{5.8}$$

It has one finite and two infinite invariant points; the latter can be real or imaginary. We shall restrict ourselves to the first case. Moreover we deal with the interesting case of equiaffine geometry; hence $\alpha\delta - \beta\gamma = 1$. If we take the invariant points as the vertices of the coordinate frame the displacement has the standard form

$$X = bx, \qquad Y = b^{-1}y. \tag{5.9}$$

Example 99. Determine the condition for α, β, γ, δ so that the invariant points at infinity are real and distinct.

If in (5.8) the six parameters are functions of t it represents an affine motion; it is an equiaffine motion if $\Delta = \alpha\delta - \beta\gamma = 1$ for any value of t. We develop instantaneous equiaffine kinematics in the following way: If the fixed and the moving frame coincide in the zero-position we have $\alpha_0 = \delta_0 = 1$, $\beta_0 = \gamma_0 = a_0 = b_0 = 0$. If furthermore the common origin is chosen at the pole (the finite fixed point) and O_X, O_Y along the (real) invariant lines, we obtain $\beta_1 = \gamma_1 = a_1 = b_1 = 0$. Then from $\Delta_1 = 0$ it follows that $\alpha_1 + \delta_1 = 0$. Hence we have up to the first order

$$\begin{aligned} X_0 &= x, \qquad X_1 = \alpha_1 x, \\ Y_0 &= y, \qquad Y_1 = -\alpha_1 y. \end{aligned} \tag{5.10}$$

The canonical frames are now determined. There is one (time-dependent) instantaneous invariant of the first order. From (5.10) it follows that the path tangent at the point $A(x, y)$ is

$$yX + xY - 2xy = 0. \tag{5.11}$$

It intersects the line at infinity at $(-x, y, 0)$; the line AO does it at $(x, y, 0)$, and O_X, O_Y at $(1, 0, 0)$ and $(0, 1, 0)$ respectively. Hence the theorem: the tangent at A is harmonic to AO with respect to the two lines through A parallel to the coordinate axes.

Example 100. Show that the tangent of a point on O_X, different from O, coincides with O_X and similar for a point on O_Y.
Example 101. Show that all points on a line through O have parallel path tangents.
Example 102. Show that the line (U_1, U_2, U_3) in the fixed plane is the tangent of the point with homogeneous coordinates $x = U_2U_3$, $y = U_1U_3$, $z = -2U_1U_2$. There exists a birational quadratic relationship between a point and the tangent to its path; determine its singular elements.

From $\Delta_2 = 0$ it follows that $\alpha_2 + \delta_2 = 2\alpha_1^2$ and we obtain the second order formulas

$$(5.12) \qquad X_2 = (\alpha_1^2 + \varepsilon)x + \beta_2 y + a_2, \quad Y_2 = \gamma_2 x + (\alpha_1^2 - \varepsilon)y + b_2,$$

with the five new (time-dependent) invariants ε, β_2, γ_2, a_2, b_2. The condition that the path has an inflection point (which is a geometrical property) reads $X_1Y_2 - X_2Y_1 = 0$; hence using (5.10) and (5.12) the equation of the inflection curve is seen to be

$$(5.13) \qquad \gamma_2 x^2 + 2\alpha_1^2 xy + \beta_2 y^2 + b_2 x + a_2 y = 0,$$

which represents a conic Γ, passing through O, the tangent at O being $b_2 x + a_2 y = 0$.

Example 103. The point of the moving plane coinciding with O passes through a cusp of its path; show that with respect to O_X and O_Y, the cuspidal tangent is harmonic with the tangent to Γ.
Example 104. Derive the condition for Γ to be an ellipse, a parabola, a hyperbola or a degenerate curve.

A parametric representation of Γ is

$$(5.14) \quad x = -(a_2 m + b_2)/N, \; y = -m(a_2 m + b_2)/N; \; N = \gamma_2 + 2\alpha_1^2 m + \beta_2 m^2;$$

(here, m is the parameter).

Hence, from (5.11), the coordinates of the path tangent of a point on Γ are

$$(5.15) \qquad U_1 = mN, \qquad U_2 = N, \qquad U_3 = 2m(a_2 m + b_2),$$

which shows that the locus of the inflection tangents is a (rational) curve C_3 of the third class.

Example 105. Show that the line at infinity is the double tangent of C_3; it is an isolated generating line of C_3 if Γ is an ellipse.
Example 106. Show that the three generating lines of C_3 through O are: O_X, O_Y and the tangent to Γ at O.
Example 107. Derive from (5.15) the equation of C_3 in terms of U_1, U_2, U_3.

The moving centrode is determined by $\dot{X} = \dot{Y} = 0$. For $t = 0$ the pole is the origin. If $x = p_1 t$, $y = q_1 t$ is the pole position up to the first order, we obtain $p_1 = -a_2/\alpha_1$, $q_1 = b_2/\alpha_1$. This implies that the pole tangent in the moving plane has the equation $b_2 x + a_2 y = 0$, which means that it coincides with the tangent at O to Γ. Moreover it is easily seen that this line is also the tangent at O to the fixed centrode.

Equiaffine plane kinematics is much more complicated than ordinary plane kinematics, although there is some similarity as we saw in the preceding. An important difference is that in the Euclidean case there is a natural geometric parameter, the rotation angle, which has no counterpart in affine geometry. Moreover a curvature theory of paths which in ordinary kinematics can be based on second order properties depends in affine geometry on higher derivatives. The affine-normal of a curve, introduced by BLASCHKE [1923] depends on third order concepts; indeed, it may be defined as the axis of the parabola which has four coinciding points in common with the curve. The center of curvature, being the intersection of two consecutive affine normals, is then a fourth order concept.

It must be mentioned that affine kinematics has been considered from a different point of view by BURMESTER [1874b, 1878, 1902] and by PELZER [1959]. Their investigations treat the case where the geometry of the moving plane is affine while that of the fixed plane is still Euclidean. Contrary to this TÖLKE [1967] considers affine kinematics proper but his "pseudo-curvatures" are not those of pure affine differential geometry, because they are not defined as intrinsic properties of the paths.

Making use of $\Delta_3 = 0$ the third order formulas are seen to be

$$X_3 = (3\alpha_1\varepsilon + \nu)x + \beta_3 y + a_3, \qquad (5.16)$$
$$Y_3 = \gamma_3 x + (3\alpha_1\varepsilon - \nu)y + b_3,$$

and we have five new invariants ν, β_3, γ_3, a_3, b_3.

The equation of the inflection curve is $\dot{X}\ddot{Y} - \ddot{X}\dot{Y} = 0$, which implies that its intersections with the next consecutive inflection curve satisfy $\dot{X}\dddot{Y} - \dddot{X}\dot{Y} = 0$, that is, for $t = 0$: $X_1 Y_3 - X_3 Y_1 = 0$, which in terms of x, y is

$$\gamma_3 x^2 + 6\alpha_1\varepsilon xy + \beta_3 y^2 + b_3 x + a_3 y = 0. \qquad (5.17)$$

This represents a conic through O. It has in general three intersections with (5.13) different from O. The conclusion is: there are instantaneously three points of the moving plane (the points of Ball) passing through an undulation point of their path; at least one of these is real.

Example 108. If $x_p(t)$, $y_p(t)$ and $X_p(t)$, $Y_p(t)$ are the centrodes in the moving and the fixed plane respectively determine $(x_p)_2$, $(y_p)_2$ and $(X_p)_2$, $(Y_p)_2$.

We shall develop a method to determine the affine normal of the path at a given point $A(x, y)$ of the moving plane. To that end we consider the pencil of conics having four consecutive points A in common with the path. One of these conics (the trivial parabola of the pencil) is the tangent counted twice; its equation is $K_1 \equiv (yX + xY - 2xy)^2 = 0$. As a second conic of the pencil we take the one through O, its equation reads

$$K_2 \equiv B_{11}X^2 + 2B_{12}XY + B_{22}Y^2 + 2B_{13}X + 2B_{23}Y = 0, \tag{5.18}$$

B_{ij} being as yet unknowns. The equation of the pencil is $K_2 + \lambda K_1 = 0$. Its two parabolas follow from the discriminant of the quadratic terms of $K_2 + \lambda K_1 = 0$: $(B_{11} + \lambda y^2)(B_{22} + \lambda x^2) - (B_{12} + \lambda xy)^2 = 0$; for the non-trivial parabola we have $(x^2B_{11} - 2xyB_{12} + y^2B_{22})\lambda + (B_{11}B_{22} - B_{12}^2) = 0$.

Substituting this value of λ into the pencil, it follows that the quadratic terms of the non-trivial parabola are: $[(B_{11}x - B_{12}y)X - (B_{22}y - B_{12}x)Y]^2$, and if the point at infinity on its axis is $(X_N, Y_N, 0)$, the affine normal to the trajectory is given by

$$X_N : Y_N = (B_{22}y - B_{12}x) : (B_{11}x - B_{12}y). \tag{5.19}$$

To determine B_{ij} we substitute

$$X = X_0 + X_1t + \tfrac{1}{2}X_2t^2 + (1/6)X_3t^3,$$

$$Y = Y_0 + Y_1t + \tfrac{1}{2}Y_2t^2 + (1/6)Y_3t^3$$

into (5.18) and write out the conditions that the terms t^k $(k = 0, 1, 2, 3)$ vanish; X_K and Y_K are given by (5.10), (5.12) and (5.16). We obtain four linear equations for the five homogeneous unknowns B_{ij}. It is easy to verify that (in all four) the coefficient of B_{11} has the factor x and that of B_{22} the factor y, from which follows a set of four homogeneous linear equations for $B_{11}x$, B_{12}, $B_{22}y$, B_{13}, B_{23}, the coefficients of B_{12} being polynomials of x, y of order two (at most) while all others are linear. If we solve the set, $B_{11}x$ and $B_{22}y$ appear as quintic polynomials and B_{12} as a quartic. Hence, from (5.19): the direction numbers of the affine normal at $A(x, y)$ are quintic polynomials of x, y; they depend on the eleven instantaneous invariants which define the time-dependent motion up to the third order. The relationship between a point and its affine normal seems too cumbersome to write explicitly.

Example 109. Show that this procedure does not hold for a point A on O_X or on O_Y. Consider these cases separately.

The affine curvature center of a path (defined as the center of the conic which has five coinciding points in common with the path) may be determined from a development along similar lines. We need then the fourth order expressions.

Example 110. Determine X_4 and Y_4 in terms of x, y; they contain five new invariants.

Affine differential geomtry defines also the number k, the affine curvature of a plane curve. It may be expressed (Blaschke [1923]) as

$$k = (12D_{12}D_{23} - 5D_{13}^2 + 3D_{12}D_{14})/(9D_{12}^{8/3}), \tag{5.20}$$

where D_{ij} stands for $X_iY_j - X_jY_i$, which is a quadratic polynomial in x, y. We have $D_{12} = 0$ for a point on the inflection curve and also $D_{13} = 0$ for a Ball point; $k = 0$ means that five consecutive points of the path are on a parabola. It follows from (5.20) that the locus of those points is a quartic curve of the moving plane.

We give some examples of special continuous plane equiaffine motions. A simple case appears if the three (instantaneous) invariant points are fixed during the motion. If those at infinity are real we have

$$X = cx, \qquad Y = c^{-1}y. \tag{5.21}$$

The paths are in general hyperbolas, and straight lines for points on O_X or O_Y; the inflection curve is degenerate. A more interesting case is the following (Bottema [1964b]): The equiaffine plane has five degrees of freedom. Hence the paths in the fixed plane Σ may be prescribed for four points of the moving plane E. We obtain a generalization of the Euclidean Cardan motion if we suppose that the paths of four points are straight lines. The main results are: There is a conic K in E all points of which describe straight lines. K is obviously the inflection curve, and the locus of the straight lines (called the *station-yard*) is the curve of the third class C_3 obtained from (5.15). The four original points may be replaced by any quadruple of points on K. At every instant K is tangent to C_3 at three points. The path of an arbitrary point of E is a conic. The center of K also describes a conic. K is the moving centrode; the fixed centrode is a quartic curve. The motion is not determined by the *yard*: there are ∞^1 different conics whose points may move along the *rails* of a given yard. There are two main cases: K can be an ellipse or a hyperbola, the parabola being a special case. In the aforementioned paper standard representations of the motion are determined.

A two-parameter affine plane motion is obtained from

$$X = \alpha x + \beta y + a, \qquad Y = \gamma x + \delta y + b, \tag{5.22}$$

if α, β, γ, δ, a, b are functions of two parameters u, v. For $u = v = 0$, let (5.22) represent the unit transformation. Consider a one-parameter motion $u(t)$, $v(t)$ belonging to the set, with $u(0) = v(0) = 0$. If for $f(u, v)$ we denote $(\partial f/\partial u)(0,0)$ by f_u and $(\partial f/\partial v)(0,0)$ by f_v, the finite pole of the one-parameter motion follows from

$$(5.23) \qquad \begin{aligned} 0 &= (\alpha_u u_1 + \alpha_v v_1)x + (\beta_u u_1 + \beta_v v_1)y + (a_u u_1 + a_v v_1), \\ 0 &= (\gamma_u u_1 + \gamma_v v_1)x + (\delta_u u_1 + \delta_v v_1)y + (b_u u_1 + b_v v_1), \end{aligned}$$

and it is seen to be at

$$(5.24) \qquad x = Q_1(m)/Q_3(m), \qquad y = Q_2(m)/Q_3(m),$$

where Q_i are quadratic expressions in $m = u_1/v_1$. Hence the locus of possible poles is, in general, a conic. (H. R. Müller [1958]).

Example 111. Show that this statement holds also for equiaffine motions.

6. Projective kinematics

Although projective differential geometry has been developed extensively (see Bol [1950]) and use has been made there of kinematical concepts, projective kinematics proper (in the sense of the present book, considering sets of transformations of the entire space) has not drawn much attention. If we remember the complications arising in the more special case of affine kinematics this is quite understandable. The general projective transformation in 3-space is represented by

$$(6.1) \qquad \boldsymbol{P} = \mathbf{A}\boldsymbol{p},$$

where $\boldsymbol{P}$ and $\boldsymbol{p}$ are column vectors in respectively the fixed and the moving space, the elements of which are four homogeneous point coordinates, and $\mathbf{A}$ is a 4×4-matrix with $|\mathbf{A}| \neq 0$ if we exclude singular phenomena. In general, that is if the eigenvalues of $\mathbf{A}$ are all distinct, there are four invariant points, the vertices of a tetrahedron. Using suitable frames (6.1) may be written

$$(6.2) \qquad X = \alpha x, \qquad Y = \beta y, \qquad Z = \gamma z, \qquad W = \delta w.$$

Example 112. Show that in the general two positions theory the locus of the joins of homologous points is a tetrahedral complex; determine its equation from (6.2).

If the eigenvalues of $\mathbf{A}$ in (6.1) are not all different several cases must be considered, depending on the so-called invariant factors of the matrix. Not only the multiplicity of an eigenvalue is important but moreover the rank of the matrix $\mathbf{A}-\lambda\mathbf{I}$ when the eigenvalue is substituted for λ. The theory gives rise to 13 types of projective transformations, each with its standard representation and with its configuration of invariant points, lines and planes.

Obviously the projective space has 15 degrees of freedom. A projective motion is defined if the 16 (homogeneous) elements of $\mathbf{A}$ are given as functions of t. For the instantaneous case there are in general four different poles; with respect to suitably chosen canonical frames we have

$$X_1 = \lambda_1 x, \quad Y_1 = \lambda_2 y, \quad Z_1 = \lambda_3 z, \quad W_1 = \lambda_4 w, \tag{6.3}$$

which means there are essentially three instantaneous invariants of the first order.

Example 113. Show that the complex of path tangents is a tetrahedral complex.

The canonical frames are completely determined by first order properties. Each additional order introduces 15 more invariants. However, X_n, Y_n, Z_n, W_n are still homogeneous linear functions of x, y, z, w and this gives rise to some general theorems.

Example 114. Show that the inflection curve is of the sixth order.
Example 115. Show that the coordinates of the osculating plane of a path are cubic functions of those of the moving point.
Example 116. A path may have a stationary osculating plane, which means that four consecutive points are coplanar. Show that the locus of those points is a quartic surface.

The formulas (6.3) must be modified if the first order transformation is not of the general type (with four different eigenvalues).

The projective plane has eight degrees of freedom. A displacement has in general three invariant points, the vertices of a triangle.

Example 117. Determine X_1, Y_1, Z_1 for canonical frames; there are two instantaneous invariants of the first order.
Example 118. Show that there are eight invariants for each order $n > 1$.
Example 119. Show that the inflection curve is a cubic.

The different types of plane projective displacements (being six in number) have been investigated from the view point of kinematics by Frank [1968], with special attention to first order properties in each case.

A kinematics theory may be associated with any geometry. This can be done for instance for (plane) Möbius geometry, which can be defined as dealing with those properties which are invariant under Euclidean displacements and inversions. Its region is the Euclidean plane to which one point ("at infinity") is added and it is therefore identical with the so-called "complex plane", introduced by Gauss and others to visualize the set of complex numbers. A displacement in the Möbius geometry corresponds to a linear transformation of these numbers and it depends therefore on six (real) parameters. Möbius kinematics has been studied by some authors (LEHMANN [1967]).

Outside the scope of this book lie such subjects as topological kinematics and kinematics in metric spaces. They have developed into autonomous branches of the science of kinematics and make use of much more sophisticated mathematical tools than are applied in this work.

CHAPTER XIII

SPECIAL MATHEMATICAL METHODS IN KINEMATICS

1. Plane kinematics by means of complex numbers

We consider a Euclidean plane E with the Cartesian frame O_{xy} and we let the point (x, y) correspond to the complex number $x + iy$. This is since Gauss' time the well-known method to visualize complex numbers. E is often denoted as "the complex plane" but as it is essentially real we shall call E the *Gaussian plane*. It is the image of the set of all points, real and imaginary, of a line.

Plane Euclidean geometry has been treated by means of complex numbers, considering the plane as Gaussian (COOLIDGE [1940], ZWIKKER [1950], MORLEY and MORLEY [1954]), and some interesting and elegant results have been found. The (finite) Euclidean plane and the Gaussian plane are essentially identical but an important difference appears if infinite elements are introduced. In the preceding chapters of this book we extended our spaces to become *projective* spaces. In particular our plane has been closed by a *line* at infinity; the Gaussian plane, however, is made complete by *one point* at infinity. Moreover our treatment has often made use of imaginary points; so, for instance, the isotropic points play an important part in our theory. The Gaussian plane has no imaginary points at all.

Euclidean plane kinematics has been systematically developed in the Gaussian plane (BEREIS [1958]). Some results are elegant but (in addition to the fact that the method is essentially restricted to two dimensions) it cannot be said that its use has per se advanced the subject very much. For the sake of completeness we give here some basic relationships.

A *displacement* in the plane is given by (1.1) of Chapter VIII,

$$(1.1) \qquad X = x\cos\phi - y\sin\phi + a, \quad Y = x\sin\phi + y\cos\phi + b.$$

Introducing the complex numbers $Z = X + iY$, $z = x + iy$, $c = a + ib$ the formulas (1.1) can be written as *one* relation

$$(1.2) \qquad Z = e^{i\phi}z + c.$$

ϕ is a real and c a complex number; hence the general displacement (1.2) depends on three real parameters.

(1.2) represents a time-dependent *motion* if ϕ and c are given as functions of t, the (real) time. Differentiation with respect to t gives us

$$\dot{Z} = i\,\mathrm{e}^{i\phi}(\dot{\phi}z) + \dot{c} \tag{1.3}$$

for the velocity of the point z. The pole P follows from $\dot{Z} = 0$ and it is seen to be

$$z_p = i\dot{c}\,\mathrm{e}^{-i\phi}\dot{\phi}^{-1}, \tag{1.4}$$

a finite point if $\dot{\phi} \neq 0$. For variable t (1.4) represents the moving centrode. The fixed centrode follows from (1.4) and (1.2):

$$Z_p = i\dot{c}\dot{\phi}^{-1} + c. \tag{1.5}$$

If we deal with geometrical (time-independent) kinematics we may take $\phi = t$ (and therefore $\dot{\phi} = 1$, $\ddot{\phi} = 0$). If, moreover, we suppose that for $t = 0$ the two frames coincide we have $c_0 = 0$. And if we take the origin at the pole for the zero-position we have $c_1 = 0$.

Example 1. Show that for $t = 0$ we obtain $\dot{Z}_P = \dot{z}_p = ic_2$, which means that the two centrodes are tangent at O.
Example 2. If our canonical frames are introduced show that c_2 is an imaginary number.

Differentiating (1.3) we obtain for $t = 0$

$$Z_2 = -z + c_2, \tag{1.6}$$

which if we set $Z_2 = 0$ gives us the point we call the "*geometrical* acceleration pole" (also called the "second geometrical pole" or simply the "second-order pole") P_2:

$$z = c_2, \tag{1.7}$$

a point on the Y-axis.

Example 3. Show that P_2 is the inflection pole and $|c_2|$ the diameter of the inflection circle.
Example 4. Derive for the zero-position "the nth geometrical pole" P_n.

2. Isotropic coordinates

Another mathematical tool, applied to some advantage in kinematics but again restricted to the plane, is the use of the so-called isotropic coordinates

(HAARBLEICHER [1933]). Here we introduce complex numbers for real concepts, but in contradiction to the method of Section 1 we suppose the Euclidean plane to be extended to a projective one. If O_{xy} is a Cartesian frame the isotropic coordinates $\boldsymbol{x}, \boldsymbol{y}$ of a point (x, y) are defined by

$$\boldsymbol{x} = x + iy, \qquad \boldsymbol{y} = x - iy. \tag{2.1}$$

The idea is that the isotropic points $I_1\ (1, i, 0)$ and $I_2\ (1, -i, 0)$ are transformed into $(0, 1, 0)$ and $(1, 0, 0)$ which simplifies some equations, such as those of circular curves (especially circles), and some analytical relationships, such as those expressing orthogonality. Furthermore (2.1) are linear (and also affine) transformations so that the degree of a curve remains unaltered. But there is, of course, a considerable loss because real points may have complex coordinates (and conversely too).

Example 5. Show that the coordinate axes have the equations $\boldsymbol{x} + \boldsymbol{y} = 0$ and $\boldsymbol{x} - \boldsymbol{y} = 0$.
Example 6. Show that the circle $(O; R)$ has the equation $\boldsymbol{xy} = R^2$.
Example 7. Show that the lines $\boldsymbol{y} = m_1\boldsymbol{x}$ and $\boldsymbol{y} = m_2\boldsymbol{x}$ are perpendicular if $m_1 + m_2 = 0$.
Example 8. Show that the general displacement is given (in terms of the symbols defined by (1.1) and (2.1)) by

$$\boldsymbol{X} = (e^{i\phi})\boldsymbol{x} + \boldsymbol{a}, \qquad \boldsymbol{Y} = (e^{-i\phi})\boldsymbol{y} + \boldsymbol{b},$$

where $\boldsymbol{a} = a + ib$ and $\boldsymbol{b} = a - ib$.
Example 9. Show that on canonical frames (with $\phi = t$) we have up to the third order

$$\boldsymbol{X}_0 = \boldsymbol{x}, \qquad \boldsymbol{X}_1 = i\boldsymbol{x}, \qquad \boldsymbol{X}_2 = -\boldsymbol{x} + ib_2, \qquad \boldsymbol{X}_3 = -i\boldsymbol{x} + a_3 + ib_3,$$

$$\boldsymbol{Y}_0 = \boldsymbol{y}, \qquad \boldsymbol{Y}_1 = -i\boldsymbol{y}, \qquad \boldsymbol{Y}_2 = -\boldsymbol{y} - ib_2, \qquad \boldsymbol{Y}_3 = i\boldsymbol{y} + a_3 - ib_3,$$

a_k and b_k are the ordinary instantaneous invariants.
Example 10. Show that the equation of the inflection circle is $2\boldsymbol{xy} + ib_2(\boldsymbol{x} - \boldsymbol{y}) = 0$.
Example 11. Determine the equation of the circling point curve in isotropic coordinates.

The exercises give an impression of the pertinent formulas for isotropic coordinates. The method has been used to advantage for some special problems in elementary plane geometry and, as far as kinematics is concerned, for the study of the four-bar sextic and its various special cases (GROENMAN [1950]).

3. Dual numbers

In the two preceding sections use was made of common complex numbers. We introduce now a less known type of algebra, developed in the mid-nineteenth century by Clifford and systematically applied to kinematics by STUDY [1891, 1903] and KOTEL'NIKOV [1895], and after them by many others.

The dual number system is a "complex" system with two units just as in ordinary complex numbers. An element of the latter is $a + ib$, a and b are real numbers and i is defined by $i^2 = -1$. A dual number is given by $a + \varepsilon b$, with a and b real and ε defined by $\varepsilon^2 = 0$. Addition, subtraction and multiplication exist for any pair and they are commutative, associative and distributive. The following rules are valid

$$0\varepsilon = \varepsilon 0 = 0, \qquad a\varepsilon = \varepsilon a, \qquad (a_1 + \varepsilon b_1) + (a_2 + \varepsilon b_2) = a_1 + a_2 + \varepsilon(b_1 + b_2),$$

$$(a_1 + \varepsilon b_1)(a_2 + \varepsilon b_2) = a_1 a_2 + \varepsilon(a_1 b_2 + a_2 b_1).$$

If $a_1 + \varepsilon b_1 = a_2 + \varepsilon b_2$ then $a_1 = a_2$, $b_1 = b_2$.

The system, however, is not a field (as is the case for real and for complex numbers) because a product may be zero without any of the factors being zero; indeed $(\varepsilon a)(\varepsilon b) = 0$, for any a and b. Hence division is not always possible: it has no sense if the divisor is of the type εb, in which case it is a "pure dual" number. Division by $a + \varepsilon b$, $a \neq 0$, however, is possible and unambiguous:

$$\begin{aligned}(a_1 + \varepsilon b_1)/(a + \varepsilon b) &= (a_1 + \varepsilon b_1)(a - \varepsilon b)/a^2 \\ &= (a_1/a) + \varepsilon(ab_1 - a_1 b)/a^2.\end{aligned} \tag{3.1}$$

We can define the function of a dual number $\mathrm{f}(a + \varepsilon b)$ by expanding it formally in a Taylor series with ε as variable. Because $\varepsilon^n = 0$ if $n > 1$ we obtain

$$\mathrm{f}(a + \varepsilon b) = \mathrm{f}(a) + \varepsilon \mathrm{f}'(a) \tag{3.2}$$

where $\mathrm{f}(a)$ is $\mathrm{f}(a + \varepsilon b)$ at $\varepsilon = 0$ and $\mathrm{f}'(a)$ is $(\mathrm{d}\mathrm{f}(a + \varepsilon b)/\mathrm{d}\varepsilon)$ evaluated at $\varepsilon = 0$.

Example 12. Show

$$\sin(a + \varepsilon b) = \sin a + \varepsilon b \cos a,$$

$$\cos(a + \varepsilon b) = \cos a - \varepsilon b \sin a,$$

$$\cot(a + \varepsilon b) = \cot a - \varepsilon b/\sin^2 a, \qquad \sin a \neq 0,$$

$$(a + \varepsilon b)^{1/2} = a^{1/2} + \varepsilon b/(2a^{1/2}), \qquad a > 0.$$

If $\boldsymbol{p}$ and $\boldsymbol{q}$ are two vectors, say in Euclidean 3-space, such that $\boldsymbol{p} = (p_1, p_2, p_3)$, $\boldsymbol{q} = (q_1, q_2, q_3)$ we define a *dual vector* by $\boldsymbol{p} + \varepsilon \boldsymbol{q}$, its components being three dual numbers:

$$\boldsymbol{p} + \varepsilon \boldsymbol{q} = (p_1 + \varepsilon q_1, p_2 + \varepsilon q_2, p_3 + \varepsilon q_3).$$

We mention the following rules. A dual vector may be multiplied by a (dual) scalar

(3.3) $$(a + \varepsilon b)(\boldsymbol{p} + \varepsilon \boldsymbol{q}) = a\boldsymbol{p} + \varepsilon(a\boldsymbol{q} + b\boldsymbol{p}).$$

For the scalar product of two dual vectors we have

(3.4) $$(\boldsymbol{p} + \varepsilon \boldsymbol{q}) \cdot (\boldsymbol{r} + \varepsilon \boldsymbol{s}) = \boldsymbol{p} \cdot \boldsymbol{r} + \varepsilon(\boldsymbol{p} \cdot \boldsymbol{s} + \boldsymbol{q} \cdot \boldsymbol{r}),$$

and for the vectorial product

(3.5) $$(\boldsymbol{p} + \varepsilon \boldsymbol{q}) \times (\boldsymbol{r} + \varepsilon \boldsymbol{s}) = \boldsymbol{p} \times \boldsymbol{r} + \varepsilon(\boldsymbol{p} \times \boldsymbol{s} + \boldsymbol{q} \times \boldsymbol{r}).$$

Example 13. Express the right-hand side of (3.4) and (3.5) in terms of the components of the four ordinary vectors $\boldsymbol{p}, \boldsymbol{q}, \boldsymbol{r}, \boldsymbol{s}$.

Certain results in spatial kinematics can be obtained directly from the analogous spherical case using a process we call *dualization.* It is simply necessary to replace all the quantities in a "spherical equation" by their dual analogs. To illustrate this we consider equation (3.1) of Chapter V, which gives an expression for the angle between the lines i_1, k_1. If we replace the vectors by their duals, so that $\boldsymbol{s}_{ik} \to \boldsymbol{s}_{ik} + \varepsilon \boldsymbol{s}'_{ik}$, etc., and the angle by its dual, i.e., $\phi_{ik} \to \phi_{ik} + \varepsilon d_{ik}$, then the expression (5.10) of Chapter V, for the distance between i_1, k_1 follows.

Example 14. Show that dualizing (3.1) of Chapter V, results in two ordinary vector equations. The "proper" (or real) part is simply V, (3.1) while the dual part is (5.10) of Chapter V.

In Section 5 we will employ a similar procedure to map points from the elliptic plane into lines in 3-space. The notion of dualizing takes its validity from a general principle known as the *transfer principle*, which was apparently formulated independently by E. Study and A. P. Kotel'nikov. The principle and its realms of application are discussed by DIMENTBERG [1965, 1971].

4. The motion of a line in three-space

Rather unexpectedly dual numbers have been applied to study the motion of a line in space; they seem even to be the most appropriate apparatus for this end. It was first done by STUDY [1891] and since his time dual numbers have an established place in kinematics as a tool with which to attack problems dealing with lines in space. There exists a vast literature on the subject (for example: BLASCHKE [1958], KELER [1959], YANG and FREUDENSTEIN [1964], DIMENTBERG [1965, 1971], ROTH [1967], YUAN et al. [1971], TÖLKE [1976], VELDKAMP [1976]).

The application of dual numbers to the lines of 3-space is carried out by a mapping of the (finite) lines on a set of points. This has been done by two different although related methods. In one of them oriented lines, called *spears*, are considered. A spear is defined as a line along which there is a given direction; any line corresponds to two *opposite* spears. A line l is determined by its Plücker vectors $\boldsymbol{q}, \boldsymbol{q}'$, satisfying the fundamental condition $\boldsymbol{q} \cdot \boldsymbol{q}' = 0$. For any line not at infinity we have $\boldsymbol{q} \neq 0$. A spear is given a unit vector in the line's Plücker direction, hence $\boldsymbol{q}^2 = 1$; that for the opposite spear is then $-\boldsymbol{q}$. We represent the spear by the dual vector $\hat{\boldsymbol{q}} = \boldsymbol{q} + \varepsilon \boldsymbol{q}'$, a vector in a dual 3-space, $\Sigma(d)$. From $\boldsymbol{q}^2 = 1$ and $\boldsymbol{q} \cdot \boldsymbol{q}' = 0$ it follows that $\hat{\boldsymbol{q}}^2 = 1$, $\hat{\boldsymbol{q}}$ is a unit vector. Now if one considers the given spear as being represented by (i.e., mapped onto) the endpoint of $\hat{\boldsymbol{q}}$, it follows that the spears are mapped onto the points of the unit sphere S in $\Sigma(d)$. Opposite spears correspond to diametrical points on S. Any point on S is the image of a spear.

Under any displacement in 3-space there is a linear transformation of the Plücker coordinates of l and it induces in $\Sigma(d)$ a linear transformation which leaves S invariant, which means S undergoes a Euclidean rotation about the origin. Hence kinematics in 3-space with spears as moving elements corresponds to spherical kinematics in Σ, as dealt with in Chapter VII.

We shall not follow this idea here but make use of a somewhat different mapping, not of spears but of non-oriented lines. Therefore we introduce a projective dual plane U. A point of U is defined as a triplet of three dual numbers $(\hat{x}, \hat{y}, \hat{z})$, with the provision that *not all three are pure duals*; furthermore $(\hat{x}, \hat{y}, \hat{z})$ is identical with the point $(\hat{\rho}\hat{x}, \hat{\rho}\hat{y}, \hat{\rho}\hat{z})$, $\hat{\rho}$ being any non pure-dual factor. Hence $\hat{x}, \hat{y}, \hat{z}$ are homogeneous coordinates. The coordinates of a point $(\hat{x}, \hat{y}, \hat{z})$ will be called *normalized* if $\hat{x}^2 + \hat{y}^2 + \hat{z}^2$ is a real number. If $\hat{x} = x + \varepsilon x'$, $\hat{y} = y + \varepsilon y'$, $\hat{z} = z + \varepsilon z'$ and $\hat{\rho} = \rho + \varepsilon \rho'$, $\rho \neq 0$, we have $\Sigma (\hat{\rho}\hat{x})^2 = \Sigma [\rho x + \varepsilon (\rho x' + \rho' x)]^2 = \rho^2 \Sigma x^2 + 2\varepsilon\rho[\rho \Sigma xx' + \rho' \Sigma x^2]$ which implies that the coordinates may always be normalized by a suitable choice of ρ'/ρ. They remain normalized if multiplied by a non-zero real factor. Using the dual plane U the mapping is defined as follows: If the line l in finite 3-space is determined by the Plücker vectors (q_1, q_2, q_3), (q'_1, q'_2, q'_3) it is mapped on the point $L(\hat{q}_1, \hat{q}_2, \hat{q}_3)$ of U, with $\hat{q}_i = q_i + \varepsilon q'_i$ and therefore with normalized coordinates in view of $\Sigma q_i q'_i = 0$. For a line l, not at infinity, not all q_i are zero and therefore L is indeed a point of U. On the other hand any point L may be given normalized coordinates and hence it corresponds to a unique line l located in finite 3-space. The relationship between this mapping and that of oriented lines onto the points of the unit sphere S comes to this: two opposite spears, represented by diametrically opposite points of S correspond to the same diameter and the direction of the diameter corresponds to a point of U.

Example 15. Show that a point L whose normal coordinates are all real corresponds to a line l through the origin of the 3-space and conversely.
Example 16. Determine the points L corresponding to the axes O_X, O_Y and O_Z respectively.

A displacement in 3-space corresponds to a transformation in U which we shall now derive. The transformation for the line coordinates follows from the general displacements for points: $\boldsymbol{X} = \mathbf{A}\boldsymbol{x} + \boldsymbol{d}$, where $\mathbf{A} = \| a_{ij} \|$ is an orthogonal matrix and $\boldsymbol{d} = (d_1, d_2, d_3)$ is the displacement vector of the origin. Here we take $\boldsymbol{x}$ and $\boldsymbol{x} + \boldsymbol{q}$ as the two points which define the line, using $\boldsymbol{q}' = \boldsymbol{q} \times \boldsymbol{x}$ and recalling that each element of $\mathbf{A}$ equals its minor we get:

$$Q_1 = a_{11}q_1 + a_{12}q_2 + a_{13}q_3, \quad Q_2 = a_{21}q_1 + a_{22}q_2 + a_{23}q_3,$$
$$Q_3 = a_{31}q_1 + a_{32}q_2 + a_{33}q_3,$$

(4.1)
$$\begin{aligned} Q_1' &= a_{11}q_1' + a_{12}q_2' + a_{13}q_3' + (a_{21}d_3 - a_{31}d_2)q_1 \\ &\quad + (a_{22}d_3 - a_{32}d_2)q_2 + (a_{23}d_3 - a_{33}d_2)q_3, \\ Q_2' &= a_{21}q_1' + a_{22}q_2' + a_{23}q_3' + (a_{31}d_1 - a_{11}d_3)q_1 \\ &\quad + (a_{32}d_1 - a_{12}d_3)q_2 + (a_{33}d_1 - a_{13}d_3)q_3, \\ Q_3' &= a_{31}q_1' + a_{32}q_2' + a_{33}q_3' + (a_{11}d_2 - a_{21}d_1)q_1 \\ &\quad + (a_{12}d_2 - a_{22}d_1)q_2 + (a_{13}d_2 - a_{23}d_1)q_3. \end{aligned}$$

If the line $(\boldsymbol{q}, \boldsymbol{q}')$ is mapped on the point $(\hat{x}, \hat{y}, \hat{z})$ of U and $(\boldsymbol{Q}, \boldsymbol{Q}')$ on $(\hat{X}, \hat{Y}, \hat{Z})$ we have

(4.2)
$$\begin{aligned} \hat{x} &= q_1 + \varepsilon q_1', & \hat{y} &= q_2 + \varepsilon q_2', & \hat{z} &= q_3 + \varepsilon q_3', \\ \hat{X} &= Q_1 + \varepsilon Q_1', & \hat{Y} &= Q_2 + \varepsilon Q_2', & \hat{Z} &= Q_3 + \varepsilon Q_3'. \end{aligned}$$

From (4.1) and (4.2) it follows that the relation between the vectors $\hat{\boldsymbol{X}} = (\hat{X}, \hat{Y}, \hat{Z})$ and $\hat{\boldsymbol{x}} = (\hat{x}, \hat{y}, \hat{z})$ is represented by

(4.3)
$$\hat{\boldsymbol{X}} = \hat{\mathbf{A}}\hat{\boldsymbol{x}},$$

where

(4.4)
$$\hat{\mathbf{A}} = \left\| \begin{matrix} a_{11} + \varepsilon(a_{21}d_3 - a_{31}d_2) & a_{12} + \varepsilon(a_{22}d_3 - a_{32}d_2) & a_{13} + \varepsilon(a_{23}d_3 - a_{33}d_2) \\ a_{21} + \varepsilon(a_{31}d_1 - a_{11}d_3) & a_{22} + \varepsilon(a_{32}d_1 - a_{12}d_3) & a_{23} + \varepsilon(a_{33}d_1 - a_{13}d_3) \\ a_{31} + \varepsilon(a_{11}d_2 - a_{21}d_1) & a_{32} + \varepsilon(a_{12}d_2 - a_{22}d_1) & a_{33} + \varepsilon(a_{13}d_2 - a_{23}d_1) \end{matrix} \right\|$$

It is easy to verify that this matrix $\hat{\mathbf{A}}$, with dual elements, is *orthogonal.* This may be done by proving that the scalar product of any column by itself is *one* and that of two different columns is *zero.*

Example 17. Show that $\hat{\mathbf{A}}\hat{\mathbf{A}}^T = \mathbf{I}$.
Example 18. Show that any element of $\hat{\mathbf{A}}$ is equal to its minor.
Example 19. Show that $|\hat{\mathbf{A}}| = 1$.
Example 20. Determine $\hat{\mathbf{A}}$ if the displacement in 3-space is a rotation about O; ($d_1 = d_2 = d_3 = 0$).
Example 21. Determine $\hat{\mathbf{A}}$ if the displacement in 3-space is a translation ($a_{ij} = \delta_{ij}$).

From (4.4) it follows that (4.3) leaves the quadratic form $\hat{x}^2 + \hat{y}^2 + \hat{z}^2$ invariant. This implies that (4.3) is an elliptic displacement in U in the sense of the non-Euclidean metrics developed in Chapter XII, Section 4. *Euclidean geometry in 3-space with the lines as elements corresponds to elliptic geometry (of the points) in the dual plane* U.

In ordinary elliptic geometry, in a real plane, the distance ψ between two points (x_1, y_1, z_1) and (x_2, y_2, z_2) is defined by

$$\cos\psi = (x_1x_2 + y_1y_2 + z_1z_2)/((x_1^2 + y_1^2 + z_1^2)^{1/2}(x_2^2 + y_2^2 + z_2^2)^{1/2}), \tag{4.5}$$

which gives us (because the right-hand side is a number a satisfying $-1 \leq a \leq 1$) always real values for ψ. If for instance the homogeneous coordinates (x_1, y_1, z_1) are multiplied by a negative number, a is replaced by $-a$. This implies that ψ is not unambiguously defined; even if we restrict ψ to $0 \leq \psi \leq \pi$ it may be changed into $\pi - \psi$. This is natural in elliptic geometry and it is related to the circumstance that in this geometry the total length of a line is finite and equal to π.

In our dual plane we define the distance between two points $(\hat{x}_1, \hat{y}_1, \hat{z}_1)$ and $(\hat{x}_2, \hat{y}_2, \hat{z}_2)$ in an analogous way. We suppose that both points are given normalized coordinates which as we know is always possible. Making use of (4.2) we obtain for the dual distance $\hat{\psi} = \psi + \varepsilon\psi'$:

$$\begin{aligned} \cos(\psi + \varepsilon\psi') &= (\hat{x}_1\hat{x}_2 + \hat{y}_1\hat{y}_2 + \hat{z}_1\hat{z}_2)/((\hat{x}_1^2 + \hat{y}_1^2 + \hat{z}_1^2)^{1/2}(\hat{x}_2^2 + \hat{y}_2^2 + \hat{z}_2^2)^{1/2}) \\ &= [(q_1r_1 + q_2r_2 + q_3r_3) \\ &\qquad + \varepsilon(q_1r_1' + q_2r_2' + q_3r_3' + q_1'r_1 + q_2'r_2 + q_3'r_3)] \\ &\quad /[(q_1^2 + q_2^2 + q_3^2)^{1/2}(r_1^2 + r_2^2 + r_3^2)^{1/2}], \end{aligned} \tag{4.6}$$

where (q_i, q_i'), (r_i, r_i') are the Plücker coordinates of the two lines represented by these two points. Keeping in mind that $\cos(\psi + \varepsilon\psi') = \cos\psi - \varepsilon\psi'\sin\psi$ and comparing (4.6) with the formulas for the distance and the sine and cosine of the angle between the two lines, we have the following interpretation of the dual distance between two image points: *ψ means the angle between the two lines and* (provided $\psi \neq 0$) *ψ' is their distance*. Neither one is uniquely defined: the angle may be replaced by its supplement and the distance by its opposite value. (For some problems it may be desirable to define ψ and ψ' unambiguously, for instance by $0 < \psi \leqslant \pi/2$, $\psi' \geqslant 0$.)

We mention some special cases of (4.6). The image points of two parallel lines have no well-defined distance ($\cos(\psi + \varepsilon\psi') = 1$, $\psi = 0$, $\sin \psi = 0$, ψ' is arbitrary). If $\cos(\psi + \varepsilon\psi')$ is real, but $\neq 1$, the two lines intersect; if $\cos(\psi + \varepsilon\psi')$ is a pure dual the two lines are orthogonal. The most important case is $\cos(\psi + \varepsilon\psi') = 0$, i.e., $\hat{x}_1\hat{x}_2 + \hat{y}_1\hat{y}_2 + \hat{z}_1\hat{z}_2 = 0$: in this case *the two lines intersect orthogonally* and their image points are conjugate with respect to the quadratic form $\Omega \equiv \hat{x}^2 + \hat{y}^2 + \hat{z}^2 = 0$.

Example 22. Show that the vertices of a polar triangle of Ω are the images of the three axes of a Cartesian frame in 3-space.

We can now give an interpretation of a straight line in the dual plane U. It is defined as the locus of points satisfying a homogeneous linear equation $\hat{u}\hat{x} + \hat{v}\hat{y} + \hat{w}\hat{z} = 0$, with dual coefficients, not all three pure duals. Each point on the line is a conjugate of the pole $(\hat{u}, \hat{v}, \hat{w})$ of the line. Hence *the points of a line are the images of those lines in three-space intersecting a fixed line orthogonally*. These lines constitute a metrically special hyperbolic congruence; we shall call the fixed line its basis. This implies that there exists a (1, 1) correspondence between the lines in U and the finite lines in 3-space. A line $(\hat{u}, \hat{v}, \hat{w})$ is the image of that basis which has the point $(\hat{u}, \hat{v}, \hat{w})$ as its image.

Example 23. Consider the line $\hat{x} = 0$ in U. It corresponds to the O_X-axis in 3-space. Points on the line are the mappings of the hyperbolic congruence with O_X as its basis. Any point on the line may be given the (normalized) coordinates $(0, \sin\psi + \varepsilon d \cos\psi, \cos\psi - \varepsilon d \sin\psi)$; it is the mapping of the line $\boldsymbol{q} = (0, \sin\psi, \cos\psi)$, $\boldsymbol{q}' = (0, d\cos\psi, -d\sin\psi)$ which intersects O_X orthogonally at $(d, 0, 0)$ the angle with O_Z being ψ.

Two points $L(\hat{x}_1, \hat{y}_1, \hat{z}_1)$ and $M(\hat{x}_2, \hat{y}_2, \hat{z}_2)$ have in general a join LM whose coordinates are

$$(4.7) \qquad \hat{y}_1\hat{z}_2 - \hat{y}_2\hat{z}_1, \qquad \hat{z}_1\hat{x}_2 - \hat{z}_2\hat{x}_1, \qquad \hat{x}_1\hat{y}_2 - \hat{x}_2\hat{y}_1,$$

but if the lines l and m corresponding to L and M are parallel, the numbers (4.7) are all three pure duals and they do not define a line in U. If LM exists it corresponds to a line in 3-space which is the basis of a hyperbolic congruence to which l and m belong. Hence the *line LM corresponds to the common perpendicular of* l *and* m. The operation of joining two points in U is translated in 3-space as the operation of determining the common perpendicular of the corresponding lines in 3-space; each point on the join corresponds to a line which intersects the perpendicular orthogonally.

Example 24. Even if $(\hat{x}_1, \hat{y}_1, \hat{z}_1)$ and $(\hat{x}_2, \hat{y}_2, \hat{z}_2)$ are normalized coordinates the same does not hold for the coordinates (4.7) of their join. Show that this is only the case if L and M are conjugate with respect to Ω (or, what is the same, if l and m are orthogonally intersecting lines).

Example 25. Two lines in U have in general a point of intersection. Determine its meaning in three-space.

Example 26. Apart from its application in kinematics the mapping may be used to prove geometrical theorems about lines in space. Let L, M, N be three non-collinear points of U, and L' the pole with respect to Ω of MN, etc. As is well-known the triangles LMN and $L'M'N'$ are perspective: LL', MM', NN' are concurrent at a point H (or in the terminology of elliptic geometry: the heights of a triangle pass through one point, its orthocenter). The translation of this theorem in 3-space is the following. l, m, n are three arbitrary lines, l′ is the common perpendicular of m and n etc., (ll′) is the common perpendicular of l and l′, etc. Show that (ll′), (mm′) and (nn′) have a common perpendicular. This is the theorem of Petersen–Morley. The configuration in space consists of ten lines, each being the common perpendicular of three others. The mapping on U is a Desargues configuration of ten points (and ten lines, it is self-dual).

A curve k in U is the locus of points satisfying an equation $\hat{F}(\hat{x}, \hat{y}, \hat{z}) = 0$, which can be written as $F(x, y, z) + \varepsilon F'(x, y, z, x', y', z') = 0$, which implies $F = F' = 0$. Since the two equations $F = 0$, $F' = 0$ correspond in 3-space to equations of two different line complexes, *the curve* k *is the mapping of a congruence of lines.*

$F = 0$ is a condition for the direction of the line, so that the first complex consists of lines intersecting a given curve in the plane at infinity; the second is of a more general character. An important example will be discussed in the next section.

Example 27. Consider the case where k is a straight line.

Example 28. Determine the congruence of lines corresponding to a circle in U. Let its equation be $\hat{x}^2 + \hat{y}^2 - \hat{a}^2\hat{z}^2 = 0$. Derive the equations of the two complexes $F = 0$ and $F' = 0$. Show that the congruence determined by their intersection is degenerate: it consists of all lines in the plane at infinity, all lines through the isotropic points of the plane O_{XY} and a (2, 2) congruence. Show that the latter is built up of ∞^1 pencils of parallel lines.

Our mapping may be used to study spatial kinematics considering lines as the moving elements. The standard formulas for a displacement in 3-space are

$$X = x\cos\phi - y\sin\phi, \quad Y = x\sin\phi + y\cos\phi, \quad Z = z + d. \tag{4.8}$$

From this it follows that the corresponding displacement formulas in U given by means of (4.3) and (4.4), are:

$$\begin{aligned} \hat{X} &= (\cos\phi + \varepsilon d\sin\phi)\hat{x} - (\sin\phi - \varepsilon d\cos\phi)\hat{y} \\ &= \cos(\phi - \varepsilon d)\hat{x} - \sin(\phi - \varepsilon d)\hat{y}, \\ \hat{Y} &= (\sin\phi - \varepsilon d\cos\phi)\hat{x} + (\cos\phi + \varepsilon d\sin\phi)\hat{y} \\ &= \sin(\phi - \varepsilon d)\hat{x} + \cos(\phi - \varepsilon d)\hat{y}, \\ \hat{Z} &= \hat{z}, \end{aligned} \tag{4.9}$$

which represents a rotation with center $(0, 0, 1)$ and (dual) rotation angle $\hat{\phi} = \phi - \varepsilon d$.

(4.9) can be used to study two-positions theory in 3-space with emphasis on orthogonally intersecting lines.

Example 29. Show that in the elliptic dual plane U a birational quadratic relationship exists between a point and its join to the homologous point. Formulate what this means for 3-space. Show that a line in 3-space is in general the common perpendicular of two homologous lines.

In principle the mapping may be used to study n-positions theory in space as far as lines are concerned. For the case $n = 3$ we have introduced in Chapter IV the "screw triangle" consisting of the screw axes s_{23}, s_{31}, s_{12} and their common perpendiculars. This configuration is mapped on the vertices of a triangle and its sides. To study this configuration by more advanced methods, three-positions theory in elliptic geometry (Chapter XII, Section 4) is an appropriate tool.

In Chapter V, for $n = 4$ we have introduced the notion of a "complimentary-screw quadrilateral" consisting of four screw axes and the normals between adjacent sides. This configuration is mapped on the vertices of a quadrilateral and its sides. This too could be studied by the methods of this section. In fact the screw cone (Chapter V, (3.3)), the cubic complex (Chapter V, (5.15)) and their ((9, 3) congruence) intersection represent examples of configurations which we would obtain in U as $F = 0$, $F' = 0$, and k respectively. We shall, however, not develop the subject further. In the next section we restrict ourselves to instantaneous kinematics.

5. The instantaneous case

We have derived for plane elliptic kinematics (in Chapter XII, (4.13)) the formulas for instantaneous kinematics based on the introduction of canonical frames. Restricting ourselves to second order properties we found the following scheme

$$\begin{aligned} &X_0 = x, \quad X_1 = -\omega y, \quad X_2 = -\omega^2 x - \beta_3 y + \beta_2 z, \\ &Y_0 = y, \quad Y_1 = \omega x, \qquad Y_2 = \beta_3 x - \omega^2 y, \\ &Z_0 = z, \quad Z_1 = 0, \qquad\ \ Z_2 = -\beta_2 x, \end{aligned} \tag{5.1}$$

where ω, β_3, β_2 are the (time-dependent) instantaneous invariants up to the second order.

On the other hand, again up to the second order, the pertinent development for a moving line in space has been given in Chapter VI, (8.1):

$$(5.2)\quad \begin{aligned} &Q_1 = q_1 - q_2 t - \tfrac{1}{2}(q_1 - eq_3)t^2, \quad Q_2 = q_2 + q_1 t - \tfrac{1}{2}q_2 t^2, \quad Q_3 = q_3 - \tfrac{1}{2}eq_1 t^2,\\ &Q_1' = q_1' + (\sigma_0 q_2 - q_2')t + \tfrac{1}{2}(2\sigma_0 q_1 + \lambda q_2 - \mu e q_3 + eq_3' - q_1')t^2,\\ &Q_2' = q_2' + (-\sigma_0 q_1 + q_1')t + \tfrac{1}{2}(2\sigma_0 q_2 - \lambda q_1 - q_2')t^2,\\ &Q_3' = q_3' + \tfrac{1}{2}(-eq_1' + \mu e q_1)t^2. \end{aligned}$$

Here σ_0, e (written instead of ε to avoid confusion with the dual unit), μ and λ are the (time-independent) spatial instantaneous invariants up to the second order.

If in the set (5.2) we substitute the mapping formulas (4.2) we obtain after some algebra

$$(5.3)\quad \begin{aligned} &\hat{X}_0 = \hat{x}, \quad \hat{X}_1 = -(1-\varepsilon\sigma_0)\hat{y}, \quad \hat{X}_2 = -(1-2\varepsilon\sigma_0)\hat{x} + \varepsilon\lambda\hat{y} + e(1-\varepsilon\mu)\hat{z},\\ &\hat{Y}_0 = \hat{y}, \quad \hat{Y}_1 = (1-\varepsilon\sigma_0)\hat{x}, \qquad \hat{Y}_2 = -\varepsilon\lambda\hat{x} - (1-2\varepsilon\sigma_0)\hat{y},\\ &\hat{Z}_0 = \hat{z}, \quad \hat{Z}_1 = 0, \qquad\qquad\quad \hat{Z}_2 = -e(1-\varepsilon\mu)\hat{x}. \end{aligned}$$

Comparing this with (5.1) we see that the canonical frame in three-space corresponds to an analogous canonical frame for the elliptic geometry of the dual plane U and that the respective instantaneous invariants are related:

$$(5.4)\qquad \omega = 1 - \varepsilon\sigma_0, \quad \beta_3 = -\varepsilon\lambda, \quad \beta_2 = e(1-\varepsilon\mu).$$

Example 30. Obtain (5.3) by direct differentiation of (4.3). (KIRSON [1975].)

We now consider some first order properties. The moving line l in 3-space describes a ruled surface R. The images in U of its generators are the points of a certain curve r. It is not identical with a general curve in U (defined by an equation F = 0 and corresponding to a congruence of lines) for the coordinates of the points of r are functions of the time t and this parameter has only real values.

Related to a point L, the image of l, are two lines in U: the tangent t′ and the normal n′ of r at L. The first is the join of L and the consecutive point on r, the second passes through L and is perpendicular to t′, it passes also through the pole O. The interpretation in 3-space comes to this: t′ corresponds to the line t, the common perpendicular of l and the consecutive generator intersecting l orthogonally at the striction point S of l; the normal n′ is the image of the line n, the common perpendicular of l and screw axis s, n also passes through S. The configuration consisting of s, l, n and t is given in Fig. 105. The plane

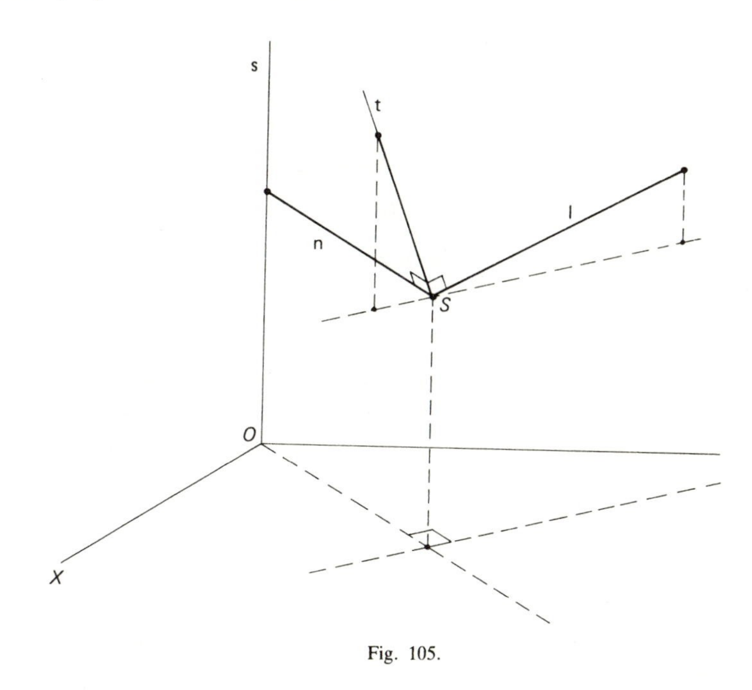

Fig. 105.

(l, t) is the tangent plane at S to R, n is the normal to R at S. The mapping confirms the first order properties we have found in Chapter VI.

Example 31. Show that the equation of t′ is $\hat{x}\hat{z}\hat{X}+\hat{y}\hat{z}\hat{Y}-(\hat{x}^2+\hat{y}^2)\hat{Z}=0$; determine the normalized coordinates of t′; show that the Plücker coordinates of t are quartic polynomials of those of l.

New results arise if we consider second order properties. In U the moving point L may pass through an inflection point of its path; this takes place if three consecutive positions of L are collinear. The locus of these points is the inflection curve. The corresponding situation in 3-space is that three consecutive positions of a moving line l have a common perpendicular. The condition for three lines to have a common perpendicular is expressed by two equations. Hence we may expect a congruence of lines l with the said property, we will call it the inflection congruence. Obviously it is mapped on the inflection curve in U.

The equation of the inflection curve in the real elliptic plane is (Chapter XII, (4.14))

$$-\beta_2 x(x^2+y^2+z^2)+\omega^2(x^2+y^2)z=0, \tag{5.5}$$

and by means of (5.4) it is in U:

(5.6) $$-e(1-\varepsilon\mu)\hat{x}(\hat{x}^2+\hat{y}^2+\hat{z}^2)+(1-2\varepsilon\sigma_0)(\hat{x}^2+\hat{y}^2)\hat{z}=0.$$

Substituting in (5.6) the mapping expressions (4.2) and equating to zero the real and the dual part of the left-hand side we obtain

(5.7) $$\mathrm{F}=-e(q_1^2+q_2^2+q_3^2)q_1+(q_1^2+q_2^2)q_3=0,$$

(5.8) $$\begin{aligned}\mathrm{F}'=&-e(-\mu q_1+q_1')(q_1^2+q_2^2+q_3^2)+q_3'(q_1^2+q_2^2-2q_3^2)\\&-2\sigma_0(q_1^2+q_2^2)q_3=0.\end{aligned}$$

Hence the inflection congruence C appears as the intersection of two cubic complexes, which is in general a (9, 9) congruence, but in our special case it is degenerate. F contains only the coordinates q_i which means that the lines in the plane V_0 at infinity are triple lines; they are moreover double lines of (5.8); hence the intersection contains the line field V_0 six times. Furthermore any line through an isotropic point (I_1 or I_2) of O_{XY} satisfies $q_1^2+q_2^2=q_3^2=0$ and all these lines belong to both complexes. Hence C is essentially a (7, 3) congruence.

If a line with direction numbers q_1, q_2, q_3 passes through the point (x, y, z) then $q_1'=q_2z-q_3y$, $q_2'=q_3x-q_1z$, $q_3'=q_1y-q_2x$. Substituting this in (5.8) we obtain

(5.9) $$\begin{aligned}&-q_2(q_1^2+q_2^2-2q_3^2)x+[eq_3(q_1^2+q_2^2+q_3^2)+q_1(q_1^2+q_2^2-2q_3^2)]y\\&\quad-eq_2(q_1^2+q_2^2+q_3^2)z+e\mu q_1(q_1^2+q_2^2+q_3^2)-2\sigma_0(q_1^2+q_2^2)q_3=0.\end{aligned}$$

Hence the lines of F′ with a given direction lie in a plane. (5.7) implies that the lines of C intersect V_0 at the points of a cubic curve K_3. This implies that C is built up of ∞^1 pencils of parallel lines; the vertex of a pencil is a point of K_3 and its plane is determined by (5.9).

An arbitrary plane W has three intersections S_1, S_2, S_3 with K_3, each being the vertex of such a pencil. Hence in W there lie three lines of C, its intersections with the planes of the pencils associated with S_i $(i=1,2,3)$. This verifies that the class of C is indeed equal to three.

If in (5.9) we consider x, y, z as fixed, the equation represents the intersection with V_0 of the cubic-complex cone with (x, y, z) as its vertex. This intersection is a cubic curve K_3' in V_0, which has 9 points of intersection with K_3, two of which are the isotropic points $q_1^2+q_2^2=q_3^2=0$. This verifies that the degree of C is indeed seven.

Example 32. Determine K_3' if $x=y=z=0$. Show that the seven lines of C through O are: $\mathrm{O}_Y, \mathrm{O}_Z$ (counted twice), OI_1 and OI_2 (each counted twice).

Example 33. Determine the lines of C through the point $(0, 0, z_0)$.
Example 34. Show that $(0, 1, 0, 0)$ and $(0, 0, 1, 0)$ are on K_3 and determine the lines of C through each of these points.
Example 35. Determine the lines of C in the planes $x = 0$, $x = a$ $(\neq 0)$, $y = 0$, $z = 0$.
Example 36. Show that any finite line in space is orthogonally intersected by three lines of C.
Example 37. Show that the lines of C in V_0 are the tangents of a curve of class seven.
Example 38. Compare the foregoing discussion and results with the finite positions case (i.e., three homologous lines with a common normal, Chapter V, Section 7).

Another second order concept in the dual plane U is the center of curvature M of the path of the moving point L; it may be defined as the intersection of two consecutive normals n′.

n′ is the image of the normal n of the ruled surface R described by l, at the striction point S of l. Hence the point M is the mapping of a line m which can be defined as the common perpendicular of n and the consecutive normal. It intersects n at the striction point of n with respect to the ruled surface described by n; this surface is the counterpart of the evolute of the path in U.

The line m depends on the moving line l in a manner similar to the way the center of curvature of the path in U depends on the moving point L, it may be defined in more than one way independent of the mapping; it was first studied in spatial kinematics by DISTELI [1914] and it has been named after him: the *Disteli axis* of l.

In Chapter XII (4.16a) we have given the coordinates of the center of curvature M of the point x, y, z for canonical frames:

$$X_M = \beta_2 x^2, \quad Y_M = \beta_2 xy, \quad Z_M = \beta_2 xz - \omega^2(x^2 + y^2). \tag{5.10}$$

Writing these expressions in terms of dual numbers, and using (5.4), these coordinates in U are seen to be

$$\begin{aligned} \hat{X}_M &= ex[x + \varepsilon(2x' - \mu x)], \qquad \hat{Y}_M = e[xy + \varepsilon(xy' + x'y - \mu xy)], \\ \hat{Z}_M &= exz - (x^2 + y^2) + \varepsilon[-e\mu xz + e(x'z + xz') - 2(xx' + yy') \\ &\qquad + 2\sigma_0(x^2 + y^2)]. \end{aligned} \tag{5.11}$$

To determine the line in 3-space corresponding to M the coordinates of the latter must be normalized. This can be done by multiplying (5.11) by a function of the form $A + \varepsilon B$ where $A : B = G_1(x, y, z) : G_2(x, y, z, x', y', z')$, and G_1 and G_2 are quartic polynomials of their arguments. From (4.2) it now follows that the coordinates (m_i, m_i') of the Disteli axis associated with the moving line (q_i, q_i') are sextic polynomials of q_i, q_i'.

Example 39. Show that $G_1 = e^2(x^2 + y^2 + z^2)x^2 - 2e(x^2 + y^2)xz + (x^2 + y^2)^2$.
Example 40. Determine G_2.
Example 41. Show that the direction of m is given by $m_1 : m_2 : m_3 = eq_1^2 : eq_1q_2 : eq_1q_3 - (q_1^2 + q_2^2)$.

Example 42. Verify that m is perpendicular to the normal n; determine its angle with the moving line and with the screw axis s.

If in U the moving point L passes through an inflection point, the two consecutive normals intersect at the point M conjugate to the tangent of the path. The conclusion is: if a moving line belongs to the inflection congruence C then its Disteli axis coincides with the line t (through the striction point S, perpendicular to both l and the normal n).

The Disteli axis may also be discussed by means of the Euler–Savary equation in the elliptic plane, given in Chapter XII, (4.17b):

$$(\cot \rho - \cot \rho_1) \sin \eta = \omega^2/\beta_2. \tag{5.12}$$

In this formula ρ and ρ_1 are the distances from the origin to, respectively, the moving point and its center of curvature; η is the angle to their ray from the pole tangent $x = 0$.

Applying this to the dual plane U the variables in (5.12) are interpreted as follows: ρ and ρ_1 stand for, respectively, the dual distances OL and OM, η for the dual angle between the pole tangent $\hat{x} = 0$ and OL, and

$$\omega^2/\beta_2 = (1 - 2\varepsilon\sigma_0)/(e(1 - \varepsilon\mu)) = [1 + \varepsilon(\mu - 2\sigma_0)]/e.$$

We note that the pole tangent $\hat{x} = 0$, that is the line $(1,0,0)$, corresponds in three-space to the O_X-axis.

We introduce in three-space a coordinate system for lines, similar to polar coordinates for points in U (Fig. 106). The striction point S is given the cylindrical coordinates ψ, h, r; the line l is determined by S and by the angle α. The Disteli axis m intersects $S'S$ orthogonally and is determined by ψ, h, r_1, α_1. The dual distances $\hat{\rho}$ and $\hat{\rho}_1$ from s to l and s to m respectively are $\alpha + \varepsilon r$ and $\alpha_1 + \varepsilon r_1$. Furthermore the dual angle $\hat{\eta}$ between O_X and $S'S$ is $\psi + \varepsilon h$.

From the trigonometry of dual angles

$$\begin{aligned} \cot \hat{\rho} &= \cot(\alpha + \varepsilon r) = \cot \alpha - \varepsilon r/\sin^2 \alpha, \\ \cot \hat{\rho}_1 &= \cot \alpha_1 - \varepsilon r_1/\sin^2 \alpha_1, \\ \sin \hat{\eta} &= \sin(\psi + \varepsilon h) = \sin \psi + \varepsilon h \cos \psi. \end{aligned} \tag{5.13}$$

Substituting all this into the modified formula (5.12) and separating the real and the dual parts we obtain

$$(\cot \alpha - \cot \alpha_1) \sin \psi = e^{-1}, \tag{5.14}$$

$$(\cot \alpha - \cot \alpha_1) h \cos \psi - ((r/\sin^2 \alpha) - (r_1/\sin^2 \alpha_1)) \sin \psi = (\mu - 2\sigma_0)/e. \tag{5.15}$$

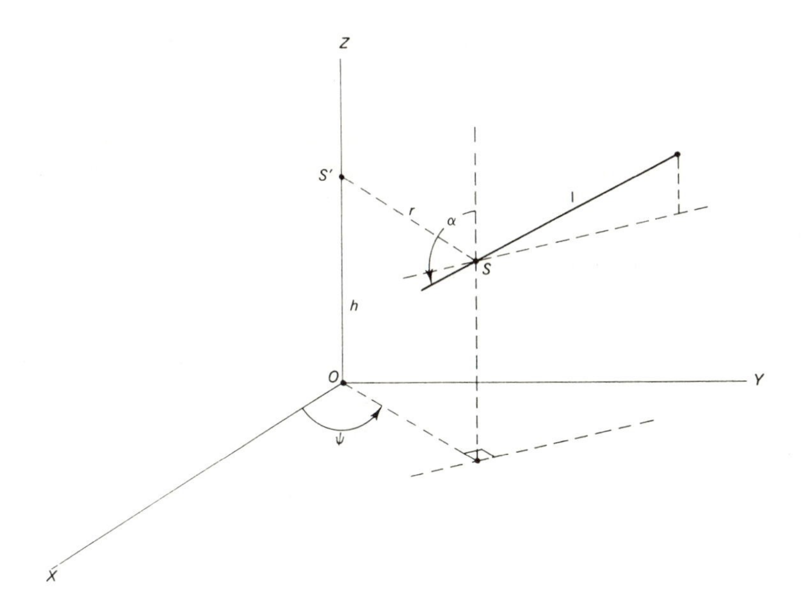

Fig. 106.

(5.14) deals only with the direction of l and m, it has the same simple form as the Euler–Savary equation in plane elliptic geometry. (5.15) is more complicated; by means of (5.14) it may be simplified to

$$(5.16)\quad (h/e)\cot\psi - [r(\cot^2\alpha + 1) - r_1(\cot^2\alpha_1 + 1)]\sin\psi = (\mu - 2\sigma_0)/e.$$

If (besides ψ and h) r and α are known then α_1 follows from (5.14) and r_1 from (5.16). Inversely, if we start with r_1 and α_1 they determine r and α. Hence there exists a birational relationship between l and m, which is, as we found before, of degree six in the coordinates of the two lines.

If in U the point L is on the pole tangent (which implies $\hat{\eta} = 0$) we have $\hat{\rho}_1 = 0$ for all $\hat{\rho}$: the center of curvature of L coincides with the pole (as in the case in plane Euclidean kinematics). The corresponding theorem in three-space reads: the Disteli axis of each line intersecting O_X orthogonally coincides with the screw axis.

Example 43. Show that l and m are never parallel (for $\alpha \neq 0$).

Example 44. If l belongs to the inflection congruence C we have $r = r_1$, $\alpha_1 = \alpha + \pi/2$. Show that, h and ψ being given, there are two lines l, different from s, intersecting SS' orthogonally and belonging to C; this is in accordance with the theorem that any line in space is intersected orthogonally by three lines of C.

It is interesting to compare (5.16) with Chapter VI, (9.16). Clearly the Disteli axis of a line l is the same as the line, l_c, which to the second order is at a fixed distance and angle from l. This may seem surprising since the Disteli axis is associated with the ruled surface generated by l while l_c is associated with a congruence which is defined by the motion of l. It should also be noted that the striction axis defined in Chapter VI, Section 8 also coincides with the Disteli axis. (In fact, what we have called d in Chapter VI, (8.26) is exactly equal to r_1 in the present development.)

We restrict ourselves to these remarks on second order properties of moving lines in space. The mapping of lines on the points of the dual plane U enables one in principle to study higher order concepts. For instance, formulas for the curvature of the centrodes, in U, correspond to properties of the Disteli axes of the generators of the axodes.

6. Quaternions

There exists a species of algebra by means of which spatial kinematics and especially spherical kinematics may be treated in an elegant way.

A *quaternion* $\mathbf{Q}$ is defined as a complex number depending on four units $1, i, j, k$:

$$\mathbf{Q} = c_0 + c_1 i + c_2 j + c_3 k, \tag{6.1}$$

c_i $(i = 0, 1, 2, 3)$ are real numbers called the *components* of $\mathbf{Q}$. The addition of quaternions is defined by

$$\begin{aligned} \mathbf{Q} + \mathbf{Q}' &= (c_0 + c_1 i + c_2 j + c_3 k) + (c_0' + c_1' i + c_2' j + c_3' k) \\ &= (c_0 + c_0') + (c_1 + c_1')i + (c_2 + c_2')j + (c_3 + c_3')k. \end{aligned} \tag{6.2}$$

The multiplication of two quaternions is distributive with respect to summation and is defined by the following rules for the multiplication of the units:

$$\begin{aligned} &1i = i1 = i, \qquad 1j = j1 = j, \qquad 1k = k1 = k, \\ &i^2 = j^2 = k^2 = -1, \\ &jk = -kj = i, \qquad ki = -ik = j, \qquad ij = -ji = k. \end{aligned} \tag{6.3}$$

Hence

$$\begin{aligned} \mathbf{Q}\mathbf{Q}' &= (c_0 + c_1 i + c_2 j + c_3 k)(c_0' + c_1' i + c_2' j + c_3' k) \\ &= (c_0 c_0' - c_1 c_1' - c_2 c_2' - c_3 c_3') + (c_0 c_1' + c_1 c_0' + c_2 c_3' - c_3 c_2')i \\ &\quad + (c_0 c_2' + c_2 c_0' + c_3 c_1' - c_1 c_3')j + (c_0 c_3' + c_3 c_0' + c_1 c_2' - c_2 c_1')k. \end{aligned} \tag{6.4}$$

From (6.3) it follows that the multiplication is *not commutative.*

Example 45. Prove that the multiplication is associative: $(\mathbf{QQ}')\mathbf{Q}'' = \mathbf{Q}(\mathbf{Q}'\mathbf{Q}'')$; hence $\mathbf{QQ}'\mathbf{Q}''$ has an unambiguous meaning.
Example 46. Show that the system of quaternions, with respect to addition and multiplication is a *skew field* (also called a "division ring"); contrary to the set of dual numbers it has no zero-divisors unequal to zero.

If (c_0, c_1, c_2, c_3) is a quaternion $\mathbf{Q}$ the *conjugate* quaternion $\tilde{\mathbf{Q}}$ is defined by $(c_0, -c_1, -c_2, -c_3)$. From (6.4) it follows that $\mathbf{Q}\tilde{\mathbf{Q}} = \tilde{\mathbf{Q}}\mathbf{Q} = c_0^2 + c_1^2 + c_2^2 + c_3^2$, a non-negative number called the *norm* $\mathrm{N}(\mathbf{Q})$ of $\mathbf{Q}$. If $\mathrm{N}(\mathbf{Q}) = 1$ then $\mathbf{Q}$ is called a *unit quatcrnion.*

Example 47. Determine from (6.4) the conjugate of $\mathbf{QQ}'$ and show that it is equal to $\tilde{\mathbf{Q}}'\tilde{\mathbf{Q}}$.
Example 48. Determine from (6.4) the norm of $\mathbf{QQ}'$ and prove $\mathrm{N}(\mathbf{QQ}') = \mathrm{N}(\mathbf{Q})\mathrm{N}(\mathbf{Q}')$.
Example 49. Show $\mathbf{Q}^{-1} = \mathrm{N}^{-1}(\mathbf{Q})\tilde{\mathbf{Q}}$.
Example 50. If $\mathbf{Q}$ is a unit quaternion then $\tilde{\mathbf{Q}}$ is a unit quaternion.

For a quaternion with $c_0 = 0$ the components (c_1, c_2, c_3) may be considered as those of a Euclidean vector; such a quaternion is called a *vector quaternion.*

Example 51. $\mathbf{Q}$ is a vector quaternion if and only if $\mathbf{Q} + \tilde{\mathbf{Q}} = 0$.
Example 52. The vector corresponding to a unit vector quaternion is a unit vector.
Example 53. Two conjugate vector quaternions correspond to opposite vectors.

Any quaternion may be written formally as the sum of a real number and a vector

$$\mathbf{Q}(c_0, c_1, c_2, c_3) = c_0 + \boldsymbol{c}(c_1, c_2, c_3). \tag{6.5}$$

Example 54. The product of two vector quaternions $\mathbf{Q} = (0, \boldsymbol{c})$ and $\mathbf{Q}' = (0, \boldsymbol{c}')$ is $\mathbf{QQ}' = -\boldsymbol{c} \cdot \boldsymbol{c}' + \boldsymbol{c} \times \boldsymbol{c}'$.
Example 55. The product of two quaternions $\mathbf{Q}$ and $\mathbf{Q}'$ may be written

$$\mathbf{QQ}' = (c_0 + \boldsymbol{c})(c_0' + \boldsymbol{c}') = c_0c_0' - \boldsymbol{c} \cdot \boldsymbol{c}' + c_0\boldsymbol{c}' + c_0'\boldsymbol{c} + \boldsymbol{c} \times \boldsymbol{c}'.$$

7. Application to spherical kinematics

Let a point in space be given in terms of position vector $\boldsymbol{x} = (x_1, x_2, x_3)$ and let $\mathbf{x}$ stand for the corresponding vector quaternion $(0, x_1, x_2, x_3)$. Let $\mathbf{Q}$ be a unit quaternion (c_0, c_1, c_2, c_3), $\Sigma c_i^2 = 1$. Then from (6.4)

$$\begin{aligned}\mathbf{Qx} = [&-(c_1x_1 + c_2x_2 + c_3x_3), (c_0x_1 + c_2x_3 - c_3x_2),\\ &(c_0x_2 + c_3x_1 - c_1x_3), (c_0x_3 + c_1x_2 - c_2x_1)].\end{aligned} \tag{7.1}$$

If we calculate $\mathbf{Qx\tilde{Q}}$, $\mathbf{\tilde{Q}} = (c_0, -c_1, -c_2, -c_3)$, again by means of (6.4), it is seen that its first component vanishes which means that the product is a vector quaternion, we denote it as $\mathbf{X} = (0, X_1, X_2, X_3)$. The result is

$$\begin{aligned} X_1 &= (c_0^2 + c_1^2 - c_2^2 - c_3^2)x_1 + 2(-c_0c_3 + c_1c_2)x_2 + 2(c_0c_2 + c_1c_3)x_3, \\ X_2 &= 2(c_0c_3 + c_2c_1)x_1 + (c_0^2 - c_1^2 + c_2^2 - c_3^2)x_2 + 2(-c_0c_1 + c_2c_3)x_3, \\ X_3 &= 2(-c_0c_2 + c_3c_1)x_1 + 2(c_0c_1 + c_3c_2)x_2 + (c_0^2 - c_1^2 - c_2^2 + c_3^2)x_3. \end{aligned} \tag{7.2}$$

These formulas are seen to be identical with those of Chapter VI, (2.1). The conclusion is: *the operation* $\mathbf{Qx\tilde{Q}}$ *on the point* $\mathbf{x}$, $\mathbf{Q}$ *being a unit quaternion, transforming* $\mathbf{x}$ *into the point* $\mathbf{X}$, *is a displacement in space, leaving the origin invariant*; *the components of* $\mathbf{Q}$ *are the Euler parameters of the displacement.*

In order to prove directly that $\mathbf{R} = \mathbf{Qx\tilde{Q}}$ represents a displacement we remark that in view of Example 47 we have $\mathbf{\tilde{R}} = \mathbf{Q\tilde{x}\tilde{Q}}$; hence $\mathbf{R} + \mathbf{\tilde{R}} = \mathbf{Qx\tilde{Q}} + \mathbf{Q\tilde{x}\tilde{Q}} = \mathbf{Q(x + \tilde{x})\tilde{Q}}$, which is equal to zero because $\mathbf{x} + \mathbf{\tilde{x}} = 0$. Hence $\mathbf{Qx\tilde{Q}}$ is a vector quaternion (which we call $\mathbf{X}$). Furthermore $\mathrm{N}(\mathbf{X}) = \mathrm{N}(\mathbf{Qx\tilde{Q}}) = \mathrm{N}(\mathbf{Q})\mathrm{N}(\mathbf{x})\mathrm{N}(\mathbf{\tilde{Q}})$, in view of Example 48. But $\mathrm{N}(\mathbf{Q}) = \mathrm{N}(\mathbf{\tilde{Q}}) = 1$ and therefore $\mathrm{N}(\mathbf{X}) = \mathrm{N}(\mathbf{x})$ or $X_1^2 + X_2^2 + X_3^2 = x_1^2 + x_2^2 + x_3^2$ which concludes the proof.

We have now derived a (1, 1) relationship between spherical displacements and unit quaternions. Let D_1 and D_2 be two rotations about the origin, and $\mathbf{Q}_1, \mathbf{Q}_2$ the corresponding unit quaternions. If the point (x_1, x_2, x_3) is displaced by D_1 into (y_1, y_2, y_3) and the latter by D_2 into (z_1, z_2, z_3), we have

$$\mathbf{y} = \mathbf{Q}_1\mathbf{x}\mathbf{\tilde{Q}}_1, \quad \mathbf{z} = \mathbf{Q}_2\mathbf{y}\mathbf{\tilde{Q}}_2, \tag{7.3}$$

and therefore

$$\mathbf{z} = \mathbf{Q}_2\mathbf{Q}_1\mathbf{x}\mathbf{\tilde{Q}}_1\mathbf{\tilde{Q}}_2 = \mathbf{Q}_3\mathbf{x}\mathbf{\tilde{Q}}_3, \quad \mathbf{Q}_3 = \mathbf{Q}_2\mathbf{Q}_1, \quad \mathbf{\tilde{Q}}_3 = \mathbf{\tilde{Q}}_1\mathbf{\tilde{Q}}_2. \tag{7.4}$$

Hence *the resultant rotation* $\mathrm{D}_2\mathrm{D}_1$ *corresponds to the quaternion which is the product*, $\mathbf{Q}_2\mathbf{Q}_1$, *of the two corresponding unit quaternions.*

The Euler parameters c_i $(i = 0, 1, 2, 3)$ have a geometrical meaning and the same holds therefore for the components of the corresponding unit quaternion: c_1, c_2, c_3 are direction numbers of the rotation axis and $c_0 = \cos\phi/2$, ϕ being the rotation angle.

Example 56. Show that $\mathbf{X} = \mathbf{Qx\tilde{Q}}$ implies $\mathbf{x} = \mathbf{\tilde{Q}XQ}$; if $\mathbf{Q}$ corresponds to a displacement D then $\mathbf{\tilde{Q}}$ corresponds to the inverse displacement.

Example 57. Show that the corresponding quaternion of the standard rotation $X_1 = x_1\cos\phi - x_2\sin\phi$, $X_2 = x_1\sin\phi + x_2\cos\phi$, $X_3 = x_3$ is $(\cos\phi/2, 0, 0, \sin\phi/2)$.

Example 58. Show that the rotation about the axis with direction angles α, β, γ and with the rotation angle ϕ corresponds to the unit quaternion $(\cos\phi/2, \cos\alpha\sin\phi/2, \cos\beta\sin\phi/2, \cos\gamma\sin\phi/2)$.

Example 59. Show that a vector quaternion corresponds to a half-turn.
Example 60. Determine by means of (6.4) the product D_2D_1 of the two rotations given by $(\alpha_i, \beta_i, \gamma_i; \phi_i)$, $i = 1, 2$. Verify that the product is not commutative.

Quaternions are an elegant tool to describe spherical displacements. A certain school of kinematicians (BLASCHKE [1960], H. R. MÜLLER [1962]) has developed this apparatus to study spherical motions (with one and with two parameters) from a mathematical viewpoint. However, for the main subjects of this book—n-positions theory, instantaneous kinematics, special motions—the method seems less useful. Moreover, if the components of **Q** must be taken into account one might as well make use of the Euler parameters and their geometrical interpretation. Therefore we restrict ourselves here to some remarks on the velocity distribution: A spherical motion can obviously be described by

$$\mathbf{X} = \mathbf{Q}(t)\mathbf{x}\tilde{\mathbf{Q}}(t), \tag{7.5}$$

the unit quaternion **Q** being a function of the scalar t. From this it follows (the ordinary rules for differentiation are valid for this kind of function)

$$\dot{\mathbf{X}} = \dot{\mathbf{Q}}\mathbf{x}\tilde{\mathbf{Q}} + \mathbf{Q}\mathbf{x}\dot{\tilde{\mathbf{Q}}},$$

or eliminating **x** by means of $\mathbf{x} = \tilde{\mathbf{Q}}\mathbf{X}\mathbf{Q}$:

$$\dot{\mathbf{X}} = \dot{\mathbf{Q}}\tilde{\mathbf{Q}}\mathbf{X} + \mathbf{X}\mathbf{Q}\dot{\tilde{\mathbf{Q}}}. \tag{7.6}$$

If **R** stands for $\dot{\mathbf{Q}}\tilde{\mathbf{Q}}$ then $\tilde{\mathbf{R}} = \mathbf{Q}\dot{\tilde{\mathbf{Q}}}$; since $\mathbf{Q}\tilde{\mathbf{Q}} = 1$ we have $\mathbf{R} + \tilde{\mathbf{R}} = 0$, which implies that **R** is a vector quaternion: $\mathbf{R} = (0, \boldsymbol{r})$. We obtain

$$\dot{\mathbf{X}} = \mathbf{R}\mathbf{X} - \mathbf{X}\mathbf{R}, \tag{7.7}$$

or, in view of Example 54,

$$\dot{\mathbf{X}} = 2\boldsymbol{r} \times \boldsymbol{X}, \tag{7.8}$$

which shows that $2\boldsymbol{r}$ is equal to the angular velocity vector $\boldsymbol{\Omega}$.

Example 61. If $\mathbf{Q} = (c_0, c_1, c_2, c_3)$, $c_i(t)$, $\Sigma c_i^2 = 1$, determine **R** in terms of c_i and $\dot{c}_i$.
Example 62. If for $t = 0$ we have $c_0 = 1$, $c_i = 0$ $(i = 1, 2, 3)$ show that $\dot{c}_0(0) = 0$; express $\mathbf{R}(0)$ in terms of $\dot{c}_i(0)$, $i = 1, 2, 3$; determine the scalar angular velocity $\omega(0)$.

8. Dual quaternions

In the preceding section we have developed by means of quaternions a certain short-hand method to describe spherical motions. In Section 4, considering lines as the moving elements and making use of dual numbers, we derived a mapping of the general motions in space onto motions of the unit dual sphere (or elliptic motions in the dual plane). In this section we combine the two results in order to describe general displacement in space by means of (dual) quaternions.

A dual quaternion $\hat{\mathbf{Q}}$ is defined as a quaternion whose components are dual numbers: $\hat{c}_i = c_i + \varepsilon c'_i$, $i = 0, 1, 2, 3$, c_i and c'_i being real numbers. $\hat{\mathbf{Q}}$ is a unit quaternion if $\Sigma \hat{c}_i^2 = 1$, which implies $\Sigma c_i^2 = 1$, $\Sigma c_i c'_i = 0$; $\hat{\mathbf{Q}}$ is a vector quaternion if $\hat{c}_0 = 0$, hence $c_0 = c'_0 = 0$.

If $\hat{\mathbf{x}}$ is a dual vector quaternion, $\hat{\mathbf{Q}}$ the unit dual quaternion $(\hat{c}_0, \hat{c}_1, \hat{c}_2, \hat{c}_3)$, then following a procedure analogous to that in Section 7,

$$\hat{\mathbf{X}} = \hat{\mathbf{Q}}\hat{\mathbf{x}}\tilde{\hat{\mathbf{Q}}} \tag{8.1}$$

represents a spherical displacement of $\hat{\mathbf{x}}$ onto the vector quaternion $\hat{\mathbf{X}}$; (8.1) is explicitly given by the formulas (7.2) if $\hat{x}_i$, $\hat{X}_i$ and $\hat{c}_i$ are written instead of x_i, X_i and c_i. Here $\hat{\mathbf{X}}$ and $\hat{\mathbf{x}}$ are the mapping of the lines as given by (4.2), and our aim will be to determine $\hat{c}_i$ in such a way that these formulas are identical with those expressed by the dual orthogonal matrix (4.4). Substituting into the latter from (7.2) so that the a_{ij} are in terms of the (real) Euler parameters c_i we obtain nine equations, the first of which reads

$$\begin{aligned} c_0^2 + c_1^2 - c_2^2 - c_3^2 + \varepsilon[2(c_0c_3 + c_2c_1)d_3 - 2(-c_0c_2 + c_3c_1)d_2] &= \\ = (c_0^2 + c_1^2 - c_2^2 - c_3^2) + 2\varepsilon(c_0c'_0 + c_1c'_1 - c_2c'_2 - c_3c'_3)&, \end{aligned}$$

which implies in view of

$$\sum c_i c'_i = 0, \tag{8.2}$$

$$(c_0c_3 + c_2c_1)d_3 + (c_0c_2 - c_3c_1)d_2 = 2(c_0c'_0 + c_1c'_1). \tag{8.3}$$

The two analogous equations are

$$\begin{aligned} (c_0c_1 + c_3c_2)d_1 + (c_0c_3 - c_1c_2)d_3 &= 2(c_0c'_0 + c_2c'_2), \\ (c_0c_2 + c_1c_3)d_2 + (c_0c_1 - c_2c_3)d_1 &= 2(c_0c'_0 + c_3c'_3). \end{aligned} \tag{8.4}$$

(8.2), (8.3) and (8.4) are four linear equations for c'_i $(i = 0, 1, 2, 3)$, with the solutions

$$\begin{aligned} c'_0 &= \tfrac{1}{2}(c_1d_1 + c_2d_2 + c_3d_3), \quad & c'_1 &= \tfrac{1}{2}(-c_0d_1 - c_3d_2 + c_2d_3), \\ c'_2 &= \tfrac{1}{2}(c_3d_1 - c_0d_2 - c_1d_3), \quad & c'_3 &= \tfrac{1}{2}(-c_2d_1 + c_1d_2 - c_0d_3). \end{aligned} \tag{8.5}$$

We have obtained the following result: *the general displacement in space with the rotational part determined by the Euler parameters* c_i (with $\Sigma c_i^2 = 1$) *and with the origin of coordinates having the displacement vector* (d_1, d_2, d_3), *may be written as* (8.1). $\hat{\mathbf{Q}}$ *is the dual unit quaternion with components* $c_i + \varepsilon c'_i$, *where the* c'_i *are given by* (8.5).

Example 63. The formulas (8.5) have been derived by identifying only the diagonal elements of (4.4) and (7.2); show that this is sufficient to guarantee that two (direct) orthogonal matrices are identical; verify that the non-diagonal elements are also equal.

Example 64. Show that the dual unit quaternion $\hat{\mathbf{Q}}$ corresponding to the standard spatial displacement (of a point (x_1, x_2, x_3))

$$X_1 = x_1 \cos\phi - x_2 \sin\phi, \quad X_2 = x_1 \sin\phi + x_2 \cos\phi, \quad X_3 = x_3 + d,$$

has the components

$$[\cos(\phi/2) + \tfrac{1}{2}\varepsilon d \sin(\phi/2), 0, 0, \sin(\phi/2) - \tfrac{1}{2}\varepsilon d \cos(\phi/2)],$$

and therefore (8.1) yields (for a line)

$$Q_1 = q_1 \cos\phi - q_2 \sin\phi, \quad Q'_1 = q'_1 \cos\phi - q'_2 \sin\phi + q_1 d \sin\phi + q_2 d \cos\phi, \text{ etc.}$$

Example 65. Show that the (point) translation $X_1 = x_1 + d_1$, $X_2 = x_2 + d_2$, $X_3 = x_3 + d_3$ corresponds to $\hat{\mathbf{Q}} = (1, -\tfrac{1}{2}\varepsilon d_1, -\tfrac{1}{2}\varepsilon d_2, -\tfrac{1}{2}\varepsilon d_3)$, and therefore (8.1) yields $Q_1 = q_1$, $Q'_1 = d_2 q_3 - d_3 q_2$, etc.

Example 66. Show that the displacement, which is the resultant of the rotation with angle ϕ about the line l (with direction angles α, β, γ and passing through the origin) and the translation d along l, corresponds to the dual unit quaternion

$$\hat{\mathbf{Q}} = [\cos(\phi/2) + \tfrac{1}{2}\varepsilon d \sin(\phi/2), \cos\alpha(\sin(\phi/2) - \tfrac{1}{2}\varepsilon d \cos(\phi/2)),$$
$$\cos\beta(\sin(\phi/2) - \tfrac{1}{2}\varepsilon d \cos(\phi/2)), \cos\gamma(\sin(\phi/2) - \tfrac{1}{2}\varepsilon d \cos(\phi/2))].$$

Consider the special case where the rotational part is a half-turn.

Example 67. If the dual unit quaternion $\hat{\mathbf{Q}} = (c_0 + \varepsilon c'_0, c_1 + \varepsilon c'_1, c_2 + \varepsilon c'_2, c_3 + \varepsilon c'_3)$ is given, the rotational part of the corresponding displacement follows immediately from c_i $(i = 0, 1, 2, 3)$; d_i may be determined from (8.5). Show that we obtain $d_1 = 2(c'_0 c_1 - c'_1 c_0 + c'_2 c_3 - c'_3 c_2)$ and for d_2 and d_3 the values that follow by cyclic substitution of the indices.

Example 68. Verify that the dual parts c'_i of the components of the dual unit quaternion $\hat{\mathbf{Q}}$ are related to Study's soma coordinates g_i (Chapter VI, (2.2)), of the displacement corresponding to $\hat{\mathbf{Q}}$, by the equations $c'_i = \tfrac{1}{2} g_i$.

Example 69. The displacement corresponding to the dual unit quaternion $\hat{\mathbf{Q}} = (\hat{c}_0, \hat{c}_1, \hat{c}_2, \hat{c}_3)$ is a rotation if $c_1 d_1 + c_2 d_2 + c_3 d_3 = 0$. Prove by means of Example 67 that this condition is identical to $c'_0 = 0$.

Example 70. Two spatial displacements D_1 and D_2 correspond to the dual unit quaternions $\hat{\mathbf{Q}}_1$, with components $(\hat{c}_i)$, and $\hat{\mathbf{Q}}_2$ with components $(\hat{C}_i)$. Show that the displacement $\mathrm{D}_2^{-1}\mathrm{D}_1$ is a rotation if $\Sigma(c_i C'_i + c'_i C_i) = 0$. Show that this condition is identical to equation (2.9) of Chapter VI, which is the expression for the condition that two Study soma's have a point (and by implication a line) in common.

Example 71. Show that a general screw displacement can also be represented by a dual quaternion $\hat{\boldsymbol{\theta}}$ such that $\hat{\boldsymbol{\theta}} = \hat{\boldsymbol{s}} \tan(\hat{\phi}/2)$ where $\hat{\boldsymbol{s}}$ is the dual unit vector along the screw axis, and the dual angle $\hat{\phi} = \phi + \varepsilon d$ (with its positive sense defined according to $\hat{\boldsymbol{s}}$ using the right-hand-rule) specifies the rotation and translation along the screw axis. Show that an arbitrary screw $\hat{\boldsymbol{x}}$ is transformed by the screw $\hat{\boldsymbol{\theta}}$ into $\hat{\boldsymbol{X}}$ according to

$$\hat{\boldsymbol{X}} = \hat{\boldsymbol{x}} + (2\hat{\boldsymbol{\theta}}/(1+(\tan\hat{\phi}/2)^2)) \times (\hat{\boldsymbol{x}} + \hat{\boldsymbol{\theta}} \times \hat{\boldsymbol{x}})$$

which is analogous to Chapter III, (12.10) (DIMENTBERG [1950, 1965]). Here, $\times$ is the ordinary vector cross product.

Example 72. Show that two successive screw displacements

$$\hat{\boldsymbol{\theta}}_{12} = \hat{\boldsymbol{s}}_{12} \tan(\hat{\phi}_{12}/2) \quad \text{and} \quad \hat{\boldsymbol{\theta}}_{23} = \hat{\boldsymbol{s}}_{23} \tan(\hat{\phi}_{23}/2)$$

can be replaced by a single screw displacement specified by the displacement screw $\hat{\boldsymbol{\theta}}_{13} = \hat{\boldsymbol{s}}_{13} \tan(\hat{\phi}_{13}/2)$ where (using the ordinary vector cross and dot products)

$$\hat{\boldsymbol{\theta}}_{13} = (\hat{\boldsymbol{\theta}}_{12} + \hat{\boldsymbol{\theta}}_{23} - \hat{\boldsymbol{\theta}}_{12} \times \hat{\boldsymbol{\theta}}_{23})/(1 - \hat{\boldsymbol{\theta}}_1 \cdot \hat{\boldsymbol{\theta}}_2).$$

(DIMENTBERG [1950, 1965].)

Example 73. If we use the dual quaternion operator (8.1) where

$$\hat{\mathbf{Q}} = \cos(\hat{\phi}/2) + \hat{\boldsymbol{s}} \sin(\hat{\phi}/2), \quad \tilde{\hat{\mathbf{Q}}} = \cos(\hat{\phi}/2) - \hat{\boldsymbol{s}} \sin(\hat{\phi}/2)$$

then the result of the previous example can be written more simply as $\hat{\mathbf{Q}}_3 = \hat{\mathbf{Q}}_2\hat{\mathbf{Q}}_1$ where $\hat{\mathbf{Q}}_i = \cos(\hat{\phi}_i/2) + \hat{\boldsymbol{s}}_i \sin(\hat{\phi}_i/2)$ $i = 1, 2, 3$. (MCAULAY [1898], YANG [1963].)

It is possible to use the properties of quaternions and exponentials to advantage in obtaining expressions for the geometry between various screw axes. The Taylor expansion of a function of the dual number $\hat{x} = x + \varepsilon x'$ is

$$\mathrm{f}(\hat{x}) = \mathrm{f}(x + \varepsilon x') = \mathrm{f}(x) + \varepsilon x' \mathrm{df}(x)/\mathrm{d}x.$$

Therefore $\mathrm{e}^{\hat{x}}$ may be written

$$\mathrm{e}^{\hat{x}} = \mathrm{e}^x + \varepsilon x' \mathrm{e}^x = \mathrm{e}^x(1 + \varepsilon x').$$

But we know $\mathrm{e}^{\hat{x}} = \mathrm{e}^{(x+\varepsilon x')} = \mathrm{e}^x \mathrm{e}^{\varepsilon x'}$ and so we have the identity

$$(1 + \varepsilon x') = \mathrm{e}^{\varepsilon x'}. \tag{8.6}$$

With the aid of (8.6) we may express a dual number $\hat{x}$ in exponential form

$$\hat{x} = x + \varepsilon x' = x(1 + \varepsilon p_x) = x\mathrm{e}^{\varepsilon p_x} \tag{8.7}$$

where $p_x = x'/x$ is called the pitch, which characterizes the dual number $\hat{x}$.

Example 74. Show that if $\hat{a} = a + \varepsilon a'$, $\hat{b} = b + \varepsilon b'$,

$$\hat{a}\hat{b} = ab + \varepsilon(a'b + ab') = ab\mathrm{e}^{\varepsilon(p_a+p_b)}$$

$$\hat{a}/\hat{b} = (a/b) + \varepsilon((a' - b')/b^2) = (a/b)\mathrm{e}^{\varepsilon(p_a-p_b)}$$

$$\hat{a}^n = a^n \mathrm{e}^{\varepsilon n p_a}.$$

A screw can be represented by a dual vector

$$\hat{\boldsymbol{\theta}} = \hat{\phi}\hat{\boldsymbol{s}} = \phi\mathrm{e}^{\varepsilon p_\phi}\hat{\boldsymbol{s}}$$

where the unit screw $\hat{\boldsymbol{s}}$ is the dual unit line-vector along the screw axis, i.e., $\hat{\boldsymbol{s}} = \boldsymbol{s} + \varepsilon \boldsymbol{s}_0 \times \boldsymbol{s}$ where $\boldsymbol{s}_0$ is the position vector from the origin to any point on the screw. $\hat{\phi}$ is the dual modulus of the screw, i.e., $\hat{\phi} = \phi + \varepsilon d$. Hence the "screw" $\hat{\boldsymbol{\theta}}$ is a vector dual quaternion: its scalar part is zero.

The scalar and "screw" products of two screws $\hat{\boldsymbol{\theta}}_{12} = \hat{\phi}_{12}\hat{\boldsymbol{s}}_{12}$ and $\hat{\boldsymbol{\theta}}_{23} = \hat{\phi}_{23}\hat{\boldsymbol{s}}_{23}$ are respectively

$$\hat{\boldsymbol{\theta}}_{12} \cdot \hat{\boldsymbol{\theta}}_{23} = \hat{\phi}_{12}\hat{\phi}_{23} \cos \hat{\alpha} = \phi_{12}\phi_{23}e^{\varepsilon(p_{\phi_{12}}+p_{\phi_{23}})} \cos \hat{\alpha},$$

$$\hat{\boldsymbol{\theta}}_{12} \times \hat{\boldsymbol{\theta}}_{23} = (\hat{\phi}_{12}\hat{\phi}_{23} \sin \hat{\alpha})\hat{\boldsymbol{n}}_2 = [\phi_{12}\phi_{23}e^{\varepsilon(p_{\phi_{12}}+p_{\phi_{23}})} \sin \hat{\alpha}]\hat{\boldsymbol{n}}_2,$$

where $\hat{\alpha} = \alpha + \varepsilon a$ is the dual angle subtended by the two screw axes, and $\hat{\boldsymbol{n}}_2$ is the unit screw along the common perpendicular in the direction of $\boldsymbol{s}_{12} \times \boldsymbol{s}_{23}$.

Example 75. Show that if the scalar product is zero the axes of $\hat{\boldsymbol{\theta}}_{12}$ and $\hat{\boldsymbol{\theta}}_{23}$ intersect orthogonally; if the screw product is zero the screws are coaxial.

Example 76. Show that if $\hat{\boldsymbol{\theta}}_{34} = \hat{\phi}_{34}\hat{\boldsymbol{s}}_{34}$ then

$$\begin{aligned}\hat{\boldsymbol{\theta}}_{12} \times \hat{\boldsymbol{\theta}}_{23} \cdot \hat{\boldsymbol{\theta}}_{34} &= \hat{\phi}_{12}\hat{\phi}_{23}\hat{\phi}_{34} \sin \hat{\alpha}\, (\hat{\boldsymbol{n}}_2 \cdot \hat{\boldsymbol{s}}_{34}) \\ &= \hat{\phi}_{12}\hat{\phi}_{23}\hat{\phi}_{34} \sin \hat{\alpha} \cos \hat{\beta},\end{aligned}$$

where $\hat{\beta} = \beta + \varepsilon b$ is the dual angle subtended by the axis of $\hat{\boldsymbol{\theta}}_{34}$ and the common perpendicular between the axes of $\hat{\boldsymbol{\theta}}_{12}$ and $\hat{\boldsymbol{\theta}}_{23}$.

9. Displacement matrices

The general finite displacement has been described in this book by equations of the form

$$\boldsymbol{p}' = \mathbf{A}\boldsymbol{p} + \boldsymbol{d}, \quad \boldsymbol{P} = \mathbf{A}\boldsymbol{p} + \boldsymbol{d}, \quad \text{or} \quad \boldsymbol{P}_2 = \mathbf{A}\boldsymbol{P}_1 + \boldsymbol{d}.$$

The particular form depended on which properties we were interested in discussing. We have already given the most common expressions for **A** and $\boldsymbol{d}$. These are in terms of:

i) The screw parameters; using a special coordinate system (Chapter I, (4.4)), and general ones (Chapter III, (12.9), (12.10), and (12.11)).

ii) Euler parameters (Chapter VI, (2.1)), including Cayley's formula (Chapter I, (5.7)) and Study's soma (Chapter VI, Section 2).

iii) Eulerian angles (Chapter VI, (3.1)).

iv) Dual quaternions (Section 8 of this chapter).

In regard to $\boldsymbol{d}$: we point out that $\boldsymbol{d}$ always represents the displacement of the origin of coordinates and can always be given by (12.13) of Chapter III, regardless of the form of the elements of **A**.

In this section we briefly describe some additional expressions for the elements of **A** and ***d***.

There is a modification of the Euler parameter formulation which uses the so-called Cayley–Klein parameters $\alpha, \beta, \gamma, \delta$. The Euler parameters can be replaced by the Cayley–Klein parameters by substituting into (2.1) of Chapter VI,

$$(9.1)\quad c_0 = (\gamma + \beta)/(2i), \quad c_1 = (\delta + \alpha)/2, \quad c_2 = (\delta - \alpha)/(2i), \quad c_3 = (\gamma - \beta)/2,$$

where $i = \sqrt{-1}$, and $\alpha, \beta, \gamma, \delta$ are restricted so that $\alpha\delta - \beta\gamma = 1$. The result is

$$(9.2)\quad \mathbf{A} = \left\| \begin{matrix} \frac{1}{2}(\alpha^2 - \gamma^2 + \delta^2 - \beta^2) & (i/2)(\alpha^2 + \gamma^2 - \beta^2 - \delta^2) & \beta\delta - \alpha\gamma \\ (i/2)(\gamma^2 - \alpha^2 + \delta^2 - \beta^2) & \frac{1}{2}(\alpha^2 + \gamma^2 + \beta^2 + \delta^2) & i(\alpha\gamma + \beta\delta) \\ \gamma\delta - \alpha\beta & -i(\alpha\beta + \gamma\delta) & \alpha\delta + \beta\gamma \end{matrix} \right\|$$

The Cayley–Klein formulation has been used mainly in quantum kinematics and in studying gyroscopic motions.

Example 77. Show that in terms of the Euler angles (Chapter VI, Section 3) the Cayley–Klein parameters are:

$$\alpha = e^{i(\psi+\xi)/2}\cos(\theta/2), \qquad \beta = ie^{i(\psi-\xi)/2}\sin(\theta/2)$$

$$\gamma = ie^{-i(\psi-\xi)/2}\sin(\theta/2), \qquad \delta = e^{-i(\psi+\xi)/2}\cos(\theta/2).$$

Example 78. Show that the parameters $(\alpha_{13}, \beta_{13}, \gamma_{13}, \delta_{13})$ associated with the equivalent of rotation $(\alpha_{12}, \beta_{12}, \gamma_{12}, \delta_{12})$ followed by rotation $(\alpha_{23}, \beta_{23}, \gamma_{23}, \delta_{23})$ are given by

$$\alpha_{13} = \alpha_{23}\alpha_{12} + \gamma_{23}\beta_{12}, \quad \beta_{13} = \alpha_{12}\beta_{23} + \beta_{12}\delta_{23},$$

$$\gamma_{13} = \gamma_{12}\alpha_{23} + \delta_{12}\gamma_{23}, \quad \delta_{13} = \gamma_{12}\beta_{23} + \delta_{12}\delta_{23}.$$

A modification of the Euler angle formulation which uses dual angles has been applied to the study of the kinematics of mechanical linkwork and gyroscopic systems (YANG [1969]). In this method we simply replace the angles in the Euler angle form of **A** (i.e., Chapter VI, (3.1)) by their dual equivalents. Hence $\psi \to \hat{\psi}$, $\theta \to \hat{\theta}$, and $\xi \to \hat{\xi}$, where

$$(9.3)\qquad \hat{\psi} = \psi + \varepsilon u, \quad \hat{\theta} = \theta + \varepsilon v, \quad \hat{\xi} = \xi + \varepsilon w.$$

Here $\varepsilon^2 = 0$; and u, v, w are linear displacements.

The total transformation is then of the form

$$(9.4)\qquad \hat{\boldsymbol{P}} = \hat{\mathbf{A}}\hat{\boldsymbol{p}}$$

where $\hat{\boldsymbol{p}}$ is the matrix representation of a dual line vector fixed in E and $\hat{\boldsymbol{P}}$ is the same dual line vector measured in Σ after transformation by $\hat{\mathbf{A}}$. The dual line vectors are described in terms of their Plücker coordinates

$$\hat{p} = \begin{Vmatrix} p_{14} + \varepsilon p_{23} \\ p_{24} + \varepsilon p_{31} \\ p_{34} + \varepsilon p_{12} \end{Vmatrix} \tag{9.5}$$

similarly for $\hat{\boldsymbol{P}}$.

The transformation $\hat{\mathbf{A}}$ represents the total screw displacement. Its form is exactly the same as in Chapter VI, (3.1) except now dual angles replace the ordinary ones. From this it follows that the displacement may be considered as the ordered product of three screw displacements $\hat{\mathbf{A}}_3, \hat{\mathbf{A}}_2, \hat{\mathbf{A}}_1$ about screw axes fixed in Σ:

i) with $\hat{\mathbf{A}}_3$, E is displaced using the O_Z-axis as screw axis and (ψ, w) as screw parameters;

ii) under $\hat{\mathbf{A}}_2$, E is then displaced about the O_X-axis using screw parameters (θ, v);

iii) finally, under $\hat{\mathbf{A}}_1$, E is (again) screwed about the O_Z-axis using parameters (ξ, u).

Hence we have $\hat{\mathbf{A}} = \hat{\mathbf{A}}_1\hat{\mathbf{A}}_2\hat{\mathbf{A}}_3$ such that

$$\hat{\mathbf{A}} = \begin{Vmatrix} \cos\hat{\xi} & -\sin\hat{\xi} & 0 \\ \sin\hat{\xi} & \cos\hat{\xi} & 0 \\ 0 & 0 & 1 \end{Vmatrix} \begin{Vmatrix} 1 & 0 & 0 \\ 0 & \cos\hat{\theta} & -\sin\hat{\theta} \\ 0 & \sin\hat{\theta} & \cos\hat{\theta} \end{Vmatrix} \begin{Vmatrix} \cos\hat{\psi} & -\sin\hat{\psi} & 0 \\ \sin\hat{\psi} & \cos\hat{\psi} & 0 \\ 0 & 0 & 1 \end{Vmatrix} \tag{9.6}$$

which yields the dual form of Chapter VI, (3.1).

Example 79. In analogy to Chapter VI, Section 3, Example 8, show that if the displacement of E is a screw displacement $S_{(\xi,u)}$ about O_Z followed by a screw displacement $S_{(\theta,v)}$ about the common perpendicular from O_Z to o_z, and then by a screw $S_{(\psi,w)}$ about o_z, it follows that $S_{(\xi,u)} = \hat{\mathbf{A}}_1$, $S_{(\theta,v)} = \hat{\mathbf{A}}_1\hat{\mathbf{A}}_2\hat{\mathbf{A}}_1^{-1}$, $S_{(\psi,w)} = (\hat{\mathbf{A}}_1\hat{\mathbf{A}}_2)\hat{\mathbf{A}}_3(\hat{\mathbf{A}}_1\hat{\mathbf{A}}_2)^{-1}$ and therefore $\hat{\mathbf{A}} = S_{(\psi,w)}S_{(\theta,v)}S_{(\xi,u)} = \hat{\mathbf{A}}_1\hat{\mathbf{A}}_2\hat{\mathbf{A}}_3$. Alternatively, show if we screw about axes fixed in Σ the order is: (ψ, w) about O_Z, (θ, v) about O_X, (ξ, u) about O_Z.

Example 80. Using the results of Example 79 and referring to Fig. 107, show that (9.6) can be modelled in the following way: We introduce two fictitious systems E', E'' and then attach E to E' with a screw axis such that it coincides with o_z, we attach E' to E'' with a screw axis which coincides with the common normal between O_Z and o_z (in the sense from O_Z toward o_z), and attach E'' to Σ with a screw along O_Z.

It is useful to express u, v, w in terms of the displacement of a point in the moving system. If origins O (in Σ) and o (in E) coincide before the displacement and if o_X, o_Y, o_Z are the coordinates of o after the displacement it follows from the geometry (Fig. 107) of the three transformations (9.6) that

$$\begin{aligned} u &= o_Z - (o_X \sin\xi - o_Y \cos\xi)\cot\theta, \\ v &= o_X \cos\xi + o_Y \sin\xi, \\ w &= (o_X \sin\xi - o_Y \cos\xi)/\sin\theta. \end{aligned} \tag{9.7}$$

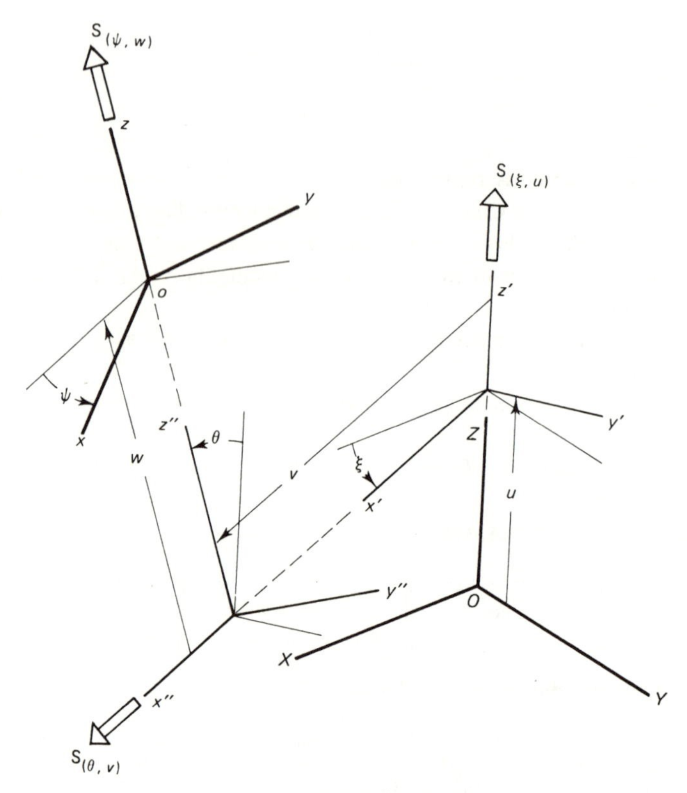

Fig. 107.

Hence if the six parameters (o_X, o_Y, o_Z), (ψ, θ, ξ) are known u, v, w follow from (9.7), and the elements of (9.6) are completely determined.

Example 81. Show that the $\hat{a}_{11}$ element of $\hat{\mathbf{A}}$ is

$$\cos\psi\cos\xi - \cos\theta\sin\xi\sin\psi + \varepsilon(-w\sin\psi\cos\xi - u\sin\xi\cos\psi + v\sin\theta\sin\xi\sin\psi - u\cos\theta\cos\xi\sin\psi - w\cos\theta\sin\xi\cos\psi).$$

The dual elements $\hat{a}_{ij}$ of $\hat{\mathbf{A}}$ have the following physical interpretation: $\hat{a}_{ij} = a_{ij} + \varepsilon\, e_{ikl} o_k a_{lj}$ where ($\varepsilon^2 = 0$, we sum over k and l according to e_{ikl}—the Levi–Civita epsilon) a_{ij} are the direction cosines describing the orientation of E relative to Σ, and o_k are the coordinates of the origin of E in Σ. Which is to say the dual part is the ith component of the moment relative to Σ of the jth coordinate axis of E.

Example 82. Using this, obtain (9.7) by showing that the $\hat{a}_{33}$ element yields $-v\sin\theta = -o_X\sin\theta\cos\xi - o_Y\sin\theta\sin\xi$, the $\hat{a}_{31}$, $w\sin\theta\cos\psi + v\cos\theta\sin\psi = o_X(\cos\psi\sin\xi + \sin\psi\cos\xi\cos\theta) - o_Y(\cos\psi\cos\xi - \sin\psi\sin\xi\cos\theta)$, and the $\hat{a}_{13}$, $v\cos\theta\sin\xi + u\sin\theta\cos\xi = o_Y\cos\theta + o_Z\cos\xi\sin\theta$.

There is an isomorphic correspondence between 3×3 real orthogonal matrices and 2×2 complex unitary matrices. This means any relation among matrices of one set is also satisfied by matrices of the other set. Hence complex 2×2 transformations can be defined which are analogous to our 3×3's. We take $\boldsymbol{p}$ and $\boldsymbol{P}$ as respectively

$$\boldsymbol{p} = \begin{Vmatrix} z & x - iy \\ x + iy & -z \end{Vmatrix}, \qquad \boldsymbol{P} = \begin{Vmatrix} Z & X - iY \\ X + iY & -Z \end{Vmatrix}$$

where $i = \sqrt{-1}$. Using the Cayley–Klein parameters one defines the unitary matrix

$$\mathbf{U} = \begin{Vmatrix} \alpha & \beta \\ \gamma & \delta \end{Vmatrix}$$

with elements satisfying the auxiliary condition $\alpha\delta - \beta\gamma = 1$. It is easy to verify that

$$\mathbf{U}^{-1} = \begin{Vmatrix} \delta & -\beta \\ -\gamma & \alpha \end{Vmatrix}$$

and that

$$\boldsymbol{P} = \mathbf{U}^{-1}\boldsymbol{p}\mathbf{U} \tag{9.8}$$

yields the same result as (9.2).

Example 83. Show that if $\boldsymbol{p}$ is rotated to position $\boldsymbol{P}_2$ by $\mathbf{U}_{12}$ and then to position $\boldsymbol{P}_3$ by $\mathbf{U}_{23}$ the direct rotation to $\boldsymbol{P}_3$ will be given by using $\mathbf{U}_{13}$ in (9.8) where $\mathbf{U}_{13} = \mathbf{U}_{23}\mathbf{U}_{12}$.

Example 84. Show that the equivalent complex 2×2 matrix in terms of the Euler angles is the result of the following product

$$\mathbf{U} = \begin{Vmatrix} e^{i\xi/2} & 0 \\ 0 & e^{-i\xi/2} \end{Vmatrix} \begin{Vmatrix} \cos(\theta/2) & i\sin(\theta/2) \\ i\sin(\theta/2) & \cos(\theta/2) \end{Vmatrix} \begin{Vmatrix} e^{i\psi/2} & 0 \\ 0 & e^{-i\psi/2} \end{Vmatrix}$$

and that if the Euler angles are replaced by their dual-angle equivalents the result is the same as from (9.6).

Example 85. Show that if we define the three matrices

$$\boldsymbol{\sigma}_x = \begin{Vmatrix} 0 & 1 \\ 1 & 0 \end{Vmatrix}, \qquad \boldsymbol{\sigma}_y = \begin{Vmatrix} 0 & -i \\ i & 0 \end{Vmatrix}, \qquad \boldsymbol{\sigma}_z = \begin{Vmatrix} 1 & 0 \\ 0 & -1 \end{Vmatrix},$$

(these are the so-called Pauli spin matrices) then

$$P = X\sigma_x + Y\sigma_y + Z\sigma_z, \quad p = x\sigma_x + y\sigma_y + z\sigma_z,$$

and $\mathbf{U}$ of the previous example becomes

$$\mathbf{U} = (\mathbf{I}\cos(\xi/2) + i\sigma_z \sin(\xi/2))(\mathbf{I}\cos(\theta/2) + i\sigma_x \sin(\theta/2))(\mathbf{I}\cos(\psi/2) + i\sigma_z \sin(\psi/2))$$

where

$$\mathbf{I} = \begin{Vmatrix} 1 & 0 \\ 0 & 1 \end{Vmatrix}.$$

If we use exponentials it is possible to describe general spatial displacements in a rather efficient way (ÖZGÖREN [1976]). We use the components (s_X, s_Y, s_Z) of the unit vector s parallel to the screw to define the matrices

$$\mathbf{S} = \begin{Vmatrix} 0 & -s_Z & s_Y \\ s_Z & 0 & -s_X \\ -s_Y & s_X & 0 \end{Vmatrix}$$

and

$$\mathbf{U} = \begin{Vmatrix} s_X^2 & s_X s_Y & s_X s_Z \\ s_X s_Y & s_Y^2 & s_Y s_Z \\ s_X s_Z & s_Y s_Z & s_Z^2 \end{Vmatrix}.$$

Then, clearly, the following vector and matrix operations can be equated

$$s \times (P_1 - S_0) = \mathbf{S}(P_1 - S_0), \quad s(s \cdot (P_1 - S_0)) = \mathbf{U}(P_1 - S_0),$$

and so we can write for the vector transformation equation (12.9) of Chapter III, the matrix form

$$P_2 = [\mathbf{I}\cos\phi + \mathbf{S}\sin\phi + \mathbf{U}(1 - \cos\phi)](P_1 - S_0) + D_0, \tag{9.9}$$

where $S_0(S_{0_X}, S_{0_Y}, S_{0_Z})$ is the position vector of a point on the screw axis, d is the translation,

$$\mathbf{D}_0 = \begin{Vmatrix} S_{0_X} + ds_X \\ S_{0_Y} + ds_Y \\ S_{0_Z} + ds_Z \end{Vmatrix}$$

and $\mathbf{I}$ is the unit identity matrix.

We note (9.9) has the following form

$$P_2 = \mathbf{F}(s, \phi)(P_1 - S_0) + D_0. \tag{9.10}$$

Since $s_X^2 + s_Y^2 + s_Z^2 = 1$, the following identities exist

$$\mathbf{S}^2 = \mathbf{U} - \mathbf{I}; \quad \mathbf{SU} = \mathbf{US} = 0. \tag{9.11}$$

Hence

$$\mathbf{F}(s,\phi)=\mathbf{I}+\mathbf{S}\sin\phi+\mathbf{S}^2(1-\cos\phi), \tag{9.12}$$

from which a Taylor series expansion yields

$$\mathbf{F}(s,\phi)=\mathbf{I}+\mathbf{S}(\phi-(1/3!)\phi^3+\cdots)+\mathbf{S}^2((1/2!)\phi^2-(1/4!)\phi^4+\cdots). \tag{9.13}$$

If we multiply $\mathbf{S}^2=\mathbf{U}-\mathbf{I}$ by successive powers of $\mathbf{S}$ we get from (9.11)

$$\mathbf{S}^3=-\mathbf{S},\quad \mathbf{S}^4=-\mathbf{S}^2,\quad \mathbf{S}^5=\mathbf{S},\quad \mathbf{S}^6=\mathbf{S}^2,\quad \mathbf{S}^7=-\mathbf{S},\cdots.$$

Hence (9.12) can be written as

$$\mathbf{F}(s,\phi)=\mathbf{I}+(\mathbf{S}\phi)+(1/2!)(\mathbf{S}\phi)^2+(1/3!)(\mathbf{S}\phi)^3+(1/4!)(\mathbf{S}\phi)^4+\cdots. \tag{9.14}$$

But this is nothing more than the Taylor series expansion of $e^{\mathbf{S}\phi}$, hence

$$\mathbf{F}(s,\phi)=e^{\mathbf{S}\phi}, \tag{9.15}$$

which is very similar in form to the planar rotation operator $e^{i\phi}$.

Example 85. Show that for a series of rotations ϕ_i, $i=1,2,\cdots,m$, the rotation angle from position 1 to m is $\Sigma\phi_i$ if all the screw axes are parallel; i.e., $e^{\mathbf{S}_1\phi_1}e^{\mathbf{S}_2\phi_2}e^{\mathbf{S}_3\phi_3}\cdots$ equals $e^{\mathbf{S}(\Sigma\phi_i)}$ if $\mathbf{S}_1=\mathbf{S}_2=\mathbf{S}_3=\cdots=\mathbf{S}$.

Example 86. Show that $e^{\mathbf{S}\phi}=\mathbf{I}+(\sin\phi)\mathbf{T}$, where $\mathbf{T}=\mathbf{S}(\mathbf{I}+\tan(\phi/2)\mathbf{S})$.

Example 87. Use the previous example to show that

$$(\mathbf{I}+(\sin\phi_{13})\mathbf{T}_{13})=(\mathbf{I}+(\sin\phi_{23})\mathbf{T}_{23})(\mathbf{I}+(\sin\phi_{12})\mathbf{T}_{12})$$

and using the fact that $\mathbf{S}_i^T=-\mathbf{S}_i$ and $\mathbf{T}_i-\mathbf{T}_i^T=2\mathbf{S}_i$ show that

$$(\sin\phi_{13})\mathbf{S}_{13}=(\sin\phi_{23})\mathbf{S}_{23}+(\sin\phi_{12})\mathbf{S}_{12}$$
$$+\tfrac{1}{2}\sin\phi_{23}\sin\phi_{12}(\mathbf{T}_{23}\mathbf{T}_{12}-\mathbf{T}_{12}^T\mathbf{T}_{23}^T).$$

The result is that the general spatial displacement can be written as

$$\boldsymbol{P}_2=e^{\mathbf{S}\phi}(\boldsymbol{P}_1-\boldsymbol{S}_0)+\boldsymbol{D}_0. \tag{9.16}$$

It has recently become fashionable to use 4×4 matrices in describing spatial displacements. In doing this the position vector is described in homogeneous coordinates and the fourth coordinate is usually taken as unity. Hence

$$\left\|\begin{matrix}P_{2x}\\P_{2y}\\P_{2z}\\1\end{matrix}\right\|=\left\|\begin{array}{ccc|c} & \mathbf{A} & & \boldsymbol{d}\\ & & & \\ \hline 0&0&0&1\end{array}\right\|\left\|\begin{matrix}P_{1x}\\P_{1y}\\P_{1z}\\1\end{matrix}\right\| \tag{9.17}$$

represents the total transformation with $\mathbf{A}$ being 3×3 and $\boldsymbol{d}$ being 3×1. The value of $\boldsymbol{d}$ is always

$$\boldsymbol{d} = (\mathbf{A} - \mathbf{I})(-\boldsymbol{S}_0) + d\boldsymbol{s}.$$

In terms of an arbitrary point R, not necessarily on the screw axis, it is of course $\boldsymbol{d} = \boldsymbol{R}_2 - \mathbf{A}\boldsymbol{R}_1$.

Example 88. A very useful form of the 4×4 is due to HARTENBERG and DENAVIT [1964] who have used it to solve 3-dimensional linkage analysis problems. Show that its elements

$$\left\|\begin{matrix} \cos\theta & -\sin\theta\cos\alpha & \sin\theta\sin\alpha & a\cos\theta \\ \sin\theta & \cos\theta\cos\alpha & -\cos\theta\sin\alpha & a\sin\theta \\ 0 & \sin\alpha & \cos\alpha & s \\ 0 & 0 & 0 & 1 \end{matrix}\right\|$$

can be obtained from (9.6) if we set $\psi = w = 0$, and $v = a$, $u = s$, $\theta = \alpha$, $\xi = \theta$.

Example 89. From the foregoing it follows that the Hartenberg and Denavit matrix is the product of two screw displacements. Show that this same result follows from Chapter III, (12.11) if we take the first screw as $S_{(\alpha,a)}$ (i.e., $\boldsymbol{s}(1,0,0)$, $\boldsymbol{S}_0(0,0,0)$, $\theta = \alpha$, $d = (a,0,0)$) and the second as $S_{(\theta,s)}$ (i.e., $\boldsymbol{s}(0,0,1)$, $\boldsymbol{S}_0(0,0,0)$, $\theta = \theta$; $d = (0,0,s)$).

Example 90. Verify that the inverse of the 4×4 transformation matrix given in (9.17) is obtained by replacing $\mathbf{A}$ by its transpose, $\mathbf{A}^T$, and $\boldsymbol{d}$ by the column vector $-\mathbf{A}^T\boldsymbol{d}$.

Although we have given some of the most useful forms, there are of course many other possible forms for the spatial displacement.

We end this discussion by pointing out that the displacement equations for lines and planes can also be written in terms of the matrices $\mathbf{A}$ and $\boldsymbol{d}$. If the point with coordinate matrix $\boldsymbol{p}$ is displaced into:

$$\boldsymbol{P} = \mathbf{A}\boldsymbol{p} + \boldsymbol{d},$$

then a line with Plücker vectors $\boldsymbol{q}, \boldsymbol{q}'$ (in column matrix form) is displaced into:

$$\boldsymbol{Q} = \mathbf{A}\boldsymbol{q}, \qquad \boldsymbol{Q}' = \mathbf{A}\boldsymbol{q}' + \mathbf{A}\mathbf{d}^*\boldsymbol{q}$$

where

$$\mathbf{d}^* = \left\|\begin{matrix} 0 & -d_3 & d_2 \\ d_3 & 0 & -d_1 \\ -d_2 & d_1 & 0 \end{matrix}\right\|,$$

d_1, d_2, d_3 are the components of $\boldsymbol{d}$. A plane with coordinates u_i, $i = 1,2,3,4$, is displaced into:

$$\left\|\begin{matrix} U_1 \\ U_2 \\ U_3 \\ U_4 \end{matrix}\right\| = \left\|\begin{array}{c|c} & 0 \\ \mathbf{A} & 0 \\ & 0 \\ \hline -\boldsymbol{d}^T\mathbf{A} & 1 \end{array}\right\| \left\|\begin{matrix} u_1 \\ u_2 \\ u_3 \\ u_4 \end{matrix}\right\|$$

BIBLIOGRAPHY

1. Authors cited in text

ALT [1921]
ALT, H., Zur Synthese der ebenen Mechanismen, *Z. Angew. Math. u. Mech.*, 1 (1921), 373–398.
ALT [1932 a]
ALT, H., Koppelgetriebe als Rastgetriebe, *Z. VDI.*, 76 (1932), 456–462 and 533–537.
ALT [1932 b]
ALT, H., Zur Geometrie der Koppelrastgetriebe, *Ing.-Archiv.*, III (1932), 394–411.
APPELL [1900]
APPELL, P., Propriété caractéristique du cilindroïde, *Bull. Soc. Math.*, 28 (1900), 261–265.

BALL [1876]
BALL, Robert S., *The Theory of Screws, A Study in the Dynamics of a Rigid Body*, 1876.
BALL [1900]
BALL, Robert S., *A Treatise on the Theory of Screws*, Cambridge (England) 1900, 544 pp.
BALTZER [1881]
BALTZER, R., *Theorie und Anwendung der Determinanten*, Leipzig, 1857; fifth edition, 1881.
BENNETT [1913]
BENNETT, G. T., The skew isogram-mechanism, *Proceedings of the London Mathematical Society*, 13, 2nd Series (1913–1914), 151–173.
BEREIS [1958]
BEREIS, R., Die Kinematik in der Gauss'schen Zahlenebene, *Wissenschaftliche Zeitschrift der Technischen Universitaet Dresden*, 8 (1958), 1–8.
BETH [1937]
BETH, H. J. E., Die Bewegungen eines starren Koerpers, die bei der Studyschen Abbildung algebraischen Bildkurven entsprechen, *Christiaan Huygens*, 16 (1937) 6, 226–272.
BETH [1938]
BETH, H. J. E., *Idem*, 16 (1938) 3, 145–154.
BETH [1949]
BETH, H. J. E., *Kinematica in het platte vlak*, Gorinchem, 1949.
BEYER [1953]
BEYER, Rudolf, *Kinematische Getriebesynthese*, Springer-Verlag, 1953, (English language translation, *The Kinematic Synthesis of Mechanisms*, McGraw-Hill Book Co., 1963).
BLASCHKE [1911]
BLASCHKE, W., Euklidische Kinematik und nichteuklidische Geometrie, *Zeitschr. Math. Phys.*, 60 (1911), 61–91 and 203–204.
BLASCHKE [1923]
BLASCHKE, Wilhelm, *Vorlesungen über Differentialgeometrie II*, Berlin, 1923, 259 pp.
BLASCHKE AND MÜLLER [1956]
BLASCHKE, W., AND MÜLLER, H. R., *Ebene Kinematik*, München, 1956, 269 pp.
BLASCHKE [1958]
BLASCHKE, W., Anwendung dualer Quaternionen auf die Kinematik, *Annales Academiae Scientiarum Fennicae*, (1958), 1–13.
BLASCHKE [1960]
BLASCHKE, W., *Kinematik und Quaternionen*, Berlin, 1960.

Bobillier [1870]
Bobillier, E. E., *Cours de Géométrie*, 14th edition, Paris, 1870, (see pp. 232–3), 403 pp.
Bol [1950]
Bol, Gerrit, *Projektive Differentialgeometrie*, Göttingen, 1950, Part 1, 365 pp. (See also Part 2, 1954, 372 pp., and Part 3, 1967, 527 pp.).
Bottema [1944 a]
Bottema, Oene, Over bewegingen der elliptische ruimte, waarbij alle punten congruente vlakke krommen beschrijven, *Ned. Akad. Wet., Verslag gewone vergaderingen Afd. Natuurkunde*, 53 (1944), 25–30.
Bottema [1944 b]
Bottema, Oene, De bijzondere bewegingen van Darboux in de elliptische ruimte, (*Ibid.*) 58–65.
Bottema [1949]
Bottema, O., On Cardan positions for the plane motion of a rigid body, *Proceedings Koninklijke Nederlandsche Ak. van Wetenschappen*, 52 (1949), 643–651.
Bottema [1954]
Bottema, O., On Alt's special three-bar sextic, *Proceedings Koninklijke Nederlandse Akademie van Wetenschappen*, Series A, 57 (1954), 498–504.
Bottema [1964 a]
Bottema, Oene, On the determination of the Burmester points for five positions of a moving plane, *Proceedings Koninklijke Nederlandse Akademie van Wetenschappen*, Series A, 67 (1967), 3, 310–318.
Bottema [1964 b]
Bottema, Oene, Ein Problem der affinen Kinematik, *Indagationes Mathematicae*, XXVI, *Fasciculus* 3, 290–300.
Bottema [1965]
Bottema, Oene, Acceleration axes in spherical kinematics, *Transactions of the ASME, J. Eng. for Ind.*, 87B (1965), 150–153.
Bottema [1966]
Bottema, Oene, On Ball's Curve, *Journal of Mechanisms*, 1 (1966), 3–8.
Bottema [1967]
Bottema, Oene, On some loci in plane instantaneous kinematics, *Journal of Mechanisms*, 2 (1967), 141–146.
Bottema, et al. [1970]
Bottema, O., Koetsier, T., and Roth, B., On the smallest circle determined by three positions of a rigid body, *Transactions of the ASME*, Series B, 93 (1971), 328–333.
Bottema [1970]
Bottema, Oene, On some loci of lines in plane kinematics, *Journal of Mechanisms*, 5 (1970), 541–548.
Bottema [1971]
Bottema, Oene, Instantaneous kinematics for spatial two-parameter motion, *Koninkl. Nederl. Akademie van Wetenschappen-Amsterdam*, Series B, 74 (1971), 53–62.
Bottema [1975 a]
Bottema, Oene, Cardan motion in elliptic geometry, *Can. J. Math.*, XXVII (1975), 37–43.
Bottema [1975 b]
Bottema, Oene, On the instantaneous binormals to the paths of points in a moving body, *Mechanism and Machine Theory*, 10 (1975), 11–15.
Bottema [1976]
Bottema, O., Line-envelope coupler curves of hinged four bars (in Russian), *Teoriya mashin i mechanizmov* (Dedicated to the 70th Birthday of Acad. I. I. Artobolevskii), Moscow, 1976, 156–161.

BOULAD [1916]
BOULAD, Farid, Sur la détermination du centre de courbure des trajectoires orthogonales d'une famille quelconque de courbes planes, *Bulletin des Sc. Math.*, XL (1916), 292–295.
BRICARD [1926]
BRICARD, Raoul, *Leçons de Cinématique*, I, Paris, 1926, 334 pp.; II, Paris, 1927, 352 pp.
BRISSE [1875]
BRISSE, M. C., Sur le déplacement fini quelconque d'une figure de forme invariable (Suite), *Liouville Journal*, 1, 3e serie (1875), 141–180.
BURMESTER [1874 a]
BURMESTER, L., Kinematisch-geometrische Untersuchungen der Bewegung ähnlich-veränderlicher ebener Systeme, *Z. für Mathematik und Physik*, 19 (1874), 145–69.
BURMESTER [1874 b]
BURMESTER, L., Kinematisch-geometrische Untersuchungen der Bewegung affin-veränderlicher und collinear-veränderlicher ebener Systeme, (*Ibid.*), 465–91.
BURMESTER [1876]
BURMESTER, Ludwig, Geradführung durch das Kurbelgetriebe, *Der Civilingenieur*, 22 (1876), 597–606.
BURMESTER [1877]
BURMESTER, Ludwig, Geradführung durch das Kurbelgetriebe, *Der Civilingenieur*, 23 (1877), 227–250 and 319–342.
BURMESTER [1878]
BURMESTER, Ludwig, Kinematisch-geometrische Theorie der Bewegung der affin-veränderlichen, der ähnlich veränderlichen und starren räumlichen oder ebenen Systeme, *Z. für Mathematik und Physik*, 23 (1878), 103–131.
BURMESTER [1888]
BURMESTER, Ludwig, *Lehrbuch der Kinematik*, Leipzig, 1888, 941 pp.
BURMESTER [1902]
BURMESTER, Ludwig, Kinematisch-geometrische Theorie der Bewegung der affin-veränderlichen und starren räumlichen oder ebenen Systeme, *Z. für Mathematik und Physik*, 47 (1902), 128–156.

CAYLEY [1875]
CAYLEY, A, On three-bar motion, *Proceedings of the London Mathematical Society*, VII (1875–6), 136–166.
CHASLES [1831]
CHASLES, Michel, Note sur les propriétés générales du système de deux corps semblables entre eux, placés d'une manière quelconque dans l'espace; et sur le déplacement fini, ou infiniment petit d'un corps solide libre, *Bulletin des Sciences Mathématiques de Férussac*, XIV (1831), 321–336.
CHASLES [1837]
CHASLES, Michel, *Aperçu historique sur l'origine et le développement des méthodes en géométrie*, (Note 34), Paris, 1837; 3ième edition, Paris, 1889, 851 pp.
CHEN AND ROTH [1969 a]
CHEN, P., AND ROTH, B., A unified theory for the finitely and infinitesimally separated position problems of kinematic synthesis, *Transactions of the ASME, Series B., J. of Engineering for Industry*, 91 (1969), 203–208.
CHEN AND ROTH [1969 b]
CHEN, P. AND ROTH, B., Design equations for the finitely and infinitesimally separated position synthesis of binary links and combined link chains, *Ibid*, 209–219.
COOLIDGE [1940]
COOLIDGE, Julian Lowell, *A History of Geometrical Methods*, Oxford, 1940, (also reissued by Dover Publications, N.Y., 1963, 451 pp.).

DARBOUX [1881]
DARBOUX, G., Sur le déplacement d'une figure invariable, *Comptes rendus de l'Académie des sciences*, XCII (1881), 118–121.
DARBOUX [1887]
DARBOUX, Gaston, *Leçons sur la théorie générale des surfaces et les applications géométriques du calcul infinitesimal*, I–IV. Paris, 1887–1896.
DARBOUX [1897]
DARBOUX, Gaston, Notes, in KOENIGS [1897].
DARBOUX [1916]
DARBOUX, Gaston, Remarque sur la note de M. Farid Boulad, *Bulletin des Sc. Math.*, XL (1916), 292–295.
DIMENTBERG [1950]
DIMENTBERG, F. M., *Determination of the motion of spatial mechanisms*, (in Russian), *Akad. Nauk.*, Moscow, 1950.
DIMENTBERG [1965]
DIMENTBERG, F. M., *The screw calculus and its applications in mechanics*, (in Russian), Moscow, 1965, (English translation: AD680993, Clearinghouse for Federal Technical and Scientific Information, Virginia).
DIMENTBERG [1971]
DIMENTBERG, F. M., *Method of Screws in Applied Mechanics*, (in Russian), Moscow, 1971, 264 pp.
DISTELI [1914]
DISTELI, Martin, Über des Analogon der Savaryschen Formel und Konstruktion in der kinematischen Geometrie des Raumes, *Zeitschrift für Mathematik und Physik*, 62 (1914), 261–309.
DIZIOĞLU [1967]
DIZIOĞLU, Bekir, *Getriebelehre*, (Band 2) *Massbestimmung*, Braunschweig, 1967, 254 pp.
DIZIOĞLU [1974]
DIZIOĞLU, B., Einfache Herleitung der Euler–Savaryschen Konstruktion der räumlichen Bewegung, *Mechanism and Machine Theory*, 9 (1974), 247–254.

EULER [1770]
EULER, L., Problema algebraicum ob affectiones prorsis singulares memorabili (1770), *Opera Omnia*, I, 6 (1921), 287–315.

FRANK [1968]
FRANK, H., *Ebene projektive Kinematik*, Diss. Univ. Karlsruhe, 1968.
FRANK [1971]
FRANK, H., Zur ebenen hyperbolischen Kinematik, *Elem. Math.*, 26 (1971), 121–131.
FREUDENSTEIN [1960]
FREUDENSTEIN, Ferdinand, The cardan positions of a plane, *Trans. of the Sixth Conference on Mechanisms*, Penton Publishing Co., 1960, 129–133.
FREUDENSTEIN [1965]
FREUDENSTEIN, Ferdinand, Higher path-curvature analysis in plane kinematics, *Journal of Engineering for Industry*, *Trans. ASME*, Series B, 87 (1965), 184–190.
FREUDENSTEIN AND WOO [1968]
FREUDENSTEIN, F., AND WOO, L. S., On the curves of synthesis in plane instantaneous kinematics, *Proceedings of the 12th International Congress of Theoretical and Applied Mechanics*, Springer-Verlag, 1969, 400–414.
FREUDENSTEIN et al. [1969]
FREUDENSTEIN, F., BOTTEMA, O., AND KOETSIER, T., Finite conic-section Burmester theory, *Journal of Mechanisms*, 4 (1969), 359–373.

GARNIER [1951]
GARNIER, René, *Cours de Cinématique*, (Tome III) *Géométrie et Cinématique cayleyennes*, Paris, 1951, 376 pp.
GARNIER [1954]
GARNIER, René, *Cours de Cinématique*, (Tome I, 3e édition) *Cinématique du point et du solide. Composition des mouvements*, Paris, 1954, 244 pp.
GARNIER [1956]
GARNIER, René, *Cours de Cinématique*, (Tome II, 3e édition) *Roulement et viration. La formule de Savary et son extension à l'espace*, Paris, 1956, 341 pp.
GOLDSTEIN [1950]
GOLDSTEIN, Herbert, *Classical Mechanics*, Addison-Wesley Pub. Co., 1950, 399 pp.
GROENEVELD [1954]
GROENEVELD, B., *Geometrical considerations on space kinematics in connection with Bennett's mechanism*, Dissertation Technical University of Delft, 1954, 112 pp.
GROENMAN [1950]
GROENMAN, J. T., *Behandeling van de Koppelkromme met behulp van isotrope coördinaten*, Dissertatie Technische Hogeschool Delft, 1950, 104 pp.
GRÜNWALD [1906]
GRÜNWALD, A., Darstellung der Mannheim–Darbouxschen Umschwungbewegung eines starren Körpers, *Zeitschrift für Mathematik und Physik*, 54 (1906), 154–221.
GRÜNWALD [1911]
GRÜNWALD, J., Ein Abbildungsprinzip, welches die ebene Geometrie und Kinematik mit der räumlichen Geometrie verknüpft, *Sitzber. Ak. Wiss. Wien*, 120 (1911), 677–741.

HAARBLEICHER [1933]
HAARBLEICHER, A., Application des coordonnées isotropes à l'étude de la courbe des trois barres, *Journal de l'Ecole Polytechnique*, 31 (1933), 13–40.
HACKMÜLLER [1938 a]
HACKMÜLLER, E., Zur Konstruktion der Burmesterschen Punkte, *Maschinenbau/Betrieb*, 6 (1938), 648–649.
HACKMÜLLER [1938 b]
HACKMÜLLER, E., Eine analytisch durchgeführte Ableitung der Kreispunkts- und Mittelpunktskurve, *Z. angew. Math. Mech.*, 18 (1938), 252–254.
HALPHEN [1882]
HALPHEN, M., Sur la théorie du déplacement, *Nouvelles Annales de Math.*, 3 (1882) 1, 296–299.
HAMEL [1912]
HAMEL, G., *Elementare Mechanik*, Leipzig-Berlin, 1912, 404 pp.
HARTENBERG AND DENAVIT [1964]
HARTENBERG, R. S., AND DENAVIT, J., *Kinematic Synthesis of Linkages*, McGraw-Hill Book Co., 1964, 435 pp.
HARTMANN [1893]
HARTMANN, W., Ein neues Verfahren zur Aufsuchung des Krümmungskreises, *Zeitschrift VDI*, 37 (1893), 95–102.
HIPPISLEY [1920]
HIPPISLEY, R. L., A new method of describing a three-bar curve, *Proc. London Math. Soc.*, 18 (1920), 136–140.
HOLDITCH [1858]
HOLDITCH, A., *Lady's and gentleman's diary for the year 1858.*
HUNT [1968]
HUNT, K. H., Note on complexes and mobility, *J. of Mechanisms*, 3 (1968), 199–202.
HUNT [1970]

HUNT, K. H., *Screw Systems in Spatial Kinematics*, MMERS3, Dept. of Mech. Eng., Monash University, 114 pp.

JOHNSON [1960]
JOHNSON, R. A., *Advanced Euclidean Geometry*, Dover Publications, New York, 1960, 356 pp., (see p. 222).

KELER [1959]
KELER, M. K., Analyse und Synthese der Raumkurbelgetriebe mittels Raumliniengeometrie und dualer Grössen, *Forschung ing.-Wes.*, 25 (1959), 26–63.
KIRSON [1975]
KIRSON, Yoram, *Higher Order Curvature Theory in Space Kinematics*, Ph.D. Dissertation, Univ. of Calif. at Berkeley, 1975, 140 pp.
KOENIGS [1897]
KOENIGS, Gabriel, *Leçons de Cinématique, Avec des notes par G. Darboux, E. et F. Cosserat*, Paris, 1897, 499 pp.
KOENIGS [1917]
KOENIGS, G., Recherches sur les mouvements plans à deux paramètres, *Bulletin des Sc. Math.*, XLI (1917), 120–127, 153–164, 181–196.
KOTEL'NIKOV [1895]
KOTEL'NIKOV, A. P., *Vintovoe Schislenie i Nikotoriya Prilozheniya evo k geometrie i mechaniki*, (in Russian), Kazan, 1895.
KRAMES [1937 a]
KRAMES, J., Über Fusspunktkurven von Regelflächen und eine besondere Klasse von Raumbewegungen (Über symmetrische Schrotungen I), *Monatsh. Math. Phys.*, 45 (1937), 394–406.
KRAMES [1937 b]
KRAMES, J., Zur Bricardschen Bewegung, deren sämtliche Bahnkurven auf Kugeln liegen (Über symmetrische Schrotungen II), *Monatsh. Math. Phys.*, 45 (1937), 407–417.
KRAMES [1937 c]
KRAMES, J., Zur aufrechten Ellipsenbewegung des Raumes (Über symmetrische Schrotungen III), *Monatsh. Math. Phys.*, 46 (1937), 38–50.
KRAMES [1937 d]
KRAMES, J., Zur kubischen Kreisbewegung des Raumes (Über symmetrische Schrotungen IV), *Sitz.-Berichte der Ak. der Wissenschaften, Wien*, 146 (1937), 145–158.
KRAMES [1937 e]
KRAMES, J., Zur Geometrie des Bennett'schen Mechanismus (Über symmetrische Schrotungen V), *Sitz.-Berichte der Ak. der Wissenschaften, Wien*, 146 (1937), 159–173.
KRAMES [1937 f]
KRAMES, J., Die Borel–Bricard Bewegung mit punktweise gekoppelten orthogonalen Hyperboloiden (Über symmetrische Schrotungen VI), *Monatsh. Math. Phys.*, 46 (1937), 172–195.
KRAMES [1940]
KRAMES, J., Über die durch aufrechte Ellipsenbewegung erzeugten Regelflächen, *Jbr. D. Math. Ver.*, 50 (1940), 58–65.
KRAUSE [1910]
KRAUSE, M., Zur Theorie der ebenen ähnlich veränderlichen Systeme, *Jahresbericht der deutschen Mathematiker-Vereinigung*, 19 (1910), 327–329.
KRAUSE [1920]
KRAUSE, Martin, *Analysis der ebenen Bewegung*, Berlin and Leipzig, 1920, 216 pp.

LEHMANN [1967]
LEHMANN, H., *Zur Möbius-Kinematik*, Diss. Univ. Freiburg/Br., 1967.

MANNHEIM [1875]
MANNHEIM, A., Sur les surfaces trajectoires des points d'une figure de forme invariable dont le déplacement est assujetti à quatre conditions, *J. de Math.*, (3), 1 (1875), 57–74.
MANNHEIM [1889]
MANNHEIM, A., Etude d'un déplacement particulier d'une figure de forme invariable, *Rendic. Circ. Math. Palermo*, 3 (1889), 131–144.
MANNHEIM [1894]
MANNHEIM, A., *Principes et Développements de Géométrie Cinématique*, Paris, 1894, 589 pp.
MAYER [1937]
MAYER, A. E., Koppelkurven mit drei Spitzen und spezielle Koppelkurvenbüschel, *Z. Math. Phys.*, 43 (1937), 389. (Also see *Z. VDI*, Vol. 82 (1939), 124.)
MCAULAY [1898]
MCAULAY, A., *Octonions — a development of Clifford's bi-quaternions*, Cambridge University Press, 1898.
MORLEY AND MORLEY [1954]
MORLEY, Frank and MORLEY, F. V., *Inversive Geometry*, Chelsea Publishing Co., 1954, 273 pp.
MÜLLER [1889]
MÜLLER, Reinhold, Über die Doppelpunkte der Koppelkurve, *Z. Math. Phys.*, 34 (1889), 303–305 and 372–375. (This and 11 other of Müller's papers have been translated into English by D. Tesar under the title: (Translations of) Papers on geometrical theory of motion applied to approximate straight line motion, *Kansas State University Bulletin*, 46 (1962), No. 6, special report no. 21.)
MÜLLER [1891 a]
MÜLLER, Reinhold, Über die Krümmung der Bahnevoluten bei starren ebenen Systemen, *Zeitschrift für Mathematik und Physik*, 36 (1891), 193–205.
MÜLLER [1891 b]
MÜLLER, Reinhold, Konstruktion der Krümmungsmittelpunkte der Hüllbahnevoluten bei starren ebenen Systemen, *Zeitschrift für Mathematik und Physik*, 36 (1891), 257–266.
MÜLLER [1892]
MÜLLER, Reinhold, Über die Bewegung eines starren ebenen Systems durch fünf unendlich benachbarte Lagen, *Zeitschrift für Mathematik und Physik*, 37 (1892), 129–150.
MÜLLER [1898]
MÜLLER, Reinhold, Über die angenäherte Geradführung mit Hilfe eines ebenen Gelenkvierecks, *Zeitschrift für Mathematik und Physik*, 43 (1898), 36–40.
MÜLLER [1903]
MÜLLER, Reinhold, Über einige Kurven, die mit der Theorie des ebenen Gelenkvierecks im Zusammenhang stehen, *Zeitschrift für Mathematik und Physik*, 48 (1903), 224–248.
MÜLLER [1907]
MÜLLER, Reinhold, Polbestimmung für Verzweigungslagen bei der Bewegung eines ebenen ähnlich-veränderlichen Systems in seiner Ebene, *Jrb. D. Math. Ver.*, 16 (1907), 242–243.
MÜLLER [1910 a]
MÜLLER, Reinhold, Über die Momentanbewegung eines ebenen ähnlich-veränderlichen Systems in seiner Ebene, *Jahresbericht der deutschen Mathematiker-Vereinigung*, 19 (1910), 29–89.
MÜLLER [1910 b]
MÜLLER, Reinhold, Über die Momentanbewegung eines ebenen ähnlich-veränderlichen Systems bei unendlich fernem Pol., (*Ibid.*), 147–154.
MÜLLER [1910 c]
MÜLLER, Reinhold, Erzeugung der Koppelkurve durch ähnlich-veränderliche Systeme, *Z. Math. Phys.*, 58 (1910), 247–251.
MÜLLER [1953]

MÜLLER, Hans Robert, Zur Kinematik des Rollgleitens, I, II, *Arch. Math.*, (Part I), 4 (1953), 239–246; (Part II), 6 (1955), 471–480.
MÜLLER [1958]
MÜLLER, H. R., Zur Kinematik der ebenen affin-veränderlichen Felder, *Mathem. Nachr.*, 18 (1958), 136–140.
MÜLLER [1962]
MÜLLER, Hans Robert, *Sphärische Kinematik*, Berlin, 1962, 121 pp.
MÜLLER [1970]
MÜLLER, Hans Robert, Kinematische Geometrie, *Jbr. D. Math. Ver.*, 72 (1970) 143–164.
MÜLLER AND KRUPPA [1923]
MÜLLER, E., AND KRUPPA, E., *Vorl. über Darstellende Geometrie I*, Leipzig and Wien, 1923.

ÖZGÖREN [1976]
ÖZGÖREN, Kemal, Optimization of Manipulator Motions, *Preprints, Second CISM-IFToMM Symposium, On Theory and Practice of Robots and Manipulators*, Warsaw, 1976, 27–36.

PELZER [1959]
PELZER, W., *Über die Kinematik affin-veränderlicher ebener Systeme*, Diss. Techn. Univ. Berlin, 1959.
PRIMROSE et al. [1964]
PRIMROSE, E. J. F., FREUDENSTEIN, F., AND SANDOR, G. N., Finite Burmester theory in plane kinematics, *Transactions of the ASME, Series E, Journal of Applied Mechanics*, 86 (1964), 683–693.

ROBERTS [1870]
ROBERTS, Samuel, On the motion of a plane under certain conditions, *Proceedings London Mathematical Society*, III (1869–71), 286–319.
ROBERTS [1875]
ROBERTS, Samuel, On three-bar-motion in plane space, *Proceedings of the London Mathematical Society*, VII (1875), 14–23.
ROBERTS [1876]
ROBERTS, Samuel, Further note on the motion of a plane under certain conditions, *Proceedings London Mathematical Society*, VII (1875–76), 216–225.
RODRIGUES [1840]
RODRIGUES, Olinde, Des lois géométriques qui régissent les déplacements d'un système solide dans l'espace, et de la variation des coordonnées provenant de ces déplacements considérés indépendamment des causes qui peuvent les produire, *Journal De Mathématiques Pures et Appliquées*, 5, 1st Series (1840), 380–440.
ROOM [1938]
ROOM, T. G., *The Geometry of Determinantal Loci*, Cambridge, (England), 1938.
ROTH [1967 a]
ROTH, B., On the screw axes and other special lines associated with spatial displacements of a rigid body, *Transactions of the ASME, Ser. B*, 89 (1967), 102–110.
ROTH [1967 b]
ROTH, B., The kinematics of motion through finitely separated positions, *Trans. of the ASME, Series E, Journal of Applied Mechanics*, 34 (1967), 591–598.
ROTH [1967 c]
ROTH, B., Finite position theory applied to mechanism synthesis, (*Ibid.*), 599–605.
ROTH [1968]
ROTH, B., The design of binary cranks with revolute, cylindric, and prismatic joints, *Journal of Mechanisms*, 3 (1968), 61–72.

SALMON [1954]
SALMON, George, *A Treatise on Conic Sections*, 6th ed., Chelsea Pub. Co., 1954 (see pp. 275, 277).
SANDOR AND FREUDENSTEIN [1967]
SANDOR, George N., and FREUDENSTEIN, F., Higher-Order plane motion theories in kinematic synthesis, *Transactions of the ASME, Series B, J. of Engrg. for Industry*, 89 (1967), 223–230.
SCHOENFLIES [1886]
SCHOENFLIES, Arthur, *Geometrie der Bewegung in Synthetischer Darstellung*, Leipzig, 1886, 194 pp. (See also the French translation: *La Géométrie du Mouvement*, Paris, 1893, which is a slightly revised, and augmented version.)
SCHOENFLIES [1892]
SCHOENFLIES, Arthur, Über Bewegung starrer Systeme im Fall cylindrischer Axenflächen, *Mathem. Annalen*, 40 (1892), 317–331.
SCHÖNEMANN [1855]
SCHÖNEMANN, Th., Construction von Normalebenen gewisser Krummen u. Linien, *Bericht über die Verhandlungen der Königl. Preufs. Akademie der Wissenschaften zu Berlin*, (1855), 255–260.
SEMPLE AND ROTH [1949]
SEMPLE, J. G., AND ROTH, L., *Introduction to Algebraic Geometry*, Oxford Univ. Press, 1949, 446 pp.
SHARIKOV [1961]
SHARIKOV, V. I., The theory of screws in structural and kinematic analysis of pairs of mechanisms (in Russian), *Seminar po teorii Mashin i Mekhanizmov*, 22 (1961), 108–136.
SOMMERVILLE [1934]
SOMMERVILLE, D. M. Y., *Analytical Geometry of Three Dimensions*, Cambridge University Press, 1934, 416 pp.
STEINER [1840]
STEINER, Jakob, *Ges. Werke, II*, Berlin, 1881-1882, 99–159.
STICHER [1972]
STICHER, F. C. O., On the principal normals to the paths of points in a moving body, *Mechanism and Machine Theory*, 7 (1972), 355–361.
STIELTJES [1884]
STIELTJES, T. J., Note sur le déplacement d'un système invariable dont un point est fixe, *Archives Néerlandaises des sciences exactes et naturelles*, 19 (1884), 372–390.
STUDY [1891]
STUDY, E., Von Bewegungen und Umlegungen, *Mathem. Annalen*, 39 (1891), 441–564.
STUDY [1903]
STUDY, E., *Die Geometrie der Dynamen*, Leipzig, 1903, 437 pp.
STURM [1893]
STURM, R., *Die Gebilde ersten und zweiten Grades der Liniengeometrie in synthetischer Behandlung, II*, Leipzig, 1893, 471 pp.
TESAR [1967]
TESAR, Delbert, The generalized concept of three multiply separated positions in coplanar motion, *Journal of Mechanisms*, 2 (1967), 461–474.
TESAR [1968]
TESAR, Delbert, The generalized concept of four multiply separated positions in coplanar motion, *Journal of Mechanisms*, 3 (1968), 11–23.
TESAR AND SPARKS [1968]
TESAR, Delbert and SPARKS, J. W., The generalized concept of five multiply separated positions in coplanar motion, *Journal of Mechanisms*, 3 (1968), 25–33.
TÖLKE [1967]
TÖLKE, J., *Affine Kinematik der Ebene*, Diss. Univ. Karlsruhe, 1967.
TÖLKE [1974]

TÖLKE, J., Kinematik der hyperbolischen Ebene, I, II, III; I. *J. reine angew. Math.*, 265 (1974), 145–153; 11. *Idem*, 267 (1974), 143–150; III. *Idem*, 273 (1975), 99–108.
TÖLKE [1976]
TÖLKE, J., Contributions to the theory of the axes of curvature, *Mechanism and Machine Theory*, 11 (1976), 123–130.
TSAI AND ROTH [1972]
TSAI, L. W., AND ROTH, B., Design of dyads with helical, cylindrical, spherical, revolute and prismatic joints, *Mechanism and Machine Theory*, 7 (1972), 85–102.
TSAI AND ROTH [1973]
TSAI, L. W., AND ROTH, B., A Note on the Design of Revolute-Revolute Cranks, *Mechanism and Machine Theory*, 8 (1973), 23–31.
TSAI [1977]
TSAI, Lung-wen, Instantaneous Kinematics of a Special Two-Parameter Motion, *Transactions of the ASME*, Series B, 99 (1977), 336–340.

VELDKAMP [1963]
VELDKAMP, G. R., *Curvature Theory in Plane Kinematics*, Dissertation Technical University of Delft, 1963.
VELDKAMP [1967 a]
VELDKAMP, G. R., Canonical systems and instantaneous invariants in spatial kinematics, *Journal of Mechanisms*, 2 (1967), 329–388.
VELDKAMP [1967 b]
VELDKAMP, G. R., Some remarks on higher curvature theory, *Trans. of the ASME, Ser. B, Journal of Engineering for Industry*, 84 (1967), 84–86.
VELDKAMP [1976]
VELDKAMP, G. R., On the use of dual numbers, vectors and matrices in instantaneous, spatial kinematics, *Mechanism and Machine Theory*, 11 (1976), 141–156.
VOINEA AND ATANASIU [1962]
VOINEA, R. P., AND ATANASIU, M. C., Théorie géométrique des vis et quelques applications à la théorie des mécanismes, *Revue Méc. appl. Buc.*, 7 (1962), 845–860.

WALDRON [1969]
WALDRON, K. J., *The Mobility of Linkages*, Ph.D. Dissertation Stanford University, 1969.
WEISS [1942]
WEISS, E. A., Die Koppelkurve als Laguerresches Bild einer Hesseschen Korrespondenz, *Mathematische Zeitschrift*, 47 (1942), 187–198.
WILLS [1931]
WILLS, A. P., *Vector Analysis with an Introduction to Tensor Analysis*, Dover Publications, 1958, 285 pp.
VAN DER WOUDE [1926]
VAN DER WOUDE, W., On the motion of a plane fixed system with two degrees of freedom, I. *Proc. Kon. Ned. Ak. Wet. Amsterdam*, XXIX (1926), 652–663; II. *Idem*, XXXI (1928), 519–530; III. *Idem*, XXXIV (1931), 948–950.

YANG [1963]
YANG, A. T., *Application of Quaternion Algebra and Dual Numbers to the Analysis of Spatial Mechanisms*, Doctoral Dissertation, Columbia University, 1963.
YANG AND FREUDENSTEIN [1964]
YANG, A. T., AND FREUDENSTEIN, F., Application of dual-number quaternion algebra to the analysis of spatial mechanisms, *Transactions of the ASME, Series E, Journal of Applied Mechanics*, 86 (1964), 300–308.

YANG [1969]
YANG, An Tzu, Displacement analysis of spatial five-link mechanisms using (3 × 3) matrices with dual-number elements, *Transactions of the ASME, Ser. B, Journal of Eng. for Industry*, 91 (1969), 152–157.
YUAN et al. [1971]
YUAN, M. S. C., FREUDENSTEIN, F., AND WOO, L. S., Kinematic analysis of spatial mechanisms by means of screw coordinates: Part I–Screw coordinates; Part II–Analysis of spatial mechanisms, *Trans. of the ASME, Series B, J. of Eng. for Industry*, 93 (1971), I, 61–66; II, 67–73.

ZWIKKER [1950]
ZWIKKER, C., *Advanced Plane Geometry*, Amsterdam, 1950, (also reissued as *The Advanced Geometry of Plane Curves and Their Applications*, Dover Publications, 1963, 299 pp.).

2. Review articles and bibliographies

BOTTEMA, O., Recent work on kinematics, *Applied Mechanics Reviews*, 6 (1953), 169–170.
BOTTEMA, O. AND FREUDENSTEIN, F., Kinematics and the theory of mechanisms, *Idem*, 19 (1966) 4, 287–293.
DE GROOT, J., *Bibliography on Kinematics*, I and II, Eindhoven University of Technology, 1970. (This contains about 7000 items.)
HAIN, Kurt, *Applied Kinematics*, McGraw-Hill Book Co. 1967, 727 pp. (This is the English language translation of *Angewandte Getriebelehre*, Düsseldorf, 1961.) (Although mainly dealing with mechanisms, this book contains over 2000 references, many of which deal with kinematic theory.)
Linkage, *Tôhoku Mathematical Journal*, 37 (1933), 294–319. (A bibliography on the theory of linkages.)
SCHOENFLIES, A. AND GRÜBLER, M., Kinematik, *Encyklopädie der Mathematischen Wissenschaften*, IV 3 (1902), 190–277. (This article contains many references and much historical information.)

3. Additional related references

ARTOBOLEVSKII, I. I., LEVITSKII, N. I., AND CHERKYDINOV, S. A., *Cintez ploskich mechanizmov*, (in Russian), Moscow, 1959, 1084 pp.
BARRAU, J. A., Mouvements algébriques dans le plan, *Journal de Mathématiques pures et appliquées*, 7ᵉ Ser., 3 (1917).
BEREIS, R., Aufbau einer Theorie der ebenen Bewegung mit Verwendung komplexer Zahlen, *Österr. Ing. Arch.*, 5 (1951), 246–266.
BIEZENO, C. B., Fläche vierten und achten Grades, welche bei der Bewegung einer mit vier festen Punkten in vier festen ebenen bleibenden Geraden entstehen, *N. Arch. v. Wsk.*, (2) XI (1915), 329–393.
BOUMAN, J. N., *Kinematische Projectie*, Dissertation, Utrecht 1937.
DARBOUX, Gaston, Sur une nouvelle définition de la surface des ondes, *Comptes rendus de l'Académie des Sciences*, XCII (1881), 446–448.
DEGEN, W., Projektive Kinematik, *Abh. Math. Sem. Hamburg*, 27 (1964), 231–249, (with many references).
DITTRICH, Günter, *Über die momentane Bewegungsgeometrie eines sphärisch bewegten starren Systems*, Dissertation, Aachen, 1964, 101 pp.
EVERETT, J. D., On a new method in statics and kinematics, *Messenger Math.*, 45 (1875), 36–37.
VAN HAASTEREN, A., *Over de formule van Euler–Savary en haar uitbreidingen in de cinematische*

meetkunde van de euclidische ruimte en van het niet-euclidische vlak, Dissertation, Leiden, 1947, 125 pp.

HILBERT, D., AND COHN-VOSSEN, S., *Geometry and the Imagination*, Chelsea Publishing Co., 1956.

JULIA, G., *Cours de cinématique*, Paris, Gauthier-Villars, 1936, 161 pp.

LOCHS, G., Die Affinnormalen der Bhan- und Hüllkurven bei einer ebenen Bewegung, *Mh. Math. Phys.*, 38 (1931), 39–52.

MANNHEIM, M. A., Sur une droite qui se déplace de façon que trois de ses points restent sur les faces d'un trièdre trirectangle, *Bulletin des Sciences Mathématiques*, 2e série, IX (1885), 137.

MÜLLER, Hans Robert, Die kinematischen Abbildungen im dreidimensionalen Raum, *Monatshefte für Mathematik*, 65 (1961), No. 3, 252–258.

MÜLLER, Hans Robert, Zur Bewegungsgeometrie in Räumen höherer Dimension, *Monatshefte für Mathematik*, 70 (1966), 47–57.

MÜLLER, Hans Robert, Über eine infinitesimale kinematische Abbildung, *Idem.*, 54 (1950), 108–129.

MÜLLER, Hans Robert, Flächenläufige Bewegungsvorgänge im elliptischen Raum I, II, *Idem.*, 57 (1953), I, 29–43; II, 129–133.

MÜLLER, Hans Robert, Zur Ermittlung von Hüllflächen in der räumlichen Kinematik, *Idem.*, 63 (1959), 231–240.

MÜLLER, Hans Robert, Die Formel von Euler und Savary in der affinen Kinematik, *Archiv der Mathematik*, X (1959), 71–80.

MÜLLER, Hans Robert, Die Bewegungsgeometrie auf der Kugel, *Monatshefte für Mathematik*, 55 (1950), 28–42.

MÜLLER, Hans Robert, Verallgemeinerung der Bresseschen Kreise für höhere Beschleunigungen, *Archiv der Mathematik*, IV (1953), 337–342.

MÜLLER, Hans Robert, Über Integrale bei mehrgliedrigen Bewegungsvorgängen, *Mathematische Nachrichten*, 7 (1952), 159–164.

PELECUDI, Christian, *Teoria mecanismelor spatiale*, Bucharest, 1972, 511 pp.

ROTH, B., AND YANG, A. T., Application of instantaneous invariants to the analysis and synthesis of mechanisms, *Trans. of the ASME, Ser. B, J. of Eng. for Industry*, 99 (1977), 97–103.

SANDOR, George N., Principles of a general quaternion-operator method of spatial kinematic synthesis, *Transactions of the ASME, Series E, J. of Applied Mechanics*, 40 (1968), 40–46.

SICARD, H., *Traité de cinématique théorique. Avec de notes par A. Labrousse*, Paris, 1902, 185 pp.

SKREINER, M., On the points of inflection in general spatial motion, *Journal of Mechanisms*, 2 (1967) 4, 429–434.

STÜBLER, E., Das Beschleunigungssystem bei der Bewegung des starren Körpers, *DMV-Berichte*, 19 (1910), 177–185.

TÖLKE, J., Zur Strahlkinematik I, *Sitzber. Oester. Ak. Wiss.*, 182 (1973), 177–202.

TÖLKE, J., Zur Konstruktion der Krümmungsmittelpunktes einer sphärischen Bahnkurve, *Monatsh. Math.*, 80 (1975), 61–65.

TÖLKE, J., *Projektive kinematische Geometrie*, Berichte der Mathematisch-Statistischen Sektion im Forschungszentrum Graz., Berichte nr. 39 (1975), 51 pp. (contains 51 references).

TSAI, L. W., AND ROTH, Bernard, Incompletely specified displacements: Geometry and spatial linkage synthesis, *Transactions of the ASME, Series B, J. of Eng. for Industry*, 95 (1973), 603–611.

WEISS, E. A., *Einführung in die Liniengeometrie und Kinematik*, Leipzig, 1935, 122 pp.

WOLFF, J., *Dynamen, beschouwd als duale Vectoren*, Dissertatie, Universiteit van Amsterdam, 1907, 154 pp.

VAN DER WOUDE, W., On the motion of a fixed system, *Proc. Kon. Ak. Wet. Amsterdam*, XXIII (1920) 4, 589–602.

WUNDERLICH, W., *Ebene Kinematik*, Mannheim-Wein-Zurich, 1970.

WUNDERLICH, W., Zur Schraubung im vierdimensionalen euklidischen Raum, *J. reine angew. Math.*, 285 (1976), 79–99.

INDEX

A CATALOG OF SELECTED

DOVER BOOKS

IN SCIENCE AND MATHEMATICS

NUMERICAL METHODS FOR SCIENTISTS AND ENGINEERS, Richard Hamming. Classic text stresses frequency approach in coverage of algorithms, polynomial approximation, Fourier approximation, exponential approximation, other topics. Revised and enlarged 2nd edition. 721pp. 5⅜ × 8½.
65241-6 Pa. $14.95

THEORETICAL SOLID STATE PHYSICS, Vol. I: Perfect Lattices in Equilibrium; Vol. II: Non-Equilibrium and Disorder, William Jones and Norman H. March. Monumental reference work covers fundamental theory of equilibrium properties of perfect crystalline solids, non-equilibrium properties, defects and disordered systems. Appendices. Problems. Preface. Diagrams. Index. Bibliography. Total of 1,301pp. 5⅜ × 8½. Two volumes. Vol. I 65015-4 Pa. $12.95
Vol. II 65016-2 Pa. $12.95

OPTIMIZATION THEORY WITH APPLICATIONS, Donald A. Pierre. Broad-spectrum approach to important topic. Classical theory of minima and maxima, calculus of variations, simplex technique and linear programming, more. Many problems, examples. 640pp. 5⅜ × 8½. 65205-X Pa. $12.95

THE MODERN THEORY OF SOLIDS, Frederick Seitz. First inexpensive edition of classic work on theory of ionic crystals, free-electron theory of metals and semiconductors, molecular binding, much more. 736pp. 5⅜ × 8½.
65482-6 Pa. $14.95

ESSAYS ON THE THEORY OF NUMBERS, Richard Dedekind. Two classic essays by great German mathematician: on the theory of irrational numbers; and on transfinite numbers and properties of natural numbers. 115pp. 5⅜ × 8½.
21010-3 Pa. $4.95

THE FUNCTIONS OF MATHEMATICAL PHYSICS, Harry Hochstadt. Comprehensive treatment of orthogonal polynomials, hypergeometric functions, Hill's equation, much more. Bibliography. Index. 322pp. 5⅜ × 8½. 65214-9 Pa. $8.95

NUMBER THEORY AND ITS HISTORY, Oystein Ore. Unusually clear, accessible introduction covers counting, properties of numbers, prime numbers, much more. Bibliography. 380pp. 5⅜ × 8½. 65620-9 Pa. $8.95

THE VARIATIONAL PRINCIPLES OF MECHANICS, Cornelius Lanzcos. Graduate level coverage of calculus of variations, equations of motion, relativistic mechanics, more. First inexpensive paperbound edition of classic treatise. Index. Bibliography. 418pp. 5⅜ × 8½. 65067-7 Pa. $10.95

MATHEMATICAL TABLES AND FORMULAS, Robert D. Carmichael and Edwin R. Smith. Logarithms, sines, tangents, trig functions, powers, roots, reciprocals, exponential and hyperbolic functions, formulas and theorems. 269pp. 5⅜ × 8½. 60111-0 Pa. $5.95

THEORETICAL PHYSICS, Georg Joos, with Ira M. Freeman. Classic overview covers essential math, mechanics, electromagnetic theory, thermodynamics, quantum mechanics, nuclear physics, other topics. First paperback edition. xxiii + 885pp. 5⅜ × 8½. 65227-0 Pa. $17.95

THE FOUR-COLOR PROBLEM: Assaults and Conquest, Thomas L. Saaty and Paul G. Kainen. Engrossing, comprehensive account of the century-old combinatorial topological problem, its history and solution. Bibliographies. Index. 110 figures. 228pp. 5⅜ × 8½. 65092-8 Pa. $6.00

CATALYSIS IN CHEMISTRY AND ENZYMOLOGY, William P. Jencks. Exceptionally clear coverage of mechanisms for catalysis, forces in aqueous solution, carbonyl- and acyl-group reactions, practical kinetics, more. 864pp. 5⅜ × 8½. 65460-5 Pa. $18.95

PROBABILITY: An Introduction, Samuel Goldberg. Excellent basic text covers set theory, probability theory for finite sample spaces, binomial theorem, much more. 360 problems. Bibliographies. 322pp. 5⅜ × 8½. 65252-1 Pa. $7.95

LIGHTNING, Martin A. Uman. Revised, updated edition of classic work on the physics of lightning. Phenomena, terminology, measurement, photography, spectroscopy, thunder, more. Reviews recent research. Bibliography. Indices. 320pp. 5⅜ × 8¼. 64575-4 Pa. $7.95

PROBABILITY THEORY: A Concise Course, Y.A. Rozanov. Highly readable, self-contained introduction covers combination of events, dependent events, Bernoulli trials, etc. Translation by Richard Silverman. 148pp. 5⅜ × 8¼. 63544-9 Pa. $4.50

THE CEASELESS WIND: An Introduction to the Theory of Atmospheric Motion, John A. Dutton. Acclaimed text integrates disciplines of mathematics and physics for full understanding of dynamics of atmospheric motion. Over 400 problems. Index. 97 illustrations. 640pp. 6 × 9. 65096-0 Pa. $16.95

STATISTICS MANUAL, Edwin L. Crow, et al. Comprehensive, practical collection of classical and modern methods prepared by U.S. Naval Ordnance Test Station. Stress on use. Basics of statistics assumed. 288pp. 5⅜ × 8½. 60599-X Pa. $6.00

WIND WAVES: Their Generation and Propagation on the Ocean Surface, Blair Kinsman. Classic of oceanography offers detailed discussion of stochastic processes and power spectral analysis that revolutionized ocean wave theory. Rigorous, lucid. 676pp. 5⅜ × 8½. 64652-1 Pa. $14.95

STATISTICAL METHOD FROM THE VIEWPOINT OF QUALITY CONTROL, Walter A. Shewhart. Important text explains regulation of variables, uses of statistical control to achieve quality control in industry, agriculture, other areas. 192pp. 5⅜ × 8½. 65232-7 Pa. $6.00

THE INTERPRETATION OF GEOLOGICAL PHASE DIAGRAMS, Ernest G. Ehlers. Clear, concise text emphasizes diagrams of systems under fluid or containing pressure; also coverage of complex binary systems, hydrothermal melting, more. 288pp. 6½ × 9¼. 65389-7 Pa. $8.95

STATISTICAL ADJUSTMENT OF DATA, W. Edwards Deming. Introduction to basic concepts of statistics, curve fitting, least squares solution, conditions without parameter, conditions containing parameters. 26 exercises worked out. 271pp. 5⅜ × 8½. 64685-8 Pa. $7.95